**W9-BKT-677**

# BASIC ENVIRONMENTAL TECHNOLOGY

Water Supply, Waste Management,
and Pollution Control

*Fourth Edition*

JERRY A. NATHANSON, M.S., P.E.

*Union County College
Cranford, New Jersey*

Prentice
Hall

Upper Saddle River, New Jersey
Columbus, Ohio

**Library of Congress Cataloging-in-Publication Data**

Nathanson, Jerry A.

   Basic environmental technology: water supply, waste management, and pollution control / Jerry A. Nathanson. — 4th ed.

     p. cm.

   Includes bibliographical references and index.

   ISBN 0-13-093085-7

   1. Environmental engineering. 2. Sanitary engineering. 3. Pollution. I. Title

TD145.N38  2003

628—dc21                                     2001059125

**Editor in Chief:** Stephen Helba
**Executive Editor:** Ed Francis
**Production Editor:** Christine M. Buckendahl
**Production Coordination:** TechBooks
**Design Coordinator:** Diane Ernsberger
**Cover Designer:** Ali Mohrman
**Production Manager:** Matt Ottenweller
**Marketing Manager:** Mark Marsden

This book was set in ITC Century Book by TechBooks and was printed and bound by Courier Kendallville, Inc. The cover was printed by Phoenix Color Corp.

Pearson Education Ltd.
Pearson Education Australia Pty. Limited
Pearson Education Singapore Pte. Ltd.
Pearson Education North Asia Ltd.
Pearson Education Canada, Ltd.
Pearson Educación de Mexico, S.A. de C.V.
Pearson Education—Japan
Pearson Education Malaysia Pte. Ltd.
Pearson Education, *Upper Saddle River, New Jersey*

Reprinted with corrections April, 2003.

10 9 8 7 6 5 4
ISBN: 0-13-093085-7

# FOR GINGER AND ADAM

*Sailing up my dirty stream*
*Still I love it and I'll dream*
*That some day, though maybe not this year*
*My Hudson River and my country will run clear*

PETE SEEGER

# PREFACE

*Basic Environmental Technology* offers a pragmatic introduction to the topics of municipal water supply, waste management, and pollution control. The book is designed primarily for use by students in civil/construction technology programs and related disciplines in community colleges and technical institutes. It can also be useful in baccalaureate engineering and technology programs when a practical but elementary course of study is desired, or for independent study by individuals who wish to explore the rudiments of environmental quality control and public health protection. Experienced technicians, engineers, scientists, and others in different disciplines who may become involved in environmental work for the first time will also find this book of value as an initial reference.

The qualities that continue to distinguish this book in its fourth edition are its clear, easy-to-read style and its logical and systematic treatment of the subject. Since the field of environmental technology is multidisciplinary and very broad in scope, review or primer sections are included so that readers with little or no experience in biology, chemistry, geology, and hydraulics can comprehend and use the book. Mathematical topics are presented at a relatively basic level; to understand all the numerical examples in the book, some knowledge of algebra and geometry will be useful.

Hundreds of example problems, diagrams, and photographs are used throughout to illustrate and clarify important topics. Numerous review questions and practice problems follow each chapter; answers to the practice problems are presented in Appendix G. SI metric as well as U.S. Customary units are used, since students and practitioners in the United States must still be familiar with both systems. A separate Instructor's Manual is available with worked-out solutions for the end-of-chapter practice problems and with supplementary problems that can be used for additional homework assignments or test questions.

The first chapter of the book provides an overview of environmental technology, including elements of public health, ecology, geology, and soils. The next nine chapters focus on water and wastewater topics, including hydraulics and hydrology, water quality and water pollution, drinking water treatment and distribution, sewage collection, sewage treatment and disposal, and stormwater management. Municipal solid waste, hazardous waste, air pollution, and noise pollution are covered in Chapters 11 through 14. Finally, appendixes covering environmental impact statements and audits; the employment of technicians; technologists, and engineers; basic mathematics; units and conversions; selected references; an extensive glossary; and a color photo insert (at the back of the book) are included.

There is more than ample material in this book for a typical one-semester course. Chapters 1 through 10 should suffice for introductory courses that focus mostly on water and wastewater topics. In courses where air quality, solid and hazardous waste, and noise pollution are also part of the syllabus, the instructor will probably find it necessary to be selective in coverage of topics from the first ten chapters to allow time for discussion and study of the last four chapters. In such circumstances, less time could be spent on the quantitative parts of the text (for example, hydraulics) and more time spent on the descriptive and qualitative aspects of environmental technology. Another option could be to focus in lectures on the first ten chapters for most of the semester, and allow students to select topics of special interest to them from those among the last four chapters for a term paper and/or oral presentation to the whole class. In this way, students get some exposure to those topics as well as practice in communication skills.

In this fourth edition, the text has been updated where necessary and some new topics have been added. These topics include nonuniform open channel flow, the

iv

rainfall severity index, mass balance, sewer plan and profile details, Geographic Information System (GIS) applications, description of software applications such as EPANET and HydroCAD, best management practices for stormwater quality control, and new water/wastewater treatment technologies. The book now includes a primer of basic mathematics as well as an expanded discussion of units and unit conversions in Appendix C. The number of case studies has been increased by about 50 percent, the number of relevant Web sites has been increased by about 30 percent, and many new terms have been added to the glossary in Appendix E.

This textbook addresses a wide range of environmental subjects. Every effort has been made to maintain a balance between thoroughness and practicality in covering the material to ensure that the book will continue to be a useful learning tool for students. The topics included here are covered in greater depth and detail in other, more narrowly specialized and advanced texts; they are presented here in a form that is more readily accessible to undergraduates and others who may have occasion to use the book. It is hoped that this book will motivate as well as prepare readers to study the discipline of environmental engineering or technology at a more advanced level.

## ACKNOWLEDGMENTS

For reviewing the manuscript of this fourth edition and for offering many helpful suggestions for its improvement, I would like to thank the following people: Alan B. Chace, Mohawk Valley Community College; Francis J. Hopcroft, Wentworth Institute of Technology; Jim Callison, Utah Valley State College; Douglas P. Macdonald, Florence–Darlington Technical College; and Ron Newton, Chemeketa Community College. I also would like to thank Robert St. Amand, Union County College, for his many helpful suggestions regarding the sections on fundamental concepts in chemistry and chemical parameters of water quality.

I am also indebted to many people for reviewing the manuscripts of one or more of the previous three editions of this book and for offering many helpful comments and suggestions regarding its content. I would like to thank them all here: Louis Chanin, United Water Resources; Leo Ebel, Washington University; Jerry Haimowitz, Boro of North Plainfield; Keith Hancock, Larimer County Vocational–Technical Center; Gayle Huges, Nashville State Technical Institute; Paul Klopping, Environmental Training Consultants, Inc.; Paul Mazur, Columbus Technical Institute; Andrew Potter, Monroe Community College; Karl Schnelle, Jr., Vanderbilt University; Paul Trotta, Northern Arizona University; Paul Cheremisinoff, New Jersey Institute of Technology; Roger Hlavek, Indiana University; Charles Ballou, Jr., Mohawk Valley Community College; Francis Hopcroft, Wentworth Institute of Technology; and Douglas Macdonald, Florence–Darlington Technical College.

I am grateful to Albert Mellini and Kevin Koch, Killam Associates, for their contributions to the chapters on municipal and hazardous waste management; David Fenster, URS Greiner Woodward Clyde, for his advice with the section on geology and soils; and Thomas Ombrello, Union County College, for his advice with the section on ecology. For help in preparing the color photo insert, I thank Russell Shallieu and Ken Zippler, Killam Associates; James Kircher, Public Works Journal Corporation; and Scott Edwards and Jim Force, USFilter. Finally, I would like to thank all those (too numerous to list) who provided many of the other photographs and diagrams used throughout the book.

I have tried to keep errors and inaccuracies in the text to a minimum. Of course, I remain fully responsible for any mistakes that may be found, and I welcome constructive comments and suggestions for the book's improvement from those who use it.

*Jerry A. Nathanson, P.E.*
Union County College
Cranford, New Jersey
nathanson@ucc.edu

# CONTENTS

## Chapter Outline

CHAPTER

1

# Basic Concepts

Environmental technology involves the application of engineering principles to the *planning, design, construction,* and *operation* of the following systems:

- Drinking water treatment and distribution
- Sewage disposal and water pollution control
- Stormwater drainage and control
- Solid and hazardous waste management
- Air and noise pollution control
- General community sanitation

The structures and facilities that serve these functions, including pipelines, pumping stations, treatment plants, and waste disposal sites, make up a major portion of society's *infrastructure*—the public and private works that allow human communities to thrive and function productively.

The practice of environmental technology encompasses two fundamental objectives:

1. *Public health protection* to help prevent the transmission of diseases among human beings.

1

**2.** *Environmental health protection* to preserve the quality of our natural surroundings, including water, land, air, vegetation, and wildlife.

Actually, there is considerable overlap of these two objectives because of the relationship between the quality of environmental conditions and the health and well-being of people. In fact, the terms *public health* and *environmental health* are often used synonymously.

Public health includes more than just the absence of illness. It is a condition of physical, mental, and social well-being and comfort. The cleanliness and esthetic quality of our surroundings—the atmosphere, rivers, lakes, forests, and meadows, as well as towns and cities—have a direct impact on this condition of human well-being and comfort, and *sanitation*, that is, the promotion of cleanliness, is a basic necessity in the effort to protect public and environmental health.

Environmental technology is usually considered to be a part of the *civil engineering* profession,* which has traditionally been called on to plan, design, build, and operate the facilities required for environmental health protection. Until fairly recently, this particular specialty field within civil engineering had several different names. It was also called

- Sanitary engineering
- Public health engineering
- Pollution control engineering
- Environmental health engineering

Whatever the profession is called, a knowledgeable and skilled team of engineers, technologists, and technicians is needed to accomplish its fundamental objectives.

Environmental technology is an *interdisciplinary field* because it encompasses several different technical subjects. In addition to such traditional civil engineering topics as hydraulics and hydrology, these include biology, ecology, geology, chemistry, and others. This variety makes the field interesting and challenging.

Fortunately, it is not necessary to be an expert in all these subjects to understand and apply the basic principles of environmental technology. This particular text has been designed so that a student with little academic background in some or all of the supporting subjects can still use it productively.

---

*Visit the Web site of the American Society of Civil Engineers at http://www.asce.org.

This chapter is a review of basic and pertinent topics in public health, ecology, and geology. Practical hydraulics is covered in Chapter 2 and the fundamentals of hydrology are presented in Chapter 3. The essential concepts and terminology from chemistry and microbiology are presented in sections of Chapter 4, on water quality. The remaining chapters of the book build on these subjects by presenting principles and applications of environmental technology. Each chapter includes a list of relevant Web sites where the student can find additional and timely information.

## 1.1   OVERVIEW OF ENVIRONMENTAL TECHNOLOGY

Before beginning a study of the many different topics that make up environmental technology, it would be helpful to have an understanding of the overall goals, problems, and alternative solutions available to practitioners in this field.

To present an overview of such a broad subject, we can consider an engineering project involving the subdivision and development of a tract of land into a new community, which will include residential, commercial, and industrial centers. Whether the project owner is a governmental agency or a private developer, a wide spectrum of environmental problems will have to be considered and solved before construction of the new community can begin. Usually, the project owner retains the services of an independent environmental consulting firm to address these problems. (See Case Study on page 5.)

### Water Supply

One of the first problems project developers and consultants must consider is the provision of a *potable* water supply, one that is clean, wholesome, safe to drink, and available in adequate quantities to meet the anticipated demand in the new community. Some of the questions that must be answered are as follows:

1. Is there an existing public water system nearby with the capacity to connect with and serve the new development? If not,
2. Is it best to build a new centralized treatment and distribution system for the whole community, or would it be better to use individual well supplies? If a centralized treatment facility is selected,
3. What types of water treatment processes will be required to meet federal and state

drinking water standards? (Water from a river or a lake usually requires more extensive treatment than groundwater does, to remove suspended particles and bacteria.) Once the source and treatment processes are selected,

4. What would be the optimum hydraulic design of the storage, pumping, and distribution network to ensure that sufficient quantities of water can be delivered to consumers at adequate pressures?

Illustrating the importance of water supply in new community development and environmental planning is a new California law (implemented in October 2001) which forces builders to prove that there will be adequate water to supply their new developments. This law imposes strict requirements for cities and counties when issuing permits for new subdivisions of 500 or more homes. The local water agencies must verify that water quantities are ample enough to serve the project for at least 20 years, including periods of drought. California is the first state to pass such strict legislation linking new development to water supply.

## Sewage Disposal and Water Pollution Control

When running water is delivered into individual homes and businesses, there is an obvious need to provide for the disposal of the used water, or *sewage*. Sewage contains human wastes, wash water, and dishwater, as well as a variety of chemicals if it comes from an industrial or commercial area. It also carries microorganisms that may cause disease and organic material that can damage lakes and streams as it decomposes.

It will be necessary to provide the new community with a means for safely disposing of the sewage, to prevent water pollution and to protect public and environmental health. Some of the technical questions that will have to be addressed include the following:

1. Is there a nearby municipal sewerage system with the capacity to handle the additional flow from the new community? If not,

2. Are the local geological conditions suitable for on-site subsurface disposal of the wastewater (usually *septic systems*), or is it necessary to provide a centralized sewage treatment plant for the new community and to discharge the treated sewage to a nearby stream? If treatment and surface discharge are required,

3. What is the required degree or level of wastewater treatment to prevent water pollution? Will a *secondary* treatment level, which removes at least 85 percent of biodegradable pollutants, be adequate? Or will some form of advanced treatment be required to meet federal and state discharge standards and stream quality criteria? (Some advanced treatment facilities can remove more than 99 percent of the pollutants.)

4. Is the flow of industrial wastewater an important factor?

5. Is it possible to use some type of *land disposal* of the treated sewage, such as spray irrigation, instead of discharging the flow into a stream?

6. What methods will be used to treat and dispose of the *sludge*, or *biosolids*, that is removed from the wastewater?

7. What is the optimum layout and hydraulic design of a sewage collection system that will convey the wastewater to the central treatment facility with a minimum need for pumping?

## Stormwater Management

The development of land for human occupancy and use tends to increase the volume and rate of stormwater runoff from rain or melting snow. Basically, this is due to the construction of roads, pavements, or other impervious surfaces, which prevent the water from seeping into the ground. The increase in surface runoff may cause flooding, soil erosion, and water pollution problems both on the site and downstream. The following are some of the questions the developer and consultant have to consider:

1. What is the optimum layout and hydraulic design of a surface drainage system that will prevent local flooding during wet weather periods?

2. What intensity and duration of storm would the system be designed to handle without *surcharging*, or overflowing?

3. Do local municipal land-use ordinances call for facilities that keep postconstruction runoff rates equal to or less than the amount of runoff from the undeveloped land? If so,

4. What are the "best management practices" (BMP) for reducing the peak runoff flows and protect water quality during wet weather periods?

5. What provisions can be made, during and after construction, to minimize problems related to soil erosion from runoff?

6. What is the best way to manage combined sewer overflows (CSOs) in older sewer systems?

## Solid and Hazardous Waste Management

The development of a new community (or growth of an existing community) will certainly lead to the generation of more municipal refuse and industrial waste materials. Ordinarily, the collection and disposal of solid wastes is a responsibility of the local municipality. However, some of the wastes from industrial sources may be particularly dangerous, requiring special handling and disposal methods.

There is a definite relationship between public and environmental health and the proper handling and disposal of solid wastes. Improper garbage disposal practices can lead to the spread of diseases such as *typhus* and *plague* due to the breeding of rats and flies.

If municipal refuse is improperly disposed of on land in a "garbage dump," it is also very likely that surface and groundwater resources will be polluted with *leachate* (leachate is a contaminated liquid that seeps through the pile of refuse into nearby streams as well as into the ground). On the other hand, incineration of the refuse may cause significant air pollution problems if proper controls are not applied or are ineffective.

Hazardous wastes, such as poisonous or ignitable chemicals from industrial processes, must receive special attention with respect to storage, collection, transport, treatment, and final disposal. This is particularly necessary to protect the quality of groundwater, which is the source of water supply for about half the population in the United States. In recent years, an increasing number of water supply wells have been found to be contaminated with synthetic organic chemicals, many of which are thought to cause cancer and other illnesses in humans. Improper disposal of these hazardous materials, usually by illegal burial in the ground, is the cause of the contamination.

Some of the general questions related to the disposal of solid and hazardous wastes from the new community include the following:

1. Is there a *materials recycling facility* (MRF, or "murf") serving the area? What will be the waste storage, collection, and recycling requirements (for example, will source separation of household refuse be necessary)?

2. Will a waste processing facility (such as one that provides for shredding, pulverizing, baling, composting, or incineration) be needed to reduce the waste volume and improve its handling characteristics?

3. Is there a suitable *sanitary landfill* serving the area, and will it have sufficient capacity to handle the increased amounts of solid waste for a reasonable period of time? (Despite the best efforts to recycle solid waste or reduce its volume, some material will require final disposal in the ground in an environmentally sound manner.) If not,

4. Is there a suitable site for construction and operation of a new landfill to serve the area? (A modern sanitary landfill site must meet strict requirements with respect to topography, geology, hydrology, and other environmental conditions.)

5. Will commercial or industrial establishments be generating hazardous waste, and, if so, what provisions must be made to collect, transport, and process that material? Is there a *secure landfill* for final disposal available, or must a new one be constructed to serve the area?

## Air and Noise Pollution Control

Major sources of air pollution include fuel combustion for power generation, certain industrial and manufacturing processes, and automotive traffic. Project developers can exercise the most control over traffic. Private industry will have to apply appropriate air pollution control technology at individual facilities to meet federal and state standards.

The volume of traffic in the area will obviously increase, leading to an increase in exhaust fumes from cars and other vehicles. Proper layout of roads and traffic-flow patterns, however, can minimize the amount of stop-and-go traffic, thus reducing the amount of air pollution in the development.

Usually, the developer's consultant will have to prepare an *environmental impact statement (EIS)*, which will describe the traffic plan and estimate the expected levels of air pollutants. It will have to be shown that air quality standards will not be violated, for the project to gain approval from regulatory agencies. (In addition to air pollution, the completed EIS will address all other environmental effects related to the proposed project.)

Noise can be considered to be a type of air pollution in the form of waste energy—sound vibrations. Noise pollution will result from the construction activity, causing a temporary or *short-term impact*. The builders may have to observe limitations on the types of construction equipment and the hours of operation to minimize this negative effect on the environment. A *long-term impact* with respect to the generation of noise will be caused by the increased amount of vehicular traffic. This is another environmental factor that the consultants will have to address in the EIS.

## Other Environmental Factors

Not to be overlooked as an environmental factor in any land development project is the potential impact on local vegetation and wildlife. The destruction of woodlands and meadows to make room for new buildings and roads can lead to significant ecological problems, particularly if there are any rare or endangered species in the area. Cutting down trees and paving over meadows can cause short-term impacts related to soil erosion and stream sedimentation. On a long-term basis, it will cause the displacement of wildlife to other suitable habitats, presuming, of course, that such habitats are available nearby. Otherwise, several species may disappear from the area entirely.

Human activity in wetland areas, including marshes and swamps, can be very damaging to the environment. Coastal wetlands are habitats for many different species of organisms, and the tremendous biological productivity of these wetland environments is a very important factor in the food chain for many animals. When wetlands are drained, filled in, or dredged for building and land development projects, the life cycle of many organisms is disrupted. Many species may be destroyed as a result of habitat loss or loss of a staple food source. Wetlands also play important roles in filtering and cleansing water and in serving as a reservoir for floodwaters. There is a definite need to control or restrict construction activities in wetland environments and to implement a nationwide wetlands protection program.

Environmental concerns related to general sanitation in a new community include food and beverage protection, insect and rodent control, radiological health protection, industrial hygiene and occupational safety, and the cleanliness of recreation areas such as public swimming pools. These concerns are generally the responsibility of local health departments.

CASE STUDY

### Development of a Master-Planned Community

Anthem Community Park, one of the largest *master-planned communities* in Maricopa County, Arizona,

is undergoing development on approximately 2400 hectares (ha) [5800 acres (ac)] located north of Phoenix. Zoning densities on the property allow for the construction of approximately 14,000 residential units, with about 240 ha (600 ac) set aside for mixed commercial uses. The year 2001 population of 2500 residents is expected to reach its ultimate design population of 30,000 residents in 10 years.

Existing and planned features for the expanding Anthem community include school sites, a community center, two golf courses, a water park, single family and multifamily housing, as well as mixed commercial uses. The planned Anthem community is a good example of a project for which the developer must consider a wide range of environmental factors; this case study will focus only on the water supply and wastewater effluent systems.

As part of the engineering plans for this project, a consulting engineering firm has been hired by the developer to construct computer models of Anthem's water supply and wastewater systems. The initial purpose of the computer models was to establish design parameters and construction phasing for the community's future infrastructure. However, in addition to use as a planning tool, the models also serve to maintain, operate, and update the existing system on an ongoing basis. The computer modeling software is used to analyze the existing water system (made up of over 550 pipes), predict future system characteristics, and design the most efficient layout to meet interim and future needs for the Anthem community. (Computer modeling software applications are discussed in more detail in later chapters.)

The growing Anthem community must meet the guidelines of the Arizona Department of Water Resources, which requires that surface water be used to provide for any new development and that a 100-year water supply be assured. Groundwater cannot be used as the sole source of water in the Phoenix Active Management Area in which the project is situated due to overpumping of the aquifers within the area. Wells can be used, but the volume of groundwater withdrawn must be equal to or less than the recharge volume. (Surface water, groundwater, and wells are topics covered in more detail in Chapter 3.) Since there is no permanent source of surface water supply at Anthem, it was necessary for the developer to obtain an assured 100-year supply from Lake Pleasant on the Central Arizona Project (CAP) canal, a long distance away. A 750-mm (30-in.)-diameter ductile iron pipeline more than 13 km (8 mi) long was built to transport CAP water to the Anthem community. (Water transmission and distribution topics are discussed in Chapter 7.)

The task of providing water to the growing community is further complicated by the fact that the Anthem property is located in two different governmental jurisdictions. On the west side, it is within the Phoenix city limits, and on the east, it is in Maricopa County. Each of these political entities has different engineering criteria for planning and design. The public infrastructure designed for Anthem must meet the design criteria for both jurisdictions.

The required fire flows vary in the community; a fire flow of 1500 gallons per minute (gpm) is required in residential areas and a fire flow of 3000 gpm is required in all commercial areas.

Minimum and maximum water pressures in the distribution system also vary. These and other variables are used in the computer model of the system, which is analyzed to ensure that the minimum and maximum pressures are maintained under all water demand scenarios, and that maximum flow velocities are not exceeded. Analyses are performed for average day, maximum day, and peak hour demand conditions. The system model is also analyzed for different fire flow alternatives. The ability to analyze many alternatives or scenarios with one hydraulic model is a key benefit provided by the computer software. (A *scenario* refers to a model run for a given set of water demand and system operating conditions, which are stored as various alternative datasets in the computer. The alternative datasets can be reused in many scenarios.)

Wastewater is collected and treated to allow the reuse of the effluent for irrigation of landscaping in roadway medians, community parks, and golf courses. Treatment processes include rotary drum screens as well as biological purification and microfiltration. (Wastewater treatment is discussed in Chapter 10.) Effluent (treated wastewater) in excess of irrigation needs will be allowed to percolate into the groundwater aquifer, using a network of recharge trenches. In the initial stages of the project, recycled wastewater quantities will not be sufficient to meet irrigation needs; CAP canal water will be purchased to meet the balance of those needs, and will also be stored for emergencies at the recharge facility. Raw CAP water, potable water, sewage, and treated effluent will be managed by using automated radio telemetry systems to optimize the eventual total reuse of treated wastewater in the planned community.

## Environmental Interrelationships

In the preceding overview of environmental technology, we have briefly considered many factors that are very much interrelated and overlapping, as illustrated in Figure 1.1. In a textbook, it is necessary to organize these factors into chapters and sections. But this is only for academic convenience. The interrelationships should always be kept in mind. Water, land, and air pollution are part of a single problem.

Sometimes, due to unanticipated interrelationships and overlaps, a solution of one environmental problem inadvertently causes a different problem to arise. For example, the use of catalytic converters since the mid-1970s to reduce smog caused by automobile exhaust gases has been found to contribute to a different air pollution problem—*global warming* (or the "greenhouse effect"). Catalytic converters can form significant quantities of nitrous oxide ("laughing gas"), which is a potent gas that can trap heat energy and warm the atmosphere. (The greenhouse effect and atmospheric warming are discussed in more detail later, in Section 13.4, Global Air Pollution.)

Another example involves the contamination of groundwater and surface water in some cities by MTBE (methyl tertiary butyl ether), an organic chemical added to gasoline to reduce air pollution. MTBE has been used as a fuel additive since the early 1990s to increase gasoline octane levels and help reduce carbon monoxide and ozone concentrations in the air. It can contaminate water sources, largely as a result of leaking underground

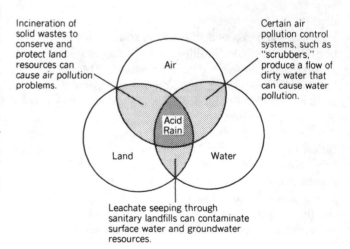

**FIGURE 1.1**

*Most environmental problems pertaining to air, water, and land quality are interrelated. A problem called* acid rain, *for example, is caused by air pollution, and it damages both aquatic and terrestrial ecosystems.*

storage tanks (see Section 12.3) and the use of motorized watercraft on lakes and reservoirs. MTBE may be a *carcinogen* (cancer-causing agent), and it can give water a bad taste and odor even at low levels; scientific research is underway to further understand its adverse health effects and to find effective methods to remove it from contaminated water sources.

As more is learned in the future about the potential interrelationships among environmental phenomena, engineers and technologists will be better able to create pollution control systems that will not have any unexpected harmful effects on other components of the environment, and will be able to avoid situations like the foregoing.

## 1.2   PUBLIC HEALTH

Preventing the spread of disease and thereby protecting the health of human populations is a fundamental goal of environmental technology. Public health protection is, of course, a primary concern of doctors and other medical professionals. But engineering technology also plays a significant role in this effort. In fact, the high standard of health enjoyed by citizens of the United States and other developed nations is largely due to the construction and operation of modern water treatment and pollution control systems. The spread of diseases in countries with inadequate sanitary facilities is a major problem for millions of people.

Diseases are classified into two broad groups: *communicable diseases* and *noninfectious diseases*. Communicable diseases are those that can be transmitted from person to person, commonly referred to as being infectious or contagious. Noninfectious diseases, as the name implies, are not contagious; they cannot be transmitted from one person to another by any means. The kinds of noninfectious diseases of concern in environmental technology are associated with contaminated water, air, or food. The contaminants are usually toxic chemicals from industrial sources, although biological toxins can also cause disease.

### Communicable Diseases

Communicable diseases are usually caused by *microbes*. These microscopic organisms include bacteria, protozoa, and viruses (see Section 4.4). Most microbes are essential components of our environment and do not cause disease. Those that do are called pathogenic organisms, or simply *pathogens*.

The ways in which diseases are spread from one person to another vary considerably. They are called

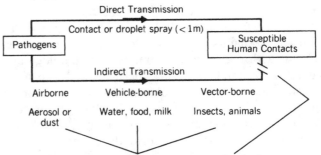

Modes of Disease Transmission

FIGURE 1.2
*Communicable diseases are spread in several ways, many of which can be controlled or intercepted by applications of modern environmental technology.*

*modes of transmission* of disease and are summarized in Figure 1.2. It is important to make distinctions among the various modes of transmission to be able to apply suitable methods of control. *Direct transmission* involves an immediate transfer of pathogens from a carrier (infected person) to a susceptible contact, that is, a person who has had direct contact with the carrier and is liable to acquire the disease. Clearly, control of this mode of transmission is not within the scope of environmental technology; it is in the province of personal hygiene and the medical profession (who provide immunization and quarantine infected persons).

Environmental technology can be applied to intercept many of the modes of *indirect transmission*. The three indirect modes of disease transmission are *airborne*, *vector-borne*, and *vehicle-borne*. Airborne transmission involves the spread of microbes from carrier to contact in contaminated mists or dust particles suspended in air. It is the least common of the indirect modes. (This should not be confused with the noninfectious public health problems associated with chemical air pollution, which will be discussed later.)

*Vectors* of disease include insects, rodents, and other animals that can transport pathogens to susceptible human contacts. The animals that carry the pathogenic microbes are also called *intermediate hosts* if the microbes have to develop and grow in the vector's body before becoming infective to humans. Vector-borne disease can be controlled to some extent by proper sanitation measures.

A *vehicle* of disease transmission is any nonliving object or substance that is contaminated with pathogens. For example, forks and spoons, handkerchiefs, soiled clothes, or even children's toys are potential vehicles of transmission. They can physically transport and transfer the pathogens from carrier to contact.

Water, food, and milk are also potential vehicles of disease transmission; these are perhaps the most significant with regard to environmental technology and sanitation. Water, in particular, plays a major role in the transmission of communicable diseases, but it is most amenable to engineering and technological controls. Water and wastewater treatment facilities effectively block the pathway of waterborne diseases.

## Types of Communicable Diseases

Waterborne and food-borne diseases are perhaps the most preventable types of communicable diseases. The application of basic sanitary principles and environmental technology have virtually eliminated serious outbreaks of these diseases in technologically developed countries.

Water- and food-borne diseases are also called *intestinal diseases* because they affect the intestinal tract of humans. The pathogens are excreted in the feces of infected people. If these pathogens are inadvertently ingested by others in contaminated food or water, the cycle of disease can continue, possibly in *epidemic* proportions, that is, when the number of occurrences of a disease in a community is far above normal.

Symptoms of intestinal disease include diarrhea, vomiting, nausea, and fever. Intestinal diseases can incapacitate large numbers of people in an epidemic and sometimes result in the deaths of many infected individuals. Water contaminated with untreated sewage (domestic wastewater) is generally the most common cause of this type of disease.

The most prevalent waterborne diseases include *typhoid fever, dysentery, cholera, infectious hepatitis,* and *gastroenteritis* (common diarrhea and cramps). These can also be transmitted by contaminated food or milk products. Diseases caused by bacterial toxins include *botulism* and *Staphylococcus* food poisoning. Refrigeration, as well as proper cooking and sanitation at food-processing facilities and restaurants, are important for control of these food-borne diseases.

Although cholera and dysentery have not generally been a problem in the United States, they are prevalent diseases in India and Pakistan and in many of the technologically underdeveloped countries of southeast Asia. In fact, they are considered to be *endemic* (habitually present) in these areas. Typhoid fever is more common in occurrence than cholera or dysentery. Until the beginning of the 1900s, typhoid mortality rates in some urban areas of the United States were as high as 650 deaths per 100,000 population. The beginning of mod-

ern water purification technology at about that time helped to lower the typhoid death rate to considerably less than 1 per 100,000 people per year. (Immunization and improvements in food and milk sanitation also played a role in reducing the incidence of typhoid.)

Amoebic dysentery, caused by a single-cell microscopic animal called an amoeba, occurred in epidemic proportions in Chicago during the early 1930s. About 100 of the approximately 1000 people who contracted the disease died from it. The cause of this epidemic was traced to sewage that contaminated the water supplies of two hotels in the city. Although epidemics of intestinal disease like this one are not at all common in the United States, when they do occur they are usually very localized and can be traced to contaminated water supplies in hotels, restaurants, schools, or camps. Generally, the contamination is caused by *cross-connections* in the water distribution system, which may allow backflow of wastewater into the drinking water supply.

*Giardiasis* and *cryptosporidosis* are two waterborne diseases that can cause gastrointestinal illness and serious public health problems. They are both caused by single-celled microscopic animals called *protozoa* (see Section 4.4 for a discussion of microorganisms) that can contaminate drinking water supplies. A very large outbreak of cryptosporidosis, for example, occurred in Milwaukee, Wisconsin, in 1993. The city's water supply comes from Lake Michigan. An unusual combination of circumstances during a period of heavy rainfall and runoff allowed the protozoan *Cryptosporidium* to pass through the water treatment plant. More than 40,000 people became ill, about 4000 people were hospitalized, and more than 50 deaths were attributed to this outbreak. The original source of the contamination is uncertain. Since that incident, improved water quality standards and treatment rules make a repetition of this type of outbreak unlikely.

Insect-borne diseases include those transmitted by the bites of mosquitoes, lice, and ticks. *Malaria, yellow fever,* and *encephalitis* are typical diseases spread by certain species of mosquitoes. Flies also transmit disease, but not by biting; the contact of their germ-laden bodies, wings, and legs with food consumed by humans spreads diseases such as typhoid fever and gastroenteritis.

The elimination of the breeding places of insects is one of the most important control measures. Proper garbage disposal reduces fly breeding places, and elimination of standing water is one of the methods available for eliminating mosquito breeding areas. Chemical control with insecticides is usually a last resort because of the environmental and potential health problems associated with the use of toxic substances.

In addition to insects, other vectors of disease transmission are vertebrate animals such as dogs and rats. Rabies is a familiar example of a disease spread by the bite of an infected dog or other mammal, but it is not generally related to environmental conditions. Rodent-borne diseases, such as *typhus* and *bubonic plague*, are more readily controlled by applications of environmental technology. Rat populations can be controlled by good community sanitation practices; rodent access to garbage and water should be prevented. Modern building codes include specifications for rodent-proof building construction.

## Noninfectious Diseases

It is a well-documented fact that the overall death rate for people residing in heavily polluted urban areas is significantly higher than the mortality rate in areas that are relatively pollution-free. This is not necessarily because of the incidence of sewage pollution and the spread of infectious diseases. In fact, many current public health problems related to environmental pollution are considered to be the result of contamination of water, food, and air with toxic chemicals. The resulting diseases are noninfectious.

Some noninfectious illnesses associated with toxic chemical pollution have a relatively sudden and severe onset, and the acute or immediate health effects can be readily traced to a specific contaminant. A group of substances known as the *heavy metals* is particularly notorious in this regard. Other noninfectious diseases may take years to develop and can involve chronic or long-lasting health problems. Generally, various synthetic organic substances cause this type of problem, even in extremely small concentrations. Some organics are considered to be carcinogenic, having the potential to cause cancer in humans.

Lead is one of the heavy metals involved in noninfectious disease. The public health problems related to lead poisoning have long been associated primarily with ingestion by children of peeling lead-based paint. Lead poisoning can lead to blindness, kidney disease, and mental retardation (particularly in children).

The evidence against lead as a dangerous environmental pollutant is overwhelming. It is a cumulative poison; that is, it accumulates in human tissue and can build up to toxic levels over time. As a result, environmental agencies in Europe and the United States have banned the use of lead additives in gasoline.

Mercury is another heavy metal associated with environmental pollution and noninfectious illness. It was first noted as such when it afflicted large numbers of people living in the Minamata Bay region of Japan in the 1950s. Mercury compounds, discharged into the bay in wastewater from a local factory, were ingested by people who ate contaminated fish. A severe epidemic of disease, resulting in blindness, paralysis, and many deaths, was the result. Less severe symptoms included hand tremors, irritability, and depression.

At the time of the Minamata Bay incident, mercury vapor was known to be harmful, although metallic mercury itself was not considered hazardous (it has long been used in dental fillings). Research after the poisoning episode in Japan, however, led to the discovery that certain microorganisms can cause the metallic mercury to combine with other substances in the water, forming harmful mercury compounds, such as *methylmercury*. This substance was ingested by microscopic organisms in the water, called plankton, and entered the food chain. People who ate the contaminated fish were made ill by the methylmercury.

The episode of mercury poisoning in Japan is one example of a relatively sudden and acute illness related to environmental pollution. The concentration of the pollutant was relatively high and the harmful effects were noticed within a short time. Questions remain as to the chronic or long-term effects of lower concentrations of mercury compounds. It is common to detect small amounts of mercury in fish and wildlife even in rivers and lakes far from industrial centers.

Discarded batteries and dry cells are a major source of mercury. This is becoming a very serious problem due to the difficulty in properly disposing of the many batteries generated by the growing electronics industry and the use of calculators, cameras, portable CD players, and watches.

Unfortunately, mercury and lead are not the only harmful chemical substances that become environmental pollutants when poorly managed or controlled. For example, the pesticide *Kepone* has seriously polluted the James River and Chesapeake Bay. The Hudson River is known to be contaminated with the toxic industrial chemical *PCB* (polychlorinated biphenyl) (see page 16). This oily substance was widely used in electrical transformer fluids, coolants, paints, and other products. It persists in the environment because it is nonbiodegradable; that is, it does not readily decompose and dissipate by natural processes. PCBs have accumulated in the bottom deposits of rivers, and many species of fish are contaminated with them.

Like the pesticide DDT, PCB has been banned from manufacture and most uses in the United States, but because these substances are extremely persistent in the environment, they remain potential dangers to public health for many years after their initial discharge. Traces of DDT and PCB are still found in the body tissues of animals far removed from the sources of pollution. Both of these chemicals are considered potential human carcinogens.

Environmental pollution with harmful chemicals and the resulting incidences of noninfectious disease are part of a problem now commonly referred to as *hazardous waste disposal*. This will be discussed in more detail in Chapter 12.

Perhaps one of the most publicized environmental disasters in the United States that was related to improper disposal of hazardous wastes occurred in the late 1970s at Love Canal in Niagara Falls, New York. Waste chemicals in steel drums were buried in the unused canal over a period of several years. The land was sold and many homes were built on top of the site. Eventually, the chemicals leaked out of the drums and into the soil, water, and air in the vicinity of the old dump site. Soon it was evident that residents in the area of Love Canal were suffering from unusually high rates of cancer, miscarriage, birth defects, kidney disease, and other illnesses. In incidents like this, it is difficult to tie a particular chemical to a specific health problem, but the fact that the noninfectious illnesses suffered by the residents of the area were associated with environmental contamination with chemical wastes is beyond question. Research is being conducted by many universities and governmental agencies to determine some of the long-term effects of heavy metals and synthetic organic chemicals on human health.

Finally, several noninfectious diseases are specifically associated with air pollution. Air pollution and its control is discussed in Chapter 13. Briefly, common diseases related to air pollution include *bronchial asthma*, *bronchitis*, and *emphysema*. *Lung cancer* also occurs more frequently among people who live in congested industrial and urban areas, and poor air quality is considered to play a role in this. Again, it is difficult to prove a direct cause-and-effect relationship between a specific pollutant and these illnesses, but the overall negative effect of dirty air on public health is obvious: The incidence of respiratory ailments and increased mortality rates are directly related to the severity of air pollution.

## 1.3 ECOLOGY

Ecology is the branch of biological science concerned with the relationships and interactions between living organisms and their physical surroundings or environment. Living organisms and the environment with which they exchange materials and energy together make up an *ecosystem*, which is the basic unit of ecology. An ecosystem includes *biotic components*—the living plants and animals—and *abiotic components*—the air, water, minerals, and soil that constitute the environ-

ment. A third and essential component of most natural ecosystems is *energy*, usually in the form of sunlight.

Familiar examples of land-based, or *terrestrial*, ecosystems include forests, deserts, jungles, and meadows. Water-based, or *aquatic*, ecosystems include streams, rivers, lakes, marshes, and estuaries. There is no specific limitation on the size or boundaries of an ecosystem. A small pond can be studied as a separate ecosystem, as can a desert comprising hundreds of square kilometers. Even the entire surface of Earth can be viewed as an ecosystem; the term *biosphere* is often used in this context.

If Earth is imagined to be about the size of an apple, then the layer of air surrounding it would not be much thicker than the skin of that apple. This thin envelope of air and the shallow crust of land and water just beneath it provide the abiotic components that support life in the biosphere. It is a *closed* ecosystem because there is essentially no transfer of material into or out of it. Only the constant flow of energy from the sun provides power to sustain the life cycles within the biosphere. Nutrients are continually recycled and reused.

The biosphere seems so big that it is sometimes difficult to believe that humans can affect or disrupt its natural balances. But global problems related to environmental pollution, such as *acid rain*, the *ozone hole*, and the *greenhouse effect*, are significant and must be controlled before irreversible environmental changes occur. These and other pollution problems are discussed later in the text.

In addition to natural ecosystems, such as lakes or forests, several types of artificial ecosystems are of particular importance in environmental technology. For example, one of the most common methods of wastewater treatment is based on a biological system called the *activated sludge process*. This is an *engineered* ecosystem comprising a steel or concrete tank, a suspended population of microorganisms in wastewater, and a constant input of air. The microbes are the biotic component; the tank, wastewater, and air are the abiotic components. The system removes organic pollutants from the wastewater. This method is discussed in more detail in Chapter 10.

### Food Chains and Metabolism

There are two basic principles or *laws* of ecology: they involve the *oneway flow of energy* and the *circulation of materials*.

Energy is the capacity to do work. It can be transformed from one form to another, such as from mechanical to electrical energy or from energy in the form of sunlight to potential energy stored in food molecules,

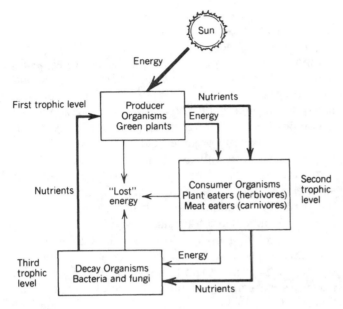

**FIGURE 1.3**
*Simplified diagram of a* food chain. *Nutrients are recycled, but energy must be continuously supplied by the sun. The efficiency of energy transfer from one trophic level to the next is less than 10 percent.*

but it cannot be created or destroyed. No energy transformation is 100 percent efficient; some is always lost to the environment. Because of this, energy cannot be recycled in an ecosystem; it can only flow one way.

On the other hand, nutrient materials needed to sustain life can be reused over and over again. They are constantly recycled or circulated through the ecosystem. The one-way flow of energy and the circulation of nutrients is illustrated in Figure 1.3. This is a very simplified diagram of a *food chain*, showing three broad groups or types of organisms: the *producers*, the *consumers*, and the *decomposers*.

The biological and chemical process by which an organism sustains its life is called *metabolism*. Two fundamental metabolic processes of living organisms are *photosynthesis* and *respiration*, which will be discussed shortly. Living organisms require energy and, as shown in Figure 1.3, the original or primary source of energy for all natural ecosystems is the sun.

In addition to energy, living organisms need certain chemicals from the environment, called *nutrients*, in sufficient quantities. All organisms need water, and most require gaseous oxygen. In addition, plants and animals require carbon, hydrogen, phosphorus, potassium, iodine, nitrogen, sulfur, calcium, iron, and magnesium, as well as other elements in smaller amounts. For animals, some of these elements must be in the form of organic molecules, such as carbohydrates or

proteins. (A brief review of basic inorganic and organic chemistry is given in Chapter 4.)

**Photosynthesis and Respiration**

The food chain shown schematically in Figure 1.3 begins with what ecologists call the first *trophic level* of organisms—the producers. These are the green plants. Green plants are *autotrophic*, which simply means that they are self-nourishing. They have the unique ability to convert carbon dioxide, water, and some basic nutrients into organic compounds that store the sun's energy.

This natural process, called *photosynthesis*, is illustrated in Figure 1.4. The plants utilize solar energy to form carbohydrates from carbon dioxide and water. The carbohydrates can also combine with nitrogen, phosphorus, sulfur, and other elements, forming other organic compounds that are the building blocks of living organisms. *Chlorophyll*, the pigment that gives plants their characteristic green color, plays a key role in trapping solar energy and converting it into chemical energy. A portion of the energy-rich organic compounds stored in the plant tissue are available for use by other organisms that consume the plants at the next trophic level.

During the process of photosynthesis, gaseous oxygen is released into the atmosphere. Oxygen is essential for the metabolism of the next trophic level in the food chain—the consumers. Actually, the consumer organisms include several intermediate trophic levels, including the *herbivores*, the *carnivores*, and the *omnivores*. Herbivores are plant-eating animals, carnivores are meat-eaters, and omnivores eat both plants and animals.

The consumer organisms are *heterotrophic*. Unlike the autotrophic plants, which manufacture their own food from simple inorganic chemicals, the herbivores must utilize the energy-rich compounds synthesized by the plants. In turn, the carnivores obtain energy for their metabolism when they consume the herbivores. The process by which the consumers obtain energy from the

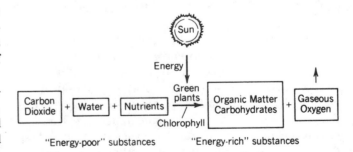

**FIGURE 1.4**
*Schematic diagram of* photosynthesis. *Energy from the sun is stored in organic molecules and is available for use by the next trophic level.*

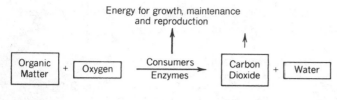

**FIGURE 1.5**

*Schematic diagram of respiration, the opposite of photosynthesis. Organic matter is metabolized or "burned," thereby releasing the stored energy for use by consumer organisms. Enzymes are chemicals that help the metabolism reactions occur in the living cell.*

organic material stored in plants and animals they eat is called *respiration*.

Respiration, illustrated in Figure 1.5, may be viewed as a process of slow combustion or *oxidation* of organic material, in which energy is released. Essentially, respiration is the opposite of photosynthesis. Photosynthesis builds energy-rich organic substances and gives off oxygen; respiration breaks down the organics and gives off carbon dioxide. Photosynthesis requires carbon dioxide, and respiration requires oxygen. This is one of the fundamental balances in nature.

The simplified food chain shown in Figure 1.3 is completed or closed by the decomposers, or *decay organisms*. These are primarily microscopic organisms, such as bacteria and fungi. During their own metabolism, microorganisms break down the waste products and the remains of dead organisms into simpler inorganic substances, which are then readily usable by the autotrophs. For example, nitrogen in ammonia is not available in plants as a nutrient until it is broken down and converted to inorganic nitrates by certain bacteria. The nitrates can be absorbed by the plants. Not only are decomposers essential for all natural ecosystems, they are the workhorses of engineered water pollution control systems.

## Aerobic and Anaerobic Decomposition

Decomposition that occurs in the presence of free oxygen is called *aerobic decomposition*, and the microorganisms that thrive in oxygen are called *aerobes*. Aerobic decomposition results in the oxidation of the carbon, hydrogen, sulfur, nitrogen, and phosphorus that are tied up in complex organic molecules. These elements become combined with oxygen, forming carbon dioxide, water, sulfates, nitrates, and other simple substances that can be taken up by green plants for photosynthesis. The energy released from the organic molecules in this process is used by the microbes for growth and reproduction. Aerobic decomposition is an efficient

and "clean" biochemical process and does not produce the offensive odors often associated with decay.

Certain species of microorganisms are able to decompose organic material in the absence of freely available oxygen. These organisms are called *anaerobes*, and the process is called *anaerobic decomposition*. As illustrated in Figure 1.6, the end products of anaerobic decomposition include methane, ammonia, hydrogen sulfide, and volatile organic acids, many of which are responsible for the unpleasant odors associated with *putrefaction* (the anaerobic decay of proteins). Hydrogen sulfide, with the chemical formula $H_2S$, causes the familiar rotten-egg odor. (See Section 4.1 for a review of chemical symbols and formulas.)

Anaerobic decomposition is an inefficient biochemical process. Although the anaerobes get energy from it for their growth and reproduction, the end products are still relatively unstable and can decompose further. In effect, anaerobic decay is similar to incomplete combustion. It plays a key role in some wastewater treatment processes. Methane, $CH_4$, one of the few odorless products of anaerobic decomposition, has a high enough energy value to be useful as a fuel; it is collected for that purpose at some sewage treatment plants and sanitary landfills (see page 367).

A type of anaerobic decomposition that is useful in producing certain foods and beverages is called *fermentation*, the decomposition of carbohydrates by microbes without free oxygen. Although it is used to produce cheese and alcohol, for example, some kinds of fermentation are not desirable, such as those that sour milk or produce acetic acid in wine.

## Biogeochemical Cycles

Although an ecosystem needs a constant source of energy from outside, the nutrients upon which life depends can be recycled indefinitely. The pathways in

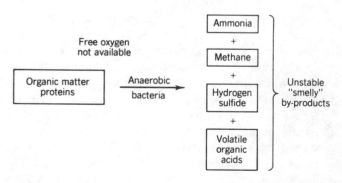

**FIGURE 1.6**

*Anaerobic decomposition of proteins in the absence of free or molecular oxygen is called* putrefaction.

which the chemical nutrients move through the biotic and abiotic components of the ecosystem are called *biogeochemical cycles* or *nutrient cycles.*

A cycle of particular concern in the field of environmental technology is the *hydrologic (water) cycle.* This will be discussed in some detail in Chapter 3. The important nutrient cycles considered here are the *carbon cycle,* the *nitrogen cycle,* and the *phosphorus cycle.*

Carbon, nitrogen, and phosphorus are considered to be among the *macronutrients* because they are needed in relatively large amounts in *protoplasm,* the fundamental substance of which a living cell is made. Other macronutrients essential to life include hydrogen, oxygen, potassium, calcium, magnesium, and sulfur. The many *micronutrients,* required only in very small quantities, include iron, manganese, copper, zinc, and sodium.

## Carbon Cycle

Carbon dioxide ($CO_2$) in the air and dissolved in water is the primary source of the element carbon. Through the process of photosynthesis, the carbon is removed from the $CO_2$ and incorporated with other chemical elements in complex organic molecules. The $CO_2$ eventually finds its way back into the atmosphere when the organics are broken down during respiration. A schematic diagram of this cycle is shown in Figure 1.7.

The combustion of fossil fuels (oil and gasoline) for energy is a human activity that increases the con-

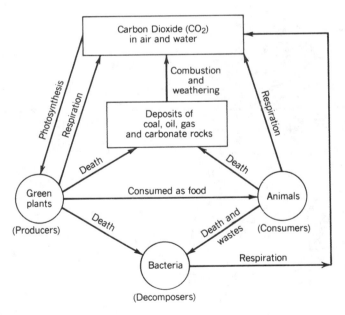

**FIGURE 1.7**
*Simplified diagram of the carbon cycle. The arrows show the various directions of carbon transfer through the biosphere.*

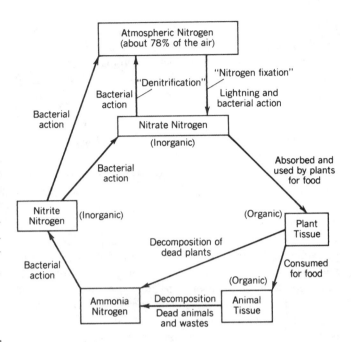

**FIGURE 1.8**
*Simplified diagram of the nitrogen cycle. Molecular nitrogen must first be* fixed *(combined with oxygen) into the form of nitrate nitrogen before it can be used by plants as a nutrient.*

centration of $CO_2$ in the atmosphere. Carbon dioxide plays a role in absorbing radiated heat and in regulating global atmospheric temperatures. A rise in $CO_2$ levels in the atmosphere will tend to cause the average temperature to increase. This problem is called the *greenhouse effect* (see Section 13.4).

## Nitrogen Cycle

About 78 percent of the atmosphere is nitrogen gas, $N_2$, but in this molecular form it is not active in biological systems. The nitrogen must first be *fixed* in the form of nitrates, $NO_3^-$, in which form it can be utilized by plants during photosynthesis. Eventually, it is combined with other substances and converted into proteins, consumed by heterotrophs, and broken down again in the process of decay. This cycle is illustrated in Figure 1.8. *Nitrification,* the process in which nitrogen in the form of ammonia, $NH_3$, is converted to nitrate nitrogen, is of particular significance in water pollution control (see page 105).

## Phosphorus Cycle

Phosphorus is another nutrient that plays a central role in aquatic ecosystems and water quality. Unlike carbon and nitrogen, which come primarily from the atmosphere, phosphorus occurs in large amounts as a mineral in phosphate rocks and enters the cycle from erosion and mining activities. This is the nutrient considered to be the main cause of excessive growth of rooted and free-floating microscopic plants in lakes (algal blooms).

## Stability, Diversity, and Succession

Each species of living organism occupies a particular habitat and serves a particular function in an ecosystem. The function and habitat constitute the organism's *ecological niche*. A basic characteristic of a healthy or well-balanced ecosystem is an overlapping of niches occupied by different species. The more complex the ecosystem is, in terms of the numbers and interrelationships among different species, the more stable it will be. A stable ecosystem can withstand some external stress, such as pollution, construction, or hunting, without being completely disrupted or damaged.

In a stable ecosystem, if any one species disappears because of natural or artificial causes, other species are available to occupy its niche and take over its role in the food chain. Actually, the term *food web* is more appropriate for a healthy ecosystem because of the overlapping nature and complexity of the eat-and-be-eaten-by relationships. A tropical rain forest is a good example of a stable ecosystem because of the tremendous number of plant and animal species thriving in it. The loss of one species of tree or one species of animal is not likely to have a significant impact on the whole ecosystem.

In an ecosystem with little diversity, that is, only a few different species of organisms, the situation is more unstable and susceptible to the effects of stress. The disappearance of a group of organisms from the food web is more likely to break the chain of trophic levels and severely disrupt the ecosystem. Diversity of species, then, provides a factor of safety or buffer against ecological disruptions by increasing the likelihood of adaptation to changing environmental conditions. *The greater the diversity of species, the healthier is the ecosystem.*

Although aquatic ecosystems such as streams and lakes are generally stable, they are sensitive to disruption from human activity. A diagram of an aquatic system is shown in Figure 1.9. Most desirable organisms, from the fish down to the microscopic plankton and bacteria (see Section 4.4), need oxygen to survive.

One effect of water pollution is the reduction of the dissolved oxygen level in the water. This type of pollution changes the ecological balance, favoring a smaller number of species of organisms that are tolerant of low oxygen levels. In heavily polluted water, only maggots and sludge worms may survive.

In studying the health or quality of a stream or lake, ecologists may use a formula to compute a *diversity index* for the ecosystem. In a field survey the

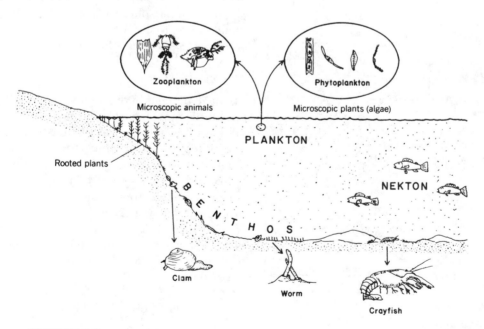

**FIGURE 1.9**

*An aquatic ecosystem showing the various biological components of a freshwater or marine habitat.* (Courtesy of the U.S. Environmental Protection Agency)

number of different species is counted and the population of each species is estimated by sampling limited areas. These data are used in the diversity index formula, and a single number or index is determined to characterize the condition of the ecosystem.

Generally, a low diversity index is indicative of a polluted ecosystem, and the pollution-tolerant species are readily identified. In a clean stream, for example, many different species of fish may be found, including trout. But in a polluted stream, only a few species of more tolerant organisms, such as catfish, may be found.

It is important to realize that even healthy or well-balanced ecosystems change over time in a process called *natural succession*. For example, a lake will eventually become shallower as silt and organic material accumulate in bottom sediments. As time goes on, the lake will eventually turn into a marsh and finally a meadow. These natural changes in a lake, called *eutrophication*, are discussed in more detail in Chapter 5. As shown there, this natural process can be affected by human activity and pollution.

Although the lake, marsh, and meadow may be stable and healthy ecosystems during their individual lifetimes, natural geological and biological processes will cause the succession from one stage to another. If geological and weather conditions are suitable, the process of natural succession will continue until a *climax stage* is reached. For example, the meadow, once a lake, will eventually become a hardwood forest in many temperate ecosystems. Natural succession, though, takes place over very long periods of time, and the changes are not ordinarily visible during a human life span.

## Biological Monitoring in Lakes and Streams

Small insects and other organisms that live on the bottom of streams and lakes form an important part of the aquatic food web. Ecologists call them *benthic* (which means "bottom-dwelling") *macroinvertebrates*. They are sensitive to many factors in their environment and are useful as indicators of the condition or "health" of streams and lakes. Routine macroinvertebrate monitoring or sampling (for example, about six times a year) can indicate problems that may not easily be detected by chemical testing, and can detect pollution problems that may no longer be evident in water samples (for example, from a chemical spill that washes downstream).

Macroinvertebrates depend on adequate water quality for survival. The time required for insect communities to return to their natural state after disturbances, such as those from point-source industrial pollutants, can be on the order of many years for streams

and decades for lakes. As a result, changes in their numbers and species can indicate pollution from various sources. Biological sampling and monitoring of these communities provides an effective method for determining if a watercourse has been impacted by pollution. (It is best to sample either in the spring, when late-stage larval forms are present, but have not yet begun their final maturation, or in late fall, after most species have mated.)

More than 4000 species of aquatic insects have been reported. These benthic macroinvertebrates are therefore a highly diverse group, which makes them excellent candidates for studies of changes in biodiversity. Changes in population numbers or behavior of these organisms can indicate that the physical or chemical conditions are outside their preferred limits. Also, the presence of numerous families of highly tolerant organisms usually indicates poor water quality.

## Biological Magnification

When a living organism cannot metabolize or excrete an ingested substance, that substance gradually accumulates in the organism. This phenomenon, called *biological accumulation* (or *bioaccumulation*), refers to the process by which a substance first enters a food chain. The extent to which bioaccumulation will occur depends on an organism's metabolism and on the solubility of the substance. If the substance is soluble in fat, it will typically accumulate in the fatty tissues of the organism. Bioaccumulation is of particular concern when the substance being concentrated is a toxic environmental pollutant and the organism is of a relatively low trophic level in a food chain.

When many contaminated organisms are consumed by a second organism that can neither metabolize nor excrete the substance, the concentration of the substance will build to even higher levels in the second organism. This effect is magnified at each successive trophic level, and the process is called *biological magnification* (or *biomagnification*). In other words, biomagnification is the steadily increasing concentration of a substance as it moves from one level of a food chain to the next (for example, from plankton to fish to birds or to humans). Biomagnification is of particular importance when chemicals are concentrated to harmful levels in organisms higher up in the food chain. Even very low concentrations of environmental pollutants can eventually find their way into organisms in high enough doses to cause serious problems.

Biomagnification occurs only when the pollutants are environmentally persistent (last a long time before breaking down into simpler compounds), mobile, and soluble in fats. If they are not persistent, they will not last long enough in the environment to be concentrated in the food chain. (Persistent substances are generally not biodegradable.) If they are not mobile, that is, not easily transported or moved from place to place in the environment, they are not likely to be consumed by many organisms. Finally, if they are soluble in water rather than fatty tissue, they are much more likely to be excreted by the organism before building up to dangerous levels.

## Impact of DDT

The incidence of mercury poisoning in people who consumed contaminated fish in the Minamata Bay region of Japan in the 1950s is just one example of the detrimental effects of biomagnification (see page 9). Another classic example involves *DDT*, an abbreviation for the organic chemical dichlorodiphenyltrichloroethane. It is a type of chemical known as a *chlorinated hydrocarbon*, and it takes a long time to break down in the environment. With a "half-life" of 15 years, if 10 kg of DDT were released into the environment in the year 2000, 5 kg would still persist in the year 2015, about 2.5 kg would remain in 2030, and even after 100 years had elapsed, in the year 2100, more than 100 g of the substance would still be detected in the environment. Of course, long before that time span elapsed, some of the DDT could be inadvertently consumed by living organisms as they forage for food, and thereby enter a food chain.

DDT is toxic to insects, but not very toxic to humans. It was much used in World War II to protect U.S. troops from tropical mosquito-borne malaria as well as to prevent the spread of lice and lice-borne disease among civilian populations in Europe. After the war, DDT was used to protect food crops from insects as well as to protect people from insect-borne disease. As one of the first of the modern pesticides, it was overused, and by the 1960s, the problems related to biomagnification of DDT became very apparent. Rachel Carson's book *Silent Spring*, published in 1962, raised the public awareness of the environmental dangers involved in its continued use (see page 23). DDT has since been banned from most applications in the United States.

If DDT is not very toxic to humans, why was there a problem with its use? The difficulty lies in the adverse ecological impact of DDT on bird populations, particularly the thinning of egg shells and the detrimental effect on egg hatching and brood survival. Ospreys and bald eagles were severely impacted, as were other species (the title of Rachel Carson's book alluded to the disappearance of songbirds). Since the ban on DDT in the United States in 1972, many bird populations have recovered, but exposure in other countries may still be a problem for some species of birds. (*Note:* In addition to the impact on bird populations, before its ban, DDT had also been found in human mothers' milk at seven times the level permitted for milk sold in stores, and may have had adverse impacts on human health.)

Many other substances in addition to mercury and DDT exhibit bioaccumulation and biomagnification in an ecosystem. These include copper, cadmium, lead, and other heavy metals, pesticides other than DDT, and cyanide, selenium, and PCBs.

## PCBs in the Hudson River

PCBs are a group of organic industrial chemicals that become very persistent contaminants when released into the environment. (They were used as insulating materials in transformers and other electrical equipment before being banned from such use in 1977.) One example of PCB contamination involves the Hudson River in New York State, into which large amounts of PCBs were discharged (legally, at the time) from two General Electric plants between 1947 and 1977; over 136,000 kg (300,000 lb) of PCBs remain concentrated in bottom sediments of the Hudson River. The extensive spread of PCBs throughout the food chain of the Hudson River is considered to be one of the most significant hazardous waste pollution problems in the nation. [In fact, about 320 km (200 miles) of the Hudson River bottom are considered to be a *Superfund* site.]

Small amounts of PCBs are consumed by microorganisms in the riverbed and are passed up and through the food chain in ever-increasing concentrations. People who consume fish contaminated with PCBs are in jeopardy. PCB levels in most fish found in the Hudson exceed the 2-ppm limit set by the federal government; commercial fishing is banned in most of the Hudson River because of this. [PCB was classified as a carcinogen by the Environmental Protection Agency (EPA)].

The danger of exposure to PCB will remain as long as that substance remains in the river. In spite of the ban on commercial fishing in the Hudson and periodic issuance of recreational fishing advisories, many people continue to eat contaminated fish from the river. Many scientists believe the best way to reduce PCB levels in Hudson River fish (as well as in the people who eat the fish) is to remove the contaminated sediments from the river. Studies have been conducted to determine if dredging the contaminated sediments from the riverbed is the optimum solution to the problem. The EPA has recommended the removal of 2 million m$^3$ of sediment from hundreds of so-called *PCB hotspots* along a 64-km stretch of the river north of Albany, a project that would take about 5 years to complete and may begin by 2003. This plan involves dredging only the most contaminated portions of the river rather than extensive bank-to-bank dredging. The sediment would be transported away from river communities by railroad for proper disposal. General Electric would be responsible for the cleanup under the Superfund law (see page 388). Updated information on the progress of the EPA plan can be found at http://www.epa.gov/hudson.

## Endangered Species Act

Although extinction is a natural process, a number of species of fish, wildlife, and plants in the United States have become extinct as a consequence of economic growth and development that took place without adequate ecological concern and conservation measures. Some species of fish, wildlife, and plants have been depleted in numbers to the point that they are now threatened with extinction. These species are of esthetic, ecological, educational, historical, recreational, and scientific value to the nation and its people. Because of this, the *Endangered Species Act* was drafted. As defined in the Act, *endangered species* means "any species which is in danger of extinction throughout all or a significant portion of its habitat." The term *threatened species* means "any species which is likely to become an endangered species within the foreseeable future throughout all or a significant portion of its range."

Enacted by Congress in 1973, the law provides for the protection and conservation of threatened or endangered plants and animals and their habitats. A list of more than 600 endangered species and about 200 threatened species is maintained by the U.S. Fish and Wildlife Service (FWS), a branch of the Department of the Interior. This list includes a wide variety of insects, crustaceans, birds, fish, reptiles, and mammals, as well as flowers, grasses, and trees. The law prohibits any action that would adversely affect or reduce the numbers of the listed organisms. Anyone can petition the FWS to include a particular species on the list. The final determination of a species' listing by the FWS is based on available scientific data.

Protecting endangered and threatened species and restoring them to a secure status in the wild is the basic objective of the endangered species program; administration of the program includes the following responsibilities:

- Listing, reclassifying, and delisting species
- Providing biological opinions to federal agencies on their activities that may affect listed species

- Overseeing recovery activities for listed species
- Providing for the protection of important habitat
- Providing grants to states to assist with their endangered species conservation efforts

(For more details about this law, see the *Endangered Species Home Page* on the World Wide Web; the URL address is given in Section 1.6.)

## 1.4 GEOLOGY AND SOILS

To a large extent, both liquid and solid wastes are disposed of on top of or below the ground surface. An important concern in environmental technology is the interaction between such waste materials and naturally occurring bodies of water. The protection of groundwater quality is of particular concern. This involves soil types and characteristics. Since soil comprises unconsolidated rock particles, it will be helpful for the student who has little or no background in the subject of geology or soils to first review the brief discussion in this section.

## Types of Rock

Rocks are composed of inorganic substances called *minerals*. Some common minerals are quartz, mica, feldspar, calcite, magnetite, and kaolinite. The fundamental chemical elements making up these minerals include silicon, potassium, aluminum, calcium, iron, oxygen, and many others.

The three major types of rocks are igneous, sedimentary, and metamorphic. *Igneous rocks*, which compose most of the solid crust of the planet, have cooled and solidified from a hot molten state. *Granite*, composed primarily of the minerals quartz and feldspar, is a common type of igneous rock.

Even the hardest and most durable igneous rocks that are exposed at Earth's surface are subject to physical disintegration and chemical decay. Changes in the composition and structure of the rock are constantly occurring because of the action of wind, water, temperature changes, carbon dioxide, and oxygen. This gradual process is called *weathering*. The solid rock made up of consolidated minerals is broken down into relatively small unconsolidated fragments called *soil*.

When the soil particles are moved by wind or water and deposited elsewhere, they form *sediments*.

These sediments may be covered under additional deposits of material; eventually, they are compacted and consolidated under the load of overlying layers. With time, the rock fragments can become cemented together, forming a second type of rock called *sedimentary rock. Sandstone,* formed from cemented sand grains, and *shale,* formed from consolidated and lithified mud (silt and clay), are common sedimentary rocks. *Limestone,* consisting primarily of the mineral calcite (calcium carbonate, $CaCO_3$) crystallizing in a marine environment, is also a widely occurring sedimentary rock type.

Under conditions of excessive heat and pressure caused by environmental conditions, both the igneous and sedimentary rock types can be changed from their original forms and mineral structures. The newly formed rock is called *metamorphic rock. Marble* and *slate* are familiar examples of this third fundamental type of rock. Marble is formed as calcite in limestone recrystallizes due to heat and pressure. Slate is the metamorphic equivalent of shale.

Both sedimentary and metamorphic rocks can again be subject to the weathering, transportation, and deposition process in a continual cycle of rock formation. This is illustrated in Figure 1.10.

The physical properties of rock that are of primary interest in environmental technology include those that are related to the underground storage and flow of water. Two terms are of significance in this regard: *porosity* and *permeability.* Rock is not entirely solid through and through. Porosity refers to the percentage of total rock volume that is occupied by *voids,* or *pore spaces.* Permeability refers to the characteristic of the rock that enables water to flow through the pore spaces. This is called the *matrix permeability.* Note that porous rocks are not necessarily highly permeable, particularly if the pore spaces are very small or are not interconnected.

Sedimentary rocks generally are porous and relatively permeable, whereas most igneous and metamorphic rocks are impermeable. Carbonate sedimentary rocks, such as limestone, are relatively soluble. In this type of rock, *solution cavities* may form as water slowly flows through the pore spaces and dissolves the mineral calcite. The solution cavities increase the rock's permeability.

Rock formations have characteristic structural features that may also affect permeability. For example, a feature called *layering* or *stratification* is usually present in sedimentary rocks. This comes about because the unconsolidated soil particles are deposited as sediments by water or wind in horizontal layers; these layers may differ in particle size or mineral composition. As a consequence, permeability is generally higher in the direction parallel to the layers in sedimentary rocks.

Rock masses can gradually bend and fold because of environmental changes and earth movements. When the rocks shatter and crack from excessive stresses, fractures or fissures are formed; these are called *joints.* Joints in which one side of the rock mass moves or is displaced relative to the other side, as happens during an earthquake, are called *faults.* Unless cemented or tightly closed, both joints and faults serve as convenient pathways for the flow of water and add *fracture permeability* to the rock mass. In soluble rocks like limestone, the joints can be enlarged by solution to form caves.

## Types of Soil

Soil, the unconsolidated rock fragments formed from weathering, may be classified or grouped on the basis of *texture,* or the size and shape of the soil particles. There are four major textural classifications of soil:

1. *Gravel:* rock fragments between 4.75 and 75 mm in size.
2. *Sand:* rock particles larger than 0.075, but less than 4.75 mm in size.
3. *Silt:* fine, powderlike particles larger than 0.002, but less than 0.075 mm.
4. *Clay:* very small particles, less than 0.002 mm in size.

The term *loam* is also used for a combination of silt and sand that also contains organic material and that is suitable for the growth of plants or crops. Clay differs from gravel, sand, and silt not only in size, but also in shape and mineral composition. Gravel, sand,

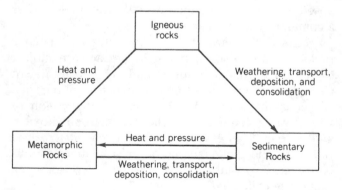

**FIGURE 1.10**
*Schematic diagram of the* rock cycle.

and silt consist of relatively coarse grained, bulky particles; the very finely divided clay particles are plate-like in shape and have a strong affinity for water.

The permeability of soil decreases as the particle size decreases. Gravel and sand are porous and highly permeable, readily allowing the flow of water through the spaces between the soil grains. Silt is considerably less permeable because of the small particle and void size, and clay, although a porous material, is virtually impervious to the flow of water. This is basically because the water in the clay is held by molecular forces on the platelike clay particles. The impermeability of clay soil can be used to advantage in building a sanitary landfill for solid waste disposal, but it is a serious disadvantage for subsurface disposal of wastewater. This will be discussed in more detail in subsequent chapters.

In addition to *porosity* and *permeability*, other terms related to the flow of water in soil include *infiltration* and *percolation*. Infiltration refers to the penetration of the water through the ground surface layer of soil or rock, and percolation refers to the continuing movement or flow of water through the pore spaces under the force of gravity.

Generally, the *tight soils*, which contain a significant percentage of silt and clay, have low infiltration and percolation rates. However, coarse-textured soils, containing mostly sand or gravel, have high infiltration and percolation rates. Tight soils are not suitable for on-site subsurface disposal of sewage. The percolation or "perc" test, used when designing on-site septic systems, is discussed in Chapter 10.

## Soil Gradation

Although soils can be classified according to particle size as gravel, sand, silt, or clay, naturally occurring soil deposits are not usually found in such distinctive groupings. They are generally mixed, containing a variety of soil particle sizes. The distribution and percentage of different particle sizes in the soil is referred to as *soil gradation*.

Soil gradation can be easily determined mechanically by separating the different soil particles in sieves or screens with varying sizes of mesh openings. The result of this mechanical analysis is usually expressed graphically in a *gradation curve*. A typical gradation curve is shown in Figure 1.11, where the vertical axis shows the percentage of the soil particles that are finer than a specific particle size, and the horizontal axis shows the particle size. This particular curve shows that 60 percent of the soil is less than 0.7 mm in size and 10 percent of the soil is less than about 0.12 mm.

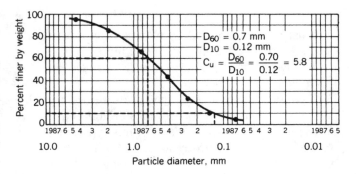

**FIGURE 1.11**

*Typical soil gradation curve. The uniformity coefficient $C_u$ can be used to classify the soil.*
(From D.F. McCarthy, *Essentials of Soil Mechanics and Foundations*, 2nd ed., Reston Pub. Co., Reston, VA. Copyright 1982 Reston Publishing Co. Used with permission.)

The particle size for which 10 percent of the soil is smaller is called the *effective size* of the soil because there is a reasonably consistent relationship between the size and soil permeability. The effective size, designated $D_{10}$, is used as a factor when specifying sand for filters used in water or wastewater treatment systems. Another gradation factor, called the *uniformity coefficient*, is also used in this regard. The uniformity coefficient is the ratio of two particle sizes: the 60 percent finer size and the effective size, or $D_{60}/D_{10}$. The soil depicted in the gradation curve in Figure 1.11 has an effective size of 0.12 mm and a uniformity coefficient of 0.7 mm/0.12 mm = 5.8.

Mixed soils can be described as being well-graded or poorly graded, depending on the distribution of particle sizes. Good gradation is represented by curve *A* in Figure 1.12. There is a wide variation of particle sizes

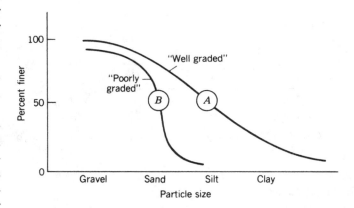

**FIGURE 1.12**

*Well-graded soil (curve A) has lower porosity and permeability than poorly graded soil (curve B).*

in this kind of soil; the smaller particles fit into and fill up most of the pore spaces between the larger particles. As a result, a well-graded soil tends to have relatively low porosity and permeability. It is also resistant to erosion and scour.

Gradation curve *B* is typical of a poorly graded soil, which consists primarily of soil grains in a very narrow range of sizes. In a soil of this type, the porosity and permeability are relatively high because there are not enough small particles to fill the voids between the larger particles. Poorly graded soils generally have uniformity coefficients of less than 10.

Mixed soils are ordinarily classified or grouped in order to help predict their behavior and characteristics. Several classification systems are used in soil mechanics technology, most of which are based upon sieve analysis and other laboratory tests. One of the simplest methods for classifying soil makes use of a *triangular classification chart*, as illustrated in Figure 1.13.

The triangular classification chart relies only on the relative amounts of sand, silt, and clay in the soil sample. For example, a soil that contained 25 percent sand, 50 percent silt, and 25 percent clay would be characterized as *clay–silt* on the chart. Generally, when the proportion of clay exceeds 20 percent of the total soil sample, the clay will tend to dominate the soil characteristics and behavior, and the primary soil type or designation will be clay. A classification scheme more common than the triangular chart, called the *unified system*, makes use of additional lab test data to classify the soil.

## Soil Survey Maps

A group of related soils that has developed from similar parent rock formations is called a *soil series*. Soils within a specific soil series are essentially alike in all basic characteristics. They are designated on the basis of their textural classification and the name of the geographic location that is particularly representative of the soil type.

The Natural Resource Conservation Service (NRCS), formerly called the Soil Conservation Service (SCS), of the U.S. Department of Agriculture has surveyed local areas of the country and has prepared countywide *soil survey maps*. Soil survey maps show the location of different soil series in an area and are readily available from most county NRCS offices. A part of a typical NRCS soil survey map is illustrated in Figure 1.14.

Soil survey maps are superimposed on top of aerial photographs of the area. In addition to the soil series distribution, they show lakes, roads, and other physical land features. They are particularly useful as an aid for good land-use planning and development. The letter symbols, such as HnB and RrD, designate the soil series. For example, HnB stands for "Hibernia very stony loam, 3 to 8 percent slopes." RrD stands for "Rockaway rock outcrop association, sloping and moderately steep." Hibernia and Rockaway are the geographic locations in Sussex County, New Jersey, where these soils have been identified.

The slope of the ground is of importance in environmental planning and land development. Slopes expressed in percent (%) represent the change in elevation of the ground per 100 m or 100 ft of horizontal distance. For example, a hill with a 20 percent slope changes 20 m in elevation in a distance of 100 m. This is illustrated in Figure 1.15.

Slopes of less than 5 percent are usually considered to be *gentle*, whereas those over 15 percent are considered to be *steep*. Steep slopes are much more susceptible to high rates of stormwater runoff and soil erosion than are gentle slopes. On the other hand, areas of very flat topography (0 to 2 percent slopes) may suffer from poor stormwater drainage and flooding problems. Modern municipal land-use ordinances may limit the extent of home building and other development allowed on steep slopes, and proper hydraulic design of storm drainage systems is important for the very flat areas.

The descriptive material in the NRCS soil survey reports contains a good deal of information regarding

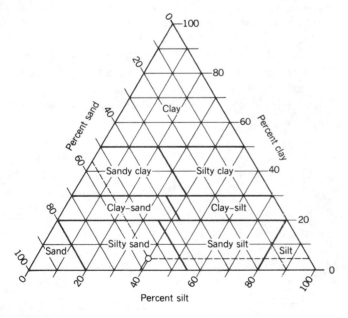

**FIGURE 1.13**
*Triangular soil classification chart.* (From B.K. Hough, *Basic Soils Engineering*, Ronald Press, New York, 1957. Copyright © 1957 Ronald Press. Used with permission of John Wiley & Sons, Inc.)

Soil series boundary line

**FIGURE 1.14**
*Typical NRCS soil series map. The symbols identify the different soil series in an area. The characteristics of each soil series are described in the NRCS publications. (Courtesy of the Soil Conservation Service, United States Department of Agriculture.)*

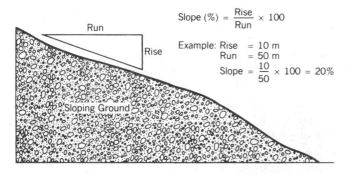

$$\text{Slope (\%)} = \frac{\text{Rise}}{\text{Run}} \times 100$$

Example: Rise = 10 m
Run = 50 m
$$\text{Slope} = \frac{10}{50} \times 100 = 20\%$$

**FIGURE 1.15**
*The slope of the ground surface is important in environmental planning. Slopes of over 15 percent are generally considered to be steep.*

soil characteristics and behavior. Data on depth to *bedrock* and depth to the *water table* are given. Bedrock is the unweathered rock formation underlying the surface soils. The water table represents the depth at which the soil is saturated with water, as is discussed in more detail in Chapter 3. Depth to bedrock and the water table are important with respect to designing solid or liquid waste disposal facilities. Other NRCS data include soil index properties, gradation, permeability, erosion potential, and suitability of the land for various types of development. For example, the HnB series is described as having a seasonal high water table that severely limits the use of the ground for a sanitary landfill or septic tank absorption fields.

## 1.5    HISTORICAL PERSPECTIVE

Water was an important factor in the location of the earliest settled communities, and evolution of public water supply systems is the first focus of *environmental technology* (which, of course, is a modern expression.) In the development of water resources beyond their natural condition in local rivers, lakes, and springs, the digging of shallow wells was perhaps the first innovation. As the need for water increased and tools were developed, the wells were made deeper. The need to move water supplies from distant sources was an outcome of the growth of urban communities.

Piped water supply systems have been in existence for well over 2000 years. Archeological findings show that pipes were made out of hollow logs, clay, and other materials that could not withstand much in the way of pressure. Even the most sophisticated of the ancient systems, the masonry aqueducts and tunnels constructed by the ancient Romans (many of which are still standing), were not capable of carrying water under pressure.

Ancient Rome had sewers as well as aqueducts, but these were used primarily as storm drains rather than for sanitary disposal of human wastes. At that time in history there was no knowledge of the relationship between improper waste disposal and the spread of disease.

Not until the middle of the 1800s did people realize that there was a direct connection between contaminated drinking water and disease. The cause of a localized cholera epidemic in London was traced by Dr. John Snow to a polluted community well, known as the Broad Street well. People who drew their water from the well were affected by the epidemic, whereas people with other water sources were not affected at all. When the well was taken out of service by simply removing the pump handle, the outbreak of cholera subsided.

The Broad Street well incident of 1849 pointed to the need for proper disposal of human wastes and for clean water supplies. Today, it is common knowledge that sewage carries "germs" that can spread disease, but in 1849 the germ theory of disease had not even been postulated. The only evidence of the need for sanitation was the coincidence of the dirty well water and the epidemic. It was not until the late 1800s that Robert Koch proved that the presence of certain microscopic living organisms caused disease in humans. Koch's famous postulates state that (1) a specific organism can always be found in association with a given disease. (2) The organism can be isolated and grown in pure culture in the laboratory. (3) The pure culture will produce the disease when inoculated into a susceptible animal. (4) It is possible to recover the organism in pure culture from the experimentally infected animal.

In the United States, a piped water supply system was used for the first time in 1801 when a steam-powered waterworks station was put into operation in Philadelphia. By the end of the 1800s, about 25 percent of all urban households in the country had running water. During the mid-1800s, sewers began to replace cesspools for waste disposal because of the rapid growth of urban populations and the provision of running water in individual homes. Construction of the first large sewage collection systems in the United States was completed in Chicago and Brooklyn in the late 1850s. Serious outbreaks of cholera and other epidemics of waterborne disease were frequent in American cities throughout the 1800s.

Water filtration plays an important role in disease prevention. The first slow sand filters in the United States were built in Poughkeepsie, New York, in 1872 to treat water from the Hudson River, called "one of the most polluted and potentially dangerous water sources in the world" by engineers and scientists of that time. A filter plant was also constructed in Lawrence, Massachusetts, in 1892, dramatically reducing the number of typhoid outbreaks caused by the polluted Merrimac River water source. The Lawrence filter plant proved to be a turning point in establishing the practice of "sanitary engineering" on a scientific basis.

By the beginning of the 20th century, newly applied water purification techniques, mostly by disinfection using chlorine, drastically reduced the incidence of waterborne diseases in the United States. The dramatic relationship between the number of American water supplies being chlorinated and the decline in waterborne disease is illustrated in Figure 1.16. Chlorine was used for the first time in the United States for disinfecting the Jersey City, New Jersey, water supply in 1908.

In the first half of the 20th century, many public health authorities believed that "the solution to pollution is dilution." They felt that the best and most economical way to protect public health was to purify drinking water, not wastewater. Most sewage was discharged untreated into streams and rivers. The first Imhoff tank facility for primary sewage treatment was built in New Jersey, in 1911, and the first activated sludge secondary sewage treatment plant was built in Texas, in 1916. A large-scale municipal sewage treatment plant was built in Milwaukee, Wisconsin, in 1919. But it was not until the late 1950s that it really became apparent that wastewater treatment was important for water pollution control and public health protection.

Earth Day in 1970 heralded a new environmental era in the United States, when air quality and solid and hazardous waste disposal joined with drinking water supply and water pollution control as major concerns.

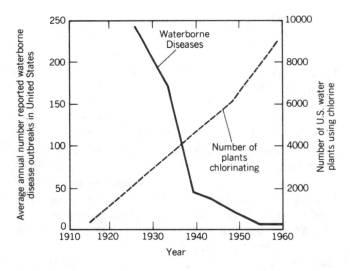

**FIGURE 1.16**

*Fewer outbreaks of waterborne disease are the result of increasing use of chlorine for disinfecting water supplies.* (From G.L. Culp and R.L. Culp, *New Concepts in Water Purification*, Litton Educational, New York, 1974. Copyright © 1974 by Litton Educational Publishing Co. Used with permission of Van Nostrand Reinhold Company, Inc.)

The more comprehensive profession called *environmental engineering* replaced the field called *sanitary engineering*. This new environmental era is discussed below. (The historical development of solid and hazardous waste disposal methods, and air pollution control requirements, are discussed in Chapters 11, 12, and 13, respectively.)

## An Era of Environmental Awareness

In the 1960s, a broad awareness of environmental pollution problems developed among the general public. Many people came to realize the value and importance of protecting environmental quality. Clean air and clean water were worth rallying about, and public demonstrations were held. People wanted streams and lakes that could be used for swimming and fishing as well as for safe drinking water supplies.

The words *ecosystem* and *biosphere* became popular buzzwords, and newspaper articles about local pollution problems became more common. Educational programs that focused on environmental issues were developed for grade school through the university level and grew in popularity.

In addition to stopping air and water pollution, solving problems related to refuse disposal, radiation, noise, pesticides, and wildlife preservation became im-

portant in the modern quest for environmental quality. Although infectious diseases like typhoid and cholera had virtually disappeared in the United States, people became aware of other types of problems caused by human and ecosystem exposure to industrial toxic chemical substances.

The book *Silent Spring* focused attention on the environmental damage caused by improper use of pesticides. DDT and other chemicals used to protect agricultural crops were seriously disrupting the natural balance of ecosystems on a wide scale. After DDT was finally banned for most uses in the United States, the concentrations of this chemical found in human and animal tissues was observed to decline significantly, and several endangered species, such as the bald eagle, began to thrive once again.

The emergence of an environmental awareness on the part of the general public in the 1960s was apparently more than just a passing fad. It was a genuine concern that served to focus the attention of politicians, lawmakers, and governmental agencies on the need for an appropriate legal and regulatory framework for environmental quality control.

In 1970 the *National Environmental Policy Act (NEPA)* was signed into law in the United States. NEPA established a national policy to "maintain conditions under which humans and nature can exist in productive harmony, and fulfill the social, economic, and other requirements of present and future generations of Americans." The concept of using *environmental impact statements (EIS)* as a planning tool to minimize the harmful effects of land development and urban growth was developed for the first time under NEPA. In 1970, the *Environmental Protection Agency (EPA)* was also created, consolidating into one independent federal agency a means for establishing and enforcing pollution control and environmental quality standards.

The EPA is basically a regulatory agency, but it has other responsibilities in addition to establishing and enforcing environmental standards. It is a research organization with a scientific and technical staff that collects environmental data and studies the causes, effects, and methods of control for all types of pollution. The EPA also provides technical and financial assistance to state governments. Individual states have set up their own environmental protection agencies to provide similar assistance to county and municipal governments.

## Environmental Regulations

In the years since 1970, many laws have been enacted by Congress to establish and implement standards of environmental quality for water, land, and air. Many of these laws have been revised and amended several times since their enactment, and some remain under congressional review. A few of the key environmental laws are listed next and are discussed in pertinent sections of this book:

- *Clean Air Act* (CAA)
- *Clean Water Act* (CWA)
- *Comprehensive Environmental Response, Compensation, and Liability Act* (CERCLA)
- *Resource Conservation and Recovery Act* (RCRA)
- *Safe Drinking Water Act* (SDWA)

Implementing and enforcing these and other environmental laws is very expensive. To some people, the objectives of environmental protection seem to be at odds with other goals—industrial growth and economic "progress." But in the long run, the *external diseconomies* caused by environmental degradation are likely to be far greater than the costs of regulation. In other words, if the environment is damaged now in the interest of profits, society will have to "pay the piper" in the future, either in the form of expensive cleanup operations and increased health problems and medical costs or in a generally lower quality of life.

Although there is often controversy on just what steps to take and how much to spend for environmental protection, most people agree that the problem cannot be ignored altogether. A certain amount of governmental regulation will always be necessary. Eventually, a balance will be reached in which environmental, economic, energy, and social problems will be solved without one preempting or overshadowing the others. Meanwhile, there is a need for technical personnel at all levels of training and education to plan, design, build, and operate environmental control systems (See Appendix B).

 ## 1.6 RELEVANT WEB SITES

A vast amount of information related to environmental technology is now accessible on the World Wide Web. In fact, there is so much information, and so many links among the profusion of Web pages, it is easy for a beginner to be quickly overwhelmed by it all. It is best to start exploring environmental topics on the Web gradually, getting familiar with some informative sites a few at a time.

One of the advantages of using a textbook such as this is that the information is organized and pre-sented in a way that allows the student to read, study, and understand underlying technical concepts before delving into the details of specific environmental subjects. Not all Web sites follow this approach; most assume that the reader has prior scientific or technical knowledge. As the topics in this book are studied, the student will be better prepared to mine the wealth of environmental information in cyberspace.

Because there are so many environmentally related Web sites, it is not practical or even possible to provide a complete list of site addresses or URLs. Instead, an abbreviated list of Web sites relevant to the chapter topics is given at the end of each chapter, along with a brief description of the site. The Web sites listed here will lead the student to information related to many of the topics introduced in this first chapter and represent excellent starting points for the exploration of environmental Web sites.

Not all Web sites are permanent, and their Internet addresses (or URLs) may sometimes change. It can be expected that a few of the sites listed in the chapters of this textbook may not be readily accessible in the future. In addition, although most of the sites listed are from governmental or educational sources and are reliable, users of the Internet must use personal discretion in determining the validity of the information presented. The author would welcome comments and suggestions for additional relevant Web sites to be added to the list. Send comments by e-mail to *nathanson@ucc.edu.*

The first two sites with which a student should become familiar are the EPA and the U.S. Geological Survey (USGS) sites:

### UNITED STATES ENVIRONMENTAL PROTECTION AGENCY (EPA)

http://www.epa.gov

This Home Page of the EPA provides links to much useful information regarding all aspects of environmental protection activities, including laws and regulations, projects and programs, research, publications, databases and software, and a wide variety of other environmental information.

### UNITED STATES GEOLOGICAL SURVEY (USGS)

http://www.usgs.gov

The Home Page of the USGS provides links to much useful information regarding earth science and geology, remotely sensed data and pictures, charts, maps related to many aspects of water, and a wide variety of other environmental information.

Additional Web sites relevant to the topics in this chapter include the following:

## ENDANGERED SPECIES HOME PAGE

http://endangered.fws.gov

This Web site provides information related to the U.S. endangered species program, including listings, reclassifications, and delistings, recovery activities, regulatory changes, changes in species' status, research activities, new ecological threats, and a variety of other related topics.

## ENVIRONMENT NEWS SERVICE (ENS)

http://www.ens-news.com

This online newspaper focuses exclusively on current environmental issues, events, and activities.

## ENVIROSOURCES

http://www.envirosources.com

This search engine focuses specifically on the fields of environmental science and engineering, occupational health and safety, industrial hygiene, and the civil engineering disciplines related to wastewater, water supply, and other areas of environmental infrastructure development.

## NATIONAL INSTITUTE OF ENVIRONMENTAL HEALTH SCIENCES (NIEHS)

http://www.niehs.nih.gov/

This Web site provides information on environmental-related diseases and health risks.

## NATIONAL SOIL SURVEY CENTER

http://www.statlab.iastate.edu/soils/nssc/

This Web site provides information about the distribution and properties of soils, and factors affecting the soil environment.

## NATURAL RESOURCE CONSERVATION SERVICE (NRCS)

http://www.nrcs.usda.gov

This is the Home Page of the Natural Resource Conservation Service (NRCS), a federal agency that sets conservation goals and provides technical assistance for rural and urban communities to reduce erosion, conserve and protect water, and solve other natural resource problems. Links to a teachers' question and answer page, and helpful technical resources such as the Natural Resources Inventory, are also provided.

# REVIEW QUESTIONS

1. Give a brief definition of *environmental technology*, including mention of basic activities and objectives.

2. List at least 15 technical factors or options related to the environmental aspects of a land development project.

3. Define the term *pathogen*.

4. Briefly discuss the modes of transmission of communicable disease.

5. List five common intestinal diseases. What is the most common mode of transmission of these diseases?

6. Briefly discuss the environmental aspects of non-infectious diseases with reference to some specific illnesses.

7. What constitutes an *ecosystem?* Give examples of five different types and sizes of ecosystems.

8. Make a sketch that illustrates the two basic principles of ecology.

9. Briefly describe two fundamental metabolic processes of living organisms.

10. What is the difference between a *heterotrophic* and an *autotrophic* organism?

11. What is the difference between *aerobic* and *anaerobic* decay?

12. What is the difference between *putrefaction* and *fermentation?*

13. Sketch a biogeochemical cycle for two different macronutrients.

14. What are *plankton*, and what role do they play in the aquatic food web?

15. Why is species diversity important for an ecosystem?

16. What is meant by *natural succession* of an ecosystem?

17. Briefly describe three types of rock and give one example of each type. Briefly compare their permeabilities. What role do structural features such as layering, joints, and faults play in permeability?

18. What is the difference between the terms *porosity* and *permeability?* What is the difference between *infiltration* and *percolation?*

19. What is *soil?* List four basic types of soil and compare their permeability characteristics.

20. What is meant by *soil gradation?* What is meant by the *effective size* and *uniformity coefficient* of soil? Why is soil gradation important in environmental technology?

21. What is the difference between a well-graded soil and a poorly graded soil? Illustrate the difference by sketching typical gradation curve sharpes for each. If a soil sample had a uniformity coefficient of 50, would it be considered well-graded or poorly graded?

22. A soil sample is determined to contain 25 percent sand, 20 percent silt, and 55 percent clay. Using the triangular classification chart in Figure 1.13, how would you classify this soil? What do you think its permeability will be, high or low?

23. What information does an NRCS soil survey map convey, and what other information is given in NRCS publications?

24. During what period of history did people first begin to recognize the connection between contaminated drinking water and disease? About when was the germ theory of disease proved?

25. Approximately when were filtration and chlorination applied to drinking water supplies in the United States? What was the effect on public health?

26. The first major challenge for environmental technology was the control of communicable disease. Has this challenge been met? How? What would you say is the present-day challenge in this field?

27. What major piece of environmental legislation was passed by Congress in 1970? What did it accomplish?

28. List at least five other federal environmental laws. What do SDWA and RCRA stand for?

29. Briefly describe what is meant by the term *biomagnification*. What is the difference between biomagnification and *bioaccumulation?* Briefly describe two examples of biomagnification and its effects.

30. Briefly describe the purpose and scope of the *Endangered Species Act*. Visit the *Endangered Species Home Page* and write a brief description of the contents of that Web site.

31. At the EPA Web site, click on "Education Resources," and visit the "Student Center." Click on the link "Laws and Regulations." Select and read the brief overviews of the *Clean Air Act*, the *Clean Water Act*, and one or more other laws of your choice. Write a brief summary describing what you learned.

32. At the USGS Web site, click on the "Water" link and then visit and explore the Web pages that describe the water resources of the state in which you live. Write a brief summary describing what you learned.

33. To organize all the links you explore, create an Environmental Technology Folder in the "Bookmarks" or "Favorites" list of your browser. Using EnviroSources (see Section 1.6) or one of the popular Internet search directories or search engines, locate at least one additional Web site relevant to one of the topics in this chapter. Add the link(s) to your Environmental Technology Folder. Write a brief description of what the Web site(s) contain.

## Chapter Outline

CHAPTER

2

# Hydraulics

The study of water at rest and in motion is called *hydraulics*. Applied hydraulics is concerned primarily with the computation of flow rates, pressures, and forces in water or wastewater storage and conveyance systems. For practical purposes, water and wastewater are considered to be incompressible liquids, each with a unit weight of 62.4 pounds per cubic foot (62.4 lb/ft$^3$). The physical and hydraulic behavior of wastewater (sewage) is so similar to that of clean water that there is generally no difference in the design or analysis of systems involving these two liquids.

In the SI metric system, the unit for weight (force due to gravity) is called a newton (N), and the unit weight of water is 9800 newtons per cubic meter (9800 N/m$^3$). More appropriately, this is expressed as 9.8 kilonewtons per cubic meter (9.8 kN/m$^3$), where the prefix *kilo* stands for 1000. (See Appendix C for a discussion of the SI metric system and the U.S. Customary, or inch–pound, system of units.)

Hydraulics is a very important aspect of environmental technology. A knowledge of basic hydraulic principles is particularly necessary for technical personnel working on the design or analysis of water supply, drainage, and water pollution control systems. The purpose of this chapter is to present only the fundamental concepts of hydraulics, which will be necessary for an understanding of the environmental topics covered subsequently in the text. It can serve as a primer for students who have not yet been exposed to the subject. For those who have previously studied hydraulics or fluid mechanics in other courses, this chapter may be useful for a quick review.

## 2.1 PRESSURE

Water or wastewater exerts forces against the walls of its container, whether it is stored in a tank or flowing in a pipeline. We can also say that it exerts a *pressure*. There is a difference between force and pressure, although they are closely related. Specifically, pressure is defined as a *force per unit area*. In equation form, this can be expressed as

$$P = \frac{F}{A} \qquad (2\text{-}1)$$

where $P$ = pressure
  $F$ = force
  $A$ = area over which the force is distributed

In U.S. Customary units, pressure is usually expressed in terms of pounds per square inch (lb/in.$^2$ or psi). In SI metric units, pressure is expressed in terms of newtons per square meter (N/m$^2$). For convenience, the unit N/m$^2$ is called a *pascal*, abbreviated as Pa. Since a pressure of 1 Pa is relatively small (1 Pa = 0.000 145 psi), the term *kilopascal* (kPa) is used in most practical hydraulics applications: 1 kPa = 1000 Pa = 0.145 psi.

### Hydrostatic Pressure

The pressure that water at rest exerts is called *hydrostatic pressure*. The following very important principles always apply for hydrostatic pressure.

1. The pressure depends only on the height of water above the point in question (not on the water surface area).
2. The pressure increases in direct proportion to the depth.
3. The pressure in a continuous volume of water is the same at all points that are at the same depth.

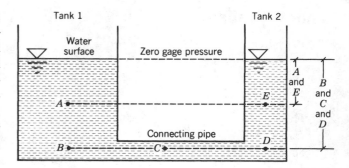

**FIGURE 2.1**
*The pressure at point* A *equals the pressure at point* E, *since these points are at the same depth in the water. Likewise, the hydrostatic pressures at points* B, C, *and* D *are equivalent.*

4. The pressure at any point in the water acts in all directions at the same magnitude.

Consider the two tanks connected by a horizontal pipe shown in Figure 2.1. The water surfaces in both tanks are at the same elevation. We can consider the pressure at the water surface to be equal to zero. Actually, there is some pressure at the free surface because of the weight of the column of air above. This pressure is called *atmospheric* or *barometric pressure*.

Atmospheric pressure at sea level is approximately 101 kPa or 14.7 psi. For most practical applications, we neglect the atmospheric pressure in hydraulic computations. In other words, we consider atmospheric pressure to be a zero reference or starting point. When we do this, we are working in terms of *gage pressure* as opposed to *absolute pressure*.

A total vacuum would have a pressure of absolute zero. Pressures less than atmospheric, but greater than absolute zero, are called *partial vacuums*. Partial vacuums expressed in terms of gage pressure have a negative sign; absolute pressures are always positive. For example, an absolute pressure of 61 kPa (8.9 psi) is equivalent to a gage pressure of −40 kPa (−5.8 psi). This is illustrated in Figure 2.2.

The second principle of hydrostatic pressure listed is one that can be appreciated from personal experience when diving under water in a pool or lake. One can feel the pressure on one's body (especially eardrums) increase when descending deeper into the water. The pressure at point $B$ near the bottom of tank 1 (Figure 2.1) is greater than the pressure at point $A$; also, the pressure at point $A$ is greater than zero. Actually, if point $B$ were exactly twice as deep as point $A$, the gage pressure at point $B$ would be exactly twice the pressure at point $A$ because the pressure varies in direct proportion to the depth.

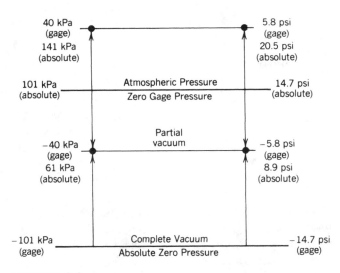

**FIGURE 2.2**
*Pressure measured with reference to standard or normal atmospheric pressure is called* gage pressure. *Gage pressure can have a negative sign when it is less than atmospheric pressure. Absolute pressure is always positive.*

Consider point E in tank 2. Since point E is at the same depth below the water surface as point A, the pressure at point E is the same as the pressure at point A. It makes no difference that tank 2 is narrower than tank 1. Hydrostatic pressure depends only on the height of water above the points and not on the volume or surface area of the water.

Even though point C in the connecting pipe does not have water directly above it, it still has the same pressure as points B and D. This is in accordance with the third principle of hydrostatics listed. Another way of expressing this is to say that *pressure in a continuous fluid at rest is transmitted undiminished at the same depth throughout the fluid.*

**Computation of Pressure**

Consider the tank shown in Figure 2.3a, with a bottom area of $1\,m^2$. If the tank is filled with water to a height of 1 m, the volume of water would be $1\,m^3$, and its weight would be 9.8 kN. The pressure at the bottom of the tank can be computed from Equation 2-1 as $P = F/A = 9.8\ kN/1\ m^2 = 9.8\ kN/m^2 = 9.8\ kPa$.

If the height of water in the tank were increased to 2 m, the total weight of water would be $2 \times 9.8 = 19.6$ kN, and the pressure at the bottom would be $P = 19.6\ kN/1\ m^2 = 19.6\ kPa$, as shown in Figure 2.3b. It can also be seen, in Figure 2.3c, that for a tank with a bottom area of $4\ m^2$ filled with water to a height of 2 m, the bottom pressure is still 19.6 kPa. This is because the additional weight of water is spread over a propor-

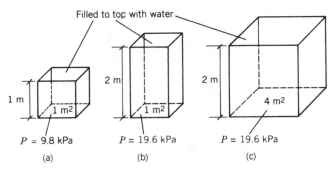

**FIGURE 2.3**
*Hydrostatic pressure at a point depends on the depth of water above the point, but not on the water surface area or volume. Note that the pressure at the bottom of tank (b) is the same as the pressure at the bottom of (c).*

tionally greater area. This again demonstrates one of the basic principles of hydrostatics: Pressure at a point in water depends only on the height of water above the point. Expressed in the form of an equation, this becomes

$$P = 9.8 \times h \qquad \textbf{(2-2a)}$$

where $P$ = hydrostatic pressure, kPa
$h$ = water depth from surface, m

In Customary units, $1\ ft^3$ of water weighs 62.4 lb. For a 1-ft height of water on an area of $1\ ft^2$, or $144\ in.^2$, the pressure at the bottom is $P = 62.4\ lb/144\ in.^2 = 0.43$ psi. Following the same reasoning as given earlier, we can say that in water

$$P = 0.43 \times h \qquad \textbf{(2-2b)}$$

where $P$ = hydrostatic pressure, psi
$h$ = water depth from surface, ft

The following examples illustrate the use of Equations 2-2a and 2-2b.

**EXAMPLE 2.1**

The tank shown in Figure 2.4 has a total depth of 25 ft of water in it. What pressure would be recorded on gage A at the tank bottom? What pressure would gage B, at a height of 15 ft from the bottom, indicate?

*Solution* Using Equation 2-2b, we can compute the pressure at the tank bottom as

$$P_A = 0.43 \times 25 = 11\ psi \qquad \text{(rounded to two significant figures)}$$

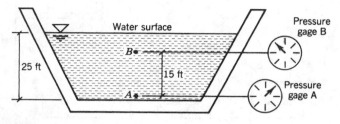

**FIGURE 2.4**
*Illustration for Example 2.1.*

To compute the pressure that would be recorded by gage B, it is first necessary to determine the depth of point B below the water surface. This is $25 - 15 = 10$ ft. Do not use the 15-ft height above the tank bottom! The pressure is computed as

$$P = 0.43 \times 10 = 4.3 \text{ psi}$$

## EXAMPLE 2.2

An elevated water storage tank and connecting pipeline are shown in Figure 2.5. Compute the hydrostatic pressures at points $A$, $B$, $C$, $D$, and $E$.

*Solution*   Point A: The height of water above point $A$ in the tank is equal to the difference in elevation between the water surface and the tank bottom, or $100.00 - 95.00 = 5.00$ m. Using Equation 2-2a, we get

$$P_A = 9.8 \times 5 = 49 \text{ kPa}$$

Point B: The total height of water above point $B$ is $100.00 - 70.00 = 30.00$ m. The pressure at that point is

$$P_B = 9.8 \times 30.00 = 290 \text{ kPa} \qquad \text{(rounded to two significant figures)}$$

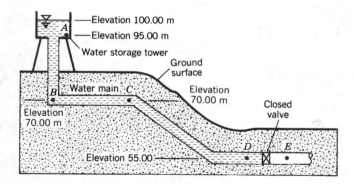

**FIGURE 2.5**
*Illustration for Example 2.2.*

Point C: The pressure at point $C$ is equal to the pressure at point $B$ because these points are at the same elevation. Thus,

$$P_C = P_B = 290 \text{ kPa}$$

Point D: The total height of water above point $D$ is $100.00 - 55.00 = 45.00$ m. The pressure at that point is

$$P_D = 9.8 \times 45.00 = 440 \text{ kPa} \qquad \text{(rounded to two significant figures)}$$

Point E: There is not enough information to determine the pressure at point $E$ because it is isolated from the system above it by the closed valve. Remember, pressures are transmitted only in a continuous fluid; the water at point $E$ is not continuous with the water on the other side of the valve.

## Pressure Head

It is often convenient to express pressure in terms of the height of a column of water, in meters or feet, instead of in terms of kPa or psi. This *pressure head*, as it is called, is the actual, or equivalent, height of water above the point in equation.

For example, in the tank shown in Figure 2.4, we could simply say that the pressure head at point $A$ is 25 ft of water instead of 11 psi. Likewise, the pressure head at point $B$ is 10 ft of water. In Figure 2.5, the pressure head at point $D$ is 45 m of water. If a tall vertical tube were inserted into the pipe at point $D$, the water in the tube would rise 45 m, to the original water surface elevation of 100.00 m.

In some instances, we might know the pressure in terms of kPa or psi, but would like to use units of pressure head instead. This is done, for example, when evaluating water distribution systems. Rearranging the terms in Equations 2-2a and 2-2b, we get, respectively,

$$h = \frac{P}{9.8} = 0.1 \times P \qquad \textbf{(2-3a)}$$

and

$$h = \frac{P}{0.43} = 2.3 \times P \qquad \textbf{(2-3b)}$$

where Equation 2-3a is for SI units and Equation 2-3b is for Customary units.

## EXAMPLE 2.3

A pressure gage on an open tank of water at a point 5 ft above the tank bottom registers a pressure of 13 psi. What is the pressure head at that point? What is the total depth of water in the tank?

*Solution*  Using Equation 2-3b, we get

$$h = 2.3 \times 13 = 30 \text{ ft} \qquad \text{(rounded to two significant figures)}$$

The pressure head is equivalent to the height of water above the gage. The total depth of water in the tank, therefore, is $30 + 5 = 35$ ft.

## EXAMPLE 2.4

A sealed tank, shown in Figure 2.6, has a pocket of air trapped above the water, which is 1 m deep. A pressure gage at the bottom of the tank reads 30 kPa. Determine (a) the pressure head of water at the tank bottom, (b) the height that water will rise in the vertical tube if the valve is opened, and (c) the pressure in the trapped air.

*Solution*
(a) Using Equation 2-3a, we can compute the pressure head as

$$h = 0.10 \times 30 = 3 \text{ m}$$

Notice that the pressure head is greater than the depth of water in the tank. This means that the air in the tank must be exerting additional pressure, pushing downward on the water.

(b) The water would rise 3 m in the vertical tube, a height equal to the pressure head at the bottom of the tank.
(c) If the tank were open to the atmosphere, then 1 m of water depth would cause a pressure of only 9.8 kPa to register on the gage. The difference, or $30 - 9.8 = 20$ kPa (rounded to two significant figures), must be exerted by the pressurized air in the sealed tank. Therefore, the air pressure in the tank is 20 kPa. (The pressure in a small volume of gas is considered to be uniform and does not depend on the height or depth of gas.)

## Measurement of Pressure

Pressure measurement is important in the operation of environmental control facilities. The operating pressure of pumps in water and sewage treatment plants must be monitored, and the pressures throughout a water distribution system must be determined to ensure adequate service. Often, pressure measurements are made and recorded automatically with electromechanical instrumentation, but it is sometimes necessary for technical personnel to measure pressures in the field with other devices.

The simplest way to determine pressure is to use a *piezometer tube*. For example, if a narrow transparent tube is attached to a pipeline under pressure, as shown in Figure 2.7, the water in the pipe will rise in the tube until the pressure head caused by the column of water is equal to the pressure in the pipe. By measuring the height of the column in meters or feet and using a simple computation (Equation 2-2a or Equation 2-2b), one can find the pressure in kPa or psi.

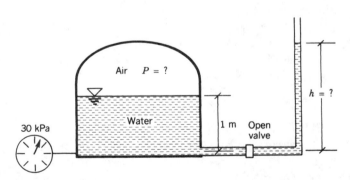

**FIGURE 2.6**
*Illustration for Example 2.4. When the valve is open, the water will rise in the vertical pipe to a certain height* h *that depends on the pressure in the tank.*

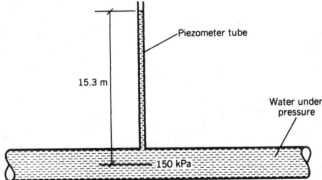

**FIGURE 2.7**
*A piezometer tube offers a simple means for determining pressure by direct measurement of the corresponding pressure head.*

Although they are simple, piezometer tubes are not very practical for field use. As seen in Figure 2.7, for a pressure of 150 kPa (22 psi), the piezometer tube would have to be about 15 m (50 ft) high. These tubes are used primarily in laboratory situations to measure very low pressures. In Section 3.7, on groundwater, the term *piezometric surface* is used to indicate pressure in water that is confined under the ground.

The *manometer* is a somewhat more practical device for measuring pressure using the height of a column of liquid. Here the liquid in the manometer tube is different from the liquid in the system being measured. A well-type manometer, using mercury as the manometer fluid, is illustrated in Figure 2.8. Mercury is a heavy metal that is liquid at room temperature; it is 13.6 times as heavy as water. In a well-type mercury manometer, the equivalent pressure head of water in the system is 13.6 times the measured height of the column of mercury. For example, if the column of mercury shown in Figure 2.8 is 16 in. high, the pressure in the system is $P = 13.6 \times 0.43 \times (16/12)$ ft = 7.8 psi.

One of the most commonly used pressure-measuring devices is the *Bourdon tube gage*. It works on the principle that a flattened hollow metal tube, curved in the form of a spiral or circular arc, tends to uncurl as pressure is applied inside the tube. As the tube uncurls, a pointer linked to it indicates the pressure on a calibrated scale. This is illustrated schematically in Figure 2.9. Bourdon tubes are calibrated to indicate gage pressure, that is, pressure above atmospheric pressure, which is taken as 0 psi.

*Pressure transducers,* devices that sense changes in pressure and convert them to pneumatic or electrical signals, are installed in water and wastewater treat-

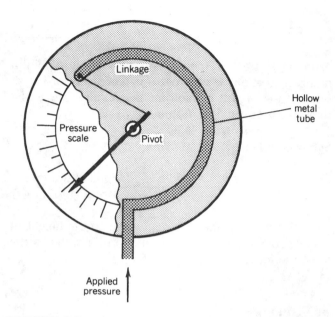

**FIGURE 2.9**
*Simplified cutaway view of a Bourdon pressure gage; the internal hollow tube uncurls as pressure is applied, thereby moving a pointer on the scale.*

ment plants and in pumping stations. They transmit signals to a central control panel, where the operator can see the pressure readings conveniently displayed. One example of a pressure transducer is illustrated in Figure 2.10. In this device, a change in pressure causes the vertical cantilever arm to bend. The displacement changes the wire resistance in attached strain gages, and an appropriate signal proportional to the pressure change is transmitted.

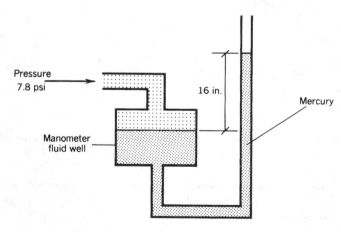

**FIGURE 2.8**
*A well-type mercury manometer is more practical than a piezometer tube for measuring pressures in most hydraulic systems.*

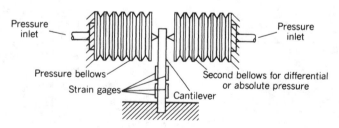

**FIGURE 2.10**
*An example of a pressure transducer that converts pressure into a proportional electrical signal.* (From *Operation of Wastewater Treatment Plants,* Manual of Practice No. 11, Water Pollution Control Federation, Alexandria, Virginia, 1976. Copyright © 1976 by the Water Pollution Control Federation. Used with permission)

## 2.2 FLOW

Most applications of hydraulics in environmental technology involve water in motion—in pipes under pressure or in open channels under the force of gravity. The volume of water flowing past any given point in the pipe or channel per unit time is called the *flow rate* or *discharge*.

In the SI metric system, the unit for flow rate is cubic meters per second ($m^3/s$). The term *liters per second* (L/s) is also used, particularly for relatively small flow rates. Other SI expressions for flow rate are megaliters per day (ML/d) (*note:* $1 m^3 = 1000 L$ and $1 ML = 10^6 L$) and cubic meters per day ($m^3/d$ or CMD).

In U.S. Customary units, flow rate may be expressed as cubic feet per second ($ft^3/s$ or cfs), gallons per minute (gpm), or million gallons per day (mgd). An approximate, but convenient conversion to remember is that 1 mgd = 1.55 cfs = 700 gpm. Also, 1 mgd = 3.79 ML/d = 44 L/s.

### EXAMPLE 2.5

Convert a flow of 50 $m^3/s$ to its equivalent value expressed in terms of L/s and ML/d.

*Solution*

$$50 \ m^3/s \times 1000 \ L/m^3 = 50\ 000 \ L/s$$

$$50\ 000 \ L/s \times 3600 \ s/h \times 24 \ h/d \times 1 \ ML/10^6 \ L$$
$$= 4320 \ ML/d$$

### EXAMPLE 2.6

Convert a flow rate of 50 $ft^3/s$ to its equivalent value expressed as gpm and mgd.

*Solution*  Using the conversion $1 \ ft^3 = 7.48$ gal (see Appendix C), we find

$$50 \ ft^3/s \times 7.48 \ gal/ft^3 \times 60 \ s/min$$
$$= 22{,}440 \ gal/min \quad or \quad 22{,}440 \ gpm$$

and

$$22{,}440 \ gal/min \times 60 \ min/h \times 24 \ h/d$$
$$= 32.3 \times 10^6 \ gal/d \ or \ 32 \ mgd$$

*Alternative Solution*  Using the approximate conversions and setting up appropriate ratios, we obtain

$$\frac{flow}{50 \ cfs} = \frac{700 \ gpm}{1.55 \ cfs} \quad and \quad flow = 50 \times \frac{700}{1.55}$$
$$= 22{,}580 \ gpm$$

To the nearest 500 gpm, this is for practical purposes the same as the first answer, or 22,500 gpm, and for mgd we get

$$\frac{flow}{50 \ cfs} = \frac{1 \ mgd}{1.55 \ cfs} \quad and \quad flow = 50 \times \frac{1}{1.55}$$
$$= 32 \ mgd$$

Many students confuse *flow rate* with *velocity of flow*, but there is a distinct difference between those two terms. Flow rate represents volume per unit time, whereas velocity represents distance per unit time. There is a relationship among flow rate, flow velocity, and the flow area, expressed by the formula

$$Q = A \times V \tag{2-4}$$

where $Q$ = flow rate or discharge
   $A$ = cross-sectional flow area
   $V$ = velocity of flow

In SI units, area $A$ is expressed in terms of $m^2$ and velocity $V$ is expressed in terms of m/s, resulting in units of $m^3/s$ for $Q$. In U.S. Customary units, $V$ is usually expressed in terms of ft/s; therefore, $A$ should be in units of $ft^2$ and the units for $Q$ are $ft^3/s$. It is important to use the appropriate units in Equation 2-4 so that the results are dimensionally correct. The following examples illustrate the use of the basic flow equation, $Q = A \times V$.

### EXAMPLE 2.7

Water is flowing with an average velocity of 4.0 ft/s in an 18-in.-diameter storm drain. The pipe is flowing full. Compute the flow rate in cfs.

*Solution*  Since the pipe is flowing full, the flow area is the same as the cross-sectional area of the pipe. The formula for the area of a circle is $A = \pi D^2/4$, where $D$ is the diameter and $\pi$ can be approximated as 3.14.

To keep units consistent in applying Equation 2-4, the area must be expressed in terms of $ft^2$. First, convert $D$ from inches to feet:

$$18 \ in. \times 1 \ ft/12 \ in. = 1.5 \ ft$$

Now compute the flow area:

$$A = \frac{\pi \times 1.5^2}{4} = 1.77 \text{ ft}^2$$

Now, applying Equation 2-4, we find

$$Q = A \times V = 1.77 \text{ ft}^2 \times 4 \text{ ft/s}$$
$$= 7.1 \text{ ft}^3/\text{s or } 7.1 \text{ cfs}$$

## EXAMPLE 2.8

Determine the required diameter of a pipe that will carry a discharge of 50 ML/d of water at a velocity of 3 m/s.

*Solution*   The pipe diameter $D$ can be determined from the required flow area $A$. Rearranging the terms in Equation 2-4 to solve for $A$, we get $A = Q/V$. In this problem, both $Q$ and $V$ are given, but the proper units must be used so that the equation is dimensionally correct. First, convert the flow rate of 50 ML/d to an equivalent value in terms of $m^3/s$, as follows:

$$Q = 50 \times 10^6 \text{ L/d} \times 1 \text{ d/24 h} \times 1 \text{ h/3600 s}$$
$$\times 1 \text{ m}^3/1000 \text{ L}$$
$$= 0.58 \text{ m}^3/\text{s}$$

Now apply Equation 2-4 to get

$$A = \frac{0.58 \text{ m}^3/\text{s}}{3 \text{ m/s}} = 0.19 \text{ m}^2$$

Rearranging the terms in the formula $A = \pi D^2/4$ gives

$$D = \left[\frac{4 \times A}{\pi}\right]^{1/2} = \left[\frac{4 \times 0.19}{\pi}\right]^{1/2} = 0.49 \text{ m}$$

or

$$D = 0.49 \text{ m} \times 1000 \text{ mm/m} = 490 \text{ mm}$$

## Continuity of Flow

Water is considered to be an incompressible fluid. In other words, its volume does not change significantly with changing pressure. Thus, for a steady discharge in a pipe, the flow rate $Q$ must be constant at any section in the pipe, no matter how the flow area or velocity may change.

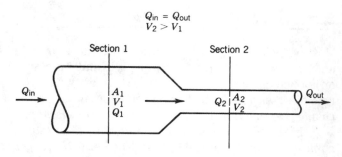

**FIGURE 2.11**
*For an incompressible fluid, such as water or sewage, the volume flow rate Q is constant at any section of the pipeline. Since Q = A × V, when the flow area A is constricted, the velocity V of flow must increase.*

Referring to Figure 2.11, it can be said that the flow rate $Q_1$ at section 1 must equal the flow rate $Q_2$ at section 2, since water is neither added to nor removed from the pipe between those two sections. But the path of flow is constricted at section 2 of the pipe. Common sense says that something must be happening to the water because of the reduced area of flow.

One of the things that is happening is that the flow velocity is increasing as the water moves from section 1 to section 2 of the pipe. Since $Q$ is constant and $Q = A \times V$, when $A$ gets smaller, $V$ must get larger; the product $A \times V$ must always equal $Q$. Conversely, if the area of flow increases, the velocity of flow must decrease.

This principle is sometimes referred to as *continuity of flow*. A common formula used to express this is

$$Q = A_1 \times V_1 = A_2 \times V_2 \qquad \text{(2-5)}$$

Equation 2-5 is sometimes called the *continuity equation*. The product of area and velocity is constant anywhere in the pipeline. The following examples illustrate the concept of continuity of flow.

## EXAMPLE 2.9

In the pipeline shown in Figure 2.11, the area at section 1 is 0.50 m² and the area at section 2 is 0.25 m². For $Q_{in} = 1000$ L/s, determine the velocities at sections 1 and 2.

*Solution*   First convert 1000 L/s to 1.0 m³/s for use in Equation 2-5:

$$Q = A_1 \times V_1 = A_2 \times V_2$$
$$1 \text{ m}^3/\text{s} = 0.50 \times V_1 = 0.25 \times V_2$$

and

$$V_1 = \frac{1 \text{ m}^3/\text{s}}{0.50 \text{ m}^2} = 2.0 \text{ m/s}$$

$$V_2 = \frac{1 \text{ m}^3/\text{s}}{0.25 \text{ m}^2} = 4.0 \text{ m/s}$$

Note that because the area decreased by a factor of $\frac{1}{2}$, the velocity increased by a factor of 2. The velocity is inversely proportional to the area. Also, the velocity is inversely proportional to the square of the diameter. If the diameter of a pipe is reduced by a factor of 3, for example, the velocity increases by a factor of $3^2$ or 9.

## EXAMPLE 2.10

For the branching pipe section shown in Figure 2.12, compute the velocity of flow at section C in the 6-in.-diameter branch. The velocity in the 12-in. branch at section A is 1.0 ft/s, and the velocity at section B in the 4-in. branch is 5.0 ft/s.

*Solution* Since water is incompressible, the total volume of flow entering the system at branch A must equal the total volume of flow leaving the system in branches B and C. This can be stated mathematically as $Q_A = Q_B + Q_C$.

First, compute the cross-sectional flow areas using the unit of feet for diameter:

$$A_A = \frac{\pi(1 \text{ ft})^2}{4} = 0.785 \text{ ft}^2$$

$$A_B = \frac{\pi(\frac{1}{3} \text{ ft})^2}{4} = 0.087 \text{ ft}^2$$

$$A_C = \frac{\pi(\frac{1}{2} \text{ ft})^2}{4} = 0.196 \text{ ft}^2$$

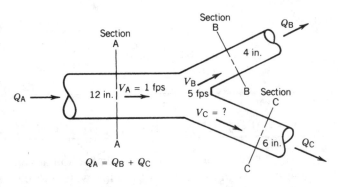

**FIGURE 2.12**
*Illustration for Example 2.10.*

Now compute the flow rates in branches A and B as

$$Q_A = A_A \times V_A = 0.785 \text{ ft}^2 \times 1.0 \text{ ft/s} = 0.79 \text{ ft}^3/\text{s}$$

$$Q_B = A_B \times V_B = 0.087 \text{ ft}^2 \times 5.0 \text{ ft/s} = 0.44 \text{ ft}^3/\text{s}$$

The flow rate in branch C is the difference between that in branch A and that in branch B, or $Q_C = Q_A - Q_B$:

$$Q_C = 0.79 \text{ cfs} - 0.44 \text{ cfs} = 0.35 \text{ cfs}$$

and

$$V_C = \frac{Q_C}{A_C} = \frac{0.35}{0.196} = 1.8 \text{ ft/s}$$

## Conservation of Energy

It is a basic principle in physics that energy can neither be created nor destroyed, but it can be converted from one form to another. In a given closed system, the total energy is constant. This is the *law of conservation of energy*. Applied to problems involving the flow of water, it proves to be a most useful principle.

In hydraulic systems, there exist three forms of mechanical energy: potential energy due to elevation, potential energy due to pressure, and kinetic energy due to velocity. Energy has the units of foot-pounds (ft-lb) or newton-meters (N-m). It is convenient to express hydraulic energy in terms of *energy head*, in meters or feet of water. This is equivalent to foot-pounds per pound of water (ft-lb/lb = ft) or newton-meters per newton of water (N-m/N = m).

In a hydraulic system, then, there are elevation head, pressure head, and velocity head. (Pressure head has already been discussed in Section 2.1.) The total energy head in a hydraulic system is equal to the sum of these individual energy heads. This can be expressed mathematically as

$$E = z + \frac{p}{w} + \frac{v^2}{2g} \quad \text{(2-6)}$$

total = elevation + pressure + velocity
head     head      head      head

where $E$ = total energy head
$z$ = height of the water above a reference plane, m (ft)
$p$ = pressure, kPa (psf)
$w$ = unit weight of water, 9.8 kN/m³ (62.4 lb/ft³)
$v$ = flow velocity, m/s (ft/s)
$g$ = acceleration due to gravity, 9.8 m/s² (32.2 ft/s²)

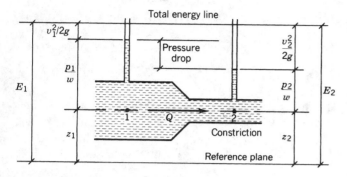

**FIGURE 2.13**
*Since the velocity and kinetic energy of the water flowing in the constricted section must increase, the potential energy must decrease (from the law of energy conservation). This is observed as a pressure drop in the constriction.*

Consider the constricted section of pipe shown in Figure 2.13. From the law of energy conservation, the total energy head at section 1, $E_1$, must equal the total energy head at section 2, $E_2$. Setting $E_1 = E_2$ and using Equation 2-6, we get

$$z_1 + \frac{p_1}{w} + \frac{v_1^2}{2g} = z_2 + \frac{p_2}{w} + \frac{v_2^2}{2g} \qquad \text{(2-7)}$$

This equation is called *Bernoulli's equation* and is one of the most useful formulas in hydraulics. As written here, it applies to *ideal* fluids because viscosity and energy loss due to friction are neglected.

For the system shown in Figure 2.13, we can simplify Bernoulli's equation because the pipeline is horizontal and $z_1 = z_2$. Since they are equal, the elevation heads cancel out from both sides, leaving

$$\frac{p_1}{w} + \frac{v_1^2}{2g} = \frac{p_2}{w} + \frac{v_2^2}{2g} \qquad \text{(2-8)}$$

Now consider what happens as water passes through the constricted section of the pipe, section 2. From continuity of flow, the velocity at section 2 must be greater than the velocity at section 1 because of the smaller flow area at section 2. This means that the velocity head in the system increases as the water flows into the constricted section.

But the total energy must remain constant. For this to happen, the pressure head, and therefore the pressure, must drop. In effect, pressure energy is converted into kinetic energy in the constriction. The fact that the pressure in the narrower pipe section is less than the pressure in the bigger section contradicts what many beginning students often "feel" about the system.

But it follows logically from continuity of flow and conservation of energy. As shown in the next section, the fact that there is a pressure difference will allow measurement of flow rate in the closed pipe.

**EXAMPLE 2.11**

For the system illustrated in Figure 2.13, the diameter at section 1 is 12 in. and that at section 2 is 4 in. The flow rate through the pipe is 2.0 cfs, and the pressure at section 1 is 100 psi. What is the pressure in the constriction at Section 2?

*Solution*    First, compute the flow area at each section, as follows:

$$A_1 = \frac{\pi(1 \text{ ft})^2}{4} = 0.785 \text{ ft}^2$$

$$A_2 = \frac{\pi(0.333 \text{ ft})^2}{4} = 0.087 \text{ ft}^2$$

Now, from $Q = A \times V$ or $V = Q/A$,

$$V_1 = \frac{2.0 \text{ ft}^3/\text{s}}{0.785 \text{ ft}^2} = 2.5 \text{ ft/s}$$

$$V_2 = \frac{2.0}{0.087} = 23 \text{ ft/s}$$

Applying Equation 2-8, we find

$$\frac{100 \times 144}{62.4} + \frac{2.5^2}{2 \times 32.2} = \frac{p_2 \times 144}{62.4} + \frac{23^2}{2 \times 32.2}$$

Note that the pressures are multiplied by 144 in.$^2$/ft$^2$ to convert from psi to lb/ft$^2$ to be consistent with the units for $w$; the energy head terms are in feet of head. Continuing, we obtain

$$231 + 0.1 = 2.3p_2 + 8.2$$

and

$$p_2 = \frac{231.1 - 8.2}{2.3} = \frac{222.9}{2.3} = 97 \text{ psi}$$

## 2.3  FLOW IN PIPES UNDER PRESSURE

When water flows in a pipeline, there is friction acting between the flowing water and the pipe wall, and between the layers of water moving at different velocities in the pipe. This is because of the *viscosity* of the water. The flow velocity is actually zero at the pipe wall and maximum along the centerline of the pipe. When the term *velocity of flow* is used in this text, it means the average velocity over the cross section of flow.

The frictional resistance to flow causes a loss of energy in the system. This loss of energy is manifested as a continuous pressure drop along the path of flow. It is often necessary to be able to compute the expected pressure drop in a given system or to design a new system with a specified maximum pressure loss.

In Figure 2.14a, a straight section of pipe filled with water under pressure is shown attached to a tank. There is no flow in the system and therefore no pressure loss when the valve in the pipe is closed. It can be seen that the pressure head at section 1 equals the pressure head at section 2.

When the valve is opened, flow begins to occur with corresponding energy loss due to friction. This loss can be seen by measuring the pressures along the pipeline. In Figure 2.14b, the difference in pressure

heads between sections 1 and 2 can be seen in the piezometer tubes attached to the pipe. A line connecting the water surface in the tank with the water levels at sections 1 and 2 shows the pattern of continuous pressure loss along the pipeline. This is called the *hydraulic grade line (HGL)* of the system. It is a very useful graphical aid when analyzing pipe flow problems.

The HGL is actually a graph of the pressure head along the pipe, plotted above the pipe centerline. It is not necessary to draw the piezometer tubes, as in Figure 2.14. *The HGL always slopes downward in the direction of flow* unless additional energy is added to the system by a pump. The vertical drop in the HGL between two sections separated by a distance $L$ is called the *head loss* $(h_L)$. The ratio of $h_L$ to $L$ is the slope *(S)* of the HGL or *hydraulic gradient*. In equation form, $S = h_L/L$.

The HGL always passes through the free water surface of any storage tank in the system, since that elevation is equivalent to the system's pressure head at that point. The greater is the flow rate in a given pipeline, the greater is the rate of pressure loss, and the steeper is the slope of the HGL.

## Hazen–Williams Equation

To be able to design new water distribution pipelines or sewage force mains or to analyze existing pipe networks, it is necessary to be able to compute head losses, pressures, and flows throughout the system. There are several formulas in hydraulics to do this, but one of those most commonly used is the Hazen–Williams equation:

$$Q = 0.28 \times C \times D^{2.63} \times S^{0.54} \qquad \textbf{(2-9)}$$

where $Q$ = flow rate, m³/s or gpm
    $C$ = pipe roughness coefficient
    $D$ = pipe diameter, m (in.)
    $S$ = slope of HGL, dimensionless

(*Note:* Equation 2-9 is not the original form of the Hazen–Williams equation, but with the constant 0.28, this equation can be used with sufficient accuracy for both SI units and U.S. Customary units, as long as the appropriate units noted here for $Q$ and $D$ are used. For practical purposes, computed values may be rounded off to two significant figures.)

## Hazen–Williams Nomograph

In engineering practice, sophisticated computer programs are used for hydraulic analysis of large water or sewerage systems. But engineers and technicians often

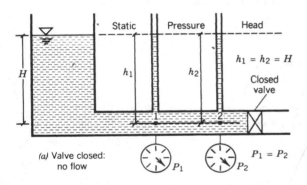

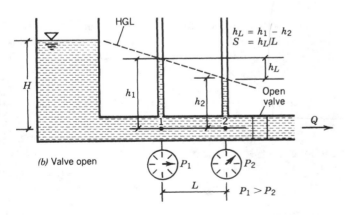

**FIGURE 2.14**
*The hydraulic grade line, or HGL, is a graph of the pressure head above the pipe centerline. Its downward slope in the direction of flow shows pressure loss due to friction.*

have occasion to do computations "by hand," that is, with a hand-held electronic calculator. Programmable calculators are useful for solving formulas such as the Hazen–Williams equation. Many practitioners, though, make use of charts, tables, or graphs that provide a quick and easy solution to a specified equation. These design aids provide numerical accuracy that is sufficient for most engineering purposes. One such chart, called a *nomograph*, is shown in Figure 2.15. A ruler or

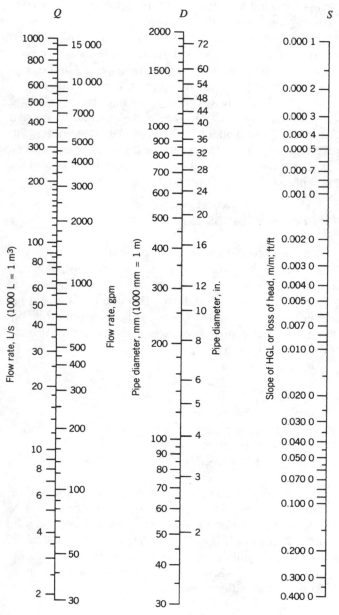

**FIGURE 2.15**

*A nomograph that provides a graphical solution to the Hazen–Williams equation for water flowing in circular pipes under pressure, with C = 100.*

straightedge is used to line up known data and to read a solution for the unknown quantity where the ruler intersects the appropriate axis.

The Hazen–Williams nomograph presented in Figure 2.15 has been prepared for a pipe roughness coefficient of 100. A value of $C = 100$ could represent the friction of an unlined iron pipe that is about 20 years old. Most designers would use this value when preparing plans for a new unlined pipeline to account for the inevitable aging and deterioration of the pipe. Newer pipes are smoother and allow more flow ($C > 100$); older pipes offer more resistance to flow ($C < 100$). Concrete pipes generally have a value of $C = 130$, and smooth plastic pipes may have a $C$ value as high as 150. In this text, all problems will have an assumed value of $C = 100$.

To use the Hazen–Williams nomograph, two of the three variables ($Q$, $D$, and $S$) must be known; the unknown variable can then be determined. A straightedge is placed across the axes so that the two known variables are intersected by a straight line. The third variable is found where the line crosses its corresponding nomograph axis. The following examples illustrate this procedure.

## EXAMPLE 2.12

A 12-in.-diameter pipe carries water with a head loss of 10 ft per 1000 ft of pipeline. Determine the flow rate in the pipe using the nomograph in Figure 2.15. Check the solution using Equation 2-9.

*Solution*   Compute the value of $S = h_L/L = 10$ ft/1000 ft $= 0.010$.

Place a straightedge across the nomograph intersecting 12 in. on the $D$ line and 0.010 on the $S$ line. Read a flow rate of about 1650 gpm on the $Q$ line.

Applying Equation 2-9, we find

$$Q = 0.28 \times 100 \times 12^{2.63} \times 0.010^{0.54} = 1605 \text{ gpm}$$

A nomograph solution is sufficiently accurate for most practical applications.

## EXAMPLE 2.13

An 8-in.-diameter pipe carries a flow of 1.0 cfs. Compute the pressure drop per mile of pipeline, in psi.

*Solution*   First, it is necessary to convert the flow to units of gpm in order to use the nomograph:

$$1.0 \text{ ft}^3/\text{s} \times 7.48 \text{ gal/ft}^3 \times 60 \text{ s/min} = 450 \text{ gpm}$$

Using a straightedge to connect $Q = 450$ gpm and $D = 8$ in. on the nomograph, read a corresponding value of 0.0064 on the $S$ axis.

Since $S = h_L/L$, $h_L = S \times L$. In this problem, $L$ is specified to be 1 mi, or 5280 ft. Therefore,

$$h_L = 0.0064 \times 5280 = 34 \text{ ft}$$

In other words, the HGL would drop a vertical distance of 34 ft over each mile of pipeline. To convert this pressure head loss to its equivalent value in terms of psi, use Equation 2-2b:

$$\text{pressure drop } P = 0.43 \times h_L = 0.43 \times 34$$
$$= 15 \text{ psi per mile}$$

---

## EXAMPLE 2.14

What minimum pipe diameter is required to carry a flow of 30 L/s without causing the pressure to drop more than 10 kPa per kilometer of pipeline?

*Solution*   Convert the pressure drop to an equivalent pressure head, using Equation 2-3a, as follows:

$$h_L = 0.10 \times P = 0.10 \times 10 = 1.0 \text{ m/km}$$

Since $S = h_L/L$ and $L = 1$ km $= 1000$ m, $S = 1.0$ m/1000 m $= 0.001$.

Entering the nomograph with $Q = 30$ L/s and $S = 0.001$, read 310 mm on the $D$ or diameter axis.

---

The energy loss due to viscosity and friction along the straight length of the pipeline accounts for most of the pressure drop. This loss is called the *major loss*. As the water flows through valves, bends, and other pipe fittings, there are additional losses due to turbulence. These losses are called *minor losses*.

In larger water distribution systems, the combined minor losses are very small compared to the major loss, and they are neglected. In pumping stations, where there may be many valves and bends in a confined space, the minor losses must be accounted for by using appropriate loss coefficients or equivalent length factors.

## Flow Measurement

The rate at which water is pumped into a distribution system or sewage is pumped in a force main must be known for proper control and operation of the system. One of the most common types of flow meters used to measure the discharge in a closed pipe under pressure is the *venturi meter*. This is actually a differential pressure meter; the flow rate is related to the pressure difference caused by the meter, using the formulas from continuity of flow and the Bernoulli equation.

A section through a venturi meter is shown in Figure 2.16. As seen in the discussion on energy conservation in Section 2.2, the pressure in the constricted section, called the *throat* of the venturi tube, must be lower than the pressure just upstream of the converging section.

Using the continuity equation and Bernoulli's equation, the following formula can be derived to relate the discharge $Q$ to the measured pressure difference $p_1 - p_2$:

$$Q = C \times A_2 \times \left[ \frac{2g(p_1 - p_2)/w}{1 - (A_2/A_1)^2} \right]^{1/2} \quad \textbf{(2-10)}$$

In Equation 2-10, $C$ is a discharge coefficient that accounts for a small amount of head loss in the venturi meter; it is usually about 0.98. All the other

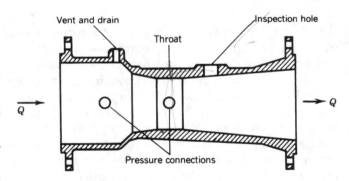

**FIGURE 2.16**

*A venturi meter can be installed in a pipeline to measure the flow rate. The difference between the pressure in the throat of the venturi and that in the upstream section can be converted to discharge, using the Bernoulli equation. (From* Operation of Wastewater Treatment Plants, *Manual of Practice No. 11, Water Pollution Control Federation, Alexdandria, Virginia, 1976. Copyright © 1976 by the Water Pollution Control Federation. Used with permission)*

terms are as previously defined in the continuity and Bernoulli equations. Care must be taken to use appropriate units so that the equation is dimensionally consistent.

## EXAMPLE 2.15

A venturi meter in a 100-mm pipe has a throat diameter of 50 mm. A pressure difference of 75 kPa is measured in the meter. What is the flow rate under these conditions?

*Solution*   First, compute the flow areas in terms of m², as follows:

$$A_1 = \frac{\pi \times (0.1 \text{ m})^2}{4} = 0.007\ 85 \text{ m}^2$$

$$A_2 = \frac{\pi \times (0.05 \text{ m})^2}{4} = 0.001\ 96 \text{ m}^2$$

The ratio $A_2/A_1 = 0.00196/0.00785 = 0.25$, and $1 - (0.25)^2 = 0.9375$. Applying Equation 2-10, we obtain

$$Q = 0.98 \times 0.00196 \times \left[ \frac{2 \times 9.8 \times 75/9.8}{0.9375} \right]^{1/2}$$

$$= 0.024 \text{ m}^3/\text{s} = 24 \text{ L/s}$$

In most systems where flow rates must be monitored continuously, the pressure difference in the venturi tube is sensed by pressure transducers and the flow rate is recorded automatically on a rotating chart in the control room. Venturi meters are available in a wide range of sizes from several manufacturers. They must have the correct shape and proportions in the converting section, throat, and diverging section to maintain streamline flow and accurate measurements. Other pressure differential meters, such as orifice or nozzle meters, are available. They are shorter in overall length, but they obstruct the flow and cause greater head loss than the venturi meter.

Another device that is used to measure flow rates in closed pipelines is the *magnetic flow meter*. It has the advantage of not causing any constriction at all in the path of flow. The operating principle is based on the fact that water is a slight conductor of electricity, and when it moves through a magnetic field, it induces a voltage. The meter produces the magnetic field around the pipe and also senses the induced voltage. The greater the flow rate, the greater is the voltage. The voltage signals are

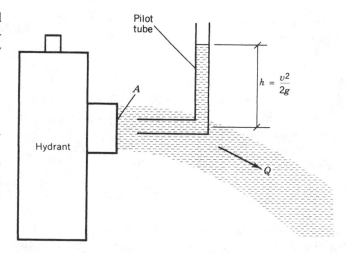

**FIGURE 2.17**
*The Pitot tube is a simple device that can be used to measure flow velocity. The velocity is proportional to the square root of the height of water in the tube.*

transmitted to a recording chart calibrated in units of flow rate.

A device called a *Pitot static tube*, illustrated schematically in Figure 2.17, can be used to measure the flow discharged by an open hydrant in a water distribution system. It consists basically of a tube open at both ends and bent so that one end can be pointed into the flowing water while the other end is vertical. The vertical part of the tube fills with water to a height that is proportional to the flow velocity. If the area of the hydrant opening is known, the discharge can be computed by a formula derived from the continuity and Bernoulli equations.

The common household water meter records the total volume of water that passes through it. That provides a means for the water utility company to bill customers on the basis of actual water use instead of at a flat rate. Water conservation is encouraged when users must pay for actual metered consumption.

A common type of meter for small water service connections is a positive-displacement or *nutating-disk meter*. In this device the water passes through a small chamber of known volume. An inclined hard-rubber disk undergoes a wobbling rotation as the water flows through the chamber. The number of rotations on the disk, which is proportional to the volume of water, is transmitted to a recording register. In modern installations, a digital register can be mounted outside the customer's house to facilitate meter reading by the utility company. Computerized billing systems based on automatic remote meter readings are also used.

## 2.4 GRAVITY FLOW IN PIPES

When water flows in a pipe or channel with a *free surface* exposed to the atmosphere, it is called *open channel* or *gravity flow*. Gravity provides the moving force, while friction resists the motion and causes energy loss. Stream or river flow is open channel flow. Flow in storm and sanitary sewers is also open channel flow, except when the water is pumped through a pipe under pressure (a *force main*).

In most routine problems in the design or analysis of storm or sanitary sewer systems, a condition called *steady uniform flow* is assumed. Steady flow means that the discharge is constant with time. Uniform flow means that the slope of the water surface and the cross-sectional flow area are also constant. A length of a stream, channel, or pipeline that has a relatively constant slope and cross section is called a *reach*.

Under steady uniform flow conditions, the slope of the water surface is the same as the slope of the channel bottom. The HGL lies along the water surface and, as in pressure flow in pipes, it slopes downward in the direction of flow. Energy loss is manifested as a drop in elevation of the water surface. A typical profile view of uniform steady flow is shown in Figure 2.18. The slope of the water surface represents the rate of energy loss. It may be expressed as the ratio of the drop in elevation of the surface in the reach to the length of the reach.

Typical cross sections of open channel flow are shown in Figure 2.19. In Figure 2.19a, the pipe is only partially filled with water and there is a free surface at atmospheric pressure. It is still open channel flow, even though the pipe is a closed conduit underground. The important factor is that gravity, not a pump, is moving the water.

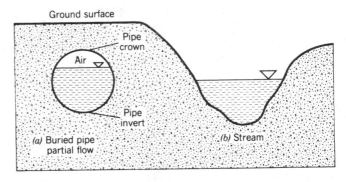

**FIGURE 2.19**
*Any flow that occurs with a free surface exposed to atmospheric pressure is open channel flow, whether it occurs in a surface stream or in an underground pipe.* (*Note:* ▽ indicates a free surface.)

The top of the inside pipe wall is called the *crown* and the bottom of the pipe wall is called the *invert*. One of the basic objectives of sewer design is to establish appropriate invert elevations along the pipeline. The length of wetted surface on the pipe or stream cross section is called the *wetted perimeter*. The size of the channel, as well as its slope and wetted perimeter, are important factors related to its discharge capacity.

### Manning's Formula

A common formula for solving open channel flow problems is called *Manning's formula*, written as

$$Q = \frac{1.0 \text{ or } 1.5}{n} \times A \times R^{2/3} \times S^{1/2} \quad \textbf{(2-11)}$$

It is an empirical or experimentally derived equation, where

$Q$ = channel discharge capacity, m³/s (ft³/s)
1.0 = constant for SI metric units
1.5 = constant for U.S. Customary units (rounded from the traditional value of 1.486)
$n$ = Manning channel roughness coefficient
$A$ = cross-sectional flow area, m² (ft²)
$R$ = hydraulic radius of the channel, m (ft)
$S$ = slope of the channel bottom, dimensionless

The *hydraulic radius* of a channel is defined as the ratio of the flow area to the wetted perimeter $P$. In formula form, $R = A/P$. The roughness coefficient $n$ depends on material and age for a pipe or lined channel and on topographic features for a natural stream bed. It can range from a value of 0.01 for a smooth clay pipe to 0.1 for a small natural stream. A value of $n$ commonly assumed for concrete pipes or lined channels is 0.013.

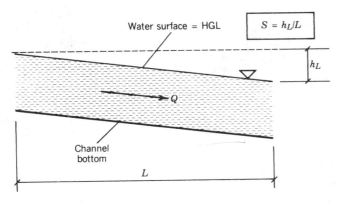

**FIGURE 2.18**
*In* steady uniform open channel flow, *the slope of the water surface, or HGL, is equal to the slope of the channel bottom.*

The following example illustrates the application of Manning's formula for a channel with a rectangular cross section.

## EXAMPLE 2.16

A rectangular drainage channel is 3 ft wide and is lined with concrete, as illustrated in Figure 2.20. The bottom of the channel drops in elevation at a rate of 0.5 ft per 100 ft. What is the discharge in the channel when the depth of water is 1.5 ft? Assume $n = 0.013$.

*Solution*   Since the data use U.S. Customary units, the constant 1.5 is used in Equation 2-11. Refer to Figure 2.20; the cross-sectional flow area $A = 3.0$ ft $\times$ 1.5 ft = 4.5 ft$^2$ and the wetted perimeter $P = 1.5$ ft + 3.0 ft + 1.5 ft = 6.0 ft. The hydraulic radius $R = A/P = 4.5$ ft$^2$/ 6.0 ft = 0.75 ft. The slope $S = 0.5/100 = 0.005$.

Applying Manning's formula, we obtain

$$Q = \frac{1.5}{0.013} \times 4.5 \times 0.75^{2/3} \times 0.005^{1/2} = 30 \text{ cfs}$$

$$\text{(rounded off)}$$

### Circular Pipes Flowing Full

Most sanitary and storm sewer systems are built with sections of circular pipe. In addition to the roughness coefficient, the important factors related to the design or analysis of these pipelines are discharge $Q$, flow velocity $V$, pipe diameter $D$, and pipe slope $S$. Limitations on slope and pipe velocity are discussed in Chapters 8 and 9.

In a circular pipe carrying water such that the pipe is just full to the crown (but still under atmospheric pressure and gravity flow), the flow area $A$ is $\pi D^2/4$, the area of the pipe. The wetted perimeter $P$ is the perimeter of the pipe, or $\pi D$. Since hydraulic radius $R$ is defined as $A$ divided by $P$, we have

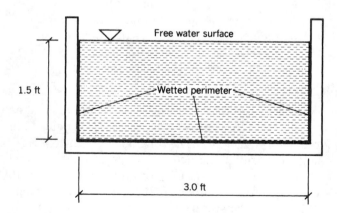

**FIGURE 2.20**
*Illustration for Equation 2-16.*

$$R = \frac{(\pi D^2)/4}{\pi D} = \frac{\pi D^2}{4} \times \frac{1}{\pi D} = \frac{D}{4}$$

For circular pipes flowing full, Manning's formula then takes the following form:

$$Q = \frac{1.0 \text{ or } 1.5}{n} \times \frac{\pi D^2}{4} \times \left(\frac{D}{4}\right)^{2/3} \times S^{1/2}$$

For a given value of $n$, only the pipe diameter and slope are needed to solve for discharge in a circular pipe flowing full. To facilitate the application of Manning's formula, particularly for routine problems with circular pipes, charts or nomographs are usually used. A nomograph for Manning's formula for circular pipes flowing full and $n = 0.013$ is shown in Figure 2.21. For pipes with values of $n$ not equal to 0.013, simply multiply the $Q$ and $V$ values obtained from the nomograph by the ratio of 0.013 to the actual value of $n$.

To use the nomograph, two of the four variables ($D$, $S$, $Q$, and $V$) must be known. A straightedge lined up across the two known variables will intersect the solution for the other two variables on their respective axes. Applications of the Manning nomograph are illustrated in the following examples.

## EXAMPLE 2.17

A 12-in.-diameter pipeline is built on a slope of 1 percent. Assuming that $n = 0.013$, determine the discharge capacity of the pipeline with full flow. What is the flow velocity?

*Solution*   Line up the values $D = 12$ and $S = 0.01$ on the nomograph in Figure 2.21. Read a value of 1580 gpm for flow rate or discharge and a value of 4.6 ft/s for velocity on the appropriate axes.

Using Manning's formula to check the accuracy of the nomograph, we obtain

$$Q = \frac{1.5}{0.013} \times \frac{\pi \times 1^2}{4} \times \left(\frac{1}{4}\right)^{2/3} \times 0.01^{1/2} = 3.6 \text{ cfs}$$

$$Q = 3.6 \text{ ft}^3/\text{s} \times 7.48 \text{ gal/ft}^3 \times 60 \text{ s/min} = 1600 \text{ gpm}$$

and

$$V = \frac{Q}{A} = \frac{3.6}{(\pi \times 1^2)/4} = 4.6 \text{ ft/s}$$

These values check very well with the nomograph solution.

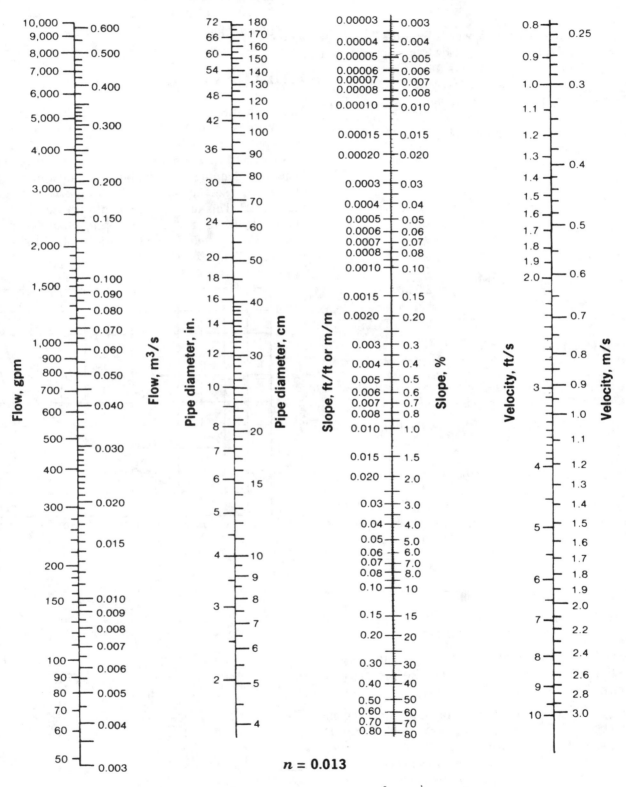

**FIGURE 2.21**

Manning's nomograph *for circular pipes flowing full, with* n = 0.013. *Manning's equation is used for open channel or gravity flow, whereas the Hazen–Williams equation is used for flow under pressure.* (Reprinted with permission from the U.S. Pipe and Foundry Company, Birmingham, Alabama, and WATER/Engineering & Management, Scranton Gillette Communications, Inc., Des Plaines, Illinois.)

## EXAMPLE 2.18

A 450-mm-diameter storm sewer is built on a grade of 2.0 percent. What are the discharge capacity and velocity of flow when the pipe is full?

*Solution* Using the Manning nomograph, line up 45 cm (1 cm = 10 mm) with 2.0 percent, or 0.02, on the appropriate axes. The nomograph solution is $Q = 0.4 \text{ m}^3/\text{s}$ and $V = 2.6 \text{ m/s}$.

Manning's formula gives

$$Q = \frac{1.0}{0.013} \times \frac{\pi \times 0.45^2}{4} \times \left(\frac{0.45}{4}\right)^{2/3} \times 0.02^{1/2} = 0.4 \text{ m}^3/\text{s}$$

and

$$V = \frac{Q}{A} = \frac{0.4}{\pi \times 0.45^2/4} = 2.5 \text{ m/s}$$

## EXAMPLE 2.19

What diameter pipe is needed to carry a peak flow of at least 500 L/s on a 0.25 percent grade?

*Solution* First, convert the flow to cubic meters per second.

$$Q = 500 \text{ L/s} \times 1 \text{ m}^3/1000 \text{ L} = 0.5 \text{ m}^3/\text{s}$$

Line up $Q = 0.5 \text{ m}^3/\text{s}$ and $S = 0.25$ percent, or 0.0025, on the appropriate axes on the nomograph in Figure 2.21. The solution obtained on the diameter axis is about 73 cm, or 730 mm. In practice, the next largest standard pipe size that is manufactured would be selected.

## EXAMPLE 2.20

On what slope should a 16-in.-diameter sanitary sewer be built if it is to carry at least 750 gpm of sewage at a velocity of not less than 2 ft/s?

*Solution* First, line up $D = 16$ in. and $Q = 750$ gpm on the Manning nomograph, and read $S = 0.00048$. But at that slope the velocity would be about $V = 1.2$ ft/s, which is less than the minimum allowable velocity of 2 ft/s. It is necessary to increase the slope to increase the flow velocity.

Line up $D = 16$ and $V = 2$ on the nomograph. Read $S = 0.0014$, or 0.14 percent. Note that the actual discharge capacity of a full 16-in. pipe at that slope is

$Q = 1250$ gpm. In this problem, the required minimum velocity is the controlling factor. If the flow rate is 750 gpm, the pipe is flowing partially full.

## Partial Flow in Pipes

Most of the time, gravity sewers flow only partially full. The free water surface is usually below the crown of the pipe. This condition is depicted in Figure 2.19a. The partial-flow hydraulics can be analyzed directly, using the Manning formula, but it is much more convenient to use the *partial-flow diagram* shown in Figure 2.22. This diagram takes into account the variation of hydraulic radius with depth, which otherwise would require tedious computations. The following examples illustrate its use.

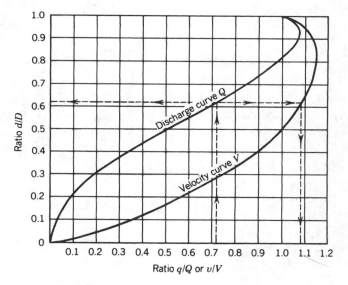

Nomenclature:
$d$ = partial depth
$D$ = full depth or pipe diameter
$q$ = partial discharge
$Q$ = full-flow discharge
$v$ = velocity, partially full
$V$ = velocity, full

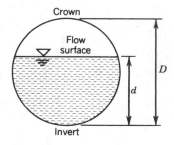

**FIGURE 2.22**

*A partial-flow diagram for a circular pipe that carries flow with the water surface below the pipe crown.*

## EXAMPLE 2.21

A 300-mm-diameter pipe is constructed on a slope $S = 0.02$. What is the depth of flow in the pipe when it carries a flow of 96 L/s? What is the velocity of flow at that depth?

*Solution*   First, using the Manning nomograph, it is found that a 300-mm or 30-cm pipe on a 0.02 slope carries $Q = 0.135$ m³/s = 135 L/s = 135 L/s when flowing full (which is the condition the nomograph is set up for). Its full-flow velocity is $V = 1.9$ m/s.

The discharge under partial-flow conditions is $q = 96$ L/s, and the ratio of partial flow to full flow is $q/Q = 96/135 = 0.71$.

Now enter the partial-flow diagram, Figure 2.22, on the horizontal or $x$-axis with the value $q/Q = 0.71$. Move straight up to an intersection with the "Discharge curve $Q$"; from that point on the $Q$ curve, move horizontally to the left and read $d/D = 0.62$. Since $D = 300$ mm, solve for the partial depth $d$ as follows:

$$d/300 = 0.62 \quad \text{and} \quad d = 300 \times 0.62 = 186 \text{ mm}$$

To compute the velocity at the depth of flow, reenter Figure 2.22 on the vertical or $y$-axis with $d/D = 0.62$ and move horizontally to the right to an intersection with the "Velocity curve $V$"; from that point on the $V$ curve, move straight down and read $v/V = 1.08$ on the horizontal axis. Since the full-flow velocity $V = 2$ m/s, compute the partial-flow velocity as follows:

$$v/1.9 = 1.08 \quad \text{and} \quad v = 1.9 \times 1.08 = 2.1 \text{ m/s}$$

Notice that the partial-flow velocity is actually greater than the velocity under full-flow conditions. As can be seen from Figure 2.22, there is a range of depths at which discharge as well as velocity exceeds full-depth values. The maximum velocity in the pipe occurs when the depth is 82 percent of the diameter, and the maximum discharge occurs when the depth is 93 percent of the diameter of the pipe. The basic reason for this is the reduction in friction at the pipe wall. At the shallower depths, however, the smaller flow area outweighs the effect of the reduced friction, and the discharge ratio drops below 1.0.

## EXAMPLE 2.22

What is the maximum possible discharge capacity of a 900-mm-diameter pipe built on a slope of 0.1 percent?

*Solution*   The Manning nomograph gives the full-flow capacity as $Q = 550$ L/s. From the partial-flow diagram, the maximum discharge occurs when $d/D = 0.93$; at that depth, $q/Q = 1.08$. Therefore, the maximum discharge $q = 1.08 \times 550 = 590$ L/s. It occurs at a depth of $d = 900 \times 0.93 = 840$ mm.

## EXAMPLE 2.23

An 18-in.-diameter sewer lines drops 1.6 ft in elevation over a 400-ft distance. Determine the discharge and velocity in the pipe when the depth of flow is 6 in.

*Solution*   The slope $S = 1.6/400 = 0.004$. Using the Manning nomograph, we obtain $Q = 2900$ gpm and $V = 3.8$ fps. The partial-flow depth ratio is $6/18 = 0.33$, and from the partial-flow diagram, $q/Q = 0.22$ and $v/V = 0.82$. Therefore, $q = Q \times 0.22 = 2900 \times 0.22 = 640$ gpm and $v = V \times 0.82 = 3.8 \times 0.82 = 3.1$ ft/s.

## Open Channel Flow Measurement

An approximate, but very simple method for determining open channel discharge is to measure the velocity of a floating object moving in a straight uniform reach of the channel. If the cross-sectional geometry of the channel is known and the depth of flow is measured, then the flow area can be computed. From the relationship $Q = A \times V$, the discharge $Q$ can be estimated.

This is a useful way to get a ballpark estimate for the flow rate as part of a preliminary field study, but it would not be suitable for routine measurements. The average velocity of flow in a reach is approximated by timing the passage of the floating object along a measured length of the channel.

## EXAMPLE 2.24

A floating object is placed on the surface of water flowing in a stormwater drainage ditch and is observed to travel a distance of 10 m downstream in 20 s. The ditch is 1.5 m wide, and the average depth of flow is estimated to be 0.5 m. Estimate the discharge under these conditions.

*Solution*   The flow velocity is computed as distance over time, or

$$V = D/T = 10 \text{ m}/20 \text{s} = 0.5 \text{ m/s}$$

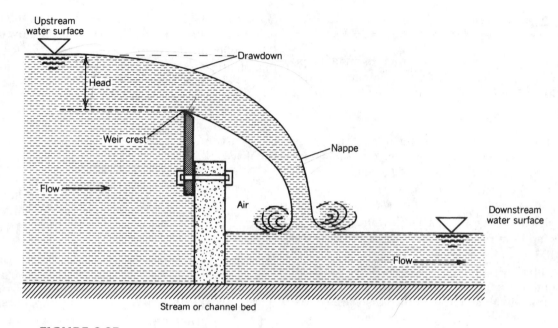

**FIGURE 2.23**
*Side view of a* sharp-crested weir, *a simple device used for measuring open channel or stream flow.*

The channel area is $A = 1.5\ \text{m} \times 0.5\ \text{m} = 0.75\ \text{m}^2$.
The discharge $Q = A \times V = 0.75\ \text{m}^2 \times 0.5\ \text{m/s} = 0.38\ \text{m}^3/\text{s}$ (rounded off).

## Weirs

A widely used device to measure open channel flow is the *sharp-crested weir*. A weir is simply a dam or obstruction placed in the channel so that the water backs up behind it and then flows over it. The sharp crest or edge allows the water to spring clear of the weir plate and to fall freely in the form of a *nappe*, as shown in Figure 2.23.

When the nappe discharges freely into the air, there is a hydraulic relationship between the height or depth of water flowing over the weir crest and the flow rate. This height, the vertical distance between the crest and the water surface, is called the *head on the weir*; it can be measured directly with a meter or yardstick or automatically by float-operated recording devices (see Section 3.4 on stream gaging stations). The head on the weir should be measured a short distance upstream of the weir plate to avoid the effect on the surface drawdown as the flow passes over the crest.

The part of the weir plate over which the water flows can have one of several different shapes, depending on the particular application. The most common shapes are rectangular, triangular, and trapezoidal. A *contracted rectangular weir* is one that does not

extend across the full width of the channel, as shown in Figure 2.24a. Triangular weirs, also called V-notch weirs, can have notch angles ranging from 22.5° to 90°, but right-angle notches are the most common; this is

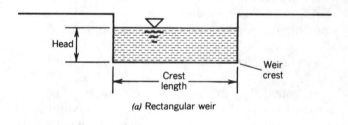

*(a) Rectangular weir*

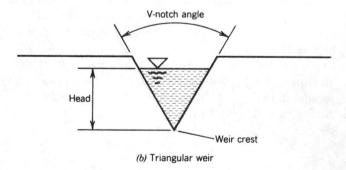

*(b) Triangular weir*

**FIGURE 2.24**
*Commonly used weir shapes include* (a) *the contracted rectangular weir and* (b) *the triangular V-notch weir.*

illustrated in Figure 2.24*b*. (A trapezoidal or *Cipolletti* weir is shown in Figure 3.18, installed as part of a stream gaging station.)

V-notch weirs allow accurate measurement of much lower discharges than rectangular or trapezoidal weirs. They are commonly used in small sewage treatment plants to monitor the sewage effluent flow rate. V-notch weirs can also be inserted into sewer lines to measure sewer discharges, and they are used frequently during infiltration–inflow studies (see Section 8.4). These weirs, manufactured to fit a range of different pipe sizes, can be held in the end of a pipe in a manhole; the flow rate can be read directly from a calibrated scale on the weir plate.

For larger sewer lines, or when continuous flow data over 24 h are needed, a temporary weir installation can be set up in the manhole. An installation that includes an automatic level recorder is shown in Figure 2.25. The disadvantage of this type of flow-metering installation is that sewage solids settle out and accumulate behind the weir plate.

There are many equations, tables, and charts in hydraulics textbooks and handbooks that relate discharge to the head on a weir. Some of the formulas account for end contractions, approach velocities, and other factors. A simple formula for a 90° V-notch weir is

$$Q = 2.5 \times H^{2.5} \qquad \textbf{(2-12)}$$

where $Q$ = discharge, ft$^3$/s
$H$ = head on weir, ft

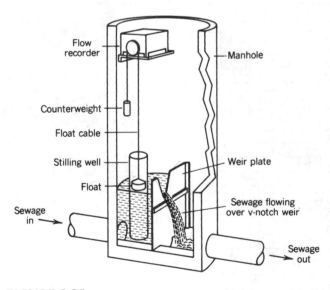

**FIGURE 2.25**
*A temporary weir installation used to measure wastewater flow rates in a sewer manhole.*
(Courtesy of Leupold & Stevens, Inc.)

## EXAMPLE 2.25

Estimate the discharge in L/s over a 90° V-notch weir when the head on the weir $H = 100$ mm.

*Solution*   It is necessary to use some conversion factors in this problem, since Equation 2-12 is given only for U.S. Customary units. We have

$$100 \text{ mm} \times 1 \text{ in.}/25.4 \text{ mm} \times 1 \text{ ft}/12 \text{ in.} = 0.33 \text{ ft}$$

Applying Equation 2-12, we obtain

$$Q = 2.5 \times 0.33^{2.5} = 0.156 \text{ ft}^3/\text{s}$$

and

$$Q = 0.156 \text{ ft}^3/\text{s} \times 28.32 \text{ L/ft}^3 = 4.4 \text{ L/s}$$

A formula used to compute discharges flowing over a Cipoletti weir with steep side-slopes is

$$Q = 3.4 \times b \times h^{1.5} \qquad \textbf{(2-13)}$$

where $Q$ = discharge, ft$^3$/s
$b$ = base width of weir, ft
$h$ = head on weir, ft

A graphical solution in metric units can be obtained using a chart like the one shown in Figure 2.26.

## EXAMPLE 2.26

Estimate the discharge flowing over a Cipoletti weir with a base width of 18 in. when the head on the weir is 9 in.

*Solution*   Applying Equation 2-13, after converting units from inches to feet, we get

$$Q = 3.4 \times 1.5 \times 0.75^{1.5} = 3.3 \text{ ft}^3/\text{s}$$

After converting from inches to metric units, compare this result with that obtained from Figure 2.26: 18 in. = 1.5 ft × (0.3048 m/ft) = 0.46 m, and 9 in. × (2.54 cm/in.) = 23 cm. Entering the chart with $h$ = 23 cm on the right, moving left to the sloping line marked 0.45, then down to the $Q$-axis, obtain a discharge of about 90 L/s, which is equivalent to about 3.2 ft$^3$/s. Note that calculations of this type are only approximate at best.

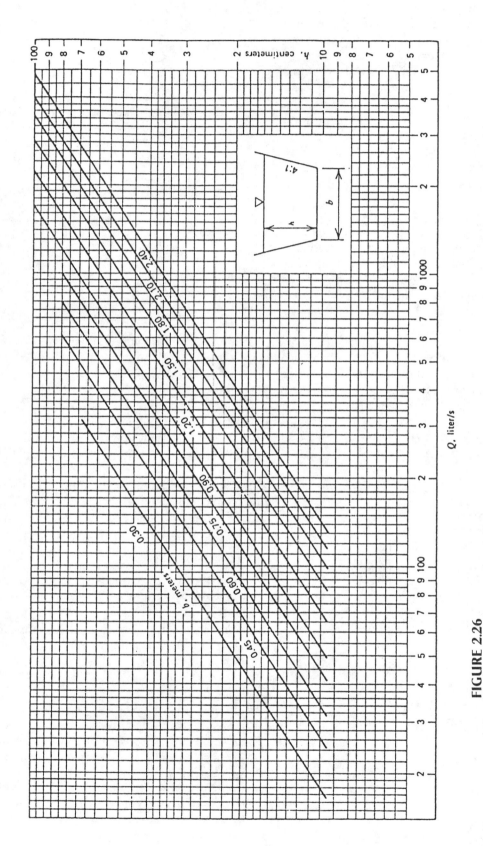

**FIGURE 2.26**
*Discharge nomograph for Cipoletti weirs.* (Courtesy of *Public Works Magazine.*)

## Flumes

A flume is a specially shaped constricted section in an open channel (similar to the venturi tube in a pressure conduit). The geometry of the flume causes a *free-fall* condition in the channel, which allows a correlation to be made between the discharge and the depth of flow. Although they are more expensive to install than weirs, flumes offer the advantage of a self-cleansing action that prevents deposits of sewage solids. Also, there is a little head loss and no significant backup of sewage upstream of the meter.

One flume often used for a permanent sewage flow-metering installation is the *Parshall flume,* shown in Figure 2.27. For small channels, prefabricated fiberglass flumes can be installed, but for larger systems, the flumes are built of cast-in-place concrete.

A set of tables or a nomograph can be used to relate the water level in the flume to the flow rate. Usually, automatic recording devices provide a continuous record of discharge once the instrument has been calibrated with the flume and level-sensing device. Another type of flume, called a *Palmer–Bowlus flume,* can be placed in existing circular channels or sewer pipes, as illustrated in Figure 2.28.

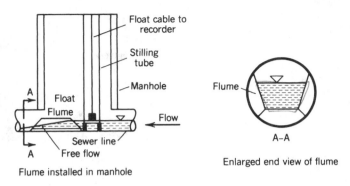

**FIGURE 2.28**
*A typical Palmer–Bowlus flume installation.*

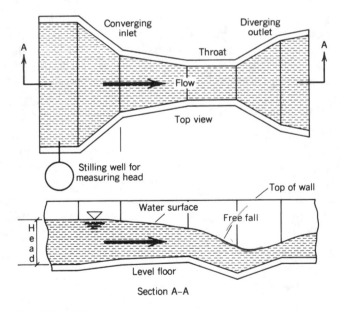

**FIGURE 2.27**
*A Parshall flume is often used to measure flow rate in an open channel that carries sewage; discharge is related to the head or depth of water just upstream of the constricted flume section.*

## 2.5 NONUNIFORM OPEN CHANNEL FLOW

Section 2.4 focuses on steady uniform gravity flow, primarily in circular pipes. There are many other applications of open channel flow hydraulics in environmental technology. Open channel flow that is not always steady over time or uniform in cross-sectional area is frequently encountered. Such *gradually varied flow* occurs in natural streams and rivers as well as urban sanitary and storm sewer systems. The physical principles and mathematical methods needed to fully analyze nonuniform flow problems are covered in more advanced engineering textbooks. This section provides a brief descriptive overview of nonuniform and gradually varied flow, to introduce students to hydraulic conditions that are often encountered in practice.

As described in Section 2.4, steady uniform flow refers to the hydraulic condition where the discharge (flow rate) and cross-sectional area of flow remain constant throughout the entire length the of the channel. Under these conditions, the velocity and the depth of flow must also be constant, and the hydraulic grade line must be parallel to the actual channel slope. The depth of flow is called the *normal depth.* In prismatic (constant-cross-section) channels that are sufficiently long, the flow conditions naturally approach uniform flow and the depth naturally approaches the normal depth, with steady discharge.

## Specific Energy

Channels do not always flow at normal depth, particularly in the vicinity of changes of slope or channel cross-sectional changes. An important concept related to the type of flow in a channel is called the *specific energy,* which is defined as the sum of the depth of flow and the velocity head in a channel. In equation form, this can be expressed as

$$E = y + v^2/2g$$

where $E$ = specific energy, m (ft)
    $y$ = depth of flow, m, (ft)
    $v$ = average velocity, m/s (ft/s)
    $g$ = acceleration due to gravity, m/s$^2$ (ft/s$^2$)
    $v^2/2g$ = "velocity head"

If friction losses and change in elevation are neglected for a very short section of a channel, the specific energy must be constant for two adjacent channel cross sections, or $E_1 = E_2$, and thus

$$y_1 + v_1^2/2g = y_2 + v_2^2/2g$$

From the principle of continuity of flow, the velocity of flow is inversely proportional to the area of flow, with constant discharge. Also, area of flow is a function of channel depth. Therefore, for a given discharge the *specific energy is solely a function of depth at each point in a channel*, and there may be more than one depth having the same specific energy. This is depicted in Figure 2.29, which is a plot of channel depth $y$ versus specific energy $E$ for a constant flow rate.

## Critical Flow

It can be seen from Figure 2.29 that there is one flow depth at which the specific energy reaches a minimum value. This is called the *critical depth*, and the velocity at which critical depth occurs is called the *critical velocity*. If the actual depth of flow is higher than the critical depth, the type of flow is characterized as *subcritical*; flow velocity in subcritical flow is slower than the critical velocity. If the actual depth of flow is lower

than the critical depth, the flow is characterized as *supercritical*; flow velocity in supercritical flow is faster than the critical velocity. Typically, water flowing rapidly down a steeply sloped, shallow channel is undergoing supercritical flow, whereas water flowing slowly in a relatively deep channel on a gentle or mild slope is in subcritical flow conditions.

## Gradually Varied Flow

When the flow is steady and the change in water depth along a reach of a channel is negligible, the constant water depth is the normal depth of flow. In many cases, though, in sanitary and in storm sewer systems and other hydraulic structures, as well as in natural streams or rivers, there may be some condition that prevents water depth from equaling normal depth throughout the full length or reach of the channel. For example, an obstruction at the end of a storm drain may force the depth to be above normal at the downstream outlet of the pipe. Under such conditions, the engineer must be able to calculate the upstream water elevations, or *hydraulic grade*, to make sure flooding or pipeline surcharging will not occur.

When difference in water depth from one end of an open channel reach to the other is significant, the flow is characterized as *gradually varied flow*, and an advanced form of engineering analysis is required. The channel is divided into a series of smaller lengths or segments, and each segment is analyzed separately. The first step is to determine the type of flow that is expected to occur in the channel (or gravity pipe) segment. This determination is based upon the slope of the channel, the calculated normal depth and critical depth, and controlling downstream (or upstream) hydraulic conditions (or boundary conditions).

### Slope Classification

The normal depth and critical depth can be computed for the section, using appropriate engineering formulas. If the normal depth is higher than the critical depth, the slope is said to be *mild*. If the normal depth is equal to the critical depth, the slope is said to be *critical*. If the normal depth is lower than the critical depth, the slope is said to be *steep*.

### Flow Profile Classification

There are three zones of gradually varied flow, as follows:

1. *Zone 1*: The actual flow depth is higher than both normal depth and critical depth.
2. *Zone 2*: The actual flow depth is between normal depth and critical depth.
3. *Zone 3*: The actual flow depth is lower than both normal depth and critical depth.

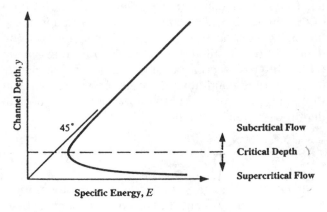

**FIGURE 2.29**
*A graph of the flow in an open channel versus the* specific energy *of flow. The critical depth occurs when the specific energy is minimum.*

Once the slope classification and flow zone have been determined, the flow profile type can be defined. (The flow profile is a "side view" of the flow, showing the straight channel bottom and the curved water surface, as contrasted with a cross-section view.) There are several flow profile classifications. It is necessary to know the profile classification in order to be able to proceed with hydraulic grade computations. A selection of some gradually varied flow profile types is depicted in Figure 2.30. In Figure 2.30, the solid lines represent the actual water surfaces, the lines with small dashes represent the calculated normal depths, and the lines with large dashes represent the calculated critical depths. Knowing the profile type and the location of the actual water depth, the engineer can determine whether the flow is subcritical or supercritical.

In order to prevent excessive velocities that could cause pipe scour or channel erosion, most storm sewers are designed with mild slopes to carry subcritical flows. This means that the hydraulic control is at the downstream end of the section and computations proceed in the upstream direction. When the flow depth is above the normal depth (Zone 1), on a mild slope, the flow profile is characterized as an M1 profile. The computation of upstream water elevations (the hydraulic grade) for an M1 profile is called a *backwater analysis*.

When the flow depth is between critical and normal depth (Zone 2), on a mild slope, the profile is characterized as an M2 profile. The computation of the hydraulic grade for an M2 profile is called a *drawdown analysis*. When supercritical flows occur, the controlling section is at the upstream end of the conduit and the computations proceed in the downstream direction. This type of calculation is called a *frontwater analysis*.

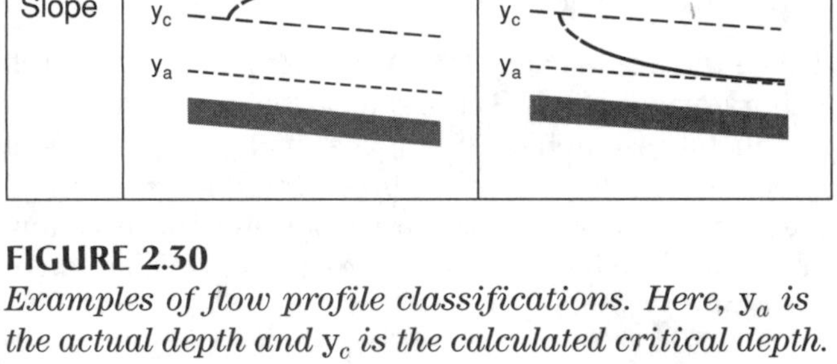

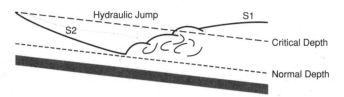

**FIGURE 2.31**
*In a hydraulic jump, the flow suddenly changes from supercritical to subcritical and the depth of flow increases over a short segment of the channel.*

## Rapidly Varied Flow

In some cases involving gravity flow, the profile types are mixed within a channel segment. Rapidly varied turbulent flow may result from a sudden change in upstream and downstream slopes. One common example of rapidly varied flow is the *hydraulic jump*, which occurs when the flow passes from supercritical to subcritical conditions. Figure 2.31 illustrates a hydraulic jump on a steep slope channel, with an S2 profile making an abrupt transition to an S1 profile.

In a storm sewer system, for example, a hydraulic jump will occur when there is a discharge from a steeply sloping pipe into a high *tailwater* condition. (The term tailwater refers to the downstream hydraulic grade condition, which may be caused by a channel constriction, an obstruction like a weir, or another condition that backs up the flow.) Figure 2.31 illustrates this situation. There are significant energy losses associated with hydraulic jumps due to the amount of turbulence (rapid mixing) that occurs. The forces caused by the jump can cause significant erosion, so engineers may try to prevent hydraulic jumps from occurring in sewer systems. As an alternative, they can calculate the expected location of the jump in order to provide adequate channel, pipe, or structure protection. For instance, when a hydraulic jump occurs just downstream of a dam spillway (Figure 2.32),

| | Zone 1 Flow Profiles | Zone 2 Flow Profiles |
|---|---|---|
| Mild Slope | M1 $y_a$ $y_c$ | M2 $y_a$ $y_c$ |
| Steep Slope | S1 $y_c$ $y_a$ | S2 $y_c$ $y_a$ |

**FIGURE 2.30**
*Examples of flow profile classifications. Here, $y_a$ is the actual depth and $y_c$ is the calculated critical depth.*

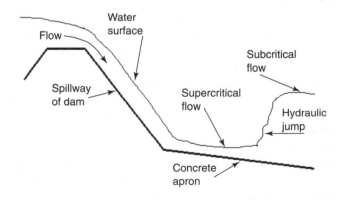

**FIGURE 2.32**
*The location of the hydraulic jump below a dam must be determined in order to design the apron.*

a suitable apron must be designed and built so that excessive channel bottom erosion does not occur.

## 2.6 COMPUTER APPLICATIONS IN HYDRAULICS

It is essential for students to fully understand the underlying physical and mathematical concepts when doing hydraulic computations. This reduces the chance for errors in any hydraulic design or analysis problem. Once the underlying concepts are understood, however, the solution processes can become repetitive and tedious, even with computational aids such as nomographs. Repetitive computational problems are very well suited to computer analysis. Although some hand-held calculators can be programmed to do the calculations automatically, the use of commercial software packages avoids the need to program the calculator, and generally offers a more extensive set of options for doing the calculations.

There are several advantages of using computerized solutions for performing common hydraulic calculations. First and foremost, the amount of time to do the calculations is greatly reduced. As long as there are no "bugs" in the program, the solution process will be less prone to algebraic and unit conversion errors; conversions between SI metric and U.S. Customary units can be done automatically, without mistakes. Because of the speed and accuracy of computer calculations, a wide range of design options can be evaluated in less time than a single computation can be done by "hand." Last, but not least, computer solutions are readily documented and reproducible.

Using computers to perform engineering calculations does not relieve the user of the software from liability for the accuracy of the results. In other words, an engineer who uses software with bugs and produces an erroneous analysis or poor-quality design cannot blame the software for the mistakes. The engineer is still legally responsible for the work. Before using any software for the first time, it is best to test it out first by comparing the computer solutions to "hand" calculated solutions for a few typical problems. This is one reason why it is important to know the underlying theoretical principles for the solutions.

One example of a powerful, easy-to-use software program for hydraulic calculations is called FlowMaster® (Haestad Methods, Inc., Waterbury, Connecticut; see their Web site at www.haestad.com). This is an easy-to-use hydraulics program that rapidly solves design and analysis problems involving pipes and open channels, like the example problems in the preceding sections of this chapter. FlowMaster computes such quantities as flows, pressures, slopes, and diameters, using a wide choice of well-known hydraulic equations, such as the Manning and Hazen–Williams formulas. The user simply inputs known values and the program automatically computes and displays the desired quantity. Tabular and graphical output can be viewed on the screen, saved to a file, or printed. An unlimited number of worksheets can be created for analyzing uniform sections of pressure pipe or open channels, including irregular sections (such as natural streams).

 ## 2.7 RELEVANT WEB SITES

### FLOW MEASUREMENT TECHNOLOGY

http://www.usbr.gov/wrrl/fmt/index.htm

This Web site, maintained by the Water Resources Research Laboratory of the Bureau of Reclamation, provides much useful information on flow measurement methods, devices, and instrumentation, including a Comprehensive Water Measurement Manual.

### HYDROLOGIC ENGINEERING CENTER (HEC)

http://www.hec.usace.army.mil/

This is the Home Page of the HEC, an office of the U.S. Army Corps of Engineers. The HEC conducts hydrologic engineering and water resources research, training, planning analysis, and technical assistance. The HEC incorporates state-of-the-art procedures and techniques into manuals and comprehensive computer programs that are available to the public.

### IOWA INSTITUTE OF HYDRAULIC RESEARCH

http://www.iihr.uiowa.edu/

This is the Home Page of the Iowa Institute of Hydraulic Research (IIHR), a unit of the University of Iowa's College of Engineering, one of the nation's oldest fluids research and engineering laboratories. The IIHR seeks to educate students and to conduct research in the broad fields of hydraulics and fluid mechanics.

### WATER RESOURCES RESEARCH LABORATORY

http://www.usbr.gov/wrrl

This is the Home Page of the Water Resources Research Laboratory of the U.S. Bureau of Reclamation, which conducts research and applies hydraulic modeling expertise to solve water resources, hydraulics, and fluid mechanics problems in the areas of water conservation, environmental hydraulics, and testing and evaluation of hydraulic systems.

## REVIEW QUESTIONS

1. Why is hydraulics an important aspect of environmental technology?

2. What is the definition of *pressure?*

3. List four important characteristics of hydrostatic pressure.

4. What is the difference between *absolute pressure* and *gage pressure?*

5. What is *pressure head?*

6. Briefly describe three different ways to measure pressure.

7. What is the simple formula that relates flow rate to flow velocity and area?

8. Briefly describe what is meant by *continuity of flow.*

9. Briefly describe the principle of conservation of energy as it relates to hydraulic systems.

10. Explain why the pressure in a pipeline drops in a constricted section of the pipe.

11. What is a *hydraulic grade line?*

12. Briefly describe the operating principle of a venturi meter.

13. Give a definition of *uniform, steady, open channel flow.*

14. Briefly describe two methods for measuring discharge in an open channel.

15. Briefly define *normal depth* in open channel flow. Under what conditions does the depth of flow in an open channel differ from the normal depth?

16. What is *specific energy?* Sketch and label a graph depicting depth of flow in a channel versus specific energy. What is *critical depth* and how is it related to specific energy?

17. Briefly describe the difference between *subcritical* flow and *supercritical* flow in an open channel.

18. What is *gradually varied flow?* Briefly describe three zones or classifications of gradually varied flow.

19. Briefly describe the conditions for an *M1* flow profile type to occur in an open channel. Do the same for an *M2* flow profile.

20. Give an example of *rapidly varied flow* and briefly explain its significance in the design of hydraulic facilities.

21. If, when using a computer software package to perform hydraulic calculations, erroneous results are obtained, who is legally liable for any damages that may occur—the software provider or the engineer who uses the software?

22. Visit the *Iowa Institute of Hydraulic Research* Web site. From the "Projects" hyperlink, select a research report in the Environmental Hydraulics category. Write a brief summary of the report.

23. Visit the *Water Resources Research Laboratory* Web site and click on the link to the Water Measurement Manual. From the Table of Contents, select a measurement device not discussed in this chapter and write a brief description of the operation and application of the device.

24. Using the EnviroSources search engine (see Section 1.6) or another Internet search directory or search engine, locate at least one additional Web site relevant to one of the topics in this chapter. Add the link(s) to your Environmental Technology Folder. Write a brief description of what the Web site(s) contain.

## PRACTICE PROBLEMS

1. A water supply reservoir is 50 ft deep. Compute the pressure at the bottom of the reservoir and at a point 30 ft above the bottom.

2. The pressure at the bottom of an open tank is measured to be 50 kPa. How deep is the water in the tank?

3. A water storage tank is situated on a hill, as shown in Figure 2.33. The water main is 2 m below the ground surface at all points. The elevations above sea level of the ground and the water surface are given. Compute the hydrostatic pressure at the closed valve at the bottom of the hill.

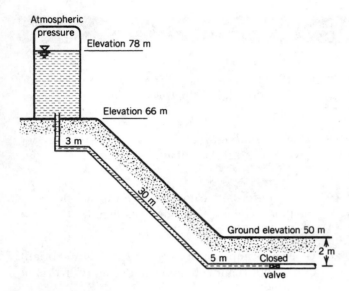

**FIGURE 2.33**
*Illustration for Problem 3.*

4. The pressure in a water main is 50 psi. What is the pressure head in the main? What is the hydrostatic pressure at a customer's tap that is 40 ft above the main?

5. If the air pressure in the sealed tank of Figure 2.6 were 30 kPa, what would the pressure gage read at the tank bottom and how high would the water rise in the vertical tube?

6. A 300-mm-diameter pipe carries a flow of 100 L/s. Compute the velocity of flow.

7. What diameter pipe is needed to carry a flow of 500 gpm at a velocity of 1.4 fps?

8. Water is flowing at a velocity of 2 m/s in a 200-mm-diameter pipe. The pipe diameter is reduced to 100 mm at a constriction in the pipe. Determine the flow velocity in the constricted pipe section.

9. For the branching pipe section shown in Figure 2.34, determine the flow velocity in pipe A if the velocity of flow in pipe B is 2 m/s and that in pipe C is 1 m/s.

10. For the system illustrated in Figure 2.13, the diameter at section 1 is 16 in. and that at section 2 is 8 in. The flow rate through the pipe is 6.0 cfs, and the pressure at section 1 is 50 psi. What is the pressure in the constriction at section 2?

11. For the system illustrated in Figure 2.13, the diameter at section 1 is 300 mm and that at section 2 is 100 mm. The flow rate through the pipe is 50 L/s, and the pressure at section 1 is 700 kPa. What is the pressure in the constriction at section 2?

12. A 600-mm-diameter pipe carries a flow of 0.20 m³/s. Compute the pressure drop per kilometer of pipeline.

13. What minimum pipe diameter is needed to carry a flow of 1000 gpm without exceeding a pressure loss of 20 psi per mile of pipeline?

14. A 300-mm-diameter pipe carries water with a head loss of 10 m/km of pipeline. Determine the flow rate in the pipe using the Hazen–Williams nomograph and check the solution using Equation 2-9.

15. A venturi meter has a pipe diameter of 6 in. and a throat diameter of 3 in. A pressure difference of 10 psi is measured in the meter. What is the flow rate under these conditions?

16. A venturi meter has a pipe diameter of 150 mm and a throat diameter of 75 mm. A pressure difference of 100 kPa is measured in the meter. What is the flow rate under these conditions?

17. An 800-mm-diameter sanitary sewer is built on a slope of 0.2 percent. What is the full-flow discharge capacity of the sewer and what is the flow velocity?

18. An 18-in.-diameter sewer is placed on a grade of 1.5 ft/1000 ft. What are the discharge and the flow velocity when the pipe is full?

19. What diameter pipe is needed to carry a peak flow of 200 L/s on a grade of 0.007?

20. What grade is required for a 36-in.-diameter pipe to carry at least 7 mgd of sewage at a velocity of not less than 2 ft/s?

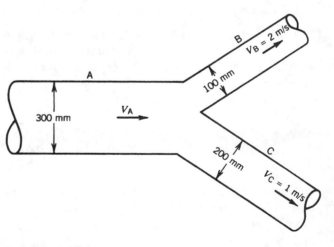

**FIGURE 2.34**
*Illustration for Problem 9.*

21. A 300-mm-diameter sewer is built at a grade of 2 percent. What is the depth of sewage flowing in the pipe when it carries a flow of 50 L/s? What is the velocity?

22. What is the highest discharge capacity of an 18-in.-diameter pipe on a slope of 0.15 percent, and what is the depth of flow in the pipe?

23. What is the highest flow velocity in a 900-mm pipe on a 0.1 percent slope? What are the discharge and the depth of flow?

24. The invert elevation of a 600-mm sewer drops 0.5 m over a 100-m distance. Determine the discharge and the flow velocity in the sewer when the depth of flow is 200 mm.

25. A float placed on the surface of water flowing in a stormwater drainage ditch is seen to move a distance of 25 m downstream in 75 s. The ditch is 2 m wide and the average depth of flow is estimated to be 0.75 m. Estimate the channel discharge.

26. Estimate the discharge in cfs over a 90° V-notch weir when the head on the weir is 4 in.

27. Estimate the discharge in L/s over a 90° V-notch weir when the head is 150 mm.

28. Estimate the discharge over a Cipolleti weir with a base width of 20 in. when the head on the weir is 10 in. Compare this result to that obtained from Figure 2.26.

## Chapter Outline

CHAPTER

# 3

# Hydrology

Hydrology is a branch of earth science that is concerned with the distribution and movement of water on and under Earth's surface. The science of hydrology is of great importance in environmental technology for many reasons. Two extreme hydrologic conditions, droughts (not enough water where needed) and floods (too much water in the wrong place), are well known for the environmental problems they cause. But droughts and floods are not the only aspects of hydrology that are important. In general, the presence and quantity of water must be estimated in order to plan, design, and operate water supply, pollution control, and stormwater management facilities.

The purpose of this chapter is to present fundamental hydrologic concepts for measuring present conditions and estimating future variations in water availability. Applications of these concepts are discussed in further detail in Chapter 9 on stormwater management and water pollution and in other sections of the book.

## 3.1   WATER USE AND AVAILABILITY

Everyone knows that water is essential for sustaining life. It also plays a central role in the growth and environmental health of cities and towns. People depend on water for more than just drinking, cooking, and personal hygiene. Vast quantities are often required for industrial and commercial uses. In some parts of the country, large quantities of water for irrigation are necessary to support agriculture. Water resources are also essential for power generation, recreation, fish and wildlife conservation, and navigation.

*Water use* refers to the withdrawal of water from its source, which may be a river, lake or well, and the transport of that water to a specific location. For example, water used for cooling purposes in a power plant may be diverted from a nearby river, passed through the power plant, and then discharged back into the river without significant loss in quantity. (The water would have to be cooled down before discharge to prevent thermal pollution, which is discussed in Chapter 5.) Navigation and recreation are other examples of *nonwithdrawal use*. However, it is necessary to make a distinction between water *use* and water *consumption*. Water that is used for drinking or combined with a product and is not directly available for use again is consumed water.

More than 100 million cubic meters ($m^3$) of water per day is withdrawn for public water supplies in the United States. More than 500 million $m^3$ is withdrawn each day for irrigation. Industrial use accounts for the largest share of water demand, almost 1 billion $m^3$ per day. Most of this is used as cooling water at electric power utilities. These approximate figures are presented to give an appreciation of the tremendous quantities of water needed.

Water is present in abundant quantities on and under Earth's surface, but only less than 1 percent of Earth's water is actually available for use in economically satisfying the needs mentioned. Most of Earth's water is salt water or is frozen in the polar ice caps.

Many freshwater lakes and rivers have been deteriorating in quality because of land development and pollution, limiting the availability of water for use, particularly for public water supplies. Even groundwater is affected by pollution in some areas, although much of it is just too deep to pump out of wells economically.

In 1998, a UN conference of 84 nations was held to discuss management of the world's limited supply of fresh water. It was estimated at the conference that about one quarter of the world's 5.9 billion people have no access to clean drinking water. The conference delegates agreed that water should be paid for as a commodity rather than considered an essential staple to be supplied virtually free of cost. Water shortages are so important that governments may need to rely on private funds for the large investments needed for water networks and treatment systems.

## The Distribution of Water

In addition to the limited availability of usable water, another basic problem in managing water resources is that it is not evenly distributed geographically. In some regions there is ample precipitation, including rain, snow, hail, sleet and dew, and water is readily available for use. On the average, about one third of this precipitation becomes available in lakes and rivers, and some makes its way into the groundwater. But where there is little precipitation, water is scarce. The fact that there is a close relationship between the amount of rain or snow and the amount of water available for use should be self-evident. This is discussed in more detail later in this chapter.

Figure 3.1 illustrates the different annual precipitation amounts across the United States. Except for the extreme northwestern corner of the country, where the total annual amount of rainfall may exceed 2500 mm (100 in.), it can be seen that the eastern half of the country gets significantly more rainfall than the western half. In some areas of the Southwest, less than 100 mm (4 in.) of rain may fall in any one year. In the Northeast, an annual rainfall of about 1000 mm (40 in.) is moderate compared to the two previously mentioned extremes.

The amount of rainfall and the availability of water can vary considerably even within a relatively small area. California is an example of a state with a very uneven distribution of water. Although southern California is very dry, the growing population there generates a large demand for water. Most of the needed water must be transported to the south from the northern part of the state, where water is more readily available. A huge system of reservoirs, open channels, pumping stations, and tunnels is used to accomplish this transfer of water.

Part of the system, called the California Aqueduct, can convey about 2800 $m^3$ (100,000 $ft^3$) of water per second. The aqueduct is an open channel, about 40 m (130 ft) wide at the surface and about 9 m (30 ft) deep. At one point in the system, the water is pumped up about 600 m (2000 ft) to get over a mountain, quite an engineering undertaking.

The uneven distribution of water from one geographic location to another is only part of the problem in hydrology and water resources management. The occurrence and availability of water also vary with time.

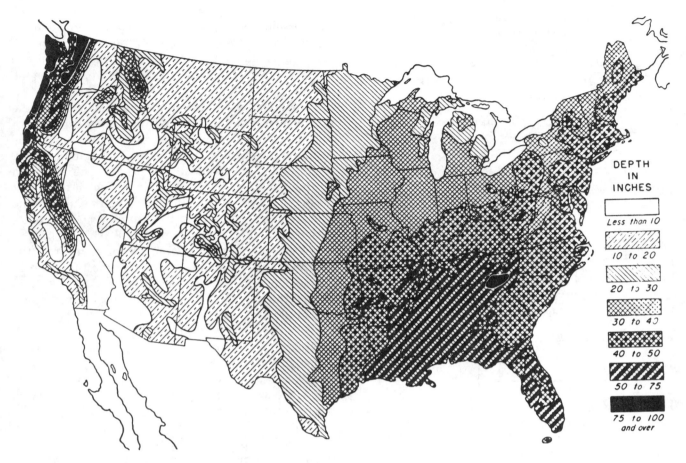

**FIGURE 3.1**
*The distribution of annual precipitation, and therefore the availability of water, is not uniform over the continental United States.* (From American Society of Civil Engineers, Committee on Hydrology, *Hydrology Handbook*, ASCE, New York, 1949. Copyright © 1949 by the American Society of Civil Engineers. Used with permission)

In any given location there may be occasional periods of little rainfall or drought, and severe water shortages may result as water in storage reservoirs is used up during these dry periods.

On the other hand, the same area may sometimes experience periods of above-average rainfall. Serious flooding problems may result, with accompanying loss of lives and property, as well as environmental pollution problems. In any given area, then, there can be too little water or too much water, depending on natural climatic conditions.

## 3.2   THE HYDROLOGIC CYCLE

Water is in constant motion on, under, and above Earth's surface. Even in what appears to be a stagnant pond, the water is evaporating, changing into a vapor and moving into the atmosphere. Powered by energy from the sun and from gravity, there is a constant circulation of water and water vapor. This natural process is called the *hydrologic cycle*. It is illustrated in schematic form in Figure 3.2.

Surprisingly, there was a time when people did not have an understanding of the cyclical motion of water through the environment and had misconceptions about the origin of water in streams or lakes. Even today, some people still have misconceptions, particularly with respect to groundwater.

Although the hydrologic cycle looks simple when sketched in schematic form as in Figure 3.2, there is more to it than initially meets the eye. The science of hydrology gets quite complicated, applying a good deal of statistics and higher mathematics. The basic objective is to measure and analyze the relationships controlling the form, quantity, and distribution of water.

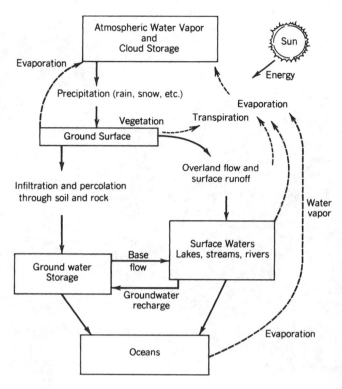

**FIGURE 3.2**

*Schematic diagram of the natural* hydrologic cycle. *The constant circulation of water is powered by energy from the sun and by gravity.*

When these relationships are understood, reliable predictions may be made concerning the occurrence of future floods or droughts. It is important that technicians involved in environmental control have an appreciation of the basic structure of the hydrologic cycle.

Precipitation begins when atmospheric moisture (water vapor) is cooled and condensed into water droplets. The precipitation can follow three different paths after it reaches the ground. First, some of it may be intercepted by vegetation or small surface depressions. In other words, it is temporarily stuck on the surfaces of leaves or grass or it is retained in puddles. Second, a portion of the water can *infiltrate* through the earth's surface and seep (or *percolate*, as it is called) downward into the ground. Third, a portion of the water can flow over the ground's surface. Measuring and predicting the relative amounts of water that follow each of these paths is of importance in hydrology.

Some of the intercepted water soon evaporates, and some of it is absorbed by the vegetation. A process called *transpiration* takes place as water is used by the vegetation and passes through the leaves of grass, plants, and trees, returning to the atmosphere as vapor.

The combined process of evaporation and transpiration is called *evapotranspiration*. Overall, more than half of the precipitation that reaches the ground is returned to the atmosphere by this process before reaching the oceans.

Overland flow and surface runoff occur when the rate of precipitation exceeds the combined rates of infiltration and evapotranspiration. Eventually, the overland flow finds its way into stream channels, rivers, and lakes, and finally the oceans. The ocean can be thought of as the final "sink" to which the water flows. As previously mentioned, about one third of the average annual rainfall in the United States becomes surface runoff in streams and rivers. This, of course, varies from region to region. In some areas of the Southwest, for example, there is no runoff for years at a time, since the rate of precipitation does not often exceed the rate of infiltration and evapotranspiration in that area.

The water that infiltrates the ground surface will percolate into saturated soil and porous rock layers, forming vast groundwater reservoirs. A "groundwater reservoir" should not be visualized as an underground lake—the water actually fills the tiny voids or spaces between the soil particles and fractures in the rock in what may be called an *aquifer*. (Aquifers are discussed further in Section 3.7.) The groundwater may later seep out onto the ground surface in springs or into streams. (Groundwater flowing into streams is referred to as the *base flow*, which may be the sole source of streamflow during dry weather periods.) Eventually, the groundwater makes its way to the ocean, either directly or via surface streams. Evaporation from the ocean surface substantially replenishes the water vapor in the atmosphere, winds carry the moist air over land, and the hydrologic cycle continues.

**Urban Hydrologic Cycle**

This description of the hydrologic cycle is only a brief summary of a complex natural phenomenon. Some of the details of this natural cycle are discussed in the following sections of this chapter. But one water cycle should be mentioned here—the *urban* water cycle, illustrated in Figure 3.3.

In human communities there is a constant circulation of water. Water is withdrawn from its source in the natural hydrologic cycle—surface waters or groundwater—and is pumped through treatment and distribution systems. After use, the wastewater is collected in sewer systems, treated to reduce the effect of pollution, and finally disposed of back into surface water or groundwater. A most significant aspect of environmental technology is the maintenance of this

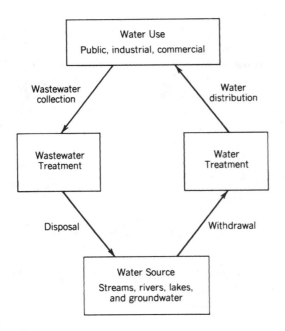

**FIGURE 3.3**
*The urban hydrologic cycle.*

urban water cycle while protecting public health and environmental quality. Much of this textbook focuses on this topic.

## 3.3 RAINFALL

Water in streams, rivers, and lakes, as well as water in the ground, is the residue of precipitation. It is possible, and often necessary, to refer to records of rainfall in order to estimate the quantity of water that will be found on and under the ground. Other factors, such as topography and land use, play a role in the relationship between rainfall and water availability. These will be considered later. In this section, basic concepts related to rainfall intensity and volume are discussed.

### Depth, Volume, and Intensity

The collection of rainfall data is the responsibility of the U.S. National Weather Service, a government agency that maintains rain-gage stations throughout the country. Rainfall amounts are expressed in terms of the depth of water accumulated in the rain gage during a storm. The units can be expressed in millimeters or inches. It is usually necessary to compute weighted averages of rainfall amounts over a region, using the data from several rain gages. The data may be weighted in proportion to the area covered by each gage.

Sometimes it is necessary to compute the total volume of water that falls on an area during a storm. The volume is computed by multiplying the land area by the rainfall depth, as follows:

$$\text{volume} = \text{depth} \times \text{area} \qquad \textbf{(3-1)}$$

In SI metric units, the volume is usually expressed in terms of cubic meters, but rainfall depth is expressed in terms of millimeters. To keep the units consistent when applying Equation 3-1, area should be expressed in square meters and rainfall depth should be converted to meters. Relatively large areas that are expressed in units of hectares (ha) should first be converted to $m^2$ (1 ha = 10 000 $m^2$).

### EXAMPLE 3.1

During a 20-min rain storm, a depth of 25 mm of rainfall was recorded for an area of 2.5 ha. Compute the total volume of water that fell on that area during the storm.

*Solution*    First, convert the rainfall depths from mm to m:

$$25 \text{ mm} \times \frac{1 \text{ m}}{1000 \text{ mm}} = 0.025 \text{ m}$$

Now convert ha to $m^2$, as follows:

$$2.5 \text{ ha} \times \frac{10\,000 \text{ m}^2}{1 \text{ ha}} = 25\,000 \text{ m}^2$$

Applying Equation 3-1, we get

$$\text{volume} = \text{depth} \times \text{area}$$
$$= 0.025 \text{ m} \times 25\,000 \text{ m}^2 = 625 \text{ m}^3$$

We can round this up to 630 $m^3$, because the given data have only two significant figures. (In this problem, we do not need the data regarding storm duration. We will use that information in a subsequent example.)

Often more important than total volume of rain is the rate at which the rain falls. This is called *rainfall intensity*. As discussed later, the rainfall intensity that occurs during a storm is of particular significance in civil and environmental technology, particularly in the design of urban drainage facilities.

Rainfall intensity is expressed in terms of depth per unit time, as in./h, mm/min, or mm/h. The National Weather Service gathers this kind of data using automatic rain gages that record rainfall duration as well as depth; a continuous record of rainfall amount and intensity is plotted on a revolving drum. It is generally observed that short-duration storms have higher average rainfall intensities than longer duration storms. This will be of significance when we consider problems in stormwater control.

## EXAMPLE 3.2

For the storm described in Example 3.1, compute the rainfall intensity.

*Solution*  Even though the storm lasted only 20 min, its intensity can still be computed in terms of mm/h. In effect, this is the same as computing how much rain would have fallen if the storm had lasted for 1 h at a steady intensity:

$$\text{intensity} = \frac{25 \text{ mm}}{20 \text{ min}} \times \frac{60 \text{ min}}{1 \text{ h}} = 75 \frac{\text{mm}}{\text{h}}$$

When using U.S. Customary units, *cubic feet* is a common unit for volume. But in hydrologic applications, large volumes of water are usually expressed in terms of *acre–feet* (ac–ft). This may seem to be a strange term at first, but, as illustrated in Figure 3.4, 1 ac–ft can easily be visualized as the volume required to cover 1 ac of land to a depth of 1 ft. Since 1 ac is equivalent to 43,560 $\text{ft}^2$, 1 ac–ft is equal to 43,560 $\text{ft}^2 \times 1$ ft, or 43,560 $\text{ft}^3$ (325,900 gal).

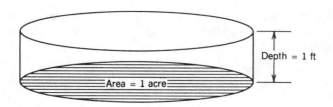

## FIGURE 3.4
*One acre–foot of water is equivalent to the volume that would cover 1 ac of land at a depth of 1 ft, or 43,560 cubic feet ($ft^3$) of water. (Not to scale.)*

## EXAMPLE 3.3

During a storm, a total of 4.0 in. of rain fell on an area of 120 ac. The storm duration was 8.0 h. What was the average rainfall intensity during the storm? Determine the total volume of rain that fell on the area in the 8-h period. Express the volume in acre–feet and cubic feet.

*Solution*  The average intensity is determined by dividing the total depth of rain by the storm duration, as follows:

$$\text{intensity} = \frac{4 \text{ in.}}{8 \text{ h}} = 0.5 \frac{\text{in.}}{\text{h}}$$

To apply Equation 3-1, the depth should be expressed in feet. The computation can be done in one step, as follows:

$$\text{volume} = 120 \text{ ac} \times 4.0 \text{ in} \times 1 \text{ ft./12 in.}$$
$$= 40 \text{ acre–feet}$$

To convert from acre–feet to $\text{ft}^3$, we use

$$\text{volume} = 40 \text{ ac–ft} \times 43{,}560 \text{ ft}^3/\text{ac–ft}$$
$$= 1{,}700{,}000 \text{ ft}^3$$

(The answer is rounded off to two significant figures.)

It is important to consider that the rainfall intensity is not constant over the duration of a storm, although the average intensity is a very useful number in a wide variety of hydrology problems and applications. In some hydrologic analyses, though, it is necessary to have more detailed information regarding the rainfall intensity. These data can be depicted in a *hyetograph*, which is a graph of rainfall intensity (or volume) versus time. An example hyetograph is depicted in Figure 3.5. Note that the average rainfall intensity over the 60-min duration of the storm is about 2.2 in./h, whereas the peak intensity is about 8 in./h.

## Recurrence Interval

Common experience shows that hydrologic events, such as rain storms, do not occur with any definite regularity. The time span or period between storms is not constant. The occurrence of rainfall, its intensity, and its duration are random natural events. Consider, for

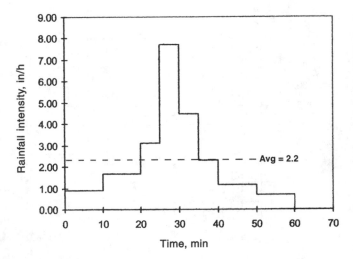

**FIGURE 3.5**
*Example of a* hyetograph.

instance, the storm described in Example 3.1. It dropped 25 mm of rain in 20 min. Despite the random nature of precipitation events, it is possible to determine average frequencies of occurrence of storms having specific intensities and durations. It would be convenient if the exact dates on which identical storms would occur in the future could be predicted, but obviously that is not possible. For instance, even though the date of the next 20-min, 25-mm storm cannot be determined in advance, it is possible to predict how many times a similar storm can be expected to occur over the next year or several years. In addition, a prediction of the likelihood or probability of observing that storm again in any given period can be made.

By examining many years of rainfall records and applying statistical analyses, the average number of years between storms of specific intensities and durations can be determined. This time span between identical storms is called the *recurrence interval* or *return period* of the storm. These return periods are determined and reported by the National Weather Service, and designers of environmental facilities must know how to interpret and use the data.

When applying these data, the expression *N-year storm* is used, where *N* stands for the recurrence interval in years. For example, a storm with a return period of 5 years is called a 5-year storm. This means that over a long period of time, the average time span between storms of that particular intensity and duration is 5 years. It does *not* mean that a similar storm will occur once exactly every 5 years. In fact, it is possible that more than one of these 5-year storms could occur within a shorter time span, even within a single year, but the

chances for this are slim. Note, too, that the probability of the 5-year storm occurring in any given 5-year period is not quite 100 percent. In other words, no one can say for sure that what is called a 5-year storm will actually take place within, say, the next 5 years. But over a long time span, 500 years, for example, it is a good bet that there will be about 100 of these 5-year storms.

## Probability of Occurrence

Data on storm intensity, duration, and return period are important in the design of urban drainage structures and for predicting peak flows in rivers. On the other end of the hydrologic spectrum, knowing the severity of droughts and their frequency of occurrence is of importance in designing water supply reservoirs.

Because of the uncertain and irregular nature of hydrologic events, there is always some risk of failure when designing a structure or facility involving water resources. For example, a river used for water supply may not provide enough water for a growing community during dry periods. Even if a small reservoir were built to overcome this deficiency, there would remain the risk that a more severe (though less frequent) drought would cause the reservoir to run dry. This risk can be reduced by building a larger reservoir, but this would be more expensive. Designers must be able to balance the economics and the risks, using probability concepts.

The probability or chance that a given event will occur can be expressed as a fraction, a decimal, or a percent. For example, the probability of a tossed coin coming up heads is one chance out of two, or $^1/_2 = 0.5 = 50$ percent. In the long run, 50 of 100 tosses can be expected to come up heads. A probability of 1 or 100 percent represents a certainty, and a probability of 0 represents an impossibility.

There is a simple relationship between the return period of a hydrologic event and the probability of occurrence of that event. If *N* is the recurrence interval of the event (in years), then the probability *P* of that event being equaled or exceeded in any given year is the reciprocal of *N*. Expressed as a formula, this is

$$P = \frac{1}{N} \tag{3-2}$$

For example, the probability of a 5-year storm occurring in any single year is $P = {}^1/_5 = 0.2$ or 20 percent. In effect, this also means that there is less than a 20 percent chance that a worse or more intense storm will occur in any given year.

Relying on common experience again, it can be seen that the really intense storms are few and far between. In other words, the more extreme the hydrologic event, the larger is its recurrence interval. And the larger the recurrence interval *N*, the lower is the probability of occurrence *P*, because of the inverse relationship between the two. For example, there is only a 1 percent chance that a 100-year storm will occur in a given year. It is much less likely to observe a severe 100-year storm than a 5-year storm. (Although in many regions of the country rainfall records do not go as far back as 100 years, statistics and probability theory can be used to extrapolate or extend the existing data beyond the actual period of record.)

To summarize, the larger the recurrence interval *N*, the less likely it is for a hydrologic event to be equaled or exceeded in a given year. This is an important concept. Generally, the more critical a project is in terms of potential loss of life, economic damage, or adverse environmental effects, the larger is the value of *N* used in design computations.

A dam, for instance, may be designed to accommodate a 100-year flood, whereas a local storm drain may be designed to handle only the flow from a 2-year storm. In the former case, designing the dam for the big flow will reduce the chance of failure or breach of the dam and ensure the protection of human lives and property downstream. In the latter case, a trade-off is made between saving money for construction and taking more of a chance on the storm drain backing up or overflowing once every 2 years or so.

## Intensity–Duration–Frequency Relationships

In these discussions, terms such as *storm intensity*, *storm duration*, and *recurrence interval* have been examined as if they were independent quantities. But these three factors are related to each other and must be considered together. The term *frequency* is often used instead of return period. The frequency of a storm or other hydrologic event varies inversely with its return period. A 10-year storm, for example, will occur less frequently than a 5-year storm.

The rainfall data collected by the National Weather Service are compiled, analyzed, and published in various forms. The relationships among rainfall intensity, duration, and frequency may be shown graphically in curves or maps, or they may be expressed as formulas. As shown in Chapter 9 on stormwater management, these data are used by designers to estimate storm runoff and peak streamflow or discharge.

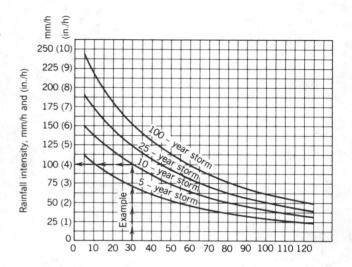

**FIGURE 3.6**
*Typical rainfall intensity–duration–frequency curves. Curves like these are prepared from rainfall statistics by the U.S. National Climatic Data Center.*

### Rainfall Curves

A typical set of rainfall intensity–duration–frequency curves is illustrated in Figure 3.6. Rainfall patterns vary significantly with geographic location and climate. For an actual application of rainfall data to a real design problem, the appropriate rainfall curves for the specific location under study should be obtained from the National Weather Service or from appropriate state or county agencies.

Rainfall curves of this form are generally used by entering the horizontal axis with a preselected storm duration, moving vertically to an intersection with a specific storm return period (the curved lines), and then moving horizontally to the vertical axis, where an expected rainfall intensity is read. For example, it can be seen in Figure 3.6 that a 10-year storm with a 30-min duration would have an intensity of 100 mm/h (or about 4 in./h). The shape of these rainfall curves reflects the fact that storms of shorter durations have higher average intensities than do longer storms. Also, for a given duration, the higher intensities correspond to storms with longer recurrence intervals.

### EXAMPLE 3.4

A storm of 40-min duration drops 50 mm (2 in.) of rain. Using the rainfall curves in Figure 3.6, estimate the probability of observing a similar storm in the next year.

*Solution*  First, compute the storm intensity, as follows:

$$\text{intensity} = \frac{50 \text{ mm}}{40 \text{ min}} \times \frac{60 \text{ min}}{1 \text{ h}} = 75 \text{ mm/h}$$

Now enter Figure 3.6 on the horizontal axis with 40 min and on the vertical axis with 75 mm/h. The intersection of horizontal and vertical lines extended from those points falls about halfway between the 5-year storm and the 10-year storm. From this it can be concluded that the return period for the storm in question is about 7.5 years. The probability of observing a similar or more intense storm in the next year is computed using Equation 3-2, as follows:

$$P = \frac{1}{7.5} = 0.13 \quad \text{or} \quad 13 \text{ percent}$$

### Rainfall Formulas

Rainfall intensity–duration–frequency relationships may be expressed in equation form instead of in the form of curves on a graph. One of the equations that may be used is

$$i = \frac{A}{t + B} \qquad \textbf{(3-3)}$$

where $i$ = rainfall intensity, mm/h  (in./h)
  $t$ = rainfall duration, min
$A$ and $B$ = constants that depend on the recurrence interval and geographic locale

Values of the constants $A$ and $B$ have been derived from data for various sections of the country. For example, for a 10-year storm in the eastern Middle Atlantic states, $A$ and $B$ are reported as 5840 and 29, respectively. $A$ and $B$ are reported as 1520 and 13, respectively, for the western states. (These values of $A$ and $B$ are for use in the SI metric system; intensity $i$ in this case is in mm/h.)

### EXAMPLE 3.5

Using the rainfall formula given by Equation 3-3, determine the expected rainfall intensity for a 10-year storm of 60-min duration (*a*) in California and (*b*) in Delaware.

*Solution*  (*a*) For California, use the constants $A = 1520$ and $B = 13$ for the western states. Applying Equation 3.3 with $t = 60$, we find

$$i = \frac{1520}{60 + 13} = \frac{1520}{73} = 21 \text{ mm/h}$$

(*b*) For Delaware, use the constants $A = 5840$ and $B = 29$ for the Middle Atlantic states. Applying Equation 3-3 with $t = 60$, we find

$$i = \frac{5840}{60 + 29} = \frac{5840}{89} = 66 \text{ mm/h}$$

The rainfall formula presented in Equation 3-3 covers large regions of the country for given values of $A$ and $B$ and can be rather insensitive to more local variations in rainfall patterns. It is preferable to use more local information regarding rainfall data if they are available.

### Rainfall Maps

Rainfall data can be depicted on a map, as illustrated in Figure 3.7. The lines seen crisscrossing the United States may look like ground contour lines at first glance. But, instead of contours, they represent lines of equal rainfall depth for a 24-h storm with a 100-year return period. It can be seen, for example, that a 100-year, 24-h storm would cause 5 in. of rain in northern Maine, western Texas, and many other locations in between. A storm with the same frequency and duration would cause 14 in. of rain in southern Florida. Maps of this nature are available for a wide range of durations and frequencies from the National Weather Service.

### Rainfall Classification System

Phrases using return periods, such as "5-year rainfall" or "100-year storm," are commonly used to describe rainfall events. At first glance, it might be assumed that a 5-year event is small compared to a 100-year event. But this is not necessarily the case because the duration of a storm also plays an important role in characterizing the *severity* of the storm. For example, it is possible that a 2-year, 24-h storm would have a larger rainfall depth that a 50-year, 1-h storm. That would result in more runoff from the 2-year storm, which may have a more severe environmental impact and cause more flood damage. In order to more fully characterize the severity of a storm, then, it is necessary to account for both return periods and durations. (Another factor that perhaps should be taken into account is the size of the watershed over which the storm occurs.)

One suggested rainfall classification system is the Haestad Severity Index™ (HSI), which uses ten levels of magnitude defined by a logarithmic relationship between return period and duration and five categories of severity similar to those used for hurricanes and

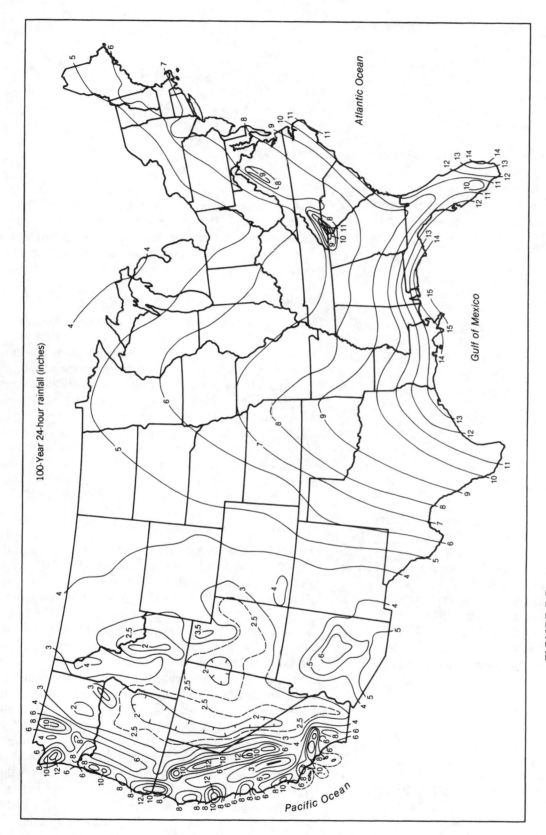

**FIGURE 3.7**

Rainfall maps, which show lines of equal rainfall depth across the nation, are available for a range of recurrence intervals and storm durations. The lines of equal rainfall depth are called isohyetal lines. These maps are updated periodically as the hydrologic database increases. (Courtesy of the National Weather Service; copies available for purchase)

100-Year 24-hour rainfall (inches)

Atlantic Ocean

Gulf of Mexico

Pacific Ocean

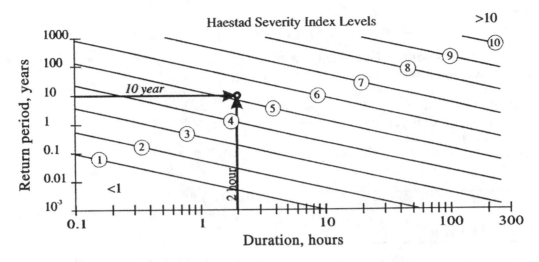

**FIGURE 3.8**
*A logarithmic plot of rainfall return period versus rainfall duration for determining*
*the Haestad Severity Index (HIS) for a storm. For example, a 10-year, 2-h storm has*
*an HSI level of 5.1.* (Reprinted with permission from Haestad Methods, Inc.)

tornados. A graph depicting the HSI levels is shown in Figure 3.8 and a chart depicting a relationship among the HSI levels and the categories of severity is shown in Table 3.1. (A full description of the Haestad Severity Index and Rainfall Classification System can be found at http://www.severityindex.com.)

## EXAMPLE 3.6

Compare the HSI level of two rainfall events with a return period of 10 years, and determine the category of each event. The first storm lasts for 2 h and the second storm lasts for 48 h.

*Solution*   From Figure 3.8, the intersection of the horizontal and vertical lines for a 10-year, 2-h storm yields

a severity level or HIS of about 5. From Table 3.1, it is seen that such a storm would be a Category 2 event, a level that would approach the design capacity of typical storm sewers and inlets. Again from Figure 3.8, for the 10-year, 48-h storm, the HSI is about 7. From Table 3.1, it would be a Category 4 storm, which would cause major flood damage.

Both storms have the same return period or recurrence interval, yet the longer duration storm has a greater severity or magnitude. This may seem obvious for storms with the same return period—the longer the duration, the greater is the overall severity. But it is not as obvious when comparing storms of different return periods, as illustrated in the following example.

**TABLE 3.1**
**Haestad Severity Index™ (HSI) levels and rainfall category levels**

| HSI | Category | General description |
|---|---|---|
| <2.5 | None | Insignificant in terms of flooding; still may be applicable to water quality concerns |
| 2.5 to 4 | 1 | Minor flooding in parking lots, clogged inlets, and poorly drained areas |
| 4 to 5.5 | 2 | Approaches design capacity of typical storm sewers and inlets; street flooding expected |
| 5.5 to 7 | 3 | Most storm sewer and channel capacities exceeded; significant rise in channel water surface elevations |
| 7 to 8.5 | 4 | Approaches or exceeds typical maximum design for all drainage conveyance systems; floodplains inundated; major flood damage |
| ≥8.5 | 5 | Catastrophic rainfall event; all storm sewers, channels, streams, and rivers exceeded; massive flooding; extreme flood damage |

*Source:* Reprinted with permission from Haestad Methods, Inc.

## EXAMPLE 3.7

Which storm has a greater severity, a 5-year, 24-h storm or a 100-year, 20-min storm event?

*Solution* From Figure 3.8 and Table 3.1, it is seen that the 5-year, 24-h storm is a Category 3 event, whereas the 100-year, 20-min (0.33 h) storm is a Category 2 event, which is less severe. In this case the 5-year storm could do more damage than the 100-year storm.

## 3.4 SURFACE WATER

Water that flows over the ground is often called *runoff*. Runoff that has not yet reached a definite stream channel is called *overland flow* or *sheet flow* (on a smooth surface, such as pavement). This type of surface water is important in the discussion of stormwater drainage systems. For the most part, the term *surface water* refers to water flowing in streams and rivers as well as water stored in natural or artificial lakes.

### Watersheds

As mentioned in Section 3.2, runoff occurs when the rate of precipitation exceeds the rate of interception and evapotranspiration. The total land area that contributes runoff to a stream or river is called a *watershed*. It may also be called a *drainage basin* or *catchment area*, particularly if the water flows toward or in an urban drainage system. Generally, engineers are interested in determining the amount of runoff at a specific point in the natural stream or engineered drainage system. This point is called the *basin outlet* or *point of concentration*.

The natural boundary or perimeter of the watershed may be determined from a topographic map, using the ground elevation contour lines. Viewed on a topographic map, water flowing freely over the ground's surface would move in a direction perpendicular to the contour lines, which is the direction of the steepest slope at any given point. By examining the contour map and visualizing the pattern of overland flow, it is possible to locate the boundary of the watershed. This boundary is called the *drainage divide line* or *ridge line*; it separates adjacent watersheds.

A simplified picture of a watershed is that of a funnel (Figure 3.9). The wide rim at the top of the funnel represents the ridge line and the circular area encompassed by the rim represents the catchment area. As water falls within the rim, it flows downward toward the narrow outlet at the bottom, which represents the

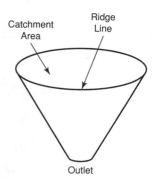

**FIGURE 3.9**
*Simplified view of a watershed or drainage basin.*

point of concentration. In practical applications, the ridge line must be located and drawn on a topographic (topo) map by the engineer or technician. Invariably, the ridge line forms an irregular shape rather than a circle like the rim of a funnel, and the point of concentration lies on the line rather than in the center of the area, because the plan view of a watershed is depicted.

To draw a drainage divide line on a topo map, the following procedure may be followed:

1. Start at the point of concentration. This might be at the intersection of two streams, at a point where a stream flows through a highway culvert, or at the location of a dam. The divide line will begin and end at this point.

2. Examine the contours to determine flow patterns. Imagine a drop of water on the ground at any given point and visualize which way it will flow. Start to sketch sections of the divide line that clearly separate the watershed from an adjacent watershed. These sections of the line will follow ridges and pass through topographic saddles. Remember, the natural drainage divide line is always perpendicular to the contour lines.

3. Fill in any gaps that may be left in the line being sketched. Occasionally, the divide line will turn sharply on the top of a ridge to pass through one of the saddles on the line.

A perspective sketch depicting flow patterns and a drainage divide line is shown in Figure 3.10*a*, and a plan view of the same area is shown in Figure 3.10*b*.

The sharp turns that may characterize a divide line as it passes through adjacent ridges and saddles are illustrated in Figure 3.11. (Another drainage divide line is shown in Figure 3.22 as the dot–dash line.)

The point at which two streams converge or intersect is called a point of *confluence*. As small streams

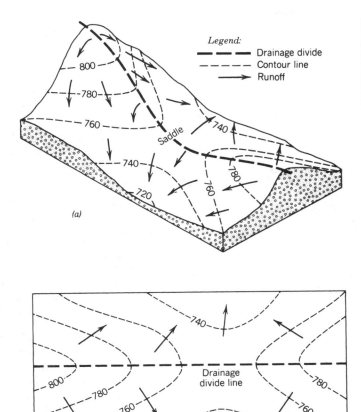

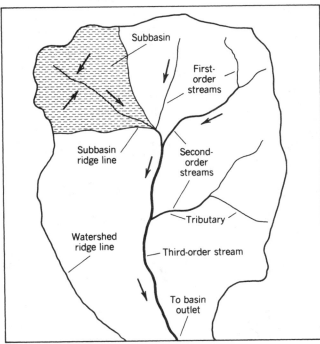

**FIGURE 3.12**

*A large watershed usually comprises several smaller catchment areas or subbasins.*

**FIGURE 3.10**

*(a) A perspective view of runoff patterns; arrows show the direction of sheet flow. The drainage divide line passes through the ridges and the saddle, separating two adjacent watersheds. The direction of sheet flow is perpendicular to the contour lines.*

*(b) A plan view topographic map that shows the same drainage divide line and contours as depicted in (a).*

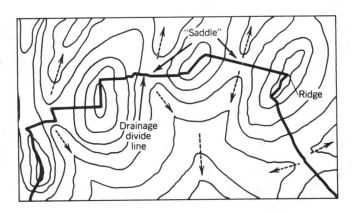

**FIGURE 3.11**

*A drainage divide line will sometimes turn sharply on a ridge or saddle, as shown here. The dashed arrows show the direction of overland flow.*

converge, larger streams and eventually rivers are formed. The catchment area for a particular stream may be only a part of a larger watershed; the smaller area is called a *subbasin* of the watershed. A typical drainage network is shown in Figure 3.12. The streams may be classified by their position in the overall network. Typical classifications are first-order streams, second-order streams, and so on. A first-order stream does not have any *tributaries* or smaller streams flowing into it.

A watershed for a large river may encompass thousands of square miles and include many smaller tributaries. These large watersheds are also called *river basins*. The Raritan River Basin in New Jersey, for example, encompasses about 2850 km² (1100 mi²). The Raritan basin is illustrated in Figure 3.13. The USGS has divided the United States into 2149 basic watershed units, the smallest of which encompasses about 1800 km² (700 mi²).

The size of a drainage basin refers to its total horizontal surface area. Relatively small basins may be expressed in terms of acres or hectares. A mechanical device called a *planimeter* is often used to measure the area by simply tracing the boundary of the watershed. Modern electronic planimeters can be calibrated to display the area digitally, based on the scale of the map being used.

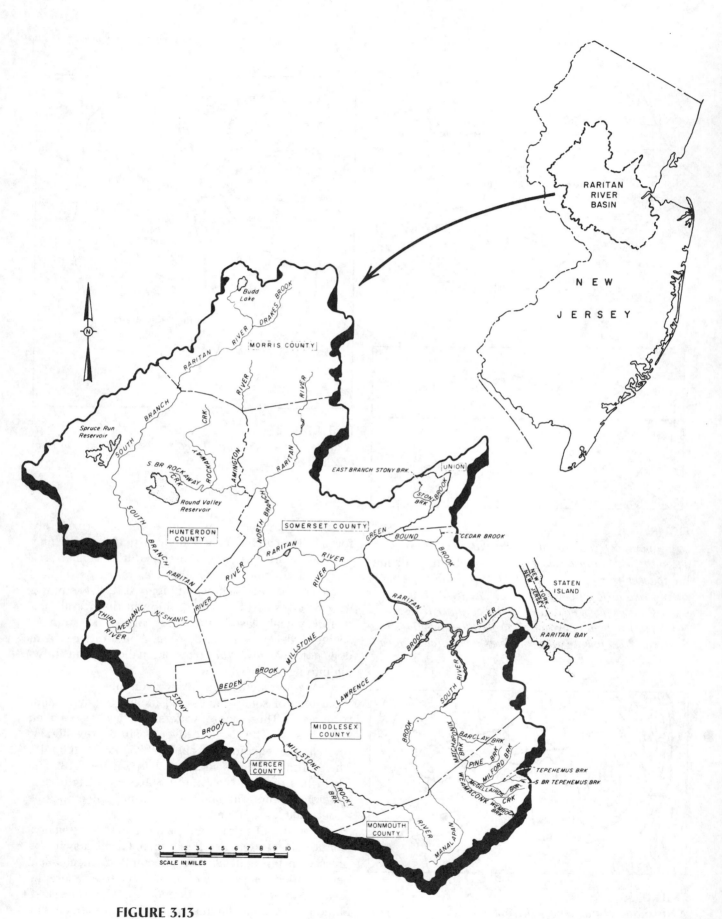

**FIGURE 3.13**
*The Raritan River Basin. This 2850-km² watershed includes a total of 536 km of streams.* (Courtesy of the Division of Water Resources, New Jersey Department of Environmental Protection)

The volume and rate of runoff in a watershed are functions of many variables. The basin area and the intensity and duration of rainfall have a direct effect on the amount and rate of runoff. Other factors include the slope of the ground, the type of soil and vegetative cover, and the type of land use. For example, a flat area with sandy soil would produce less runoff than a sloping area with clay soil. More of the water would infiltrate the ground surface through the porous sand in the former case, leaving a smaller fraction of the rain to become surface flow. Also, densely populated urban areas generate more runoff than suburban or rural areas. The relationships among these various factors and runoff are discussed in more detail in Chapter 9.

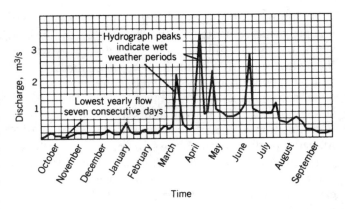

**FIGURE 3.14**

*Typical* annual hydrograph *for a small stream.*

## Streamflow

The amount or volume of water that flows in a stream is called the flow rate or *discharge* of the stream. The discharge is expressed in terms of volume per unit time passing any given point in the stream. The SI units for discharge are usually cubic meters per second ($m^3$/s), cubic meters per hour ($m^3$/h), or megaliters per day (ML/d). In U.S. Customary units, discharge may be expressed as cubic feet per second ($ft^3$/s or cfs), gallons per minute (gpm), million gallons per day (mgd), or acre–feet per day (ac–ft/d). A section of a stream that has a relatively constant slope, cross section, and discharge may be called a *reach* of the stream.

Stream discharge varies with time. Generally, higher flow rates are observed in the spring and summer months, whereas lower discharges occur in the fall and winter. This is particularly the case for the northeastern United States. Snowmelt can contribute significantly to streamflow. Variations in discharge that occur on a weekly, daily, or even hourly basis are directly associated with rainfall events. In some streams, an extremely wide variation in discharge can occur, from a raging torrent during wet weather to hardly a trickle of water during dry periods.

Low flow rates can cause environmental problems in streams receiving discharges from wastewater treatment plants because there is less water in the stream to dilute the wastewater. In Chapter 5, the relationship between the discharge and the *waste-assimilative capacity* of the stream is discussed. (Under certain conditions, a stream can *assimilate* or absorb biodegradable wastes without excessive environmental damage.)

Low stream discharges also cause problems if the stream is used as a source for water supply. On the other end of the spectrum, excessively high discharges usually necessitate the construction of flood control facilities.

### Hydrographs

A graph of discharge versus time is called a *hydrograph*. The vertical axis represents stream discharge and the horizontal axis represents time. Time intervals may span several years or several hours, depending on the type of hydrograph and its use.

Streamflow over a span of 1 year may be depicted in an *annual hydrograph*, as illustrated in Figure 3.14. The isolated peaks on the hydrograph correspond to periods of heavy rainfall.

A *flood hydrograph* or *storm hydrograph* represents the flow response of a stream to one particular rainfall event. The time interval on the horizontal axis is generally hours or days, rather than months as in the annual hydrograph. A typical storm hydrograph is shown in Figure 3.15.

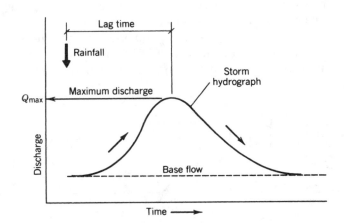

**FIGURE 3.15**

*A storm or* flood hydrograph *shows the direct effect of a specific rainfall event on streamflow. A sharply rising limb, a peak flow after a lag time, and a gradually receding limb are typical characteristics.*

Shortly after the storm begins and overland flow reaches the stream channel, the discharge in the stream starts to increase. This is depicted as the rising limb of the hydrograph. After a time interval called the *lag time*, which depends on the physical characteristics of the watershed, the maximum discharge occurs. This peak streamflow can occur many hours after the rain stops. (Computation of peak flow is discussed in Section 9.1.) After the peak is reached, the streamflow gradually recedes toward the *base flow*. The base flow is the normal dry-weather flow in the stream; it is sustained by groundwater seeping out of the ground into the stream channel. A stream that has a base flow throughout the year is called a *perennial stream*. A stream that dries up completely during periods of very little rainfall is called an *intermittent* or *ephemeral stream*. Perennial stream channels penetrate the groundwater table, whereas ephemeral streams lie above the water table. Groundwater is discussed in more detail in Section 3.7.

## Gaging Stations

The U.S. Geological Survey (USGS) is a government agency that measures streamflows and publishes records of discharges for most of the large streams or rivers in the United States as one of its responsibilities. In many cases, a permanent structure called a *gaging station* is constructed along the river to provide a continuous record of flow versus time. A sketch of a typical gaging station is shown in Figure 3.16.

The basic measurement made at a gaging station is of the depth of water in the stream or river. The elevation or height of the water surface above a reference level is called the *stage* of the stream. The stage changes as the discharge changes; as one would expect, higher stages correspond to higher discharges.

The stage may be measured and recorded graphically on a rotating chart by a float-operated device. A cable with a float on one end and a counterweight on

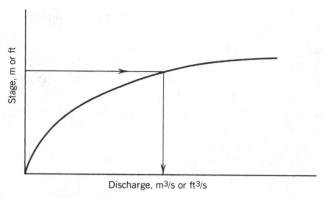

**FIGURE 3.17**
*A typical stage–discharge curve for a stream or river, showing the relationship between flow rate and depth of water in that particular river.*

the other end is hung over a pulley, as shown. The float moves up or down as the stage changes, rotating the pulley and thus changing the position of a pen on the chart. The stilling well, connected to the stream channel by a pipe, prevents excessive fluctuations of the water level due to wind or other disturbances. Modern digital recording and telemetry technology is also used.

Before a gaging station can provide data on streamflow, it is necessary to determine the actual relationship between the stage and the discharge. This relationship is often expressed graphically in a rating curve or stage–discharge curve, as illustrated in Figure 3.17. Once a rating curve is established for the stream, it is necessary only to measure the stage in order to know what the discharge is in terms of a volume flow rate.

One of the methods used to correlate stage with discharge is to construct a low dam or *weir* in the stream channel. As discussed in more detail in Section 2.4, a weir is an obstruction in the stream over which the water must flow. The height of water flowing over the weir, called the *head* on the weir, is related hydraulically to the volume flow rate. Figure 3.18 shows the installation of a weir and unsheltered gaging station in a small stream.

In larger streams or rivers, it may be impractical to obstruct the flow of water with a weir. The increase in water depth behind the weir can cause excessive flooding upstream. Instead, devices called *current meters* can be submerged at various points in the river to measure velocity of flow at various stages. A typical current meter includes a small propeller that rotates in the water at a rate proportional to the water velocity. Knowing the depth and cross-sectional area of the stream where the current measurement is made, one can compute the discharge. (The hydraulic relationship between velocity and area of flow is explained in Section 2.2.) Since the depth and shape of the stream bed may change gradually because of erosion or sedimentation, the rating curve must be checked and revised from time to time.

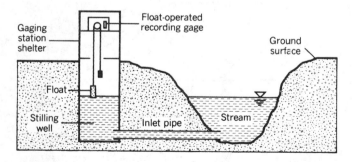

**FIGURE 3.16**
*Cross section of a typical stream gaging station. The water elevation in the stilling well is the same as the stage of the river or stream.*

**FIGURE 3.18**
*An unsheltered gaging station and a trapezoidal weir for a small stream.* (Courtesy of Leupold & Stevens, Inc.)

Streamflow statistics in areas without gaging stations can now be estimated using an automated equation-based system recently developed by the USGS. (A pilot system for Massachusetts can be accessed on the Internet at http://ma.water.usgs.gov/streamstats.) At a user-selected site, the physical characteristics of the watershed that drains to the site are determined from the computerized database, a system of equations is solved, and the estimated streamflow statistics along with a topographic location map are provided to the user. These estimates are useful for carrying out river basin planning studies, creating flood insurance rate maps, and other applications.

## 3.5  DROUGHTS

In everyday terms, a *drought* is a long period of dry weather that causes a lack of available water. At the other extreme, a *flood* is what happens when a stream or river overflows its banks, shortly after periods of excessive rainfall or snowmelt. Both of these events are hydrologic extremes that are notorious for the environmental problems, in addition to possible loss of life and property damage, that they cause.

To reduce or mitigate the problems caused by floods or droughts, designers of hydraulic structures and water management facilities must be able to quantify the severity and frequency of these events. The magnitude of the *N*-year flood for a particular watershed must be determined if flood control efforts are to be effective. The low flow in a stream due to a drought must

be estimated if the problems associated with prolonged periods of dry weather are to be avoided.

To a large extent, the occurrence and severity of floods or droughts can be related to precipitation. Since precipitation records are more commonly available than streamflow data, designers often have little choice but to make estimates of the occurrence of droughts or floods from correlations with rainfall data. The computation of peak streamflows from rainfall data is discussed in more detail in Chapter 9. It is assumed that the return period of the maximum stream discharge is the same as the return period of the storm from which the discharge is computed.

The low flows that occur in perennial streams during droughts are of importance for two reasons. If a stream is to be used for water supply, it must be determined whether a storage reservoir must be built to ensure adequate supply during a drought; and if the stream is receiving wastewater discharges from a sewage treatment plant, it must be determined whether the low streamflow will still be adequate to dilute the sewage or if some type of advanced wastewater treatment is necessary.

### MA7CD10 Flow

In water pollution studies, a drought flow is commonly defined as the lowest average discharge over a period of 1 week with a recurrence interval of 10 years. This is called the *minimum average 7-consecutive-day 10-year flow*, or the *MA7CD10 flow*. Since the value of *N* for the MA7CD10 flow is 10 years, there is only a 10 percent probability that there will be a more severe drought in any given year. In other words, the probability is 90 percent that the minimum weekly discharge in the stream will be greater than the MA7CD10 flow. This is generally considered to provide an acceptable risk for water pollution control projects, and the MA7CD10 flow is used for design computations.

When many years of records of stream discharge are available, a statistical procedure called *frequency analysis* may be used to estimate return periods or frequencies of droughts. The same method may be used to determine flood frequencies or recurrence intervals of storms from precipitation records. It is a good idea for designers or others who make use of return period data to have some understanding of how the data are determined.

To illustrate the procedure, a simplified example for determining a drought flow in a stream is presented here. In Example 3.8, only 5 years' worth of discharge records is used. In practical applications, much longer periods of record are required for meaningful results, but for the sake of clarity, as well as to illustrate extrapolation beyond the period of record, this is a useful example.

## EXAMPLE 3.8

Given the following record of streamflow data, estimate the MA7CD10 for the stream:

| Year | Lowest 7-d average flow, $m^3/s$ |
|------|------|
| 1980 | 4.4 |
| 1981 | 2.8 |
| 1982 | 4.0 |
| 1983 | 3.4 |
| 1984 | 5.2 |

*Solution*  First, rearrange the flow data in decreasing order of magnitude and assign a *rank* or *m* value to each flow, beginning with 1 and increasing by 1 sequentially.

The probability of observing an equal or higher flow in any given year is estimated by dividing the rank $m$ by the number of years of record plus 1 $(n + 1)$; in this example, $n = 5$. In formula form, the probability $P = m/(n + 1)$. We have

| Low flow, $m^3/s$ | Rank | Probability |
|------|------|------|
| 5.2 | 1 | 1/6 = 0.167 |
| 4.4 | 2 | 2/6 = 0.333 |
| 4.0 | 3 | 3/6 = 0.500 |
| 3.4 | 4 | 4/6 = 0.667 |
| 2.8 | 5 | 5/6 = 0.833 |

Hydrologic data are often plotted on a special type of graph paper called logarithmic probability paper. The points usually plot as a straight line or close to it. The low flows and their corresponding probabilities in this problem are plotted in Figure 3.19. A

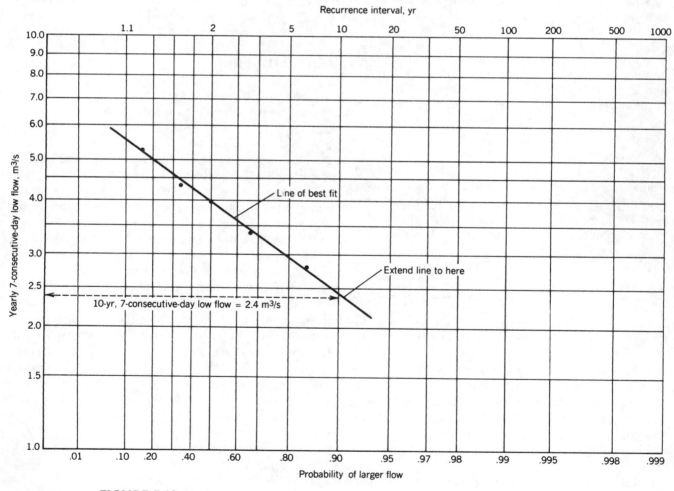

**FIGURE 3.19**

*Logarithmic probability paper is used to estimate the MA7CD10 drought flow in a stream or river. (See Example 3.8.)*

straight line of best fit has been drawn through the plotted points and extended, or extrapolated, to the 90 percent probability value. This identifies a flow rate on the vertical axis of the graph that would be exceeded nine times out of ten in any given subsequent year. Conversely, the probability of observing a lower flow (a more severe drought) is 10 percent. This flow, therefore, represents the MA7CD10 flow. As seen in Figure 3.19, the MA7CD10 flow for this stream (based on the very limited 5 years of record) is estimated at $2.4 \ \text{m}^3/\text{s}$.

## 3.6 RESERVOIRS

When streamflow is not sufficient for a dependable water supply, particularly during dry spells, a reservoir can be built to overcome this problem. A reservoir equalizes the flow in a stream, storing excessive wet weather flows for use during periods of low streamflow. A reservoir serving primarily for water supply is called a *conservation reservoir*. A reservoir of this type is built on a natural site with suitable topography by blocking the stream with a dam, allowing an artificial lake to be formed.

Conservation reservoirs are usually large and provide capacity for long periods of dry weather. The flooding of land by the artificial lake can have significant environmental as well as social effects; these must be considered in addition to the purely technical or economic aspects of the project.

Because of economic and environmental factors, it is often not feasible to build a dam or reservoir for only one purpose, such as water supply. Reservoirs that simultaneously serve this and other needs, such as for flood control, hydroelectric power, and recreation, are called *multipurpose reservoirs*. Other types of reservoirs include *distribution storage reservoirs* for water distribution and *detention reservoirs* for stormwater control. These are discussed in subsequent chapters.

The storage capacity of a large reservoir is usually expressed in units of megaliters or acre–feet. The *yield* of a reservoir represents the amount of water the reservoir can supply in a specific time interval without going dry. The relationship between reservoir yield and storage capacity is a key factor in its design.

According to a recent report by the World Commission on Dams, there are about 800,000 dams worldwide. Of these, roughly 45,000 are considered *large dams*—more than 15 m (50 ft) high or more than 3 million $\text{m}^3$ (2500 ac–ft) in reservoir volume. (About half the world's large dams were built primarily for land irrigation.)

The Three Gorges Dam on the Yangtze River in China, designed primarily for flood control and electric power, is considered the world's largest infrastructure project. Unfortunately, when the reservoir is full, the water behind the 180-m (600-ft)-high dam will submerge many cities, towns, and heritage sites, and displace more than 1 million people from their homes.

## Summation Hydrograph

To determine the required volume of a conservation reservoir, records of streamflow spanning many years must be used. Conservation reservoirs are often designed to provide the needed yield during a drought equal to the worst drought on record. A *summation hydrograph*, also called a *mass diagram*, is a convenient graphical tool for determining the required storage volume. This technique is illustrated in Example 3.9.

### EXAMPLE 3.9

A conservation reservoir is to provide a uniform withdrawal or yield of 60 ML/month without being depleted. The streamflow records for the year of lowest flows are summarized on a monthly basis as follows:

| Month | Jan | Feb | Mar | Apr | May | Jun |
|---|---|---|---|---|---|---|
| Streamflow, ML/month | 60 | 100 | 180 | 20 | 15 | 15 |

| Month | Jul | Aug | Sep | Oct | Nov | Dec |
|---|---|---|---|---|---|---|
| Streamflow, ML/month | 5 | 15 | 115 | 200 | 180 | 100 |

Determine the required reservoir volume.

*Solution*  First, determine the cumulative streamflow entering the reservoir on a monthly basis. For example, in February the cumulative flow would be $60 + 100 = 160$ ML, and in March it would be $160 + 180 = 340$ ML. Just keep adding the flows for each month. Prepare a table of cumulative monthly flows as follows:

| Month | Jan | Feb | Mar | Apr | May | Jun |
|---|---|---|---|---|---|---|
| Cumulative flow, ML | 60 | 160 | 340 | 360 | 375 | 390 |

| Month | Jul | Aug | Sep | Oct | Nov | Dec |
|---|---|---|---|---|---|---|
| Cumulative flow, ML | 395 | 410 | 525 | 725 | 905 | 1005 |

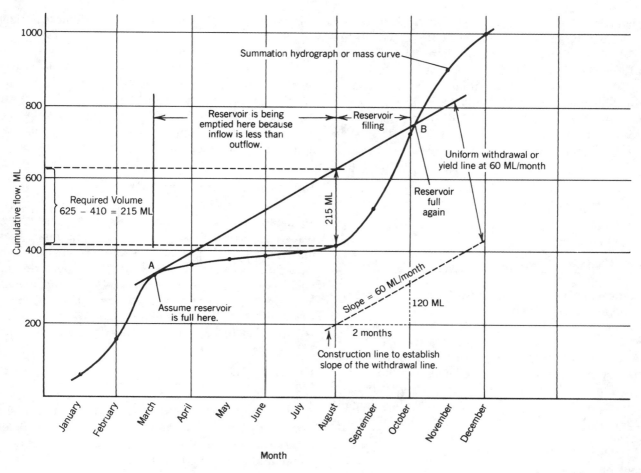

**FIGURE 3.20**
*Summation hydrograph. (For Example 3.9.)*

Now the cumulative monthly flows can be plotted on a graph as shown in Figure 3.20. This graph is the summation hydrograph; it is a plot of flow versus time, but the flows are cumulative with time.

The slope of the summation hydrograph or mass curve represents the rate of inflow into the reservoir. Notice that the slope is very flat during the summer months because of the low streamflows during that period.

A uniform yield or withdrawal can be represented as a straight line on the graph; in this case the withdrawal line has a slope of 60 ML/month, as shown. Where the mass curve slope is flatter than the withdrawal line slope, more water is leaving the reservoir than is flowing in; the reservoir is emptying. Where the mass curve is steeper than the withdrawal line, more water is flowing in than is flowing out of the reservoir; the reservoir is filling up.

Draw a line parallel to the withdrawal line and tangent to the mass curve at the point labeled A in

Figure 3.20. Point A, in general, represents a peak of the mass curve where it is concave downward. Assume that the reservoir is just full at this point. It would immediately start to decrease in water volume since, just past point A, the rate of withdrawal exceeds the rate of inflow. But after several months, the slope of the mass curve increases and the rate of inflow surpasses the rate of withdrawal; the water volume begins to increase. At point B, where the line crosses the mass curve, the reservoir will be at full capacity once again.

The vertical distance between the yield line AB and the mass curve represents the volume of water taken out of storage to satisfy the yield or withdrawal. In this example, the maximum vertical distance is measured to be 215 ML, as shown in Figure 3.20. This is the minimum storage volume needed to ensure that the specified yield can be satisfied.

Since this volume of 215 ML was determined for the worst drought year on record, it is reasonable to assume that during years of normal precipitation and streamflow the reservoir will be more than adequate to provide the required yield. But it is still possible that a more severe drought could occur. A frequency analysis could be done to provide estimates of recurrence intervals and probabilities of more serious droughts.

## Reservoir Capacity

The maximum volume of water that can be stored in a reservoir depends on the elevation of the spillway of the dam forming the reservoir and on ground topography upstream of the dam. In addition to this total volume, it is important to know the relationship between the volume and the elevation of the reservoir surface. A graph of water elevation versus volume is called a *reservoir capacity curve* or *elevation-storage curve*. A typical capacity curve is illustrated in Figure 3.21. Using a curve like this, it is possible to determine the volume of water in the reservoir at any given time by simply measuring the elevation of the water supply.

## Measuring Reservoir Volume

The volume of a reservoir can be estimated from a topographic map. For example, in Figure 3.22, if the spillway of a dam at *AB* is to have an elevation of 100 ft, the water in the reservoir will cover the area enclosed by the 100-ft contour, as shown by the cross-hatching. Since the contour interval is 10 ft on this map, the total body of water contained in the reservoir can be viewed as a stack of layers 10 ft thick separated by level surfaces at each contour line. Each layer would then be a solid figure bounded at the top and bottom by parallel plane surfaces. The area at each of these surfaces could be measured with a planimeter by tracing each of the contour lines. Multiplying the average area for each pair of surfaces by the 10-ft thickness of the layer can then approximate the volume of each layer. The sum of the volumes, plus an approximation of the volume below the lowest contour, provides an estimate of the total reservoir volume.

All streams and rivers carry suspended soil particles to some degree. These particles tend to settle out by gravity in the reservoir, forming deposits of *sediment*. All reservoirs ultimately become filled with sediment and therefore have limited design lives or periods during which they can fulfill their intended purposes.

Figure 3.23 illustrates the accumulation of sediment behind a dam. Although reservoir sedimentation cannot be prevented, it can be controlled and slowed down. Sluice gates below the crest of the dam permit the occasional discharge of sediment before it has a chance to settle to the bottom.

## Environmental Impacts

In addition to reducing the design life of a reservoir, the accumulation of sediment behind a dam can cause unwanted environmental impacts on the downstream ecosystem. A notable example of this is the Aswan High Dam in Egypt, built on the Nile River to control floods and provide hydroelectric power. The loss of silt and plant nutrients that were previously deposited on the downstream fields after a flood disrupted agricultural yields in the Nile Valley.

Other environmental impacts of dams include deleterious effects on the water quality and water temperature. In the United States, a federal court ruled in 2001 that the operation of four Snake River dams violated the *Clean Water Act*, and endangered salmon and trout in eastern Washington State. The U.S. Army Corps of Engineers, a federal agency responsible for operating the dams, was ordered to find ways to lower water temperatures and reduce levels of nitrogen in the reservoirs.

Another example of the harmful environmental impacts of a dam and its reservoir, and an attempt to mitigate those impacts, is described in the following case study.

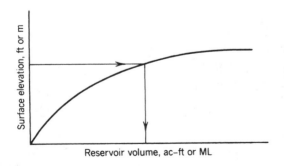

**FIGURE 3.21**
*Typical* reservoir capacity curve.

## CASE STUDY

### Environmental Impact of a Reservoir: Artificial Flood Created to Rejuvenate the Grand Canyon

PAGE, Ariz., March 26 (AP)—Four huge arcs of foamy white Colorado River water shot out of a dam with a roar today as the Federal Government began a weeklong

**FIGURE 3.22**
*The spillway of a dam at AB would form a reservoir with a surface elevation of 100 ft, shown by the cross-hatching. The watershed ridge line for the reservoir is shown by the dot–dash line.* (From P. Kissam, *Surveying for Civil Engineers*, 2nd ed., McGraw-Hill, New York, 1981. Copyright © 1981 by McGraw-Hill. Used with permission)

flood designed to turn back the clock on the Grand Canyon (see Figure 3.24).

As the Colorado River below the dam crept higher up the salmon-colored, sandstone canyon walls, several dozen scientists in hard hats looked on at their effort to bring the canyon closer to its natural state.

"The roar of the water is like what Mother Nature would've been doing naturally this time of year," said

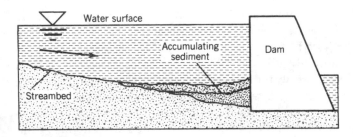

**FIGURE 3.23**
*Sedimentation in a reservoir reduces its capacity to store water. Over time, a reservoir can eventually become completely filled in with sediment.*

David Wegner, program manager for the Federal Bureau of Reclamation, the agency that manages the nation's dams.

Because sediment settles out of the water as it sits behind the Glen Canyon Dam, the once warm and muddy river downstream now runs cold and clear green.

The flood should stir up sediment and redistribute it through the Grand Canyon, which starts about 15 miles downstream from the dam, creating hundreds of new sandy shores where vegetation can take root to feed birds and fish.

Interior Secretary Bruce Babbitt called it "a new beginning" as he pushed a button, cranked a lever and turned a wheel to open the first of four huge valves, releasing millions of gallons of the river from behind the dam.

It is "a new era for ecosystems, a new era for dam management, not only for the Colorado but for every river system and every watershed in the United States," Mr. Babbitt said.

Today at the dam, water shot hundreds of feet out of the four, eight-foot steel tubes, filling the normally quiet quarter-mile Glen Canyon with the thundering sounds of a waterfall.

"Woo-hoo! Check that out!" exclaimed a grinning Clay Bravo, assistant director for natural resources for the Hualapai Indians, one of several tribes living along

**FIGURE 3.24**
*Water erupted on March 26, 1996, from an 8-ft tube into the Colorado River. The flooding is a way to manage the Grand Canyon downstream.* (Copyright © 1996 by the Associated Press. Used with permission)

the river. "But this is nothing compared to the days before the dam."

The scientists behind the $2.7 million experiment, the Government's first scientifically documented artificial flood, said it was intended to mimic seasonal flows restricted by the 33-year-old dam.

The scientists have warned that the flood may wash away fragile fish eggs and some plant life, but they expect flora and fauna to return in greater abundance. Before the dam was built, floods three to four times the strength of the current release came through with each spring's snow melt.

After the dam was built, the cold water made the river a premier fishing spot for rainbow trout—a breed exotic to the area. Leafy tamarisk and cottonwood trees—also foreigners—now thrive in the canyon.

In addition, the cold water wiped out some native warm-water fish. Of the seven endangered species of fish that lived in the canyon before the dam was built in 1963, only three survive. Scientists hope the flood will leave warmer, safer water in backwater canyons for endangered fish like the humpback chub and razorback sucker.

Spectators atop the 710-foot-high dam watched the gates open and release water fast enough to fill the 110-story Sears Tower in Chicago in 17 minutes. The river is expected to rise 10 to 15 feet inside the Grand Canyon. More than 117 billion gallons of water will be sent into the canyon over the week.

The water level behind the dam was especially high because it was a wet winter and because water releases were reduced in the days leading up to the flood.

The newly churning river forced officials to ban motorless boats for 15 miles below the dam. Only experienced river runners were expected to notice the flow's effects inside the canyon, and several tourist trips were on the water today.

*Source:* From the *New York Times*, March 27, 1996. For a follow-up article (A Dam Opens, Grand Canyon Roars Again, *New York Times*, February 25, 1997), search the *New York Times* archive (at the Web site www.nytimes.com) with key words "Glen Canyon Dam."

## 3.7   GROUNDWATER

As discussed in Section 3.2, part of the precipitation that falls on the land may infiltrate the surface, percolate downward through the soil under the force of gravity, and become what is known as *groundwater*.

The groundwater is an extremely important part of the hydrologic cycle. Almost half of the people in the United States obtain their public water supply from groundwater. Overall, there is more groundwater than surface water in the United States, including the waters in the Great Lakes. It sometimes is uneconomical to pump it to the surface for use, however, and in recent years the pollution of groundwater supplies from improper disposal of wastes has become a significant problem.

After water first infiltrates the ground surface, it seeps downward through a layer of soil called the *zone of aeration* or *vadose* zone. This is a layer of soil in which the small spaces between the solid soil particles are partially filled with air as well as with water. As the water continues to percolate downward, it eventually reaches the *zone of saturation*, a layer of soil or rock in which all the pore spaces or rock fissures are completely filled with water. Even though the individual pore spaces and rock crevices are relatively small, the total volume of groundwater is large because the geological formations that can hold water are so vast. Groundwater can be considered to be a huge subsurface reservoir.

The dividing line between the zone of aeration and the zone of saturation is called the *water table* (or *phreatic surface*). An excavation or a well that is deep enough to penetrate the zone of saturation will fill up with water to the height or elevation of the water table. This is illustrated in Figure 3.25.

The elevation of the water table is not constant. It depends on weather conditions and varies seasonally. The water table is generally closer to the ground surface in the spring or during rainy periods and deeper during dry spells. The water table can also be lowered by pumping, as is seen in the discussion on wells later in this section.

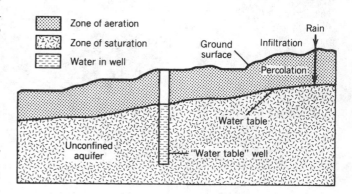

**FIGURE 3.25**

*Soil below the* water table *is saturated with water.*

## Aquifers

An *aquifer* is a layer of soil or rock in which groundwater can move relatively freely. It is, in other words, a geological stratum that can transmit water in sufficient quantities to permit economical use of the groundwater for supply purposes. Porous sand and gravel aquifers yield more water than do more impermeable silt or clay deposits. Rock formations may contain enough cracks or fissures to yield significant quantities of water.

An aquifer that is sandwiched between two impermeable layers that block the flow of water is called a *confined* or *artesian* aquifer. The water in a confined aquifer is under hydrostatic pressure. It does not have a free water table. An imaginary line, representing the *piezometric surface*, can be used to represent the height to which water would rise in a well that penetrated the aquifer. This is illustrated in Figure 3.26.

The *recharge area* for an aquifer is where precipitation infiltrates the ground to replenish the water flowing through the aquifer. As seen in Figure 3.26, the recharge area may be remote from the point of actual water use. This is an important factor in land-use planning and urban development. Covering recharge areas with pavements and parking lots blocks the infiltration process and reduces the amount of water that can be withdrawn from the aquifer.

## Groundwater Flow

Groundwater is in a constant state of motion through the pores and crevices of the aquifer in which it occurs. The water table is rarely level; it generally follows the shape of the ground surface. The groundwater flows in

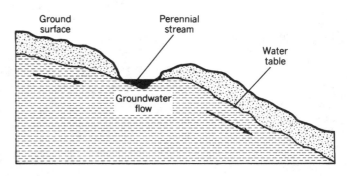

**FIGURE 3.27**
*Groundwater flows slowly through the soil. Sometimes it seeps out of the ground in a* spring *or into the channel of a* perennial stream, *resulting in a relatively stable dry-weather base flow in the stream.*

the downhill direction of the sloping water table, as illustrated in Figure 3.27. The water table sometimes intersects low points of the ground, where it seeps out into springs, lakes, or streams. As previously discussed, the base flow of a perennial stream is actually sustained by groundwater flow. Such streams are also called *influent* streams.

The rate of movement of groundwater due to gravity is usually very slow. It is limited by the frictional resistance to flow in the soil and rock openings. A velocity of about 18 m/d (60 ft/d) is considered high, even in porous sand and gravel deposits. In more impervious clay soils, the velocity of flow may be as low as a fraction of a meter per year.

The velocity of groundwater flow is a function of the slope of the water table and the permeability of the soil. This relationship is expressed in a formula known as *Darcy's law*, as follows:

$$V = K \times S \qquad (3\text{-}4)$$

where $V$ = flow velocity, mm/s
$\quad K$ = permeability coefficient, mm/s
$\quad S$ = slope of the water table

Permeability is a characteristic of a porous material that allows it to transmit water; it has the same units as velocity. The slope of the water table is the drop in elevation divided by the horizontal distance; it is a dimensionless number.

Darcy's law is the basis of more complicated mathematical analyses of groundwater hydraulics. If aquifer conditions are known, it is possible to predict such things as how much water can be pumped out

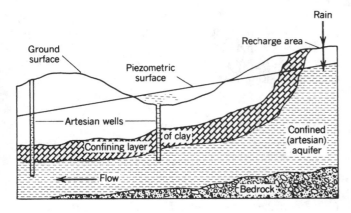

**FIGURE 3.26**
*Water in a confined or* artesian aquifer *is under hydrostatic pressure. Water in an artesian well rises to the level of the* piezometric surface.

**TABLE 3.2**
**Typical permeability coefficients**

| Soil type | K, mm/s |
|-----------|---------|
| Gravel | 10 to 40 |
| Sand | 0.01 to 10 |
| Silty sand | 0.001 to 0.02 |
| Silt | 0.0001 to 0.005 |
| Clay | $10^{-6}$ to $10^{-8}$ |

of a well. But the limitations of mathematics, including Darcy's law, for evaluating groundwater conditions need to be mentioned. To apply the formulas, it is necessary to assume that the aquifer is a uniform or homogeneous material that can be described by a single coefficient $K$. In reality, natural soil deposits or rock aquifers are rarely uniform over large areas. Nevertheless, Darcy's law is of value in making initial estimates of the rate of groundwater movement. Typical values of $K$ are presented in Table 3.2, and the application of Darcy's law is illustrated in the following example.

## EXAMPLE 3.10

Compute the velocity of groundwater flow in an aquifer that has a coefficient of permeability $K = 0.01$ mm/s. The water table slopes at a rate of 1 m over a distance of 200 m.

*Solution*   The slope of the water table is computed as

$$S = \frac{1 \text{ m}}{200 \text{ m}} = 0.005$$

and

$$V = K \times S = 0.1 \times 0.005 = 0.0005 \text{ mm/s}$$

or

$$V = 0.0005 \text{ mm/s} \times 3600 \text{ s/h} \times 24 \text{ h/d}$$
$$= 432 \text{ mm/d} \approx 0.43 \text{ m/d}$$

## Wells

The most common method for withdrawing groundwater is to penetrate the aquifer with a vertical well and then pump the water up to the surface. Other methods include using natural springs or infiltration galleries. An infiltration gallery consists basically of several horizontal perforated pipes radiating outward from a large-diameter central shaft.

Wells may be constructed in a variety of ways, depending on the depth and nature of the aquifer. A *dug well* is a shallow excavation, up to about 10 m (30 ft) deep, that penetrates an unconfined aquifer. It is generally lined with stone or masonry to support the side walls. Dug wells are not dependable sources of water because of the seasonal variation in the depth of the water table and the well's susceptibility to pollution. They still may be observed in use in some agricultural and rural areas, but modern environmental sanitation standards generally prohibit their construction for public water supply.

Wells up to about 20 m (65 ft) deep may be constructed in soft soils by driving a *well point* into the ground. A well point is a section of perforated pipe with an internal screen and a point on the lower end. The required depth is reached by coupling additional sections of pipe to the well point as it is driven down. This type of well is more commonly used to dewater construction excavations than to provide a water supply.

Deep wells, those more than 30 m (100 ft) deep, are most commonly used for public water supplies. They can penetrate extensive aquifers with more dependable yields of water and better water quality than shallower wells. Deep wells are typically 100 to 300 mm (4 to 6 in.) in diameter. They are drilled using percussion or rotary drilling techniques.

Deep wells are permanently lined with a metal pipe, called a *casing*. [Plastic, for example, polyvinyl chloride (pvc), casings are used in some instances.] The annular space around the casing is filled with cement grout. The casing and the grout serve to seal off poor-quality water coming from the surface and the upper soil layer, protecting the well from contamination. A sanitary seal is installed at the top of the casing to further protect the water quality. In unconsolidated aquifers, a slotted well screen is usually attached to the bottom of the casing to strain silt and sand out of the well water. These basic features of deep well construction are illustrated in Figure 3.28.

Multistage vertical turbine pumps (a type of centrifugal pump) are commonly used in deep wells to lift the water. (The operating characteristics of centrifugal pumps are discussed in Chapter 7.) The well pump may be driven by an electric motor at ground level connected to the submerged pump by a shaft or by a special submersible motor directly connected to the pump.

The elevation of the water table in the well before pumping begins is called the *static level*. When the well is pumped, the water level in the well drops below the

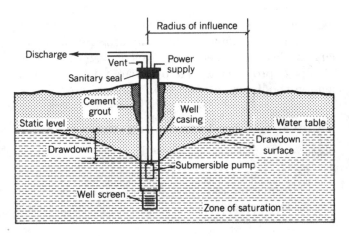

**FIGURE 3.28**
*Schematic diagram of a water table well showing the* drawdown *that occurs during water withdrawal by pumping.*

static level, as seen in Figure 3.28. The elevation difference between the static level and the pumping level is called the *drawdown*. A *drawdown surface* of the water table, or *cone of depression*, as it is also called, is formed around the well during pumping.

As the distance from the well increases, the slope of the drawdown curve flattens out, eventually merging with the undisturbed static water table. The horizontal distance from the well to the area where the water table has not been appreciably affected by the pumping is called the *radius of influence* of the well. These terms are illustrated in Figure 3.28.

The drawdown and the circle of influence will increase as the pumping rate is increased. From the discussion of groundwater flow, it should be clear that for the same rate of pumping, a very permeable aquifer will have a smaller drawdown than an aquifer with less permeability. Likewise, the radius of influence will be larger in the more porous aquifer. Mathematical formulas allow for the computation of all the terms defined here, but a full discussion of groundwater hydraulics is beyond the scope of this book.

The *safe yield* of a well is the rate at which water can be withdrawn without pumping the well dry. The larger the drawdown, the greater is the yield. The relationship between the yield and drawdown of a well is called its *specific capacity*. For example, if the drawdown in a well is 50 m when the withdrawal rate is $500 \text{ m}^3/\text{h}$, the specific capacity is expressed as 500/50 or $10 \text{ m}^3/\text{h}$ per meter of drawdown. If the drawdown is 25 m instead of 50 m, the output from the well can be estimated to be $10 \text{ m}^3/\text{h/m} \times 25 \text{ m} = 250 \text{ m}^3/\text{h}$. The diameter of a well has very little effect on the yield; doubling the diameter increases the yield by only about 10 percent.

Wells must be *developed* before they are put into use, in order to remove silt and fine sand adjacent to the well screen. Well development unplugs the aquifer, producing a natural filter of coarser particles around the well screen and allowing silt-free water to flow freely into the well. In one method of developing a well, called *surging*, a plunger is moved rapidly up and down in the well. In aquifers consisting of very fine uniform sand and silt, a filter must be constructed around the well screen. This is called *gravel packing* and is accomplished by filling the annular space around the well screen with gravel.

After a well is developed, a pump test is conducted to determine if it can supply the required amount of water. The well is generally pumped for at least 6 h at a rate equal to or greater than the desired yield. A stabilized drawdown should be obtained at that rate, and the original static level should be recovered within 24 h after pumping stops. During this test period, samples are taken and tested for bacteriological and chemical quality.

More than 15 million U.S. households obtain their drinking water from their own well, according to the 1990 U.S. Census. These private wells serve about 16 percent of the U.S. population. Another 36 percent get their drinking water from centralized public well systems.

##  3.8   RELEVANT WEB SITES

### AMERICAN WATER RESOURCES ASSOCIATION

http://www.uwin.siu.edu/~awra/index.html

This is the Home Page of the AWRA, which promotes understanding of water resources and related issues by providing a multidisciplinary forum for education, professional development, and information exchange.

### BUREAU OF RECLAMATION

http://www.usbr.gov

This is the Home Page of the Bureau Reclamation, best known for the dams (including the Hoover and Grand Coulee dams), canals, and power plants it constructed in the western United States. It now provides water to more than 31 million people and is the second largest producer of hydroelectric power in the western United States. This site provides information on water conservation, water recycling, and water reuse.

### EPA's SURF YOUR WATERSHED

http://www.epa.gov/surf/

This multifaceted, information-packed site provides comprehensive data about all watersheds, including the Index of Watershed Indicators, a compilation of information on the "health" of aquatic resources in the United States.

## Hydrologic Conditions—Delaware River Basin

http://www.state.nj.us/drbc/hydro.htm

This site presents comprehensive information about the hydrology of the Delaware River Basin, including stream flow, flood, and drought information.

## Hydrologic Engineering Center (HEC)

http://www.hec.usace.army.mil/

The HEC, an office of the U.S. Army Corps of Engineers, is involved in hydrologic engineering and water resources planning and management, and in improving hydrologic analytical methods for water resources planning. The HEC incorporates state-of-the-art procedures and techniques into manuals and computer programs.

## Hydrology and Related Internet Resources

http://terrassa.pnl.gov:2080/EESC/resourcelist/
hydrology.html

This site is a gateway to a wide variety of hydrologic information and Internet resources, including publications, computer modeling software, government agencies, electronic discussion groups, and more.

## Hydrogeologists Home Page

http://www.thehydrogeologist.com/index.htm

This page is a collection of hundreds of links to hydrogeological organizations, software and data repositories, publications, and other resources of potential use to hydrogeologists.

## Hydrology Laboratory

http://hydrolab.arsusda.gov/

The Hydrology Laboratory, a research unit of the U.S. Department of Agriculture, specializes in the improve-

ment of methods for predicting water yield, evaluating the impact of management practices, and assessing large-scale environmental changes on water resources. This page provides access to research data and information about current research.

## National Ground Water Association

http://www.ngwa.org/

This Home Page is a gateway to professional and technical information about the groundwater industry and groundwater resources.

## National Weather Service Office of Hydrology

http://hsp.nws.noaa.gov/oh/

The Office of Hydrology provides administrative, policy, and technical guidance for hydrologic matters at a national level within the National Weather Service and coordinates interagency matters in the field of operational hydrology. Activities include developing hydrologic procedures (computer models) and collecting and processing hydrologic data for river and flood forecasts and warnings, and for water-supply forecasts.

## USGS—Water Resources of the United States

http://h2o.er.usgs.gov/

The U.S. Geological Survey has the principal responsibility within the federal government to provide the hydrologic information and understanding needed to achieve the best use and management of the nation's water resources. This site provides a wealth of information about the quantity, quality, and use of the nation's water resources.

## USGS Hydrology Primer

http://wwwdutslc.ws.usgs.gov/
infores/hydrology.primer.html

This site provides an informative and interesting overview of what hydrology is and what hydrologists do.

# REVIEW QUESTIONS

1. Why is the science of hydrology of importance in environmental technology?

2. List two uses of water other than for public supplies. Which use requires the greatest amount of water?

3. Is there a difference between water withdrawal and water consumption? Name two nonwithdrawal uses of water.

4. Briefly discuss the relative availability of water across the United States.

5. Briefly outline the basic features of the hydrologic cycle.

6. What is the origin of subsurface water (ground-water)?

7. What is the meaning of *rainfall intensity?* How is it measured?

8. What is an *acre–foot?*

9. Briefly discuss the meaning of *N-year storm.*

10. If a 5-year storm occurred today, when would you next expect to observe a similar storm? Explain.

11. Is a 50-year storm more likely to be observed than a 20-year storm? Explain.

12. What is the difference between the expressions *storm recurrence interval* and *storm frequency?*

13. On average, are the intensities of long-duration storms less than, equal to, or greater than the intensities of short-duration storms? Sketch a graph of intensity versus duration to illustrate your answer.

14. Which rainstorm is likely to cause more environmental damage, a 1-year, 24-h storm or a 100-year, 8-h storm? What are the likely environmental consequences of each storm? (Use Figure 3.8 and Table 3.1.)

15. List three basic characteristics of a drainage divide line.

16. List three general characteristics of a watershed that may affect the volume and rate of runoff.

17. What is a hydrograph?

18. What is meant by *base flow* of a perennial stream?

19. Briefly explain the basic operation of a stream gaging station.

20. What is a stage–discharge curve?

21. What is the MA7CD10 flow of a stream?

22. What is a summation hydrograph and how is it used?

23. Sketch a typical reservoir capacity curve. Why do you think it is shaped concave downward?

24. What is the difference between the zone of aeration and the zone of saturation?

25. What is a *water table?* What may cause it to change position?

26. What is an *aquifer?*

27. What is the difference between a water table well and an artesian well?

28. Does groundwater constantly flow through the ground or is it stationary? Explain.

29. What is the most common method of groundwater withdrawal for public water supply? Briefly discuss construction and operation details.

30. For a given rate of groundwater withdrawal, will the drawdown occurring in a sand and gravel aquifer be any different from that in a less permeable, silty aquifer? Will there be any difference in the radius of influence?

31. Visit the *Hydrology Laboratory* Web site. Briefly describe some of the research activities of the Hydrology Laboratory.

32. Explore the *EPA's Surf Your Watershed* Web site. Research the Index of Watershed Indicators (IWI) for the watershed in which your home is located. Briefly describe what the IWI is, how it is calculated, and what the results are for your watershed.

33. Using the EnviroSources search engine (see Section 1.6) or another Internet search directory or search engine, locate at least one additional Web site relevant to one of the topics in this chapter. Add the link(s) to your Environmental Technology Folder. Write a brief description of what the Web site(s) contain.

# PRACTICE PROBLEMS

1. A total of 500 mm of rain fell on a 75-ha watershed in a 10-h period. Compute the average rainfall intensity and the total volume of rain that fell on the watershed.

2. A total of 1 in. of rain fell on a 96-ac area in 30 min. Compute the average rainfall intensity and the total volume of rain that fell.

3. Using the rainfall curves in Figure 3.6, determine the expected rainfall intensity for *(a)* a 5-year storm of 10-min duration, *(b)* a 10-year storm of 1.5-h duration, and *(c)* a 100-year storm of 2-h duration.

4. A storm of 30 min duration causes 75 mm of rainfall. Using the rainfall curves in Figure 3.6, estimate

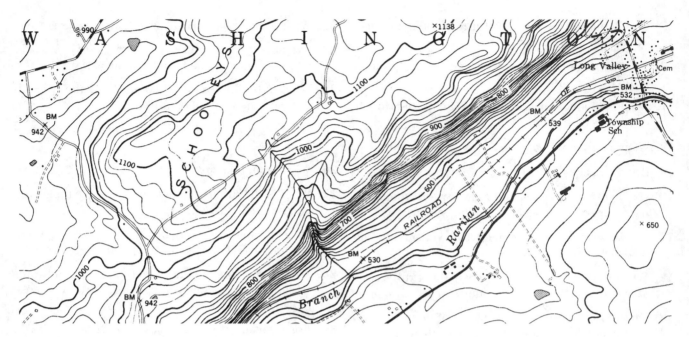

**FIGURE 3.29**
*Illustration for Problem 8(a).* (Courtesy of the U.S.
Geological Survey)

the probability of observing a similar storm in the next year.

5. Using Equation 3-3, determine the expected rainfall intensity for a storm of 1.5-h duration if $A = 3000$ and $B = 20$.

6. What is the probability of a 20-year storm being equaled or exceeded in any given year?

7. Based on the following record of weekly low flows in a river, determine the MA7CD10 drought flow. Use the probability graph of Figure 3.19; multiply vertical axis values by 10.

| Year | Discharge, $m^3$/s | Year | Discharge, $m^3$/s |
|------|------|------|------|
| 1970 | 50 | 1978 | 40 |
| 1971 | 47 | 1979 | 45 |
| 1972 | 57 | 1980 | 50 |
| 1973 | 42 | 1981 | 33 |
| 1974 | 36 | 1982 | 45 |
| 1975 | 39 | 1983 | 48 |
| 1976 | 53 | 1984 | 50 |
| 1977 | 44 | 1985 | 41 |

8. (a) In Figure 3.29, a stream tributary to the south branch of the Raritan River is shown on a USGS topographic map. (The point of confluence is near BM 530 on the railroad line.) Sketch the drainage divide line for the stream, beginning at its point of confluence.
(b) In Figure 3.30, Clyde Potts Reservoir is shown on a USGS topographic map. Sketch the drainage basin boundary for the reservoir.

9. A conservation reservoir is needed to provide a uniform withdrawal or yield of 0.5 mgd without being depleted. The streamflow records for the year of lowest flows are summarized on a monthly basis as follows:

| Month | Jan | Feb | Mar | Apr | May | Jun |
|-------|-----|-----|-----|-----|-----|-----|
| Streamflow, mil. gal/month | 20 | 20 | 10 | 2 | 10 | 38 |

| Month | Jul | Aug | Sep | Oct | Nov | Dec |
|-------|-----|-----|-----|-----|-----|-----|
| Streamflow, mil. gal/month | 8 | 3 | 2 | 6 | 30 | 50 |

Determine the required reservoir volume.

10. Compute the velocity of groundwater flow in soil that has a coefficient of permeability of 0.05 mm/s if the water table drops 0.5 m in elevation over a distance of 100 m.

**FIGURE 3.30**
*Illustration for Problem 8(b).* (Courtesy of U.S. Geological Survey)

**11.** The slope of a water table is determined to be 0.035 and the average velocity of groundwater flow is determined to be 0.5 m/h. In what type of soil deposit is the flow probably occurring?

**12.** The specific yield of a 100-mm-diameter well is 2 m³/h per meter. What is the yield of the well when the drawdown is 15 m? If the well diameter is doubled to 200 mm, what do you expect the yield to be at the same drawdown of 15 m?

## Chapter Outline

# CHAPTER 4

# Water Quality

The topic of water quality focuses on the presence of foreign substances in water and their effects on people or the aquatic environment. Water of good quality for one purpose may be considered to be of poor quality for some other use. For example, water suitable for swimming may not be of good enough quality for drinking. But even drinking water may not be suitable for certain industrial or manufacturing purposes that require pure water.

What exactly is pure water? Just how pure does it have to be for drinking or for other uses? Obviously, it is not enough to simply describe water quality as being "good" or "poor." Some quantitative measures for determining and describing the condition of the water are needed. It is necessary to determine what substances are in the water and in what concentrations they are present. Some knowledge of the effects of those substances on public and environmental health is also needed. Finally, some yardsticks or standards against which to compare the results of our analysis and thereby judge the suitability of the water for a particular use are needed.

Water has a remarkable tendency to dissolve other substances. Because of this, it is rarely found in nature in a pure condition. Even water in a mountain stream, far from civilization, contains some natural impurities in solution and in suspension.

Changes in water quality begin with precipitation. As rain falls through the atmosphere, it picks up dust particles and such gases as oxygen and carbon dioxide. In some industrialized regions, the quality of rain water is altered significantly before it ever touches the ground. *Acid rain*, a prime example of this, is discussed in Chapter 13.

Surface runoff picks up silt particles, bacteria, organic material, and dissolved minerals. Groundwater usually contains more dissolved minerals than surface water because of its longer contact with soil and rock. Finally, water quality is very much affected by human activities, including land use (such as agriculture) and the direct discharge of municipal or industrial wastewaters to the environment.

Protecting water quality and modifying it for a particular purpose are major objectives in the field of environmental technology. It is therefore necessary to make use of technical terms in discussing the various aspects of water quality and pollution. In particular, reference to the different *parameters* of physical, chemical, and biological quality needs to be made.

This chapter begins with an overview of chemical concepts and terminology. In the discussion of specific water quality parameters that follows, only brief reference is made to actual laboratory analysis procedures. These are thoroughly described in *Standard Methods for the Examination of Water and Wastewater* (published by the American Water Works Association) and other professional publications. Discussion of water quality standards is included in subsequent chapters as they relate to drinking water, surface water, or wastewater treatment plant effluents.

Portable field test kits, as illustrated in Figure 4.1, are particularly useful for conducting preliminary water quality surveys. But for most water quality analyses to be official and able to stand up to legal scrutiny if challenged, they must be done by qualified personnel in certified laboratories, following *Standard Methods*.

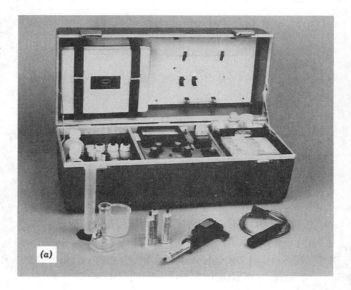

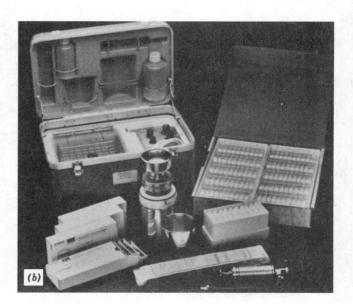

**FIGURE 4.1**
*Portable water testing kits are used in the field to measure (a) chemical and physical quality* (Courtesy of HACH Company, Loveland, Colorado) *and (b) microbiological or sanitary quality.* (Courtesy of Millipore Corporation, Bedford, Massachusetts)

## 4.1 FUNDAMENTAL CONCEPTS IN CHEMISTRY

The study of water quality and pollution control requires a basic knowledge of *chemistry*, a science that focuses on the composition and properties of substances. For those students with little or no previous training in chemistry, this section will provide a foundation for an understanding of the environmental topics covered later. It may serve as a quick review for others.

## Elements and Compounds

All matter is composed of basic substances, called *elements*, which cannot be subdivided or broken down into simpler substances by ordinary chemical change. The smallest part of an element that can exist and still retain the same chemical characteristics of that substance is called an *atom*.

There are over 100 known elements. Some of the more common elements along with the symbols used to represent them in chemical formulas and equations are listed in Table 4.1.

The science of chemistry is also concerned with how the elements react and combine with each other, forming *compounds*. Compounds are substances made up of various combinations of the basic elements. The smallest part of a chemical compound that can exist

**TABLE 4.1**
**Common elements and their symbols and atomic weights**

| Element | Symbol | Atomic weight |
|---------|--------|---------------|
| Aluminum | Al | 27 |
| Arsenic | As | 75 |
| Barium | Ba | 137 |
| Cadmium | Cd | 112 |
| Calcium | Ca | 40 |
| Carbon | C | 12 |
| Chlorine | Cl | 35 |
| Chromium | Cr | 24 |
| Copper | Cu | 64 |
| Fluorine | F | 19 |
| Hydrogen | H | 1 |
| Iron | Fe | 56 |
| Lead | Pb | 207 |
| Magnesium | Mg | 24 |
| Manganese | Mn | 55 |
| Mercury | Hg | 201 |
| Nitrogen | N | 14 |
| Oxygen | O | 16 |
| Phosphorus | P | 31 |
| Potassium | K | 39 |
| Selenium | Se | 79 |
| Silicon | Si | 28 |
| Silver | Ag | 108 |
| Sodium | Na | 23 |
| Sulfur | S | 32 |
| Zinc | Zn | 65 |

and still retain the same chemical properties of that compound is called a *molecule*.

Molecules can be represented using combinations of the symbols for the atoms in the molecule; such a combination is called a *chemical formula*. For example, a single molecule of water is composed of two atoms of hydrogen H and one atom of oxygen O. Its chemical formula is $H_2O$, pronounced "H-two-O." The subscript 2 after the H indicates that two atoms of hydrogen are in a water molecule. The formula for iron oxide, commonly called rust, is $Fe_3O_4$, indicating that there are three atoms of iron and four atoms of oxygen in one molecule of this compound.

There are hundreds of thousands of known compounds. Chemists have traditionally separated them into two broad groups, called *organic* compounds and *inorganic* compounds. Organic compounds are typically complex molecules of carbon in combination with other elements, such as hydrogen and oxygen.

Inorganic compounds usually do not contain carbon, although there are exceptions to this. In a very general sense, organic compounds are closely related to living organisms, whereas inorganic compounds are more a part of the inanimate world. A list of common inorganic compounds is presented in Table 4.2. Many of these compounds are used in water or wastewater treatment operations and will be mentioned again later. The list includes the chemical name, the common name, and the formula of each compound, as well as its physical state (solid, liquid, or gas) at ordinary room temperature and pressure.

### Atomic Structure

The way in which elements combine with each other to form compounds depends on their atomic structure. A simplified model of atomic structure includes a dense center or *nucleus* of positively charged particles called *protons* and uncharged or neutral particles called *neutrons*. Very light, negatively charged particles called *electrons* spin around the atomic nucleus in concentric shells or orbitals. To illustrate this basic model of atomic structure, schematic diagrams of a hydrogen atom and an oxygen atom are shown in Figure 4.2.

The unique identity of an element is established by the number of protons in its nucleus, called its *atomic number*. Each element has a different atomic number. For example, the atomic number of hydrogen is 1 and the atomic number of oxygen is 8. The atoms themselves are electrically neutral because the number of negatively charged electrons orbiting the nucleus is the same as the number of positively charged protons inside the nucleus; the opposite charges cancel or balance each other.

**TABLE 4.2**
**Common inorganic compounds**

| Chemical name | Common name | State | Formula |
|---|---|---|---|
| Aluminum sulfate | Alum | Solid | $Al_2(SO_4)_2$ |
| Ammonia | — | Gas | $NH_3$ |
| Calcium carbonate | Limestone | Solid | $CaCO_3$ |
| Calcium hydroxide | Slaked lime | Solid | $Ca(OH)_2$ |
| Calcium hypochlorite | — | Solid | $Ca(ClO)_2$ |
| Calcium oxide | Lime | Solid | $CaO$ |
| Carbon dioxide | — | Gas | $CO_2$ |
| Carbon monoxide | — | Gas | $CO$ |
| Chlorine | — | Gas | $Cl_2$ |
| Copper sulfate | Blue vitriol | Solid | $CuSO_4$ |
| Iron oxide | Rust | Solid | $Fe_3O_4$ |
| Hydrogen | — | Gas | $H_2$ |
| Hydrogen sulfide | — | Gas | $H_2S$ |
| Hydrochloric acid | Muriatic acid | Liquid | $HCl$ |
| Hypochlorous acid | — | Liquid | $HClO$ |
| Nitric acid | — | Liquid | $HNO_3$ |
| Nitrogen | — | Gas | $N_2$ |
| Nitrogen oxide | — | Gas | $NO$ |
| Nitrogen dioxide | — | Gas | $NO_2$ |
| Oxygen | — | Gas | $O_2$ |
| Ozone | — | Gas | $O_3$ |
| Sodium carbonate | Soda ash | Solid | $Na_2CO_3$ |
| Sodium chloride | Table salt | Solid | $NaCl$ |
| Sodium hydroxide | Lye | Solid | $NaOH$ |
| Sodium hypochlorite | — | Solid | $NaClO$ |
| Sulfur dioxide | — | Gas | $SO_2$ |
| Sulfuric acid | Oil of vitriol | Liquid | $H_2SO_4$ |

The total number of protons plus neutrons in a nucleus is called the mass number of the element and is approximately equal to its *atomic weight*. The electron has very little mass or weight compared to the proton. For example, the atomic weight of hydrogen is 1 (since it has no neutrons in its nucleus) and the atomic weight of oxygen is 16 (8 protons plus 8 neutrons).

The way an element behaves chemically depends primarily on the number of electrons in the atom's outermost shell or orbital. The orbital closest to the nucleus is most stable when it has two electrons, which is the maximum it can contain. The second orbital is most stable when it contains its maximum of eight electrons. Larger atoms have additional electron orbitals.

### Formation of Molecules

Compounds are formed by either the transfer or the sharing of electrons among two or more atoms. For example, sodium chloride, NaCl, is formed by the transfer of one electron from the outermost shell of the Na atom to the outermost shell of the Cl atom. As a result, the Na has a positive charge and the Cl has a negative

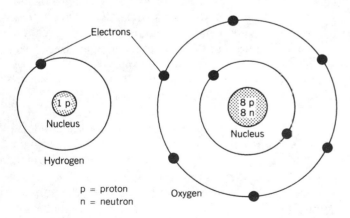

**FIGURE 4.2**
*The solar system model of atomic structure for hydrogen and oxygen. Hydrogen has one proton in its nucleus; oxygen has eight protons and eight neutrons.*

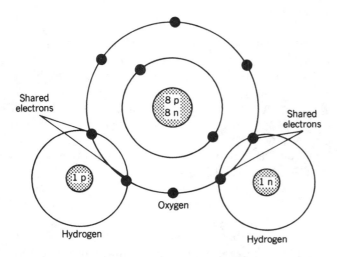

**FIGURE 4.3**
*The water molecule, written as $H_2O$, is formed by the sharing of electrons between a hydrogen and an oxygen atom. This is an example of* covalent bonding.

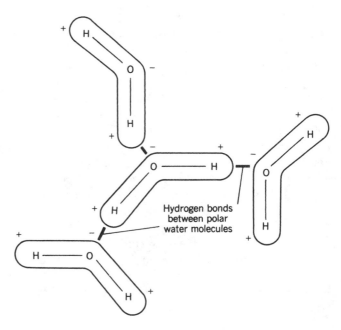

**FIGURE 4.4**
*Water molecules stick together because of attractive forces in what are called* hydrogen bonds.

charge. Since unlike charges attract each other, the compound NaCl is formed as the $Na^+$ and $Cl^-$ ions stick together. This is an example of what is called *ionic bonding* between atoms. (Ions are discussed further below.)

In the case of the water molecule, $H_2O$, the two atoms of hydrogen tend to share their electrons with the oxygen atom, as illustrated in Figure 4.3. This is called *covalent bonding*. In ionic bonding there is a complete transfer of electrons, but in covalent bonding, the outer orbitals are stabilized by a sharing of electrons.

The atoms in a water molecule are arranged at an angle instead of along a straight line, and the shared electrons are pulled closer to the oxygen atom than to the hydrogen atoms. This results in what is called a *polar molecule*, in which the positive and negative charges are not evenly distributed. The oxygen end of the molecule is negatively charged and the hydrogen ends are positively charged. It is this polarity of the water molecule that accounts for most of its properties, including its ability to dissolve many other substances. A schematic drawing of polar water molecules and the hydrogen bonds that hold them together in a volume of water is shown in Figure 4.4.

## Solutions

A *solution* is a uniform mixture of two or more substances existing in a single phase, that is, as a solid, a liquid, or a gas. Solutions in water are called *aqueous solutions* and are the most familiar to people. Carbon monoxide, CO, can be considered to be dissolved in the air; this is a gaseous solution. Carbon, C, and man-

ganese, Mn, are dissolved in iron, forming steel, an example of a solid solution.

In these mixtures, the substance present in the largest amount is called the *solvent* and the substances present in smaller amounts are called *solutes*. The properties of solutions differ from the properties of the solvent. For example, although water freezes at 0°C, the presence of a dissolved salt, such as NaCl, in water lowers the freezing point of the solution to below 0°C. Most chemical changes or reactions take place in solution.

Consider the aqueous solution of sugar shown schematically in Figure 4.5. The individual sugar molecules

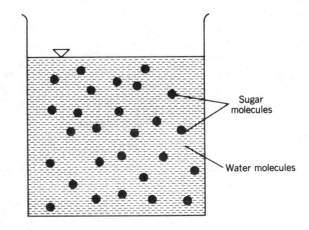

**FIGURE 4.5**
*Schematic representation of an aqueous sugar solution. The sugar molecules remain uniformly dispersed in the volume of water.*

are uniformly dispersed throughout the water and do not settle out to the bottom. The solute molecules will always remain evenly mixed in the solvent because of the kinetic energy and constant motion of all the molecules.

If more sugar is added, the solution will eventually reach a state where the sugar molecules no longer dissolve. At this point, the solution is called a *saturated solution*. Temperature has a great effect on the saturation point, that is, on the amount of solute a solution can hold before it becomes saturated. Most solid substances are more soluble in warm water than in cold water (for example, increasing the temperature of the sugar–water solution will allow more sugar molecules to dissolve).

Some liquid substances can dissolve in water without limit. Alcohol, for example, can be mixed with water in all proportions. If there is more alcohol than water, the alcohol is then considered to be the solvent and water, the solute.

The solubility of various gases in water, such as oxygen, carbon dioxide, and chlorine, is of particular concern in environmental technology. As with solids, the solubility of gases depends on temperature, but the relationship is just the opposite of that for solids. *The solubility of a gas decreases with increasing temperature.*

The solubility of oxygen is of particular importance in water quality, as discussed in more detail later in this chapter and in Chapter 5. Typical saturation values of dissolved oxygen in fresh water, at selected temperatures, are summarized in Table 4.3. This illustrates the very limited solubility of oxygen as well as the pronounced effect of temperature on solubility.

The term *mg/L* used in the table for the concentration of oxygen in the water is an abbreviation for *milligrams per liter.* This and other ways of expressing concentrations are discussed on pages 96 to 97.

Factors other than temperature also affect the solubility of gases. At higher altitudes and lower atmospheric pressures, the solubility of a gas is less than it is at sea level. Also, increasing the salinity, or salt content, decreases the solubility of gases. For example, less oxygen can be dissolved in seawater than in fresh water under the same conditions of temperature and pressure.

**TABLE 4.3**
**Solubility of oxygen in fresh water**

| Temperature, °C | Saturation solubility, mg/L |
|:---:|:---:|
| 0 | 14.6 |
| 10 | 11.3 |
| 20 | 9.2 |
| 30 | 7.6 |

## Ionization

In the previous illustration of an aqueous sugar solution, the sugar molecules retained their identity. In other words, they did not break apart into fragments smaller than sugar molecules. The uncharged or neutral sugar molecules remained dispersed in the solution, surrounded by the water molecules.

There are many substances, however, that *dissociate*, or break apart, as they dissolve, forming smaller electrically charged particles called *ions*. This process is called *ionization.*

Sodium chloride, NaCl, was previously described as an ionic compound because of the nature of the chemical bond between the $Na^+$ and $Cl^-$ ions. Sodium chloride dissociates in water, as shown in the following chemical equation:

$$NaCl \rightarrow Na^+ + Cl^-$$

sodium      positive      negative
chloride   sodium ion   chloride ion

The sodium ion, $Na^+$, has a positive charge because it has given up its outermost electron to the chlorine atom and therefore has more protons than electrons. The chloride ion, $Cl^-$, is negative because it has more electrons than protons; its outer shell is stable with the extra electron from the sodium atom. Ionic solutions are neutral; the total positive charges must equal the total negative charges in the solution. Figure 4.6 illustrates schematically the solution of NaCl in water.

## Polyatomic Ions

In some cases, molecules dissociate into charged particles consisting of groups of atoms that act together

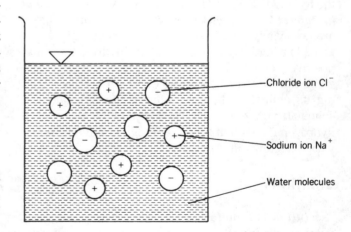

**FIGURE 4.6**
*A schematic representation of an ionic solution of sodium chloride, NaCl, in water.*

as a unit. These charged fragments are called *polyatomic ions*, and they have special names that are used quite frequently in water chemistry. For example, the water molecule can dissociate into a hydrogen ion, $H^+$, and a *hydroxyl ion*, $OH^-$, as follows:

$$H_2O \rightarrow H^+ + OH^-$$

Sulfuric acid is an example of a compound that dissociates readily in water, as follows:

$$H_2SO_4 \rightarrow 2H^+ + SO_4^{2-}$$

The $SO_4^{2-}$ is called a *sulfate ion;* it has a double negative charge, balancing the total positive charge of the two hydrogen ions. The complex sulfate ion behaves like a simple ion, in that it does not dissociate into smaller fragments.

Other complex ions of interest in environmental applications include *nitrate*, $NO_3^-$, *phosphate*, $PO_4^{2-}$, *ammonium*, $NH_4^+$, and *hypochlorite*, $OCl^-$. There are many others, some of which appear in later discussions of water treatment and pollution.

## Suspensions and Colloids

Although many substances occur in solution in molecular or ionic form (true solution), some substances may be suspended in the mixture in fragments significantly larger than the size of molecules. The properties of these mixtures differ from those of true solutions and are of particular significance in environmental technology.

Perhaps one of the most important characteristics of a true solution is that the solute particles do not settle out or separate from the mixture, no matter how long the solution remains under quiescent or still conditions. Furthermore, solutes cannot be physically separated from the solvent by filters.

In contrast to this, most suspended particles, if allowed enough time, will settle out from the water, due to gravity. Also, suspended particles can be removed from the water by filters. Because of this, suspensions of silt, organic material, and even microbes are among the first of the impurities to be removed in conventional water and sewage treatment systems.

A suspension of relatively large particles is called a *coarse suspension*. In quiescent or still water, large particles settle out of the water in a matter of minutes. Finer particles, however, take many hours to settle out. Even some bacteria, about 0.001 mm, or 1 $\mu$m in size, eventually settle to the bottom (a $\mu$m, pronounced "micrometer," is one millionth, or $10^{-6}$, of a meter; see

Appendix C for further discussion of metric terms and symbols). It can be predicted mathematically that bacteria will settle at a rate of about 1 m in 175 h, but this is hardly an efficient way to remove bacteria from water.

Extremely fine particles, those less than about 0.1 $\mu$m, are generally too small to settle out because of the force of gravity or to be removed by most filters. These particles, smaller than those in coarse suspensions, but larger than those in true solutions, are called *colloids*. Colloids occur in air as well as in water. In water, for example, clay particles or tiny fragments of decaying vegetation and organic wastes may form colloidal suspensions. In air, colloidal suspensions of tiny solid particles (smoke) or liquid particles (fog) are often encountered.

Like the particles in a true solution, the particles in a colloidal suspension cannot be seen under a microscope. In a true solution, a beam of light can pass through the solution without any scattering of the light. However, in a colloidal suspension, the colloidal particles will scatter the light, allowing the beam to be seen. This is called the *Tyndall effect*, and it is one of the characteristics distinguishing colloids from solutions. The Tyndall effect is illustrated in Figure 4.7.

Colloidal particles may have either all positive or all negative electrical charges of various magnitudes, depending on the nature of the substance. Since like charges repel each other, colloidal particles keep their distance from each other. There is a force of repulsion between them. Because colloidal particles repel each other, they very rarely collide, so they have no chance

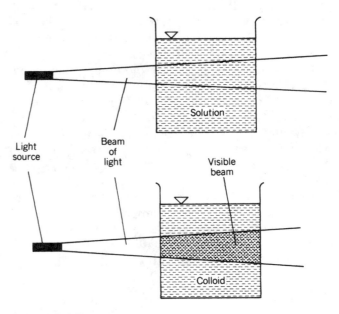

**FIGURE 4.7**

*The* Tyndall effect: *Light is scattered by the colloidal particles, and the light beam is visible in the liquid.*

to stick together to form larger, heavier particles. In addition to their small size, this is a basic reason that colloids are stable and do not settle out of suspension. A very common water treatment process involves the addition of certain chemicals to neutralize the effect of the colloidal charges. This allows the particles and chemicals to collide and form *flocs*, which can settle out or be separated from the water by filters. This is discussed in more detail in Chapter 6.

## Expressing Concentrations

The properties of solutions, suspensions, and colloids depend to a large extent on their concentrations. A *dilute* or weak solution has a relatively small amount of solute dissolved in the solvent. It has characteristics different from those of a *concentrated* or strong solution of the same substance, in which a relatively large amount of solute is present. Since concentrations need to be expressed quantitatively, instead of qualitative terms like *dilute* or *strong*, concentrations are usually expressed in terms of mass per unit volume, parts per million or billion, or percent.

### Mass per Unit Volume

One of the most common terms for concentration is *milligrams per liter (mg/L)*. For example, if a mass of 10 mg of oxygen is dissolved in a volume of 1 L of water, the concentration of that solution is expressed simply as 10 mg/L. If 0.3 g of salt is dissolved in 1500 mL of water, then the concentration is expressed as 300 mg ÷ 1.5 L = 200 mg/L, where 0.3 g = 300 mg and 1500 mL = 1.5 L (1 g = 1000 mg; 1 L = 1000 mL).

Very dilute solutions are more conveniently expressed in terms of *micrograms per liter (μg/L)*. For example, a concentration of 0.004 mg/L is preferably written as its equivalent 4 μg/L. Since 1000 μg = 1 mg, simply move the decimal point three places to the right when converting from mg/L to μg/L. Move the decimal three places to the left when converting from μg/L to mg/L. For example, a concentration of 1250 μg/L is equivalent to 1.25 mg/L.

In air, concentration of particulate matter or gases are commonly expressed in terms of *micrograms per cubic meter ($\mu g/m^3$)*. (Computations related to the concentrations of air pollutants are discussed in Section 13.6.)

### Parts per Million

One liter of water has a mass of 1 kg. But 1 kg is equivalent to 1000 g or 1 million mg. Therefore, if 1 mg of a substance is dissolved in 1 L of water, we can say that there is 1 mg of solute per 1 million mg of water. In other words, there is *one part per million (1 ppm)*.

Neglecting the small change in the density of water as substances are dissolved in it, we can say that, in general, a concentration of 1 mg per liter is equivalent to one part per million: *1 mg/L = 1 ppm*. Conversions are very simple; for example, a concentration of 17.5 mg/L is identical to 17.5 ppm.

The expression *parts per million* is useful in conveying a picture of just how small most of the concentrations encountered in environmental technology actually are. If 1 lb of salt were dissolved in a 50 ft × 50 ft × 50 ft tank of water, the concentration of salt would be about 1 ppm. One part per million is the same as 1 in. in about 16 mi, or 1 s in about 12 d.

Very dilute concentrations can be expressed in terms of *parts per billion (ppb)*, instead of parts per million. For example, a concentration of 0.005 ppm is better written as its equivalent of 5 ppb. Even such tiny concentrations of some substances can significantly affect environmental quality and human health. Modern analytical instruments are capable of detecting these very low concentrations. (One part per billion corresponds roughly to 1 s in 33 years, or a single penny in 10 million dollars.)

The expression *mg/L* is preferred over *ppm*, just as the expression *μg/L* is preferred over its equivalent of *ppb*. But both types of units are still used, and the student should be familiar with each.

### Percentage Concentration

Concentrations in excess of 10 000 mg/L are generally expressed in terms of percent, for convenience. For practical purposes, the conversion of 1 percent = 10 000 mg/L can be used even though the densities of the solutions are slightly more than that of pure water (10 000 mg/L = 10 000 mg/1 000 000 mg = 1 mg/100 mg = 1 percent).

The concentration of salts in seawater is about 35 000 mg/L. To convert to percent salts, divide by 10,000, obtaining 3.5 percent. The concentration of wastewater sludge may be about 3 percent solids. To convert this to mg/L, multiply by 10,000, getting 30 000 mg/L solids.

A concentration expressed in terms of percent may also be computed from the following equation:

$$\text{percent} = \frac{\text{mass of solute (mg)}}{\text{mass of solvent (mg)}} \times 100 \quad \textbf{(4-1)}$$

### U.S. Customary Units

The expression *grains per gallon (gpg)* is sometimes used for the concentrations of certain substances in water. One grain per gallon is equivalent to a concentration of 17.1 milligrams per liter: *1 gpg = 17.1 mg/L.*

The expression *pounds per million gallons* is also used in U.S. Customary units of concentration for water treatment applications. Since 1 gal of water weighs 8.34 lb, 1 gal/mil gal is the same as 8.34 lb/mil gal. Or we can say that *1 mg/L = 8.34 lb/mil gal.* To convert from mg/L to lb/mil gal, multiply by 8.34; to go from lb/mil gal to mg/L, divide by 8.34.

## EXAMPLE 4.1

A 500-mL aqueous salt solution has 125 mg of salt dissolved in it. Express the concentration of this solution in terms of *(a)* mg/L, *(b)* ppm, *(c)* gpg, *(d)* percent, and *(e)* lb/mil gal.

*Solution*
*(a)* 125 mg/500 mL × 1000 mL/L = 250 mg/L.
*(b)* 250 mg/L = 250 ppm.
*(c)* 250 mg/L × 1 gpg/17.1 mg/L = 14.6 gpg.
*(d)* Applying Equation 4-1 and the fact that 500 mL of water has a mass of 500 g gives

> percent = 0.125 g/500 g × 100 = 0.025 percent

Or divide 250 mg/L by 10,000 to get 0.025 percent.
*(e)* 250 mg/L × 8.34 = 2090 lb/mil gal.

## EXAMPLE 4.2

How many pounds of chlorine gas should be dissolved in 8 mil gal of water to result in a concentration of 0.2 mg/L?

*Solution*

$$0.2 \text{ mg/L} \times 8.34 = 1.67 \text{ lb/mil gal}$$

and

$$1.67 \text{ lb/mil gal} \times 8 \text{ mil gal} \approx 13 \text{ lb}$$

## Acids, Bases, and pH

An *acid* is a substance that causes an increase of the hydrogen ion ($H^+$) concentration in an aqueous solution. A substance that causes the hydroxyl ($OH^-$) concentration to increase is called a *base*. Acids and bases may be characterized as *strong* or *weak*, depending on the degree to which they increase the relative concentrations of $H^+$ or $OH^-$.

Hydrochloric acid, HCl, is an example of a strong acid that readily dissociates in water, forming $H^+$ and $Cl^-$ ions; the $H^+$ concentration in the water is greatly increased. Sodium hydroxide, NaOH, is an example of a strong base that dissociates into $Na^+$ and $OH^-$, greatly increasing the $OH^-$ concentration. A substance that is basic, like NaOH, is also called an *alkaline* substance.

The chemical reaction between an acid and a base is called *neutralization*. The two products of a neutralization reaction are water and a salt. Table salt, NaCl, for example, is a product of the neutralization reaction between hydrochloric acid, HCl, and sodium hydroxide, NaOH, as follows:

$$HCl + NaOH \rightarrow NaCl + H_2O$$

### The pH Scale

The pH is a dimensionless number that indicates the strength of an acidic or a basic solution. The pH scale ranges from 0 to 14. The middle of the range, pH = 7, represents a neutral solution, or one that is neither acidic nor basic. Pure water is neutral because it contains the same number of hydrogen ions, $H^+$, as hydroxyl ions, $OH^-$.

Solutions with pH values less than 7 are acidic; those with pH values greater than 7 are basic or alkaline. Figure 4.8 illustrates this scale, along with the relative positions of some familiar substances.

In more technical terms, pH is defined as the negative logarithm of the hydrogen ion concentration. For

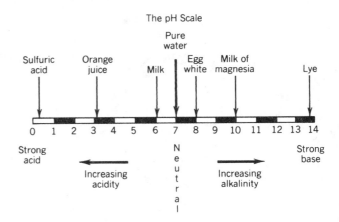

**FIGURE 4.8**
*The pH scale is used to indicate the intensity or strength of an acidic or basic solution. A pH of 7 is neutral, neither acidic nor basic.*

example, in pure water the numerical value of the hydrogen ion concentration is $10^{-7}$. The logarithm (or exponent) is $-7$, and the negative of that is 7.

Because the pH scale is based on logarithms to the base 10, each unit change in pH actually represents a tenfold change in the degree of acidity or alkalinity of a solution. For instance, a solution with a pH = 5 is ten times more acidic than a solution with a pH = 6. Likewise, a solution with a pH = 4 is 100 times more acidic than the solution with pH = 6.

## Organic Substances

The defining characteristic of an organic compound is that it contains carbon in combination with other elements, such as hydrogen, oxygen, nitrogen, phosphorus, and sulfur. All living organisms are composed of organic compounds, some of which are so complex in molecular structure that they are still not fully understood by scientists.

Hundreds of thousands of organic chemicals are known to exist. Many of these occur naturally, in animal and plant tissues and wastes. Other organic compounds are synthetic substances that never occur in nature outside the chemist's laboratory. The basic reason for there being so many organic compounds, both natural and synthetic, is that a carbon atom combines readily with other carbon atoms and other elements, linking together in long chains or rings.

Organic compounds containing carbon and hydrogen are called *hydrocarbons*. The simplest hydrocarbon is methane, $CH_4$, which is a gas at ordinary temperature and pressure. It is produced naturally during the decay of other organic compounds, such as those found in sewage sludge or garbage.

The simplest *ring hydrocarbon* is benzene, $C_6H_6$, in which the carbon atoms link together to form a hexagon-shaped ring. Butane is an example of a *straight-chain hydrocarbon* molecule. Schematic diagrams showing the molecular structure of methane, benzene, and butane are shown in Figure 4.9.

In addition to the ability of carbon atoms to bond to each other, forming rings and chains of various lengths, different groups of atoms can readily replace the hydrogen atoms in the hydrocarbons. This is another reason for the existence of the extremely large number of organic substances.

| NAME | CHEMICAL FORMULA | MOLECULAR STRUCTURE |
|---|---|---|
| METHANE | $CH_4$ | 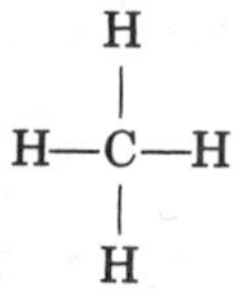 |
| BENZENE | $C_6H_6$ | 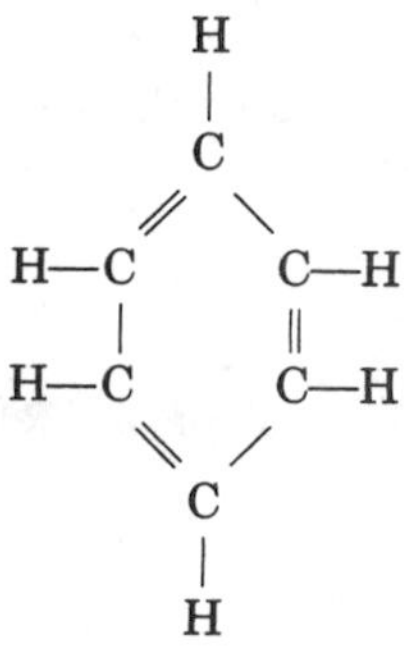 |
| BUTANE | $C_4H_{10}$ | 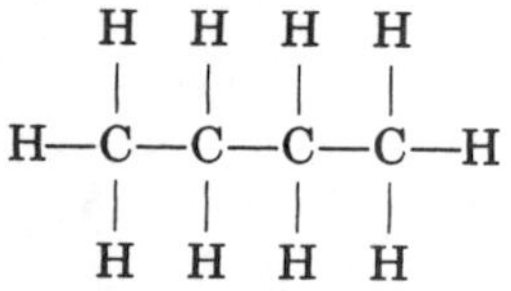 |

**FIGURE 4.9**
*Schematic representations of the molecular structures for three common organic compounds, called* hydrocarbons.

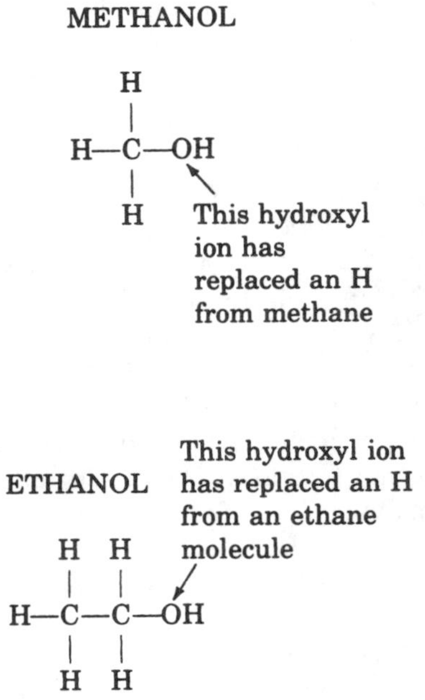

METHANOL

$$
\begin{array}{c}
\text{H} \\
| \\
\text{H—C—OH} \\
| \\
\text{H}
\end{array}
$$

This hydroxyl ion has replaced an H from methane

ETHANOL

This hydroxyl ion has replaced an H from an ethane molecule

$$
\begin{array}{c}
\text{H} \quad \text{H} \\
| \quad | \\
\text{H—C—C—OH} \\
| \quad | \\
\text{H} \quad \text{H}
\end{array}
$$

**FIGURE 4.10**
*Molecular structure for two different types of alcohol: methanol and ethanol. The $OH^-$ group replaces an H atom in both types.*

The classification of different organic chemicals depends on which particular group of atoms replaces the hydrogen. *Alcohols*, for example, are formed when hydrogen atoms are replaced by hydroxyl groups. Methanol, $CH_3OH$, is an example of an alcohol used in solvents and fuel additives. Its molecular structure is illustrated schematically in Figure 4.10. Ethanol, $C_2H_5OH$, is an alcohol resulting from the fermentation of sugar and is found in alcoholic beverages. Oxidation of alcohol may result in organic compounds called *aldehydes* and *ketones*.

Organic acids are formed when the carboxyl group, —COOH, replaces hydrogen in a hydrocarbon. Acetic acid, commonly found in vinegar, is an example of an organic acid. It is illustrated in Figure 4.11.

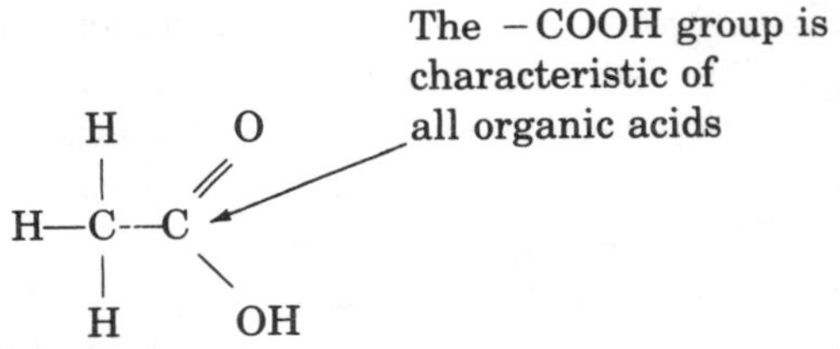

Many organic substances are *biodegradable*. This is a popular term used in reference to substances that can be used by microbes as food; biodegradable organic molecules are readily broken down into smaller, simpler molecules by biological action.

*Carbohydrates* are the most abundant group of biodegradable organic compounds and are sometimes called the "fuel of life." They are the basic products of photosynthesis in green plants. Photosynthesis is the process by which the sun's energy is converted into a form that can be used by living organisms (see Section 1.3).

Carbohydrate molecules are formed from the elements carbon, hydrogen, and oxygen; the hydrogen and oxygen always occur in the same proportion as in water, that is, two to one. The sugar called *glucose* is an example of a simple carbohydrate. Its molecular structure is illustrated schematically in Figure 4.12. *Sucrose*, common table sugar, is a carbohydrate formed by the combination of glucose and another sugar called fructose. Starch and cellulose are larger and more complex carbohydrates that are not as biodegradable as the simpler sugars. Cellulose is the primary material in plants.

*Fats* are also biodegradable organic compounds, and are composed of carbon, oxygen, and hydrogen atoms. Although they are important energy storage molecules in living organisms, they are not very soluble in water, and decompose at a slow rate. *Proteins* (formed from amino acids) are much more complex than carbohydrates or fats, and they form the primary substance of animal tissue. In addition to carbon, oxygen, and hydrogen, proteins contain nitrogen and sulfur.

## Concept of Mass Balance

A fundamental concept in science is the *law of conservation of matter*. This means that when there is no appreciable conversion of mass into energy, the sum of the masses of substances entering into a reaction must always equal the sum of the masses of the products of the reaction. Even if there is no chemical reaction occurring, the law of conservation underlies the

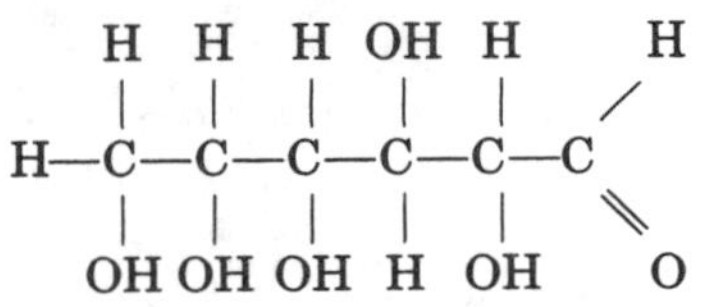

The —COOH group is characteristic of all organic acids

$$
\begin{array}{c}
\text{H} \quad\quad \text{O} \\
| \quad\quad \parallel \\
\text{H—C—C} \\
| \quad\quad \backslash \\
\text{H} \quad\quad \text{OH}
\end{array}
$$

**FIGURE 4.11**
*Molecular structure of acetic acid, commonly found in vinegar.*

$$
\begin{array}{c}
\text{H} \quad \text{H} \quad \text{H} \quad \text{OH} \quad \text{H} \quad\quad \text{H} \\
| \quad | \quad | \quad | \quad | \quad\quad \diagup \\
\text{H—C—C—C—C—C—C} \\
| \quad | \quad | \quad | \quad | \quad\quad \diagdown \\
\text{OH} \quad \text{OH} \quad \text{OH} \quad \text{H} \quad \text{OH} \quad\quad \text{O}
\end{array}
$$

**FIGURE 4.12**
*Molecular structure of the sugar called glucose, $C_6H_{12}O_6$.*

concept of *mass balance* (also called *material balance*), and is useful in environmental technology.

Mass balance calculations play an important role in the design and operation of water, sewage, air, and solid waste treatment processes. In treatment systems, the physical, chemical, and biochemical processes usually occur in vessels or tanks called *reactors*, and the particular reactions or processes are referred to as *unit processes*. In the simplest case, it can be said that the *input must equal the output*, or, in other words, "what goes in must go out." If this does not occur, there must be an accumulation (or depletion) of the material in the reactor equal to the difference between the input and output, or *accumulation = input − output*. Since, in this kind of situation, the composition of material in the reactor changes with time, it is referred to as an *unsteady-state* operation. In a *steady-state* operation, it can be assumed that the rates of input and output are constant, as is the composition of the completely mixed reactor.

Suppose, for example, two pipes containing salt solutions discharge into a tank in which the two solutions are completely mixed, and a third pipe carries the mixture out of the tank (as shown in Figure 4.13. The solution in the first pipe has a concentration of $c_1$ mg/L and that in the second pipe has a concentration of $c_2$ mg/L. The flow rates in the pipes are $Q_1$ and $Q_2$, respectively. The concept of mass or material balance can be applied to determine the concentration of the mixed solution discharged from the tank because under steady-state conditions, the total amount of salt entering the tank must be equal to the total amount leaving the tank. In other words, since the salt neither decays nor reacts with other substances (in this example), the concentration of salt in the mixture in the tank stays constant over time.

The product of concentration and volume flow rate equals the mass flow rate because mg/L × L/d = mg/d, where the volume flow rate in this example is expressed in terms of liters per day, or L/d. (See Appendix C for a discussion of unit cancellation.) For convenience here, consider that the time interval is 1 d. Then the product of $c_1 \times Q_1$ must equal the mass of salt entering the vessel in 1 d from the first pipe. Similarly, $c_2 \times Q_2$ equals the mass of salt entering the tank from the second pipe. The total mass of salt entering the tank in 1 d, that is, the *input*, must be equal to the sum from the two pipes, or *input* $= c_1 \times Q_1 + c_2 \times Q_2$.

The total mass of salt leaving the tank equals the product of the concentration in the mixture $c_3$ and the volume flow rate leaving the tank. Because water is virtually incompressible, however, that flow rate must be $Q_1 + Q_2$. Therefore, the *output* of salt is $c_3 \times (Q_1 + Q_2)$. Because the concept of mass or material balance applies here and *output = input*, the following relationship is obtained:

$$c_3 \times (Q_1 + Q_2) = c_1 \times Q_1 + c_2 \times Q_2$$

Solving the above equation for $c_3$ by dividing both sides by $(Q_1 + Q_2)$, we obtain the following mass balance equation:

$$c_3 = \frac{c_1 \times Q_1 + c_2 \times Q_2}{Q_1 + Q_2}$$

Mass balance calculations can also be applied to natural environmental systems, such as streams, rivers, lakes, and even the atmosphere. Examples of the application of the mass balance equation derived above are included in Section 5.4, Stream Pollution.

## 4.2 PHYSICAL PARAMETERS OF WATER QUALITY

The parameters that are commonly used to describe the physical quality of water include *turbidity*, *temperature*, *color*, *taste*, and *odor*.

### Turbidity

When small particles are suspended in water, they tend to scatter and absorb light rays. This gives the water a murky or *turbid* appearance, and this effect is called *turbidity*. Clay, silt, tiny fragments of organic matter, and microscopic organisms are some of the substances that cause turbidity. They occur in water naturally or because of human activities and pollution.

Turbidity is a particularly important parameter of drinking water quality. Suspended particles can provide hiding places for harmful microorganisms and thereby shield them from the disinfection process in a water treatment plant. Because of this shielding effect, the

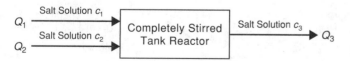

**FIGURE 4.13**

Mass balance *sketch. The total mass of salt entering the tank on the left must equal the total mass of salt leaving the tank on the right (input = output). If the salt concentrations and volume flow rates in pipes 1 and 2 are known, the concentration of salt in pipe 3 can be calculated using a mass balance equation. Note that* $Q_3 = Q_1 + Q_2$.

microbes can be consumed by people who drink the water, and the spread of disease may result.

Turbidity in drinking water is also unacceptable for esthetic reasons—it makes the water look very unappetizing. Most people find even a slight degree of turbidity in their water objectionable. Even when told that the water is safe to drink in spite of its turbidity, people tend to seek alternative water supplies (which could possibly be of poorer quality).

Turbidity is measured in units that relate the clarity of the water sample to that of a standardized suspension of silica. The interference in the passage of light caused by a suspension of 1 mg/L of silica is equivalent to one *turbidity unit* (TU). For example, a water sample that has the same degree of cloudiness as a 10 mg/L suspension of silica has a turbidity of 10 TU.

To interpret turbidity data, it is useful to be familiar with the typical ranges that occur. Turbidity in excess of 5 TU is just noticeable to the average person; most people do not complain about the clarity of the water at TU values less than 5. Turbidity in what most people would consider to be a relatively clear lake may be as high as 25 TU. In muddy water, turbidity generally exceeds 100 TU. Modern water treatment plants can routinely produce crystal clear water with turbidities of less than 1 TU.

Groundwater normally has very low turbidity because of the natural filtration that occurs as the water percolates through the soil. Most streams and rivers, though, have relatively high turbidities. This is particularly true during and just after rain storms, which cause soil erosion. The treatment of turbid stream water for drinking supplies can be an expensive process; the greater the turbidity, the greater is the amount of chemicals needed and the more frequently must the filters be cleaned.

For drinking water, instruments called *nephelometers* are used to measure the turbidity after purification. These devices measure the amount of scattered light electronically and do not depend on human vision or judgment in making comparisons to standard suspensions. Measurements made with nephelometric turbidimeters may be expressed in terms of *NTU* instead of just TU, to indicate how the measurement was made. Turbidity in filtered drinking water in the United States must be equal to or less than 0.5 NTU in at least 95% of the samples tested each month.

A conventional *Jackson candle turbidimeter* is illustrated in Figure 4.14. It may be used to measure raw (untreated) water turbidities. The water is added to a vertical glass tube until the candle flame is just obscured from view. The glass tube is graduated with turbidity units; the higher the water column required to obscure the flame, the less is the turbidity. Turbidity

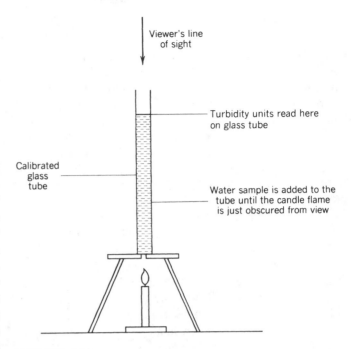

**FIGURE 4.14**
*The candle turbidimeter.*

values obtained using the candle turbidimeter may be expressed as *JTU*.

Excessive turbidity in a lake reduces the depth to which sunlight penetrates the water. This has an effect on the photosynthesis of microscopic plants, or algae, and on the overall environmental balance of the lake. In field surveys, small, white *Secchi disks* may be lowered into the water on a line marked off in meters until the disk disappears from view. The depth of the disk at that point can be correlated with the turbidity of the lake water.

## Temperature

Fish and other aquatic organisms require certain conditions of temperature in order to live and reproduce. The optimum temperature for trout, for example, is 15°C. A temperature of about 24°C is best for perch, and carp do very well at a cozy 32°C, which is more than twice the preferred temperature for trout (a cold-water fish).

Most species can adapt to a moderate change from their optimum temperature, but if the change is excessive, the organisms will perish or migrate to a new location. Generally, a change of about 5°C can significantly alter the balance and health of an aquatic environment. Sudden drops in temperature can be harmful, but usually an increase in temperature will cause more damage than a decrease (thus, rivers must be protected from warm water discharges from power plants).

A basic reason for this, as discussed previously, is that the solubility of oxygen in water decreases markedly as the temperature of the water goes up. Fish and other organisms need the oxygen to survive, and higher temperatures increase their rate of metabolism. In other words, the rate at which the organisms use oxygen to burn food for energy increases at higher temperatures. The combined effect of there being less available oxygen and the organisms having faster metabolism rates can eventually be very damaging.

Other than the fact that most people prefer cold drinking water, temperature is of little direct significance in public water supplies. Temperature plays a more important role, however, in wastewater treatment and water pollution control. Biological wastewater treatment systems are more efficient at higher temperatures. In colder regions of the country, treatment plants may be sheltered in heated enclosures to maintain optimum temperature ranges.

## Color, Taste, and Odor

Color, taste, and odor are physical characteristics of drinking water that are important for esthetic reasons. They do not cause any direct harmful effect on health, but no matter how safe the water may be to drink, most people object strongly to water that offends their sense of sight, taste, or smell.

Color may be caused by dissolved or suspended colloidal particles, primarily from decaying leaves or microscopic plants. This tends to give the water a brownish-yellow hue. Streams or rivers with tributaries in swampy areas may have this problem. Color is measured by comparing the water sample with standard color solutions or colored glass disks. One color unit is equivalent to the color produced by a 1-mg/L solution of platinum. It is not practical to isolate and identify specific chemicals that cause the color.

Hydrogen sulfide gas, $H_2S$, is a common cause of odor in water supplies. The rotten-egg smell of this gas may be encountered in water that has been in contact with naturally occurring deposits of decaying organic matter. Groundwater supplies sometimes have this problem; the wells are called *sulfur wells*.

Odor is measured and expressed in terms of a *threshold odor number*. The threshold odor number is the ratio by which the sample has to be diluted for the odor to become virtually unnoticeable. For example, if a 50-mL volume of water sample has to be diluted to a volume of 200 mL for the odor to be just barely detectable, the threshold number would be 200/50 = 4. A similar technique may be applied in measuring the taste of the water. Taste and odor measurements are very subjective and depend on the sensitivity of the person conducting the test.

## 4.3   CHEMICAL PARAMETERS OF WATER QUALITY

Many organic and inorganic chemicals affect water quality. In drinking water, these effects may be related to public health or to esthetics and economics. In surface waters, chemical quality can affect the aquatic environment. Several chemical parameters are also of concern in wastewater. In this section, the most common chemical parameters of water quality are discussed.

## Dissolved Oxygen

Dissolved oxygen is generally considered to be one of the most important parameters of water quality in streams, rivers, and lakes. It is usually abbreviated simply as DO. Just as people need oxygen in the air they breathe, fish and other aquatic organisms need DO in the water to survive. With most other substances, the less there is in the water, the better is the quality. But the situation is reversed for DO. *The higher the concentration of dissolved oxygen, the better is the water quality.*

Oxygen is only slightly soluble in water. For example, the saturation concentration at 20°C is about 9 mg/L or 9 ppm. (Remember that this is equivalent to the relationship between 9 in. and 16 mi.) Because of this very slight solubility, there is usually quite a bit of competition among aquatic organisms, including bacteria, for the available dissolved oxygen. As discussed in some detail later, bacteria will use up the DO very rapidly if there is much organic material in the water. Trout and other fish soon perish when the DO level drops. Another factor to remember is that oxygen solubility is very sensitive to temperature. Changes in water temperature have a significant effect on DO concentrations.

Dissolved oxygen has no direct effect on public health, but drinking water with very little or no oxygen tastes flat and may be objectionable to some people. Dissolved oxygen does play a part in the corrosion or rusting of metal pipes; it is an important factor in the operation and maintenance of water distribution networks.

Dissolved oxygen is used extensively in biological wastewater treatment facilities. Air, or sometimes pure oxygen, is mixed with sewage to promote the aerobic decomposition of the organic wastes. The role of dissolved oxygen in water pollution and wastewater treatment is discussed in subsequent chapters.

The DO concentration can be determined by using standard wet chemistry methods of analysis or membrane electrode meters in the lab or in the field. Field instruments are available that have probes that can be lowered directly into a stream or treatment tank. The electrode probe senses small electric currents that are proportional to the dissolved oxygen level in the water.

## Biochemical Oxygen Demand

Bacteria and other microorganisms use organic substances for food. As they metabolize organic material, they consume oxygen. The organics are broken down into simpler compounds, such as $CO_2$ and $H_2O$, and the microbes use the energy released for growth and reproduction.

When this process occurs in water, the oxygen consumed is the DO. If oxygen is not continually replaced in the water by artificial or natural means, then the DO level will decrease as the organics are decomposed by the microbes. This need for oxygen is called the *biochemical oxygen demand*. In effect, the microbes "demand" the oxygen for use in the biochemical reactions that sustain them. The abbreviation for biochemical oxygen demand is *BOD*; this is one of the most commonly used terms in water quality and pollution control technology.

As discussed in Chapter 5, organic waste in sewage is one of the major types of water pollutants. It is impractical to isolate and identify each specific organic chemical in these wastes and to determine its concentration. Instead, the BOD is used as an indirect measure of the total amount of biodegradable organics in the water. *The more organic material there is in the water, the higher the BOD exerted by the microbes will be.*

In addition to being used as a measure of the amount of organic pollution in streams or lakes, the BOD is used as a measure of the *strength* of sewage. As seen in Chapter 10, this is one of the most important parameters for the design and operation of a water pollution control plant. A *strong* sewage has a high concentration of organic material and a correspondingly high BOD. A *weak* sewage, with a low BOD, may not require as much treatment.

The complete decomposition of organic material by microorganisms takes time, usually 20 d or more under ordinary circumstances. The amount of oxygen used to completely decompose or *stabilize* all the biodegradable organics in a given volume of water is called the *ultimate BOD*, or $BOD_L$. For example, if a 1-L volume of municipal sewage requires 300 mg of oxygen for complete decomposition of the organics, the $BOD_L$ would be expressed as 300 mg/L. One liter of wastewater from an industrial or food processing plant may require as much as 1500 mg of oxygen for complete stabilization of the waste. In this case, the $BOD_L$ would be 1500 mg/L, indicating a much stronger waste than ordinary municipal or domestic sewage. In general, then, the BOD is expressed in terms of mg/L of oxygen.

The BOD is a function of time. At the very beginning of a BOD test, or time = 0, no oxygen will have been consumed and the BOD = 0. As each day goes by, oxygen is used by the microbes and the BOD increases. Ultimately, the $BOD_L$ is reached and the organics are completely decomposed. A graph of the BOD versus time has the characteristic shape illustrated in Figure 4.15. This is called the *BOD curve*.

The BOD curve can be expressed mathematically by the following equation:

$$BOD_t = BOD_L \times (1 - 10^{-kt}) \qquad \textbf{(4-2)}$$

where $BOD_t$ = BOD at any time $t$, mg/L
$BOD_L$ = ultimate BOD, mg/L
$k$ = a constant representing the rate of the BOD reaction
$t$ = time, d

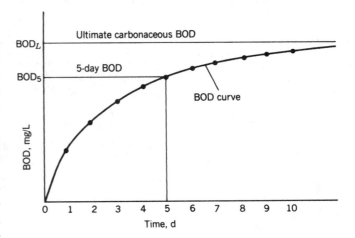

**FIGURE 4.15**
Biochemical oxygen demand, *or BOD, increases over time until all the organic pollutants are stabilized. The value of the BOD after 5 d, or $BOD_5$, is used for routine measurement and analysis.*

The rate at which oxygen is consumed is expressed by the constant $k$. The value of this rate constant depends on the temperature, the type of organic material, and the type of microbes exerting the BOD. For ordinary domestic sewage, at a temperature of 20°C, the value of $k$ is usually about 0.15/d.

## EXAMPLE 4.3

A sample of sewage from a town is found to have a BOD after 5 d ($BOD_5$) of 180 mg/L. Estimate the ultimate BOD ($BOD_L$) of the sewage. Assume that $k = 0.1/d$ for this wastewater.

*Solution*   Applying Equation 4-2 gives

$$180 = BOD_L \times (1 - 10^{-0.1 \times 5})$$
$$= BOD_L \times (1 - 10^{-0.5})$$
$$= BOD_L \times (1 - 0.316) = BOD_L \times 0.684$$

Rearranging terms to solve for $BOD_L$ gives

$$BOD_L = 180/0.684 = 260 \text{ mg/L} \qquad \text{(rounded off)}$$

The effect of different temperatures on the rate of the BOD reaction and on the shape of the BOD curve is shown in Figure 4.16. At higher temperatures, the organics decompose at a faster rate, but the $BOD_L$ remains the same.

The 20 d or so required for the ultimate BOD to develop is much too long a time to wait for lab results.

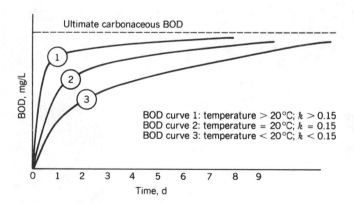

**FIGURE 4.16**
*The rate of the BOD reaction is directly proportional to temperature, but the total amount of organics in the sample and therefore the ultimate BOD do not change.*

This is particularly true when the BOD data are used to monitor the efficiency of a water pollution control plant. It has been found that more than two thirds of the $BOD_L$ is usually exerted within the first 5 d of decomposition. For instance, in the preceding example, the 5-d BOD is 180/260 = 0.69, or 69 percent of the ultimate BOD. For practical purposes, the 5-d BOD, or $BOD_5$, has been chosen as a representation of the organic content of water or wastewater. For standardization of results, the test must be conducted at a temperature of 20°C.

In summary, the parameter of $BOD_5$ is the amount of dissolved oxygen used by microbes in 5 d to decompose organic substances in water at 20°C.

### Measurement of $BOD_5$

The traditional BOD test is conducted in standard 300-mL glass *BOD bottles*. The test for the 5-d BOD of a water sample involves taking two DO measurements: an initial measurement when the test begins, at time $t = 0$, and a second measurement, at $t = 5$, after the sample has been incubated in the dark for 5 d at 20°C. The $BOD_5$ is simply the difference between the two DO measurements.

For example, consider that a sample of water from a stream is found to have an initial DO of 8.0 mg/L. It is placed directly into a BOD bottle and incubated for 5 d at 20°C. After the 5 d, the DO is determined to be 4.5 mg/L. The BOD is the amount of oxygen consumed, or the difference between the two DO readings. That is, $BOD_5 = 8.0 - 4.5 = 3.5$ mg/L.

Very clean bodies of surface water usually have a $BOD_5$ of about 1 mg/L due to the presence of naturally occurring organics from decaying leaves and animal wastes. $BOD_5$ values in excess of 10 mg/L, however, usually indicate the presence of sewage pollution.

When measuring the $BOD_5$ of sewage, it is necessary to first dilute the sample in the BOD bottle. Domestic sewage generally has a $BOD_5$ value of about 200 mg/L. If the sample were not diluted, all the DO would be very quickly depleted and it would not be possible to get a DO reading on the fifth day. Computation of the $BOD_5$, using this *dilution method* in a 300-ml BOD bottle, is done using the following equation:

$$BOD_5 = \frac{(DO_0 - DO_5) \times 300}{V} \qquad \textbf{(4-3)}$$

where $DO_0 =$ initial DO at $t = 0$
$DO_5 =$ DO at $t = 5$ d
$V =$ sample volume, mL

## EXAMPLE 4.4

A 6.0-mL sample of wastewater is diluted to 300 mL with distilled water in a standard BOD bottle. The initial DO in the bottle is determined to be 8.5 mg/L, and the DO after 5 d at 20°C is found to be 0.5 mg/L. Determine the $BOD_5$ of the wastewater and compute its $BOD_L$. Assume that $k = 0.1/d$.

*Solution*    Applying Equation 4-3 gives

$$BOD_5 = \frac{(8.5 - 5.0) \times 300}{6.0} = \frac{3.5 \times 300}{6.0}$$

$$= 180 \text{ mg/L}$$

Now applying Equation 4-2 gives

$$180 = BOD_L \times (1 - 10^{-0.1 \times 5})$$

and

$$BOD_L = \frac{180}{0.684} = 260 \text{ mg/L}$$

In some cases, particularly when analyzing industrial or food-processing wastewater that does not contain bacteria, the dilution water must be *seeded* with sewage. This provides a suitable population of microorganisms for the BOD reaction to take place. Remember that, even though there may be a lot of organic material present in the water, if there are no microbes to use oxygen and stabilize the organics, a measurement of the BOD cannot be obtained. When seeded dilution water is used, Equation 4-3 must be modified to account for the BOD added by the dilution water.

## Nitrification

In Figure 4.15, the BOD curve flattens out after about 8 d, as it approaches the ultimate BOD. This $BOD_L$ is called the ultimate *carbonaceous* BOD because, during the first week or so of decomposition, the bacteria act primarily on the carbon-containing substances.

As time goes on and the carbonaceous material is depleted, another group of bacteria become active. These are called *nitrifying bacteria*. This group of microorganisms thrives on the noncarbonaceous ammonia, $NH_3$, in the wastewater, metabolizing it for energy. In this process, called *nitrification*, the ammonia is converted into the more stable nitrite, $NO_2^-$, and nitrate, $NO_3^-$, ionic forms of nitrogen.

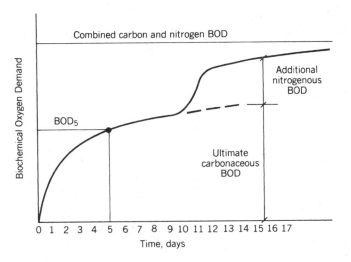

## FIGURE 4.17

*A complete BOD curve, showing the delayed effect of nitrification on the total oxygen demand. Additional oxygen is consumed by the nitrifying bacteria as they convert ammonia to nitrate.*

During this process, the nitrifying bacteria consume additional oxygen, causing a rise in the BOD curve after the first 8 to 10 d of decomposition. This is illustrated in Figure 4.17. Most sewage treatment plants remove only the carbonaceous BOD. But it is possible that the discharge from a conventional treatment plant could still deplete the DO in a small receiving stream because of nitrogenous BOD that is exerted. Sometimes, advanced treatment systems must be built to also remove the ammonia and protect sensitive aquatic environments from DO depletion due to nitrification.

## Chemical Oxygen Demand

The BOD test provides a measure of the biodegradable organic material in water, that is, of the substances that microbes can readily use for food. There also might be nonbiodegradable or slowly biodegradable substances that would not be detected by the conventional BOD test.

The *chemical oxygen demand*, or *COD*, is another parameter of water quality, which measures all organics, including the nonbiodegradable substances. It is a chemical test using a strong oxidizing agent (potassium dichromate), sulfuric acid, and heat. The results of the COD test can be available in just 2 h, a definite advantage over the 5 d required for the standard BOD test.

COD values are always higher than BOD values for the same sample, but there is generally no consistent correlation between the two tests for different wastewaters. In other words, it is not feasible to simply measure the COD and then predict the BOD. Because most wastewater treatment plants are biological in their mode of operation, the BOD is more representative of the treatment process and remains a more commonly used parameter than the COD.

## Solids

Solids occur in water either in solution or in suspension. These two types of solids are distinguished by passing the water sample through a glass-fiber filter. By definition, the *suspended solids* are retained on top of the filter and the *dissolved solids* pass through the filter with the water.

If the filtered portion of the water sample is placed in a small dish and then evaporated, the solids in the water remain as a residue in the evaporating dish. This material is usually called *total dissolved solids*, or *TDS*. The concentration of TDS is expressed in terms of mg/L. It can be calculated as follows:

$$TDS = \frac{(A - B) \times 1000}{C} \qquad (4\text{-}4)$$

where $A$ = weight of dish plus residue, mg
$B$ = weight of empty dish, mg
$C$ = volume of sample filtered, mL

### EXAMPLE 4.5

The weight of an empty evaporating dish is determined to be 40.525 g. After a water sample is filtered, 100 mL of the sample is evaporated from the dish. The weight of the dish plus the dried residue is found to be 40.545 g. Compute the TDS concentration.

*Solution*  Applying Equation 4-4, we get

$$TDS = \frac{(40\,545 \text{ mg} - 40\,525 \text{ mg}) \times 1000}{100 \text{ mL}}$$

$$= \frac{20 \times 1000}{100}$$

$$= 200 \text{ mg/L}$$

In drinking water, dissolved solids may cause taste problems. Hardness, corrosion, or esthetic problems may also accompany excessive TDS concentrations. In wastewater analysis and water pollution control, the suspended solids retained on the filter are of primary

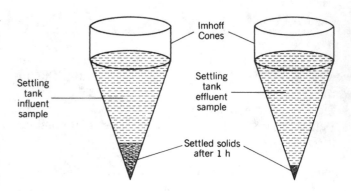

**FIGURE 4.18**
Imhoff cones *are used to measure the amount of settleable solids in raw and in treated sanitary sewage.*

importance and are referred to as *total suspended solids*, or *TSS*.

The TSS concentration can be computed using Equation 4-4, where $A$, represents the weight of the filter plus retained solids, $B$ represents the weight of the clean filter, and $C$ represents the volume of sample filtered.

One routine test used in wastewater treatment plants to determine the efficiency of the treatment process is the measurement of *settleable solids*. Settleable solids are the coarser fraction of the suspended solids that readily settle out because of gravity. A 1-L volume *Imhoff cone*, illustrated in Figure 4.18, is filled with the sewage sample. After 1 h of quiescent settling, the solids accumulate at the bottom of the cone; the cone is graduated in milliliters, and the amount of settleable solids is expressed in terms of mL/L.

Another classification of solids that is of particular significance in wastewater treatment is *volatile solids*. These are organic substances that can be burned off or volatilized at 550°C in a furnace. The residues remaining after burning at that temperature are the *fixed* or nonvolatile solids. The concentration of volatile suspended solids gives an indication of the organic loading on biological treatment units. It can be determined by measuring the loss in weight of the glass-fiber filter plus solids after burning.

## Hardness

*Hardness* is a term used to express the properties of certain highly mineralized waters (high TDS concentrations). The dissolved minerals cause problems such as scale deposits in hot water pipes and difficulty in producing lather with soap. The economic aspects of these problems, rather than any adverse health effect, are what make hard water generally unacceptable to the consumer.

Calcium, $Ca^{2+}$, and magnesium, $Mg^{2+}$, ions cause the greatest portion of hardness in naturally occurring waters. These minerals enter the water primarily from contact with soil and rock, especially limestone deposits. In general, groundwater is harder than surface water because it is in contact with mineral deposits for long periods.

Hardness is usually expressed in terms of milligrams per liter of calcium carbonate, $CaCO_3$; grains per gallon is also used to express hardness concentrations. Water with more than 300 mg/L of hardness is generally considered to be hard, and water with less than 75 mg/L is considered to be soft. Very soft water is undesirable in public supplies because it tends to increase corrosion problems in metal pipes; also, some health officials believe it to be associated with the incidence of heart disease.

## Iron, Manganese, Copper, and Zinc

Although iron, Fe, and manganese, Mn, do not cause health problems, they do impart a noticeable bitter taste to drinking water, even at very low concentrations. These metals usually occur in groundwater in solution as ferrous, $Fe^{2+}$, and manganous, $Mn^{2+}$, ions. When exposed to air, they form the insoluble ferric, $Fe^{3+}$, and manganic, $Mn^{3+}$, forms, making the water turbid and unacceptable to most people. They also cause brown or black stains on laundry and on plumbing fixtures.

Copper, Cu, and zinc, Zn, are nontoxic in small concentrations, and in fact they are both beneficial and essential for human health. They cause undesirable tastes in drinking water, however, and at high concentrations zinc imparts a milky appearance to the water.

## Fluorides

A moderate amount of fluoride ions, $F^-$, in drinking water contributes to good dental health. Extensive research over many years has demonstrated that a fluoride concentration of about 1 mg/L is effective in preventing tooth decay, particularly in children, without any harmful side effects. Fluoridation, the intentional addition of compounds containing fluoride to drinking water, is practiced by many cities and towns throughout the United States.

Fluorides occur naturally in water in some areas. When the concentrations are excessive, either alternative water supplies must be used or treatment to reduce the fluoride concentration must be applied. This is because excessive amounts of fluoride cause mottled or discolored teeth, a condition called *dental fluorosis*. There is a very small margin of error between beneficial levels of fluoride and levels that cause fluorosis. The maximum allowable levels of fluoride in public water supplies depend on local climate because climate affects the amount of water consumption. In the warmer regions of the country, the maximum allowable concentration of fluoride is 1.4 mg/L; in colder climates, up to 2.4 mg/L is allowed.

## Chlorides

Chloride ions, $Cl^-$, in drinking water do not cause any harmful effects on public health, but high concentrations can cause a salty taste that most people find objectionable. Salt levels are, of course, very high in ocean waters—about 35,000 mg/L.

Chlorides do occur naturally in groundwater, streams, and lakes, but the presence of relatively high chloride concentrations in fresh water (about 500 mg/L or more) may be an indication of sewage pollution. Salt, NaCl, used in foods, is excreted with body wastes; sanitary sewage carries these chlorides into the receiving waters. Also, chlorides from roadway deicing salts may enter the groundwater as well as streams and lakes. Saltwater intrusion into wells is a problem in some coastal areas.

## Chlorine Residual

It is important to make a distinction between chloride ions and chlorine in water; many beginning students confuse these two. Chlorine, $Cl_2$, does not occur naturally in water. It is, however, one of the most common chemicals added to water and wastewater, primarily for disinfection. This is discussed in more detail in subsequent chapters.

Although chlorine itself is a toxic gas, in dilute aqueous solutions it is not harmful to human health. One advantage of chlorine as a disinfectant is that a leftover or *residual* concentration can be maintained in the water distribution system, ensuring good sanitary quality of the water. In drinking water, a residual of about 0.2 mg/L is optimal. The measurement of chlorine residual in a water sample can be made using a color comparator test kit, illustrated in Figure 4.19.

One problem with chlorination of water supplies is that the chlorine can react with organics in the water, forming toxic compounds. The naturally occurring organics, primarily from vegetation, are called *precursors*. By themselves they are harmless. The toxic compounds,

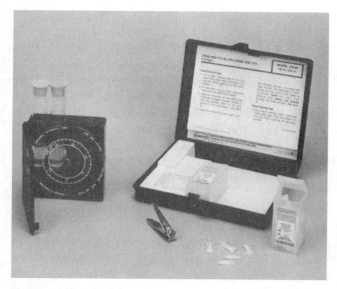

**FIGURE 4.19**
*A color comparator test kit for measuring chlorine residual.* (Courtesy of HACH Company, Loveland, Colorado)

called *trihalomethanes*, or THMs, have been identified as potential *carcinogens*, or cancer-causing substances. One of the most common THM compounds formed is chloroform, $CHCl_3$. The chlorine replaces three of the hydrogen atoms in the methane molecule (see Figure 4.9) to form chloroform.

It is ironic that the disinfection process using chlorine, designed to destroy microbes that cause disease, may be the source of a different public health hazard. Methods of control, including removal of the precursors before chlorination, are now being applied.

## Sulfates

Sulfate ions, $SO_4^{2-}$, occur in natural waters and in wastewater. If high concentrations are consumed in drinking water, there may be objectionable tastes or unwanted laxative effects, but there is no significant danger to public health from sulfates.

Sulfates in sewage can result in offensive odors from the formation of hydrogen sulfide gas, $H_2S$, with its characteristic rotten-egg odor. It also leads to a problem in sewer systems called *crown corrosion*. This is discussed in more detail in Chapter 8.

## Nitrogen

Nitrogen, $N_2$, occurs in many forms in the environment and takes part in many biochemical reactions. The four forms of nitrogen that are of particular significance in environmental technology are organic nitrogen, ammonia nitrogen, nitrite nitrogen, and nitrate nitrogen. The circulation of nitrogen in its various forms through the environment is illustrated in the nitrogen cycle in Figure 1.8.

In water contaminated with sewage, most of the nitrogen is originally present in the form of complex organic molecules (proteins) and ammonia, $NH_3$. These substances are eventually broken down by microbes to form nitrites and nitrates. This process of *nitrification* was discussed in Section 4.3 under biochemical oxygen demand.

Nitrogen, particularly in the nitrate form, is a basic nutrient that is essential to the growth of plants. Excessive nitrate concentrations in surface waters encourage the rapid growth of microscopic plants called *algae*; excessive growth of algae degrades water quality. This problem, referred to as *eutrophication*, is discussed in more detail in Chapter 5.

Nitrates can enter the groundwater from chemical fertilizers used in agricultural areas. Excessive nitrate concentrations in drinking water pose an immediate and serious health threat to infants under 3 months of age. The nitrate ions react with blood hemoglobin, reducing the blood's ability to carry oxygen; this produces a disease called *blue baby* or methemoglobinemia.

## Phosphorus

Like nitrogen, phosphorus, P, is an essential nutrient that contributes to the growth of algae and the eutrophication of lakes, although its presence in drinking water has little effect on health. Phosphorus can enter water from sewage or from agricultural runoff containing fertilizers and animal wastes. Phosphate, $PO_4^{3-}$, the inorganic form of phosphorus, had been commonly used in detergents in the past, but even with the ban on phosphate-based detergents, the amount of phosphorus occurring in water from other sources poses a significant environmental problem.

## Acidity, Alkalinity, and pH

Very high levels of either acidity or alkalinity in water may indicate the presence of industrial or chemical pollution, but acidity and alkalinity also occur naturally. Carbon dioxide from the atmosphere or from the respiration of aquatic organisms causes acidity when dissolved in water by forming carbonic acid, $H_2CO_3$. Dissolved carbonate ions, $CO_3^{2-}$, or bicarbonate ions,

$HCO_3^-$, of sodium, calcium, or magnesium cause natural alkalinity. Contact between the water and minerals in the ground is the major source of these substances.

Acidic substances yield $H^+$ ions in water, and alkaline substances yield $OH^-$ ions. The pH is a measure of the intensity of the acidity or alkalinity, as discussed in Section 4.1. The primary reason for measuring the acidity, alkalinity, and pH of water is to be able to control the water treatment process in a water purification facility. The required doses of various chemicals depend on the concentration of acidity or alkalinity, or on the pH of the water.

Water with moderate amounts of acidity or alkalinity can be consumed without adverse health effects, but excessive concentrations cause objectionable tastes; acids are sour and alkaline solutions are bitter.

The acidity and alkalinity in natural waters provide a buffering action that protects fish and other organisms from sudden changes in pH. For example, if an acidic chemical has somehow contaminated a lake that had natural alkalinity, a neutralization reaction occurs between the acid and alkaline substances; the pH of the lake water remains essentially unchanged. Most aquatic organisms can survive in a pH range of about 6 to 9.5.

## Toxic and Radioactive Substances

A wide variety of *toxic* inorganic and organic substances may be found in water in very small or trace amounts. Even in trace amounts, they can be a danger to public health. Some toxic substances are from natural sources, but many come from industrial activities and improper management of hazardous waste (see Chapter 12).

A toxic chemical may be a poison, causing death, or it may cause disease that is not noticeable until many years after exposure. A carcinogenic substance is one that causes cancer; substances that are *mutagenic* cause harmful effects in the offspring of exposed people.

Some *heavy metals* that are toxic are cadmium, Cd, chromium, Cr, lead, Pb, mercury, Hg, and silver, Ag. Arsenic, As, barium, Bar, and selenium, Se, are also poisonous inorganic elements that must be monitored in drinking water.

Many toxic organic chemicals have been identified and are currently monitored in public water supplies. Among these are the trihalomethanes formed after chlorination, as previously discussed in the section on chlorine residual. Pesticides such as endrin and toxaphene are toxic chlorinated hydrocarbons that are monitored; DDT and chlordane are not routinely checked for in drinking water because they have been banned from use.

Relatively expensive and sophisticated instruments are required for analyzing water samples for trace contaminants. Atomic absorption spectrophotometers for detecting heavy metals and gas chromotography/mass spectrometry (GC/MS) instrumentation for detecting organics are now commonly found in water quality labs. These instruments are capable of detecting substances in extremely dilute concentrations, in the parts-per-billion or micrograms-per-liter range.

To illustrate the low concentrations involved, consider that the maximum allowable concentration of the pesticide Lindane in drinking water is 0.2 $\mu$g/L or 0.2 ppb; this is equivalent to the presence of only one Lindane molecule among several billion water molecules. It is quite a technical achievement to make measurements in this range—like finding a needle in a haystack as big as a house.

### Radiation

The emission of subatomic particles or energy from the unstable nuclei of certain atoms, referred to as *radiation*, poses a serious public health hazard. Obviously, the consumption of radioactive substances in water is undesirable, and maximum allowable concentrations of radioactive materials have been established for public water supplies. Potential sources of radioactive pollutants in water include wastes from nuclear power plants, from industrial or medical research using radioactive chemicals, and from refining of uranium ores. Radon sometimes occurs naturally in groundwater. The unit of radioactivity used in water quality applications is the picocurie per liter (pCi/L); 1 pCi is equivalent to about two atoms disintegrating per minute.

## 4.4  BIOLOGICAL PARAMETERS OF WATER QUALITY

The presence or absence of living organisms in water can be one of the most useful indicators of its quality. In streams, rivers, and lakes, the diversity of fish and insect species provides a measure of the biological balance or health of the aquatic environment. A wide variety of different species of organisms usually indicates that the stream or lake is unpolluted. The disappearance of certain species and overabundance of other groups of organisms is generally one of the effects of pollution. Trout, for example, will soon disappear from a polluted stream, whereas catfish and other scavenger organisms will thrive. If the pollution is very severe, fish life will vanish altogether. Biologists can survey the fish and insect life of natural waters and assess the water quality on the basis of a computed *species diversity index* (see Section 1.3).

Microscopic plants and animals are also important in assessing the quality of water, particularly of drinking water and sewage. In this section, some basic facts about bacteria and other microbes are discussed. The main focus is on a group of organisms called *coliforms*, which are perhaps the most important of the biological parameters of water quality.

## Microorganisms

Microscopic plants and animals play an essential role in the life processes of all living organisms, including humans. Contrary to a popular misconception that microbes are harmful, the fact is that most of them are beneficial, particularly in their role as decomposers in the food chain (see the discussion on ecology in Section 1.3). Only a relatively small number of species of microbes cause disease in humans or otherwise harm the environment.

Microorganisms are ubiquitous in nature, that is, they occur everywhere. There are millions of bacteria and molds living in a single gram of rich garden soil, for example. They serve to decompose organic materials, converting them into simpler nutrients, which can be absorbed through the roots of plants. Foods also contain microorganisms, such as yeasts, which cause fermentation, producing $CO_2$ and alcohol from sugars.

Since foods are not sterile, human bodies acquire a normal population of microbes in the intestinal tract; the *coliform* group of bacteria makes up a large part of this population. Animal wastes consist primarily of microorganisms from the intestines. Although sewage contains millions of microbes per milliliter, most of them are harmless. It is only when sewage contains wastes from people infected with disease that the presence of harmful organisms in the sewage is likely.

### Bacteria

Bacteria are considered to be single-celled plants because of their cell structure and the way they take in food. They utilize soluble food taken in through a rigid cell wall, but unlike green plants, which use photosynthesis, bacteria do not produce their own food.

Bacteria are very small, typically about 2 $\mu$m in size, and can be seen only with the aid of a microscope. They occur in three basic cell shapes: rod-shaped or *bacillus*, sphere-shaped or *coccus*, and spiral-shaped or *spirellus*. In some cases, the individual cells grow together in larger groups or chains. *Sphaerotilus natans* is an example of a species of bacteria that grows in a chain or filament enclosed within a long sheath or tube. Excessive growth of these filamentous organisms is known to be one of the causes of reduced treatment efficiency in biological sewage treatment plants.

In less than 30 min, a single bacterial cell can mature and divide into two new cells. This process of reproduction is called *binary fission*. Under favorable conditions of food supply, temperature, and pH, bacteria can reproduce so rapidly that a bacterial culture may contain as many as 20 million individual cells per milliliter after just 1 d of growth. This rapid growth of visible *colonies* of bacteria on a suitable nutrient medium makes it possible to detect and count the number of bacteria in water. This is discussed in more detail in the section on coliform bacteria.

There are several distinctions among the various species of bacteria. One depends on how they metabolize their food. Bacteria that require oxygen for their metabolism are called *aerobic bacteria* or *aerobes*. Those that live only in an oxygen-free environment are called *anaerobic bacteria* or *anaerobes*. The distinction between aerobes and anaerobes is of great significance in water pollution and wastewater treatment. (Some species, called *facultative* bacteria, can live in either the absence or the presence of oxygen.)

Another distinction among species of bacteria is a function of the type of food that they require. Those that utilize simple inorganic compounds for nourishment are called *autotrophic* bacteria; those that require complex organic substances are called *heterotrophic* bacteria. The nitrifying bacteria, for example, which use ammonia as food and convert it to nitrate, are autotrophs. Other examples of autotrophs include the *iron bacteria* and the *sulfur bacteria*. Iron bacteria thrive in some water pipelines and often cause taste and odor problems in drinking water. The sulfur bacteria, which are also anaerobes, are active in sewers and speed the deterioration of concrete pipes by converting hydrogen sulfide gas to sulfuric acid.

One of the most important factors affecting the growth and reproduction of bacteria is temperature. At low temperatures, bacteria grow and reproduce slowly. As the temperature increases, the rate of growth and reproduction just about doubles for every additional 10°C (up to the optimum temperature for the species). The majority of species of bacteria are classified as *mesophilic*, having an optimum temperature of about 35°C. Those that do best at elevated temperatures of about 60°C are called *thermophilic* bacteria. Bacteria with an optimum growth temperature between 0°C and 20°C are called *psycrophilic* bacteria.

## Algae

Algae are microscopic plants, which contain photosynthetic pigments, such as chlorophyll. They are autotrophic organisms and support themselves by converting inorganic materials into organic matter by using energy from the sun. During the process of photosynthesis, they take in carbon dioxide from the air and give off oxygen.

A basic characteristic of these simple plants is their lack of roots, stems, and leaves. Free-floating algae are also called *phytoplankton*. (*Plankton* are tiny floating plants or animals that live in either fresh or salt waters. Over 90 percent of atmospheric oxygen is produced by salt water or marine phytoplankton, by the process of photosynthesis.) Even though most species of algae are microscopic, they can be easily noticed when their numbers proliferate in the water. Excessive growths of algae, called *algal blooms*, are often unsightly. Some algal species are multicellular, growing as filaments that sometimes appear as a green slime in the water.

Common species include the blue-green algae, such as *Anabaena*, green algae, such as *Spirogyra*, yellow-green algae, such as *Botrydium*, and red algae, such as *Gelidium*. Another important group of algae, called *diatoms*, produce hard shells of silica. Deposits of these shells, from dead diatoms, that have accumulated over many hundreds of years form *diatomaceous earth*, a material sometimes used for filtering water.

Algae play a role in the eutrophication (aging) of lakes (discussed in Section 5.5). They are also important for wastewater treatment in stabilization ponds (discussed in Section 10.3). In regard to public health considerations, algae are primarily nuisance organisms in water supplies because of the taste and odor problems they create and the extra expense needed to filter them out of the water. Occasionally, certain species of algae do cause serious environmental and public health problems. Blue-green algae, for example, can kill cattle and other domestic animals if the animals drink water containing those species. A toxic alga called *Pfiesteria* can kill fish and sicken people who consume tainted water. Another organism, *Chattonella verruculosa*, can kill fish and sicken beach-goers when so-called "red tides" occur in seawater.

## Protozoa

Protozoa are the simplest of animal species. These single-celled microscopic animals consume solid organic particles, bacteria, and algae for food. They are, in turn, ingested as food by higher level multicellular animals. Floating freely in water, these *zooplankton*, as they are sometimes called, are a vital part of the natural aquatic food chain. They are also of significance in biological wastewater treatment systems.

*Amoebae* are protozoa that move by projecting sections of their bodies; this mobile protoplasm of the amoebae is also used to surround and engulf food particles. Amoebae are commonly found in slimes formed in certain types of sewage treatment processes.

A group of protozoa called *flagellates* move around in water by means of a long threadlike strand, called a *flagellum*, that propels them with its whiplike action. One such organism, *Giardia lamblia*, is an intestinal parasite that causes a form of dysentery in humans. Another type of protozoa has hundreds of short hairs, called *cilia*, that propel the organism through the water and serve to direct food particles into its digestive system. The *paramecia*, for example, are ciliated protozoa commonly found in freshwater ponds and lakes.

A species of protozoa called *Cryptosporidium* has been found to be the cause of recent waterborne gastrointestinal disease outbreaks in the United States (see Sections 1.2 and 6.5). These pathogens are frequently found in lakes and streams and are very resistant to disinfection by chlorination (although they can be controlled by ozonation). The 1996 amendments to the *Safe Drinking Water Act* call for enhanced surface water treatment rules to prevent such outbreaks.

Several types of protozoa, as well as some common forms of algae and bacteria, are illustrated in Figure 4.20.

## Viruses

Viruses are extremely small pathogens, so small that they can pass through filters that do not permit the passage of bacteria; most viruses can be seen only with the aid of a powerful electron microscope. Since they are incapable of independent metabolism and reproduction, there is debate as to whether viruses should be called "living" organisms. To reproduce, viruses must invade a suitable host cell and take over the cell's metabolic processes for their own use.

Viruses can cause a variety of illnesses in humans, including chicken pox, rabies, yellow fever, polio, influenza, gastroenteritis, and the common cold. They can be transmitted among people in a variety of ways, including by ingestion of water contaminated with sewage. Viruses that can infect cells of the intestinal tract of humans are called *enteric viruses* or *enteroviruses*.

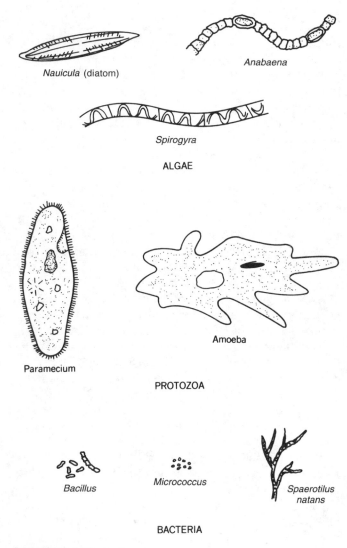

*Nauicula* (diatom)

*Anabaena*

*Spirogyra*

ALGAE

Paramecium

Amoeba

PROTOZOA

*Bacillus*

*Micrococcus*

*Spaerotilus natans*

BACTERIA

**FIGURE 4.20**
*Sketches of some typical microorganisms found in water and/or sewage.*

Many types of viral infection can be controlled by use of vaccines or by eradication of insect vectors that transmit the viruses to humans. Most waterborne viruses can be inactivated by water treatment methods, which include coagulation, filtration, and disinfection (see Chapter 6); the inactivation occurs during the disinfection process, after coagulation and filtration remove substances that can interfere with disinfection.

## Indicator Organisms

One of the most important attributes of good-quality water is that it be free of disease-causing organisms—

*pathogenic* bacteria, viruses, protozoa, or parasitic worms. Water contaminated with sewage may contain such organisms because they are excreted in the feces of infected individuals. If contaminated water is consumed by others before it is properly treated, the cycle of disease can continue in epidemic proportions.

Testing water for the presence of individual pathogens such as the *Salmonella typhosa* bacterium, which causes typhoid fever, or *Entamoeba histolytica*, the protozoan that causes dysentery, is a very difficult, time-consuming, and impractical task. The concentrations of these organisms in a contaminated water sample may be small enough to elude detection, making it necessary to test large volumes of water. Further, it would be necessary to test for a wide variety of different organisms before the water could be considered safe.

A more practical and reliable approach than testing for individual pathogens is to test for a single species that would signal the possible presence of sewage contamination. If sewage is present in the water, it can be assumed that the water may also contain pathogenic organisms and is a threat to public health. A species of organisms that serves this purpose is called an *indicator organism*.

### Coliforms

A very important biological indicator of water quality and pollution used in environmental technology is the group of bacteria called *coliforms*. Not pathogenic, coliforms are always present in the intestinal tract of humans, and millions are excreted with body wastes. Consequently, water that has been recently contaminated with sewage will always contain coliforms.

A particular species of coliforms found in domestic sewage is *Escherichia coli*, or *E. coli*. Even if the water is only slightly polluted, they are very likely to be found; there are roughly 3 million *E. coli* bacteria in a 100-mL volume of untreated sewage. Most strains of *E. coli* are generally harmless, but infected individuals also excrete pathogens along with the coliforms.

Coliform bacteria are hardy organisms and survive in water longer than most pathogens. They are also relatively easy to detect. In general, it can be stated that, if a sample of water is found not to contain coliforms, then there has not been recent sewage pollution and the presence of pathogens is therefore extremely unlikely. On the other hand, if coliforms are detected, there is a possibility of recent sewage pollution. However, additional tests would be required to prove that the coliforms are from sewage and not from other sources.

In summary, we can say

no coliforms → no sewage → no pathogens

Coliforms are actually a broadly defined group of microorganisms. They occur naturally in the soil as well as in the digestive tract of warm-blooded animals, including humans. It is necessary to make a distinction between two groups: *total coliforms* and *fecal coliforms*. Total coliforms refers to all the members of the group regardless of origin. Fecal coliforms are those from the intestines of warm-blooded animals; *E. coli* are fecal coliforms from humans.

A total coliform test is particularly applicable to the analysis of drinking water to determine its sanitary quality. Drinking water must be free of coliforms of any kind. On the other hand, a fecal coliform test is more appropriate for monitoring pollution of natural surface waters or groundwaters, since a total coliform count would be inconclusive in this case. Municipalities and industries are required to test for fecal coliforms in their wastewater treatment plant discharges to make sure that the disinfection process is working properly.

### Fecal Coliform–to–Fecal Strep Ratio

It is sometimes necessary to determine whether fecal coliforms in a tested sample originated from human wastes or animal wastes. The presence of another type of bacteria, called *fecal streptococci* (or *strep*), can provide the necessary clue. Fecal strep bacteria are also intestinal bacteria, but they predominate over coliforms in animals other than humans. When the ratio of the number of fecal coliform bacteria to fecal strep bacteria is more than 2, the contamination is likely to be of human origin. When this ratio, abbreviated FC/FS, is less than 1, then animal wastes rather than sewage are more likely to be the source of pollution. FC/FS ratios between 1 and 2 are inconclusive. In addition to the FC/FS ratio, an investigation called a *sanitary survey* is usually required to determine the source of water pollution. In a sanitary survey, factors such as the extent of agricultural activity, the location and condition of residences, and the prevalence of individual on-site sewage disposal systems are studied.

## Testing for Coliforms

Two testing procedures can be used for detecting and measuring coliforms in water—the *membrane filter method* and the *multiple-tube fermentation method*. The membrane filter method takes less time and pro-vides more of a direct count of the coliforms than the multiple-tube method. It also requires less laboratory equipment. Although the membrane filter method is gaining in use, the multiple-tube procedure is still practiced in some labs; the membrane filter method is not applicable to turbid samples. It is necessary to understand the essential differences between these two tests.

### Membrane Filter Method

In this procedure, a measured volume of sample is drawn through a special membrane filter by applying a partial vacuum. The filter, a flat, paperlike disk about the size of a silver dollar, has uniform microscopic pores small enough to retain the bacteria on its surface while allowing the water to pass through.

After the sample is drawn through, the filter is placed in a sterile container called a *petri dish*. The petri dish also contains a special *culture medium* that the bacteria use as a food source. This nutrient medium is usually available in small glass containers called *ampuls* (or *ampoules*), from which it is readily transferred into the petri dish. Its composition is such that it promotes the growth of coliforms while inhibiting the growth of other bacteria caught on the filter.

A membrane filter apparatus is shown in Figure 4.21, and the filter is shown being placed in a petri dish in Figure 4.22.

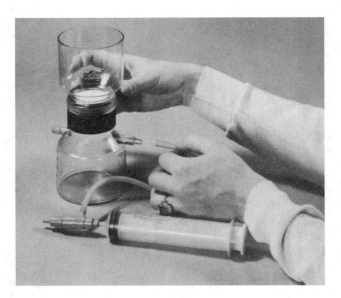

**FIGURE 4.21**
*A membrane filter apparatus for detecting and counting bacteria in water or sewage.* (Courtesy of Millipore Corporation, Bedford, Massachusetts)

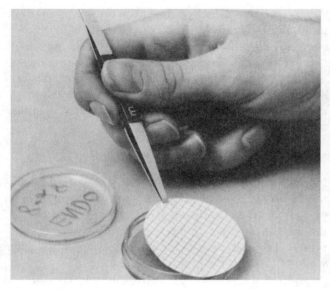

**FIGURE 4.22**
*After filtration, the membrane filter is placed in a petri dish that contains a nutrient medium. The trapped bacteria on the filter will grow into visible colonies.* (Courtesy of Millipore Corporation, Bedford, Massachusetts)

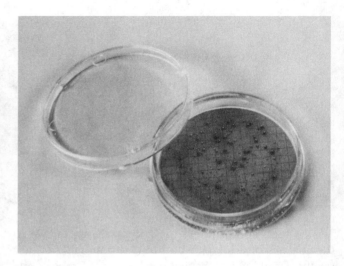

**FIGURE 4.23**
*The visible colonies have a characteristic green metallic sheen that is readily identifiable.* (Courtesy of Millipore Corporation, Bedford, Massachusetts)

The petri dish holding the filter and nutrient medium is usually placed in an incubator, which keeps the temperature at 35°C, for 24 h. After incubation, colonies of coliform bacteria, each containing millions of organisms, will be visible. The colonies form by the reproductive process of binary fission. They appear as specks or dots, with a characteristic green metallic sheen. This is illustrated in Figure 4.23.

The coliform concentration is obtained by counting the number of colonies on the filter. A basic premise for the membrane filter test is that each colony started growing from one single organism. From this it can be assumed that *each colony counted represents only one coliform in the original sample.*

The filter has a grid printed on it to facilitate counting colonies; a magnifying glass helps to obtain accurate results. Small samples of polluted water or wastewater must be diluted with sterile water before filtering so that the filter is not overgrown with colonies, making it impossible to get an accurate count.

Coliform concentrations are expressed in terms of the number of organisms per 100 mL of water. The following formula can be used to express the results of samples of various sizes:

$$\text{coliforms per 100 mL} = \frac{\text{number of colonies} \times 100}{\text{mL of sample}}$$

**(4-5)**

## EXAMPLE 4.6

A 4-mL volume of a water sample from a stream was drawn through a membrane filter. The filter was first covered with sterile water to dilute and spread the sample evenly over the filter. Sixteen coliform colonies were counted on the filter after incubation for 24 h at 35°C. Determine the coliform count per 100 mL.

*Solution*   Applying Equation 4-5 gives

$$\text{coliforms per 100 mL} = \frac{16 \times 100}{4} = 400$$

The basic procedure described here for the membrane filter test can be applied to tests for total coliforms or fecal coliforms, but different nutrient media are used, and the fecal coliform test is conducted at 44.5°C rather than at 35°C. A special water bath incubator is used to accurately maintain the higher temperature for the fecal coliform test. The membrane filter technique can also be used to test for fecal streptococci.

**Multiple-Tube Fermentation Method**

This technique is based on the fact that coliform organisms can use lactose, the sugar occurring in milk, as food and produce gas in the process. A measured volume of water sample is added to a tube that contains lactose broth nutrient medium. A small inverted vial in the lactose broth traps some of the gas that is produced as the coliform bacteria grow and reproduce. The gas bubble in the inverted vial along with a cloudy appearance of the broth provide visual evidence that coliforms may be present in the sample. But if gas is not produced within 48 h of incubation at 35°C, it can be concluded that coliforms were not present in the sample volume injected into the broth. This is illustrated in Figure 4.24.

The failure of gas to form after incubation is called a *negative* test. The appearance of gas and the accompanying cloudiness in the broth is called a *positive presumptive test*. Some bacteria other than coliforms occasionally produce gas in lactose. Because of this, it is usually necessary to perform another test, called the *confirmed test*, to prove that it was really the coliform bacteria that produced the gas in the positive presumptive tube.

The confirmed test involves transferring the nutrient medium from a positive presumptive tube to another fermentation tube that contains a different nutrient medium, called brilliant green bile. Now, if gas is again formed within 48 h of incubation at 35°C, the presence of coliforms is confirmed. In some cases, a third procedure, called the *completed test*, may have to be performed. The fermentation tube procedure can be used to test for fecal coliforms as well as total coliforms, but the higher temperature of 44.5°C is used for the fecal organisms.

### Most Probable Number (MPN)

The production of gas in a single fermentation tube may indicate the presence of coliforms, but it gives no clue as to the concentration of bacteria in the sample. A coliform count cannot be obtained directly. The gas bubble could have been caused by one bacterium or by thousands. To estimate the actual number of organisms, a multiple series of fermentation tubes with different sample volumes must be used.

As the size of the sample volume placed into a fermentation tube is increased, the probability of coliforms being present in the tube increases. Using statistics and probability theory, it is possible to analyze the combinations of positive and negative results in a multiple tube series and to determine the *most probable number* of coliforms in the original sample, referred to as the *MPN*.

The MPN is expressed in terms of the number of coliforms per 100 mL, but, as the name implies, it is more of an educated guess based on probability

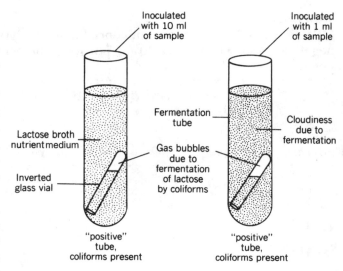

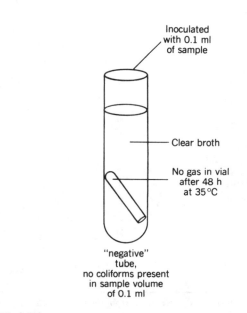

**FIGURE 4.24**

*A positive test in a fermentation tube (a trapped gas bubble and cloudiness in the lactose broth) signals the possible presence of coliform bacteria in the sample. The larger the sample volume, the more likely it is that the tube will test positive. If the number of positive tubes that occurs in a series of sample dilutions is high, then the MPN of coliforms is high.*

formulas than a direct count of organisms. Statistical analyses have been worked out for a variety of tube dilutions and combinations and conveniently summarized in references such as *Standard Methods*. To illustrate this, MPN values using a series of nine

**TABLE 4.4**
**MPN index and 95 percent confidence limits for various combinations of positive and negative results when three 10-mL portions, three 1-mL portions, and three 0.1-mL portions are used**

| *Number of tubes giving positive reaction out of:* 3 of 10 mL each | 3 of 1 mL each | 3 of 0.1 mL each | *MPN index per 100 mL* | *95% confidence limits* Lower | Upper |
|---|---|---|---|---|---|
| 0 | 0 | 1 | 3 | < 0.5 | 9 |
| 0 | 1 | 0 | 3 | < 0.5 | 13 |
| 1 | 0 | 0 | 4 | < 0.5 | 20 |
| 1 | 0 | 1 | 7 | 1 | 21 |
| 1 | 1 | 0 | 7 | 1 | 23 |
| 1 | 1 | 1 | 11 | 3 | 36 |
| 1 | 2 | 0 | 11 | 3 | 36 |
| 2 | 0 | 0 | 9 | 1 | 36 |
| 2 | 0 | 1 | 14 | 3 | 37 |
| 2 | 1 | 0 | 15 | 3 | 44 |
| 2 | 1 | 1 | 20 | 7 | 89 |
| 2 | 2 | 0 | 21 | 4 | 47 |
| 2 | 2 | 1 | 28 | 10 | 150 |
| 3 | 0 | 0 | 23 | 4 | 120 |
| 3 | 0 | 1 | 39 | 7 | 130 |
| 3 | 0 | 2 | 64 | 15 | 380 |
| 3 | 1 | 0 | 43 | 7 | 210 |
| 3 | 1 | 1 | 75 | 14 | 230 |
| 3 | 1 | 2 | 120 | 30 | 380 |
| 3 | 2 | 0 | 93 | 15 | 380 |
| 3 | 2 | 1 | 150 | 30 | 440 |
| 3 | 2 | 2 | 210 | 35 | 470 |
| 3 | 3 | 0 | 240 | 36 | 1300 |
| 3 | 3 | 1 | 460 | 71 | 2400 |
| 3 | 3 | 2 | 1100 | 150 | 4800 |

Source: Reprinted from *Standard Methods for the Examination of Water and Wastewater*, by permission. Copyright © 1981, The American Water Works Association.

tubes are presented in Table 4.4. Three of the nine tubes are inoculated with 10 mL of sample, three are inoculated with 1 mL of sample, and three are inoculated with 0.1 mL. Example 4.7 illustrates the use of the table.

## EXAMPLE 4.7

The results of a multiple-tube fermentation test on a sample of river water are as follows:

| *Dilution series* | *Results after incubation* |
|---|---|
| 10 mL | 2 positive, 1 negative |
| 1 mL | 1 positive, 2 negative |
| 0.1 mL | 0 positive, 3 negative |

Determine the MPN of coliforms from these data.

*Solution* Entering Table 4.4, locate the row with 2, 1, and 0 in the first three columns, respectively. These numbers represent the number of positive tubes in each dilution series. Under the column headed MPN index, read an MPN of 15 coliforms per 100 mL. The last two columns of the table point out the statistical nature of the MPN. The probability is 95 percent that the actual coliform concentration is at least 3, but no more than 44, per 100 mL.

## 4.5 WATER SAMPLING

Proper sampling procedures are an important part of any survey to assess water (or wastewater) quality and to check compliance with water quality standards. A sample that has been improperly collected, preserved, transported, or identified will result in invalid and useless test results, despite the precision of the analytical lab procedure. Since the results of water quality tests are the basis for decisions that affect public health, good sampling procedures must be followed. There are two basic sampling methods: *grab sampling* and *composite sampling*.

### Grab Samples

As its name implies, a grab sample is a single sample collected over a very short period of time. Most people envision this as a quick "scoop," but technically it can take up to 15 min to fill the sample container and still be considered a grab sample.

It is important to note that the test results from a grab sample only represent the conditions of the water or wastewater at the particular time and location of sample collection. Grab samples are most suitable when testing for chlorine residual, pH, coliforms, and dissolved oxygen. They are usually collected manually.

For stream or wastewater grab sampling, devices that provide easy access to the flow channel from boats, spillways, or docks are available. This is illustrated in Figure 4.25. Special containers that allow samples to be collected at specific depths below the surface, without mixing with air, are also available. This is particularly important for DO sampling. Sampling of surface water and groundwater at contaminated waste disposal sites is discussed in Section 12.4.

### Composite Samples

In many instances, grab samples alone are not enough to adequately characterize water or wastewater quality. This is particularly true for wastewater collection and treatment systems in which quality as well as quantity

**FIGURE 4.25**
*A typical grab sampler device, which allows safe access to the water or wastewater from which a sample is to be collected.* (Courtesy of Wheaton Instruments, Millville, New Jersey)

changes from hour to hour. Composite sampling is more appropriate when it is necessary to determine overall or average conditions over a certain period of time.

Composite samples are obtained by mixing individual grab samples taken at regular intervals over the sampling period. For example, a composite sample may consist of a mixture of smaller samples taken every 20 min over an 8-h period.

In wastewater studies, the volumes of the smaller grab samples that make up the composite are generally taken in proportion to the flow rate, for more meaningful results. For example, if a 100-mL grab sample is taken when the flow rate is 5 L/s, then a 200-mL sample would be taken when the flow increased to 10 L/s.

Automatic sampling equipment is usually used for composite sampling. The cost of the equipment is balanced by the savings on the labor involved in manual collection and mixing. An automatic sampler installed

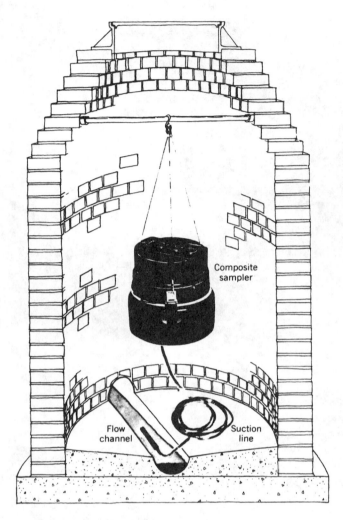

**FIGURE 4.26**
*Automatic composite samplers can be left unattended in sewer manholes, providing a sample that is representative of average flow conditions.* (Courtesy of ISCO, Inc., Environmental Division, Lincoln, Nebraska)

in a sewer manhole is shown in Figure 4.26, and a device being set up for composite stream sampling is shown in Figure 4.27.

## General Requirements

The methods for taking and preserving samples vary, depending on the specific water quality parameter and analysis to be made. The sampling frequencies and locations are stipulated in the NPDES permit for wastewater effluent standards and by the SDWA for drinking water. (The NPDES is discussed on page 284 and the SDWA on page 149.) A summary of four general considerations that apply for any type of sample follows:

**FIGURE 4.27**
*Automatic composite samplers can be used to collect stream samples in order to determine the average water quality conditions over a period of time.* (Courtesy of ISCO, Inc., Environmental Division, Lincoln, Nebraska)

1. *The sample must be truly representative of the existing conditions.* For instance, collecting a water sample from a faucet without first letting the water run for a while will not give results representative of conditions in the water main, but only of the water that was stagnant in the service line for an unknown period of time.

2. *The time between collection and analysis should be as short as possible* for the most reliable results. Certain tests, such as chlorine residual or temperature tests, must be determined immediately. Dissolved oxygen is another parameter that needs immediate analysis, although it is possible to add a chemical that fixes the DO concentration, allowing later testing in the lab.

3. *Appropriate preservation techniques should be applied* to slow down the biological or chemical changes that may occur in the time between sample collection and sample analysis. This usually involves refrigeration to cool the sample or chemical fixing (as for DO).

4. *Accurate and thorough sampling records must be kept* to avoid any confusion as to the "what, when, and where" of the sample, as well as to satisfy legal requirements. Figure 4.28 illustrates a sample bottle label that facilitates thorough record keeping.

```
                    COLIFORM BOTTLE No._____                          LAB RESULTS
    Sample Taken: Date _____ Time _____            Date Tested _____
    Sample Source □ Consumer's Faucet  □ Raw Water Supply
    Sample Location: _____              MF _____ per 100 ml
    Sampled By:_____              LB ——————  ——————
    Type of Sample  □ Initial Sample   Chlorine Residual mg/l _____         24        48
                    □ Routine Sample  WaterTemp. _____          BGB ——————  ——————
                    □ Check Sample                                       24        48
                    □ Special Purpose Sample                      SPC_____
    Water System Name _____           MPN _____
            Address _____           Sample Is:
            City _____ State _____ Zip _____          □ Safe   □ Unsafe
                                                                   □ Unsatisfactory for test
                                                                          Please Resubmit
```

**FIGURE 4.28**

*All sample bottles should be properly identified, and records of the date, time, place, and type of sampling should be kept.* (Reprinted from *Safe Drinking Water Act Self-Study Handbook,* by permission. Copyright © 1978, The American Water Works Association)

## 4.6 RELEVANT WEB SITES

### CALIFORNIA WATER QUALITY PROGRAM

http://wwwomwq.water.ca.gov/

The California State Water Project water quality program collects and publishes detailed information on the concentrations and distribution of chemical, physical, and biological parameters at more than 30 sites in the California Aqueduct and associated reservoirs.

### EPA MICROBIOLOGY HOME PAGE

http://www.epa.gov/nerlcwww/

The purpose of this EPA site is to provide access to microbiology-related information that has been developed or managed by the Agency; included are online publications, software, and images related to microbiological research.

### SOCIETY OF ENVIRONMENTAL TOXICOLOGY AND CHEMISTRY

http://www.setac.org

The Society of Environmental Toxicology and Chemistry (SETAC) provides a forum for individuals and institutions engaged in study of environmental issues, management and conservation of natural resources, environmental education, and environmental research and development.

### WATER QUALITY ASSOCIATION (WQA)

http://www.wqa.org/

The WQA is an international trade association representing the household, commercial, and industrial water quality improvement industry. Member companies manufacture and sell point-of-use/point-of-entry equipment, prefabricated water treatment plants, and customized water treatment systems. This site provides technical papers and consumer information.

### WATER QUALITY INFORMATION CENTER

http://www.nal.usda.gov/wqic/

The Water Quality Information Center at the National Agricultural Library is part of the U.S. Department of Agriculture's Agricultural Research Service. The center collects, organizes, and communicates scientific findings, educational methods, and public policy issues related to water quality and agriculture.

### WATER QUALITY MONITORING

http://www.epa.gov/OWOW/monitoring/

This EPA Web page addresses methods and tools for monitoring, assessing, and reporting on the health of U.S. water resources, and software and automated information systems for managing monitoring data.

# REVIEW QUESTIONS

1. What is the difference between an element and a compound? What is the difference between an atom and a molecule?

2. Briefly describe the difference between ionic bonding and covalent bonding. Give one example compound for each.

3. Briefly describe the difference between a suspension and a solution and between a colloid and a solution.

4. How does the solubility of solids in water change with increasing temperature? Is the situation the same for gases?

5. Match the following symbols or formulas for certain chemical substances on the left with the appropriate descriptive term(s) from the list on the right.

   _____ Mg            (a)  atom
   _____ Pb            (b)  ion
   _____ $Na^+$        (c)  molecule
   _____ $Cl^-$        (d)  complex ion
   _____ $O_3$         (e)  organic
   _____ CaO           (f)  manganese
   _____ $OH^-$        (g)  magnesium
   _____ $H^+$         (h)  silver
   _____ $NO_3^-$      (i)  lead
   _____ $CH_4$        (j)  oxygen
   _____ — COOH        (k)  lime
   _____ HCl           (l)  hydroxyl
   _____ Mn            (m)  proton
   _____ $Cl_2$        (n)  nitrate
                       (o)  chlorine
                       (p)  acid

6. What is the approximate saturation concentration of oxygen in fresh water at a temperature just above freezing?

7. Give one example of an ionic solution.

8. What is a *complex ion*? Give three examples.

9. What do *ppm*, *ppb*, and *gpg* stand for?

10. What is a *neutralization reaction*?

11. A solution has a pH of 8.5. Is it acidic or basic? Which substance is more likely to cause this pH in water, HCl or NaOH? Why?

12. The pH of a solution changes from 6 to 3. By what factor did the strength of its acidic condition increase?

13. What is an *organic compound*? List three different groups or types of organic substances.

14. Match the following parameters of water quality on the left with the possible effects listed on the right.

   _____ turbidity        (a)  causes dysentery
   _____ TDS              (b)  interferes with disin-
   _____ DO                    fection
   _____ iron             (c)  suffocates fish
   _____ fluoride         (d)  causes algal blooms
   _____ phosphorus       (e)  is toxic to humans
   _____ fecal coliforms  (f)  prevents tooth decay
   _____ lead             (g)  may cause cancer
   _____ THM              (h)  increases corrosion
                          (i)  indicates sewage
                               pollution
                          (j)  causes taste prob-
                               lems in water

15. What do the terms *JTU* and *NTU* mean?

16. Briefly describe the significance of temperature in water quality.

17. Briefly describe how odor or taste is measured in water.

18. What is DO? Why is it a significant parameter of water quality?

19. What does *BOD* stand for? What does this parameter indicate about water or wastewater quality?

20. What is *ultimate* BOD? Why is $BOD_5$ used for the standard BOD test?

21. Briefly describe how the BOD of a sewage sample is determined.

22. Briefly explain the effect of nitrification on oxygen demand.

23. What is the difference between a BOD test and a COD test?

24. What is the difference between TDS and TSS? How are they measured?

25. What is an *Imhoff cone*?

26. What is the difference between hard and soft water? Is it advisable to remove all the hardness from drinking water? Why?

27. Are high concentrations of fluoride ions in drinking water beneficial for public health? Why?

28. If a sample of stream water had an unusually high chloride ion concentration, what might you conclude about its quality? Briefly describe the difference between chloride and chlorine residual in water.

29. What are two effects of sulfate ions in water?

30. What is the significance of nitrogen and phosphorus in water quality?

31. What are two reasons for measuring acidity, alkalinity, and pH?

32. Briefly describe the basic characteristics of bacteria, algae, and protozoa. List one example of a species from each group.

33. What is the difference between an aerobe and an anaerobe?

34. Why are indicator organisms used to evaluate the sanitary quality of water?

35. If coliform bacteria are not detected in a water sample, what can you conclude about the possibility of recent sewage pollution? If coliforms are detected, will the water definitely cause disease among people who drink it? Explain your answer.

36. A stream passing through a sparsely populated agricultural area was found to have a fecal coliform count of 80 per 100 mL and a fecal strep count of 100 per 100 mL. What is the most likely source of the bacterial contamination?

37. Briefly describe the membrane filter method for determining the coliform count of a water sample. What is the basic premise or assumption underlying this method?

38. Briefly describe the multiple-tube fermentation method for obtaining a coliform count of a water sample. What does *MPN* mean?

39. Briefly discuss two different methods of collecting samples for water (or wastewater) quality testing.

40. List and briefly discuss four general requirements for good sampling procedure.

41. Visit the Water Quality Association Web site. Read and briefly summarize the latest WQA press release.

42. Explore the *EPA Microbiology Home Page*; check out the microbial "mugshots."

43. Using the EnviroSources search engine (see Section 1.6) or another Internet search directory or search engine, locate at least one additional Web site relevant to one of the topics in this chapter. Add the link(s) to your Environmental Technology Folder. Write a brief description of what the Web site(s) contain.

# PRACTICE PROBLEMS

1. Convert a concentration of 275 ppm to an equivalent value in terms of mg/L and gpg.

2. A sample of water has 4 grains per gallon of hardness as $CaCO_3$. What is the hardness in terms of mg/L? Is this level of hardness objectionable? Why?

3. A discharge from a sewage treatment plant enters a stream at a flow rate of 3 mgd. The BOD of the discharge is 50 mg/L. How many pounds of BOD are entering the stream per day?

4. A 50-lb bag of copper sulfate, $CuSO_4$, is dissolved in a lake to control algal growth. The lake volume is 30 ac–ft. If the chemical is completely dispersed throughout the lake volume, what is its concentration in mg/L?

5. How many kilograms of chlorine gas should be dissolved in 5 ML of water to result in a concentration of 2 ppm?

6. How many pounds per day of chlorine are needed to apply a chlorine concentration of 0.5 ppm in a flow of 25 mgd?

7. A 200-mL aqueous solution contains 0.005 mg of arsenic. What is the concentration of arsenic in terms of ppb?

8. A sample of water from a stream is placed into a standard 300-mL BOD bottle and is found to have a DO of 14.0 mg/L. After 5 d of incubation at 20°C, the DO in the bottle has dropped to 6.0 mg/L. What is the $BOD_5$ of the stream? What can you conclude about the quality of the stream?

9. In Problem 8, what could you say about the BOD and the quality of the stream if, after 5 d of incubation, DO was not detected in the bottle?

10. A sample of sewage is found to have a $BOD_5$ of 250 mg/L. Assuming the rate constant is 0.15/d,

estimate the ultimate carbonaceous BOD of the wastewater.

11. A 5-mL sample of sewage is diluted to 300 mL in a standard BOD bottle. The initial DO in the bottle is 9.2 mg/L, and after 5 d of incubation the DO is found to be 4.7 mg/L. Determine the 5-d and ultimate BOD values for the sewage, assuming that the reaction rate constant is 0.14/d.

12. A wastewater sample has an ultimate BOD of 280 mg/L. A 5-mL volume of this sample is diluted to 300 mL in a BOD bottle, and the initial DO is determined to be 9.0 mg/L. What is the expected DO in the bottle after 5 d of incubation if $k = 0.1$/d?

13. The weight of an empty evaporating dish is 38.820 g. A 50-mL volume of a filtered sample is evaporated from the dish. The weight of the dish plus dried residue is found to be 38.845 g. Compute the TDS of the sample.

14. The weight of a clean glass-fiber filter is 545 mg. After filtering a 100-mL sample, the weight of the filter plus retained solids is found to be 580 mg. After ignition in a furnace at 550°C, the weight of the filter and residue is 560 mg. Compute the concentration of suspended solids in the sample and the percentage of volatile solids.

15. A 10-mL water sample was tested for fecal coliforms using the membrane filter method. A total of 22 colonies of fecal coliforms was counted on the filter after incubation at 44.5°C. What was the fecal coliform count of the sample?

16. The results of a multiple-tube fermentation test of a water sample are shown to the right. What is the MPN of the sample?

| Dilution series | Results after incubation |
| --- | --- |
| 10 mL | 3 positive |
| 1 mL | 1 positive, 2 negative |
| 0.1 mL | 2 positive, 1 negative |

# Chapter Outline

C H A P T E R

# 5

# Water Pollution

Water has such a strong tendency to dissolve other substances that it is sometimes referred to as the *universal solvent*. This is largely because of its polar molecular structure. Pure water, that is, pure $H_2O$, is not found under natural conditions in streams, lakes, groundwater, or the oceans. It always has something dissolved or suspended in it. Because of this, there is not any definite line of demarcation between clean water and contaminated water.

If pure water does not exist outside of a chemist's laboratory, how can a distinction be made between polluted and unpolluted water? In fact, the distinction depends on the type and concentration of impurities as well as on the intended use of the water. In addition, the concentrations of the dissolved or suspended substances can be compared with water quality standards set for a particular use.

In general terms, water is considered to be polluted when it contains enough foreign material to render it unfit for a specific beneficial use, such as for drinking, recreation, or fish propagation. Actually, the term *pollution* usually implies that human activity is the cause of the poor water quality.

This chapter expands on some of the topics mentioned in the preceding chapter on water quality. It focuses on many of the common types and sources of water pollutants and on their effects in streams, lakes, groundwater, and the oceans.

## 5.1 CLASSIFICATION OF WATER POLLUTANTS

To understand the effects of water pollution and the technology applied in its control, it is useful to classify pollutants into various groups or categories. First, a pollutant can be classified according to the nature of its origin as either a *point source* or a *dispersed source* pollutant.

A point source pollutant is one that reaches the water from a pipe, channel, or any other confined and localized source. The most common example of a point source of pollutants is a pipe that discharges sewage into a stream or river. Most of these discharges are treatment plant *effluents*, that is, treated sewage from a water pollution control facility; they still contain pollutants to some degree.

A dispersed or nonpoint source is a broad, unconfined area from which pollutants enter a body of water. Surface runoff from agricultural areas, for example, carries silt, fertilizers, pesticides, and animal wastes into streams, but not at only one particular point. These

materials can enter the water all along a stream as it flows through the area. Acidic runoff from mining areas is a dispersed pollutant. Stormwater drainage systems in towns and cities are also considered to be dispersed sources of many pollutants, because, even though the pollutants are often conveyed into streams or lakes in drainage pipes or storm sewers, there are usually many of these discharges scattered over a large area.

The distinction between point sources and dispersed sources of pollutants is illustrated in Figure 5.1. Point source pollutants are easier to deal with than are dispersed source pollutants; those from a point source have been collected and conveyed to a single point where they can be removed from the water in a treatment plant, and the point discharges from treatment plants can easily be monitored by regulatory agencies. Under the *Clean Water Act*, a discharge permit is required for all point sources (see Section 10.1).

Pollutants from dispersed sources are much more difficult to control. Many people think that sewage is the primary culprit in water pollution problems, but dispersed sources cause a significant fraction of the water pollution in the United States. The most effective way to control the dispersed sources is to set appropriate restrictions on land use.

In addition to being classified by their origin, water pollutants can be classified into groups of substances based primarily on their environmental or health effects. For example, the following list identifies nine specific types of pollutants:

1. Pathogenic organisms
2. Oxygen-demanding substances
3. Plant nutrients
4. Toxic organics
5. Inorganic chemicals

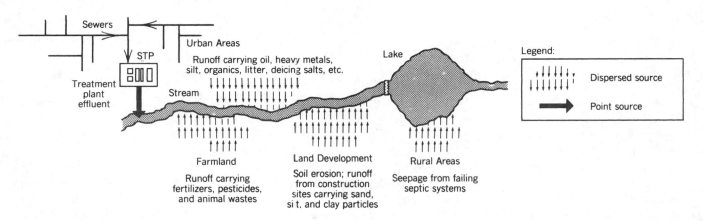

**FIGURE 5.1**
*Dispersed source pollutants are more difficult to control than are point source pollutants, which can be collected and removed from the water.*

**6.** Sediment

**7.** Radioactive substances

**8.** Heat

**9.** Oil

Domestic sewage is a primary source of the first three types of pollutants. Pathogens, or disease-causing microorganisms, are excreted in the feces of infected persons and may be carried into waters receiving sewage discharges. Sewage from communities with large populations is very likely to contain pathogens of some type.

Sewage also carries oxygen-demanding substances—the organic wastes that exert a biochemical oxygen demand as they are decomposed by microbes. This is the BOD, which is discussed in some detail in Chapter 4. BOD changes the ecological balance in a body of water by depleting the dissolved oxygen (DO) content. Nitrogen and phosphorus, the major plant nutrients, are in sewage, too, as well as in runoff from farms and suburban lawns.

Conventional sewage treatment processes significantly reduce the amount of pathogens and BOD in sewage, but do not eliminate them completely. Certain viruses, in particular, may be somewhat resistant to the sewage disinfection process. (A virus is an extremely small pathogenic organism that can only be seen with an electron microscope.) To decrease the amounts of nitrogen and phosphorus in sewage, usually some form of advanced sewage treatment must be applied.

Toxic organic chemicals, primarily pesticides, may be carried into water in the surface runoff from agricultural areas. Perhaps the most dangerous type is the family of chemicals called chlorinated hydrocarbons. Common examples are known by their common chemical names as chlordane, dieldrin, heptachlor, and the infamous DDT, which has been banned in the United States. They are very effective poisons against insects that damage agricultural crops. Unfortunately, they can also kill fish, birds, and mammals, including humans. And they are not very biodegradable, taking more than 30 years in some cases to dissipate from the environment.

Toxic organic chemicals can also get into water directly from industrial activity, either from improper handling of the chemicals in the industrial plant or, as has been more common, from improper and illegal disposal of chemical wastes. Proper management of toxic and other hazardous wastes is a key environmental issue, particularly with respect to the protection of groundwater quality. Poisonous inorganic chemicals, specifically those of the heavy metal group, such as lead, mercury, and chromium, also usually originate from industrial activity and are considered hazardous wastes.

Oil is washed into surface waters in runoff from roads and parking lots, and groundwater can be polluted from leaking underground tanks. Accidental oil spills from large transport tankers at sea occasionally occur, causing significant environmental damage. Blowout accidents at offshore oil wells can release many thousands of tons of oil in a short period of time. Oil spills at sea may eventually move toward shore, affecting aquatic life and damaging recreation areas.

## 5.2   THERMAL POLLUTION

Heat is considered to be a water pollutant because of the adverse effect it can have on the oxygen levels and the aquatic life in a river or lake. Overall, the amount of water withdrawn for cooling purposes in power plants exceeds the amount of water used for any other purpose. The cooling water carries away waste heat as it passes through the condensers in the plant. (Steam is converted back to water in the condensers.) The temperature of water used for cooling may increase by up to 15°C after it serves to condense the steam.

The discharge of warm water into a river is usually called *thermal pollution*. The warmer temperature decreases the solubility of oxygen and increases the rate of metabolism of fish. This changes the ecological balance in the river. Valuable game fish, such as trout, cannot survive in water above 25°C and will not reproduce in water warmer than 14°C. Coarser fish, such as carp or pike, can do well in water as warm as 35°C.

Because several species of coarser fish actually prefer warmer waters, some representatives of the power industry use the term *thermal enrichment* rather than thermal pollution to refer to the warm water they return to the river. Although it is true that many fish may congregate near the outfall pipe from the power plant, a problem arises if the plant is suddenly shut down for repairs. The sudden decrease in water temperature causes a fish kill of significant proportions, leaving thousands of dead fish floating belly-up in the river or washed up along the shore.

Thermal pollution may be controlled by passing the heated water through a cooling pond or a cooling tower after it leaves the condenser. The heat is dissipated into the air, and the water can then be either discharged to the river or pumped back to the plant for reuse as cooling water. This is illustrated in Figure 5.2. There is no discharge of heated water into the river, but some water will be withdrawn to make up for evaporative losses.

In locations where there is not enough room for a cooling pond, cooling towers may be built to prevent thermal pollution. These structures take up less land area than ponds. A common type is the *natural draft hyperbolic cooling tower*, in which evaporation accounts for most of the heat transfer. A photograph of a hyperbolic tower is shown in Figure 5.3*a*. Hyperbolic

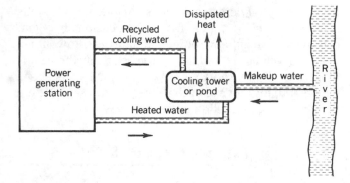

**FIGURE 5.2**
*Thermal pollution from power plants can be eliminated by using recirculating cooling towers or ponds.*

(a)

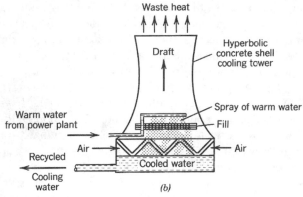

(b)

**FIGURE 5.3**
*In a natural draft cooling tower, the waste heat is dissipated into the atmosphere. The towers are typically about 100 m in diameter and about 130 m in height.* (Courtesy of Custodis–Cottrell, A Research–Cottrell Company, Somerville, New Jersey)

cooling towers are usually very tall and may dominate the landscape in their vicinity.

The operation of an evaporative cooling tower is basically a simple process, as illustrated in Figure 5.3*b*. Warm water coming from the condensers is sprayed downward over vertical sheets or baffles, called *fill*. The water flows in thin films through the fill. Cool air enters the tower through the air inlet that encircles the base of the tower and rises upward through the fill. Evaporative cooling takes place as the cool air passes over the thin films of warm water. A chimney effect or natural draft is maintained because of the density differential between the cool air outside and the warmer air inside the tower. The waste heat is dissipated into the atmosphere about 100 m above the base of the tower. The cooled water is collected in a basin at the floor of the tower and recycled back to the power plant condensers.

## 5.3 SOIL EROSION AND SEDIMENT CONTROL

The natural movement of soil particles by wind or water from one location to another is called *soil erosion*. Uncontrolled soil erosion is a significant environmental problem. Soils in agricultural regions are a precious natural resource, and the loss of these fertile soils from unwise land-use practices can be devastating. One of the most notable examples of this is the Oklahoma Dust Bowl of the 1930s. When a prolonged drought hit Oklahoma after many years of decreasing soil fertility, strong summer winds literally blew the dry topsoil away. In addition to environmental damage, this caused severe economic hardship and social dislocation.

Not all problems related to soil erosion are as dramatic as the Oklahoma Dust Bowl, but the cumulative effect of less extensive erosion episodes can still have adverse environmental effects, particularly with respect to water quality. Soil erosion has been identified as one of the most significant sources of water pollutants.

Soil particles suspended in water interfere with the penetration of sunlight. This in turn reduces photosynthetic activity of aquatic plants and algae, disrupting the ecological balance of the stream. When the water velocity decreases, the suspended particles settle out and are deposited as *sediment* at the bottom of the stream or lake. Sediment smothers *benthic*, or bottom-dwelling, organisms and disrupts the reproductive cycles of fish and other life forms.

There are two types of water-caused soil erosion: *sheet erosion* from land areas by raindrop impact and overland flow of storm runoff, and *stream erosion*, or the removal of soils from stream beds and stream banks by the swiftly moving channelized water.

The factors that affect the rate of sheet erosion include rainfall intensity, soil texture, steepness of slope, and amount of vegetative cover. The velocity of streamflow is one of the most important factors in stream erosion, although the type of soil is important, too. The quantity of eroded material carried by some of the larger streams and rivers can be enormous. The Mississippi River, for example, transports an average of 1.5 million tons of sediment per day to the ocean. Most of this material is carried as *suspended load* in the turbulent currents, but a significant portion is also carried as *bed load*, sliding or rolling along the river bottom.

A natural vegetative cover of grass and trees provides protection against sheet erosion. Land-use activities such as agriculture and construction, which temporarily remove the natural vegetation and expose the bare soils, are the main causes of serious erosion and sediment problems. Construction projects involving major land disturbances may be the more significant of these two causes. The uncontrolled erosion of soil at major construction sites can exceed the erosion from naturally vegetated areas of the same size by a factor of 100 or more.

Construction plans and specifications should describe the location and details of specific erosion control measures to conserve the soil and prevent water pollution. In some states, a *soil erosion and sediment control plan* is required to accompany the construction contract documents. The particular methods used depend primarily on the types of soil and the slope of the construction site. The Natural Resources Conservation Service (NRCS) of the Department of Agriculture sets standards for erosion and sediment control. Typical soil erosion control measures include the following:

1. *Temporary grass cover* on exposed soils can be used to reduce wind and water erosion until permanent seeding or soil stabilization is accomplished. Application of lime and fertilization should be done on the basis of soil test data, and the proper seed mixture should be applied.

2. *Mulching materials*, such as unrotted salt hay or woodchips, can be used for temporary cover on areas difficult to vegetate because of steep slopes, unsuitable soils, or winter operations.

3. *Diversion channels* can be constructed across slopes to reduce open slope length, as illustrated in Figure 5.4. These channels are constructed with a ridge on the lower side of the slope, diverting water to sites where it

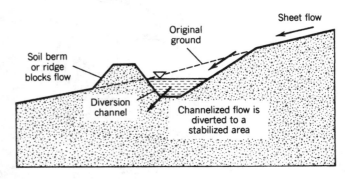

**FIGURE 5.4**

*Diversion channels reduce the distance of overland sheet flow, thereby reducing soil erosion and sedimentation of nearby streams and lakes.*

can be disposed of safely. They may be temporary or permanent.

4. *Hay bales* can be placed around stormwater inlets or at the low point on the site to intercept sediment-laden runoff and prevent the soil from entering the storm drainage system. This is illustrated in Figure 5.5.

5. *Temporary fences*, as illustrated in Figure 5.6, can be used to reduce erosion at construction sites. They are generally placed on the perimeter of the site at the lower elevations where water runs off.

6. *Sediment basins* or ponds can be built to intercept and retain water carrying suspended soil particles. The sediment is deposited in the pond, protecting streams or drainage systems downstream of the construction site. These ponds can be temporary or permanent earth structures and may be designed to reduce peak storm flows and flooding.

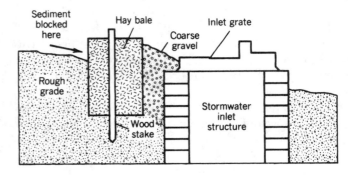

**FIGURE 5.5**

*A typical hay bale and gravel filter, which prevents sediment from entering a drainage system and then local streams; it is usually used in the vicinity of active construction sites.*

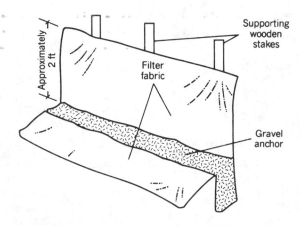

**FIGURE 5.6**
*A temporary fence can be constructed to control erosion at a construction site.*

7. *Channel stabilization* can be used to provide capacity in streams and drainage ditches for the flow of water without excessive erosion. Flow velocities should be minimized by the proper alignment and slope of the channel. Also, the channel can be protected by linings such as grass, concrete, or stone riprap.

8. *Scheduling of construction* can be done so as to minimize exposure of bare soils prior to final landscaping or paving. Drainage and soil protection facilities should be completed as early as possible.

## 5.4 STREAM POLLUTION

Rivers and streams are surface waters in which the entire water body continually moves downhill in natural channels under the force of gravity. They are shallower and narrower than lakes and have a greater proportion of water exposed to land surfaces. The flowing water carries algae rapidly downstream and tends to discourage the growth of rooted plants on the stream bed.

To a limited extent, streams and rivers have the ability to assimilate biodegradable wastes. Thus, they can recover from the effects of pollution naturally, without significant or permanent environmental damage. The capacity for self-purification depends on the strength and volume of pollutants and on the stream discharge or flow rate.

It used to be said that "the solution to pollution is dilution." The effects of dilution and the constant flushing action of the flowing water are obvious factors involved in the *waste-assimilative capacity* of a stream. Not as obvious, but equally important, is the effect of oxygen transfer between the air and the water. This is called *reaeration*.

The DO in the water is constantly replenished as atmospheric oxygen is dissolved at the water surface. Fast-flowing, shallow, turbulent streams are reaerated more effectively than slow, deep, meandering streams. This is because of the increased surface area and contact between the air and the water in the churning and well-mixed turbulent flow.

Modern-day population densities are too high for most streams and rivers to assimilate raw sewage discharges without offensive environmental conditions and public health hazards quickly developing. Some degree of treatment is required to remove enough of the BOD from the sewage so that stream dilution and reaeration can finish the job of purification. A level of treatment called *secondary treatment* is generally sufficient for this purpose; it is the minimum level of treatment required by law in the United States. This is discussed in more detail in Section 10.3.

Even when appropriate levels of treatment are applied to point sources of pollutants, a water body may remain impaired. (The EPA defines *an impaired water body* as one in which water quality standards have not been attained or maintained.) To make progress toward improving water quality in impaired waters, a new *total maximum daily load* (TMDL) rule has been promulgated and published by the EPA (as described in more detail in Section 5.9).

It is important to note that not all pollutants can be assimilated in water by natural means. This is particularly true for nonbiodegradable or persistent contaminants that do not dissipate in the environment. Even the physical process of dilution is ineffective when these persistent chemicals become trapped in river sediments. Two notable examples of this problem are the accumulation of PCB (polychlorinated biphenol, a toxic industrial chemical) in the sediments of the Hudson River in New York State and the contamination by the pesticide Kepone of the sediments of the James River in Virginia. These problems may persist unless the sediments are removed by dredging. But dredging may even increase the pollution by stirring up the contaminated deposits.

### Dilution

There are two basic steps involved in the process of waste assimilation in a stream or river: First, the *physical processes* of dilution and reaeration occur; second, *biological processes* occur, in which microorganisms in the water use dissolved oxygen to metabolize organic pollutants and convert them into harmless substances. In order to calculate the extent of assimilation, it is first necessary to account for the physical effect of dilution of the waste discharge.

When a point discharge of wastewater enters a flowing stream, the physical process of mixing and dilution begins immediately. With the exception of small turbulent streams, however, it is unlikely that the pollutants will be thoroughly mixed in the streamflow at or near the point of discharge. Instead, a *waste plume* forms, as illustrated in Figure 5.7. The length of this gradually widening mixing zone depends on the channel geometry, the flow velocity, and the design of the discharge pipe.

In water pollution control, it is often necessary to predict the BOD concentrations and DO levels downstream from a sewage discharge point. One of the first computations needed for this involves the effect of dilution. Assuming that the pollutant is completely mixed in the streamflow (at a point just below the end of the mixing zone), one can calculate the diluted concentration of any water quality parameter using the following *mass balance* equation:

$$c_d = \frac{c_s Q_s + c_w Q_w}{Q_s + Q_w} \qquad \textbf{(5-1)}$$

where $c_d$ = diluted concentration or temperature
$c_s$ = original stream concentration or temperature
$c_w$ = waste concentration or temperature
$Q_s$ = stream discharge
$Q_w$ = waste discharge

(See page 99-100 for a derivation of Eq. 5-1)

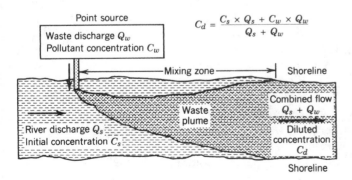

**FIGURE 5.7**
*Dilution of pollutants from a point source, such as a sewage treatment plant, occurs within the* mixing zone *of the stream.*

## EXAMPLE 5.1

The BOD$_5$ of an effluent from a municipal sewage treatment plant is 25 mg/L, and the effluent discharge is 4 ML/d. The receiving stream has a BOD$_5$ of 2 mg/L and the streamflow is 40 ML/d. Compute the combined 5-day BOD in the stream just below the mixing zone.

*Solution* Applying Equation 5-1 directly gives

$$c_d = \frac{2 \times 40 + 25 \times 4}{40 + 4} = \frac{180}{44} = 4.1 \text{ mg/L}$$

where $c_d$ represents the diluted BOD$_5$ in the combined flow.

## EXAMPLE 5.2

A river has a dry-weather discharge of 100 cfs and a temperature of 25°C. Compute the maximum discharge of cooling water at 65°C that can be discharged from a power plant into the stream. Assume the legal limit on temperature increase in the stream is 2°C.

*Solution* The maximum allowable stream temperature is

$$25 + 2 = 27°C$$

Applying Equation 5-1 gives

$$27 = \frac{25 \times 100 + 65 \times Q_w}{100 + Q_w}$$

Multiplying both sides by $(100 + Q_w)$ gives

$$2700 + 27Q_w = 2500 + 65Q_w$$

Transposing similar terms gives

$$38Q_w = 200 \quad \text{and} \quad Q_w = \frac{200}{38} = 5.3 \text{ cfs}$$

The discharge of warm water cannot exceed 5.3 ft³/s if the stream temperature is not to increase more than 2°C.

## Dissolved Oxygen Profile

When sewage is discharged into a stream, dissolved oxygen is utilized by microorganisms as they metabolize and decompose organic substances from the wastewater. The microbes exert a biochemical oxygen demand, or BOD, as discussed in Chapter 4. The BOD causes the dissolved oxygen level in the stream to gradually drop. This is illustrated in Figure 5.8 as curve A, called the stream *deoxygenation curve.*

While deoxygenation is occurring, oxygen from the air is dissolving into the water at the surface. The rate of oxygen transfer from the air into the water depends on temperature as well as on the *oxygen deficit.* The oxygen deficit is the difference between the actual DO concentration and the saturation DO value. The larger the deficit, the faster is the rate of oxygen transfer. This is illustrated as curve B in Figure 5.8, called the stream *reaeration curve.* Notice that the slope (rate of change) of the reaeration curve gradually increases as the deoxygenation curve falls.

At any given time, the DO level in the stream is a function of the combined effects of deoxygenation and reaeration. In other words, the actual DO is equal to the sum of the DO on the deoxygenation curve plus the DO on the reaeration curve. The graph of the combined DO versus time is seen as curve C, called the *dissolved oxygen sag curve.* Since the product of velocity and time equals distance $[m/s \times s = m]$, the horizontal or $x$-axis in Figure 5.8 can also be labeled "distance" for a given reach of the stream. Curve C is, in effect, a profile view of the DO concentrations along the length of the stream and is also called the *dissolved oxygen profile.*

Initially, the rate of deoxygenation exceeds the rate of reaeration, so the oxygen profile begins to sag. After most of the organics are decomposed, the rate of reaeration dominates and the oxygen profile begins to rise

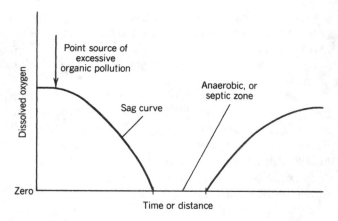

**FIGURE 5.9**
*Under heavy loads of pollution, the DO level may drop to zero. This results in obnoxious odors and very unsightly conditions in the water. With additional time and distance downstream, the water will eventually be reaerated and water quality will be restored.*

toward its original level. The minimum dissolved oxygen content in the stream occurs when the rate of reaeration equals the rate of deoxygenation. The computation of this level is of importance in water pollution studies.

In cases of extremely heavy organic pollution or very low streamflow, the oxygen in the water may be completely depleted. The sag curve intersects the horizontal axis at DO = 0, resulting in anaerobic or septic conditions. This is illustrated in Figure 5.9.

## Zones of Pollution

Most streams that are polluted by a point source of biodegradable organic substances can be described and evaluated in terms of four relatively distinct zones. These are illustrated in Figure 5.10. The first is the *zone of degradation,* which forms below the point of waste discharge. This zone is characterized by floating solids, turbidity, and other visual evidence of pollution. The DO level begins to drop rapidly in the zone of degradation.

When the DO level drops to about 40 percent of its saturation value, the *zone of active decomposition* is considered to start. This zone is characteristic of heavily polluted water. Higher forms of aquatic life and desirable species, such as trout, either die or migrate out of the area. More tolerant fish species such as carp and catfish may survive. The mixture of different species is altered because of the low DO levels. Sludge deposits of settleable solids may form in the stream. If anaerobic conditions occur (see Figure 5.9), gas bubbles, floating sludge, and obnoxious odors may be noticeable in this zone.

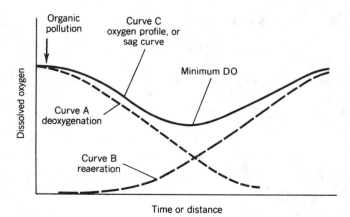

**FIGURE 5.8**
*The oxygen sag curve shows the effect of organic pollution on the DO levels in a stream or river. After the organics decompose, surface reaeration will restore the original water quality. This is called* stream self-purification.

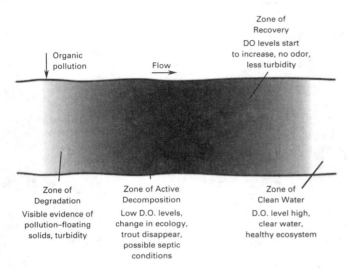

**FIGURE 5.10**

*The zones of pollution in a stream that receives biodegradable organic contaminants.*

After most of the organics have been decomposed by the microbes in the water, the rate of reaeration will exceed the rate of deoxygenation. When the DO level increases back up to 40 percent of the saturation concentration, the *zone of recovery* begins. This zone is characterized by gradually clearing water with no offensive odors; desirable aquatic species reappear. When the organic waste loading on a stream is small or when there is considerable dilution, the zone of recovery may follow directly after the zone of degradation, with no zone of active decomposition forming at all.

Following the zone of recovery is the *zone of clean water*. This zone is characterized by clear water, high in DO; diverse species of aquatic organisms thrive, utilizing the stable inorganic nutrients remaining in the water. In effect, the stream has recovered its original quality through a process of natural self-purification. Of course, additional point discharges or dispersed sources of pollutants along the stream can alter this model of pollution zones in a stream. Nevertheless, this model is of value in understanding stream pollution and in creating technical solutions to the problem.

## Computation of Minimum DO

It is important to be able to predict the minimum dissolved oxygen level in a polluted stream or river. For example, if a new sewage treatment plant is to discharge its effluent into a trout stream, it is possible that conventional (secondary) treatment levels will not remove enough BOD to prevent excessively low DO downstream. To determine if some form of advanced treatment is required to preserve the stream for trout

spawning and survival, it is necessary to compute the minimum DO caused by the sewage effluent and to compare it to the allowable value for trout streams.

One technique used to describe and predict the behavior of a polluted stream uses the so-called Streeter–Phelps equation. This equation is based on the assumption that the only two processes taking place are deoxygenation from BOD and reaeration by oxygen transfer at the surface, as previously discussed. Two key formulas from the Streeter–Phelps model of stream pollution and oxygen sag follow. Figure 5.11 illustrates some of the variables in these equations. The minimum DO in the stream is the difference between the saturation DO level and the *critical oxygen deficit*. The formulas are

$$t_c = \frac{1}{k_2 - k_1}$$
$$\times \log\left[\frac{k_2}{k_1} \times \left(1 - D_i \times \frac{k_2 - k_1}{k_1 \times BOD_L}\right)\right]$$

$$\tag{5-2}$$

$$D_c = \frac{k_1 \times BOD_L}{k_2 - k_1}$$
$$\times (10^{-k_1 t_c} - 10^{-k_2 t_c}) + D_i \times (10^{-k_2 t_c}) \tag{5-3}$$

where $t_c$ = time it takes for the critical oxygen deficit or minimum DO to develop, d

$D_c$ = critical oxygen deficit, mg/L

$D_i$ = initial oxygen deficit at time $t = 0$ just below the point of waste discharge into the stream, mg/L

$BOD_L$ = ultimate BOD in the stream just below the point of waste discharge, mg/L

$k_1$ = deoxygenation rate constant, $d^{-1}$

$k_2$ = the reaeration rate constant, $d^{-1}$

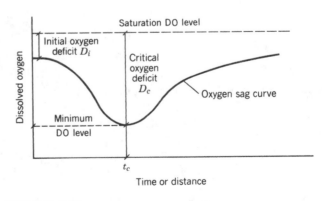

**FIGURE 5.11**

*The critical time* $t_c$ *and the critical oxygen deficit* $D_c$ *can be computed using appropriate equations. The minimum DO is the difference between the saturation DO level and the computed deficit.*

The value of $k_1$ is generally taken to be the same as the rate constant for the BOD reaction in Equation 4-2; it can be determined in the laboratory. The value of $k_2$ depends on the velocity and the depth of the flow and can be determined from field studies or by an appropriate formula. The reaeration rate constant $k_2$ can vary from about 0.1 for a sluggish river to about 4.0 for a shallow turbulent stream. Both rate constants, $k_1$ and $k_2$, depend on temperature.

Equations 5-2 and 5-3 look complicated, and they are. They are presented here to illustrate the power of mathematics as a tool for modeling the environment and helping solve water pollution problems. But as complicated as they appear, the Streeter–Phelps equations are not completely accurate representations of the oxygen profile in a polluted stream or river. Other factors that affect the oxygen balance include photosynthesis and respiration of rooted plants and algae and the oxygen demand of benthic (bottom) deposits. Equations that have been developed to include these factors are even more complicated than Equations 5-2 and 5-3.

### EXAMPLE 5.3

The $BOD_L$ in a stream is 3 mg/L and the DO is 9.0 mg/L. Streamflow is 15 mgd. A treated sewage effluent with $BOD_L = 50$ mg/L is discharged into the stream at a rate of 5 mgd. The DO of the sewage effluent is 2 mg/L. Assuming that $k_1 = 0.2$, $k_2 = 0.5$, and the saturation DO level is 11 mg/L, determine the minimum DO level in the stream. For a stream velocity of 0.5 ft/s, how far downstream does the minimum DO occur?

*Solution* First, it is necessary to compute the diluted $BOD_L$ and DO using Equation 5-1, as follows:

$$BOD_L = \frac{15 \times 3 + 5 \times 50}{15 + 5} = \frac{295}{20} = 14.8 \text{ mg/L}$$

$$DO = \frac{15 \times 9 + 5 \times 2}{15 + 5} = \frac{145}{20} = 7.3 \text{ mg/L}$$

Now compute the initial oxygen deficit as

$$D_i = \text{saturation DO} - \text{initial DO}$$
$$= 11.0 - 7.3$$
$$= 3.7 \text{ mg/L}$$

Applying Equation 5-2 gives

$$t_c = \frac{1}{0.5 - 0.2} \times \log\left[\frac{0.5}{0.2} \times \left(1 - 3.7 \times \frac{0.5 - 0.2}{0.2 \times 14.8}\right)\right]$$

$$= \frac{1}{0.3} \times \log[2.5 \times (1 - 0.375)]$$
$$= (3.33) \log 1.56$$
$$= 0.64 \text{ d}$$

It will take about 0.64 d (roughly 15 h) for the minimum DO to occur.

Now applying Equation 5-3 gives

$$D_c = \frac{0.2 \times 14.8}{0.5 - 0.2} \times (10^{-0.2 \times 0.64} - 10^{-0.5 \times 0.64}) + 3.7 \times 10^{-0.5 \times 0.64}$$

$$= 9.87 \times (0.745 - 0.479) + 3.7 \times 0.479$$
$$= 2.63 + 1.78$$
$$= 4.4 \text{ mg/L}$$

The minimum DO in the stream is the difference between the saturation DO and the critical oxygen deficit, or $11.0 - 4.4 = 5.6$ mg/L. At a velocity of 0.5 ft/s, in 0.64 d the distance downstream for the minimum DO is 0.64 d $\times$ 24 h/d $\times$ 3600 s/h $\times$ 0.5 ft/s = 27,650 ft $\approx$ 5 mi (approximately).

## 5.5 LAKE POLLUTION

The pollution of natural lakes or conservation reservoirs poses problems that are different from the problems caused by pollution of streams or rivers. This is primarily because of physical characteristics. Water in a stream is constantly moving and providing a flushing action for incoming pollutants, but in lakes the water is not moving very much at all and is detained for a relatively long period of time. In some cases, pollutants discharged into a lake can remain there for many years. Lakes are also significantly affected by seasonal temperature changes.

In streams, organic pollutants affect the oxygen profile. In lakes, water quality may be more dependent on plant nutrients than on organics from sewage. As discussed in Chapter 4, phosphorus and nitrogen are the most critical plant nutrients. When pollutants containing phosphorus and nitrogen compounds accumulate in a lake, rooted aquatic plants and free-floating algae may grow profusely.

The algae and aquatic weeds eventually die and settle to the bottom of the lake, where they are decomposed by bacteria and protozoa. This exerts an oxygen demand on the water and may deplete the DO in parts of the lake.

Excessive growth of algae, or *algal blooms*, forms slimy mats that float on the lake surface. They are unsightly and, along with the thick growths of weeds that develop along the shore, they interfere with boating, swimming, and fishing. A lake suffering from algal blooms is not a very good recreational resource. Furthermore, if the lake or reservoir is used for water supply purposes, the algae raise the cost of water treatment because the microscopic plant cells tend to clog the filters in the treatment facility, requiring them to be cleaned more frequently. Also, additional chemicals may be required to help control the tastes and odors imparted to the water by the algae.

Decaying plants, along with silt carried into the lake by overland runoff and feeder streams, gradually accumulate in significant amounts as sediment at the lake bottom. As the lake becomes shallower and, as a consequence, warmer, the balance of aquatic life shifts to favoring less desirable species. For example, trout give way to perch and bass and eventually to bullheads and carp as the process continues.

## Eutrophication

Actually, the process of *nutrient enrichment* and gradual filling in of a lake, as just described, is a natural process. It is called *eutrophication* and can be thought of as an inevitable and continual aging of the lake.

Lakes have a natural life cycle. Most lakes start out geologically as deep, cold, clear bodies of water. At this stage, they are called *oligotrophic* lakes. They usually have a sand or rock bottom, very few nutrients, and a scarcity of plant or fish life. Over the years, nutrients slowly accumulate and more organisms enter from inlet streams and the surroundings. Silty sediments begin to form at the bottom as the lake passes through a *mesotrophic stage* of existence.

The *eutrophic stage* of a lake's life cycle is characterized by a relatively shallow and warmer body of water, with enough nutrients to support large populations of plants and animals. In a eutrophic lake, there are frequent algal blooms, as previously described, and at certain times of the year the water at the bottom may be devoid of dissolved oxygen. Further aging or eutrophication leads to what is called a *senescent lake*, characterized by thick deposits of organic silts and very high nutrient levels. Senescent lakes are very shallow, with much rooted emergent vegetation growing throughout the lake. Eventually, what was once a lake will become a marsh as natural geological and ecological processes continue. The aging of a lake is illustrated in Figure 5.12.

The natural process of lake eutrophication, from the oligotrophic through the senescent stages, takes many thousands of years. It is an exceedingly slow process. But many people use the term *eutrophication* synonymously with *pollution* in reference to lakes. Perhaps a more accurate characterization of the problem would be the term *cultural eutrophication*. Cultural eutrophication is the acceleration and hastening of the natural aging process because of human activity in the drainage basin or watershed of a lake.

### Controlling Cultural Eutrophication

As many as two thirds of the lakes in the United States are significantly degraded as a result of eutrophication. About one third of the country's population lives within 5 miles of a lake. Sewage effluents and surface runoff carry large amounts of plant nutrients into these lakes, accelerating the eutrophication process.

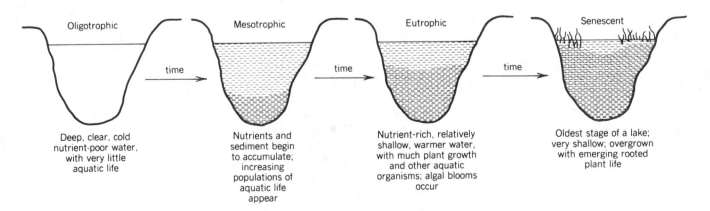

Oligotrophic — time → Mesotrophic — time → Eutrophic — time → Senescent

Deep, clear, cold nutrient-poor water, with very little aquatic life

Nutrients and sediment begin to accumulate; increasing populations of aquatic life appear

Nutrient-rich, relatively shallow, warmer water, with much plant growth and other aquatic organisms; algal blooms occur

Oldest stage of a lake; very shallow; overgrown with emerging rooted plant life

**FIGURE 5.12**
*Four stages in the life of a lake. All lakes go through a natural aging process called* eutrophication. *Human activity often accelerates this process.*

Phosphorus and nitrogen compounds are the most significant of the plant nutrients. Of these, phosphorus is generally recognized as the limiting factor and requires the greatest degree of control. It takes only a concentration of about 0.02 mg/L of inorganic phosphorus to cause algal blooms in a lake; the inorganic nitrogen concentration can be more than 10 times that level. On the other hand, even with very high nitrogen levels, if phosphorus concentrations are kept below 0.02 mg/L, excessive growths of algae usually do not occur.

Wastewater effluent and runoff containing phosphorus compounds can easily trigger algal blooms. Advanced treatment of sewage can effectively remove much of the phosphorus, as well as the nitrogen, from wastewater, but this is an expensive means of control. Advanced wastewater treatment is discussed in Section 10.4. It is being applied to control nutrient enrichment of the Great Lakes, San Francisco Bay, and many other bodies of water.

In areas where most of the nutrient input is from dispersed sources, such as surface runoff from agricultural areas, advanced sewage treatment is of little value as a control method. It has been estimated that more than 7 million tons of nitrogen and 0.5 million tons of phosphorus enter surface waters from agricultural areas each year in the United States. More efficient use of fertilizers, soil erosion control, and surface water diversion must be put into effect for lakes in agricultural areas.

Another way of reducing nutrient input is to divert wastewater effluents around the lake into some other body of water, such as a stream, that may be less sensitive to the nutrients. The city of Madison, Wisconsin, applies this method to protect a series of five lakes. Lake Washington in Seattle is another example of a lake protected by wastewater diversion. This method of controlling eutrophication requires contruction of extensive interceptor pipeline systems.

Another source of nutrient pollution in lakes is seepage from individual subsurface wastewater disposal systems (see Sections 5.6 and 10.5). In fact, many lakes, because of the nature by which the surrounding watershed becomes developed, are at first impacted by nutrient loading from septic systems. According to EPA studies, all septic systems within about 90 m (300 ft) of a lake have the potential to contribute nutrients to the lake. The extent of this pollution depends on depth to groundwater and bedrock, as well as the slope and the composition of the native soils. The ages and sizes of the septic systems are also important factors affecting nutrient loads.

The nuisances caused by excessive algal growth in lakes and reservoirs may be alleviated temporarily by the application of copper sulfate. The copper sulfate kills the algae, but its dose must be carefully controlled to prevent fish kills as well. In lieu of chemicals, harvesting of the algae and weeds can offer temporary relief from the problems related to eutrophication. Underwater weed cutters mounted on boats can be used to remove rooted aquatic plants, and dredges can be used to remove sediments, but these are not very practical measures for large bodies of water.

## Thermal Stratification

Lakes and reservoirs are affected by seasonal temperature changes. These effects include a layering or *thermal stratification* of the water as well as a mixing or *seasonal overturn* of the water because of temperature differences. Both thermal stratification and seasonal overturn can have significant impacts on pollution and the quality of the lake water. In temperate climates, the cycle of stratification and overturn occurs twice a year, whereas in warm climates where the water never freezes, the cycle occurs once.

Stratification due to temperature differences in the lake water is of most concern in the warm summer months. The lake water is warmed by the air, and the warm water forms a top layer called the *epilimnion*. Colder and therefore denser water remains at the lake bottom in a layer called the *hypolimnion*. A relatively thin layer of water with rapidly decreasing temperature from top to bottom, called the *thermocline*, separates the epilimnion and the hypolimnion. The thermocline acts as a physical barrier, which prevents mixing of water between the top and bottom layers of the lake. This is illustrated in Figure 5.13*a*.

The warm water in the epilimnion is mixed by the wind and receives energy from sunlight, allowing it to support algal growths. This relatively turbid water interferes with the penetration of sunlight to greater depths. The stagnant hypolimnion waters are relatively cool and dark. Because of this, some species of fish may prefer the hypolimnion environment, but the water at the lake bottom can often be of poor quality, particularly in a mesotrophic lake. The decaying benthic sediments exert a BOD that depletes the dissolved oxygen in that zone. Sometimes anaerobic conditions may develop at the bottom of the lake.

As the air temperature decreases during the autumn months, the epilimnion waters cool, become denser, and begin to sink toward the lake bottom. Eventually, the entire lake becomes completely mixed and the well-defined layers of the summer stratification disappear. This circulation, called the *fall overturn*, is illustrated in Figure 5.13*b*.

In the cold winter months, when ice covers the lake surface, a *winter stagnation* occurs. Then, in the spring, the ice melts, and when the water warms to 4°C (at which

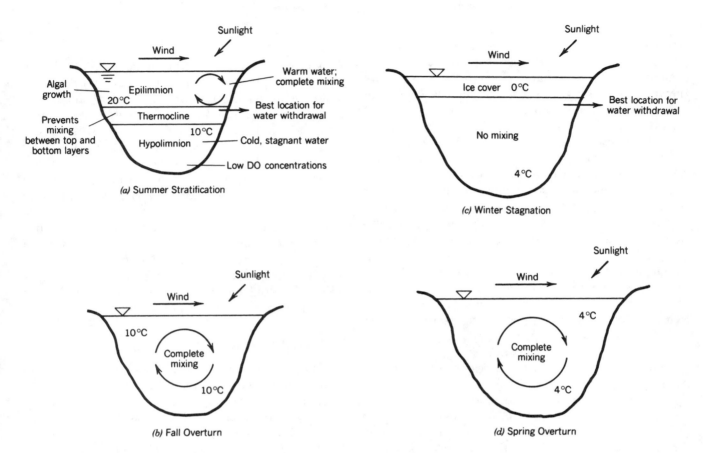

**FIGURE 5.13**
*Seasonal stratification and overturn of a lake or reservoir has an effect on
water quality.*

temperature water is densest), it starts to sink toward the bottom. Aided by the wind, the entire lake soon becomes completely mixed again. This is called the *spring over-turn.* The winter stagnation and spring overturn are illustrated in Figures 5.13*c* and 5.13*d*, respectively.

In lakes or reservoirs used for water supply, strat-ification and overturns can affect the quality of the water. During the fall overturn, for example, the poorer-quality bottom waters in the hypolimnion become mixed throughout the volume of the lake. This usually intensifies taste and odor problems in the finished wa-ter unless additional steps are taken in the treatment process. Water intake structures can be built in the lake, with inlet ports and valves at several depths. This pro-vides flexibility in operation and permits the water of optimum quality to be taken into the treatment plant.

During the winter, the best-quality water is gen-erally withdrawn from a depth just below the ice cover. In the summer, the water in the epilimnion and the wa-ter near the bottom of the hypolimnion are of the poor-est quality for reasons already discussed. At this time of the year, the best-quality water is usually withdrawn from a depth just below the thermocline.

Various methods have been tried to reduce the adverse effects of stratification. When oxygen deple-tion and anaerobic conditions become severe in the hypolimnion, compressed air is sometimes diffused through perforated pipes placed on the lake bottom to reoxygenate the water. In some cases, mechanical mixing and destratification may be effective in im-proving water quality. One method is to pump cold bottom waters up to the surface. The cooler water tends to shift algal growth to less troublesome species, thus reducing taste and odor problems in the water supply.

## 5.6  GROUNDWATER POLLUTION

Groundwater supplies about one fourth of the fresh water used for all purposes in the United States, in-cluding irrigation, industrial uses, and drinking water. About half of the U.S. population relies on groundwater for drinking water. Despite this strong reliance on it, groundwater has for many years been one of the most neglected natural resources.

Because it is underground, groundwater is less visible than other environmental resources, such as streams or lakes, and because it is less visible, it has tended to be of less public concern. In recent years, however, primarily because of well-publicized incidents of groundwater contamination, public attitude and regulation have changed.

Groundwater is usually of excellent quality. This is primarily because of the natural filtration that occurs in the layers of soil through which the water slowly moves. The distance that a pollutant can travel in the ground before being separated from the groundwater depends on both the type of soil and the type of pollutant. Deposits of fine sand, for example, may remove suspended solids or bacteria from the water in a short distance, whereas coarse gravel or fissured rock could allow those pollutants to travel considerable distances. And soluble pollutants are not affected at all by the filtering action of the soil, although other processes, such as adsorption, may take place.

Since the late 1970s, an increasing number of discoveries of contaminated groundwater in certain locations has been reported. The contaminants come from many different sources and include a variety of materials, most notably synthetic organic chemicals. These include a group of substances called chlorinated hydrocarbons, such as trichloroethylene and carbon tetrachloride. Many of these organic chemicals are toxic; some are suspected to be carcinogens or mutagens, which pose serious public health risks at concentrations as low as 10 $\mu$g/L (10 ppb).

Of all types of water pollution, this is perhaps the most insidious because at low concentrations the contaminants rarely impart any noticeable taste or odor to drinking water. The water may appear to be crystal clear, but it is far from pristine pure. In some cases, the concentrations of synthetic organics found in contaminated groundwater have exceeded by several orders of magnitude the typical levels of those compounds found in very polluted rivers.

The slow rate of flow of groundwater in an aquifer is discussed in Chapter 3. This is a significant factor in evaluating the impact of groundwater pollution. Because it moves so slowly, typically less than 30 m/yr, a contaminated aquifer will remain contaminated for hundreds of years.

If an aquifer that supplies drinking water is polluted, it may be necessary to abandon the contaminated well(s) and drill new ones some distance away or to seek alternative surface supplies. In some cases, it may be feasible to install special treatment units, such as aerators or activated carbon filters, to remove the contaminant, but this adds a permanent and often large expense to the water utility budget.

## Sources of Contamination

Even in areas far removed from human activity, groundwater is not pure. Although it is generally free of turbidity due to natural filtration, it usually contains dissolved minerals. This is to be expected because the water is in intimate contact with minerals in soil and rock deposits for long periods of time. Groundwater is usually harder (see pages 106 to 107) than surface water for this reason. But, for the most part, the natural contaminants found in groundwater pose no threat to public health.

The main problems with respect to serious groundwater pollution have been improper disposal of wastes and accidental spills of hazardous substances, especially from industrial activities. Petroleum products leaking from old underground storage tanks (USTs) are another source of groundwater pollution.

### Industrial Wastes

Industrial chemical wastes disposed of in surface impoundments, such as unlined landfills or lagoons, represent a significant source of groundwater contamination. A large fraction of the hundreds of millions of tons of industrial wastes generated each year is hazardous. Land disposal of these liquid and solid wastes is practiced, because it is the least expensive way to "get rid of" the unwanted materials. Although it seemed to be the most economical alternative for industry, in the long run it proved to be very expensive for society as a whole with respect to the health hazards and the costs of cleanup activities.

Many industrial impoundments in the United States did not have any bottom liners and did not meet new federal or state standards for land disposal of wastes. Contaminated liquids leaked out of these landfills and lagoons, percolated through the soil, and eventually reached a groundwater aquifer. This is illustrated in Figure 5.14. (The liquid from a solid waste landfill is called *leachate*.) Organic chemicals such as polychlorinated biphenol (PCB) and benzene have been found in groundwater at many industrial impoundment sites. One of the substances found most frequently is trichloroethylene (TCE), a chlorinated hydrocarbon used as a solvent and degreaser. Heavy metals such as selenium, arsenic, and cyanide have also been found.

Industrial wastes are sometimes pumped into the ground under pressure through deep wells, in a process called *deep well injection*. This is generally an acceptable method for industrial waste disposal, but the geological conditions must be suitable. At depths over 300 m (1000 ft), groundwater is often saline (high salt concentrations) and is not appropriate for other uses.

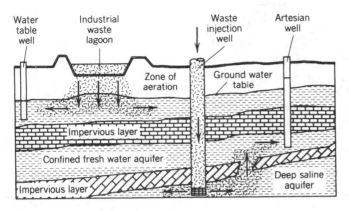

**FIGURE 5.14**
*Diagram showing two sources of groundwater pollution from industrial waste disposal—a leaky surface impoundment or lagoon and deep well wastewater injection. The arrows indicate the direction of flow of the pollutants. A bottom liner for the lagoon and thorough geological exploration of the saline aquifer can help to prevent the pollution.*

But even with deep well injection, accidental contamination of important water supply aquifers is possible, as illustrated in Figure 5.14; deep well injection is now strictly regulated (see Section 12.3).

### Subsurface Sewage Disposal Systems

Almost one third of the population of the United States is served by on-site subsurface sewage disposal systems. The most common of these is the septic tank and leaching field system, discussed in Section 10.5. Briefly, the septic tank traps and stores solids while the liquid effluent from the tank flows into a network of buried perforated pipes. The perforated pipes form what is called the *leaching* or *absorption field* and serve to spread out the sewage effluent over an area large enough for it to seep slowly into the soil and degrade naturally.

It has been estimated by the U.S. Environmental Protection Agency that more than 1 trillion gallons of sewage enter the ground each year through on-site disposal systems. Unfortunately, because of inadequate design, poor construction, or lack of maintenance, not all of these systems work properly. Septic disposal systems are frequently the sources of fecal bacteria and virus contamination in private wells. Also, the septic tank cleaning fluids that some homeowners use contain organic solvents, such as TCE. These potential human carcinogens also pollute the groundwater in areas served by septic systems. Other contaminants that can reach the groundwater from septic effluents include detergents, nitrates, and chlorides. Pollution from septic systems is illustrated in Figure 5.15.

### Municipal Landfills

Burial in the ground is one of the most common methods of disposing of municipal refuse. In the past, this practice was largely uncontrolled and the disposal sites, commonly known as garbage dumps, were literally just that—places where municipal solid wastes were simply dumped on and into the ground. These dumps were often located in low-lying areas with high groundwater tables or in abandoned sand and gravel pits. Leachate flowing through the refuse, high in BOD, chloride, nitrate, organics, heavy metals, and other contaminants, easily reaches the groundwater and enters underlying aquifers from such disposal sites. There are thousands of inactive or abandoned dumps like this throughout the United States.

Modern solid waste disposal technology can effectively prevent groundwater pollution. The disposal sites are properly called *landfills* rather than dumps. This is discussed in more detail in Section 11.6. Briefly, proper location of the landfill with respect to geological conditions and the use of bottom liners are two of the ways in which groundwater quality can be protected.

### Mining and Petroleum Production

Many of the active, as well as the abandoned, coal, metal, and other mines in the United States are a threat to groundwater quality. Surface water flowing in the vicinity of the mines can pick up dissolved metals and other solids, acidity, and even radioactive substances. As it infiltrates the earth, either in the open pits from strip mining or in underground tunnels and shafts, the polluted water easily carries these contaminants into underlying groundwater aquifers. Leaching of contaminants from tailings (residue) ponds and slag (cinder)

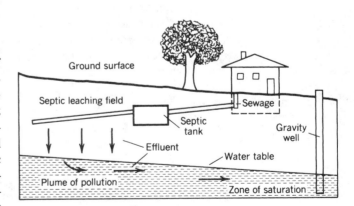

**FIGURE 5.15**
*Groundwater can be polluted from on-site sewage disposal systems. Wells located downhill from septic absorption fields are susceptible to contamination.*

piles is also a source of groundwater contamination from mining operations.

Many instances of groundwater pollution from petroleum production activities have been reported, particularly in the south central and southwestern United States. A basic cause of this has been the use of brine pits for disposal of the saline by-products of drilling.

Petroleum products such as gasoline and motor oil also are groundwater pollutants. These materials can flow through the zone of aeration and reach the groundwater table. Numerous instances of local groundwater pollution due to leaky gasoline tanks at filling stations have been reported. It is estimated by the EPA that nationwide as much as 11 million gallons of gasoline leaked into the ground each year before UST regulations were enacted. Now petroleum companies more often replace steel tanks with rustproof fiberglass tanks to eliminate leakage.

Accidental spills on the surface are also sources of oil and gasoline contamination in groundwater. Even in low concentrations, these materials cause noticeable tastes and odors in drinking water obtained from a contaminated aquifer. In addition, gasoline contains ethylene dibromide (EDB) and benzene, which may be carcinogenic in humans.

In some cases, if the plume of petroleum pollution has not traveled too far, steps can be taken to clean up the water. The contaminated water can be pumped out of the ground, put through oil separators, and then discharged back into the ground.

### Agriculture

The most significant groundwater contaminants from agricultural activities are fertilizers and pesticides. Nitrates are of particular concern in groundwater used for drinking because of the health problem called blue baby (see page 108). In the farming areas of eastern Long Island in New York State, for example, many families with infants must use bottled drinking water. The soil is very sandy, and nitrates from fertilizers are easily carried through the porous soil into the groundwater, contaminating many private wells.

### Urban Areas

The public works departments of many cities and towns often spread salt on the roads to keep them ice-free during the winter. Eventually, these salts are dissolved and carried off the pavement in sheet flow. Much of this material is carried into underlying aquifers, increasing the chloride and TDS concentrations of the groundwater. (In some communities, calcium magnesium acetate is used instead of chloride salts.)

### Saltwater Intrusion

The intrusion of salty seawater into wells is a groundwater pollution problem in many coastal cities and towns. Because of increasing population, urbanization, and industrialization, increasing quantities of groundwater are being used, and the amount of natural groundwater recharge is decreasing in these areas due to the construction of roads and parking lots. As a consequence, the elevation of the groundwater table is dropping.

In coastal areas, there is an interface or boundary between the fresh groundwater flowing from upland areas and the saline water from the sea. Because seawater is about 2.5 percent denser than freshwater, a pressure head of 40 ft of seawater is equivalent to a pressure head of 41 ft of freshwater ($1/40 = 0.025 = 2.5\%$). As a result, for each foot the water table drops in elevation, the seawater boundary rises 40 ft. This is illustrated in Figure 5.16. Wells pumping water that is salty because of seawater intrusion may have to be abandoned as sources of drinking water. In addition to water conservation, artificial recharge of the groundwater from freshwater impounds can be effective in halting saltwater intrusion.

Because of saltwater intrusion, several coastal communities in the United States require desalination plants to ensure potable water supplies. The first desalination plant in the northeast United States, for example, now serves the resort community of Cape May, New Jersey. It is a reverse osmosis facility with a treatment capacity of 7.6 ML/d (2 mgd). Reverse osmosis and other desalination methods are discussed in Section 6.6, along with a case study of the Cape May facility.

## Preventive Measures

Natural purification of chemically contaminated groundwater can take decades and perhaps centuries, and

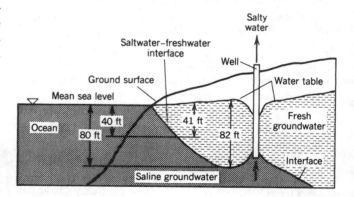

**FIGURE 5.16**
*Saltwater intrusion into coastal area wells is a significant source of groundwater contamination.*

cleanup efforts are sometimes much too expensive to be practical. The best way, then, to control groundwater pollution is to prevent it from occurring in the first place. Laws related to solid and hazardous waste disposal now significantly reduce new contamination. Not only are physical barriers between the waste and the groundwater required, but monitoring wells must be installed in some cases to provide early warning of possible leakage.

Land-use management applied on the local level by towns and cities can be effective in preventing aquifer contamination. For example, zoning ordinances that prevent residential or industrial development in areas that are known groundwater recharge zones can reduce pollution problems. Strict enforcement of regulations pertaining to the siting, design, and construction of septic systems can reduce or eliminate the incidence of sewage contamination of private wells. Prudent application of pesticides and fertilizers in agricultural areas can also be effective in this regard.

The *Safe Drinking Water Act* amendments of 1996 require states to establish a wellhead protection program (WHP) for groundwater-based public water supplies (the term "wellhead" refers to a well or the column or "head" of water within a well). Each state must develop, with public participation, a wellhead protection program plan that is to be reviewed and approved by EPA. A wellhead protection plan consists of *(a)* delineation of the wellhead protection area, *(b)* an inventory of possible sources of pollution, and *(c)* an emergency or contingency plan should a contaminant threaten the groundwater supply. The costs of completing and implementing a wellhead protection plan and monitoring and regulating some activities are far less than the costs of emergency procedures (for example, providing an emergency water supply or cleaning the groundwater, should a public well become contaminated).

## 5.7  OCEAN POLLUTION

Ocean water is naturally saline, containing about 3.5 percent dissolved solids (35 000 mg/L). This is much greater than the concentration of total solids carried in raw sewage, but the ocean is not considered to be polluted because of its natural salinity. The dissolved solids in the sea are inorganic minerals, mostly sodium chloride. The salinity, however, does make ocean water unsuitable for most uses, with the exception of instances where it is subjected to a *desalting* process (see Section 6.6) to make it potable.

For these reasons, as well as because of the tremendous amount of dilution it apparently provides, there was a natural tendency to consider the ocean as a convenient "sink" or receptacle for wastes of all kinds. Ultimately, all the sewage effluent discharged into streams and rivers makes its way to the ocean. In coastal areas, treated sewage effluent is discharged directly into the ocean.

Despite the tremendous volume of the marine environment, the natural capacity of estuaries and the ocean to assimilate wastes is limited. One of the adverse effects of ocean pollution is the destruction of marine life, particularly the phytoplankton (algae), which produces oxygen in the process of photosynthesis and serves as food for other organisms. In some instances, unsightly and perhaps dangerous waste materials are being washed back to shore.

*Estuaries* are natural transition zones between freshwater rivers and saline ocean waters. They are semienclosed bodies of water, including bays, river mouths, and salt marshes. Because they are adjacent to land, they are the first marine areas to receive wastes carried by river flow. Estuaries are considered to be one of the most biologically productive environments and are of critical importance to a variety of both terrestrial and marine organisms. Not only is pollution a threat to these vital ecosystems, poor land-use management that allows the filling in of wetland areas for residential or commercial development also takes its toll.

Beyond the estuaries and the relatively shallow coastal ocean waters is the open ocean, which composes about 70 percent of Earth's surface. The open ocean is almost entirely dependent on the estuaries for nutrients, which are transported by currents to the deeper water, and for the support of life processes.

### Diffusion of Sewage in Seawater

Treated sewage effluent from cities and towns is discharged directly into the ocean in many coastal areas. The pipes that carry the wastewater into the ocean are called *outfalls*. These are often large-diameter conduits that may extend far offshore. For example, the outfall from the Deer Island Treatment Plant in Boston is 12.5 km (7.5 mi) long.

When sewage flows out of the open end of an outfall pipe, it forms a rising column, because it is warmer and less dense than seawater. This is illustrated in Figure 5.17. When it reaches the surface, the column of sewage forms a large bubble or *boil*, which moves in the direction of the surface currents. As the current carries the boil, a plume of diluted wastewater forms, similar to a plume from a smokestack. Unfortunately, the plume is sometimes carried toward shore, raising the coliform counts near recreational areas. Occasionally,

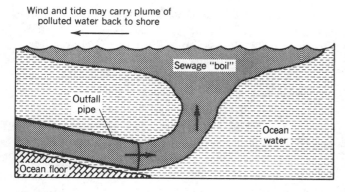

**FIGURE 5.17**
*Sewage effluent from an ocean outfall pipe forms a rising column in ocean water because of its lower density. The column reaches the surface in a boil.*

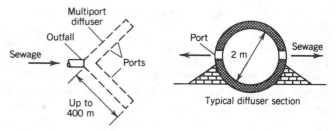

**FIGURE 5.18**
*Multiport diffusers at the end of an ocean outfall pipe increase the mixing and dilution of sewage in the seawater, thereby preventing the formation of plumes.*

beaches must be closed for swimming because of excessive coliform counts.

The effectiveness of ocean disposal of sewage effluent depends on how well the effluent is dispersed and spread out in the ocean when it exits the outfall. Sufficient dispersion of the effluent will facilitate the natural purification process, reduce bacteria concentrations, and prevent pollution at shore areas. Most outfalls are now built with *multiport diffusers* at the discharge end to maximize the amount of diffusion. The diffuser distributes the effluent over a relatively large area of the ocean bottom through many circular holes or ports. This allows a much greater degree of mixing and dilution than there is from an outfall without a diffuser. Diffusers prevent the formation of sewage boils and provide greater shore protection.

The success of a diffuser in accomplishing this objective depends on careful hydraulic design computations. It is necessary to achieve uniform flow distribution through the ports. A typical ocean outfall diffuser is illustrated in Figure 5.18. Typical dimensions are shown just to illustrate the size of these structures.

### Ocean Dumping of Sludge

In the past, many coastal cities in the United States dumped sewage sludge in the ocean. For example, for about 60 years New York City and many of the surrounding communities in New York and New Jersey used a sewage sludge dumping site in the Atlantic Ocean. It was located about 20 km (12 mi) offshore and was only about 25 m (80 ft) deep and about 36 km$^2$ (14 mi$^2$) in area. In 1984 alone, approximately 8 million tons of sewage sludge was dumped there. That area of the ocean was found to have high levels of bacteria, which prevented shellfishing, as well as high levels of toxic metals and

PCB, a known carcinogen. Normal marine life was virtually eliminated at the site, and ocean currents sometimes carried the sludge to beaches and parks along the shore.

In 1985, the EPA ordered that the 12-mi site be closed and required that sewage sludge be hauled out to a more distant site, about 175 km (106 mi) off the coast. This dumping location, just beyond the continental shelf, had an area of about 250 km$^2$ (100 mi$^2$) and was about 2500 m (8000 ft) deep. Ocean dumping of sewage sludge at this site was allowed until 1992, at which time Congress banned ocean dumping completely in all coastal areas. Now, New York City ships its sludge by rail for land disposal in locations as far away as Texas; this will continue until facilities are built to convert the sludge to fertilizer.

### Oil Spills

Accidental discharges of oil can be a serious ocean pollution problem. Oil can enter ocean waters primarily from tanker spills or from offshore well blowouts. Worldwide, in 1979 alone, about 0.75 million tons of oil from tanker accidents was spilled into the sea. In the same year almost 0.5 million tons (3 million barrels) of oil was spilled from a single offshore well off the coast of Mexico over a 9-month period. The oil and tar slick traveled about 1100 km (700 mi) north, reaching the coast of Texas.

Oil spills can cause the deaths of birds that live on the water surface and dive to obtain their food. It has been estimated that the Torrey Canyon oil spill in 1967 was responsible for the deaths of about 100,000 birds. In 1989, when the *Exxon Valdez* oil tanker spilled 42 million liters (11 million gallons) of oil into Prince William Sound, Alaska, hundreds of kilometers of shoreline was polluted and ecological damage was severe. Oil slicks are often driven toward land by wind and tide action. Oil pollution on the shore harms all forms of aquatic life and interferes with bathing and recreational uses of beach areas.

In some cases, the spread of an oil slick can be controlled by employing physical barriers, and mechanical means of collecting the spilled oil have been used with varying degrees of effectiveness. Detergents used as chemical dispersants have been used to break up the oil slick, but even these detergents can be toxic and harm the marine ecosystem.

The best way to prevent environmental damage from an oil spill is to prevent it from occurring in the first place. Stricter international standards for the design and operation of oil tankers can be effective in reducing the frequency and extent of spills. Similarly, stricter requirements for the safety, licensing, inspection, and monitoring of offshore drilling operations can protect the ocean environment.

## 5.8 WATER QUALITY STANDARDS

In the urbanized and industrialized world of today, it is necessary to have a legal basis for protecting water quality. It takes human effort, energy, and money to keep water clean enough for the many different uses for which society requires it. Without a legal framework to allow the enforcement of *water quality standards*, environmental quality and public health would be in constant jeopardy.

Water quality standards are limits on the amount of physical, chemical, or microbiological impurities allowed in water that is intended for a particular use. These are legally enforceable by governmental agencies and include rules and regulations for sampling, testing, and reporting procedures.

Within the past 35 years or so in the United States, Congress has enacted several laws that focus on the problems of water pollution and water quality protection. They require the EPA to set minimum standards; individual states have the right to adopt the same federal standards or to establish stricter standards of their own.

There are certain disadvantages in having such rigid statutory laws, including the fact that sometimes not enough scientific data are available to really confirm the validity of a particular standard. But the laws generally serve as a reasonable basis for pollution control and public health protection. As more research is done, the standards are revised so as to better balance the risk of contamination with the cost of cleanup.

There are three different types of water quality standards: *stream standards*, *effluent standards*, and *drinking water standards*. To put this in proper perspective, the relationship among these three kinds of water quality standards is shown schematically in Figure 5.19. Together, they reinforce each other and

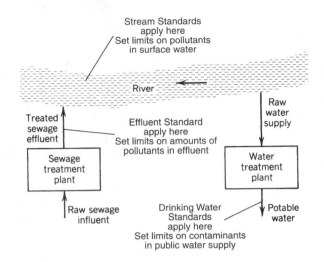

**FIGURE 5.19**

*Three different types of water quality standards are enforced by the EPA and state regulatory agencies to protect public health and environment.*

cover all bases in the ultimate goal of protecting public health and environmental quality.

Stream standards are discussed in this section, and drinking water standards are discussed in Section 6.1. Effluent standards are covered in Section 10.1.

### Stream Standards

Beginning with the *Water Quality Act* of 1965, individual states were required by federal law to classify surface waters on the basis of their "maximum beneficial use." The states also had to establish specific criteria for limiting the amount of pollutants allowed in the different classifications of surface water. These classifications and criteria are generally referred to as *stream standards*, although they have been established for lakes and coastal waters as well.

This law took into account the fact that the water quality of many streams and lakes had already deteriorated enough to prevent certain uses, but to allow others. Classification on the basis of maximum beneficial use is intended to prevent even further deterioration of the water from excessive sewage discharges. Stream classifications can be upgraded by the states as progress is made in cleaning up the water, but they cannot be lowered. In this manner, existing low-quality and high-quality waters were identified, and pollution control efforts could be focused where the maximum benefit could be obtained.

Specific classifications and criteria vary among the different states, but in general, four categories or classes of surface waters are commonly identified, as follows:

| Classification | Description |
| --- | --- |
| A | Water suitable for primary contact recreation (that is, swimming and the like) |
| B | Water suitable for the maintenance and propagation of fish, shellfish, and wildlife, and for secondary contact recreation (boating and fishing) |
| C | Water suitable for public water supply after treatment and purification |
| D | Water suitable for agricultural or industrial use |

The actual water quality criteria for the different use classifications usually include allowable limits on dissolved oxygen, coliforms, solids, turbidity, pH, and toxic wastes. For example, a minimum DO of 5 mg/L is typically set for the maintenance of fish life, but a minimum DO of only 3 mg/L may be required for class D waters just to maintain aerobic conditions. Streams to be protected as trout spawning streams, however, have minimum DO levels set at about 8 mg/L.

State standards for maximum coliform levels in class A waters may vary somewhat from the EPA recommended level of 200 fecal coliforms per 100 mL. In surface water intended for public supply, the permissible level of fecal coliforms may be as high as 2000 per 100 mL, 10 times the level for bathing waters.

At first glance, this may seem to be a mistake, but it must be kept in mind that these stream standards are for the raw or untreated water. Surface water must be treated before consumption, and modern water treatment plants can easily reduce coliform levels to an average of less than 1 coliform per 100 mL, even if the water initially contained 2000 per 100 mL.

On the other hand, swimmers do not normally swallow significant quantities of water, and even at a level of 200 coliforms per 100 mL, the probability of disease transmission is very low. Moderate levels of coliforms, then, are tolerable in primary contact recreational waters. Of course, in waters intended for public drinking supplies, it would still be best to have a source with coliform levels as low as possible so as not to overburden the treatment processes.

## Clean Water Act

The *Water Quality Act* of 1965 was amended and strengthened by the *Federal Water Pollution Control Act Amendments* (FWPCA) of 1972, which began a national permit system for all point source dischargers. This shifted the focus from stream standards, which regulate the amount of pollutants in receiving water, to effluent standards, which limit the amount of pollutants in separate discharges from point sources such as sewage treatment plants (for more on this, see Section 10.1).

The *Clean Water Act* (CWA), which now provides the national framework for water pollution control, is a 1977 amendment to the FWPCA. The CWA added control of priority toxic pollutants to the federal program. In 1987, the CWA was reauthorized. Among the provisions of the reauthorization was the requirement for the EPA to develop regulations for control of stormwater runoff and the requirement for states to prepare nonpoint-source management programs. The CWA allows for the delegation by the EPA of many permitting, administrative, and enforcement aspects of the law to state governments. In states with the authority to implement CWA programs, the EPA still retains oversight responsibilities.

A great deal of progress in cleaning up the nation's waters has been made over the last three decades under the provisions of the CWA and its predecessors. In 1972, only 30 to 40 percent of assessed waters met water quality goals such as being safe for fishing and swimming. Today, 60 to 70 percent of assessed waters meet state water quality goals. In 1972, sewage treatment plants served about 85 million people; today, less than 1 percent of the U.S. population discharges untreated wastewater. And compliance with national standards for discharges from industrial plants results in the removal of millions of kilograms of pollutants from wastewater each year.

A primary reason for the dramatic progress in reducing water pollution is the remarkable improvement in the treatment of municipal wastewater. More than $100 billion has been invested in wastewater treatment facilities since 1972. In addition to providing funding, the CWA implemented uniform sewage treatment standards (secondary treatment, see Section 10.3) for all sewage treatment systems across the country. This provision of a single sewage treatment goal helped overcome extended debates over treatment levels and became a basis for the successful construction of almost 14,000 municipal sewage treatment facilities.

## 5.9 CLEAN WATER ACTION PLAN

Under the CWA, states are required to submit a water quality report every 2 years to the EPA. The EPA must analyze the state reports and send a summary to Congress. That document, called the National Water Quality Inventory, contains information about water quality conditions in U.S. rivers, lakes, estuaries, wetlands, and groundwater. Water quality is evaluated by determining if the water is clean enough for the basic uses, such as swimming, drinking supply, and fishing. These uses are part of water quality standards set by each state (as described in the previous section.)

Despite the significant progress in reducing pollution since the early 1970s, serious water quality problems persist throughout the United States. According to the inventory data (as of 1996), about 40 percent of the nation's rivers, lakes, and estuaries remain too polluted for the basic uses of fishing or swimming. In rivers and streams, impairment of aquatic life is the most serious problem. Siltation is reported to be one of the leading causes of that problem. Silt (or sediment) clouds the fresh water, suffocates fish eggs and bottom-dwelling organisms, and interferes with water treatment processes. In the long run, siltation can profoundly alter aquatic life habitats in streams and rivers.

Agriculture is the most widespread source of pollutants entering streams and rivers: It is estimated to be responsible for about 70% of surface water pollution. Farming is a source of the silt, as well as nutrients, pesticides, and organic matter. (Livestock feeding operations, for example, confine thousands of animals in a small space and produce huge amounts of waste.) Municipal sewage treatment plants also continue to pollute rivers and streams by discharging bacteria, nutrients, and organic matter that deplete oxygen.

Fish consumption is the most frequently impaired use in ponds, lakes, and reservoirs; aquatic life and swimming are also harmed. Nutrients and metals are the most significant pollutants in these bodies of water, followed by siltation and organic matter. Lakes are most sensitive to excessive nutrients and other pollutants, because they retain their contents for long periods of time. Nutrient pollution leads to algal blooms, low DO levels, fish kills, foul odors, and aquatic weed overgrowth, which interfere with recreational activities. Farm fertilizers and manure from animals are a major source of nutrients.

In estuaries (bodies of coastal water where rivers meet the oceans), nutrients are the most widespread pollutants, followed by bacteria, toxic chemicals, and organic material. Excessive nutrients in estuaries lead to many of the same problems seen in lakes. High bacteria levels make the water unfit for swimming or harvesting shellfish, such as oysters and clams, which accumulate the bacteria. Storm runoff and industrial point sources are the primary sources of pollutants impairing estuaries, followed by municipal sewage and agriculture. The EPA now participates in a National Estuary Program with the states to manage estuarine water quality.

Loss of wetlands (see page 5) has decreased from as much as half a million acres per year between 1950 and 1970 to less than 100,000 acres per year between 1982 and 1992. Wetland losses result from commercial and residential development, agriculture, filling and draining, and road construction. Many activities degrade the remaining wetlands with pollutants such as sediment and nutrients. Although loss rates are decreasing, progress is still needed to stop the losses and increase the quality of the nation's wetlands.

As of 1996, leaking underground storage tanks (USTs) were reported to be the most widespread cause of pollution in groundwater. More than 300,000 confirmed leaks were reported up to that time (see Section 12.3 for more about USTs). Landfill leachate and flow from septic systems are also significant sources of groundwater contaminants. To protect the nation's groundwater resources, states are developing individual comprehensive state ground water protection programs.

## Watershed Management Approach

For more than two decades, the CWA has helped to steadily reduce water pollution caused by point discharges from municipal and industrial wastewater treatment plants. Now, the predominant source of the remaining water pollution is nonpoint runoff from urban and agricultural lands. The key to controlling this is the *watershed management approach*. This is a strategy that integrates water quality control activities within well-defined hydrologic drainage basins, rather than by politically defined boundaries. It is the foundation for the *Clean Water Action Plan: Restoring and Protecting America's Waters*, which was initiated in 1998.

Development and implementation of plans to restore water quality on a watershed basis should result in a significant reduction of polluted runoff. Focusing on the whole watershed balances efforts to control polluted runoff as well as point source pollution and protect drinking water sources and sensitive natural resources such as wetlands. A watershed focus also helps implement the most cost-effective pollution control strategies for meeting clean water goals.

### Index of Watershed Indicators

To describe water quality on a watershed scale, the EPA and other federal agencies have produced the *index of watershed indicators* (IWI). The IWI organizes information on several indicators of watershed health and uses them to assess the condition of the aquatic system in each watershed in the nation. The indicators include most conventional water quality data (see Chapter 4) as well as data on sediment contamination, fish consumption advisories, wetlands loss rates, soil loss, and other environmental conditions.

In 1997 the EPA released the first IWI report. At that time, 16 percent of U.S. watersheds were reported to have good water quality, 36 percent had moderate water quality problems, 21 percent had serious water quality problems, and 27 percent of the watersheds lacked sufficient information to make an overall assessment. (There are 2149 watersheds in the continental United States; see Section 3.4 for more about watersheds.)

## Total Maximum Daily Load Program

A *total maximum daily load* (TMDL) specifies the maximum amount of a specific pollutant that a water body can receive and still meet water quality standards; TMDLs also allocate the pollutant loads among point and nonpoint sources. Under the CWA, states are required to develop a list of water bodies that do not attain and maintain current water quality standards, to establish priority rankings for the list, and to calculate TMDLs for each listed water body. The list of impaired waters must be updated every 2 years. Although TMDLs have been required by the CWA since 1972, the program was not fully implemented. In an effort to speed progress toward achieving water quality standards and improving the TMDL program, the EPA published a new revised and clarified TMDL rule in July 2000 (detailed information about the new TMDL rule can be found at http://www.epa.gov/owow/tmdl/index.html).

 **5.10 RELEVANT WEB SITES**

### CLEAN LAKES PROGRAM

http://www.epa.gov/OWOW/lakescllkspgm.html

This site provides online publications and information on the water quality of U.S. lakes and the EPA Clean Lake Program, including case studies.

### CLEAN WATER ACT (SDWA)

http://water.usgs.gov/public/eap/env_guide/h2o_quality. html#HDR2

This site presents the major provisions of the *Clean Water Act* (CWA).

### DELAWARE NONPOINT SOURCE MANAGEMENT PLAN

http://www.dnrec.state.de.us/dnrec2000/Library/NPS/ NPSPlan.pdf

This site presents a report on Delaware's plan to control and manage nonpoint sources of pollution, including agriculture, construction, land disposal of waste, and urban runoff.

### GROUNDWATER PROTECTION COUNCIL (GWPC)

http://gwpc.site.net/

The GWPC is a national organization of state and federal officials promoting the safest methods and most effective regulations regarding comprehensive groundwater protection and underground injection techniques.

### INDEX OF WATERSHED INDICATORS (IWI)

http://www.epa.gov/iwi

The IWI organizes and presents U.S. water pollution information on a watershed basis. This Web site also indicates what the major pollution sources are, whether discharge permit holders are in compliance, and how one watershed compares to others in terms of water quality.

### MASSACHUSETTS BAY TUNNEL OUTFALL

http://www.mwra.com/harbor/html/outfall_update.htm

This site offers a thorough description of the 15-km (25-ft)–long, 7-m (24-ft)–diameter sewage outfall tunnel for the Deer Island Treatment Plant, serving the greater Boston area. The tunnel can convey a flow from the plant of up to 1.27 billion gallons a day. The discharge is through a diffuser at the end of the tunnel, to assure maximum dispersion and dilution and to prevent ocean pollution.

### NATIONAL ESTUARY PROGRAM

http://www.epa.gov/owow/estuaries/nep.html

This site is the Home Page for the National Estuary Program, which was established to restore and protect estuaries along the coasts of the United States. The program focuses on improving water quality in an estuary and on maintaining the integrity of the whole system— its chemical, physical, and biological properties, as well as its economic, recreational, and esthetic values. This site provides information about estuaries and current estuary restoration projects.

### NATIONAL WETLANDS RESEARCH CENTER (NWRC)

http://www.nwrc.usgs.gov/

This site offers access to NWRC publications, fact sheets, and online library. A link to the "Education/ Training" page provides educational materials for teachers and students as well as resources for professionals.

### NATIONAL WETLANDS INVENTORY CENTER (NWI)

http://www.nwi.fws.gov

This is the Home Page of the NWI, a division of the U.S. Fish and Wildlife Service, providing information on the characteristics, extent, and status of wetlands in the United States. The NWI provides a digitized wetlands database and maps for almost all wetlands.

**NONPOINT SOURCE POLLUTION CONTROL PROGRAM**

http://www.epa.gov/OWOW/NPS/

This site provides information and news and publications about nonpoint sources (NPSs) of water pollution, *Clean Water Act* NPS regulations, and control efforts.

**NORTH AMERICAN LAKE MANAGEMENT SOCIETY (NALM)**

http://www.nalms.org

This is the Home Page of the NALM, whose mission is to forge partnerships among citizens, scientists, and professionals in order to foster management and protection of lakes and reservoirs. This site provides news, a glossary, and other information about lake ecosystems and watershed management.

**OFFICE OF WETLANDS, OCEANS, AND WATERSHEDS**

http://www.epa.gov/owowwtr1/oceans/

This EPA site focuses on coastal watersheds, coastal activities, regulatory programs, dredged material management, and other related information.

**GUIDE FOR TEACHING ABOUT COASTAL WETLANDS**

http://www.nwrc.usgs.gov/fringe/ff_index.html

Produced by the National Wetlands Research Center, this site provides a framework for teaching about coastal wetlands, and includes a glossary and extensive references and activity resources.

**WETLANDS HOMEPAGE**

http://www.epa.gov/owow/wetlands/

This site provides information about monitoring wetlands and assessing water quality, laws and regulations, wetland protection, and more.

**WATERSHED ASSESSMENT, TRACKING AND ENVIRONMENTAL RESULTS (WATERS)**

http://www.epa.gov/waters

This site collects water quality information previously available only on individual state agency home pages and at several EPA and USGS Web sites. WATERS can be used to quickly identify the status of individual water bodies. It can also be used to generate summary reports on waters of a state that are impaired (not attaining water quality standards).

# REVIEW QUESTIONS

1. Give a brief definition of *water pollution*.

2. What is the difference between a point source and a dispersed source of pollutants? Give an example of each.

3. List nine different types or groups of water pollutants. Indicate a primary source of each type.

4. Briefly discuss the effects of thermal pollution.

5. How can thermal pollution be controlled or eliminated?

6. Why is suspended silt or clay considered to be a water pollutant? How does it get into streams and lakes?

7. Briefly describe eight soil erosion control methods.

8. What are two important factors affecting stream self-purification?

9. Briefly describe what a dissolved oxygen profile is. Make a sketch of a DO profile for a small, slow-moving stream receiving a raw sewage discharge.

10. Briefly describe the four zones of stream pollution.

11. Why is it important to be able to compute the minimum DO in a stream or river?

12. What is one of the basic differences between lake pollution and stream pollution?

13. Is the eutrophication of a lake a pollution problem? Why?

14. What are some methods for controlling cultural eutrophication of lakes or reservoirs?

15. Briefly explain the occurrence of thermal stratification and seasonal turnover in a lake. How does this affect the quality of the lake water?

16. Is groundwater naturally pure? Why?

17. Do you think that the contamination of an important groundwater aquifer is a serious problem? Why?

18. Briefly discuss seven different sources of groundwater contamination.

19. What measures can be taken to prevent groundwater pollution?

20. Are ocean waters immune to pollution problems? Briefly discuss your answer.

21. What is an *estuary*? Why are estuaries important ecosystems?

22. Briefly discuss the role of multiport diffusers in ocean disposal of sewage.

23. Briefly discuss ocean dumping and oil spills with regard to ocean pollution.

24. What is the function of stream classification standards? Briefly discuss four common classifications of streams.

25. What is the most widespread source of stream and river pollution in the United States? What is the most widespread source of groundwater pollution?

26. Briefly describe the main strategy of the Clean Water Action Plan.

27. Visit the *Index of Watershed Indicators* Web page. Determine the IWI for the watershed in which your home or school is located.

28. Visit and explore the EPA *Clean Lakes Program* Web page. Write a brief report about the program; describe the relationship between the Clean Lakes Program and the Nonpoint Source Management Program.

29. Visit and explore the EPA *National Estuary Program* Web site. Read the "About Estuaries" page and write a summary of that page. Link to the NEP Web page for details about the Galveston Bay Estuary Program and briefly describe that program. Do the same for one other estuary program of your choice that is on the NEP list.

30. Visit and explore the EPA *Office of Wetlands, Oceans, and Watersheds* Web site. What are the consequences of losing or degrading wetlands?

31. Using the EnviroSources search engine (see Section 1.6) or another Internet search directory or search engine, locate at least one additional Web site relevant to one of the topics in this chapter. Add the link(s) to your Environmental Technology Folder. Write a brief description of what the Web site(s) contain.

## PRACTICE PROBLEMS

1. An effluent from a sewage treatment plant has a TDS concentration of 500 mg/L and a flow rate of 1.5 mgd. The receiving stream has a TDS level of 100 mg/L and a discharge of 6 mgd. Compute the TDS concentration in the combined sewage and streamflow downstream of the mixing zone. Assume that the sewage is completely mixed in the stream water.

2. The $BOD_5$ of an effluent from a poorly operating sewage treatment plant is 100 mg/L, and the discharge is 1.5 ML/d. The receiving stream has a $BOD_5$ of 3 mg/L. What minimum streamflow is needed for a dilution such that the combined $BOD_5$ of the sewage and stream water is no greater than 10 mg/L?

3. The combined $BOD_L$ in a stream mixed with sewage effluent is 25 mg/L. The stream reaeration rate is found to be 0.4/d and the deoxygenation constant is assumed to be 0.1/d. The initial combined DO in the stream is 8.0 mg/L and the DO saturation level is 11 mg/L. Compute the minimum DO level in the stream due to the sewage discharge.

4. A sewage treatment plant discharges 4 ML/d of effluent with 28 mg/L of $BOD_5$ into a stream. The stream discharge is 16 ML/d and its initial $BOD_5$ is 6 mg/L. The initial DO in the effluent is 2 mg/L and in the stream it is 7 mg/L. Compute the minimum DO level in the stream, assuming that $k_1 = 0.1$/d and $k_2 = 0.3$/d. Assume that the saturation DO level is 10.0 mg/L. If the velocity of streamflow is 0.1 m/s, how far downstream does the minimum DO occur?

5. An effluent from a sewage treatment plant has a DO level of 3 mg/L and a flow rate of 6 ML/d. The receiving stream has a DO level of 10 mg/L and a discharge of 30 ML/d. Compute the DO concentration in the combined sewage and streamflow just downstream of the mixing zone.

6. A sewage treatment plant discharges 2 mgd of effluent with 30 mg/L of $BOD_5$ into a stream. The stream discharge is 10 mgd, and its initial $BOD_5$ is 5 mg/L. The initial DO in the effluent is 3 mg/L, and in the stream it is 9 mg/L. Compute the minimum DO level in the stream, assuming that $k_1 = 0.1$/d and $k_2 = 0.3$/d. Assume that the saturation DO level is 12.0 mg/L. If the velocity of streamflow is 20 ft/min, how far downstream does the minimum DO occur?

## Chapter Outline

C H A P T E R

# 6

# Drinking Water Purification

Water withdrawn directly from rivers, lakes, or reservoirs is rarely clean enough for human consumption if it is not treated to purify it. Even water pumped from underground aquifers often requires some degree of treatment to render it *potable*, that is, suitable for drinking.

The nature and extent of treatment required to prepare potable water from surface or subsurface sources depend on the quality of the raw (untreated) water.

Better-quality water needs less treatment. Generally, a source of raw water with a coliform count of up to 5000/100 mL and a turbidity of up to 10 units is considered good. Water with coliform counts that frequently exceed 20,000/100 mL and turbidities that exceed 250 units is considered a very poor source and requires expensive treatment to render it potable.

The primary objective of water purification is to remove harmful microorganisms or chemicals, thereby preventing the spread of disease and protecting public health. In addition to being safe to drink, the water must also be esthetically pleasing. It should be crystal clear, and it should not have any objectionable color, taste, or odor. Section 6.1 discusses the criteria and standards by which a public water supply is judged to be potable or not.

Common treatment processes that are used to prepare potable water are also discussed in this chapter. Generally, groundwater may require some degree of treatment. It is usually free of bacteria and suspended or colloidal particles because of the natural filtration that occurs as the water percolates through the soil. But because it is in direct contact with soil or rock, groundwater often contains dissolved minerals, such as calcium or iron.

As a minimum, most states require that public groundwater supplies be disinfected with chlorine to ensure the absence of pathogens. If dissolved minerals are present in excessive amounts, some combination of chemical treatment, aeration, filtration, and other processes may be needed to purify the water. Some groundwater supplies have recently been found to be contaminated with very low or trace amounts of toxic organic chemicals, usually from improper land disposal of hazardous wastes. If purification using aeration, activated carbon, or other processes is not feasible, the contaminated wells may have to be abandoned.

Surface water supplies generally require more extensive treatment than groundwater supplies because most streams, rivers, and lakes are contaminated to some extent with domestic sewage and runoff. Even in areas far removed from human activity, surface water contains suspended soil particles (silt and clay) and organics and bacteria (from decaying vegetation and animal wastes). An aerial view of a modern water treatment plant required to purify surface water is shown in Figure 6.1. (A flow diagram, showing the treatment processes used in the plant, is shown in Figure 6.22.)

The most common type of treatment for surface water includes *clarification* and *disinfection*. Clarification is usually accomplished by a combination of *coagulation–flocculation, sedimentation,* and *filtration*; the most common method for disinfection in the United States is *chlorination*. A typical flow diagram that shows the sequence of the individual treatment steps, or *unit processes*, is shown in Figure 6.2. These and other unit processes are discussed later in the chapter.

**FIGURE 6.1**
*Aerial view of a modern water treatment plant. This facility uses ozone for clarification and disinfection.* (Courtesy of United Water Resources, Harrington Park, New Jersey)

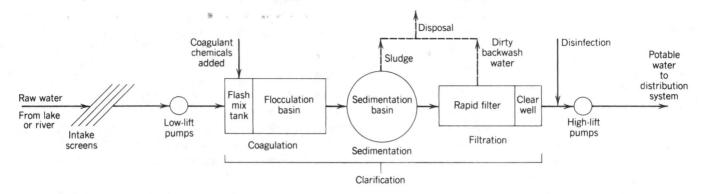

**FIGURE 6.2**
*A flow diagram of a typical surface water treatment plant. Screens keep fish
and debris out of the plant; low-pressure pumps lift the water to the flash-mix
tank; coagulation, sedimentation, and filtration remove turbidity and clarify
the water; disinfection destroys pathogenic organisms; high-pressure pumps
deliver potable water to the consumers.*

## 6.1  SAFE DRINKING WATER ACT

The *Safe Drinking Water Act* (SDWA), enacted by
Congress in 1974 and amended several times since then,
establishes minimum drinking water standards in the
United States. While *stream standards* (discussed in
Section 5.8) serve to protect surface water quality, the
SDWA standards are for water that people actually consume, and thereby serve directly to protect public
health and welfare. SDWA standards ensure that drinking water supplied to the public is safe and wholesome
by setting limits on the amounts of various substances
sometimes found in the water supply.

The SDWA applies to *public water systems*, defined as having 15 or more service connections or serving 25 or more people each day, at least 60 d per year.
A system can be owned by a private company and still
be classified as a public system if it meets this definition. Most states have been delegated the authority for
making sure that the SDWA standards are met; some
states in turn have delegated their authority to county
health departments, which routinely keep track of water quality testing results, conduct inspections, and take
enforcement actions when necessary. The EPA provides
guidance and technical assistance to the states, conducts research, and periodically revises the standards.

A major revision of the SDWA was enacted in 1996.
In addition to authorizing billions of dollars of expenditures for drinking water systems, it focuses water
program spending on contaminants that pose the greatest risk to human health and that are most likely to be
present in a public water system. The amended Act requires stricter controls on microbial contaminants as
well as on the by-products of chlorination. Health risk

reduction analyses must now include cost/benefit considerations. Additional revisions include the requirement that water utilities notify the public of water safety
violations within 24 h, that all water system operators
be certified to meet the EPA's minimum certification
standards, and that the EPA establish a database to monitor the presence of unregulated contaminants in water.

## Drinking Water Standards

Based on the results of public health research and scientific judgment, the EPA has established two types of
drinking water standards: *primary* and *secondary*.
Primary standards are designed to protect public health
by setting maximum permissible levels of potentially
harmful substances in the water. Secondary standards
are guidelines that apply to the esthetic aspects of
drinking water, which do not pose a health risk (for
example, color and odor). Primary standards are
enforceable by law; secondary standards are not.

Most primary standards are specified as *maximum contaminant levels*, or MCLs; these are the
enforceable limits. Primary standards may also be
specified as *treatment technique* (TT) requirements,
which are set for those contaminants that are difficult
or costly to measure; specific treatment processes (for
example, filtration or corrosion control) may be required in lieu of an MCL to remove those contaminants. The EPA has also issued *maximum contaminant level goals* (MCLGs). An MCLG is a level of a
contaminant not expected to cause any adverse health
effects; it is a goal, not an enforceable standard. MCLs,
which are revised periodically, are set as close to
MCLGs as current technology and economics allow.

MCLGs are set at zero for carcinogenic chemicals because there are no known safe levels for them.

## Primary MCLs

MCLs for potentially toxic or harmful substances reflect levels that can be safely consumed in water, taking into account exposure to substances from other sources. They are based on consumption of 2 L (roughly 2 quarts) of water-based fluids every day for a lifetime. The states can establish MCLs that are more stringent than those set by the EPA.

Categories of primary contaminants include *organic chemicals, inorganic chemicals, microorganisms, turbidity,* and *radionuclides.* Except for some microorganisms and nitrate, water that exceeds the listed MCLs pose no *immediate* threat to public health. However, all these substances must be controlled because drinking water that exceeds the standards over long periods of time may be harmful.

*Organic Chemicals.* Many synthetic organic chemicals (SOCs) are included in the primary regulations. Some of them (like benzene and carbon tetrachloride) readily become airborne and are known as *volatile organic compounds* (VOCs). Table 6.1 shows a partial list of maximum allowable levels for several selected organic contaminants. As more is learned from research about the health effects of various contaminants, the number of regulated organics is likely to grow. Public drinking water supplies must be sampled and analyzed for organic chemicals at least every 3 years.

It is seen from Table 6.1 that extremely small concentrations can have public health significance. Levels are expressed in terms of mg/L; 1 mg/L is equivalent to one part per million. The MCL for the insecticide *lindane,* for example, is 0.0002 mg/L; this value can also be expressed as 0.2 $\mu$g/L (micrograms per liter) and is equivalent to 0.2 parts per billion. [One part per billion is roughly proportional to the first 0.4 m (about 1.3 ft) of a trip to the moon.]

**TABLE 6.1**
**Selected primary standard MCLs and MCLGs for organic chemicals**

| Contaminant | Health effect | MCL (mg/L) | Typical source | MCLG |
|---|---|---|---|---|
| Aldicarb | Nervous system effects | 0.003 | Insecticide | 0.001 |
| Benzene | Possible cancer | 0.005 | Industrial chemicals, pesticides, paints, plastics | Zero |
| Carbon tetrachloride | Possible cancer | 0.005 | Cleaning agents, industrial wastes | Zero |
| Chlordane | Possible cancer | 0.002 | Insecticide | Zero |
| Endrin | Nervous system, liver, kidney effects | 0.002 | Insecticide | 0.002 |
| Heptachlor | Possible cancer | 0.0004 | Insecticide | Zero |
| Lindane | Nervous system, liver, kidney effects | 0.0002 | Insecticide | 0.0002 |
| Pentachlorophenol | Possible cancer, liver, kidney effects | 0.001 | Wood preservative | Zero |
| Styrene | Liver, nervous system effects | 0.1 | Plastics, rubber, drug industry | 0.1 |
| Toluene | Kidney, nervous system, liver, circulatory effects | 1 | Industrial solvent, gasoline additive, chemical manufacturing | 1 |
| Total trihalomethanes (TTHM) | Possible cancer risk | 0.1 | Chloroform, drinking water chlorination by-product | Zero |
| Trichloroethylene (TCE) | Possible cancer | 0.005 | Waste from disposal of dry cleaning materials and manufacture of pesticides, paints, waxes; metal degreaser | Zero |
| Vinyl chloride | Possible cancer | 0.002 | May leach from PVC pipe | Zero |
| Xylene | Liver, kidney, nervous system effects | 10 | Gasoline refining by-product, paint, ink, detergent | 10 |

*Source:* Environmental Protection Agency (for a complete, up-to-date list, see *http://www.epa.gov/OGWDW/mcl.html*).

*Inorganic Chemicals.* Several inorganic substances (that is, containing no carbon), particularly heavy metals, are of public health importance. Some of these inorganics are listed in Table 6.2. Treated water is sampled and tested for inorganics at least once per year in public supplies. For most inorganics, MCLs are the same as the MCLGs, but the MCLG for lead is zero; the use of lead pipe and lead solder of flux for installation or repair of public water systems is no longer allowed in the United States.

Arsenic, a well-known poison, can contaminate drinking water supplies naturally if the raw water has been in contact with certain rocks and minerals; arsenic can also enter water sources from industrial and mining activities. It is found at higher levels in groundwater than in surface waters, such as lakes and rivers. The 0.05-mg/L (50-ppb) MCL was set by the EPA in 1975, based on a U.S. Public Health Service standard originally set in 1942. In 1999, the National Academy of Sciences completed a review of updated scientific data on arsenic and recommended that the EPA lower the MCL as soon as possible. Early in 2001, the EPA

established a new standard of 0.01 mg/L (10 ppb), by 2006, all water utilities will have to comply with the new standard (10 ppb of arsenic is equivalent to one teaspoon per 5 ML, or 1.3 mil gal, of water).

Nitrate levels above 10 mg/L pose an immediate threat to children under 1 year of age. Excessive levels of nitrate can react with hemoglobin in blood to produce an anemic condition known as *blue baby*. Nitrates can enter water supplies naturally from soil and mineral deposits as well as from fertilizers and sewage pollution. The sources and health effects of other inorganic drinking water contaminants are summarized in Table 6.2.

*Lead and Copper Rule.* Treatment techniques have been set for lead and copper because the occurrence of these chemicals in drinking water usually results from corrosion of plumbing materials. All systems that do not meet the *action level* at the tap are required to improve corrosion control treatment to reduce the levels. The action level for lead is 0.015 mg/L and for copper is 1.3 mg/L.

**TABLE 6.2**
**Selected primary standard MCLs for inorganic chemicals**

| Contaminant | Health effect | MCL (mg/L) | Typical source |
|---|---|---|---|
| Arsenic | Nervous system effects | 0.05 | Geological, pesticide residues, industrial waste, smelter operations |
| Asbestos | Possible cancer | 7 MFL[a] | Natural mineral deposits, air conditioning pipe |
| Barium | Circulatory system effects | 2 | Natural mineral deposits, paint |
| Cadmium | Kidney effects | 0.005 | Natural mineral deposits, metal finishing |
| Chromium | Liver, kidney, digestive system effects | 0.1 | Natural mineral deposits, metal finishing, textile and leather industries |
| Copper | Digestive system effects | TT[b] | Corrosion of household plumbing, natural deposits, wood preservatives |
| Cyanide | Nervous system effects | 0.2 | Electroplating, steel, plastics, fertilizer |
| Fluoride | Dental fluorosis, skeletal effects | 4 | Geological deposits, drinking water additive, aluminum industries |
| Lead | Nervous system and kidney effects, toxic to infants | TT | Corrosion of lead service lines and fixtures |
| Mercury | Kidney, nervous system effects | 0.002 | Industrial manufacturing, fungicide, natural mineral deposits |
| Nickel | Heart, liver effects | 0.1 | Electroplating, batteries, metal alloys |
| Nitrate | Blue-baby effect | 10 | Fertilizers, sewage, soil and mineral deposits |
| Selenium | Liver effects | 0.05 | Natural deposits, mining, smelting |

*Source:* Environmental Protection Agency (for a complete, up-to-date list, see *http://www.epa.gov/OGWDW/mcl.html*).
[a]Million fibers per liter.
[b]Treatment technique.

*Microorganisms.* This group of contaminants includes bacteria, viruses, and protozoa. The *total coliform* group of bacteria is used to indicate the possible presence of pathogenic organisms (see Section 4.4). In testing for total coliforms, the number of monthly samples required is based on the population served and the size of the distribution system. The SDWA standards now require that coliforms not be found in more than 5 percent of the samples examined during a 1-month period. This is now known as the *presence/absence concept;* it replaces previous MCLs based on the number of coliforms detected in the sample. All coliform-positive samples have to be further tested for *fecal coliforms* (or *E. coli*), the presence of which is strong evidence of recent sewage contamination and indicates an urgent public health risk.

*Legionella* (which causes an upper respiratory disease), intestinal viruses, and *Giardia lamblia* (a protozoan cyst that causes intestinal illness) are also regulated under the SDWA, using treatment technique requirements. The *Surface Water Treatment Rule* (SWTR) requires that all public systems using surface water properly filter the water unless they can meet certain strict criteria. These systems must also disinfect the water, without exception, to kill disease-causing microorganisms.

The SWTR requires source water reductions of 99.9 percent for *Giardia* and virus concentrations, in lieu of setting MCLs. The MCLGs, though, are zero because ingestion of even very low numbers of those organisms can cause illness. Microbial standards will be improved and strengthened to control *Cryptosporidium* and other pathogens after the *Enhanced Surface Water Treatment Rule* is fully implemented, as required by the 1996 amendments to the SDWA.

*Turbidity.* The presence of suspended particles in the water is measured in *nephelometric turbidity units* (NTUs); NTUs measure the amount of light scattered or reflected from the water. Turbidity testing is not required for groundwater sources.

Turbidity affects more than just the appearance of water; it can be a health hazard in drinking water and is therefore controlled as a primary contaminant. Turbidity interferes with disinfection by shielding microorganisms. MCLs for turbidity depend on the type of treatment used to clarify the water. Conventional and direct filtration systems must be monitored at least every 4 h and have turbidity levels less than 0.5 NTU. (Beginning in 2002, the EPA will require continuous monitoring of turbidity for surface water systems serving 10,000 people or more.)

*Radionuclides.* Water can be contaminated with substances from nuclear facilities and radioactive wastes or from natural radioactive materials. The radioactive gas radon-222, for example, occurs in certain types of rock and can get into groundwater. People can be exposed to radon in water by drinking it, showering in it, or using it to wash dishes. (The primary source of exposure to radon in the home is radon seeping out of the soil and into the basement air; see Section 13.5.) Tests for radioactivity, which can cause a cancer risk, are required at least every 4 years. The MCL for radon in water is 300 pCi/L (picocuries per liter) and the MCL for radium is 20 pCi/L. Limits are also set for emitters of beta particles, photons, and alpha particles.

**Secondary MCLs**

Under the *Secondary Drinking Water Standards*, a range of concentrations is established for substances that affect only the esthetic qualities of drinking water (for example, taste, odor, and color), but have no direct effect on public health. Secondary standards are presented in Table 6.3. These standards are guidelines or suggestions related to the general acceptability of the water to consumers. States may adopt their own enforceable regulations governing these substances (most of which are discussed in Sections 4.2 and 4.3).

## Sampling Procedures

Sampling frequency requirements vary for each contaminant group as well as for individual contaminants within each group. They also depend on the population served and whether surface water or groundwater is used. The sampling frequencies can range from once every 4 h (for turbidity) to once every 9 years (for asbestos). Detection of a contaminant above a certain level sometimes triggers increased sampling requirements, even if the MCL is not exceeded.

Most samples must be collected at points representative of water quality throughout the distribution system. Generally, drinking water samples are *fully flushed.* This means that the water has run for a long enough time to represent water in the main line rather than in the household plumbing. The exception to this is monitoring for copper and lead, for which a *first draw* sample is required at the consumer's taps, where contamination is more likely to occur.

Some samples must be collected in glass containers; others must be collected in plastic. Sample volumes vary for each contaminant, ranging from 100 mL for a coliform sample to 1 L for some radionuclide samples. Certain samples must be kept cold for preservation; others can be delivered to the lab at ambient temperature. Sampling bottles for VOCs must be filled to the top with no air space. The maximum allowable time

**TABLE 6.3**
**National secondary drinking water standards**

| Contaminant or adverse effect | Suggested level | Contaminant effect |
|---|---|---|
| Aluminum | 0.05–0.2 mg/L | Discoloration of water |
| Chloride | 250 mg/L | Salty taste; corrosion of pipes |
| Color | 15 color units | Visible tint |
| Copper | 1.0 mg/L | Metallic taste; blue-green staining of porcelain |
| Corrosivity | Noncorrosive | Metallic taste; fixture staining, corroded pipes (corrosive water can leach pipe materials, such as lead, into drinking water) |
| Fluoride | 2.0 mg/L | Dental fluorosis (a brownish discoloration of the teeth) |
| Foaming agents | 0.5 mg/L | Esthetic: frothy, cloudy, bitter taste, odor |
| Iron | 0.3 mg/L | Bitter metallic taste; staining of laundry, rusty color, sediment |
| Manganese | 0.05 mg/L | Taste; staining of laundry, black-to-brown color, black staining |
| Odor | 3, threshold odor number | Rotten-egg, musty, or chemical smell |
| pH | 6.5–8.5 | Low pH: bitter metallic taste, corrosion<br>High pH: slippery feel, soda taste, deposits |
| Silver | 0.1 mg/L | Argyria (discoloration of skin), graying of eyes |
| Sulfate | 250 mg/L | Salty taste; laxative effects |
| Total dissolved solids | 500 mg/L | Taste and possible relation between low hardness and cardiovascular disease; also an indicator of corrosivity (related to lead levels in water); can damage plumbing and limit effectiveness of soaps and detergents |
| Zinc | 5 mg/L | Metallic taste |

*Source:* Environmental Protection Agency (for a complete, up-to-date list, see *http://www.epa.gov/ogwbw/mcl.html*).

between sample collection and analysis in the laboratory can range from 1 d for coliforms to 1 year for a radionuclide sample. Details of all these requirements can be obtained from the laboratory doing the testing.

## Record Keeping and Reporting

Good record-keeping procedures are important for proper operation of a public water system. They also provide data for future planning, public information, and legal protection. Under the requirements of the SDWA, each public system must maintain records of water quality test results, reports, and actions taken to correct deficiencies; depending on the type of record, it may have to be kept on file for up to 12 years.

Records of MCL analysis data for bacteria and chemicals are particularly important. Records should include the name, date, and place of sampling; the name of the technician who took the sample; the type of sample; and the place, date, method, and results of analysis.

### Public Notification

The SDWA requires that public water systems submit reports to consumers as well as to an appropriate local regulatory agency. Results of routine sampling and testing must be sent to the agency every month. Also, the state must be notified within 48 h of a violation of any of the primary regulations MCLs. The requirement for routine sample and violation reports helps to ensure that water system deficiencies and potential health hazards will be identified and corrected.

Public notification is required to advise consumers of the potential health hazards and to educate them about the importance of adequate financing and support for drinking water systems. The public is notified only of confirmed MCL violations by mail, newspaper, and radio and television broadcasts (for acute health risks). Timely public notice must describe the nature of the problem and include any steps people should take to protect their health. An illustration of a notice for violation of the nitrate MCL is shown in Figure 6.3.

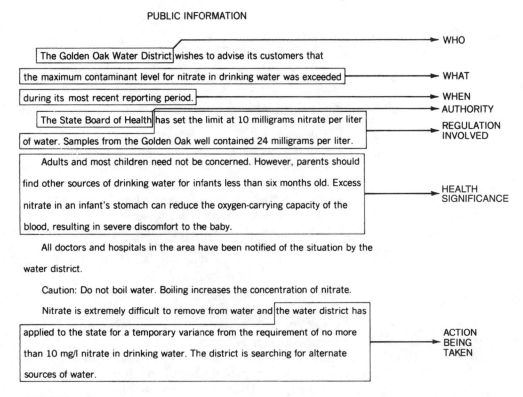

PUBLIC INFORMATION

The Golden Oak Water District wishes to advise its customers that → WHO

the maximum contaminant level for nitrate in drinking water was exceeded → WHAT

during its most recent reporting period. → WHEN
→ AUTHORITY

The State Board of Health has set the limit at 10 milligrams nitrate per liter — REGULATION INVOLVED

of water. Samples from the Golden Oak well contained 24 milligrams per liter.

Adults and most children need not be concerned. However, parents should
find other sources of drinking water for infants less than six months old. Excess — HEALTH SIGNIFICANCE
nitrate in an infant's stomach can reduce the oxygen-carrying capacity of the
blood, resulting in severe discomfort to the baby.

All doctors and hospitals in the area have been notified of the situation by the
water district.

Caution: Do not boil water. Boiling increases the concentration of nitrate.

Nitrate is extremely difficult to remove from water and the water district has
applied to the state for a temporary variance from the requirement of no more — ACTION BEING TAKEN
than 10 mg/l nitrate in drinking water. The district is searching for alternate
sources of water.

**FIGURE 6.3**
*A typical MCL violation notice that would appear in local newspapers.*
(Reprinted from *Safe Drinking Water Act Self-Study Handbook,* by permission.
Copyright © 1978, The American Water Works Association)

## 6.2  SEDIMENTATION

The impurities in water may be either dissolved or suspended. The easiest way to remove the suspended material is to let the force of gravity do the work. Under *quiescent* conditions, when flow velocities and turbulence are minimal, particles that are denser than water will be able to settle to the bottom of a tank. This process is called *sedimentation*, and the layer of accumulated solids at the bottom of the tank is called *sludge*. The tank may be called a *sedimentation tank*, a *settling tank*, or a *clarifier*.

The speed at which suspended particles settle toward the bottom of a tank depends on their size as well as on their density. The larger and heavier particles will naturally settle faster than smaller or lighter particles. The forces opposing the downward force of gravity include buoyancy and friction (drag). The temperature and viscosity of the water are additional factors that affect the particle-settling rate.

The nature of the sedimentation process also varies with the concentration of suspended particles and their tendency to interact with one another. In a dilute suspension, where the particles are free to settle without interference, the process is called free settling or *discrete settling*. As the concentration increases, the particles tend to interact and interfere with the free movement of one another; this is sometimes called *hindered settling*. In a sedimentation tank there may be up to four different zones or types of settling that occur at different depths, and exact mathematical analysis of the process can be quite complicated. This section discusses some common factors related to discrete particle settling.

### Detention Time

If a volume of water is left completely undisturbed in a tank for several days or weeks, just about all of the suspended solids have a chance to settle to the bottom. Even some bacteria, microscopic in size, eventually settle out. But this procedure is not practical for municipal water treatment plants because they generally handle large volumes of water on a continuous-flow basis. It is not feasible to simply shut a valve to stop the flow and let a fixed volume of water remain undisturbed in a tank for a long period of time. Too many very large tanks would be needed.

Instead, settling tanks for water (or wastewater) treatment are designed to operate as the flow slowly continues from the inlet to the outlet of the tank. The movement of water is slow enough to allow quiescent settling for a large percentage of the suspended particles. Generally, the water remains in the tank for only a few hours before it reaches the tank outlet. The theoretical amount of time water remains in a settling tank is called the *detention time*. It can be computed as follows:

$$T_D = \frac{V}{Q} \qquad \textbf{(6-1)}$$

where $T_D$ = detention time
$V$ = volume of water in tank
$Q$ = average flow rate (volume per unit time)

Detention time is usually expressed in terms of hours. It is important to use consistent units for $V$ and $Q$ so that Equation 6-1 will be dimensionally correct. Minimum detention times of 3 h are specified by most state health departments or environmental agencies to ensure sufficient settling; most of the settleable suspended solids will reach the sludge layer in this time period. The following examples illustrate the use of Equation 6-1.

## EXAMPLE 6.1

A sedimentation tank has a volume capacity of 15,000 $m^3$. If the average flow rate entering the tank is 120 ML/d, what is the detention time?

*Solution*  First convert cubic meters to megaliters, or vice versa, for dimensional consistency. Choosing to convert volume to megaliters gives

$$V = 15\,000 \text{ m}^3 \times 1000 \text{ L/m}^3$$

$$= 15\,000\,000 \text{ L} = 15 \text{ ML}$$

Now, applying Equation 6-1, we get

$$T_D = \frac{15 \text{ ML}}{120 \text{ ML/d}} = 0.125 \text{ d}$$

and converting to hours, we get

$$T_D = 0.125 \text{ d} \times 24 \text{ h/d} = 3 \text{ h}$$

## EXAMPLE 6.2

Water flowing at a rate of 6 mgd is to have a 3-h detention time in a sedimentation tank. Compute the required volume capacity of the tank in cubic feet. If the tank has a surface area of 10,000 $ft^2$, how deep will the water be in the tank?

*Solution*  Since it is necessary to solve for volume, first rearrange the terms of Equation 6-1 to get $V = T_D \times Q$. For dimensional consistency, convert the flow rate from units of million gallons per day to gallons per hour, as follows:

$$Q = 6,000,000 \text{ gal/d} \times 1 \text{ d/24 h} = 250,000 \text{ gal/h}$$

and

$$V = T_D \times Q = 3 \text{ h} \times 250,000 \text{ gal/h}$$

$$= 750,000 \text{ gal}$$

Converting gallons to cubic feet, we obtain

$$V = 750,000 \text{ gal} \times \frac{1 \text{ ft}^3}{7.5 \text{ gal}} = 100,000 \text{ ft}^3$$

Since the volume of a tank can be expressed as the product of its depth and surface area (volume = depth × area), the depth of water can be computed as volume ÷ area or 100,000 $ft^3$/10,000 $ft^2$ = 10 ft deep.

## Overflow Rate

Another factor or term that is of importance in the design and operation of a settling tank is the *overflow rate*, or *surface loading*, as it is called. It can be computed as follows:

$$V_o = \frac{Q}{A_s} \qquad \textbf{(6-2)}$$

where $V_o$ = overflow rate
$Q$ = average flow rate
$A_s$ = tank surface area (top view)

In SI metric units, the overflow rate is expressed in terms of cubic meters per square meter per day $(m^3/m^2 \cdot d)$; flow rate is expressed in terms of cubic meters per day $(m^3/d)$; and surface area is expressed in terms of square meters $(m^2)$. In customary units, $V_o$ is expressed in terms of gallons per day per square foot $(gpd/ft^2)$; $Q$ is expressed in terms of gallons per day

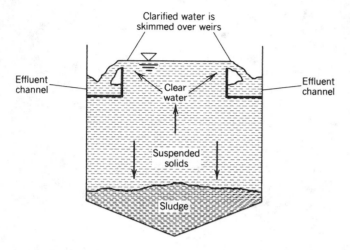

**FIGURE 6.4**
*Schematic view of the sedimentation process.*

(gpd); and $A_s$ is expressed in terms of square feet ($ft^2$). The maximum overflow rate typically allowed by state regulatory agencies is 33 $m^3/m^2 \cdot d$ or 800 gpd/ft$^2$.

Overflow rate can be visualized as an average *upflow* velocity of water in the settling tank. In fact, it can be easily seen that overflow rate is actually a velocity by canceling units in the SI metric system: $m^3/m^2 \cdot d = m/d$ (meters per day). The schematic cross section of a tank shown in Figure 6.4 illustrates the general flow pattern. The clarified water is skimmed from the surface as it flows over weirs into an effluent channel. All suspended particles that settle at a faster velocity than $V_o$ reach the sludge layer at the bottom of the tank. Only a fraction of the smaller and lighter particles that settle at velocities less than $V_o$ are removed from the water before it leaves the tank.

## EXAMPLE 6.3

What is the minimum settling velocity, in feet per hour, of suspended particles that can be completely removed in a settling tank that has an overflow rate of 700 gpd/ft$^2$?

*Solution*

$$\frac{700 \text{ gal}}{ft^2 \cdot d} \times \frac{1 \text{ ft}^3}{7.5 \text{ gal}} \times \frac{1 \text{ d}}{24 \text{ h}} = 3.9 \text{ ft/h}$$

Only a fraction of particles that settle slower than 3.9 ft/h will be removed; the slower the settling velocity, the smaller is the percentage removed. For example, only 20 percent of particles that settle at a velocity of $0.2 \times 3.9 = 0.78$ ft/h will be captured in the sludge layer.

## Settling Tank Design

By combining specified values of detention time and overflow rate, it is possible to determine the required dimensions of a settling tank. The tanks may be either rectangular or circular in shape. The actual depth of water in the tank is called the *side water depth*, or SWD. The height of the tank wall is usually about 0.45 m, or 1.5 ft, above the SWD. This is called *freeboard*, and it serves to prevent splashing of water over the tank sides. The following example illustrates how tank dimensions are determined.

## EXAMPLE 6.4

A circular sedimentation tank is to have a minimum detention time of 4 h and a maximum overflow rate of 20 $m^3/m^2 \cdot d$. Determine the required diameter of the tank and the SWD if the average flow rate through the tank is 6 ML/d.

*Solution*   Applying Equation 6-1, compute the required volume as

$$V = Q \times T_D = 6 \text{ ML/d} \times \frac{1 \text{ d}}{24 \text{ h}} \times 4 \text{ h}$$
$$= 1 \text{ ML} = 1\,000\,000 \text{ L}$$

(Note the factor 1/24 for dimensional consistency.) Converting the volume to cubic meters gives

$$V = 1\,000\,000 \text{ L} \times \frac{1 \text{ m}^3}{1000 \text{ L}} = 1000 \text{ m}^3$$

Before using Equation 6-2, it is convenient to first convert the flow rate to units of $m^3/d$ for dimensional consistency, as follows:

$$Q = 6 \text{ ML/d} \times 10^6 \text{ L/ML} \times \frac{1 \text{ m}^3}{10^3 \text{ L}} = 6000 \text{ m}^3/d$$

Now, applying Equation 6-2, compute the surface area as

$$A_s = \frac{Q}{V_o} = \frac{6000 \text{ m}^3/d}{20 \text{ m}^3/m^2 \cdot d} = 300 \text{ m}^2$$

To determine the tank diameter, use the formula for the area of a circle: $A = \pi D^2/4$. From this,

$$D = \sqrt{4A/\pi} = \sqrt{(4 \times 300)/\pi} \approx 20 \text{ m}$$

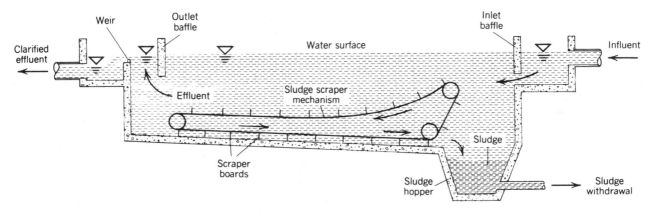

**FIGURE 6.5**
*Simplified section view of a rectangular sedimentation tank. A slowly moving endless-chain scraper mechanism pushes sludge into a hopper for removal.*

Finally, since $V = A_s \times$ SWD, rearranging terms gives

$$\text{SWD} = \frac{V}{A_s} = 1000 \text{ m}^3/300 \text{ m}^2 = 3.33 \text{ m}$$

In a rectangular settling tank, the *influent* (water flowing into the tank) is directed against a baffle that distributes the water uniformly across the width of the tank and imparts a downward velocity to the flow. This is illustrated in Figure 6.5. The *effluent* (water flowing out of the tank) is skimmed from the surface over weirs placed at the opposite end of the tank. A series of redwood boards moving on a continuous chain scrapes the sludge toward a collection hopper, from where it is pumped out of the tank.

In a circular clarifier, the water usually enters at the center of the tank and flows radially outward toward an effluent weir built along the perimeter of the tank. (See Appendix H, Figure 4.) A rotating sludge-scraper mechanism moves the sludge toward a central collection hopper. This is illustrated in Figure 6.6.

Whatever the shape of the tank, the inlets and outlets must be designed carefully to prevent currents that could resuspend the sludge. It is also necessary to avoid a condition called *short-circuiting* of the flow. In this context, short-circuiting refers to a condition that allows most of the water to flow through the tank in a period of time that is considerably less than the computed detention time. The effectiveness of sedimentation may be significantly reduced if short-circuiting occurs.

The effluent weirs in a settling tank are designed to operate at minimum head and velocity conditions, to reduce the chance for particles to be carried over in the effluent. The total weir length must be long enough so that the flow rate per foot or meter of weir

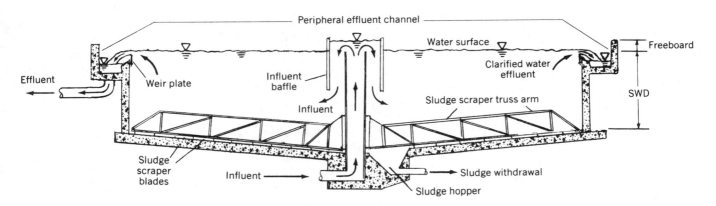

**FIGURE 6.6**
*Simplified section view of a circular sedimentation tank. Rotating scraper blades move the settled sludge to a central draw-off hopper.* (Courtesy of Envirex, Inc., a Rexnord Company, Waukesha, Wisconsin)

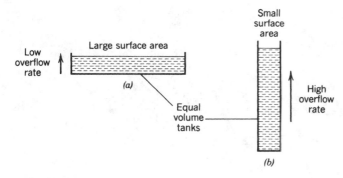

**FIGURE 6.7**

*The shallow tank* (a) *has a lower overflow rate than tank* (b) *because it has a larger surface area. It is more effective in removing suspended solids than the deeper tank.*

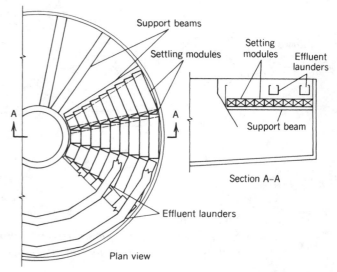

**FIGURE 6.9**

*Tube settlers can be installed in new or in existing sedimentation tanks to improve suspended solids removal efficiency.* (From G. L. Culp and R. L. Culp, *New Concepts in Water Purification*, Litton Educational, New York, 1974. Copyright © 1974 by Litton Educational Publishing Co. Used with permission of Van Nostrand Reinhold Company, Inc.)

is less than a specified maximum value, called the *weir loading rate*. The weirs usually consist of a series of uniformly spaced V-notches in a long metal plate. The effluent flows through the notches into a channel called an *effluent launder*, which directs the flow to an outlet pipe.

The effectiveness of a settling tank in removing suspended solids depends more on the surface area than on the total volume or detention time. It can be seen from Equation 6-2 that, for a given flow rate $Q$, as the surface area $A_s$ increases, the overflow rate $V_o$ decreases. Thus, the maximum particle settling velocity for complete removal also decreases with increasing surface area. Therefore, a shallow tank with a large surface area will be more effective than a deep tank with the same volume, but smaller area. This is illustrated

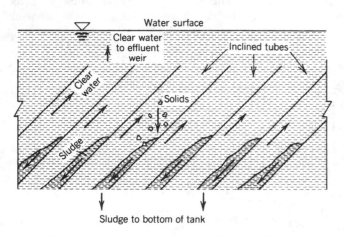

**FIGURE 6.8**

*A series of inclined tube settlers increases the effective surface area of a settling tank. Settleable solids are quickly entrapped in the downward-flowing sludge in each tube, whereas the clarified water flows upward.*

in Figure 6.7. Most settling tanks are not less than 3 m (10 ft) deep, however, to provide room for the sludge layer and the sludge scraper mechanism.

The concept of *shallow depth sedimentation*, as described here, is now often used when designing new settling tanks or for increasing the capacity and efficiency of existing tanks. Prefabricated units, or modules, comprising a series of inclined and nested tubes, can be installed near the top of the tank to increase its effective surface area. This is illustrated in Figure 6.8.

As shown in Figure 6.8, the suspended particles in the water become caught in the downward-flowing stream of sludge in each tube, whereas the clarified water flows upward toward the effluent weirs. A typical installation of these *tube settlers*, as they are called, is illustrated in Figure 6.9.

## 6.3 COAGULATION AND FLOCCULATION

Suspended particles cannot be completely removed from water by plain settling, even when they are given very long detention times and low overflow rates. Some of the very small turbidity-causing particles, called *colloids* (see Section 4.1), will not settle out of suspension by gravity without some help. If certain chemicals, called *coagulants*, are rapidly mixed in the water and then the mixture is slowly stirred before allowing sedimentation to occur, the particles will settle.

One of the properties of colloidal particles that keeps them in suspension is the small electrostatic charge that they each carry. Because of the presence of like charges, colloidal particles push each other apart and avoid collisions. The coagulant chemical, however, neutralizes the effect of the colloidal charges. Once neutralized, the colloidal particles can collide and agglomerate (stick together), forming larger and heavier particles, called *flocs*. The coagulant also reacts with the natural alkalinity in the water, forming a sticky solid *precipitate* that comes out of solution and helps in the formation of the flocs by capturing particles.

After the initial *flash-mix* of the coagulant with the water, a gentle agitation caused by slow stirring further enhances the growth of the flocs by increasing the number of particle collisions. The slow mixing or stirring process is called *flocculation*. The combined rapid mix–slow mix process is usually referred to as *coagulation*. Most of the flocs formed during coagulation are settleable and can be removed from the water in a sedimentation tank. As illustrated in Figure 6.2, coagulation generally precedes the sedimentation process in a typical water treatment plant.

Several different chemicals can be used for coagulation. The most common coagulant is aluminum sulfate, $Al_2(SO_4)_3$. It is generally referred to as *alum*. Sometimes certain synthetic organic chemicals, called *polymers* (or *polyelectrolytes*), are added along with the alum to act as *coagulant aids*. These long-chain, high-molecular-weight compounds help the formation of larger, heavier floc particles.

The success of the coagulation process depends on several factors, including the chemical dose, water temperature, pH, and alkalinity. Since the quality of a surface water supply often varies with time, it is frequently necessary to adjust the dose. The optimum coagulant dose is usually determined in the laboratory by a procedure called the *jar test.*

In the jar test, six beakers or jars are filled with a sample of the raw water, and each sample is mixed with a different amount of coagulant. A stirring apparatus, illustrated in Figure 6.10, is used to provide slow mixing, thereby simulating the flocculation process.

After the stirring paddles are stopped, the flocs are allowed to settle in the beakers. The dose in the beaker that required the least amount of coagulant to produce a clear water with well-formed, rapidly settling floc is used to compute the dose for the entire water treatment plant. Additional tests can be made with the same apparatus to determine the effects of pH or alkalinity adjustments on the formation of flocs to further optimize the process.

**FIGURE 6.10**

*A stirring apparatus for the* jar test, *which is used to determine optimum coagulant dosage.* (Courtesy of Phipps & Bird, 8741 Landmark Road, Richmond, Virginia, 23228)

The first step in the coagulation process is the flash-mix or rapid-mix of water and coagulants. This involves violent agitation to quickly spread the chemicals throughout the water and to ensure that there is a complete chemical reaction. Sometimes this is accomplished by adding the chemicals in the suction line just ahead of the centrifugal low-lift pump that brings the water into the treatment plant. The impeller of the pump provides the rapid-mix action inside the pump casing.

In most treatment plants, though, a rapidly rotating propeller is installed in a relatively small tank that provides about 1 min of detention time. These flash-mix tanks are often built immediately adjacent to the flocculation tanks in order to save on construction costs, as shown in Figure 6.11.

The size of the flocculation tank is such that it provides a detection time of up to 1 h for slow stirring. Paddle-type flocculators are the most common, using redwood slats mounted horizontally on motor-driven shafts. Rotating slowly at about one revolution per minute, the paddles provide a gentle agitation that promotes the growth of flocs.

Some relatively small water treatment plants combine chemical addition, flocculation, and sedimentation in a single tank, called a *solids-contact tank* or *upflow clarifier*. A sectional view of a typical upflow clarifier is shown in Figure 6.12. Chemical addition and rapid mixing occur where the water enters the tank, in the center. The water first flows downward under a cone-shaped hood, where flocculation occurs. Then it flows upward through the portion of the tank that serves for gravity settling. A sludge blanket of floc particles is formed at the bottom of the tank. Treatment units like this are particularly useful in plants where the water must also be softened by adding lime (discussed in Section 6.6) and where only limited space is available.

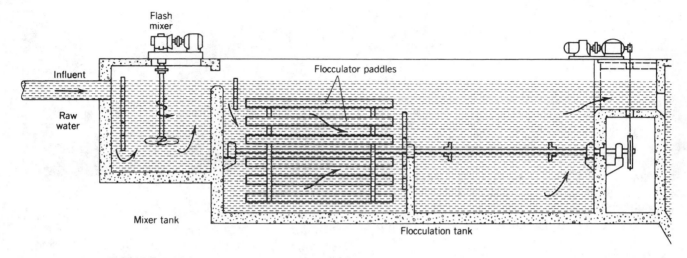

**FIGURE 6.11**
*Section view of a flash-mix and flocculation tank used in the coagulation process.* (Courtesy of FMC Corporation, Lansdale, Pennsylvania)

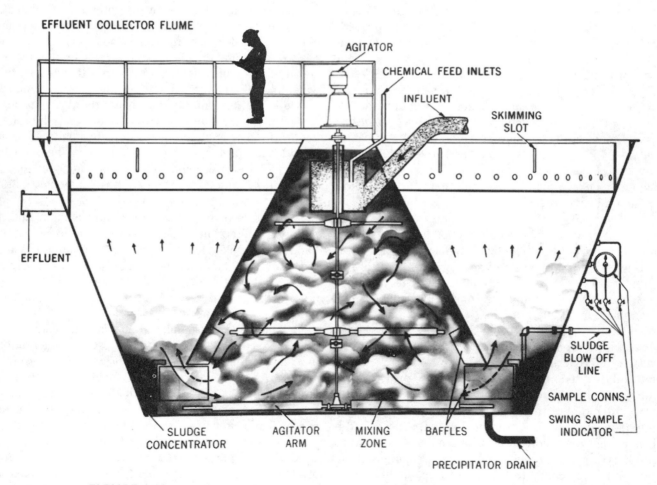

**FIGURE 6.12**
*A suspended solids contact clarifier.* (Courtesy of Permutit Company, Inc., Paramus, New Jersey)

## Ballasted Coagulation

A recent development in coagulation technology adds fine sand particles of the mix of alum and polymers. The sand particles adhere to the clumps of coagulated material and the added weight speeds up the settling process. A machine called a *hydrocyclone* separates the sand from the sludge, allowing the sand to be washed and recycled. Ballasted coagulation allows the same amount of water to be treated in a smaller tank than would otherwise be needed. There are about 25 plants currently using the method in the United States and its use is growing, especially in areas where space for a treatment plant is limited. One of the newest water treatment plants to make use of this technology is the 57-ML/d (15-mgd) Wilson, Oregon, facility, which uses water from the Willamette River.

## 6.4 FILTRATION

Even with the help of chemical coagulation, sedimentation by gravity is not sufficient to remove all the suspended impurities from water. About 5 percent of the suspended solids may still remain as nonsettleable floc particles. These remaining flocs can cause noticeable turbidity and may shield microorganisms from the subsequent disinfection process. To produce a crystal clear potable water that satisfies the SDWA requirement of 0.5 NTU (the MCL for turbidity), an additional treatment step following coagulation and sedimentation is typically needed.

This next step is a physical process called *filtration*. Filtration involves the removal of suspended particles from the water by passing it through a layer or *bed* of a porous granular material, such as sand. As the water flows through the filter bed, the suspended particles become trapped within the pore spaces of the filter material, or *filter media*, as it is called. This is shown schematically in Figure 6.13. Filtration is a very important treatment process in a surface water purification plant. In fact, many of these facilities are called *filtration plants*, even though filtration is only one step in the overall treatment sequence.

## Rapid Filters

The first filters built for water purification used very fine sand as the filter media. Because of the tiny size of the pore spaces in the fine sand, water takes a long time to flow through the filter bed, and when the surface becomes clogged with suspended particles, it becomes necessary to manually scrape the sand surface to clean the filter. These units, called *slow sand filters,*

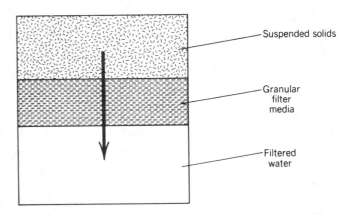

**FIGURE 6.13**
*Schematic diagram of the filtration process.*

take up a considerable amount of land area because of the slow filtration rates. Slow sand filters are still used in several existing treatment plants. They are effective and relatively inexpensive to operate.

In modern water treatment plants, the *rapid filter* has largely replaced the slow sand filter. As its name implies, the water flows through the filter bed much faster (about 30 times as fast) than it flows through the slow sand filter. This naturally makes it necessary to clean the filter much more frequently. But instead of manual cleaning by scraping of the surface, rapid filters are cleaned by reversing the direction of flow through the bed. This is shown schematically in Figure 6.14.

During filtration, the water flows downward through the bed under the force of gravity. When the filter is washed, clean water is forced upward, expanding the filter bed slightly and carrying away the accumulated impurities. This process is called *backwashing*. Cleaning by a backwash operation is a key characteristic of a rapid filter.

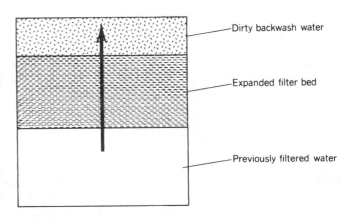

**FIGURE 6.14**
*Schematic diagram of the backwash or cleaning cycle of a rapid filter.*

Many rapid filters currently in operation use sand as the filter medium and are called *rapid sand filters*. But the sand grains (and pore spaces) are larger than those in the older, slow sand filters. In a rapid sand filter, the effective size of the sand is about 0.5 mm and the uniformity coefficient is about 1.5 (see Section 1.4). A difficulty that arises when using only sand in the rapid filter is that, after backwashing, the larger sand grains settle to the bottom first, leaving the smaller sand grains at the filter surface. This pattern of filter medium gradation is shown in Figure 6.15*a*.

Because of this small-to-larger gradation of sand grains in the direction of flow, most of the filtering action takes place in the top layer of the bed. This results in inefficient use of the filter. The filter run time (period of time between backwashes) is reduced, and frequent backwashes are required. Also, if some of the suspended material happens to penetrate the upper layer of fine sand, it is then likely to pass through the entire filter bed.

A preferable size distribution of the filter material is shown in Figure 6.15*b*. The larger-to-smaller particle gradation allows the suspended particles to reach greater depths within the filter bed. This *in-depth filtration*, as it is called, provides more storage space for the solids, offers less resistance to flow, and allows longer filter runs. The process of filtration becomes more than just a physical straining action at the surface of the bed. The processes of flocculation and sedimentation also occur within the pore spaces, and some material is adsorbed onto the surfaces of the filter medium.

To achieve the optimum gradation for in-depth filtration, it is necessary to use two or more different filter materials. For example, if a coarse layer of anthracite coal is placed above the sand, the coal grains will always remain on top after backwashing occurs. This is because the coal has a much lower density than the sand. Even though the coal grains are larger than the sand grains, they are lighter and therefore settle more slowly. The heavier sand particles settle to the filter bottom first at the end of a backwash cycle. A rapid filter that uses both coal and sand is called a *dual-media filter*. In effect, the upper coal layer acts as a rough filter, removing most of the large impurities first. This allows the sand layer to remove the finer particles without getting clogged too quickly.

The coarse-to-fine gradation shown in Figure 6.15*b* is even more closely obtained by using three filter materials: coal, sand, and garnet (a very dense material). After backwashing, the top layer of the filter bed is mostly coarse coal, the middle is mostly medium sand, and the bottom layer is mostly very fine grains of garnet. This is called a *mixed-media filter*. Filter material ranges in size from about 2 mm at the top to about 0.2 mm at the bottom. In recent years, dual- and mixed-media filters have been used to replace existing rapid sand filters in many treatment plants.

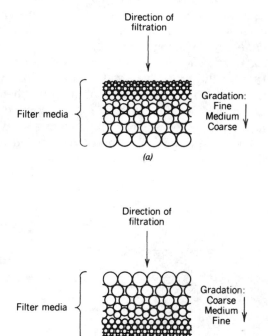

**FIGURE 6.15**

(a) *Typical gradation of a rapid sand filter bed. Solids removal occurs primarily by straining action at the top of the sand bed.* (b) *Typical coarse-to-fine gradation in a* mixed-media *filter. It is preferable to the sand bed because it provides in-depth filtration.*

## Filter Design

Rapid filters, whether of sand, dual media, or mixed media, are usually built in boxlike concrete structures, as illustrated in Figure 6.16. Multiple filter boxes or units are arranged on both sides of a central *piping gallery*, and a *clear well* used for storing filtered water is often located under the filters. Since only one unit is backwashed at a time, the filtration process can occur continuously as water flows through the treatment plant.

A typical rapid filter box is about 3 m, or 10 ft, deep, but the filter bed itself is only about 0.75 m, or 2.5 ft, deep. Located above the surface of the filter bed are *wash-water troughs*, which carry away the dirty backwash water as it flows upward through the bed and over the edge of the troughs. The filter medium is generally supported on a layer of coarse gravel. Below

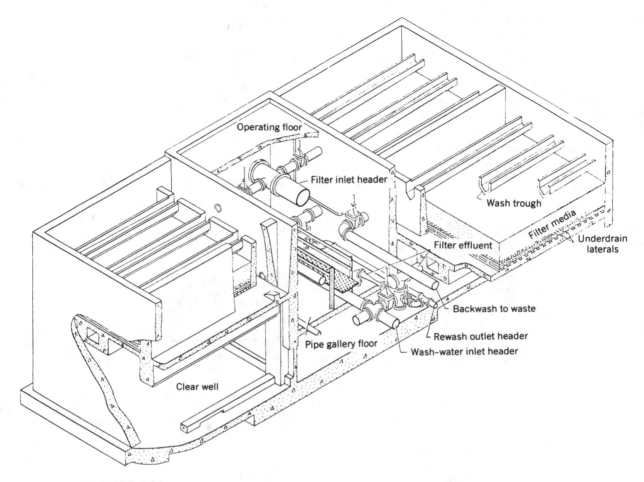

**FIGURE 6.16**
*Perspective view of a typical rapid filter facility. The filtered water is temporarily stored in the clear well. Multiple filter boxes provide operational flexibility; only one filter is backwashed at a time.* (Courtesy of Permutit Company, Inc., Paramus, New Jersey)

the gravel, which only serves to support the filter bed and does not contribute to the filtering action, is a special filter bottom or *underdrain system.*

The underdrains collect the filtered water and uniformly distribute the wash water across the filter bottom during the backwash cycle. They may consist of a grid of perforated pipes leading to a common header pipe that carries the water into the clear well. In many filters, the underdrains consist of specially manufactured porous tile blocks or steel plates with nozzles to help to distribute the backwash water. A cross section of a typical filter unit is shown in Figure 6.17.

The effectiveness of filtration and the length of a filter run depend on the *filtration rate.* Lower filtration rates generally allow longer filter runs and produce higher quality water, but they require larger filters. Filtration rate is often expressed as the flow rate of water divided by the surface area of the filter. In

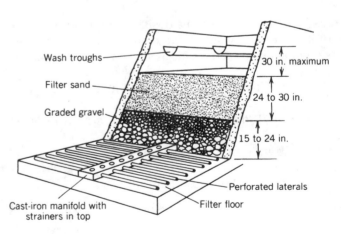

**FIGURE 6.17**
*Cross-sectional view of a typical sand filter box.* (Courtesy of National Lime Association, Arlington, Virginia)

Customary units, this is usually in terms of gallons per minute per square foot (gpm/ft$^2$ or gal/ft$^2$/min). In SI metric units, it is liters per square meter per second (L/m$^2$ · s). Rapid sand filters are usually designed to operate at an average rate of about 1.4 L/m$^2$ · s or 2 gpm/ft$^2$, whereas mixed-media filters can operate effectively at an average rate of about 3.5 L/m$^2$ · s or 5 gpm/ft$^2$. The filtration rate is proportional to the velocity of flow through the filter bed.

## EXAMPLE 6.5

A filter unit has a surface area of 50 ft$^2$. The flow rate through the filter is 0.25 mgd. Compute the filtration rate and the velocity at which the water flows through the filter bed.

*Solution*  First, convert the flow rate to gallons per minute, as follows:

$$250,000 \text{ gal/d} \times \frac{1 \text{ day}}{24 \text{ h}} \times \frac{1 \text{ h}}{60 \text{ min}} = 174 \frac{\text{gal}}{\text{min}}$$

Then

$$\text{filtration rate} = \frac{174 \text{ gpm}}{50 \text{ ft}^2} = 3.5 \text{ gpm/ft}^2$$

$$\text{velocity} = 3.5 \frac{\text{gal}}{\text{ft}^2 \cdot \text{min}} \times \frac{1 \text{ ft}^3}{7.5 \text{ gal}} = 0.47 \text{ ft/min}$$

## EXAMPLE 6.6

A square filter box is to be designed for a filtration rate of 2.8 L/m$^2$ · s. What are the required surface area and side dimension of the unit if the flow rate is 6 ML/d?

*Solution*  The flow rate in terms of liters per second is

$$6 \times 10^6 \text{ L/d} \times 1 \text{ d}/24 \text{ h} \times 1 \text{ h}/3600 \text{ s} = 69.4 \text{ L/s}$$

Since filtration rate = flow rate/area,

$$\text{area} = \text{flow rate/filter rate}$$

$$= 69.4 \text{ L/s} \div 2.8 \text{ L/m}^2 \cdot \text{s} = 25 \text{ m}^2$$

and

$$\text{side dimension} = \sqrt{25 \text{ m}^2} = 5 \text{ m}$$

## Filter Operation

Rapid filtration is usually preceded by coagulation and sedimentation. In some cases, however, depending on the quality of the raw water, *direct filtration* may be used. In direct filtration, coagulant mixing and flocculation occur, but the sedimentation step is omitted. Instead, the water flows from the flocculation basin directly to the filters. This provides a saving in treatment plant area and construction cost. (Some filtration plants also use ozone and a process called *floatation* to aid the clarification process. See Figure 6.22.)

A cross-sectional view of a typical rapid filter unit is shown in Figure 6.18. When filtration begins through

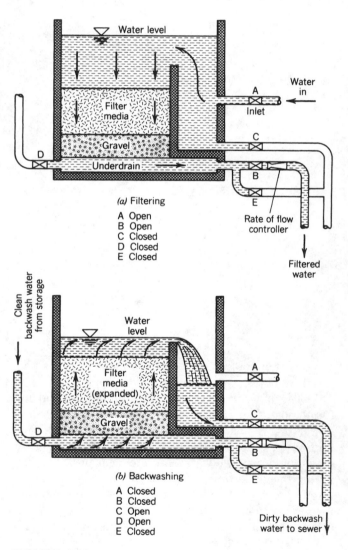

**FIGURE 6.18**

*Schematic diagrams of a rapid filter in the* (a) *filtering cycle and* (b) *backwash cycle of operation. Valves A, B, C, D, and E control the flow. Valve E is opened briefly when filtering starts.*

a clean bed, the inlet valve A is fully open and the outlet valve B is throttled (that is, only partially open). Valve B is gradually opened farther by an automatic *filter rate controller*, which operates by sensing pressure differences caused by changes in flow rate. The control device is usually a venturi meter (see Section 2.3).

As solids accumulate in the filter, the resistance to flow through the bed increases and the filtration rate tends to decrease. The reduced flow is sensed by the rate controller, which causes valve B to open farther. The gradual opening of valve B compensates for the continually increasing resistance to flow in the filter bed. In this way, the rate of flow through the filter does not vary. This type of operation is called *constant-rate filtration*. During constant-rate filtration, the water level in the filter box remains about 1 m (3 ft) above the top of the filter bed.

Eventually, the filter bed gets clogged to the extent that valve B must be wide open to maintain the desired filtration rate. At this point it becomes necessary to clean or backwash the filter. To do this, valves A and B are closed and valves C and D are opened. Water from the backwash storage tank then flows upward through the filter bed, expanding it slightly and carrying away the accumulated solids. The dirty backwash water flows into the wash-water troughs and then is either recycled or drained into a municipal sewerage system.

Filter backwash water may contain very high levels of harmful microbes, such as *Cryptosporidium* (which has caused 12 waterborne disease outbreaks in the United States since 1984). Improper backwash recycling procedures have been identified as possible causes for several of those disease outbreaks. Under the SDWA, a *Filter Backwash Recycling Rule* has been promulgated to protect public health. The rule requires that backwash water be returned to the headworks of the water treatment facility so that it undergoes all of the conventional treatment processes. Appropriate record keeping of recycle flow rates is also required.

Rapid filters are generally backwashed at a rate of about 10 L/m$^2 \cdot$ s (15 gpd/ft$^2$) for about 10 min. After the backwash flow stops, the filter material settles back in the bed and the filtration cycle begins again. For the first 5 min of filtering, however, the filtered water is discarded through valve E to ensure that any remaining solids will not be carried into the clear well.

## EXAMPLE 6.7

If the filter designed in Example 6.6 is backwashed once a day for 12 min at a rate of 10 L/m$^2 \cdot$ s, what percentage of the total flow rate is used for cleaning the filter?

*Solution*  The volume of water used for the backwash each day can be computed by multiplying the backwash rate times the filter area and the time of backwash, as follows:

$$10 \text{ L/m}^2 \cdot \text{s} \times 25 \text{ m}^2 \times 12 \text{ min} \times 60 \frac{\text{s}}{\text{min}} = 180 \text{ m}^3$$

The total daily flow is given as 6 ML = 6000 m$^3$. The percentage of water used for backwash is therefore

$$\frac{180}{6000} \times 100 = 3 \text{ percent}$$

This percentage of water used for backwashing is typical for most water treatment plants.

---

In recent years, a mode of operation called *declining-rate filtration* has been applied in some water treatment plants. In this mode of operation, rate-of-flow controllers are not used. The filtration rate is allowed to gradually decline from a maximum value at the beginning of the filter run to a minimum value when the bed is clogged. As the filter becomes clogged with accumulated solids, the water level gradually rises in the filter box. When the water level reaches a predetermined height, the filter is automatically backwashed.

Both declining-rate filtration and constant-rate filtration produce water of excellent quality. The crystal clear effluents from properly designed and operated rapid filters generally have turbidity levels less than 0.2 NTU.

## Other Types of Filters

The gravity-flow, rapid filter is the most common type of filter used for treating public water supplies, primarily because it is the most reliable. Other types of filters sometimes used to clarify water include the *pressure filter* and the *diatomaceous earth filter*.

A pressure filter is very much like a conventional rapid filter in that the water flows through a granular filter bed. But instead of being open to the atmosphere and utilizing the force of gravity, the pressure filter is enclosed in a cylindrical steel tank and the water is pumped through the bed under pressure. Because it operates under pressure, there is more of a chance that solids will get through the bed in the effluent. Since they are not as reliable as gravity filters, pressure filters are only occasionally used for treating public drinking water supplies. (See Appendix H, Figure 10.) They are more commonly used for filtering water for industrial use or for swimming pools.

The diatomaceous earth filter is also used primarily for industrial or swimming pool applications. Like the pressure filter, it is less reliable than the rapid sand or mixed-media filter. The filter medium in this type of filter is a thin layer of diatomaceous earth, a natural, powderlike material formed from the shells of microscopic organisms called diatoms. The diatomaceous earth is supported on a cylindrical metal screen or fabric, called a *septum*. A typical diatomaceous earth filter is composed of many of these small septa.

## 6.5   DISINFECTION

The unit processes described in the previous sections—coagulation, sedimentation, and filtration—together compose a type of treatment called *clarification*. Clarification removes many microorganisms from the water along with the suspended solids. But clarification by itself is not sufficient to ensure the complete removal of pathogenic bacteria or viruses. A potable water must be more than crystal clear—it must be completely free of disease-causing microorganisms. To accomplish this, the final treatment process in water treatment plants is *disinfection*, which destroys or inactivates the pathogens.

## Chlorination

Chlorine is the most commonly used substance for disinfection in the United States. The addition of chlorine or chlorine compounds to water is called *chlorination*. Chlorination is considered to be the single most important process for preventing the spread of waterborne disease. The effectiveness of chlorination is illustrated in Figure 1.16, which shows a steady decline in disease outbreaks as the number of chlorinated water supplies increased.

Molecular chlorine, $Cl_2$, is a greenish-yellow gas at ordinary room temperature and pressure. In gaseous form it is very toxic, and even in low concentrations it is a severe irritant. But when the chlorine is dissolved in low concentrations in clean water, it is not harmful, and if it is properly applied, objectionable tastes and odors due to the chlorine and its by-products are not noticeable to the average person.

Although chlorine is effective in destroying pathogens and preventing the spread of communicable disease, there may be an indirect noninfectious health problem caused by the chlorination process. Natural waters often contain trace amounts of organic compounds, primarily from natural sources such as decaying vegetation. These substances can react with the chlorine to form compounds called *trihalomethanes* (THMs), which may cause cancer in humans. Chloroform is an example of a THM compound.

The EPA has set standards that limit the maximum amount of THM compounds in drinking water. One way to prevent THM formation is to make sure that the chlorine is added to the water only after clarification and the removal of most of the organics. Also, alternative methods of disinfection are available that do not use chlorine. These are discussed later in this section.

### Chlorination Chemistry

When chlorine is dissolved in pure water, it reacts with the $H^+$ ions and the $OH^-$ ions in the water. Two of the products of this reaction are *hypochlorous acid*, HOCl, and the *hypochlorite ion*, $OCl^-$. These are the actual disinfecting agents. If microorganisms are present in the water, HOCl and $OCl^-$ penetrate the microbe cells and react with certain enzymes. This reaction disrupts the organisms' metabolism and kills them.

Hypochlorous acid is a more effective disinfectant than the hypochlorite ion because it diffuses faster through the microbe cell wall. The relative concentrations of HOCL and $OCl^-$ depend on the pH of the water. The lower the pH, the more HOCl there is relative to the $OCl^-$. In general, then, the lower the pH of the water, the more effective is the chlorination–disinfection process.

When chlorine is first added to water containing some impurities, the chlorine immediately reacts with the dissolved inorganic or organic substances and is then unavailable for disinfection. The amount of chlorine used up in this initial reaction is called the *chlorine demand* of the water. If dissolved ammonia, $NH_3$, is present in the water, the chlorine reacts with it to form compounds called *chloramines*. Only after the chlorine demand is satisfied and the reaction with all the dissolved ammonia is complete is the chlorine actually available in the form of HOCl and $OCl^-$.

Chlorine in the form of HOCl and $OCl^-$ is called *free available chlorine*, whereas chloramines are referred to as *combined chlorine*. Free chlorine is often the preferred form for disinfection of drinking water. It works faster than combined chlorine, and it does not cause objectionable tastes and odors. Combined chlorine is also effective as a disinfectant, but it is slower acting and it may cause the typical swimming-pool odor of chlorinated water. Its advantage is that it lasts longer and can maintain sanitary protection throughout the water distribution system.

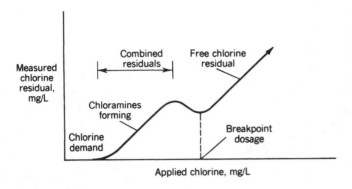

**FIGURE 6.19**
*Breakpoint chlorination curve.*

A process called *breakpoint chlorination* is sometimes used to ensure the presence of free chlorine in public water supplies. To do this, it is necessary to add enough chlorine to the water to satisfy the chlorine demand and to react with all the dissolved ammonia. When this occurs, it is said that the chlorine breakpoint has been reached. Chlorine added beyond the breakpoint will be available as a free chlorine residual in direct proportion to the amount of chlorine added. This is illustrated in Figure 6.19. The chlorine demand and the breakpoint dose vary, depending on the water quality. Sometimes, chlorine doses up to 10 mg/L are needed to obtain a free chlorine residual of 0.5 mg/L.

## Chlorination Methods

Chlorine is commercially available in gaseous form or in the form of solid and liquid compounds called *hypochlorites*. For the disinfection of relatively large volumes of water, the gaseous form of chlorine is generally the most economical, but for smaller volumes, the use of hypochlorite compounds is more common.

Gaseous chlorine is stored and shipped in pressurized steel cylinders. Under pressure, the chlorine is actually in liquid form in the cylinder; when it is released from the cylinder, it vaporizes into a gas. The cylinders may range in capacity from 45 kg (100 lb) to about 1000 kg (1 ton). Very large water (or wastewater) treatment plants may use special railroad tank cars filled with chlorine.

A device called an *all-vacuum chlorinator* is considered to provide the safest type of chlorine feed installation. It is mounted directly on the chlorine cylinder. The gaseous chlorine is always under a partial vacuum in the line that carries it to the point of application; chlorine leaks cannot occur in that line. A typical vacuum chlorine feed system is shown in Figure 6.20. The vacuum is formed by water flowing through the ejector unit at high velocity. There are other types of chlorinators, some of which have the chlorine or concentrated chlorine solutions conveyed relatively long distances under pressure. These present somewhat greater risks of chlorine leaks. In any chlorine feed

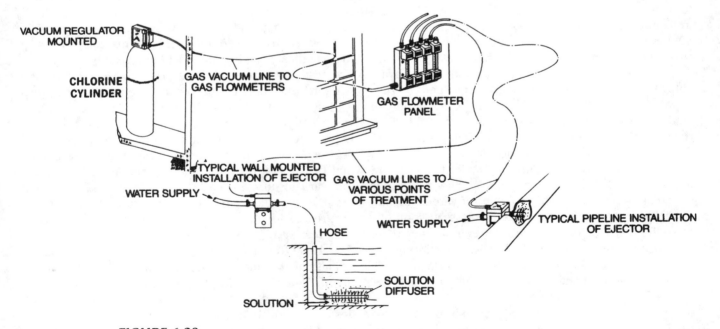

**FIGURE 6.20**
*Typical vacuum-feed chlorination system. There is little risk of a chlorine gas leak because the chlorine feed line is at less than atmospheric pressure.*
(Provided by Capital Controls Company, Inc., Colmar, Pennsylvania)

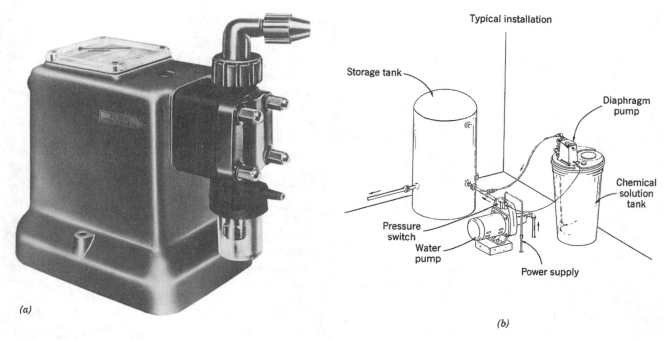

*(a)*

*(b)*

**FIGURE 6.21**
(a) *Typical hypochlorinator or solution metering pump and* (b) *a typical hypochlorinator installation.* (Courtesy of Penwalt Corporation/Wallace and Tiernan Division, Belleville, New Jersey)

installation, safety factors are very important because of the toxicity of the gas.

Hypochlorites are usually applied to water in liquid form by means of small pumps, such as the one illustrated in Figure 6.21. These are *positive-displacement*-type pumps, which deliver a specific amount of liquid on each stroke of a piston or flexible diaphragm.

Two types of hypochlorite compounds are available for disinfection: *sodium hypochlorite* and *calcium hypochlorite*. Sodium hypochlorite is available only in liquid form and contains up to 15 percent available chlorine. It is usually diluted with water before being applied as a disinfectant. (Common laundry bleach is a 5 percent solution of sodium hypochlorite.) Calcium hypochlorite is a dry compound, available in granular or tablet form; it is readily soluble in water. Calcium hypochlorite solutions are more stable than solutions of sodium hypochlorite, which deteriorate over time.

In addition to pH, the effectiveness of chlorine and chlorine compounds in destroying bacteria depends on the chlorine concentration and the *contact time*. Contact time is the time period during which the free or combined chlorine is acting on the microorganisms. At pH values close to 7 (neutral conditions), a free chlorine residual of 0.2 mg/L with a 10-min contact time has about the same disinfecting power as 1.5 mg/L of combined chlorine residual with a 1-h contact time.

The effectiveness of chlorination can be determined by the coliform test or, as mentioned in Section 4.3, by a more convenient test for chlorine residual in the treated water. The method approved for chlorine residual testing, under the SDWA regulations, is called the *DPD chlorine residual test*. Field test kits, such as the one illustrated in Figure 4.19, are readily available. The test procedure, which is based on a color comparison, takes only 5 min to complete.

In the DPD test, a chemical dye is added to the water sample. The dye turns red if chlorine residual is present, and the intensity of the red color is proportional to the chlorine concentration. It is assumed that the presence of a chlorine residual ensures that there are no surviving pathogenic organisms in the water. It is possible to measure either the total residual, the free residual, or the combined residual with the DPD test kit.

It is often necessary to compute the total weight or mass of chlorine used at a treatment plant to be able to order chlorine supplies at the appropriate time. Also, it may be necessary to determine the applied chlorine dosage or concentration if the mass or weight consumed is known. The following relationships are useful for these purposes:

$$\text{kg/d} = Q \times C \qquad \textbf{(6-3a)}$$

where $Q$ = flow rate, ML/d
$C$ = chlorine concentration, mg/L

or

$$lb/d = 8.34 \times Q \times C \qquad \textbf{(6-3b)}$$

where $Q$ = flow rate, mgd
$C$ = chlorine concentration, mg/L
8.34 = lb/gal of water

## EXAMPLE 6.8

How many pounds per day of chlorine is required to disinfect a flow of 7.5 mgd with a chlorine dose of 0.5 mg/L? How many 100-lb chlorine cylinders are needed per month?

*Solution*   Applying Equation 6-3b gives

$$lb/d = 8.34 \times 7.5 \times 0.5 = 31 \text{ lb/d}$$

and

$31 \text{ lb/d} \times 30 \text{ d/month} \times 1 \text{ cylinder/100 lb}$
$$= 9.4 \text{ cylinders}$$

At least ten chlorine cylinders should be ordered per month.

## EXAMPLE 6.9

A total of 15 kg of chlorine is used in 1 d to disinfect a volume of 50 ML of water. What is the chlorine dose?

*Solution*   Applying Equation 6-3a gives

$$15 \text{ kg/d} = 50 \text{ ML/d} \times C$$

and

$$C = \frac{15}{50} = 0.3 \text{ mg/L}$$

## Other Methods of Disinfection

Chlorination is the most widely used method for disinfection of water supplies in the United States because of its economy and its ability to maintain a protective residual. Other methods of disinfection have been receiving more attention in recent years, primarily because of the problem of THM formation and the potential effect on public health.

### Ozone

Ozone ($O_3$) is a highly reactive gas at ordinary temperature and pressures, and acts as a very potent disinfectant when mixed with water. It has been used for over 90 years in European countries as an alternative to chlorine, which sometimes leaves a noticeable taste and odor in drinking water. Ozone can be produced by passing a very high voltage electric current through air or oxygen. However, because it is very unstable and cannot be stored, it must be manufactured on site, where it is used. And because it does not leave a measurable residual in water after the initial contact time, some chlorine (although in relatively smaller amounts) must be used to ensure continued disinfection as the water flows throughout the network of water distribution pipes.

In addition to the ability of ozone to act as a disinfectant without causing taste and odor problems, it does not react to form THM (trihalomethane) compounds. Ozone is also a stronger disinfectant than chlorine and is able to inactivate most viruses in addition to bacteria. (It is approved by the EPA for disinfection.) It can assist as a coagulant when used with alum, thus reducing the amount of chemicals needed to adjust the final pH of the water to make it noncorrosive. Because it greatly aids the coagulation process, ozone can also facilitate the application of a direct filtration process and eliminate the need for large sedimentation basins.

Despite these advantages, the high cost of its production and application compared to that for chlorine has discouraged widespread use of ozone for disinfection in the United States. There are some notable exceptions to this. For example, ozone is used at the 1900 ML/d (500-mgd) Los Angeles water treatment plant. It is also used for clarification and disinfection at the 570 ML/d (150-mgd) water treatment plant in Haworth, New Jersey. Both these plants also use the direct filtration process. A flow diagram showing major treatment steps at the Haworth plant is shown in Figure 6.22.

Ozone is more effective than chlorine as a disinfectant when used to prevent outbreaks of cryptosporidiosis. Water treatment facilities in Milwaukee, Wisconsin, were recently upgraded for just this purpose. This is explained in the following case study.

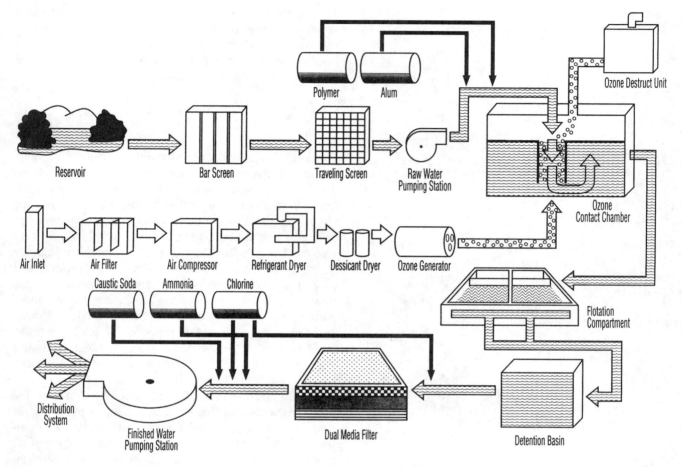

**FIGURE 6.22**

*A flow diagram of the ozone water purification process at United Water Resources Haworth water treatment plant. Major steps in the process include:*

*Step 1:* Raw reservoir water is drawn in to the treatment process by intake pumps. As the water enters the plant, bar screens remove large particles, and small amounts of chemical coagulants are added to help remove small particles and clarify the water.

*Step 2:* Outside air from which ozone is generated is first filtered, cooled, and dried.

*Step 3:* The treated air passes over many glass tubes, each individually fused with an electric filament; a lightning-like electrical charge transforms oxygen in the air to a mixture of ozone and oxygen.

*Step 4:* In the ozone contact chamber, small ozone bubbles move upward as water flows downward in a counter current. The rising bubbles help to mix the coagulants and water, carry small particles upward, and destroy bacteria and viruses. Any remaining ozone is converted back into oxygen by a catalytic converter in the ozone destruction unit.

*Step 5:* The ozone-treated water flows to flotation compartments, where skimmers remove the floating materials.

*Step 6:* The skimmed water flows into detention basins, where additional particles settle out.

*Step 7:* The water flows by gravity down through dual-media granular filters.

*Step 8:* Small amounts of chlorine are added to ensure continued disinfection of the water in the distribution system, and other chemicals are added to eliminate corrosion and the likelihood of lead dissolving into the water from household plumbing.

*Step 9:* Large electrically powered pumps push up to 200 mgd of the treated water into the distribution system.

(Courtesy of United Water Resources, Harrington Park, New Jersey)

## CASE STUDY

### Milwaukee's $89 Million Water Works Upgrade

In March 1993, after a serious outbreak of cryptosporidiosis sickened 403,000 people in Milwaukee, Wisconsin, a plan for upgrading the city's water treatment facilities to provide safer drinking water was implemented. This was done despite the fact that the Milwaukee Water Works facility was deemed by the EPA to be in full compliance with all federal and state drinking water regulations before and during the epidemic. Although the actual source of contamination was never ascertained, the city's main goal was to kill *Cryptosporidium* in raw water entering the treatment plants. In addition, the upgrade was intended to minimize levels of disinfection by-products and reduce taste and odor problems.

*Cryptosporidium* is a protozoan parasite (see Section 4.4) that can move through the aquatic environment in the form of microscopic egg-shaped capsules called *oocysts*. Oocysts, the dormant stage of the organism, are typically 1/20 the thickness of a strand of human of hair (about 5 $\mu$m) and are resistant to ordinary filtration and chlorination. Drinking water contaminated with these pathogens causes gastrointestinal illness (severe cramps and diarrhea) in humans. The effects can be fatal, particularly for people with weakened immune systems (about 100 deaths were attributed to the outbreak in Milwaukee). Cryptosporidia capable of infecting humans can come from cattle manure that washes off the land into lakes and streams, or from human waste if it is not treated before entering surface water.

The city's plan to protect public health included making improvements to its two major water treatment facilities, the 570-ML/d (150-mgd) Howard Avenue purification plant and the 1045-ML/d (275-mgd) Linwood plant, which together serve about 850,000 people. Dual-media (sand and anthracite) filters replaced the original sand filters. In addition, the upgrade plan included a 1280-m (4200-ft) extension of a 2.7-m (9-ft)–diameter prestressed concrete intake pipe, putting the end of the pipe 4 km (2.5 mi) offshore and 18.3 m (60 ft) below the surface of Lake Michigan. This improved the quality of raw source water. The main feature of the upgrade, though, is the application of *ozone* at each treatment facility, the most effective disinfectant against *Cryptosporidium*.

Ozone may be added to drinking water at one or more points in a water treatment process. Comprehensive laboratory studies at the Howard Avenue and Linwood plants indicated that pre-ozonation (adding ozone before filtration) is the preferred method for the source water in Milwaukee. Ozone is generated from liquid oxygen at the two plants and is mixed with the water using fine-bubble diffusion technology. In addition to killing *Cryptosporidium*, the pre-ozonation method reduces the need for coagulation chemicals and chemicals used for taste and odor control. Use of less coagulant also reduces sludge production and sludge disposal costs, lengthens the filter runs, and reduces the volume of backwash water.

The upgrades at the Howard Avenue and Linwood treatment plants, completed in 1998, are among the first in the United States designed specifically to inactivate *Cryptosporidium* with ozone. Not only will *Cryptosporidium* pathogens no longer be a threat to customers of the Milwaukee Water Works, there will be fewer chlorine by-products in the water, as well as enhanced taste and odor control. The project's success in Milwaukee is likely to serve as a model for the installation of additional ozone disinfection facilities nationwide.

---

### Ultraviolet Radiation

Ultraviolet (UV) light can be used for disinfection. Ultraviolet light is electromagnetic radiation just beyond the blue end of the light spectrum, outside the range of visible light. It has a much higher energy level than visible light, and in large doses it destroys bacteria and viruses. The UV energy is absorbed by genetic material in the microorganisms, interfering with their ability to reproduce and survive. UV light can be generated by a variety of lamps; submerged, low-pressure mercury lamps are best suited for use in disinfection systems because they generate a large fraction of UV energy that gets absorbed.

Ultraviolet disinfection systems do not involve chemical handling, as do chlorine or ozone systems, thereby minimizing chemical safety concerns. Like ozone, though, UV radiation leaves no measurable residual in the water. Advances in UV germicidal lamp technology are making UV disinfection a more reliable and economical option for disinfection. (A case study of UV disinfection in wastewater is presented in Section 10.3.)

## 6.6   OTHER TREATMENT PROCESSES

Clarification by coagulation, sedimentation, and filtration removes suspended impurities and turbidity from drinking water. The final step of disinfection produces potable water, free of harmful microorganisms. But other treatment processes may be required, particularly to remove some of the dissolved substances. These processes may be used in addition to clarification or applied separately, depending on the source and quality of the raw water.

Groundwater, for example, does not ordinarily require clarification because the water is filtered naturally in the layers of soil from which it is withdrawn. Disinfection of groundwater supplies, required by law for public water supply systems, is basically a precautionary step; groundwater is usually free of bacteria or other microorganisms. On the other hand, because of its contact with soil and rock, groundwater may contain high levels of dissolved minerals that must be removed. Methods to accomplish this, as well as other less common treatment processes for both surface and subsurface water, are discussed in the following paragraphs.

## Water Softening

Water that contains dissolved salts of calcium and magnesium is known as hard water, as discussed in Section 4.3. Hardness in water interferes with the lathering action of soap and causes deposits of scale in water heaters, pipes, and plumbing fixtures. This is basically an economic and esthetic problem, rather than a health problem. Generally, when the hardness exceeds about 500 mg/L in the raw water, it is best to remove the calcium and magnesium at a central municipal treatment plant. The process of removing these minerals is called *water softening*. The two most common methods of softening are the *lime–soda method* and the *ion-exchange method*.

In the lime–soda method, two chemicals are added to the water to cause what chemists call a precipitation reaction. These chemicals are lime, $Ca(OH)_2$, and soda ash, $Na_2CO_3$. A reaction takes place among these chemicals and the dissolved calcium and magnesium ions in the water, causing the formation of calcium carbonate, $CaCO_3$, and magnesium hydroxide, $Mg(OH)_2$.

Since they are very insoluble in water, the calcium carbonate and magnesium hydroxide compounds precipitate out of solution as they form during the reaction. This process is then followed by sedimentation and filtration to remove the insoluble precipitates and clarify the water. Also, carbon dioxide, $CO_2$, may be added to the water to precipitate excess calcium and to adjust the pH, which is raised by the addition of lime; this process is called *recarbonation*.

Softening by ion exchange involves passing the water through a column containing a special ion-exchange material. Several different types of ion-exchange materials are in use, including natural substances called *zeolites* and synthetic resins. When water containing calcium or magnesium ions is in contact with these materials, an exchange or trade of ions takes place. The calcium and magnesium ions are taken up by the resin, whereas sodium ions, $Na^+$, are released into the water.

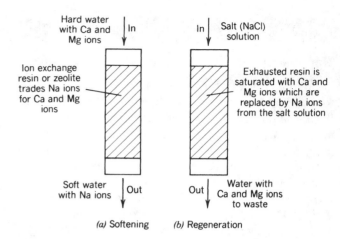

**FIGURE 6.23**

(a) *An ion-exchange column used for water softening;* (b) *the column may be regenerated and used again after washing with a strong salt solution.*

The ion-exchange process is illustrated schematically in Figure 6.23. Eventually, the exchange capacity of the zeolite or resin is used up and the ion exchangers must be regenerated for further use. This is done by washing the exchanger with a sodium chloride, NaCl, solution. Now the sodium ions replace the calcium and magnesium ions, which are discharged to a waste disposal drain. The softening process can then begin again.

Softening by ion exchange can produce water with almost zero levels of hardness, but this is not really desirable. Very soft water may be aggressive, or corrosive, causing damage to metal pipes and plumbing. Hardness levels of about 100 mg/L are considered optimum for drinking water.

There is also some evidence that the presence of moderate hardness levels in drinking water actually reduces the incidence of heart disease. Another factor that must be considered is that softened water from an ion exchanger contains sodium, which may be harmful to persons who already have heart disease. In such cases, the softened water may not be suitable for consumption. Finally, it should be noted that ion-exchange softening does not produce a precipitate or sludge and is generally less costly than lime–soda softening. But because of the disadvantages mentioned, it is usually better adapted for treating industrial water supplies or for use in individual home softening units.

## Aeration

A physical treatment process in which air is thoroughly mixed with water is called *aeration*. Thorough contact with air and oxygen can improve water quality in a number of ways. For example, one of the common uses of aeration is for *taste and odor control*. Dissolved gases that tend to cause the taste and odor problems, such as hydrogen sulfide, are transferred from the water to the air during aeration. This application is also called *air stripping*.

Aeration is also used for the *removal of iron and manganese* from the water, particularly in groundwater supplies. (See Appendix H, Figure 13.) The oxygen in the air reacts with the iron and manganese to form an insoluble precipitate (rust). Sedimentation and filtration are then necessary to clarify the water.

Several methods for aerating the water are available. The method selected depends primarily on the type and concentration of material to be removed from the water and on the available pressure head. Aeration using *spray nozzles* provides a large total air–water contact area, but relatively high pressures and much space are required. Spraying the water into the air can be followed by allowing the water to cascade and flow in thin sheets down several concrete or metal steps. Cascade structures require at least a 3-m (10-ft) drop.

Another common method for aeration makes use of *multiple-tray aerators*. These consist basically of a tall stack of perforated trays or slats with staggered openings. The water is applied at the top and trickles downward in thin films or sheets of flow. In some cases, a fan or blower might be used to force air upward through the stack to increase the contact with air.

For very large volumes of water, the use of *diffused-air aerators* is generally the most practical method. In this type of aerator, air is pumped by centrifugal blowers into a tank of water. The air enters the water at the tank bottom through special diffuser nozzles or porous fixtures, forming air bubbles that become thoroughly mixed with the water. *Mechanical aerators* consisting of a large propeller that churns the water at the surface are also available. (Mechanical and diffused-air aerators find frequent application in wastewater treatment, as discussed in Section 10.3.)

## Activated Carbon

Activated carbon derived from coal or wood has two unique properties that make it useful for water purification. First, it is a very porous material and has an extremely high ratio of surface area to unit weight—up to 100 ac of area per pound. Second, the surface of activated carbon attracts and holds many of the impurities in water, particularly the dissolved organics. This process is called *adsorption*. (Unlike *absorption*, adsorption is a surface phenomenon.)

Adsorption on activated carbon is an effective method for removing dissolved organic substances that cause taste and odor problems in drinking water. It is also effective in removing the organic *precursors* that react with chlorine to form harmful THM compounds after disinfection.

When the carbon surfaces become covered with impurities, the carbon can be cleaned or reactivated by heating to a high temperature in a special furnace. The organics are driven off by the heat, and the carbon can then be reused. But on-site reactivation, rather than complete replacement with fresh carbon, is economical only for large municipal water treatment plants.

Activated carbon is available as a very fine black powder or in granular form. The powdered carbon can be mixed with the water by a special dry-feeder device at a point in the treatment plant that precedes the filtration process; it is removed from the water by the filters. Granular carbon is sometimes used in the filter bed itself, combining both filtration and adsorption in one treatment unit.

One of the latest developments in water treatment is the use of *biologically active carbon* (BAC) media for removal of excess biodegradable dissolved organic carbon and disinfection by-products from treated drinking water. In a BAC water treatment process, heterotrophic bacteria (see Sections 1.3 and 4.4) colonize the granular carbon medium (in a supplemental tank or contactor) and metabolize the biodegradable organic matter. For certain water sources, BAC can reduce the content of organics more than that achieved by using only activated carbon. It is usually necessary to first convert nonbiodegradable organics into biodegradable forms, using ozonation before the BAC units. The BAC contactors are generally added at the end of the treatment system, before disinfection. (BAC technology is also used for advanced wastewater treatment; see Appendix H, Figure 7.)

## Corrosion Control

Corrosion or rusting of metals in water supply systems can be a serious problem. Since corrosion involves a transfer of electrons, control methods are aimed at blocking the flow of electrons between the water and the metal that is susceptible to corrosion. One way to do this is to add chemicals called *complexing agents* to the water. Complexing agents, such as sodium silicate or sodium phosphate, are added at concentrations of about 1 mg/L. They combine chemically with the metal to form a barrier that blocks corrosion reactions. Control of the pH of the treated water is also used to prevent corrosion in the distribution system.

## Fluoridation

As discussed in Section 4.3, fluorides are effective in preventing tooth decay, particularly in young children. Many communities intentionally fluoridate their water supplies as a public health measure. This is done by adding sodium fluoride, NaF, or sodium silicofluoride, $Na_2SiF_6$, to the water after filtration. It is important that the dosage of fluoride be carefully controlled; the optimum concentration is about 1 mg/L. Excessive concentrations may cause discoloration of tooth enamel, called *dental fluorosis*. Other than dental fluorosis, which only occurs if large amounts of fluoride are consumed, there are no harmful side effects of fluoridation.

## Desalination

Water with high levels of dissolved minerals or salts is unfit for most domestic, industrial, or agricultural uses. This includes both brackish water (having more than 1000 mg/L of salts) and seawater (with about 35,000 mg/L of salts). It is possible to separate freshwater from seawater or brackish water in a process called *desalination* (or *desalting*). Although expensive, desalination can be more economical than moving large quantities of freshwater over long distances. Advances in desalting technology have made it a feasible treatment option.

Desalted water now provides the primary source of municipal water in many areas of the globe, particularly in the Caribbean, the Middle East, North Africa, and other densely populated arid areas. A 180-ML/d (47-mgd) desalting facility soon to be built in the city of Ashdod, Israel, will be the largest in the world. Desalting technology for public water supply is also used in coastal areas of the United States, and its use in increasing. A 10-ML/d (2.7-mgd) desalting plant, for example, now in operation in Cape May, New Jersey, was the first to be built in the northeastern United States to desalinate brackish groundwater (see case study below).

Two basic methods are available for desalination of water: *thermal processes* and *membrane processes*. Thermal processes involve transfer of heat and a phase change of water into either vapor or ice. Membrane processes make use of thin sheets of special materials that allow freshwater to pass through, but not salt. Both of these methods require large inputs of energy to operate.

The most common thermal process in use is *multistage flash distillation*. In distillation, freshwater is separated from salt water by heating, evaporation, and condensation. Multistage flash distillation is a technique that allows the production of relatively large quantities of desalted water. It is based on the fact that the boiling temperature of water is lowered as the air pressure drops, reducing the amount of energy needed for vaporization. The process is carried out in a series of closed vessels (stages) set at progressively lower pressures. Heat is provided by steam from a boiler. When preheated salt water enters a vessel that is at low pressure, some of it rapidly boils (flashes) into vapor. The vapor is condensed into freshwater on heat-exchange tubes and then collected in trays under the tubes. The remaining salt water flows into the next stage set at even lower pressure, where some of it also flashes, thus continuing the process. Some facilities may have up to 40 stages. This process is shown schematically in Figure 6.24.

*Solar humidification* is a thermal process suitable primarily for providing desalted water to small communities where sunlight is abundant. Water evaporates from a free surface at a temperature below its boiling point.

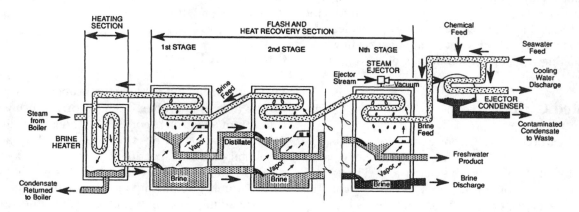

**FIGURE 6.24**
*Multistage flash distillation plant.* (From O. K. Burros, *The ABC's of Desalting*, 2nd ed., International Desalination Institute, Topsfield, Massachusetts. Used with permission)

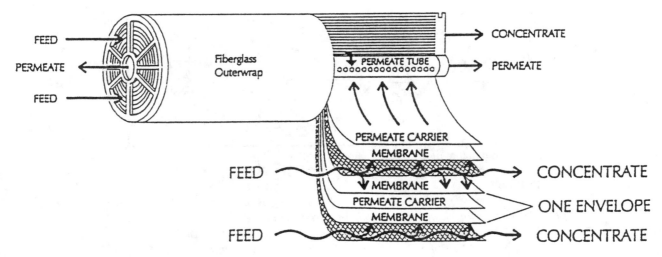

**FIGURE 6.25**

*Schematic diagram of a spiral-wound configuration for an RO semipermeable membrane unit. Brackish feedwater is pumped under high pressure into the tube. Freshwater permeate is carried to the inner permeate tube by wicklike materials sandwiched between the wound membrane.* (Provided by David H. Paul, Inc., Farmington, New Mexico. © 1995. Used with permission)

Salty water can be collected in shallow basins in a solar still (similar to a greenhouse) and warmed as sunlight passes through sloping glass or plastic covers. Vapor rises and condenses on the cooler covers; then the freshwater trickles down to a collecting channel. Although energy from the sun is free, a solar still is expensive to build, requires a large land area, and needs additional energy for pumping the water.

*Reverse osmosis* and *electrodialysis* are two membrane processes for desalination, but they are usually limited to the treatment of brackish inland or well water supplies, rather than seawater. In electrodialysis, a voltage is applied across the salty water, causing ions to migrate toward an electrode of opposite charge. Plastic membranes that are selectively permeable to either positive or negative ions are used to separate freshwater from salty water.

In a reverse osmosis (also called *ultrafiltration*), a semipermeable membrane separates salty water of two different concentrations. There is a natural tendency for the concentrations to become equalized by a flow of water from the dilute side to the concentrated side (osmosis). But high pressure applied to the high-concentration side of the membrane can reverse this direction of flow. Freshwater diffuses through the membrane, leaving a more concentrated salt solution behind. To conserve space, and to help make reverse osmosis (RO) technology an economically acceptable water treatment process, the semipermeable membranes can be packaged in multiple layers of spiral-wound units and aligned in long tubes, as illustrated schematically in Figure 6.25. (Also see the color photo in Appendix H, Figure 12.)

Next to multistage flash distillation (which is used primarily to treat seawater), reverse osmosis is the second-ranking process in the worldwide production of desalted water. However, compared to multistage flash distillation, reverse osmosis requires only about half the energy to produce potable water. No heating is necessary for reverse osmosis; the major energy required is for pressurizing the feed water. Moreover, because of recent improvements in the durability and efficiency of semipermeable synthetic membranes, the market for reverse osmosis facilities is growing much faster than that for multistage flash distillation plants.

## CASE STUDY

### Desalination in the Nation's Oldest Seashore Resort

#### The Problem

The City of Cape May at the southern end of New Jersey is a National Historic Landmark and an increasingly popular vacation destination for people from all over the globe. Its year-round population of 6800 swells to nearly 43,000 during the summer months. As the

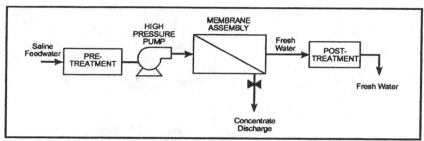

*Basic components of a reverse osmosis plant*

**FIGURE 6.26**

*In the Cape May reverse osmosis plant, pretreatment consists of fine filtration and the addition of chemicals to inhibit precipitation of chemicals with scaling potential, thereby preventing clogging of the membranes. The high-pressure pump (1700 kPa, or 250 psi) enables brackish water to pass through the membrane assembly while the salts are retained in the concentrate discharge. Post-treatment includes the addition of hypochlorite for disinfection, and lime to control the pH of the finished water.* (From O. K. Burros, *The ABCs of Desalting*, 2nd ed., International Desalination Institute, Topsfield, Massachusetts. Used with permission)

population increases, so does the daily water demand—the amount of water required each day to supply residents and businesses in the City of Cape May and the neighboring communities with potable water. The average water demand on an off-season day is about 2.6 ML (0.7 mil gal). The average daily demand is as high as 10 ML (2.7 mil gal) in the heat of the summer and at the height of the tourist season.

Through the years this water has been pumped from the "Cohansey aquifer" faster than it could be naturally replaced, causing *saltwater intrusion* (see page 138) from Delaware Bay and the Atlantic Ocean. By the early 1960s, two water supply wells had to be completely abandoned due to high salinity levels. Three more wells were drilled further inland as replacement supply wells. During the 1990s, even those new "Cohansey wells" showed signs of saltwater intrusion. In fact, one of the new wells also had to be abandoned as a potable water source. The City of Cape May was faced with a real water supply problem.

**The Solution**

As the water supply problem grew, the City of Cape May retained a firm of consulting civil and environmental engineers to evaluate the problem and to recommend a solution. The engineering study was completed in 1996. Of six water supply alternatives that were evaluated in detail, the consultant's recommendation was to construct a 7.6 ML/d (2.0-mgd) reverse osmosis (RO) desalination facility in two phases, with

two new water supply wells drilled into the so-called "Atlantic City 800-ft Sands aquifer." Even though the water in this area of the Atlantic City 800-ft Sands aquifer is brackish, the three new water supply wells were recommended so that the City's dependence on the Cohansey aquifer could be drastically reduced. By reducing the demand on the Cohansey aquifer, the intrusion of saltwater across the lower Cape Peninsula would be greatly slowed and perhaps stopped, thereby extending the water supply of neighboring communities to the north.

The City of Cape May accepted the recommended plan and awarded the consulting engineering firm a contract to design and oversee the construction of the desalination facility. The treatment plant was completed in 1999, providing a potable water supply projected to meet the needs of the City through the year 2020. The Cape May RO facility includes cartridge pressure filtration as well as chemical treatment. The cartridge filters remove particles large enough to lodge in the RO membrane. Chemicals are added to control the pH of the finished water to keep substances such as calcium carbonate from precipitating in, and clogging, the membrane and to maintain a residual disinfectant concentration in the finished water. A simplified schematic diagram of the treatment process is shown in Figure 6.26.

*Source:* Adapted from information provided by the City of Cape May Water & Sewer Utility. Used with permission.

## 6.7    RELEVANT WEB SITES

systems and provides technical assistance to America's rural water facilities.

### ABC's of Desalting

http://www.ida.bm/pages/publications/abcs.htm

This site contains a clearly written report that gives a comprehensive overview of modern desalting technology and worldwide desalination applications.

### American Water Works Association

http://www.awwa.org

This is the Home Page of the AWWA, a not-for-profit scientific and educational society dedicated to the improvement of drinking water quality and supply.

### Boca Raton Water Treatment Facility

http://www.ci.boca-raton.fl.us/services/utility/wplant.cfm

This site contains an interesting description of a groundwater treatment plant, including a process flow diagram and colorful photographs.

### Current Drinking Water Standards

http://www.epa.gov/ogwdw/mcl.html

This site contains the current National Primary and Secondary Drinking Water Regulations and Standards for inorganic chemicals, organic chemicals, radionuclides, and microorganisms.

### National Drinking Water Clearinghouse

http://www.estd.wvu.edu/ndwc/NDWC_homepage.html

The NDWC develops and maintains services and information related to small-community drinking water

### Office of Groundwater and Drinking Water

http://www.epa.gov/OGWDW/

This site is a gateway to a wide scope of up-to-date information about drinking water technology, federal regulations, activities, and drinking water standards.

### Public Drinking Water Systems Programs

http://www.epa.gov/safewater/pws/pwss.html

This is the Home Page of EPA's Public Water System Supervision Programs, providing information about public drinking water systems in each state, including facts and figures, rules, compliance reports, and contaminant occurrence in public water supplies.

### Safe Drinking Water Act

http://water.usgs.gov/public/eap/env_guide/h2o_quality.html#HDR2

The site gives the major provisions of the *Safe Drinking Water Act* (SDWA).

### Safe Drinking Water Overview

http://www.epa.gov/enviro/html/sdwis/sdwis_ov.html

This site contains information about public water systems and their violations of the EPA's regulations for safe drinking water.

## REVIEW QUESTIONS

1. Which water usually requires more extensive treatment for purification—groundwater or surface water? Why?

2. Sketch a flow diagram for a typical surface water treatment plant.

3. What is a *public water system?* What is the difference between a community system and a noncommunity system?

4. What do SDWA and MCL stand for?

5. What are the five general groups or types of contaminants that are controlled and limited under the SDWA? Briefly discuss the requirements for each group.

6. What is the difference between the SDWA primary regulations and the secondary regulations?

7. Briefly discuss the record-keeping and reporting requirements of the SDWA.

8. Describe the meaning of *detention time* and *overflow rate* with regard to the process of sedimentation.

9. Explain what is meant by *short-circuiting* in a sedimentation tank.

10. What is meant by the term *freeboard?*

11. Is a narrow and deep settling tank more effective in removing suspended solids than a wide and shallow tank? Explain your answer.

12. What is the function of a *tube settler?*

13. Briefly describe the process of *coagulation*. How does it improve the purification of drinking water?

14. What is the purpose of a *jar test?* Describe it briefly.

15. What is an *upflow clarifier?*

16. What is the purpose of *filtration?* What is a key characteristic of a *rapid filter?*

17. What is meant by *in-depth filtration?* How can it be accomplished?

18. Briefly describe the configuration of a typical rapid filter.

19. Briefly describe the operation of a typical rapid filter.

20. What is meant by *declining-rate filtration?*

21. What is meant by *direct filtration?*

22. What is considered to be the most important water treatment process with respect to preventing the spread of waterborne disease? Is there any potential harmful side effect from this process?

23. What are the disinfecting agents that kill bacteria when chlorine is added to water? What is the difference between *free chlorine* and *combined chlorine?* Compare their relative merits in disinfection.

24. What is meant by *breakpoint chlorination?*

25. Briefly describe the ways in which chlorine can be applied to water in a treatment plant.

26. What is the *DPD test?*

27. Briefly describe two methods other than chlorination that can be used to disinfect water supplies.

28. Briefly describe two methods used to soften water.

29. How can aeration improve drinking water quality? What methods are available to aerate water?

30. How can activated carbon improve drinking water quality? What are two important properties of activated carbon?

31. Why would sodium silicate be added to drinking water? Why would sodium flouride be added?

32. What is the difference between freshwater, brackish water, and seawater with respect to dissolved mineral content?

33. What are the two basic methods for desalinating water? Briefly describe a key process that is representative of each.

34. Visit and read the *ABC's of Desalting* Web page. Briefly describe the use of solar energy for desalting; describe the freezing method for desalting. Discuss both the advantages and disadvantages of these methods.

35. Visit and explore the *American Water Works Association* Web page. Read the latest news story in the *AWWA Mainstream* magazine, and write a brief summary of the article.

36. Visit the EPA *Safe Drinking Water Overview* Web page; locate your drinking water supplier and view its violations and enforcement history for the last 10 years.

37. Using the EnviroSources search engine (see Section 1.6) or another Internet search directory or search engine, locate at least one additional Web site relevant to one of the topics in this chapter. Add the link(s) to your Environmental Technology Folder. Write a brief description of what the Web site(s) contain.

38. Find out where your own household drinking water comes from (by contacting your city or township municipal utilities authority, public works department, or health department). Make an appointment to visit the facility, if possible, and tour the plant. Write a brief report on what you learned. Include data about the average and maximum plant capacities, the total population served, and the type of treatment processes used. Draw a flow diagram showing the sequence of all unit processes and treatment steps.

# PRACTICE PROBLEMS

1. A settling tank with a 50-ft diameter and a SWD of 9 ft treats a flow of 15,000 gpd. What is the detention time?

2. Compute the required volume of a sedimentation tank that provides 3 h of detention time for a flow of 10 ML/d. If the tank is 10 m by 25 m in plan dimensions, how deep is the water in the tank?

3. A clarifier operates with a surface loading of 500 gpd/ft$^2$. What is the slowest settling velocity of particles that will be completely removed in the tank?

4. A circular settling tank is to have a minimum detention time of 3 h and a maximum overflow rate of 800 gpd/ft$^2$. Determine the required basin diameter and SWD for a flow rate of 2 mgd.

5. A rectangular settling tank is to have a minimum detention time of 3.5 h and a maximum surface loading of 25 m/d. The tank length is to be twice its width. Determine the required tank dimensions, including freeboard, for a flow of 5000 m$^3$/d.

6. A rapid filter has plan dimensions of 10 ft by 15 ft, and it treats a flow rate of 1 mgd. Compute the filtration rate in terms of gpm/ft$^2$ and the velocity of flow through the filter bed.

7. Compute the required plan dimensions of a square filter box that will treat a flow of 9 ML/d at a maximum rate of 2.9 L/m$^2$ · s.

8. If the filter designed in Problem 7 is backwashed once a day for 10 min at a rate of 10.4 L/m$^2$ · s, what percentage of the total flow is used for cleaning the filter?

9. How many 100-lb chlorine cylinders should be ordered per month in order to disinfect a flow of 12 mgd of water using a 0.6-mg/L dose of chlorine? Would you recommend ordering 1-ton cylinders instead of 100-lb cylinders?

10. A mass of 150 kg/d of chlorine is used to disinfect a flow of 250 000 m$^3$/d. What is the chlorine dose?

11. A mass of 20 kg per day of chlorine is applied to water, resulting in a chlorine concentration of 0.4 ppm. What is the flow rate of the water?

## Chapter Outline

C H A P T E R

# 7

# Water Distribution Systems

A water distribution system is an interconnected network of pipelines, storage tanks, pumps, and smaller appurtenances, including valves and flow meters. The purpose of this chapter is to describe some of the practical aspects related to the design, analysis, and operation of these systems.

The chapter begins with discussion of basic design factors, materials, and appurtenances. A section on centrifugal pumps—the prime movers of a distribution system—is included. Conservation and distribution reservoirs are discussed, and the analysis of pipe network hydraulics is covered. Much of this material assumes some prior knowledge of hydraulics, so it would be best to study or review Chapter 2 before starting this chapter.

In engineering practice, sophisticated computer modeling programs are used to analyze and design water distribution networks. Many of these programs are integrated with computer-aided drafting (CAD) software and provide graphical output as well as hydraulic analysis. Application of water distribution modeling software is discussed in Section 7.5. But the best way to learn about water distribution system design and analysis is to do the computations "by hand," that is, with an electronic hand-held calculator. It is only after the student obtains a firm grasp of underlying computational methods and terminology that powerful software packages can be used effectively.

## 7.1 DESIGN FACTORS

The design of a water distribution system begins after a study of community water requirements has been completed. A water distribution system must be able to deliver adequate quantities of water for various uses in a community. Also, sufficient pressures must be maintained throughout the system.

A survey of the service area is required so that maps of streets and topographical features can be prepared. On a relatively small scale map (about 1:24,000, or 1 in. = 2000 ft), the principal elements of the system can be planned, showing the general locations of water mains, pump stations, storage tanks, and so on. On larger scale maps (about 1:600, or 1 in. = 50 ft), the exact locations of the proposed facilities are shown in detail, as are existing utilities such as sewers or gas mains. These plan drawings are accompanied by written specifications describing the materials and methods of construction.

### Required Flows and Pressures

It is convenient to classify water demands or water uses into four basic categories, as follows:

1. *Domestic* water for drinking, cooking, personal hygiene, lawn sprinkling, and the like.
2. *Public* water for fire protection and street cleaning and for use in schools or other public buildings.
3. *Commercial and industrial* water for restaurants, laundries, manufacturing operations, and the like.
4. *Loss* due to leaks in mains and house plumbing fixtures.

The total demand for water in a community varies, depending on the population, the industrial and commercial activity, the local climate, and the cost of the water. For example, in warm, dry climates, domestic use is generally a larger fraction of total consumption than it is in colder climates; lawn watering is much more common in dry climates. However, when the water bill is based on individual meter readings rather than on a flat rate, conservation is encouraged and water demand decreases.

### Per Capita Demand

If the total annual water use of a community is divided by 365 d, a value of average daily water consumption is obtained. If this value is further divided by the total population served, a *per capita* value is obtained. In SI units, this is expressed in terms of liters per day per person, in U.S. Customary units in gallons per capita per day (gpcd).

For example, if the average daily water demand is 5 megaliters per day (5 ML/d) in a system serving 10,000 people, the average per capita demand would be (5 000 000 L/d)/(10,000 people) = 500 L/d per person. Keep in mind that a figure like this includes each person's share of industrial, commercial, and public use and leakage; it is not just individual domestic use.

Since the exact water demands of a new service area may not be known, it is common to use average per capita values from similar communities in order to design the new distribution system. New systems are generally designed to accommodate populations and water demands that are anticipated 10 to 30 years in the future. Otherwise, the system would be too small soon after it was built. Table 7.1 presents overall average daily water demands in the United States.

### Variations in Water Demand

In any community, water demand will vary on a seasonal, daily, and hourly basis. For example, on a hot summer day it is not unusual for water consumption to be as much as 200 percent of the average daily demand. If the average demand is 670 L/d, then we can estimate a peak daily demand to be $2 \times 670 = 1340$ L/d per person. Generally, the pipelines and pumps of a distribution system (as well as treatment plants and wells) must be designed to accommodate peak daily flows rather than average flows. The minimum flow a system should

**TABLE 7.1**
**Estimated average water requirements in the United States**

| Type of use | L/d per person | gpcd | Percent of total |
|---|---|---|---|
| Domestic | 300 | 80 | 44 |
| Commercial/ industrial | 260 | 70 | 39 |
| Public | 60 | 16 | 9 |
| Loss | 50 | 14 | 8 |
| Total | 670 | 180 | 100 |

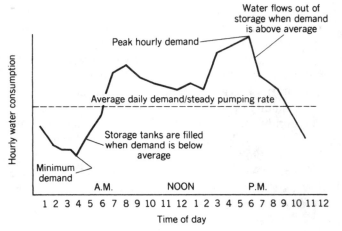

**FIGURE 7.1**

*A graph that shows the typical variation in water demand or consumption throughout the day.*

be designed for is about 1000 L/d per person (or about 250 gpcd).

Water consumption also varies hourly throughout the day, according to a somewhat predictable pattern. Peak hourly demands in residential districts usually occur in the morning and evening hours, just before and after the normal workday. In commercial or industrial districts, water consumption may be uniformly high throughout the workday. Minimum flows typically occur around 4 A.M., when almost no one is using water.

A graph illustrating typical hourly variations in water use is shown in Figure 7.1. On this graph, the peak hourly flow occurs at about 6 P.M. In extreme cases, these maximum hourly flows could be as much as 10 times the average flow, but they are usually around 3.5 times the average flow rate. As discussed later, these peak hourly demands are generally accommodated by water from storage tanks instead of by the pumps in the system. Otherwise, the pumps and pipes would have to be excessively large just to handle flows that occur for a relatively short time.

There can be a wide variation in average, peak daily, and peak hourly flow rates among different communities. As far as is practical, the specific water demands should be determined or estimated for each service area. Generally, big cities have higher per capita water use than small communities, and small service areas are noted for their very high peak rates.

**Fire Flows**

Water for fire fighting is an important part of the total demand that must be provided for in a water distribution system. Fire flows are only required once in a while, and the total amount of water used to extinguish

fires in any year is small compared to all other uses. But the rate and volume of water needed in the few hours of a fire emergency can be large in a local area. Sometimes, it can be the controlling factor affecting the size of the water mains.

Municipal insurance rates depend to a large extent on the fire protection provided by the distribution system. Factors involved in determining required fire flow capacity include type of building construction, occupancy, sprinkler protection, and so on. As a minimum, 30 L/s (475 gpm) of fire flow is required for at least 2 h. In more extreme cases, up to 760 L/s (12,000 gpm) for a 10-h duration may be necessary. The required fire flow must be added to the peak daily demand in the system when sizing pipes and pumps.

**Pressures**

Water pressures in a distribution system should not drop below 350 kPa (50 psi) in order to provide for adequate operation of home plumbing fixtures and appliances as well as for fire fighting when pumper trucks are used at fire hydrants. Maximum pressures in water mains are generally kept below 760 kPa, or 110 psi, to reduce the chances for leaks or water main breaks. Pressures of about 550 kPa (80 psi) are considered optimum. Pressure-regulating valves must be installed in the distribution system to reduce pressures in low-lying service areas; otherwise, pressure heads in the system would be too high.

**Pipeline Layout**

Water mains are generally not less than 150 mm (6 in.) in diameter. They are usually located in the street right-of-way (ROW) so as to provide water to every potential customer. The *gridiron* arrangement of pipes is preferred to a layout that has many dead-end branches. In the gridiron system, water can circulate in interconnected loops, but in the dead-end system, the water may remain relatively stagnant in sections of the system, causing taste and odor problems from bacterial growth. The two types of layouts are illustrated in Figure 7.2.

In the dead-end layout, frequent flushing of the pipes at the fire hydrants is necessary to prevent consumer complaints about taste and odor. Another disadvantage of the dead-end system is that water service could be disrupted for long periods of time while repairs are made to a broken water main. But in a gridiron system, the broken section can be isolated by valves, and water can still reach consumers from the other side of the loop. Most distribution systems combine both layouts, depending on local conditions and economic factors.

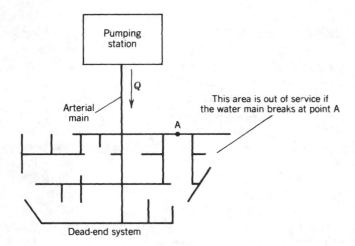

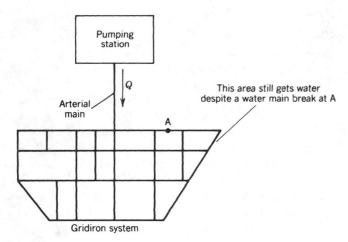

**FIGURE 7.2**
*A gridiron pattern for water mains is preferable to a dead-end type of system; gridiron networks provide greater flexibility in operation and service.*

Water mains may be referred to as *primary feeders* or *secondary feeders*. The primary feeders, also called *arterial mains*, carry large quantities of water from the treatment or pumping facility to areas of major water use. Secondary feeders are smaller pipes that provide a daily supply to local areas.

## 7.2 WATER MAINS

### Materials

The water mains in a distribution system must be strong and durable in order to resist applied forces and corrosion. The pipe is subjected to internal pressure from the water and to external pressure from the weight of the soil (backfill) and vehicles above it. Another force the pipe may have to withstand is called *water hammer*. This can occur when a valve is closed too fast, for example, causing waves of high pressure to surge through the pipe. Finally, damage due to corrosion or rusting may occur internally because of the water quality or externally because of the nature of the soil conditions. Materials commonly used in water mains to provide adequate strength and durability are discussed here.

Detailed specifications for pipe materials can be obtained from the American Water Works Association (AWWA) or from individual pipe manufacturers.

### Ductile Iron Pipe

Ductile iron is one of the most common materials used for the construction of water distribution pipelines. Because of its chemical composition, ductile iron is stronger and more elastic than gray cast iron, which was the predominant pipe material used until the mid-1900s. Cast iron (CI) is also strong and durable, with many older installations still in service after 100 years or more. Ductile iron, though, is less brittle than gray cast iron; it is less vulnerable to damage during construction and is considered to be more corrosion-resistant.

Ductile iron pipe sections are available in lengths up to about 6 m (18 ft) and in diameters up to 1200 mm (48 in.). It is manufactured in several thickness classes or groups; higher class pipe (thicker pipe walls) would be specified for deep installations or high water pressures.

Two common methods for joining individual sections of pipe together are the *push-on joint* and the *flanged joint*. In the push-on or *compression-type* joint, as it is also called, a *spigot end* of one pipe is pushed into the *bell end* of an adjacent pipe; a rubber-ring gasket in the bell end is compressed when the sections are joined, creating a watertight, but flexible, connection. Flanged connections involve the bolting together of the ends of the pipe sections; they are used for aboveground installations in treatment plants or pumping stations. Both push-on and flanged joints are illustrated in Figure 7.3.

Unprotected iron pipes are subject to a process called *tuberculation*—the formation of tubercles or small projections of rust on the inside wall of the pipe. Tuberculation significantly increases the resistance to flow, reduces pipe capacity, and increases pressure losses in the system. Iron pipes are usually coated with a thin cement–mortar lining on the inside pipe wall to prevent tuberculation and to preserve the hydraulic capacity of the pipe. (The Hazen–Williams *C* factor may

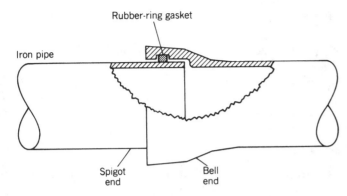

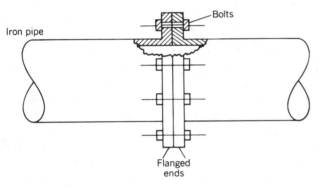

**FIGURE 7.3**
*Two common methods for joining sections of iron pipe.*

be as high as 145 for cement-lined pipes; see Section 2.3.) Tarlike coatings are also applied on the outside of the pipe to prevent external corrosion.

### Asbestos Cement (AC) Pipe

A compacted mixture of sand, cement, and asbestos fibers provides a lightweight pipe material that is smooth and corrosion-resistant. Although it is not as strong as iron pipe, the absence of tuberculation and the ease of installation make AC pipe desirable in many instances.

AC pipe has long-lasting hydraulic properties with high carrying capacity ($C = 140$). Manufactured in about 4-m (12-ft) lengths and in diameters up to about 400 mm (16 in.), it is also available in several pressure classes up to a maximum of about 1400 kPa (200 psi). The plain ends of AC pipe sections are easily joined with a coupling sleeve, as illustrated in Figure 7.4. The two rubber-ring gaskets in the sleeve provide a water-tight, yet flexible joint.

Although asbestos is a known carcinogen when inhaled (especially among smokers), there is no evidence linking ingested asbestos to any particular disease. Nevertheless, manufacturers are experimenting with substitutes for asbestos, including other minerals, carbon fibers, and plastic fibers. The *Safe Drinking Water Act* (SDWA) allows 7 million fibers per liter (whereas airborne asbestos is of concern at 100 fibers per liter).

### Plastic Pipe

Poly vinyl chloride (PVC) plastic is sometimes used as a pipe material for construction of water distribution mains. These plastic pipes are strong and durable, yet they are very lightweight and are easily handled and installed. They are resistant to corrosion and they are very smooth, providing excellent hydraulic characteristics ($C = 150$). Available in diameters up to about 600 mm (24 in.), PVC pipe sections are joined using a bell-and-spigot, compression-type joint with rubber-ring seal.

Other plastic materials used for service connections and domestic plumbing include polyethylene (PE) and acrylonitrile–butadiene–styrene (ABS) plastic. These pipes may be joined using threaded screw couplings or chemical solvent welds.

### Other Pipe Materials

Reinforced concrete pipes (RCPs) are made of welded steel cylinders wrapped with steel wire and embedded in concrete. They are used primarily in long water-transmission lines of large diameter. They can be precast in sections up to 5 m (16 ft) in length and up to about 6 m (20 ft) in diameter. RCP pipes are very strong and durable, and have excellent hydraulic characteristics. Sections may be joined using a modified bell-and-spigot

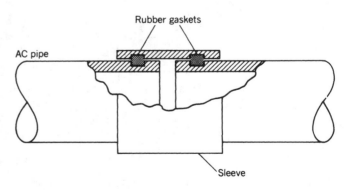

**FIGURE 7.4**
*Asbestos–cement pipes are joined with a sleeve and two rubber-ring gaskets.*

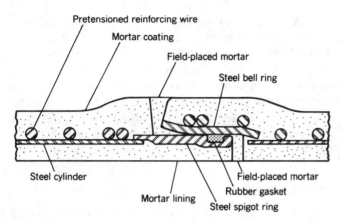

**FIGURE 7.5**
*Section of an RCP joint.* (Reprinted from *Water Distribution Operator Training Handbook*, by permission. Copyright © 1976, The American Water Works Association)

type of connection and sealed with cement mortar, as shown in Figure 7.5.

Steel pipe is sometimes used for water transmission lines, particularly for aboveground installations. It is very strong, yet it is lighter in weight than RCP. But it must be carefully protected against corrosion; this is usually done by lining the interior and painting and wrapping the exterior. Sections of steel pipe may be joined by welding or with mechanical coupling devices as shown in Figure 7.6.

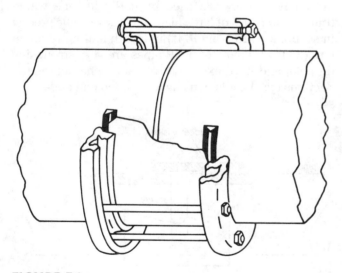

**FIGURE 7.6**
*A mechanical coupling device used to join sections of steel pipe.* (Reprinted from *Water Distribution Operator Training Handbook*, by permission. Copyright © 1976, The American Water Works Association)

## Appurtenances

Proper functioning of the water mains in a distribution system requires many different devices in addition to the sections of pipe. These devices, called *appurtenances*, include hydrants, shutoff valves, throttling valves, pressure-reducing valves, and other fittings.

### Hydrants

The primary purpose of a hydrant is to provide convenient access to water for firefighting and other emergencies. A hydrant also serves for flushing out water mains, washing debris off public streets, and providing access to the underground pipe system for pressure testing. The spacing and location of hydrants depend primarily on fire protection and insurance needs. Hydrants are also placed at dead ends and at high and low points in the pipeline.

A long valve stem inside the cast-iron barrel of the hydrant operates a shutoff valve at its base. Another valve in the pipe connecting the hydrant to the water main allows isolation of the hydrant for maintenance. The connecting pipe is usually 150 mm (6 in.) in diameter, and the hydrant has two hose connections on top. A gravel footing is provided to allow drainage from the barrel after the hydrant is used; this is particularly important in cold climates because the water can freeze and break the hydrant barrel.

### Service Connections

Water from the distribution main reaches the property line of individual consumers through a service pipe, usually made of copper or plastic, with a minimum diameter of 20 mm ($^3/_4$ in).

Service connections can be made initially when the main is installed (a dry tap) or later when the main is already in service (a wet tap). Pipe-tapping machines are available that allow wet taps to be made without affecting water service to existing users of the system. The service pipe is connected to the main by means of a special fitting called a *corporation stop*. At the user's end of the service line, there is usually a water meter and shutoff valve.

### Valves

Many different types of valves are used in water distribution systems to control the quantity and direction of flow. Many of these can be opened or closed manually by screw stems or gear train devices; large valves often are power-operated using electric or hydraulic systems.

The most common function of a valve is for complete shutoff of flow. *Gate valves* are usually used for this purpose. They are placed throughout the distribution network, allowing sections of pipeline to be shut off and isolated during repairs of broken mains, pumps, or hydrants. A gate valve consists basically of a sliding disk that is moved across the path of flow by a screw-operated stem. When the valve is in the open position, the disk is enclosed in a valve cover or housing above the pipe and is completely out of the path of flow. In the closed position, the disk is lowered and tightly wedged in a valve seat, blocking the flow. A typical gate valve is illustrated in Figure 7.7a.

Gate valves are usually either in the fully open or fully closed position; they are rarely used for throttling flow by blocking it only partially. In most distribution networks, they are placed at pipe intersections and are operated manually using an extension rod to reach the operating nut on the valve; cast-iron valve boxes, which extend from the valve up to the street surface, cover and protect the underground valve. Large gate valves may be placed in underground manhole structures or valve pits for easier access and operation.

A type of valve commonly used for throttling and controlling flow rate is the *butterfly valve.* In a butterfly valve, a movable disk rotates on an axle in the path of flow. In the closed position, the disk is tightly seated against a rubber ring in its casing. In the open position, the disk is turned 90°, allowing the water to flow around and past it. The fact that the disk is always in the flow is a disadvantage of the butterfly valve since it blocks the use of pipe-cleaning tools. Because the force of flowing water tends to close the valve, reducing gear drives are used for manual operation, and power operators are required for the large butterfly valves. A butterfly valve is illustrated in Figure 7.7b.

A device called a *check valve* is used to permit flow in only one direction in a pipe; it closes automatically when the flow stops or tends to flow in the opposite direction. A common type, called a *swing-check* valve, is illustrated in Figure 7.8. The valve disk is lifted up by the force of the flowing water and closes by gravity when the flow stops. A valve seat prevents the disk from swinging open in the opposite direction.

Check valves are usually installed in the discharge piping of a pump to prevent backflow when the pump stops. They are called *foot valves* when installed at the end of a pump suction line in a well or tank. Foot valves prevent loss of prime in the centrifugal pumps. (The operation of centrifugal pumps is discussed in the next section.) In plumbing systems, special double-check valves may be used to prevent backflow and possible contamination of a drinking water supply when a cross section with another system exists.

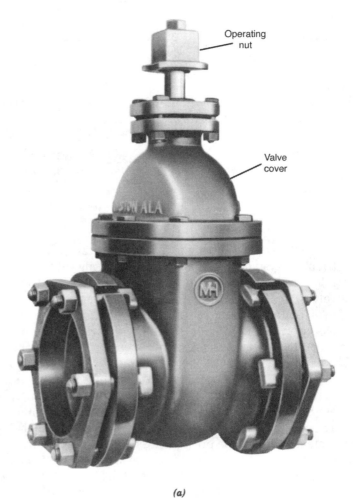

(a)

(b) BUTTERFLY VALVE

**FIGURE 7.7**

(a) *A typical gate valve* (Courtesy of M & H Valve Company, Division of McWane, Inc., Anniston, Alabama); (b) *a typical butterfly valve.* (Courtesy of American-Darling Valve, Division of American Cast Iron Pipe Co., Birmingham, Alabama)

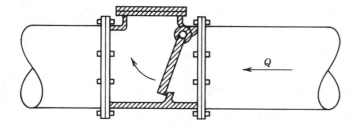

**FIGURE 7.8**
*A swing-check valve will open to allow flow in only one direction.*

Other types of valves that find use in water distribution systems include pressure-reducing valves, air-release valves, and altitude valves. Pressure-reducing valves operate automatically to lower excessive hydrostatic pressure in water mains that are at a low elevation in the system. In effect, these valves form separate networks or pressure zones in a large distribution system.

Water mains generally follow the hills and valleys of the natural topography. It is not uncommon for pockets of air to develop at the high points of the main. These pockets of air reduce the flow capacity of the system and increase pressure losses. Air-release valves are placed in the pipeline at the peaks to automatically vent the accumulated air in the system.

Another appurtenance, called an *altitude valve*, is an automatic device that controls flow into an elevated water storage tank. It automatically closes when the tank is full, preventing overflow. When there is demand for water from the tank, the lower pressure in the distribution main is sensed by the valve mechanism and the valve opens to allow flow out of the tank. In effect, the water in the tank "floats" on the water in the main and freely flows into or out of the tank, depending on pressure differentials. The hydraulics of elevated storage in water distribution systems is discussed later in this chapter.

## Installation

Water mains must be installed at sufficient depths below the ground surface to provide protection against traffic loads and to prevent freezing. Generally, these depths are in the range of 1 to 2 m (3 to 6 ft). Since flow occurs under pressure instead of by gravity, the water mains can follow the general topographic shape of the ground, uphill as well as downhill.

There are several ways of placing the pipeline in an excavated trench to provide additional strength and protection. These different pipe *bedding* methods are

illustrated in Figure 7.9. The material placed in the trench on top of the pipeline is called *backfill*. Compacting the backfill in layers around the pipe barrel increases the support for the pipe and can reduce the incidence of water main breaks. The type of bedding condition selected depends on the trench depth and the thickness class of the pipe.

Water mains should not be installed in the same trench with a sewer line. Generally, they should be at least 3 m (10 ft) away from a sewer, horizontally, and they should be at least 0.5 m (18 in.) higher than sewer lines when they cross.

| Laying Condition | Description |
|---|---|
|  Type 1* | Flat-bottom trench. † Loose backfill. |
| 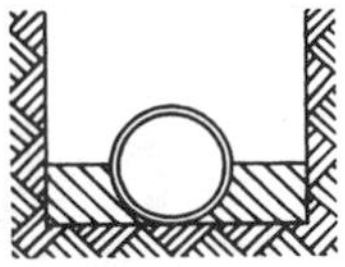 Type 2 | Flat-bottom trench. † Backfill lightly consolidated to center-line of pipe. |
| 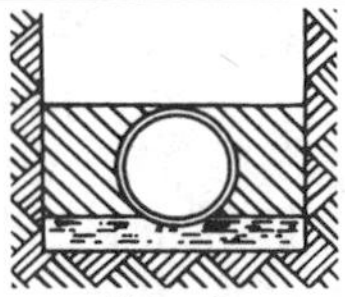 Type 3 | Pipe bedded in 4-in. minimum loose soil. †† Backfill lightly consolidated to top of pipe. |
|  Type 4 | Pipe bedded in sand, gravel or crushed stone to depth of ⅛ pipe diameter, 4-in. minimum. Backfill compacted to top of pipe. |
|  Type 5 | Pipe bedded in compacted granular material to centerline of pipe. Compacted granular or select†† material to top of pipe. |

*For 30-in. and larger pipe, consideration should be given to the use of laying conditions other than Type 1.
†"Flat-bottom" is defined as undisturbed earth.
††"Loose soil" or "select material" is defined as native soil excavated from the trench, free of rocks, foreign materials and frozen earth.

**FIGURE 7.9**
*Standard trench bedding conditions for ductile iron pipe.* (Courtesy of U.S. Pipe and Foundry Company, Birmingham, Alabama)

## Pressure Testing

No matter how well a pipeline is constructed, there may be some leakage at the joints. Within certain limits, some leakage is acceptable; the construction specifications should indicate the maximum allowable rate of leakage. A common formula used for this purpose is the following:

$$Q_L = \frac{N \times D \times P^{1/2}}{C} \qquad (7\text{-}1)$$

where $Q_L$ = allowable leakage, L/h (gal/h)
$N$ = number of joints in length of main tested
$D$ = pipe diameter, mm (in.)
$P$ = test pressure, kPa (psi)
$C$ = a constant depending on units used: for SI metric, $C = 32{,}600$ (for U.S. Customary units, $C = 1850$)

A pressure or leakage test is conducted on a newly installed water main by filling the pipe with water and maintaining a pressure of 1000 kPa (150 psi) for 1 h. If excessive leakage occurs, an amount of water greater than $Q_L$ must be pumped into the line to maintain the pressure; repairs are necessary before the pipeline can be accepted for use.

## EXAMPLE 7.1

A 300-m-long section of a newly installed 305-mm-diameter water main is pressure tested for leakage. It was observed that during the 1-h test period, a volume of 10 L of water was pumped into the pipeline to maintain the required pressure of 1000 kPa. The pipe sections are 6 m long between joints. Has the allowable rate of leakage been exceeded?

*Solution*   First, compute the number of joints in 300 m of pipeline as

$$N = \frac{300 \text{ m}}{6 \text{ m}} = 50 \text{ joints}$$

Now, applying Equation 7-1 gives

$$\frac{50 \times 305 \times 1000^{1/2}}{32\,600} = 15 \text{ L/h}$$

Since the observed leakage of 10 L/h is less than the computed allowable leakage of 15 L/h, the pipe is sufficiently watertight.

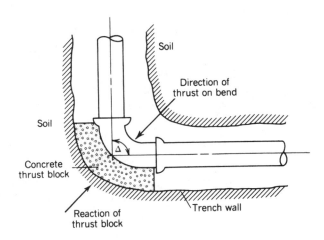

Plan (top) view

Soil

Direction of thrust on bend

Soil

Concrete thrust block

Reaction of thrust block

Trench wall

**FIGURE 7.10**
Thrust blocks *are used to anchor the pipeline and to prevent movement or possible joint opening at bends.*

## Thrust Blocks

It is usually necessary to anchor the pipeline securely in the trench at dead ends and at bends or at changes in horizontal or vertical direction. This is because of the force, or thrust, caused by the internal pressure and the kinetic energy of flow, which tends to move the pipe or fittings. Such movement can damage the joints and cause excessive leakage. One method for providing the necessary anchorage is to use concrete thrust blocks, as illustrated in Figure 7.10.

In most cases, the internal pressure causes most of the thrust, and the dynamic thrust due to the flow velocity can be neglected. For a bend in a pipeline, the thrust due to static pressure can be computed from the following formula:

$$F = 2 \times P \times A \times \sin(\Delta/2) \qquad (7\text{-}2)$$

where $F$ = force or thrust, kN (lb)
$P$ = water pressure, kPa (psi)
$A$ = cross-sectional area of pipe, m$^2$ (in.$^2$)
$\Delta$ = change in direction of the pipe, degrees

## EXAMPLE 7.2

An 18-in.-diameter pipe carries water under a pressure of 80 psi. Compute the static thrust for a 90° bend in the pipe, and compute the required bearing area of a concrete thrust block if the soil can support a bearing stress of 3000 lb/ft$^2$.

*Solution*  First, compute the area of the pipe section as

$$A = \frac{\pi D^2}{4} = \pi \times \frac{18^2}{4} = 254 \text{ in.}^2$$

Now use Equation 7-2 to compute the static thrust:

$$F = 2 \times 254 \times 80 \times \sin\left(\frac{90}{2}\right)$$

$$= 2 \times 254 \times 80 \times 0.707 \approx 30{,}000 \text{ lb}$$

If the soil can withstand 3000 lb/ft$^2$, then the required area of the thrust block that is needed to spread the force out of the supporting soil is 30,000 lb ÷ 3000 lb/ft$^2$ = 10 ft$^2$.

---

### Disinfection

A newly installed water main must be flushed clean and disinfected before being put into service. Flushing velocities of about 1 m/s (3 ft/s) are generally enough to remove dirt and debris that may accumulate in the pipe during construction. The pipe is disinfected to kill bacteria by filling it with a relatively concentrated chlorine solution for a certain period of time, as specified by local regulatory agencies. It is flushed again before being put into service.

### Rehabilitation

The use of proper material and installation methods does not guarantee trouble-free operation of a water main for an unlimited period of time. Pipeline breaks and leaks occur periodically for several reasons, and emergency repairs must be made. Most water utilities have a plan of action for dealing with these emergencies and keep spare parts, tools, and equipment readily available.

Leaks that are not readily observable from wet or sunken spots in the street can be located by using sounding rods for electronic amplification of the sound of the escaping water. Relatively small leaks can be repaired without shutting off the water pressure. This not only avoids inconvenience to utility customers, it also prevents contamination of the distribution system by backflow (backflow of dirty water into the system can occur when the water pressure is turned off). All water is pumped out of the excavated trench. Repair clamps or sleeves are installed over the leakage section of pipe and tightened until the leakage stops. The pipe should be flushed, hydrostatically retested, and disinfected after the repair has been made.

### Cleaning

Sudden water main breaks are not the only problems that can occur in a water distribution system. Loose deposits of sediment may accumulate in the pipeline, particularly in dead-end branches. These sediments, which cause taste, odor, and color problems, can be removed by periodic flushing through hydrants.

Many water mains suffer the effects of a gradual and persistent buildup of solid deposits on the inside wall of the pipe. The longer the pipeline is in service, the worse this problem becomes. These deposits, illustrated in Figure 7.11, reduce the hydraulic capacity of the pipeline and cause high pressure losses. Pumping costs increase and there is a greater chance for regrowth of bacteria in the distribution system.

The deposits may consist of tubercles or lumps of iron oxide if the pH of the water is low and the metal pipe is unlined. This problem is called *tuberculation*. When the pH of the water is high, the deposits consist of calcium carbonate scale. Maintaining a proper pH so that the water is neither corrosive nor scale-forming may be accomplished by adding chemicals at the treatment plant. Although this can minimize the formation

**FIGURE 7.11**
*View of a deteriorated water main with accumulated interior deposits that reduce the pipe capacity.* (Courtesy of Centriline Division of Raymond International Builders, Inc., Rochelle Park, New Jersey)

of additional deposits, it does not restore the lost capacity that has already occurred.

A common method for rehabilitating old water mains is to clean them using a mechanical or hydraulic scraper tool. During the cleaning operation, water supply service can be provided to customers by using small-diameter temporary bypass pipes or hoses. Thick deposits can be removed using power-driven mechanical cleaning devices. The cutting tool consists of rotating steel blades mounted on a series of body sections attached to a center rod.

Another type of cleaning device is a bullet-shaped resilient foam object called a *pig*. It is wrapped with wearing strips in a spiral or crisscross arrangement. Propelled by water pressure through the pipeline, the pig serves to scrape the deposits off the wall of the pipe. It is inserted into the line either through a fire hydrant or through a specially installed *launcher*. Several passes of the pig may be needed to remove all the deposits. Afterward, the pipe is flushed out and disinfected before being placed into service again.

### Lining

Newly installed metal pipes are supplied with a cement–mortar lining to prevent tuberculation. But many iron water mains were installed in the past without these linings. When these mains are cleaned and the deposits removed, it is necessary to install a lining that will prevent the problem of tuberculation from recurring. One such rehabilitation method, called *sliplining*, involves placing a plastic pipe inside the cleaned pipe. The plastic pipe, of slightly smaller diameter than the original pipe, is pulled through straight sections of the transmission main.

Another method for protecting the cleaned pipe wall is to apply a cement–mortar lining to the pipe in place. For pipes less than 600 mm (24 in.) in diameter, mortar can be pumped through a hose to a lining machine that is pulled through the pipeline. The mortar is sprayed centrifugally onto the pipe wall; the thickness of the lining can be controlled by the speed at which the machine is pulled through the pipe. A lining thickness of 6 mm or less is preferred, so as not to reduce the inner diameter of the pipe excessively. For pipes over 600 mm in diameter, a worker can enter the water main with the lining equipment.

Cleaning and lining can effectively rehabilitate a water main, increasing its carrying capacity and reducing pressure drops. But the pipe must not be structurally defective if this method of rehabilitation is to be used. A deteriorated pipeline that experiences frequent breaks and leaks may have to be replaced completely with a new pipeline. Large concrete water mains, however, may be lined with pipe made of steel plate to restore structural integrity with minimum loss of capacity. It is always necessary to compare the relative economics of complete replacement versus cleaning and relining. If water demands in the area are increasing, new water mains may be needed; rehabilitation alone may not be sufficient to meet the higher demands.

## 7.3  CENTRIFUGAL PUMPS

A pump is a mechanical device that adds energy to water or other liquids. In most water distribution systems, pumps are needed to raise the water in elevation and to move it through the network of water mains under pressure. One way of classifying pumps is by their application in the system. Pumps that lift the water from a river or lake and move it to a nearby treatment plant are called *low-lift pumps*. They move large quantities of water, but at relatively low discharge pressures. The pumps that discharge the treated drinking water into the transmission and distribution system are called *high-lift pumps;* they operate under relatively high heads or pressures. (See Appendix H, Figure 13).

Sometimes it is necessary to increase the pressure within the distribution system or to raise the water into an elevated storage tank; *booster pumps* can be used for this purpose. *Well pumps* lift water from an underground aquifer and often discharge directly into the distribution system.

Another way of classifying pumps is according to the mechanical principles on which they operate. The two basic types are *positive-displacement pumps* and *centrifugal pumps*. A positive-displacement pump will deliver a fixed quantity of water with each revolution of the pump rotor or piston. The water is physically pushed or displaced from the pump casing. The capacity of the pump is unaffected by changes in pressure in the system in which it operates.

Centrifugal pumps are the most common type used in water supply (as well as wastewater) systems. As discussed shortly, the capacity of the pump is very much a function of the pressure against which it operates in the system. A centrifugal pump adds energy to the water by accelerating it through the action of a rapidly rotating impeller. The water is thrown outward by the vanes of the impeller and passes through a spiral-shaped casing, where its velocity is gradually slowed down. As its velocity drops in the expanding spiral volute, the kinetic energy is converted to pressure head, called the *discharge pressure*.

Centrifugal pumps can be further classified as *radial flow* or *axial flow*. In the radial-flow type, the water discharges at right angles to the direction of flow into the pump impeller; in the axial-flow type, the water discharges in the same direction as the axis of the impeller. Centrifugal pumps with more than one impeller are called *turbine pumps.*

Centrifugal-type pumps have several advantages over positive-displacement pumps. They are simple, with only one moving part—the impeller; no internal valves are required, and there is no need for internal lubrication. Also, they operate very quietly. Disadvantages include the effect of pressure on the pump output and efficiency, and the necessity for priming the pump before it is operated. Priming involves filling the pump casing and suction line with water.

## Pump Characteristics

The performance characteristics of a centrifugal pump define the relationships among the discharge or rate of flow, the discharge pressure head against which the pump operates, the power requirements, and the efficiency of operation. These characteristics depend on the diameter, speed, and shape of the impeller(s) within the pump casing. They are different for each model of pump.

### Pump Head Curve

Pump manufacturers provide data in catalogs that describe the performance characteristics of their line of pumps. These data are usually presented graphically, for convenience. A typical graph showing the relationship between the rate of flow and the total pressure head for a centrifugal pump, for a fixed impeller speed and diameter, is shown in Figure 7.12. It is called a *pump head curve.*

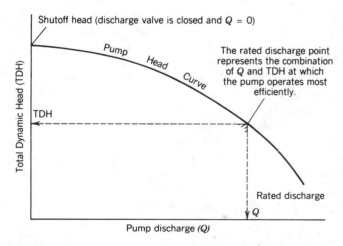

**FIGURE 7.12**

*Typical centrifugal pump head–discharge curve. The discharge decreases as the TDH or pressure head on the pump increases.*

The pressure head is plotted on the vertical axis of Figure 7.12. This is called the *total dynamic head,* or TDH, and is expressed in meters (feet) of water. The TDH depends on the configuration of the system into which the pump discharges and the flow rate; this is explained in more detail shortly. The flow rate or discharge, $Q$, is plotted on the horizontal axis and is expressed in terms of L/s or gpm.

The pump head curve clearly shows that the pump discharge rate is a function of the total pressure against which the pump works and that a given centrifugal pump can operate over a wide range of $Q$ and TDH values. *The discharge decreases as the TDH increases.* In effect, the harder the pump has to work to move the water, the less it can discharge per unit time. Specification or description of the operating conditions for a centrifugal pump must always have a $Q$ value and a corresponding TDH value.

A centrifugal pump that operates with its discharge valve completely shut will build up a maximum discharge pressure, called the *shutoff head.* At shutoff head, the discharge is zero, but the rotating impeller adds energy to the water circulating within the pump casing and this develops pressure. When the valve is gradually opened and the water begins to flow, the pressure head will decrease, as the pump head curve shows. When the valve is completely opened, the unthrottled pump will operate at a discharge and TDH that match those of the system in which the pump is working.

Pump head curves, or *head–discharge curves* as they are also called, are determined by the pump manufacturer under shop test conditions. During the test, power and efficiency characteristics of the pump are also measured. Like any other mechanical device, a centrifugal pump cannot operate at 100 percent efficiency. There is always water circulating with the rotating impeller in the casing, causing energy loss. This is called *slip,* and the more slip there is in a given pump, the lower is its efficiency.

The efficiency of a centrifugal pump depends on the discharge and TDH. At shutoff head ($Q = 0$), the efficiency is zero because the energy driving the impeller is entirely lost (as heat); no physical work is done by the completely throttled pump. As the discharge is allowed to increase, the efficiency of operation will increase to a maximum value and then begin to decrease. The combination of $Q$ and TDH at which the maximum efficiency is measured is called the *rated* or *normal discharge capacity* of the pump. The rated capacity may be indicated on the pump head curve with a mark, as shown in Figure 7.12.

Pump manufacturers often indicate the efficiency and power consumption, over a range of flows

Performance characteristics

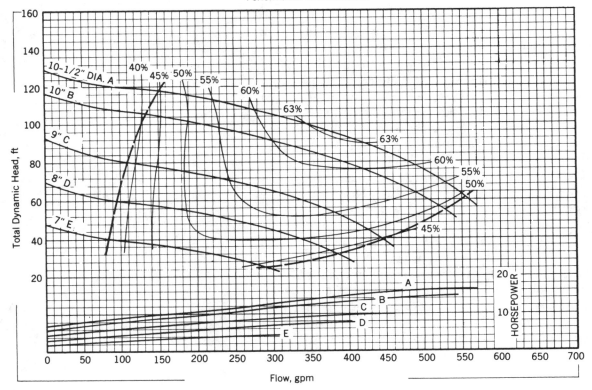

**FIGURE 7.13**

*Typical set of manufacturer's centrifugal pump characteristic curves, which
include data on head capacity, efficiency, and power requirements for various
impeller sizes.* (Courtesy of FMC Corporation, Lansdale, Pennsylvania)

and impeller speeds and/or diameters, on the graph
with the pump head curve. An example of a set
of actual pump characteristic curves is shown in
Figure 7.13.

### Impeller Speed

For a given centrifugal pump, it is sometimes desirable
to change the rotational speed of the impeller with a
variable-speed motor. The discharge of a centrifugal
pump varies directly with the impeller speed, and the
discharge head (pressure) varies directly with the square
of the impeller speed. In other words, if the impeller
speed is doubled, the discharge doubles, and the
discharge pressure head increases by a factor of
four. These relationships, called *affinity laws*, can be
expressed as follows:

$$\frac{Q_1}{Q_2} = \frac{N_1}{N_2} \qquad \textbf{(7-3)}$$

$$\frac{H_1}{H_2} = \frac{N_1^2}{N_2^2} \qquad \textbf{(7-4)}$$

where $Q_1$ and $Q_2$ = pump discharges, L/s (gpm)
$H_1$ and $H_2$ = pump discharge pressure heads,
m (ft)
$N_1$ and $N_2$ = pump impeller speeds, revolu-
tions/minute (rpm)

### EXAMPLE 7.3

Manufacturer's data for a centrifugal pump indicate that
the pump can discharge 100 L/s at a discharge pressure
of 25 m when the impeller speed is 1500 rpm. What is
the expected pump discharge and discharge pressure
head if the impeller speed is increased to 2000 rpm?

*Solution*  Applying Equations 7-3 and 7-4, we find

$$\frac{Q_1}{100} = \frac{2000}{1500} \quad \text{and} \quad Q_1 = 130 \text{ L/s}$$

$$\frac{H_1}{25} = \frac{2000^2}{1500^2} \quad \text{and} \quad H_1 = 45 \text{ m}$$

## System Characteristics

In this discussion, the term *system* refers to the network of interconnected pipes, distribution reservoirs, valves, and other appurtenances to which the pump is connected. The piping from the water source to the inlet of the pump is called the *suction line*. One possible system configuration has the pump located below the water level on the suction side, as shown in Figure 7.14a. The vertical distance between the water level and the pump centerline is called the *static suction head*. This is a good arrangement because a suction head will always keep the pump and suction line *primed* or, in other words, filled with water and ready to operate.

When the system is arranged so that the pump is above the water level on the suction side, then the vertical distance between the water surface and the pump centerline is called *static suction lift*. The maximum theoretical height to which a water column can be lifted by any pump is about 10 m (33 ft) at sea level. This is equivalent to standard atmospheric pressure. But under *dynamic* conditions, that is, when the pump is operating and there is flow in the system, the suction lift is limited to a maximum height of about 5 m (16 ft); this is because of the frictional resistance to flow in the suction line that the pump must also work against. A system configuration with static suction lift is illustrated in Figure 7.14b.

Under suction lift conditions, a *foot valve* is usually installed in the suction line to maintain prime on the pump. This is a check valve that allows water to flow into, but not out of the pipe. Without a foot valve, water would flow out of the suction line when the pump is not operating. A centrifugal pump that loses its prime for any reason must have its casing and suction line refilled with water in order to operate. This can be done manually or automatically for special self-priming pumps.

The vertical distance between the free water surfaces on the suction side and the discharge side of the pump is called the *total static head*. As illustrated in Figure 7.14, it represents the actual change in elevation of the water being pumped. But when the pump is operating and moving water through the system (a dynamic condition as opposed to a static or no-flow condition), the pump must also overcome the frictional resistance to flow.

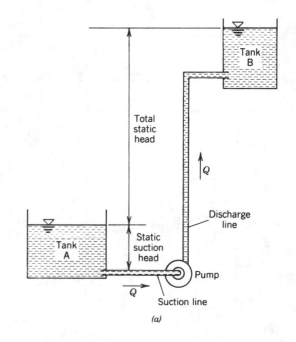

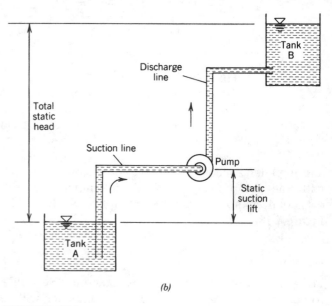

**FIGURE 7.14**

*In diagram (a) the water in tank A is above the pump, causing a condition of* suction head; *in diagram (b), the water in tank A is below the pump, causing* static suction lift. *The* total static head *is always the vertical distance between the lower and the upper water surfaces, regardless of the suction line conditions.*

### System Head Curve

The amount of energy lost because of friction in the system is primarily a function of pipe diameter and flow rate. For a given diameter pipe, the greater the flow rate, the greater is the resistance to flow and friction head loss. In addition to adding energy to lift the water and to overcome frictional resistance, the pump is adding velocity head to the system when it operates.

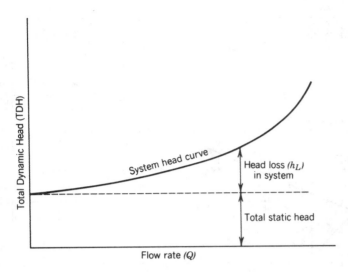

**FIGURE 7.15**

*A system head curve shows the hydraulic response of a water transmission system to various flow rates. There is greater resistance to flow and therefore a higher THD in the system for higher flow rates.*

At a given flow rate, the sum of the total static head, the total friction head, and the velocity head is called the *total dynamic head*, or TDH. (In most cases, the velocity head is small compared to the static and frictional heads, and it can be neglected.) In effect, the pump "thinks" it is lifting the water a distance equal to the TDH, which is always greater than the static lift.

This TDH is the same term used in the previous discussion of pump characteristics. But the student must keep in mind that TDH is now being considered from the perspective of the system, not the pump. A graph of TDH as a function of flow rate is called the *system head curve*. A typical system head curve is illustrated in Figure 7.15. Note that in the system *the TDH increases as the flow rate increases*. This is just the opposite of what happens from the pump's perspective: The TDH against which the pump can operate decreases as the flow increases.

## Pump Operating Point

The pump head curve gives a picture of the hydraulic response of the pump to changes in flow rate or TDH. The system head curve gives a picture of the relationship between flow rate and TDH in the system. But when an unthrottled pump operates steadily in a given system (with a fixed total static head), there is

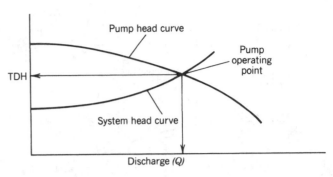

**FIGURE 7.16**

*The intersection of the head curve for a centrifugal pump and the system curve for the system in which it works represents the* operating point *for the pump in that system.*

only one point, that is, only one pair of $Q$ and TDH values, at which the pump will operate. This point can best be determined graphically by drawing both the pump and system head curves on a single graph. The intersection of the two curves represents the *operating point* of the pump in the given system. This is illustrated in Figure 7.16.

When selecting a pump for use in a particular system, the designer must examine the system head curve together with the pump head curves given in the manufacturer's data catalogs. The basic objective in pump selection is to find a pump that will provide the required flow rate at or close to its peak operating efficiency. In other words, the rated discharge for the pump should match the point at which it is expected to operate in the system. It is not a good practice to select an oversized pump for a system and then throttle the discharge to obtain the desired flow. The following example illustrates the procedure for determining a pump's operating point.

## EXAMPLE 7.4

A centrifugal pump head curve is given in Figure 7.17. Determine the pump's operating point in a system that has a total static head of 50 m and comprises 4500 m of 305-mm-diameter pipe.

*Solution*  The system head curve must be plotted on the graph in Figure 7.17 to find the operating point. To do this, several discharge values are arbitrarily selected, and friction losses at those discharges are determined using the Hazen–Williams nomograph (Figure 2.15 in Section 2.3).

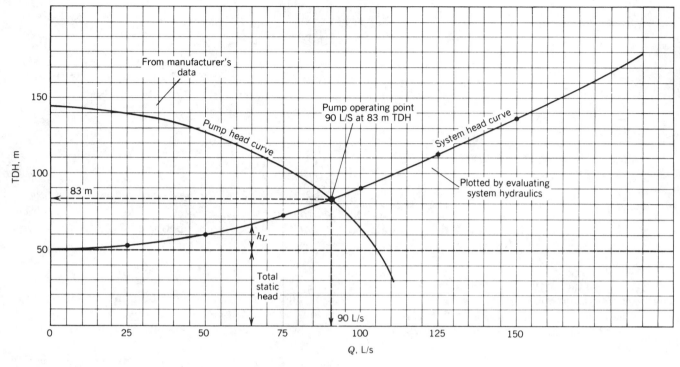

**FIGURE 7.17**
*Illustration for Example 7.4.*

Neglecting velocity head, one determines the TDH for each of the selected $Q$ values by adding the computed friction loss to the total static head. This is summarized in Table 7.2. The friction head loss is computed as $h_L = S \times L$, where $S$ is the slope of the HGL and $L$ is the pipe length of 4500 m. The TDH = 50 m + $h_L$ at each of the selected flow rates.

Plotting the values of $Q$ and TDH from Table 7.2 gives the system head curve, as shown in Figure 7.17. The intersection of the system curve with the given pump curve is the operating point for the pump in the given system. This is read from the graph to be 90 L/s at a TDH of 83 m.

**TABLE 7.2**
**Computation of TDH for Example 7.4**

| $Q$, L/s | $S$, m/m | $h_L$, m | TDH, m |
|---|---|---|---|
| 0 | 0 | 0 | 50 |
| 25 | 0.000 65 | 3 | 53 |
| 50 | 0.0025 | 11 | 61 |
| 75 | 0.005 | 23 | 73 |
| 100 | 0.009 | 41 | 91 |
| 125 | 0.014 | 63 | 113 |
| 150 | 0.019 | 86 | 136 |

## Parallel Operation

Water demand and consumption vary on an hourly as well as on a daily and seasonal basis, as discussed in Section 7.1. Although the peak hourly demands are usually satisfied by water from local storage tanks, it is still necessary to provide for varying pump outputs to satisfy the changes in daily or seasonal demand.

One way to do this is to select a pump large enough to handle the maximum expected flow in the system and then to reduce its discharge by throttling when the water demand is low. But this is not a preferred method—the pump operating efficiency will be low since the pump will rarely be working at its rated discharge. Another method for changing the pump output is to use a variable-speed motor to drive the pump. By increasing the speed of rotation of the impeller, the pump discharge can be increased. But variable-speed equipment is expensive.

A third method involving *parallel operation* of two or more pumps finds the widest application. Pumps that are connected in parallel discharge into a common header or manifold pipe; the suction and discharge pressures for each of the pumps are the same. A parallel arrangement is illustrated in Figure 7.18. Parallel operation is advantageous because one or more pumps can be shut off when water demand is low, allowing the remaining pump(s) to operate at or near peak efficiency.

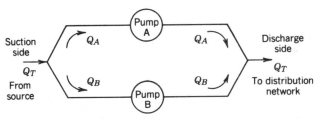

**FIGURE 7.18**

*Schematic representation of pumps connected in parallel. The total flow in the system is the sum of each individual pump discharge at the same TDH, or $Q_T = Q_A + Q_B$.*

Also, parallel connection allows maintenance work to be done on one pump without creating the need to shut down the entire pumping station.

To evaluate the performance of parallel pumps operating in a given system, it is first necessary to determine and plot the *combined head curve* of the pumps. This can be done simply by adding or combining the individual pump head curves horizontally. In other words, for selected values of TDH on the vertical axis, add the values of the discharge from each pump and plot the result along the horizontal axis. The next example illustrates this procedure for a system with two identical pumps in parallel. It is important to note that the combined discharge from the two pumps is not simply twice the discharge from one pump operating alone in the system.

## EXAMPLE 7.5

Two identical pumps with the pump head curve shown in Figure 7.17 are connected in parallel. Sketch the combined head curve for the two pumps, and determine their operating point in the system given in Example 7.4.

*Solution* First, prepare a graph with a pump head curve for a single pump, as shown in Figure 7.17. Arbitrarily select a few values of TDH, say 75 m, 100 m, and 125 m. For TDH = 75 m, the discharge of the pump is 97 L/s. For two pumps operating in parallel at the same TDH of 75 m, the combined discharge would be 97 + 97 or 194 L/s. Plot a point on the graph with coordinates $Q$ = 194 L/s and TDH = 75 m.

If the TDH is 100 m, the flow from a single pump is 80 L/s, and the combined flow from two pumps in parallel is 160 L/s. Likewise, for TDH = 125 m, the combined flow is 106 L/s. Finally, at the shutoff head of 144 m, there will be no flow from either pump and the combined $Q$ = 0 L/s. Plotting all these points on the graph and sketching a smooth line that passes through them results in the combined pump head curve for the two pumps operating in parallel. This is shown in Figure 7.19.

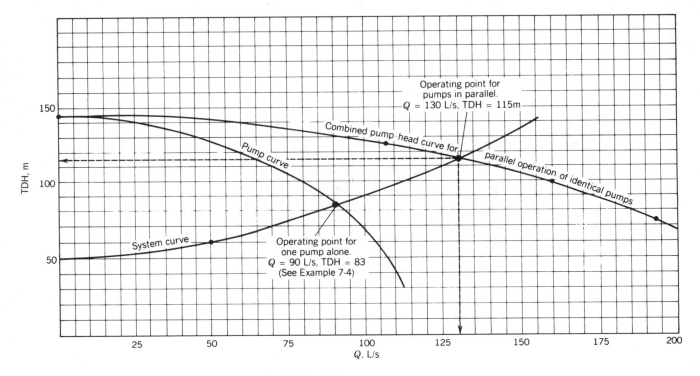

**FIGURE 7.19**

*Illustration for Example 7.5.*

The system head curve from Example 7.4 is also plotted on Figure 7.19. The intersection of the combined head curve with the system curve represents the point at which the parallel pumps operate: From the graph, we see that they will discharge a combined flow of 130 L/s at a TDH of 115 m. Each of the two pumps contributes 65 L/S of flow in parallel operation, whereas one pump operating alone in the system discharges 90 L/s.

When two or more different pumps are connected in parallel, the procedure for determining the combined head curve is similar to that just described. But the combined curve will branch off from the curve for the larger pump at the higher values of TDH. Even when the smaller pump is operating, it would not contribute any discharge when the TDH in the system is above the pump shutoff head. This is illustrated in the following example.

### EXAMPLE 7.6

Pump A and pump B are connected in parallel in a system that comprises 8000 ft of 8-in.-diameter pipe (with $C = 100$). The total static head of the system is 80 ft. The pump manufacturer has provided data that describe the pump head curves in tabular form, as shown in Table 7.3.

Determine the operating point of the parallel pumps in the given system. How much flow will each pump contribute?

**TABLE 7.3**
**Pump operating data for Example 7.6**

| Discharge, gpm | TDH, ft | |
| --- | --- | --- |
| | Pump A | Pump B |
| 0 | 240 | 200 |
| 100 | 225 | 185 |
| 200 | 200 | 155 |
| 300 | 165 | 115 |
| 400 | 125 | 60 |
| 500 | 65 | — |

*Solution*  First, plot the head discharge curves for each pump individually from the data given in Table 7.3, as shown in Figure 7.20.

Now plot the combined head curve for the two pumps operating in a parallel; add the discharges for each pump at a few selected values of TDH. Note that, when the TDH is between 200 and 240 ft, only pump A will be discharging into the system; pump B will be operating at a TDH equal to its shutoff head, and $Q = 0$. When the TDH is less than 200 ft, both pumps can contribute flow. For example, when the TDH is 100 ft, pump A discharges 450 gpm and pump B discharges 335 gpm; the combined discharge at TDH = 100 ft is therefore 785 gpm.

Now plot the system head curve. Computations for the system TDH at four selected discharge values are summarized in Table 7.4.

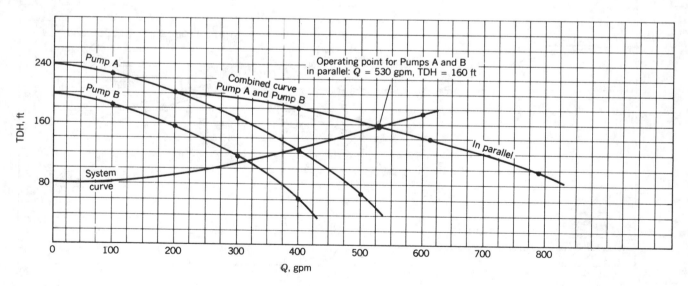

**FIGURE 7.20**
*Illustration for Example 7.6.*

**TABLE 7.4**
**Computation of TDH for Example 7.6**

| Q, gpm | S, m/m | $h_L$, ft | TDH, ft |
|--------|--------|-----------|---------|
| 0 | 0 | 0 | 80 |
| 200 | 0.0014 | 11 | 91 |
| 400 | 0.0055 | 44 | 124 |
| 600 | 0.012 | 96 | 176 |

As seen from Figure 7.20, the combined head curve and the system curve intersect at $Q = 530$ gpm and TDH = 160 ft, which represents the operating condition of the pumps in parallel. Of the total discharge of 530 gpm, pump A contributes about 330 gpm and pump B contributes about 200 gpm.

## Power and Efficiency

The rate at which a pump adds energy to the water is the power output of the pump, called the *water power*. Pumps are usually driven by electric motors (often with gasoline or diesel standby engines to be used in the event of a power failure). Since a pump (or any machine) can never operate at 100 percent efficiency, the water power is always less than the power delivered to the pump impeller by the motor. The efficiency of a pump can be expressed as follows:

$$\text{efficiency} = \frac{P_{out}}{P_{in}} \times 100 \qquad (7\text{-}5)$$

In SI metric units, power is expressed in terms of kilowatts (kW). Water power, $P_{out}$, can be computed with the following formula:

$$P_{out} = 9.8 \times Q \times \text{TDH} \qquad (7\text{-}6)$$

where  $Q$ = the pump discharge, m³/s
  TDH = total dynamic head, m

Note that 9.8 is the unit weight of water in terms of kN/m³ and power has the units kN · m/s, which is work per unit time. In the SI system, 1 kN · m/s is called one kilowatt (1 kW).

## EXAMPLE 7.7

A pump discharges 500 L/s at a TDH of 25 m. The drive motor delivers 150 kW of power to the pump. At what efficiency is the pump operating?

*Solution*  Converting 500 L/s to 0.5 m³/s and applying Equation 7-6 to compute water power gives

$$P_{out} = 9.8 \times 0.5 \times 25 = 123 \text{ kW}$$

From Equation 7-5,

$$\text{efficiency} = \frac{123}{150} \times 100 = 82 \text{ percent}$$

In U.S. Customary units, power is commonly expressed in terms of *horsepower* (hp), where 1 hp = 33,000 ft lb/min. The power input to the pump shaft is called *brake horsepower* (bhp). Water horsepower can be computed from the following formula:

$$P_{out} = \frac{Q \times \text{TDH}}{3960} \qquad (7\text{-}7)$$

where $Q$ is expressed in terms of gpm and TDH is expressed in terms of feet. Horsepower can be converted to kilowatts by multiplying by 0.746 (that is, 1 hp = 0.746 kW).

## EXAMPLE 7.8

A centrifugal pump with an efficiency of 65 percent discharges 1500 gpm into a system that includes 3000 ft of 10-in.-diameter pipe ($C = 100$). The total static head is 100 ft. Compute the required brake horsepower ($P_{in}$).

*Solution*  First, determine the TDH on the pump. Enter the Hazen–Williams nomograph with $Q = 1500$ gpm and $D = 10$ in.; read $S = 0.024$. Then $h_L = S \times L = 0.024 \times 3000 = 72$ ft, and TDH = 100 + 72 = 172 ft. Now, applying Equation 7-7 gives

$$P_{out} = \frac{1500 \times 172}{3960} = 65 \text{ hp}$$

From Equation 7-5,

$$P_{in} = \frac{P_{out}}{\text{efficiency}} \times 100 = \frac{65}{65} \times 100 = 100 \text{ hp}$$

## EXAMPLE 7.9

The overall *wire to water efficiency* of a pump and motor is 50 percent. The water horsepower is 150 kW.

If electric power costs $0.15 per kilowatt-hour (kW · h), how much does it cost to operate the pump for 8 h?

*Solution* The electric power consumption is 150/50 × 100 = 300 kW. For 8 h of operation, the energy consumed is

$$300 \text{ kW} \times 8 \text{ h} = 2400 \text{ kW} \cdot \text{h}$$

The cost of operation is

$$2400 \text{ kW} \cdot \text{h} \times \$0.15/\text{kW} \cdot \text{h} = \$360$$

## Operation and Maintenance

Proper operation and maintenance (O&M) procedures must be followed to obtain satisfactory service from centrifugal pumps. Two important aspects of O&M involve keeping the pumping station clean and keeping the pump properly lubricated. Pumps located in a dirty and messy pumping station cannot be operated properly. Oil and grease cans must be covered and kept free from dirt. Dirt in lubricating oil or grease causes excessive wear on pump bearings and shortens their life. The manufacturer's recommendations must be followed carefully for proper lubrication. It is important that the pump bearings not be overlubricated; too much grease causes damage.

The rotating impeller is the only moving part in the pump casing. Thrust bearings and guide bearings support the shaft that carries the impeller. A *packing gland* or seal is used where the pump shaft protrudes from the casing to stop air from leaking in or water from leaking out. The packing gland usually consists of several graphite-impregnated asbestos rings around the shaft. These rings can be compressed to bear against the shaft; the graphite provides some lubrication. It is very important that the packing glands be properly adjusted and not overtightened. During operation, a slight trickle of water leaking out of the casing is desirable to keep the gland cool.

A centrifugal pump must be primed or filled with water when it is started; the casing and suction piping must be free of air. When the pump is situated above the water level on the suction side, an electric or hand-operated vacuum pump can be used to prime the centrifugal pump. The foot valve is designed to keep the pump and suction line filled with water when the pump is not operating, but if the valve leaks slightly, the pump will lose its prime. The ideal arrangement is to have the pump located lower than the water surface so that it will always be primed.

When the pump is started, the suction valve should be open and the discharge valve should be closed. The discharge valve is then opened slowly when the motor is up to speed. If a slight drip or flow of water from the packing gland is not observed, the packing may be too tight and cause excessive friction on the shaft; the gland should be adjusted before allowing continued operation.

Before stopping a centrifugal pump, the discharge valve must be closed slowly. If the valve is closed too fast or if the pump is stopped suddenly, a condition called *water hammer* will occur. *Water hammer* refers to a surge or pressure wave that travels throughout the pipeline. The momentary pressure caused by water hammer is often high enough to rupture the pipe or the pump casing.

Daily inspection of operating pumps is important to check for excessive heat, noise, or gland leakage. A properly installed pump should not vibrate. Vibration is an indication of misalignment between the motor and the pump, which will cause premature wear and damage of the pumping system. If vibration is noticeable, the manufacturer's specifications for proper alignment procedures and tolerances should be checked.

## 7.4 DISTRIBUTION STORAGE

Elevated water storage tanks and towers are familiar sights in most communities. These relatively small water storage facilities serve two basic purposes: They provide *equalizing storage* and they provide *emergency storage.* They are called distribution storage tanks or reservoirs because they are part of the localized water distribution systems. Much larger storage facilities, called *conservation reservoirs*, are generally located at a considerable distance form the distribution network and are meant to store water that can be used during long dry-weather periods (see Section 3.6).

*Equalizing storage* refers to the volume of water in the tank that is available to satisfy the peak hourly demands for water use in the community. The hourly variation in water demand is illustrated in Figure 7.1. During the late night and early morning hours, when water demand is very low, the high-lift pumps move water into the distribution storage tanks. During the day, when water demand exceeds the average daily demand, water flows out of the tanks to help to meet the peak hourly needs of the community.

Distribution storage tanks are often described as *floating on the line.* This means that the flow into or out of the tank varies directly with the demand for water in the system, and the hydrostatic pressure in the water main is equivalent to the pressure head or elevation of water in the tank. A distribution storage tank is illustrated in Figure 7.21. Automatically controlled

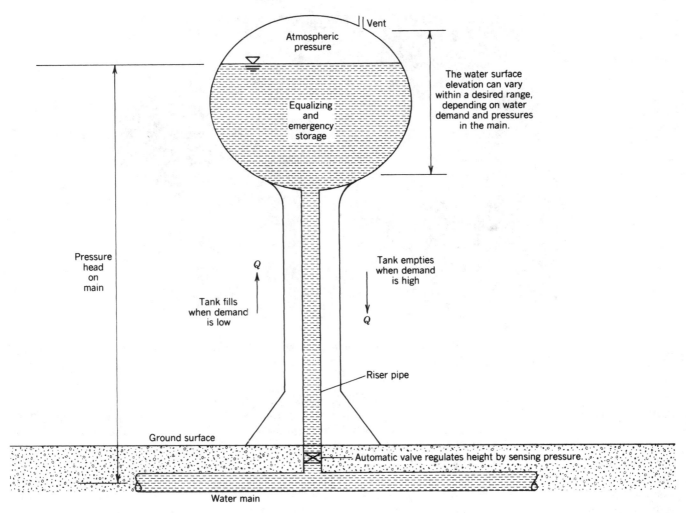

**FIGURE 7.21**
*An elevated water storage tank will* float on the line *in a water distribution system.*

altitude valves can maintain the water elevations, and therefore the water main pressure, within a desired range.

The equalizing or averaging effect on flow rates provided by the stored water allows for a uniform or steady water treatment and pumping rate. When water demand is low, the extra water being pumped fills the storage tanks, and when demand exceeds the pumping rate, the tanks are emptied to make up the difference. This has the advantage of reducing the required sizes and capacities of the pipes, pumps, and treatment facilities, resulting in reduced construction and operating costs.

Generally, the volume of water needed to balance or equalize the peak hourly flows is about 20 percent of the average daily water demand in the service area. For example, if a community has an average water demand of 1 ML/d, then at least 0.2 ML, or 200 000 L, of storage volume should be provided for equalizing purposes. In communities where adequate records of

water use and demand are available, the summation hydrograph method (see Section 3.6) can be used for a more accurate determination of the required equalizing volume in a new tank.

Distribution storage tanks are not constructed for the sole purpose of providing equalizing storage. Emergency storage volume is also provided by these tanks. This furnishes additional water for firefighting needs, or overcoming problems due to power blackouts or pump station failure.

The amount of water required for fire control varies depending on the type of service area and the capacity of the water pumping station. For example, if storage for a fire flow of 30 L/s for a 2-h duration is needed, the required volume is 30 L/s × 3600 s/h × 2 h = 216 000 L. This would be added to the volume needed for equalizing purposes to determine the total tank volume. If 200 000 L is needed for equalizing storage and

**FIGURE 7.22**
*A ground-level distribution storage reservoir.* (Courtesy of Natgun Corporation, Wakefield, Massachusetts)

216 000 L is needed for fire control, the minimum tank volume is 416 000 L, or 416 m³.

If an emergency power generator or standby diesel engine is not available at the high-lift water pumping station, it may be necessary to provide additional storage so that both domestic water demand and emergency firefighting needs can be met even during a temporary power failure. To accommodate all these distribution needs, some states simply require that the storage volume be equal to 1-d average water demand, unless it can be demonstrated with available data that a smaller volume will suffice.

**Types of Distribution Reservoirs**

Distribution reservoirs store water for relatively short periods of time (1 d or less) and are small enough to be covered to prevent contamination and reduce evaporation. In areas with flat topography, the storage tanks are elevated above the ground on towers to provide adequate pressures in the water mains. These elevated tanks are usually constructed of steel.

In hilly areas, distribution tanks can be built at ground level on hilltops higher than the service area and still float on the line while maintaining adequate pressures in the system. They may be constructed of either steel or reinforced concrete. A typical ground-level distribution reservoir is illustrated in Figure 7.22.

When the height of the storage tank is greater than its diameter, the structure is called a *standpipe*, as illustrated in Figure 7.23. Standpipes provide more

storage capacity than elevated tanks. But the storage capacity that is useful for equalizing purposes is only that volume above the elevation required for minimum pressure in the water main. The water below that elevation can be used for fire protection with pumper trucks or during other emergency conditions. Standpipes are constructed of steel, with thicker walls at the bottom to withstand the hydrostatic pressure.

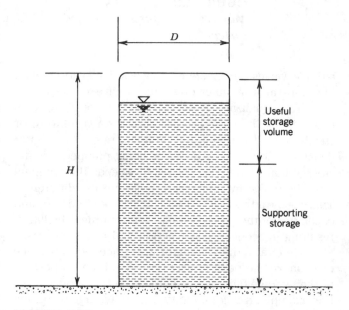

**FIGURE 7.23**
*A water storage tank is called a* standpipe *when its height is greater than its diameter* ($H > D$).

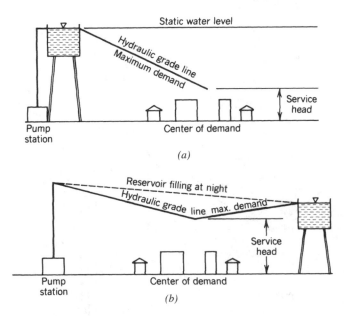

**FIGURE 7.24**

*Location of distribution storage tanks opposite the source* (b) *is preferable to location at the source* (a). (Reprinted from *Water Distribution Operator Training Handbook*, by permission. Copyright © 1976, The American Water Works Association)

### Location of Storage Tanks

Using several small storage tanks near the major centers of water withdrawal is preferable to using one large tank near the pumping station. Also, it is best to locate the tanks on the opposite side of the demand center from the pumping station. This allows for more uniform pressures throughout the distribution network, as illustrated in Figure 7.24. It also allows the use of smaller-diameter mains and pumps than would otherwise be needed. The following example illustrates the hydraulic computations involved and the effect of storage on the required discharge pressure at the pump station.

### EXAMPLE 7.10

A peak hourly flow rate of 100 L/s is required at point A, as illustrated in Figure 7.25. Pressure at that point is not to drop below 150 kPa. Determine (a) the required pressure at the pump station if the demand is satisfied without any distribution storage tank, and (b) the required pressure head at the pumps if the storage tank is used to help to meet peak demand by floating on the line at elevation 20 m, as shown.

*Solution* (a) Compute the drop in pressure head from the pump station to the point of withdrawal, as follows:

For $Q = 100$ L/s and $D = 250$ mm, read $S = 0.024$ on the Hazen–Williams nomograph (Figure 2.15). Compute $h_L = S \times L = 0.024 \times 2000 = 48$ m.

At point A, the pressure is not to drop below 150 kPa. Using Equation 2-3a, one sees that 150 kPa is equivalent to $0.1 \times 150$ or 15 m of pressure head. Since flow occurs in the direction of sloping HGL, the 48 m of head loss must be added to the 15-m minimum pressure head requirement at point A. The HGL is shown in Figure 7.25. The pressure head is then $48 + 15 = 63$ m, and using Equation 2-2a, we find that the required pressure at the pump station is $P = 9.8 \times 63 \approx 620$ kPa. (b) In this part of the problem, the storage tank is contributing flow to the withdrawal point. The head loss in the 500 m of pipeline is $20 - 15 = 5$ m. The slope of the HGL is then $5/500 = 0.01$, and from the Hazen–Williams nomograph, with $D = 200$ mm and $S = 0.01$, read $Q = 32$ L/s.

Since the total withdrawal is still 100 L/s, the flow from the pump station must be $100 - 32 = 68$ L/s. From the Hazen–Williams nomograph, with $Q = 68$ L/s and $D = 250$ mm, read $S = 0.012$. The head loss between the pump station and point A is then $h_L = 0.012 \times 2000 = 24$ m, and the required pressure head at the pump station is then $15 + 24 = 39$ m; the required pressure is $P = 9.8 \times 39 \approx 380$ kPa.

It can be seen in this problem that the required pump capacity decreases from 100 L/s at a discharge pressure of 620 kPa to a capacity of 68 L/s at a pressure of 380 kPa when peak hourly demand is partially satisfied by water from a distribution storage tank.

### Maintenance

Distribution storage reservoirs should be inspected frequently. Cracks and leaks must be repaired, and air vents should be kept clear. Blocked air vents can cause excessive pressures or vacuums to develop in the tank, which can result in structural damage.

Storage tanks are occasionally drained, cleaned, and disinfected. Accumulated silt should be removed. Sometimes the inside of a steel tank will have to be painted with an approved bituminous or vinyl coating. All traces of rust must first be removed. Disinfection can be accomplished by spraying the tank walls with a 500-mg/L chlorine solution.

Corrosion of steel tanks can be a major problem. About 10 kg (22 lb) of steel per year can be lost because of corrosion. In addition to protecting the steel with high-quality bituminous coatings, a method called *cathodic protection* is sometimes used. Corrosion involves a flow of electric current through the water in the form of positively charged metal ions. In cathodic protection, a voltage is maintained in the tank that

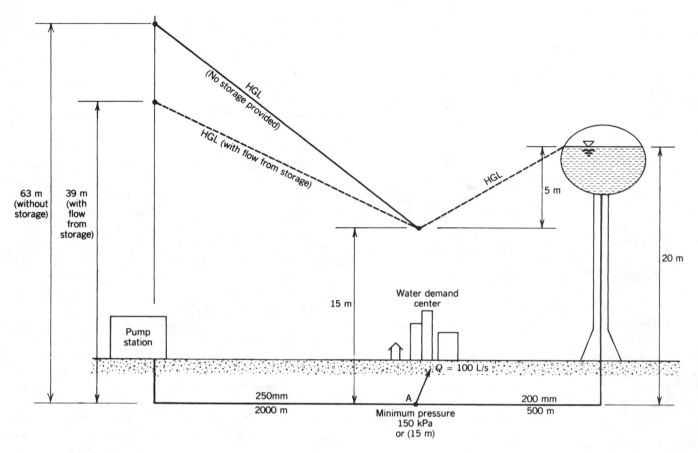

**FIGURE 7.25**
*Illustration for Example 7.10.*

tends to reverse the direction of the current. This applied voltage keeps the metal from ionizing and thus prevents corrosion. A cathodic protection system must be custom designed by specialists for each tank.

## 7.5  FLOW IN PIPE NETWORKS

Water distribution systems, particularly those serving densely populated cities, consist of a complex network of interconnected pipes and appurtenances. The hydraulic conditions in these systems must be analyzed to determine flow capacities and pressures in the main and secondary feeders and at points of significant water withdrawal.

It may be convenient to first simplify or *skeletonize* the system by replacing the many smaller water mains with *equivalent pipes* to reduce the number of loops and interconnections. This is done only conceptually, in theory; the existing pipes are not actually replaced in the streets. After the distribution system has been skeletonized, network analysis methods can be applied to study the hydraulics of the system.

Studies of this type are done to plan for possible expansion or upgrading of the water supply network.

This section presents the basic hydraulic techniques for determining hydraulically equivalent pipes and for analyzing the pipe network.

## Equivalent Pipes

An *equivalent pipe* is one that has the same hydraulic characteristics as the pipes it theoretically replaces. In other words, for any given flow rate, the pressure drop through the equivalent pipe is the same as the total pressure drop through the original pipes. And for any given pressure drop, the flow rate in the equivalent pipe is the same as the total discharge through the original pipes.

Equivalent pipes can be determined to replace pipes in series (connected end to end) or pipes in parallel (forming loops). For any given series or parallel system, there is no limit to the combinations of theoretical pipe diameters and lengths for the equivalent pipe. Usually, either the diameter (or length) is first specified, and the required length (or diameter) for hydraulic equivalence is then determined.

## Pipes in Series

The following five steps summarize the method for determining an equivalent pipe to replace several different pipes connected end to end:

**Step 1:** Assume any flow rate $Q$. The flow rate should generally be selected within the range of flows on the Hazen–Williams nomograph of Figure 2.15, but, theoretically, any flow rate can be selected.

**Step 2:** Using the nomograph, line up flow rate $Q$ and diameter $D$ for each section of the original series pipeline; read slope $S$, and compute head loss $h_L = S \times L$ for each section, where $L$ is the length of the section.

**Step 3:** Add up the head losses for all sections in the series to determine a total head loss $H_L$ for the assumed discharge $Q$ in the pipeline.

**Step 4:** Convert the total head loss $H_L$ to an overall hydraulic gradient $S' = H_L/L'$, where $L'$ is the specified total length of the equivalent pipe.

**Step 5:** Enter the nomograph again with the assumed value of $Q$ and the computed value $S'$; read the required value for the diameter $D$ of the equivalent pipe.

## EXAMPLE 7.11

A series pipeline is shown in Figure 7.26. It consists of 1500 feet of 8-in.-diameter pipe from point A to point B and 2500 ft of 12-in.-diameter pipe from point B to point C. Determine the equivalent diameter of a single 4000-ft-long pipeline from A to C that could theoretically replace the pipes AB and BC.

*Solution*

**Step 1:** Assume a flow rate $Q = 450$ gpm.

**Step 2:** For section AB, line up $Q = 450$ gpm and $D = 8$ in. on the Hazen–Williams nomograph. Read $S \approx 0.0064$. Compute $h_L = S \times L = 0.0064 \times 1500$ ft $= 9.6$ ft. Similarly for section BC, line up $Q = 450$ gpm and $D = 12$ in., then read $S \approx 0.001$ and compute

$$h_L = 0.001 \times 2500 = 2.5 \text{ ft}$$

**Step 3:** Compute the total head loss from A to C as

$$H_L = 9.6 + 2.5 = 12.1 \text{ ft}$$

**Step 4:** Compute the overall hydraulic gradient from A to C as

$$S' = \frac{12.1}{4000} = 0.003$$

**Step 5:** Now line up $Q = 450$ gpm and $S' = 0.003$ on the nomograph. Read $D = 9.5$ in.

An equivalent pipe for this problem, then, is one that would have a diameter of 9.5 in. and a length of 4000 ft. (Nomograph readings are approximate.)

Notice that the equivalent diameter of 9.5 in. in the preceding example is not simply a weighted average of the original two diameters from A to B and B to C. It is the diameter that for a 4000-ft length of pipe would have the same hydraulic characteristics as the original 8-in. and 12-in. pipes. The original pipeline could also be replaced by an equivalent 3000-ft length of an 8.9-in. pipe, a 2000-ft length of an 8.1-in. pipe, and so on; there are an unlimited number of equivalent pipes for any given system.

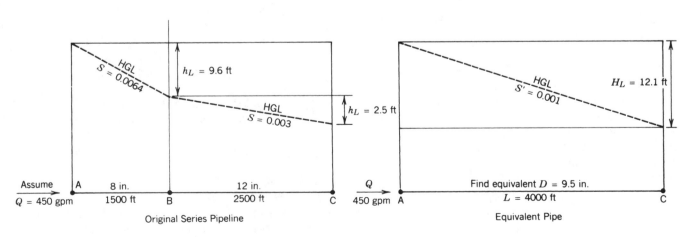

**FIGURE 7.26**
*Illustration for Example 7.11.*

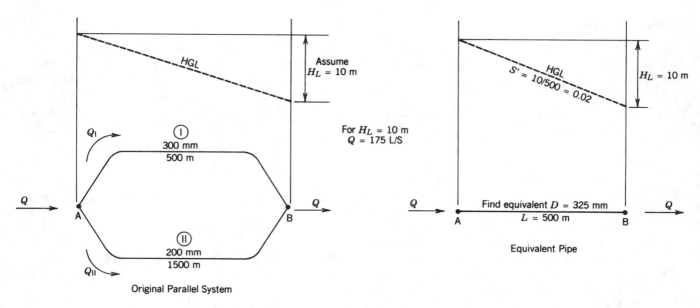

**FIGURE 7.27**
*Illustration for Examples 7.12 and 7.13.*

For Example 7.11, if a flow rate other than $Q = 450$ gpm is assumed in step 1, the same final answer of 9.5 in. will be obtained for the equivalent diameter. Try the same problem with an assumed $Q = 1000$ gpm, or any other flow rate, to verify this.

## Pipes in Parallel

The procedure for determining an equivalent pipe to replace a loop configuration of pipes differs somewhat from the preceding procedure for pipes in series. The following seven steps outline the method for a parallel or looped pipe system (see Figure 7.27):

**Step 1:** Assume a total head loss $H_L$ across the loop from the first pipe junction at A to the second pipe junction at B. This assumed head loss must be the same for both the top branch and the bottom branch of the loop, since the pressures at the ends of each pipe are the same at the junctions.

**Step 2:** For branch AIB, compute $S = H_L/L$, where $L$ is the length of AIB.

**Step 3:** Enter the Hazen–Williams nomograph with $D$ and $S$ for branch AIB; determine $Q_I$ in that branch.

**Step 4:** Repeat steps 2 and 3 for branch AIIB to determine $Q_{II}$.

**Step 5:** Compute the total flow rate $Q$ entering junction A, as $Q = Q_I + Q_{II}$.

**Step 6:** Determine an overall hydraulic gradient $S'$ as follows: $S' =$ assumed $H_L/L'$, where $L'$ is the specified length of the equivalent pipe.

**Step 7:** Enter the Hazen–Williams nomograph with $Q$ and $S'$ to determine the equivalent diameter $D$.

## EXAMPLE 7.12

Two pipelines are connected in parallel from junction A to junction B, as shown in Figure 7.27. Branch AIB consists of 500 m of 300-mm-diameter pipe, and branch AIIB consists of 1500 m of 200-mm pipe. Determine the equivalent diameter of a single 500-m-long pipeline from A to B that could replace the given loop.

*Solution*

**Step 1:** Assume that $H_L = 10$ m.
**Step 2:** For branch AIB, $S = 10/500 = 0.02$.
**Step 3:** From the Hazen–Williams nomograph with $D = 300$ mm and $S = 0.02$, read $Q_I = 143$ L/s.
**Step 4:** For branch AIIB, $S = 10/1500 = 0.0067$. From the nomograph, with $D = 200$ mm and $S = 0.0067$, read $Q_{II} = 27$ L/s.
**Step 5:** The total flow into junction A is $Q = Q_I + Q_{II} = 143 + 27 = 170$ L/s.
**Step 6:** The overall hydraulic gradient $S' = 10/500 = 0.02$.
**Step 7:** From the nomograph, with $Q = 170$ L/s and $S' = 0.02$, read an equivalent diameter of 320 mm.

The calculations in Example 7.12 show that a 500-m-long, 320-mm-diameter pipeline is hydraulically equivalent to the given loop from A to B. An unlimited number of other equivalent pipes with different combinations of length and diameter can be found. Also, if a head loss other than 10 m was assumed in step 1, the same final answer of 320 mm for a length of 500 m would result. Try the same problem with an assumed $H_L = 20$ m to verify this.

## Pipe Network Analysis

After a complex water distribution system has been simplified using equivalent pipes, it is then analyzed to determine pressures and flow rates at important points in the system. Water distribution networks can be readily modeled and analyzed using desktop computers and network modeling software. Such software allows engineers to create a mathematical model (rather than a physical model) of a network and to simulate hydraulic and water quality conditions at intervals over a specified time period. Before the development of such computer modeling software, pipe networks were analyzed "manually" (with slide rules and later with hand-held calculators), using the *Hardy Cross* method (based on the approach introduced in 1930 by the civil engineer Hardy Cross). Although the Hardy Cross method is now rarely used in modern engineering practice, it is described here so that the student can gain a real understanding of, and an appreciation for, the underlying hydraulic concepts related to "balancing" a network.

The Hardy Cross method is a controlled trial-and-error procedure; corrections are applied to assumed flow rates in a manner that leads (converges) to a *hydraulically balanced system*. A balanced system is one in which the computed flows and head losses match up at the pipe junctions of the distribution network. In other words, no matter which path or branch is followed to get to a specific junction, the computed pressure is the same. And the total flow into the junction equals the total flow out of that junction.

The corrections applied to the assumed flows are determined from the following formula, which is derived from the Hazen–Williams equation:

$$\Delta Q = -\frac{\Sigma\ h_L}{1.85 \times \Sigma(h_L/Q)} \qquad (7\text{-}8)$$

where $\Delta Q$ = flow correction
$\Sigma\ h_L$ = sum of head losses
$\Sigma\ h_L/Q$ = sum of $h_L/Q$ ratios for each pipeline in a loop

(here $\Delta Q$ is pronounced "delta $Q$" and $\Sigma\ h_L$ to pronounced "sigma $h_L$").

A sign convention is used in the Hardy Cross procedure to indicate the direction of flow in a loop: Flows in a clockwise direction (↶) are considered to be positive (+), and flows in a counterclockwise direction (↷) are considered to be negative (−). Head losses caused by clockwise flows are also considered to be positive, and head losses from counterclockwise flows are negative.

In a hydraulically balanced system, the algebraic sum of the head losses around a loop ($\Sigma\ h_L$) will add up to zero. For example, in the parallel pipe system shown in Figure 7.27, the flow rate of 143 L/s in AIB is positive (clockwise), and the 10-m head loss in that pipe is also positive ($+10$ m). The flow rate of 27 L/s in AIIB is negative (counterclockwise), and the 10-m head loss in that pipe is also negative ($-10$ m). In this simple loop, $\Sigma\ h_L = (+10) + (-10) = 0$, as it should be for a balanced system.

## EXAMPLE 7.13

Consider again the single loop of parallel pipes shown in Figure 7.27. Instead of replacing them with an equivalent pipe, the problem here is as follows: If a flow of 400 L/s enters the loop at junction A, what will be the flow rates $Q_I$ in branch AIB and $Q_{II}$ in branch AIIB?

*Solution* From the principle of continuity of flow, the flow entering a junction must equal the total discharge from the junction, or $Q_I + Q_{II} = 400$ L/s. If branches AIB and AIIB were identical in all respect, then the flow of 400 L/s would be evenly split between the two. But this is not the case here; AIIB is longer and narrower than AIB, and it will therefore offer more resistance to flow. It is reasonable to assume that the flow rate in AIIB will be less than that in AIB.

Start by assuming that the flow rate $Q_I = 300$ L/s and $Q_{II} = -100$ L/s. (The minus sign for $Q_{II}$ indicates a counterclockwise direction of flow.) The magnitudes of the flows must still add up to 400 (that is, 300 + 100 = 400).

If the assumption for flow rate is correct, then the head losses in AIB and AIIB should be equal in magnitude (but opposite in sign). Using the Hazen–Williams nomograph to check this gives the following results:

- Pipe AIB: For $Q = 300$ L/s and $D = 300$ mm,
$$S = 0.075$$
$$h_L = S \times L = 0.075 \times 500 = 37.5 \text{ m}$$
- Pipe AIIB: For $Q = -100$ L/s and $D = 200$ mm,
$$S = -0.07$$
$$h_L = S \times L = -0.07 \times 1500 = -105 \text{ m}$$

Since $\Sigma\ h_L = 37.5 + (-105) = -67.5$ m, instead of zero, the loop is not balanced and our assumed flows are incorrect. Instead of simply guessing at new flows to try, use the Hardy Cross correction formula, Equation 7-8, as follows:

$$\Delta Q = -\frac{-67.5}{1.85 \times (37.5/300 + 105/100)}$$

$$= -\frac{-67.5}{1.85 \times 1.175} = 31 \text{ L/s}$$

Adding the flow correction $\Delta Q = 31$ L/s to the assumed flows, get new flow rates as follows:

- Pipe AIB: $Q_I = 300 + 31 = 331$ L/s.
- Pipe AIIB: $Q_{II} = -100 + 31 = -69$ L/s.
- Check continuity: $331 + 69 = 400$ OK.

Note the importance of using the algebraic signs properly.

Now, use the nomograph again to see if the loop is hydraulically balanced with the new flows:

- Pipe AIB: For $Q_I = 331$ L/s and $D = 300$ mm,
  $S = 0.085$
  $h_L = S \times L = 0.085 \times 500 = 42.5$ m
- Pipe AIIB: For $Q_{II} = -69$ L/s and $D = 200$ mm,
  $S = -0.037$
  $h_L = S \times L = -0.037 \times 1500 = -55.5$ m

Now $\Sigma h_L = 42.5 + (-55.5) = -13$ m. This is still not zero, so make another flow correction as follows:

$$\Delta Q = -\frac{-13}{1.85 \times (42.5/331 + 55.5/69)}$$

$$= -\frac{-13}{1.726} = 7.53 \text{ L/s}$$

For practical purposes, round up the 7.53 to a $\Delta Q = 8$ L/s, and, for corrected flows, get the following:

- Pipe AIB: $Q_I = 331 + 8 = 339$ L/s.
- Pipe AIIB: $Q_{II} = -69 + 8 = -61$ L/s.
- Check continuity: $339 + 61 = 400$ OK.

Now check head losses again to see if the loop is balanced:

- Pipe AIB: For $Q_I = 339$ L/s and $D = 300$ mm,
  $S = 0.09$
  $h_L = S \times L = 0.9 \times 500 = 45$ m
- Pipe AIIB: For $Q_{II} = -61$ L/s and $D = 200$ mm,
  $S = -0.03$
  $h_L = S \times L = -0.03 \times 1500 = -45$ m

Since $\Sigma h_L = 45 + (-45) = 0$, the loop is balanced and the final answers for the flow rates are $Q_I = 339$ L/s and $Q_{II} = 61$ L/s (for practical purposes, round off the answers to 340 L/s and 60 L/s, respectively). The pressure drop from A to B is $9.8 \times 45 = 440$ kPa.

For networks with more than one loop, and having several points of water inflow or outflow, it is convenient to set up a table to organize and keep track of the Hardy Cross computations. This is illustrated in Example 7.14; first, the general procedure is outlined as follows:

**Step 1:** For each pipe in the network, assume a flow rate and flow direction. The only restriction on this initial assumption for flows is that the total flow going into a pipe junction must equal the total flow going out of that junction.

**Step 2:** Working with only one loop at a time, use the Hazen–Williams nomograph to determine $S$ and $h_L$ for each pipe in the loop. Also, compute $h_L/Q$ for each pipe.

**Step 3:** Compute $\Sigma h_L$ and $\Sigma h_L/Q$, using the appropriate signs (+ or −). The $h_L/Q$ terms are always positive.

**Step 4:** Compute $\Delta Q$, the flow correction for the loop, using the Hardy Cross formula (Equation 7-8). Add $\Delta Q$ to the flow in each pipe of that loop, taking care to use the appropriate sign. Do not change the flows in the other loops(s) with this $\Delta Q$.

**Step 5:** Repeat steps 2 through 4 for an adjacent loop in the network. Note that at least one of the pipes in the first loop is also part of the adjacent loop. Use the previously corrected flow(s) in the common pipe(s), but keep in mind that the algebraic signs of the flows and head losses in the common pipe(s) change, depending on which loop is being evaluated.

**Step 6:** Alternately repeat steps 2 through 4 for each loop in the network as many times as needed to arrive at a reasonably balanced system. Generally, a difference of less than 10 percent between the positive and negative head losses in a loop is acceptable, in lieu of $\Sigma h_L = 0$.

## EXAMPLE 7.14

A water distribution system has been skeletonized and reduced to the two-loop network shown in Figure 7.28. A flow rate of 60 L/s is pumped into the network at point A, and two major water withdrawal points, at C and D, discharge 20 L/s and 40 L/s, respectively. Determine the flow rates in all the pipes of the network.

*Solution* The computations outlined in steps 1 through 6 of the Hardy Cross procedure are summarized in Table 7.5.

**TABLE 7.5**
**Hardy Cross analysis for the system in Example 7.14**

| Loop | Pipe | D, mm | L, m | Q, L/s | S, m/m | $h_L$, m | $h_L/Q$ | $Q + \Delta Q$ |
|------|------|-------|------|--------|--------|----------|---------|----------------|
| I (first try) | AB | 300 | 1000 | 35 | 0.0015 | 1.5 | 0.043 | 29 |
| | BD | 300 | 3000 | 10 | 0.000 15 | 0.45 | 0.045 | 4 |
| | AD | 400 | 3500 | −25 | −0.0002 | −0.7 | 0.028 | −31 |
| | | | | | | 1.25 | 0.116 | |

$$\text{I: } \Delta Q_1 = -\frac{1.25}{1.85 \times 0.116} \approx -6 \text{ L/s}$$

| Loop | Pipe | D, mm | L, m | Q, L/s | S, m/m | $h_L$, m | $h_L/Q$ | $Q + \Delta Q$ |
|------|------|-------|------|--------|--------|----------|---------|----------------|
| II (first try) | BC | 200 | 2500 | 25 | 0.0055 | 13.75 | 0.55 | 12 |
| | CD | 200 | 600 | 5 | 0.0003 | 0.18 | 0.036 | −8 |
| | BD | 300 | 3000 | −4 | neglect | neglect | neglect | −17 |
| | | | | | | 13.9 | 0.586 | |

$$\text{II: } \Delta Q_1 = -\frac{13.9}{1.85 \times 0.586} \approx -13 \text{ L/s}$$

| Loop | Pipe | D, mm | L, m | Q, L/s | S, m/m | $h_L$, m | $h_L/Q$ | $Q + \Delta Q$ |
|------|------|-------|------|--------|--------|----------|---------|----------------|
| I (second try) | AB | 300 | 1000 | 29 | 0.001 | 1.00 | 0.034 | 24 |
| | BD | 300 | 3000 | 17 | 0.0004 | 1.20 | 0.071 | 12 |
| | AD | 400 | 3500 | −31 | −0.0003 | −1.05 | 0.034 | −36 |
| | | | | | | 1.15 | 0.139 | |

$$\text{II: } \Delta Q_2 = -\frac{1.15}{1.85 \times 0.139} \approx -5 \text{ L/s}$$

| Loop | Pipe | D, mm | L, m | Q, L/s | S, m/m | $h_L$, m | $h_L/Q$ | $Q + \Delta Q$ |
|------|------|-------|------|--------|--------|----------|---------|----------------|
| II (second try) | BC | 200 | 2500 | 12 | 0.0015 | 3.75 | 0.313 | 8 |
| | CD | 200 | 600 | −8 | −0.000 75 | −0.45 | 0.056 | −12 |
| | BD | 300 | 3000 | −12 | −0.0002 | −0.60 | 0.050 | −16 |
| | | | | | | 2.7 | 0.419 | |

$$\text{II: } \Delta Q_2 = -\frac{2.7}{1.85 \times 0.419} \approx -4 \text{ L/s}$$

| Loop | Pipe | D, mm | L, m | Q, L/s | S, m/m | $h_L$, m | $h_L/Q$ | $Q + \Delta Q$ |
|------|------|-------|------|--------|--------|----------|---------|----------------|
| I (third try) | AB | 300 | 1000 | 24 | 0.0007 | 0.7 | 0.029 | 23 |
| | BD | 300 | 3000 | 16 | 0.000 34 | 1.02 | 0.064 | 15 |
| | AD | 400 | 3500 | −36 | −0.0004 | −1.4 | 0.039 | −37 |
| | | | | | | 0.32 | 0.132 | |

$$\text{I: } \Delta Q_3 = -\frac{0.32}{1.85 \times 0.132} \approx -1 \text{ L/s}$$

| Loop | Pipe | D, mm | L, m | Q, L/s | S, m/m | $h_L$, m | $h_L/Q$ | $Q + \Delta Q$ |
|------|------|-------|------|--------|--------|----------|---------|----------------|
| II (third try) | BC | 200 | 2500 | 8 | 0.0007 | 1.75 | 0.219 | |
| | CD | 200 | 600 | −12 | −0.0015 | −0.9 | 0.075 | |
| | BD | 300 | 3000 | −15 | −0.0003 | −0.9 | 0.06 | |
| | | | | | | −0.05 | 0.354 | |

$$\text{II: } \Delta Q_3 = -\frac{-0.05}{1.85 \times 0.354} \approx 0.07 \text{ (negligible)}$$

| Pipe | Diameter, mm | Length, m |
|------|--------------|-----------|
| AB | 300 | 1000 |
| BC | 200 | 2500 |
| BD | 300 | 3000 |
| AD | 400 | 3500 |
| DC | 200 | 600 |

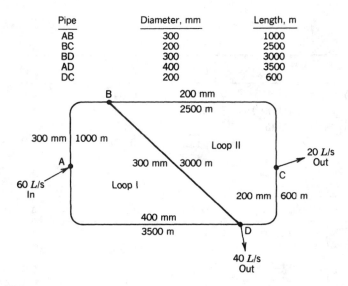

**FIGURE 7.28**
*Illustration for Example 7.14.*

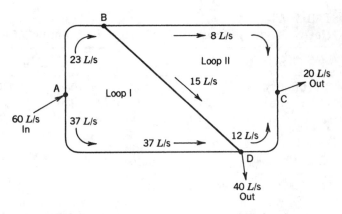

**FIGURE 7.30**
*Balanced flow rates for Example 7.14. Note that continuity of flow applies at all pipe junctions ($Q_{IN} = Q_{OUT}$). At point A, 60 = 23 + 37; at point B, 23 = 15 + 8; at point C, 8 + 12 = 20; at point D, 15 + 37 = 40 + 12 = 52.*

A running tabulation of corrected flows is shown in Figure 7.29. Notice that the corrected flow in BD ($Q + \Delta Q = 4$ L/s) from loop I/first try is used with opposite sign in the computations for loop II/first try. This is because pipe BD is common to the two loops; while the flow is clockwise in loop I, it is counterclockwise in loop II. This reversal in sign for flows in BD occurs repeatedly, as indicated in the rest of the computations.

The final flow rates in the balanced network are shown in Figure 7.30.

## Computer Modeling

As previously mentioned, the Hardy Cross method is rarely used in engineering practice. Mathematical models of distribution networks are now readily created and analyzed with desktop computers. Modern network modeling software can integrate the hydraulic calculations of water pressure and flow rates (as well as water quality) with computer-aided drafting (CAD) software and geographic information systems (GIS) programs. These generate drawings of distribution networks, pressure contour maps, graphs, and many other visual images of the network and its performance.

Network models are used for long-range planning in the water utility industry, and they serve as the basis for the preliminary design of pipe sizes and pump capacities for new facilities. Models are also used for fire flow studies, emergency response planning, energy management, and optimization of the use of existing facilities. Network models allow more time for creative thinking and providing answers to "what-if" questions, in lieu of tedious number-crunching tasks. They facilitate the planning, analysis, and design of large water distribution systems in a cost-effective manner. A case study describing the implementation of a water system modeling program is presented next, followed by a brief description of EPA's network modeling program.

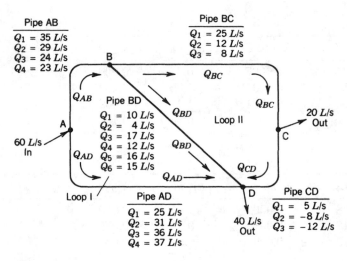

**FIGURE 7.29**
*Summary of initial flow assumptions and subsequent corrected flows for Example 7.14.*

## CASE STUDY

### Modeling Software: GCDC's "Personal" Consultant

The Genesee County Drain Commissioners (GCDC) has found an extremely useful analytical tool for managing its extensive water systems.[*] The tool is providing a rapid-response, decision-making framework in support of operations, construction, and design. Moreover, it is being maintained in-house and can grow with the water systems.

GCDC's model-development process required CAD maps that included pipes, nodes, tanks, pumps, and other elements, as well as associated data. Through the late 1980s the authority analyzed results through a numerical model and a separate CAD package.[†] This method required iterative correlation of the data from the software program with the CAD maps. Such back-and-forth shifting was manpower intensive, tedious, and error prone, making it particularly cumbersome for large systems such as GCDC's. In addition, the procedure did not transfer well between technical staff and political decision makers.

GCDC needed the potential power and flexibility of a water-distribution software package that tied the numerical model seamlessly to a CAD graphical representation of the system. It quickly seized the opportunity when such a package was released[††] in the spring of 1991.

The new interactive, visual modeling software smoothly moved the modeling methods of development, maintenance, and analysis services in-house. (They traditionally had been in the consultant's purview.) Management thus got the tools it needs to meet growing operational and design decision-making responsibilities while realizing direct-cost savings.

Using the program, GCDC quickly converted all subsystem models to the new software, using the software's import facility. GCDC thus was able to build upon its existing information database maintained in

subsystem models and CAD drawings. It thereby created a comprehensive "living model" of the entire water system (1116 pipes, 967 nodes).

The model soon proved its worth when GCDC had to do a quick analysis for a small city in the water system. This system is bisected by the Flint River. The northern half includes elevated storage with primarily older and smaller-diameter pipes. The southern half contained larger, newer water mains. The entire city was being served as a low-pressure district due to concerns about higher pressures' damaging the water mains.

Analyses using GCDC's new modeling capability showed that the city could be divided into two pressure zones. An additional loop added in the south section would improve fire flows. This division effectively protects the low-pressure area from main damage and preserves the fire-flow capabilities of the elevated storage tank. The southern half of the city, with the added fire loop, will realize enhanced service and better-than-adequate fire protection.

---

### EPANET—EPA's Network Modeling Software

EPANET is a computer program that models and simulates hydraulic and water-quality conditions within pressurized water distribution networks. Results can be viewed in a variety of formats, including color-coded network maps, data tables, time series graphs, and contour plots. The program can be used for network design (that is, determining the location and size of pipes, pumps, valves, and tanks) and analysis of existing networks. Further, it can be used for designing sampling programs, fire flow analysis, vulnerability studies, and operator training.

Developed by the Water Supply and Water Resources Division (formerly the Drinking Water Research Division) of the Environmental Protection Agency's National Risk Management Research Laboratory, EPANET is public domain software that may be downloaded for free from http://www.epa.gov/ORD/NRMRL/wswrd/epanet.html.

EPANET models a water distribution system as a collection of *links* connected to *nodes*. The links represent pipes, pumps, and control valves. The nodes represent junctions, tanks, and reservoirs. Figure 7.31 illustrates how links and nodes can be connected to one another to form a network. EPANET tracks the flow of water in each pipe, the pressure at each node, the height of water in each tank, and the concentration of a chemical species throughout the network during a simulation period comprising multiple time steps.

---

[*]The Genesee County Drain Commissioners, Division of Water & Waste Services, Flint, Michigan, provides sanitary sewer collection, sewage treatment, and potable-water distribution to 27 governing units serving some 28,000 water and 65,000 sanitary-sewer connections.

[†]In early 1988, the best tools at hand were KYPIPE and AutoCAD.

[††]CYBERNET (now called WaterCAD) from Haested Methods, Waterbury, Connecticut.

*Source*: *WaterWorld Review*, May/June 1994. Based on material provided by the Genesee County Drain Commissioners, Division of Water & Waste Services, Flint, Michigan, and Haested Methods, Waterbury, Connecticut. (Copyright © *WaterWorld Review*, May 1994)

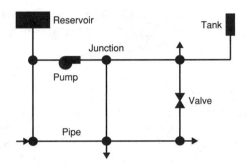

**FIGURE 7.31**
*Illustration of* links *and* nodes *in a simple distribution network.* (Courtesy of the Environmental Protection Agency)

Junctions are points in the network where links join together and where water enters or leaves the network. Tanks are nodes with storage capacity, where the volume of stored water can vary with time during a simulation, and reservoirs are nodes that represent an infinite source or sink of water to the network, such as lakes, rivers, or groundwater aquifers. Pipes, the links that convey water from one point in the network to another, are assumed to flow full at all times. The Hazen–Williams formula (see page 37) is used to compute head losses in the pipes.

 **7.6 RELEVANT WEB SITES**

**CANADIAN PIPELINER**

http://www.pipeline.ca/

This is the site of a newsletter of the Pipeline Contractor's Association of Canada, providing current news and industry information.

**DUCTILE IRON PIPE RESEARCH ASSOCIATION (DIPRA)**

http://www.dipra.org/

DIPRA is a not-for-profit association supported by all of the ductile iron pressure pipe manufacturers in the United States and Canada. This site provides specific information and publications about ductile iron pipe and its applications.

**PLASTIC PIPE INSTITUTE**

http://www.plasticpipe.org/

This is the site of the Insider Newsletter, which provides articles and information about the use of plastic pipe in water distribution systems and other applications.

**UNDERGROUND CONSTRUCTION**

http://www.undergroundinfo.com/

This site provides links to various publications and articles that focus on underground pipeline construction.

**UNI-BELL PVC PIPE ASSOCIATION**

http://www.uni-bell.org/

This is the Home Page of a nonprofit, technical, educational, and research-oriented organization focusing on providing technical literature and recommended standards for PVC pipe applications.

**WATER MAINS**

http://www.waterone.org/ds_water_mains.htm

This site provides information from Water District 1 of Johnson County, Kansas, on the installation or upgrading of water mains, requirements for water main extensions, construction scheduling, service connections, and cross connections.

**WATER TRANSMISSION AND DISTRIBUTION**

http://www.cwlp.com/Water_division/T_DWater/t&dwater.htm

This site provides a "case study" description of the water transmission and distribution system serving the City of Springfield, Illinois, including information on the transmission mains, water mains, storage tanks, service connections, fire hydrants, and system maintenance.

# REVIEW QUESTIONS

1. List four basic categories of water use. What is the effect of local climate and the use of individual water meters on water demand?

2. What does *gpcd* stand for? Explain briefly.

3. Briefly discuss variations in water demand over time. Sketch a graph that would illustrate hourly variations over a 24-h period.

4. What is the range of working pressures in a typical water distribution main?

5. What is the minimum size of a water main in a public water system?

6. Why is a gridiron pipe layout preferable to a system with many dead ends?

7. List five materials commonly used in the manufacture of water distribution pipes. Briefly discuss each type.

8. List three common types of valves used in water distribution systems. Briefly explain the use and operation of each type of valve.

9. Are water mains completely watertight?

10. What is the purpose of a thrust block?

11. Briefly discuss methods for rehabilitating a water main that has reduced flow capacity due to internal deposits on the pipe wall.

12. What is a *low-lift pump?* A *high-lift pump?*

13. Make a sketch showing the difference between static lift and static suction head for a pump. Also show total static head.

14. What happens when a centrifugal pump loses its prime?

15. What is the maximum practical height to which a column of water can be lifted by a pump under suction lift conditions?

16. What is the effect of increasing the impeller speed of a centrifugal pump on discharge and on pressure head?

17. What is *TDH?* Sketch a graph showing the relationship between TDH and the discharge of a centrifugal pump.

18. What is meant by the term *shutoff* head?

19. How does TDH vary with discharge in a water distribution system? Sketch a graph showing this relationship on the same paper that you used for

Question 17. What does the intersection between the two curves represent?

20. If two identical pumps operate in parallel in a water distribution system, would the total discharge be twice the discharge from one pump alone? Why?

21. What is meant by the term *water power* of a pump? Is it greater than, the same as, or less than the power used to operate the pump?

22. What is meant by the term *equalizing storage?* What benefits does equalizing storage provide in a water distribution system?

23. What are the basic functions of a distribution storage tank other than to provide equalizing storage?

24. Why are distribution storage tanks often elevated above the ground by a tower? What does it mean to say that the tank *floats on the line?*

25. Is it preferable to have a single large water storage tank located near the pumping station or to have several smaller tanks located near the major centers of water demand? Explain briefly.

26. What is the difference between a standpipe and a reservoir?

27. What is an *equivalent pipe?* What is the purpose of determining equivalent pipes in water distribution systems?

28. What is the first step in determining an equivalent pipe for pipes in series? For pipes in parallel?

29. What is a *hydraulically balanced* pipe network? What is the sum of the head losses around a balanced pipe loop?

30. In the Hardy Cross analysis of a pipe network, what is the only restriction on the initial assumption of flow rates in each pipe of the system?

31. Visit the *Ductile Iron Pipe Research Association* Web page and click on the link for "Ductile Iron Pipe." Write a brief report summarizing that page. Include a description of the differences between cast iron and ductile iron.

32. Using the EnviroSources search engine (see Section 1.6) or another Internet search directory or search engine, locate at least one additional Web site relevant to one of the topics in this chapter. Add the link(s) to your Environmental Technology Folder. Write a brief description of what the Web site(s) contain.

## PRACTICE PROBLEMS

1. A community has a population of 52,500 people. What is the total expected daily water demand? (Refer to Table 7.1.)

2. Referring to Problem 1, compute the peak hourly flow in liters per second for the day of maximum use. Assume that the peak daily demand is 200 percent of the average daily demand and the peak hourly flow is 400 percent of the daily demand.

3. A 240-m-long section of 600-mm water main is pressure-tested for 1 h, in which time a volume of 100 L of water is pumped into the pipeline to maintain the test pressure of 1000 kPa. Pipe sections are 6 m long. By how much is the allowable leakage exceeded?

4. Compute the static thrust on a 90° bend in a 305-mm pipe that carries water at a pressure of 600 kPa. If the soil has a bearing capacity of 100 kN/m$^2$, what area of thrust block is needed to anchor the bend?

5. Manufacturer's data for a centrifugal pump indicate that the pump can discharge 500 gpm at a discharge pressure of 100 ft when the impeller speed is 2000 rpm. What is the expected pump discharge and pressure head if the impeller speed is increased to 2500 rpm?

6. A centrifugal pump with the following characteristics is installed in a system to raise water from one reservoir to another. The water surface elevation in the first reservoir is 500 ft and in the second reservoir is 650 ft above mean sea level. The pipeline connecting the reservoirs include 2 mi of 10-in.-diameter pipe. Determine the operating point of the pump in this system.

| Q, gpm | TDH, ft |
|--------|---------|
| 0 | 210 |
| 200 | 205 |
| 400 | 190 |
| 600 | 165 |
| 800 | 125 |
| 1000 | 70 |

7. If two identical pumps with the characteristics tabulated in Problem 6 are operating in parallel in that system, what will be the operating point of the pumps?

8. Pump A and pump B are connected in parallel in a system that includes 10 km of 600-mm pipe. The following data describe the head–discharge relationship for each pump:

| | TDH, m | |
|--------|--------|--------|
| Q, L/s | Pump A | Pump B |
| 0 | 28 | 3 |
| 50 | 27 | 33 |
| 100 | 24 | 31 |
| 150 | 20 | 26 |
| 200 | 12 | 19 |
| 250 | — | 10 |

The total static head in the system is 16 m. For the pumps operating in parallel, what is the operating point? Assuming that the discharge line is throttled by a butterfly valve to a point that increases the system TDH to 30 m, describe the operating condition of the pumps.

9. For the pump described in Problem 6, compute the water horsepower. Assuming that the pump efficiency is 75 percent, what horsepower must the motor deliver to drive the pump?

10. For Problem 8, the total power input to the pumps is 75 kW. At what efficiency are the pumps operating?

11. A town has a population of 32,000 people and an average per capita water demand of 500 L/d. Assuming that the need for equalizing storage is 20 percent of the average daily demand and that storage for a fire flow of 60 L/s for a 4-h duration is required, compute the required volume of a distribution storage tank for the town.

12. Referring to Figure 7.25, determine the rate at which water is being withdrawn from the system at point A when the pump discharge pressure is 350 kPa and the pressure at point A is 250 kPa. The storage tank floats on the line with water at a height of 20 m, as shown. Sketch the HGL for the system. Determine at what rate the storage tank is being filled (or emptied) under these conditions.

13. Determine the theoretical diameter of a single 2000-m-long pipeline from A to D that would be equivalent to the series pipeline shown in Figure 7.32.

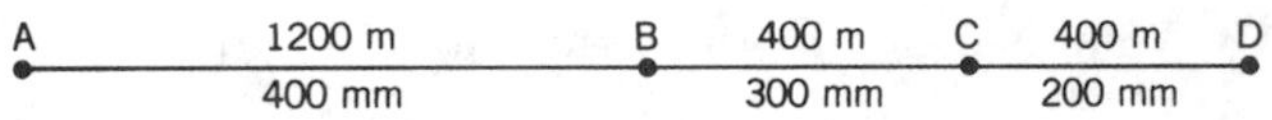

**FIGURE 7.32**
*Illustration for Problem 13.*

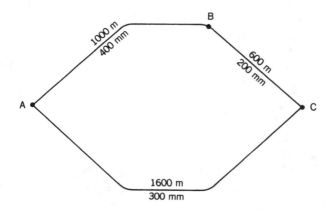

**FIGURE 7.33**
*Illustration for Problems 14 and 16.*

**14.** Determine the theoretical diameter of a single 1000-m-long pipeline from A to C that would be equivalent to the looped pipeline shown in Figure 7.33.

**15.** Determine the theoretical diameter of a single 1-mi-long pipeline from A to C that would be equivalent to the pipe system shown in Figure 7.34.

**16.** A flow of 500 L/s enters junction A of the loop shown in Figure 7.33. Determine the flow rate in each pipe if 200 L/s is withdrawn at junction B and 300 L/s is withdrawn at junction C.

**17.** Under the conditions stated in Problem 16, if the pressure at junction A is 750 kPa, what is the pressure at junction C, assuming that A and C are at the same elevation?

**18.** Consider the two-loop pipe network shown in Figure 7.28. If the flow into junction A is 100 L/s and the flows out of junctions B and C are 60 L/s and 40 L/s, respectively, compute the flows in each pipe of the network.

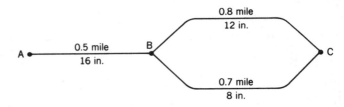

**FIGURE 7.34**
*Illustration for Problem 15.*

## Chapter Outline

CHAPTER

# 8

# Sanitary Sewer Systems

A sewage collection system consists of a network of pipes, pumping stations, channels, and appurtenances that conveys wastewater to a point of treatment, storage, or disposal. The wastewater may be *sanitary sewage, industrial sewage, storm sewage,* or a mixture of the three.

Sanitary sewage, also called domestic sewage, contains human wastes and washwater from homes, public buildings, or commercial and industrial establishments. Industrial sewage is the used water from manufacturing processes, usually carrying a variety of chemical compounds. Storm sewage, or stormwater, is the surface runoff caused by rainfall; it carries organics, suspended and dissolved solids, and other substances picked up as it travels over the ground.

Pipelines that carry a mixture of these three types of liquid wastes are called *combined sewers.* Combined sewers were commonly built in the 19th century and

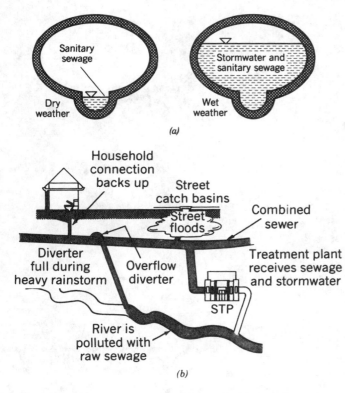

(a)

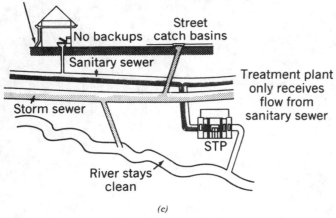

(b)

(c)

**FIGURE 8.1**

(a) *Cross section of a* combined sewer; *combined sewers, which were built in many older cities, carry both storm and sanitary sewage. A smaller channel at the bottom serves to carry the sanitary sewage at self-cleansing velocities during dry-weather periods. During rainstorms, the storm runoff and sanitary sewage become mixed together.* (b) *During very heavy rainstorms, most of the* combined sanitary and storm sewage *must be diverted directly into the receiving stream because it is too much for the treatment plant to handle; this causes water pollution.* (c) Separate sewers *are used in new construction.* (Reprinted from *Waterworld News*, by permission. Copyright © 1985, The American Water Works Association)

can be found in older cities and towns. Most of these systems, some of which are more than 100 years old, are still in use today.

Combined sewers typically consist of large-diameter pipes or tunnels. This is because the volumes of stormwater that must be carried away during wet weather periods are so large. The volume of sanitary sewage is very small when compared to stormwater flow. To keep the sanitary sewage flowing swiftly in the large-diameter conduits, a combined sewer may have the shape shown in Figure 8.1a. During dry weather, the sanitary sewage flows in the smaller channel at the bottom.

Today, combined sewers, as depicted in Figure 8.1b, are no longer built. Instead, separate sewer systems are constructed. Stormwater is carried in separate *storm sewers* to a point of storage or disposal; sanitary sewage and pretreated industrial wastewater are carried in separate *sanitary sewers*, usually to a municipal wastewater treatment plant. This is shown in Figure 8.1c.

The basic reason for building separate systems is that conventional wastewater treatment plants do not have the capacity to handle the huge volumes of stormwater that develop when it rains. Consequently, stormwater mixed with sanitary sewage must be bypassed around the treatment plant directly into receiving waters, causing water pollution. This is called *combined sewer overflow*, or CSO. In some large cities, CSO may be directed to a large storage basin and disinfected before flowing into the receiving waters. Separate sewers are designed to prevent the problem of CSO by conveying only sanitary sewage to the sewage treatment plant (STP).

In most sewer systems, the wastewater flows downhill, by gravity, in partially filled pipes that are not under pressure. Sometimes, sewage must be conveyed under pressure in pipelines called *force mains* to a treatment plant or to a point where it again can flow downhill due to the effect of gravity. In some cases, the entire sanitary sewer system for a localized area may consist of relatively small-diameter pipes under pressure, called *pressure sewer systems*. *Vacuum sewers*, which use suction to move the wastewater, are used in some situations. (See case study, page 241.)

This chapter focuses on sanitary sewer systems and includes design factors, materials, appurtenances, construction, infiltration and inflow, and rehabilitation. Stormwater management and control of CSO are covered in Chapter 9.

## 8.1 SANITARY SEWER DESIGN

Sewers that collect wastewater directly from a series of homes or buildings are called *lateral sewers*. Except for the individual house connections, these are the smallest diameter sewers in the system. Laterals carry the sewage by gravity flow into larger *submains* or *collector sewers*, which in turn tie into an even larger main sewer, called a *trunk line* or *interceptor*. The interceptor carries the sewage to the treatment plant, where most of the pollutants are removed before the sewage flows into the receiving waters. It is generally located in the lowest part of the service area or drainage basin and may be built parallel to a valley floor or river bed.

## Materials and Appurtenances

The pipes in a sanitary sewer system must be strong and durable to resist the abrasive and corrosive properties of the wastewater. They also must be able to withstand stresses caused by the soil backfill material, which is placed into the excavated trench to cover the pipe, and the effect of vehicles passing above the pipeline.

The joints between sewer pipe sections should be flexible, but tight enough to prevent excessive leakage, either of sewage out of the pipe or groundwater into the pipe. In addition to the use of appropriate materials to meet all these requirements, a variety of appurtenances is necessary for the proper operation of a sewer system. These include manholes, inverted siphons, lift stations, and flow meters. Flow meters are discussed in Section 2.4 and lift stations are discussed in Section 8.2.

### Vitrified Clay Pipe (VCP)

Vitrified clay is a good sewer material because it is very resistant to corrosion or deterioration from acids and other chemicals. It also resists erosion and scour from abrasive materials carried in the flow, but it is brittle and can break easily. Careful handling during construction is important, and proper placement in the trench, called *bedding*, will provide support to resist external loads. Vitrified clay pipe is limited in size; it is available in diameters up to about 1 m (3 ft) and in lengths up to about 2 m (6 ft). Bell-and-spigot O-ring compression joints are generally used to connect the pipe sections.

### Reinforced Concrete Pipe (RCP)

Precast sections of reinforced concrete pipe are suitable for larger sewer systems. RCP is available in diameters up to about 6 m (20 ft) and in lengths up to about 8 m (25 ft). It is a strong pipe material, but the concrete is susceptible to deterioration in the presence of hydrogen sulfide gas. This problem, called *crown corrosion*, is discussed in more detail later in this section. A variety of protective linings applied on the inside pipe wall can prevent crown corrosion. Joints are generally bell-and-spigot with O-ring rubber gaskets.

### Plastic Pipe

The use of plastic pipe for gravity sewer lines is increasing because of its lightness and ease of handling during construction. It is corrosion-resistant, and the smooth plastic pipe wall provides good hydraulic characteristics. One type of plastic commonly used is poly vinyl chloride (PVC); PVC pipe is available in diameters up to 750 mm (30 in.) and lengths up to 6 m (20 ft). Bell-and-spigot ends can be joined using either an O-ring gasket or a chemical weld joint. High-density polyethelyne (HDPE) is also used.

### Other Materials

Asbestos cement (AC), ductile iron, and steel sometimes are used in sanitary sewer systems. AC pipe is light and easy to handle during construction, but it is subject to deterioration when hydrogen sulfide or acids are present. Iron and steel pipes are generally used for force mains or in pumping stations, where the sewage is under pressure, for unusual external loading conditions, or when the sewer line is installed very close to or above a water main.

### Manholes

Manholes are structures that provide access to the sewer pipeline for cleaning, repair, sampling, and flow measurement. They are generally circular in cross section, with a diameter of at least 1.25 m (4 ft) so that a worker can move inside them without too much difficulty.

A typical manhole for a lateral sewer is shown in Figure 8.2. The bottom is made of concrete, with a semicircular channel connecting the inlet and the outlet sewer pipes. It is common practice to use precast concrete pipe sections to build the manhole up to required grade, although brick, concrete block, or plain concrete may also be used in its construction. A cast-iron frame and cover are provided to carry traffic loads and keep out surface water.

Manholes are located over the pipe centerline under the following circumstances:

1. When there is a change in pipeline diameter
2. When there is a change in pipeline slope
3. When there is a change in pipeline direction
4. At all pipe intersections
5. At the uppermost end of each lateral
6. At intervals not exceeding about 150 m (400 ft)

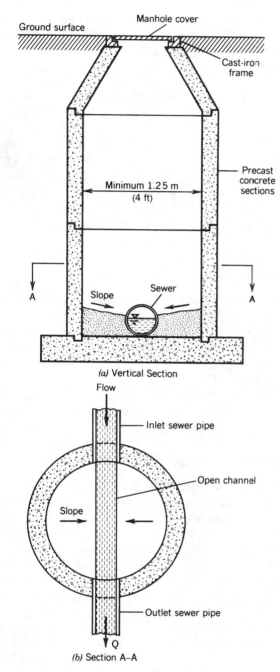

**(a) Vertical Section**

**(b) Section A–A**

**FIGURE 8.2**
*Section views of a lateral sewer manhole, which provides access for pipeline inspection, cleaning, repair, and flow sampling or measurement.*

Sometimes, when a lateral sewer joins a deeper submain sewer, the use of a *drop-manhole* will reduce the amount of excavation needed by allowing the lateral to maintain a shallow slope. This is illustrated in Figure 8.3. The wastewater drops into the lower sewer through the vertical pipe at the manhole. In this way, a worker is protected from the incoming sewage flow.

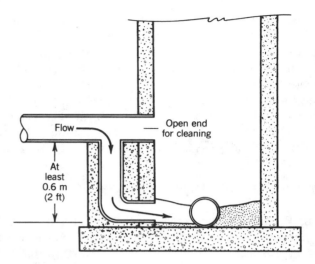

**FIGURE 8.3**
*Typical drop-manhole structure.*

Manholes should be built so as to cause minimum head loss and interference with the hydraulics of the sewer line. One way to maintain a relatively smooth flow transition through the manhole when a small sewer joins one of a larger diameter is to match the pipe crown elevations at the manhole. This is illustrated in Figure 8.4.

**Sewer Crossings**

A sewer line can be built across a stream, highway cut, or other obstruction, either below ground or above ground. When built below ground, the section of sewer that goes under the stream or road is called an *inverted siphon* or, more accurately, a *depressed sewer*. A sketch of a depressed sewer is shown in Figure 8.5.

It can be seen that this section of sewer is under the hydraulic grade line; it flows full and under pressure.

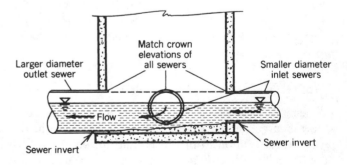

**FIGURE 8.4**
*When intersecting pipes at a manhole are of different diameters, the pipe crown elevations may be kept the same to allow for a smooth flow transition.*

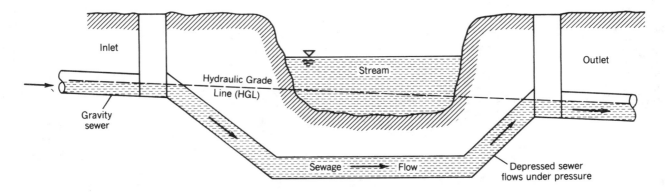

**FIGURE 8.5**
*A* depressed sewer, *or* inverted siphon *as it is called, is constructed to carry the gravity flow under an obstacle along the pipeline route.*

To maintain flow velocities high enough to prevent solids from settling out in the pipe, two or three different sizes of parallel pipes are used to carry the minimum, average, or peak flows. Since the siphon is under pressure, a concrete encasement or ductile iron pipes are provided to prevent leakage.

In some cases, it may be more economical to cross a stream or other topographic low area with the sewer above ground. For example, a pipeline may be suspended on an existing bridge. Occasionally, a shallow depression can also be crossed by supporting the pipe on concrete piers.

**CASE STUDY**

### Sinking a Sewer Line

Flood control, especially during hurricanes, is essential to San Juan, Puerto Rico, especially near the Rio Puerto Nuevo, a river that empties into San Juan Bay. The river will be dredged to widen and deepen it so that it can better accommodate floodwaters. Originally about 2 m (7 ft) deep, when dredging is complete, the river depth will be about 11 m (35 ft). The increased depth and width requires replacement of sewer and water pipelines to bring them under the channel bottom. Construction of two main sanitary sewer siphon projects, the Miramar trunk sewer close to the Bay and the San Jose trunk sewer farther upstream, began in year 2000.

HDPE pipe material was selected for the three-barrel siphons because of the substantial savings it offered compared to concrete encased ductile iron pipe. Adjustable weirs will regulate higher flows entering the two larger, 1-m-diameter siphon barrels. The ultimate design capacity is 2190 L/s (50 mgd) under a pressure of 1100 kPa (160 psi). [A complete description of the design and construction of this project is provided in John Ruhl and Steve Campbell, "Sinking a Sewer Line," *Civil Engineering Magazine* (American Society of Civil Engineers), September 2000.]

## Sewer Location and Layout

Before the detailed design of a sewer system can begin, the area to be sewered must be surveyed to obtain topographic data. A map must be prepared so that a preliminary layout can be made. Sewers are generally located in the public right-of-way (ROW), near the center of the street to conveniently serve houses on both sides. Sometimes they are located in alleys or in easements across private property.

The designer sketches the location of the sewer lines and manholes in the streets or easements on the map, with the objective of obtaining gravity (downhill) flow as much as possible. Force mains and the required pumping stations are avoided because of the added expense and potential operating problems. The slope of the ground governs the degree to which gravity or open channel flow conditions can be maintained. A typical sewer system layout is shown in Figure 8.6.

From the field survey data, detailed maps of each street are prepared at a relatively large scale showing the locations of the houses to be served by the sewers as well as the locations of other utilities, such as storm sewers and water and gas mains. In urban areas, the congestion caused by other underground utilities often complicates the design and construction of a new sewer line.

Profile views of the streets and utilities are generally drawn directly below the plan views. These plan

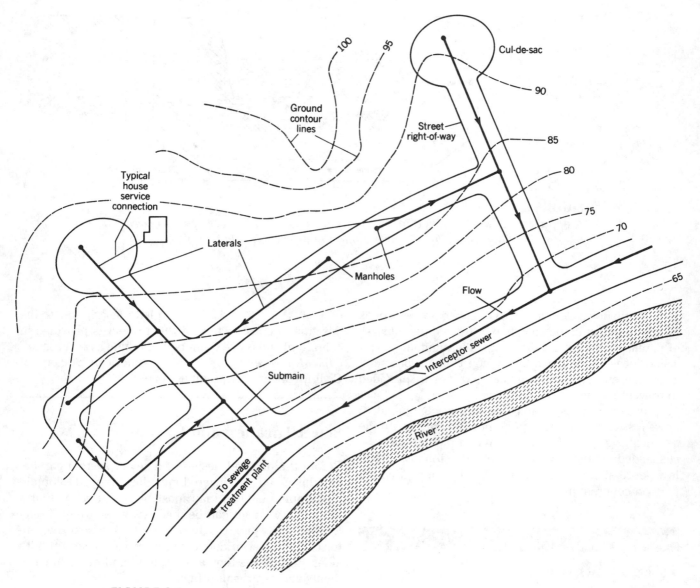

**FIGURE 8.6**
*Sanitary sewers are located so as to provide gravity (downhill) flow as much as possible. It is best to avoid the need for pumping stations.*

and profile drawings serve as working drawings during the detailed design phase. The final design of the system is then added to the drawings, which serve as part of the construction contract documents. Data from soil borings, including depth to bedrock and groundwater, are also shown on the profiles, and sometimes the boring logs or records are included with the contract documents.

The construction drawings for a sewer system must show the *location, depth, diameter,* and *slope* of the pipeline so that the construction contractor can readily excavate the trench and place the pipe sections at the proper position and grade. A typical sewer plan

and profile drawing is shown in Figure 8.7. In addition to showing the sewer diameter and slope, the drawings show manhole locations, *pipe-invert elevations,* and existing utilities. (The *invert* of a pipe is the bottom-most surface of the inner pipe wall; *elevation* refers to height, usually above mean sea level. Determination of invert elevations is discussed in Example 8.2.)

**Plan and Profile Details**

In Figure 8.7, the *plan* (top) view of the sewer line is located directly over the *profile* (side) view, and both views are fully aligned. This is not always the case, as

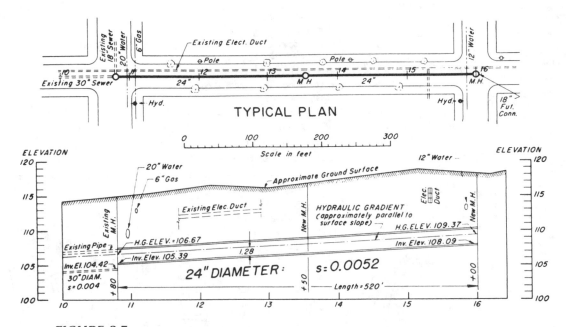

**FIGURE 8.7**

*Typical sewer plan and profile will show the location, depth, diameter, and slope of the pipeline, as well as the location of the other utility lines.* (Courtesy of the National Clay Pipe Institute, Washington, D.C.)

shown in Figure 8.8, where the sewer (called LINE B) in Florida Avenue changes direction at manhole B-2, near Lot 8. Sewer lines may change direction several times in plan view, but they are always "projected" onto a flat plane when drawn in profile view, and thus they appear as a straight line. The vertical scale of the profile is typically "magnified," normally by a ratio of 10 to 1, for clarity in depicting important elevations and slopes. Many sewer construction drawings will also contain notes and symbols similar to those shown in Figure 8.8.

A typical symbol shown on a sewer plan and profile is a *station* number, which is used to indicate distance from a starting point (often a manhole cover in a street intersection.). In U.S. Customary units, each 100 ft is called a *station*. Station numbers are labeled in two parts separated by a plus sign—first the number of hundreds of feet and then the number of feet (and decimals of a foot). For example, the starting point would be labeled 0 + 00 and the *first full* station (or distance of 100 ft) would be written as 1 + 00. A manhole at station 3 + 00 would be 300 ft from the starting point, and a manhole at station 5 + 67.89 would be 567.89 ft from the beginning of the line. In SI metric units, the meter is the base unit for length, and a full station would be 1 + 00, or 100 m. Station numbers may increase from left to right (as in Figure 8.7) or from right to the left (as in Figure 8.8).

On construction drawings, manholes may be abbreviated as M.H., or as SSMH (short for "sanitary sewer manhole"), as seen in Figure 8.8. Inverts are usually abbreviated as INV and invert elevations as IE. Other symbols shown on the plan in Figure 8.8 include S.C., S.L., and E.I., to show where each house is hooked up to the lateral sewer in the public right of way and related data. Service lines are typically installed by the house contractor and then become the responsibility of the homeowner for maintenance. A legend for abbreviations and symbols is often shown on the plan and profile drawing.

## EXAMPLE 8.1

As an exercise in reading a sewer plan and profile, provide the following information from Figure 8.8:

(*a*) What is the station number of manhole B-4?
(*b*) What is the ground elevation at manhole B-2?
(*c*) What is the "invert out" elevation of the sewer at manhole B-3?
(*d*) What is the steepest sewer slope shown on the profile?
(*e*) How far is manhole B-1 from the beginning of Line B?
(*f*) Approximately how deep is manhole B-3?

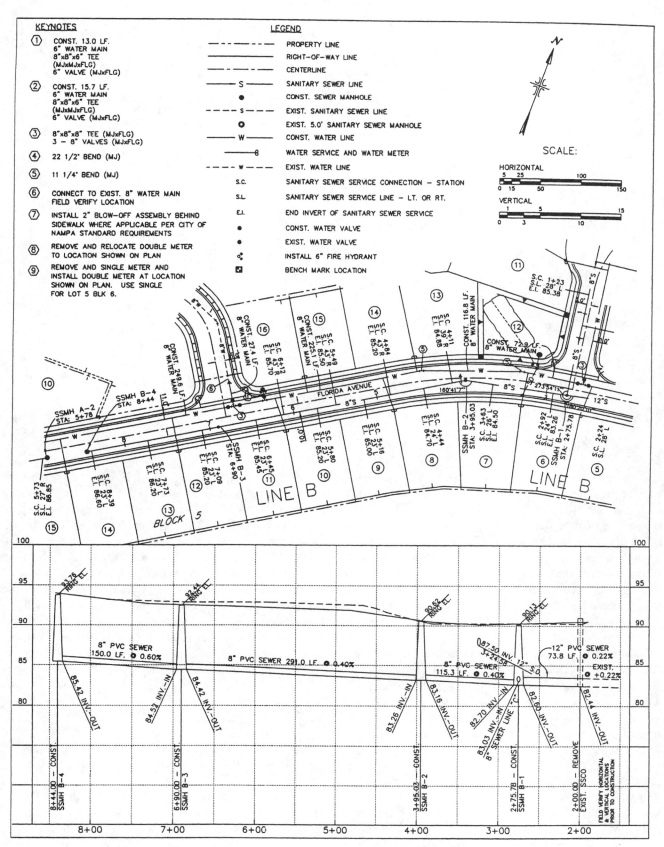

**FIGURE 8.8**

*Plan and profile view of a sanitary sewer.* (Courtesy of Capital Development, Boise, Idaho. From Madsen and Shumaker, *Civil Drafting Technology*, 3rd ed., © 1998. Reprinted by permission of Prentice-Hall, Inc., Upper Saddle River, New Jersey)

*Solution*

(a) Manhole B-4 is on the westmost end of Line B, at station 8 + 44.

(b) The ground elevation (or grade) at B-2 is 90.62 ft.

(c) The invert out elevation at B-3 is 84.42 ft.

(d) The steepest slope is 0.60% or 0.006, between B-3 and B-4.

(e) Manhole B-1 is 275.78 ft from the beginning of Line B.

(f) Manhole B-3 is about 92.44 − 84.47 = 7.97 or 8 ft deep.

## Quantity of Sewage

Before the diameter and slope of a sewer can be established, it is necessary to have an estimate of the quantity of wastewater it will carry. Actually, the quantity of sewage carried by a sewer is not constant with time. The flow rate varies throughout the day in a pattern similar to the variation in water demand, which is illustrated in Figure 7.1.

Sewers must be designed to carry the peak or maximum flow rates, not just the average flow rates. They also must maintain self-cleansing flow velocities, as is discussed shortly. State environmental agencies or local health departments generally have specific requirements for sewer design flows. For example, a typical requirement for lateral and submain sewers is that they be designed to have a full-flow capacity that is 4 times the average flow rate. Main or trunk sewers must have a capacity of 2.5 times the average flow.

The flow rate peaks are less pronounced in the larger diameter sewers because it is less likely that all the sources of wastewater in a large service area will contribute sewage into the system at the same time. In effect, this smooths out the daily flow pattern; by the time the sewage reaches the treatment plant, there is considerably less variation among peak, average, and minimum flows than there is in the laterals.

It is very important to have a good estimate of flow rates before designing the sewer system. If the quantity of sewage flow is underestimated, the selected pipe diameters and slopes may be too small to carry the peak flow rates. This is called *underdesign;* it can result in the *surcharging* of part of the sewer system.

A sewer system is surcharged when the wastewater is backed up into the manholes and the hydraulic grade line is above the pipe crown. This is illustrated in Figure 8.9. In severe cases of surcharging, the sewage may overflow out of the manholes and into the street, and it also can back up into basements of houses connected to the line. Surcharging may occur in a sewer

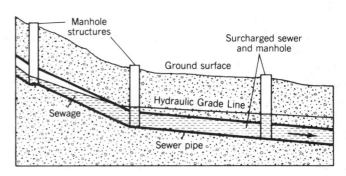

**FIGURE 8.9**
*When the sewer pipe diameter and slope are not adequate to carry the peak wastewater flows, the system will temporarily become* surcharged.

system because of poor design or because of excessive infiltration or inflow, as discussed later in this chapter.

Underdesign and surcharging are problems related to inaccurate sewage flow estimates. Another problem may occur if the amount of sewage is overestimated, resulting in *overdesign* of the system. Not only does construction of an overdesigned system become unnecessarily expensive, it also tends to attract extra land development sooner than expected. This premature development is usually undesirable, particularly because other municipal services and utilities may not have sufficient capacity to support it. In an EIS (see Appendix A), land development following overdesign of sewers is considered a secondary environmental effect of sewer construction.

Sewage flow rates depend on several factors, including population density, per capita water consumption, and commercial or industrial activity. The amount of groundwater seeping into the pipeline, called *infiltration*, must also be taken into consideration. Infiltration (and inflow) are discussed in more detail in Section 8.4.

Population densities are expressed as the number of people per hectare or acre. Sewer pipelines are generally designed to carry peak flows from the anticipated saturation (maximum possible) population for the local area to be served. These population estimates are usually available from county or state planning agencies.

Roughly 75 percent of the per capita water use (see Section 7.1) actually becomes sewage because some water is lost in lawn watering, car washing, and other uses that keep it out of the sewers. But in some cases, the quantity of infiltration and inflow into the system can more than make up for these losses. For practical purposes, then, it is often assumed that the average sewage discharge is about the same as the average water use in the community. On a nationwide basis, the average sewage discharge is presently about 400 L/d per person (roughly 100 gpcd). Of course, if more accurate data are available, such as from sewage flow studies of similar neighborhoods in the area, they should be used to estimate the flow quantities.

On the maps showing the sewer layout, the total areas that contribute sewage to segments of the system are outlined by the designer. A segment, for design purposes, is typically a length of the sewer line between manholes, called a *sewer reach*. A reach of a sewer line is constant in slope and diameter. The boundary of the tributary area is similar in concept to the drainage divide line discussed in Section 3.4, but it does not follow the same set of rules regarding contour lines and direction of flow. This is because the flow is confined in and must follow the direction of the service connections and the sewer line itself.

After the tributary boundaries have been sketched for every reach of the system, the areas can be determined. For square or rectangular shapes, computation of the areas using scaled map dimensions is easily done. For irregularly shaped areas, a planimeter is used to trace the boundary and determine the enclosed area. A high degree of precision in defining the boundary is usually not warranted because of the uncertainties in the assumed population densities and the per capita flow rates.

For a given sewer reach, the design flow is taken as the product of three factors: the tributary area, the population density for that area, and the assumed peak per capita sewage flow rate. This is illustrated in Example 8.2.

## EXAMPLE 8.2

A reach of a submain sewer is to be designed to receive flow from 100 ha of a community where the population density is estimated to average 25 people/ha. The average per capita sewage flow is estimated to be 400 L/d. Compute the design flow for the reach in liters per second.

*Solution*  First, determine the peak per capita flow rate, assuming that the submain must be designed to carry a flow rate four times the average:

$$\text{peak flow} = 4 \times 400 \text{ L/d/person}$$
$$= 1600 \text{ L/d per person}$$

The design flow rate for the reach is then computed as

$$100 \text{ ha} \times 25 \text{ persons/ha} \times 1600 \text{ L/d/person}$$
$$= 4{,}000{,}000 \text{ L/d or 4 ML/d}$$

and

$$4 \times 10^6 \text{ L/d} \times \frac{1 \text{ d}}{24 \text{ h}} \times \frac{1 \text{ h}}{3600 \text{ s}} \approx 46 \text{ L/s}$$

Design generally starts at the uppermost reach in the system and proceeds downstream, accumulating tributary areas and populations along the way. This is continued until a junction with an incoming branch sewer occurs, at which point the uppermost reach and then the rest of the branch are designed. The reach just downstream of a junction must be designed to accommodate the total population and sewage flows from the branches above it.

## Depth and Velocity Limitations

As far as is practical, sewer pipelines are placed deep enough in the ground to be able to receive wastewater flow by gravity from the houses and buildings to be served. But very deep basements or buildings located below street level because of steep topography may require individual pumping units to lift the sewage up into the public sewer. This condition is illustrated in Figure 8.10. Generally, a minimum cover of about 2 m (6 ft) above the crown of the sewer pipe is maintained, but design practice may vary in this regard. Greater depths require more excavation and increase the cost of construction. Generally, sewer depths do not exceed about 6 m (20 ft).

It is common practice to select the size and shape of a sanitary sewer so that the full-flow velocity will not be less than about 0.6 m/s (2 ft/s). This is called the

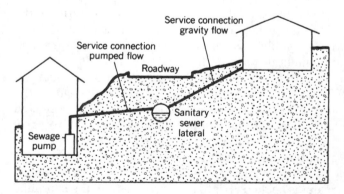

**FIGURE 8.10**

*Sewers cannot always be placed deep enough in the ground to allow gravity flow from all service connections. Individual sewage pumps or ejectors may have to be used by some homeowners.*

*minimum self-cleansing velocity* because it is the velocity that will keep sewage solids suspended in the flow. Low velocities tend to allow solids to settle out in the pipes, blocking the flow and reducing the capacity of the sewer line.

The upper limit of flow velocity is usually about 3 m/s (10 ft/s) to prevent excessive abrasion and wear of the pipe wall from sand and grit carried in the sewage. For a given diameter or slope of sewer, Manning's formula or nomograph can be used to check that the full-flow velocity is within the allowable range. For example, the minimum slope for a 300-mm (12-in.) pipe flowing full is 0.2 percent (0.002) in order to maintain the self-cleansing velocity.

### Crown Corrosion

When flow velocities in a sanitary sewer drop below the minimum self-cleansing velocity, sludge deposits form in the pipeline. As the sludge decomposes, the dissolved oxygen in the sewage is depleted, resulting in septic or anaerobic conditions. Even the remaining suspended solids have time to undergo decomposition in the pipeline, because of the long travel time, before reaching the treatment plant.

Sanitary wastewater contains sulfate ($SO_4^{2-}$) compounds, which are converted to hydrogen sulfide gas, $H_2S$, by anaerobic bacteria. This results in the characteristic rotten-egg odor of stale sewage. But a serious structural problem can also develop in concrete pipelines. In gravity sewers, which have air and moisture in the pipeline above the flowing wastewater, aerobic bacteria attached to the pipe crown oxidize the $H_2S$, converting it to sulfuric acid, $H_2SO_4$. This is illustrated in Figure 8.11. The sulfuric acid reacts with and weakens the concrete, causing the problem called *crown corrosion*. In extreme

cases, concrete sewer lines have actually collapsed from crown corrosion, necessitating immediate and expensive repairs. Proper hydraulic design of the sewer system to maintain self-cleansing flow velocities can help to avoid this problem.

## Size, Slope, and Invert Elevations

To minimize the amount of excavation, the slope or grade of the sewer should follow the slope of the ground as much as possible. A typical design procedure starts with an examination of the street profile to determine the average ground slope for a reach between two manholes. Manning's nomograph is used to determine the smallest standard pipe diameter that will carry the design flow for that reach at the same slope as the ground surface. For that diameter and slope, the velocity is then checked to make sure it is within the acceptable limits.

Appropriate adjustments are then made either in the pipe slope or diameter, as needed. For example, if the velocity is too low or if a very large diameter pipe is required, the designer might opt to increase the sewer slope. A steeper slope will increase the velocity of flow and reduce the required pipe diameter. But the designer must keep in mind that steeper slopes make it necessary to place the pipeline deeper in the ground, increasing excavation costs.

Lateral sewers are generally not less than 200 mm (8 in.) in diameter, no matter how small the design flow; this minimum size reduces the occurrence of pipe blockages and facilitates maintenance activities. But the low flow rates contributed at the upper reaches of the system lead to low flow velocities in these oversized laterals, and they need to be routinely flushed and cleaned to keep them clear of sludge deposits and blockages.

Once the pipe diameter and slopes have been established, the *invert elevations* of the pipe can be determined and the proposed sewer can be drawn on the profile. (The invert is the bottom inside wall surface of the pipe.) Usually a minimum depth of earth cover above the pipe crown is specified. The invert elevation is computed by subtracting the cover and the pipe diameter from the ground elevation at the manhole or by matching crown elevations (see Figure 8.4). This design procedure for a sewer reach is illustrated in Example 8.3.

### EXAMPLE 8.3

A 120-m reach of sewer is to be designed with a flow capacity of 100 L/s. The street elevation at the upper manhole is 90.00 m and at the lower manhole is 87.60 m,

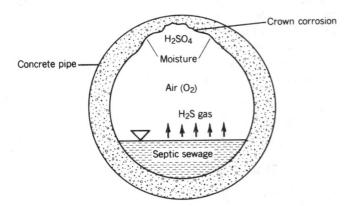

**FIGURE 8.11**
*Crown corrosion in unlined concrete sewers can eventually lead to structural collapse of the pipeline.*

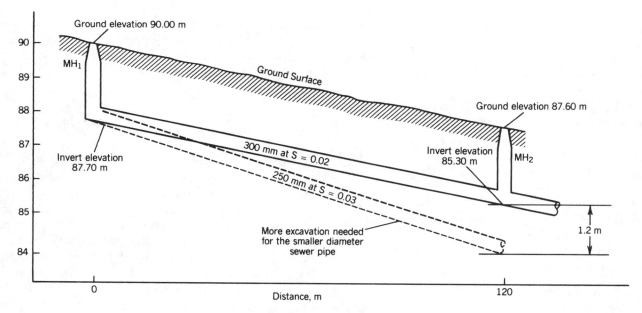

**FIGURE 8.12**
*Illustration for Example 8.3.*

as shown in Figure 8.12. Determine an appropriate pipe diameter and slope for this reach, and establish the pipe invert elevations at the upper and lower manholes. Assume a minimum earth cover of 2 m above the crown of the pipe.

*Solution*   The ground elevation drops $90.00 - 87.60 = 2.4$ m.

The ground slope is the change in elevation divided by the horizontal distance, or $S = 2.4/120 = 0.020$. Now enter Manning's nomograph (Figure 2.21) with $S = 0.02$ and $Q = 100$ L/s, or 0.1 m³/s. The straight line connecting $S$ and $Q$ crosses the diameter axis at 26 cm, or 260 mm.

It is necessary to choose a standard pipe size, one that is readily available from pipe manufacturers. A 250-mm pipe could be selected, but its slope would have to be steeper, about 0.03, in order for it to have a capacity of 100 L/s, according to the nomograph solution of Manning's formula. At a slope of 0.03, the sewer reach would drop $0.03 \times 120$ m $= 3.6$ m in elevation. Starting with 2 m of cover at the upper end, it then would have 3.2 m of cover at the lower end. In other words, the extra cover is $3.6$ m $- 2.4$ m $= 1.2$ m.

It may be preferable to select the larger, 300-mm-diameter pipe and install it at a slope of 0.02, parallel to the ground surface. This would keep the depth of cover constant at 2 m and involve less excavation than the 250-mm pipe would require. It is important to note that the savings in excavation will be more than that

one reach. If the 250-mm pipe is used, the extra depth of 1.2 m will be carried throughout the rest of the sewer line downstream of that reach. This can be seen in the sewer profile shown in Figure 8.12

Note that the full-flow capacity of the 300-mm pipe placed on a 0.02 slope is more than the required 100 L/s; from Manning's nomograph, it is seen that the actual full-flow capacity is 0.135 m³/s, or 135 L/s. Partial-flow conditions can be evaluated using Figure 2.22. If $q = 100$ L/s, then the ratio of $q/Q = 100/135 = 0.74$, and $d/D = 0.63$ is read from the partial-flow diagram. From this, $d = 0.63 \times D = 0.63 \times 300$ mm $= 190$ mm. The full-flow velocity $V = 1.95$ m/s from the nomograph. For $d/D = 0.63$, read $v/V = 1.08$ from the partial flow diagram. From this, $v = 1.08 \times V = 1.08 \times 1.95$ m/s $= 2.1$ m /s.

Let us select the 300-mm-diameter pipe and a slope of 0.02 for this reach and compute the invert elevations. The upper invert elevation is computed as follows:

upper invert elevation
$$= \text{ground elevation} - \text{cover} - \text{pipe diameter}$$
$$= 90.00 \text{ m} - 2.00 \text{ m} - 0.300 \text{ m}$$
$$= 87.70 \text{ m}$$

The fall or drop in elevation of the sewer over the length of the reach is the product of the slope and the distance, as follows:

$$\text{fall of sewer} = 0.020 \times 120 \text{ m} = 2.40 \text{ m}$$

and therefore

lower invert elevation

$$= \text{upper invert elevation} - \text{fall of sewer}$$

$$= 87.70 - 2.40$$

$$= 85.30 \text{ m}$$

This is illustrated in Figure 8.12.

---

## EXAMPLE 8.4

---

Design the two sewer reaches shown in Figure 8.13a. The design flow for reach 1 is 1 mgd, and for reach 2 it is 2 mgd. The ground elevation at the first manhole, MH 1, is 1100 ft, at MH 2 it is 1093 ft, and at MH 3 it is 1090 ft. Use a minimum cover of 8 ft.

*Solution*  For reach 1, the ground slope is computed as

$$S = \frac{1100 - 1093}{300} = \frac{7}{300} = 0.023$$

To use Manning's nomograph in Figure 2.21, first convert the flow rate from units of million gallons per day to gallons per minute, as follows:

$$Q = 1,000,000 \frac{\text{gal}}{\text{d}} \times \frac{1 \text{ d}}{24 \text{ h}} \times \frac{1 \text{ h}}{60 \text{ min}} = 690 \frac{\text{gal}}{\text{min}}$$

Now entering the nomograph with $Q = 690$ gpm and $S = 0.023$, select an 8-in.-diameter pipe. The actual full-flow capacity of an 8-in.-diameter pipe at that slope is about 840 gpm, and the velocity is about 5.3 ft/s. Invert elevations are computed as follows:

$$\text{upper invert elevation} = 1100 \text{ ft} - 8 \text{ ft} - \frac{8}{12} \text{ ft}$$

$$= 1091.33 \text{ ft}$$

$$\text{fall of sewer} = 0.023 \times 300 \text{ ft} = 6.90 \text{ ft}$$

$$\text{lower invert elevation} = 1091.33 - 6.90 = 1084.43$$

For reach 2, the ground slope is computed as

$$S = \frac{1093 - 1090}{400} = \frac{3}{400} = 0.0075$$

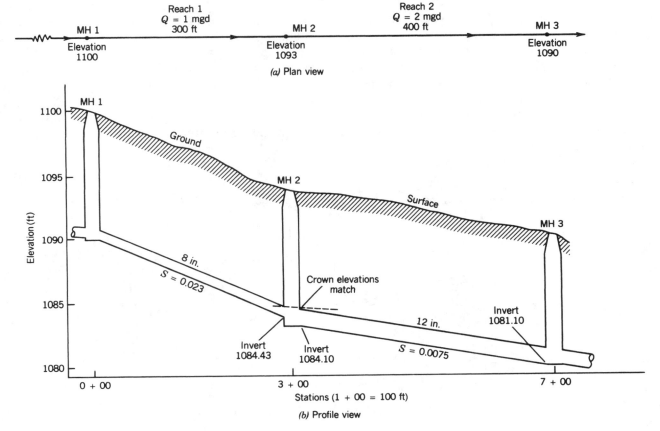

**FIGURE 8.13**
*Illustration for Example 8.4.*

From Manning's nomograph with $S = 0.0075$ and $Q = 1380$ gpm, select a 12-in.-diameter pipe for this reach. The velocity is 3.9 ft/s. To establish the upper invert elevation of the 12-in. pipe, match its crown elevation with that of the 8-in. pipe in reach 1, as follows:

The crown elevation of the 8-in. pipe is simply the sum of its invert elevation and the pipe diameter, or

$$1084.43 \text{ ft} + \frac{8}{12} \text{ ft} = 1085.10 \text{ ft}$$

The upper invert elevation for the 12-in. (1.00-ft)–diameter pipe is therefore

$$1085.10 - 1.00 \text{ ft} = 1084.10$$

The fall of sewer is $0.0075 \times 400$ ft $= 3.0$ ft, and

$$\text{lower invert elevation} = 1084.10 - 3.0 = 1081.10$$

This is illustrated in Figure 8.13*b*.

---

### Sanitary Sewer Overflows

Sanitary sewer overflows (SSOs) are discharges of raw (untreated) sewage from municipal sanitary sewer systems. SSOs can release raw sewage into basements or flow out of manholes and into public streets, playgrounds, and streams before the sewage can reach a treatment facility. They can pose a substantial public health threat and environmental challenge.

SSOs occasionally occur in almost every sewer system, even though the sewers are designed to collect and transport peak sewage flows. When SSOs happen frequently, though, poor collection system maintenance and management are often the cause. Inflow and infiltration (I&I) is often a specific factor causing chronic SSO problems. I&I may be due to excessive rainfall or snowmelt *infiltrating* into leaky or broken sanitary sewers, water *inflowing* through roof drains connected to the sewers, or poorly constructed service lines that connect homes to the pipes in the streets. I&I can be reduced by proper sewer construction and testing as well as by enforcement of laws that prohibit connections from storm drains (see Section 8.4). Other causes of SSOs include undersized systems, pipe settlement, and pipe blockages (by tree roots or sediment deposits).

## 8.2 SEWAGE LIFT STATIONS

Sanitary sewers are usually built to follow the general slope of the land so that the wastewater will flow downhill by gravity. In some cases, however, it becomes necessary to pump the sewage up from a low point to a higher elevation, either to reach a treatment plant or to reach another gravity sewer.

Special *nonclog* centrifugal pumps are available for pumping raw sewage. The impellers and casings of these pumps are designed to allow sewage solids, including rags and small sticks, to pass through without causing any damage or blockages. The structure in which these pumps operate is called a *lift station*.

A lift station may be used, for example, in an area where, because of flat topography, the sewer line would eventually become excessively deep, even at the minimum slope needed for self-cleansing velocity. This is illustrated in Figure 8.14.

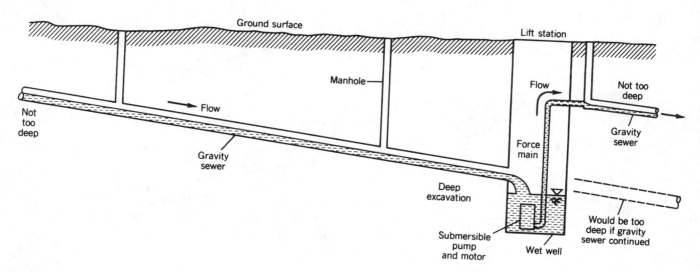

**FIGURE 8.14**
*Lift stations are used to raise the hydraulic grade line in the sewer system.*

Sewer construction costs increase with increasing trench excavation depths. When the slope of the pipeline carries it too deep, it may be more economical to install and operate a lift station in order to raise the hydraulic grade line. Another situation in which a lift station is necessary is when a new residential or industrial development is built in a topographically low area that is close to an existing municipal sewer system. A lift station will be needed to pump the sewage from the new development up into an existing submain or trunk sewer.

There are two basic types of lift stations: *dry well* installations and *wet well* installations. A wet well installation has only one chamber to receive and hold the sewage until it is pumped. One such arrangement involves the use of specially designed submersible pumps and motors, as shown in Figure 8.14. It is also possible to use a submerged pump, powered by a vertical shaft connected to a motor that is located above the wet well. In both cases, the pumps must be raised out of the wet well for maintenance.

In yet another arrangement, both the motor and the pump may be located above the wet well in a protective enclosure, but then suction lift and provision for priming the pumps are necessary. A typical prefabricated package lift station of this type is shown in Figure 8.15. The displacement switches serve to automatically shut the pumps off and on by sensing the water level in the wet well.

A dry-well installation has two separate chambers, one to receive the wastewater and another to house the pumps and controls. The protective dry chamber allows easy access to the pumps and controls for inspection and maintenance. An advantage of the dry-well installation is that the pumps may be located below the sewage level. This provides suction head instead of suction lift, eliminating the need to prime the pumps. A dry-well installation is sketched in Figure 8.16. The back pressure sensed by the air bubbler tube serves to indicate the sewage level and start or stop the pumps automatically.

The hourly variation in sewage flow rates must be known or estimated in order to design a lift station or to select and specify a suitable prefabricated package unit. Even the smallest lift station should have at least two pumps so that one can operate while the other is removed from service for maintenance or repair. Use of these two pumps can be automatically alternated so that one will not sit idle for very long periods of time.

The volume of the wet well should be large enough so that the pumps do not cycle on and off too frequently, yet it should be small enough so that the sewage does not remain stagnant for too long. If the wet well is too small, the pumps will run continuously

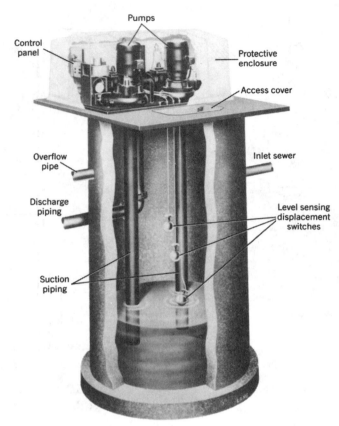

**FIGURE 8.15**
*A wet-well mounted pump station.* (Courtesy of Smith and Loveless, Inc., Lenexa, Kansas)

or cycle on and off so frequently that they will soon be worn out. If the wet well is too large, the long detention time of sewage before pumping starts will lead to anaerobic decomposition and septic conditions in the lift station. The designer must have reliable flow rate data to determine an appropriate wet-well volume.

## CASE STUDY

### Downsizing First Stage Pump Saves Energy

By adding a small booster pump to its Welches Point sewage lift station in 1997, the City of Milford, Connecticut, reduced energy consumption at the station by more than 3000 kW per month, resulting in an annual savings of almost $3000. With a total cost of $16,000, this project will pay for itself in a little more than 5 years. The lessons learned from this project will be applied to other sewage stations throughout the city.

The city of Milford, located just south of New Haven, operates 37 sewage lift stations that serve more than 48,000 people. Its Welches Point pump station was built in 1963 and handles about 2 mgd of raw sewage.

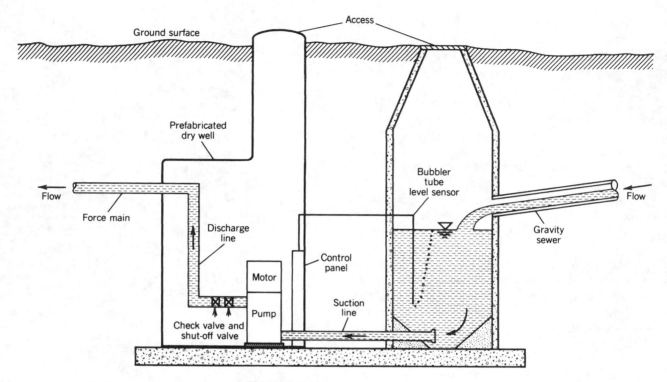

**FIGURE 8.16**
*A dry-well type of lift station.*

The station had three identical 75-hp pumps, which were vertically mounted 40 ft below ground level and driven by electric motors positioned directly above at ground level. Each pumped raw sewage to a common header from which the sewage flowed through a gravity feed header (shared by several sewage stations) to the main treatment plant. To evaluate the pumping station's efficiency, an analysis of flow volume, operating time, and energy consumption was performed by city engineers and consultants.

The old pump system was designed to operate with only one pump under normal conditions. One of the pumps would begin operating when the water level reached a set high-water level and remained on until the water dropped to a designated low-water level. During periods with very heavy inflow rates, two pumps operated simultaneously. The third pump served as a backup, operating if another pump needed repair. The pumps rarely operated for more than 15 min during each cycle.

The sewage station was designed to handle a peak inflow of 3000 gpm. The average inflow rate of sewage is 1700 gpm. Average flow rates from the station to the local treatment plant were estimated at 3350 gpm during normal conditions and 4250 gpm with two pumps operating. Each year, the old pumping system consumed about 212,000 kW of electric energy, with an overall system efficiency rated at 73 percent. To increase efficiency, several alternatives were considered, each of which involved pumping water out of the station more slowly. This would reduce head loss due to friction in the piping system, thereby reducing energy consumption.

Engineers considered installing variable-speed controls on the pumps, which can save energy in applications involving fast or frequent changes in flow rates. Because the station sump acted as a buffer in this application, the outflow rate did not need to be changed frequently, so variable-speed controls were not the answer. Other options involved trimming the impeller or replacing the original pumps. After analyzing tests performed on the original pumps, the engineers concluded that the best solution would be to install a smaller pump to operate at lower outflow rates for longer running periods.

The new system includes a 35-hp pump that replaced one of the original three pumps. It operates for longer periods, 1 to 2 h on average, but at a lower outflow rate. The lower outflow reduces friction in the piping system, which reduces energy consumption. The original two pumps will no longer operate under normal conditions, but only during periods with heavy inflow rates. The optimized system now delivers sewage at an average flow rate of 1930 gpm under normal conditions. Energy consumption is estimated to be 176,000 kW/yr, about 17 percent lower than the old system.

In addition to the energy savings, the new design has increased equipment life and reduced downtime and repair. Frequent starting and stopping of the pumps in the old system contributed to wear and tear of the equipment and increased maintenance needs. With the new system, less stress is placed on the system.

*Source*: Adapted from *WaterWorld*, July/August 1997. Used with permission.

## 8.3   SEWER CONSTRUCTION

After the engineering plan drawings and specifications for a sewer system are completed, they must be approved by the local regulatory agencies before construction can begin. The most common type of sewer construction, discussed here, uses open trenches and prefabricated circular pipe sections. Larger sewer systems or unusual construction situations sometimes require tunneling, jacking of pipe through the soil, or cast-in-place concrete sewers.

To ensure that there is quality construction and a well-built system, careful and continuous surveillance of the project by trained inspectors is required. The basic responsibility of an inspector during sewer construction is to ensure that there is compliance with the project plans and specifications. Some specific tasks of the inspector include the following:

1. Making sure that each pipe section is uncracked and fully usable
2. Checking for proper placement (bedding) of the pipe sections in the open trench
3. Checking for proper joining of pipe sections
4. Checking for proper alignment (direction and slope) of the pipeline
5. Making sure that the pipe is covered (backfilled) properly with clean fill material
6. Determining the need for trench dewatering

### Structural Requirements

Sewers must be able to support the load caused by the backfill soil above them, called the *dead load*, and the force due to vehicular traffic, called *live load*. This is illustrated in Figure 8.17. The depth of cover, the width of trench, and the type of backfill material are the key factors that affect the dead load. The key factors that affect the load-carrying capacity of the pipe are the *pipe crushing strength* and the type or class of *pipe bedding*.

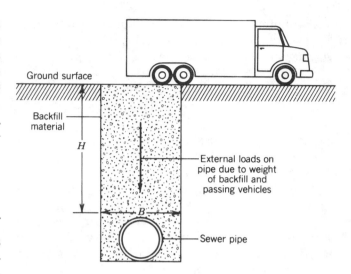

**FIGURE 8.17**
*A buried pipeline must be able to resist external forces without excessive deflection or cracking. The load on the pipe due to backfill depends on the type of soil, the depth of cover over the pipe crown (H), and the width of the trench (B).*

The crushing strength of a pipe is determined by a standard laboratory procedure, and it is specified in terms of load or force per unit length. The procedure is called a *three-edge bearing test* because the load is applied to the test sections only along three "edges" of the pipe barrel: one on the top and two on the bottom. Minimum required crushing strengths for various pipe materials and sizes are published by the American Society for Testing and Materials (ASTM), and pipe manufacturers must meet these requirements. For example, typical values for the crushing strength of vitrified clay pipe (VCP) are presented in Table 8.1.

*Bedding* refers to the way in which the pipe is placed on the bottom of the trench. Proper bedding always increases the actual supporting strength of the installed pipe above the reported crushing strength value

**TABLE 8.1**
**Selected minimum VCP crushing strengths**

| Nominal size | Standard strength | | Extra strength | |
|---|---|---|---|---|
| mm (in.) | kN/m | lb/ft | kN/m | lb/ft |
| 200   (8) | 20.4 | 1400 | 32.0 | 2200 |
| 250 (10) | 23.2 | 1600 | 35.0 | 2400 |
| 300 (12) | 26.3 | 1800 | 37.9 | 2600 |
| 380 (15) | 29.2 | 2000 | 42.3 | 2900 |
| 460 (18) | 32.0 | 2200 | 48.1 | 3300 |

by distributing the load over the pipe circumference. The ratio of the actual field supporting strength to the crushing strength is called the *load factor*.

$$\text{load factor} = \frac{\text{field supporting strength}}{\text{crushing strength}} \quad \text{(8-1)}$$

Another way of expressing this is

field supporting strength
$$= \text{load factor} \times \text{crushing strength}$$

Four types or classes of bedding are illustrated in Figure 8.18. Class D bedding is the weakest and least desirable type and is not recommended for sewer construction in most circumstances. The bottom of the trench is left flat, and the barrel of the pipe is not fully supported because of the protruding bell ends; backfill is placed loosely over the pipe without proper compaction.

Class C, called *ordinary bedding*, has compacted granular material placed under the pipe and extending partially up the pipe barrel. This provides good support, with a load factor of 1.5. In other words, the field supporting strength is 1.5 times greater than the crushing strength.

Class B, or *first-class bedding*, has the compacted granular material extending halfway up the pipe barrel, and the backfill is carefully compacted over the top of the pipe; the load factor is 1.9. In class A bedding, the pipe barrel is cradled in concrete and the backfill is carefully compacted, providing a load factor of 2.8.

In addition to the load factor provided by the pipe bedding, a *safety factor* is applied to the computations to arrive at the *safe supporting strength* of the pipeline, as follows:

safe supporting strength
$$= \frac{\text{field supporting strength}}{\text{safety factor}} \quad \text{(8-2)}$$

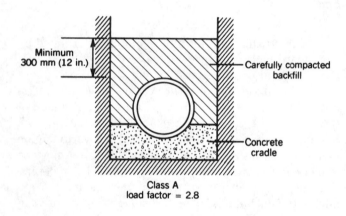

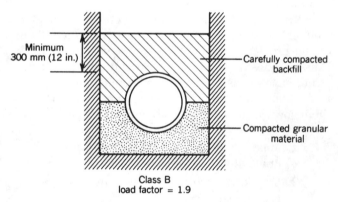

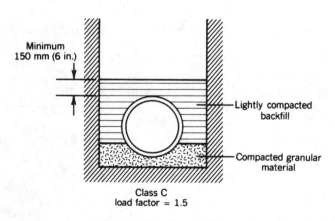

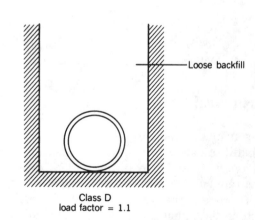

**FIGURE 8.18**
*Different types of pipe bedding conditions affect the safe supporting strength of the pipe.*

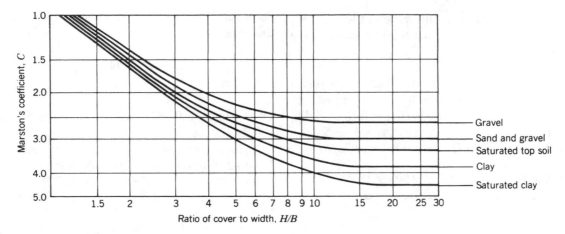

**FIGURE 8.19**
*Values of the C coefficient used in Marston's formula.* (Adapted from *Design and Construction of Sanitary and Storm Sewers*, with permission of Civil Engineering/American Society of Civil Engineers and the Water Pollution Control Federation)

and substituting from Equation 8-1 yields

safe supporting strength

$$= \frac{\text{load factor} \times \text{crushing strength}}{\text{safety factor}} \quad \textbf{(8-3)}$$

A safety factor of 1.5 is commonly used for clay or unreinforced concrete sewers to compensate for the possibility of use of poor-quality materials or for faulty construction.

**Marston's Formula**

To select an appropriate bedding condition for a pipeline, the total live load and dead load on the pipe must first be determined. The class of bedding is selected so that the safe supporting strength is equal to or greater than the computed total load on the pipe.

For pipes in shallow trenches, such as storm sewers and some water mains, the vehicular traffic load may be a significant part of the total load; tables are available to help the designer estimate these live loads. For pipes in relatively deep trenches, however, such as sanitary sewers, the live traffic loads are often insignificant when compared to the dead load due to backfill. For purposes of illustration, we will consider only the dead load on the pipe.

An equation commonly used to estimate the dead load due to backfill is known as Marston's formula and is expressed as follows:

$$W = CwB^2 \quad \textbf{(8-4)}$$

where $W$ = dead load due to backfill, kN/m (lb/ft)
  $C$ = a dimensionless coefficient
  $w$ = unit weight of the backfill soil, kN/m³ (lb/ft³)
  $B$ = trench width at the pipe crown, m (ft)

The value of the coefficient $C$ depends on the depth of cover, the trench width, and the type of backfill material. A chart for determining values of $C$ is presented in Figure 8.19. The horizontal axis represents the ratio of cover $H$ to trench width $B$.

Typical values of unit weights for a few selected soil types are presented in Table 8.2. The following simplified examples will illustrate the use of Marston's formula and the analysis of pipe bedding conditions.

**EXAMPLE 8.5**

A 300-mm-diameter pipe is placed in a 3-m-deep rectangular trench that is 0.60 m wide. The trench is

**TABLE 8.2**
**Typical values of soil unit weights**

| Soil type | Unit weight | |
| --- | --- | --- |
| | *kN/m³* | *lb/ft³* |
| Sand and gravel | 17.2 | 110 |
| Clay | 18.8 | 120 |
| Saturated clay | 20.4 | 130 |

backfilled with clay that has a unit weight of 18.8 kN/m³. Compute the dead load due to backfill that the pipe must support.

*Solution* The cover $H$ is equal to the total trench depth minus the pipe diameter, or $H$ = 3 m − 0.3 m = 2.7 m.

The ratio of cover to width is $H/B$ = 2.7/0.6 = 4.5. From Figure 8.19, read $C$ = 2.6 for clay soil. Now, applying Marston's formula (Equation 8-4), we have

$$W = CwB^2 = 2.6 \times 18.8 \times 0.60^2 \approx 18 \text{ kN/m}$$

## EXAMPLE 8.6

If the pipe in the previous example is standard strength VCP, what class of bedding should be specified for construction, using a safety factor of 1.5?

*Solution* From Table 8.1, the crushing strength of a standard-strength 300-mm VCP is 26.3 kN/m. From the solution to Example 8.5, the safe supporting strength of the pipeline must be equal to (or greater than) the dead load of 18 kN/m. Applying Equation 8-3 gives

safe supporting strength
$$= \frac{\text{load factor} \times \text{crushing strength}}{\text{safety factor}}$$

$$18 \text{ kN/m} = \frac{\text{load factor} \times 26.3 \text{ kN/m}}{1.5}$$

and from this

$$\text{load factor} = \frac{1.5 \times 18}{26.3} = 1.0$$

According to these computations, class D bedding, with a load factor of 1.1, will be adequate. But use of class D bedding is generally not considered good construction practice, and preferably class C bedding will be specified instead.

## EXAMPLE 8.7

An 8-in.-diameter VCP sewer is to be placed in a 2-ft-wide trench with 11 ft of cover. Backfill is saturated clay. Determine the required bedding condition for standard-strength as well as for extra-strength pipe, using a safety factor of 1.5.

*Solution* First, compute the cover-to-width ratio and determine $C$:

$$\frac{H}{B} = \frac{11}{2} = 5.5$$

From Figure 8.19, $C$ = 3.2. From Table 8.2, read $w$ = 130 lb/ft³.

Now, applying Marston's formula, we get $W = 3.2 \times 130 \times 2.0^2$, and the dead load $W$ = 1700 lb/ft (rounded to two significant figures). The safe supporting strength must be equal to or greater than this load.

For standard-strength 8-in. VCP, the crushing strength is equal to 1400 lb/ft. Applying Equation 8-3, We obtain

$$1700 = \frac{\text{load factor} \times 1400}{1.5}$$

and

$$\text{load factor} = \frac{1.5 \times 1700}{1400} = 1.8$$

Figure 8.18 shows that class B bedding, which has a load factor of 1.9, is needed for the standard-strength VCP in this problem; class C does not have a high enough load factor. For extra-strength VCP, the crushing strength is 2200 lb/ft, and recomputing the required load factor yields

$$\text{load factor} = \frac{1.5 \times 1700}{2200} = 1.2$$

Therefore, if extra-strength pipe rather than standard-strength pipe is used, class C bedding with a load factor of 1.5 will be sufficient. The additional cost for the stronger pipe may be justified because it could offset the time, material, and money that would be spent for class B bedding.

## Field Layout and Installation

A basic consideration in sewer construction is the accurate field layout of the *line and grade* of the pipeline, as shown in the plan drawings. The line, or horizontal alignment, defines the location and direction of the pipeline within the right-of-way. The grade, or pipe slope, must also be accurately established to provide the required hydraulic capacity of the system.

In the construction survey, the location of the sewer trench is usually established and laid out as an *offset line*, which runs parallel to the proposed sewer centerline. The offset line is marked by wooden stakes driven into the ground at uniform intervals, usually about 15 m (50 ft) long. The offset line is far enough away from the pipe centerline so that it will not be disturbed during construction, yet it is close enough so that transfer of measurements to the excavated trench can readily be done by the builder. The offset stakes may be set so that their tops are a specific height above the required trench bottom. They can then be used periodically to check the depth of the trench during excavation.

Two methods are used to set the pipe sections properly in the open trench: *batter boards* and *lasers*. The batter board method is the older of the two, but the application of low-power laser instruments in surveying and construction activities is becoming more popular. Placing the pipe using lasers is more accurate, quicker, and less labor-intensive than using the traditional batter board method. Nevertheless, batter boards are still sometimes used in sewer construction.

### Batter Boards

After the trench has been excavated, batter boards are set across the trench at uniform intervals, as shown in Figure 8.20. The tops of the boards are usually set at some even height above the required sewer invert elevation. The sewer centerline is marked on the boards by extending a line of sight with a transit or theodolite; a string is stretched from board to board along this line.

The centerline is transferred down into the trench with a plumb-bob line held against the top centerline string. Invert elevations are set and checked using a vertical rod marked off in even increments. The lower end of the rod is placed on the pipe invert, and the batter board string is checked to see if it matches with the proper elevation mark on the rod. If it does not match with the required mark, an appropriate adjustment of the pipe invert elevation is made by a worker in the trench.

### Lasers

A *laser* is an instrument that can project an intense but narrow beam of light for a long distance. This pencil-thin light beam is aimed through a pipe and can be seen on a target placed in the other end of the pipe, as shown in Figure 8.21.

The laser is securely mounted in the manhole, and the slope of the light beam is accurately set to match

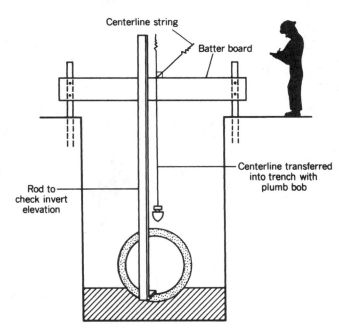

**FIGURE 8.20**
*Batter board method of construction for a sewer line.*

the required slope of the pipe. A transit mounted above the manhole is used to establish pipe alignment from field reference points and to transfer the alignment

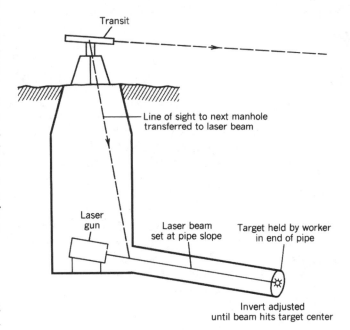

**FIGURE 8.21**
*In modern sewer construction, laser beams are used to establish the specified slope of the pipe; the laser beam is set in the proper horizontal direction with a transit or electronic theodolite.*

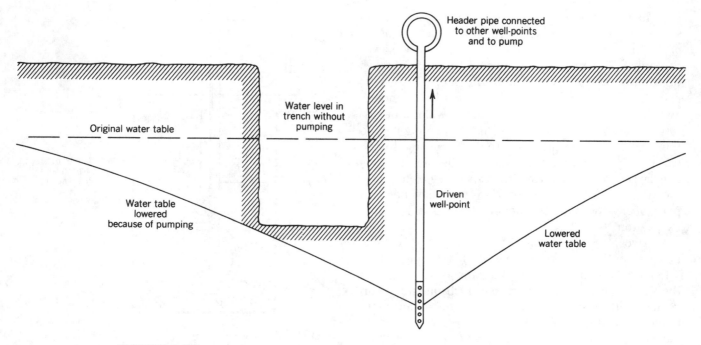

**FIGURE 8.22**
*A series of well points attached to the main header pipe and a pump can be used to dewater a trench that contains groundwater.*

down to the laser instrument. Lasers can maintain accuracies in the pipe slope of 0.01 percent over a distance of up to 300 m (1000 ft). In other words, the invert elevations can be set accurately to within 30 mm in a 1000-m length of pipeline (0.1 ft in a 1000-ft-long pipeline).

### Excavation

The most common type of equipment used for digging a sewer trench is the backhoe, although this may vary depending on the depth and type of material (soil or rock) being excavated. The trench width should be kept as narrow as possible while allowing enough room for a worker. Keeping the trench narrow not only reduces excavation costs, it also reduces the backfill load on the pipe. Generally, a minimum working room allowance of at least 300 mm (1 ft) from each side of the pipe is required.

For safety, the sides of trenches more than 1.5 m (5 ft) deep must be supported with sheeting and bracing to prevent collapse and to protect workers. Sheeting is wood planking or other material that is in contact with the trench sides; the braces extend across the trench from one side to the other.

If the trench is flooded with groundwater that seeps in through the bottom and sides, dewatering with

a pump or well-point system may be necessary. The effect of a well-point system on lowering the groundwater table in the vicinity of the trench is illustrated in Figure 8.22.

### Placement and Backfill

The importance of proper bedding design was discussed in the previous section. Improper bedding can significantly reduce the supporting strength of the pipe, causing pipe deflection, cracking, and excessive infiltration. Of prime concern in bedding is that the entire length of pipe barrel be uniformly supported by the soil or the gravel bed under the pipe. The pipe sections should be handled carefully and placed with minimum disturbance of the supporting material on the trench bottom. Care must be taken in joining pipe sections so that excessive infiltration through the joints will not be a problem.

Backfilling of the sewer trench should be done immediately after the pipe has been placed at the proper grade. The backfill must not contain boulders or large cobbles, frozen material, tree stumps, or other debris. It should be placed and compacted in uniform layers about 150 mm (6 in.) deep to a height of about 300 mm (1 ft) above the top of the pipe; after that, backfilling can usually proceed more rapidly.

## 8.4 INFILTRATION AND INFLOW

When a gravity sewer line lies below the groundwater table, the groundwater will seep into the sewer through poorly constructed pipe joints, cracked pipe sections, and leaky manhole structures. This flow of groundwater into the system is called *infiltration.*

Control of infiltration depends primarily on the quality of construction of the pipeline. Even in well-constructed sewer systems, there will be some infiltration; building a completely watertight gravity flow pipeline is neither feasible nor economical. Generally, up to 45 L/d per millimeter of pipe diameter per kilometer of pipe length (45 L/d/mm/km) is allowed in the specifications for a sewer construction project. This is roughly equivalent to 500 gal/d per inch of pipe diameter per mile of pipe length.

## EXAMPLE 8.8

What is the total allowable rate of infiltration in a 1500-m-long, 200-mm-diameter lateral sewer if 45 L/d/mm/km is acceptable? How does it compare to the minimum pipe capacity?

*Solution* The allowable infiltration is computed as follows:

$$45 \text{ L/d/mm/km} \times 200 \text{ mm} \times 1.5 \text{ km} = 13\,500 \text{ L/d}$$

The minimum capacity of a 200-mm pipe is the discharge that would have a self-cleansing velocity of 0.6 m/s. Using Manning's nomograph in Figure 2.21, read a discharge of 0.019 m³/s, or 19 L/s. Converting to L/d gives

$$19 \text{ L/s} \times 3600 \text{ s/h} \times 24 \text{ h/d} \approx 1\,600\,000 \text{ L/d}$$

The ratio of infiltration to pipe capacity is therefore

$$\frac{13\,500}{1\,600\,000} = 0.008 \quad \text{or} \quad 0.8 \text{ percent}$$

This amount of infiltration is negligible compared to the pipe capacity.

Surface water that enters a sewer system through poorly sealed manhole covers or that comes from intentional connections to roof drains and basement sump pumps on private property is called *inflow.* Most communities have sewer-use ordinances that prohibit roof or cellar drain connections to the sanitary sewer, in an attempt to control the inflow problem.

It is very important to prevent excessive infiltration or inflow in a sanitary sewer for two basic reasons:

1. Too much infiltration and inflow can cause surcharging of the sewer system during wet-weather periods.
2. Sewage treatment plants are not designed to handle the extra volume of water from infiltration and inflow. During wet weather, the rapid increase of flow through the treatment plant causes hydraulic overloading and failure of the treatment process.

Newly constructed sewer lines must be tested to ensure compliance with infiltration specifications before they are put into service. Older systems may also be tested to determine the extent of leakage. The objective is to ensure that tax dollars will not be spent on expensive new treatment plants or treatment plant additions that would have operating problems due to hydraulic overloads. This process of examining the integrity of older systems is called an *I/I survey.*

There are several methods of infiltration and inflow testing, including *direct measurement, exfiltration testing, smoke testing,* and *low-pressure air testing.*

Direct measurement for infiltration in a new sewer is done when the pipeline lies below the groundwater table at the time of the test and before service connections are made. A V-notch weir can be placed in the end of a known length of the sewer line to gage the flow; any flow in the line must be infiltration. Direct measurement can also be done as part of an I/I survey of an existing sewer system. But in existing systems, the flow measurements must be made at about 3 A.M., when little or no sanitary flow would be expected.

If a new sewer must be tested for watertightness when it lies above the water table, as is often the case, an exfiltration test is conducted in lieu of direct measurement. This is, in effect, just the opposite of measuring infiltration: It measures the flow going out of the pipeline instead of the flow coming in. After plugging the pipes as shown in Figure 8.23, the reach of sewer being tested is filled with water. The reach is kept flooded for a few hours so that entrapped air can be removed and the pipeline material becomes saturated with water.

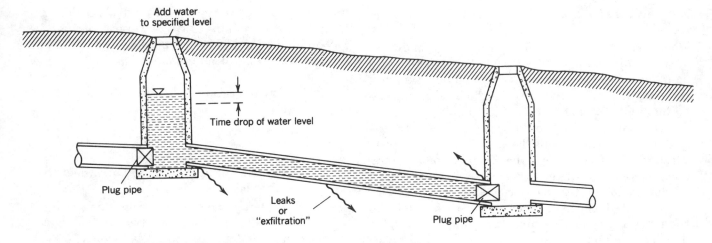

**FIGURE 8.23**
*An* exfiltration test *is one method used to judge the watertightness of a new sanitary sewer.*

If water is leaking out of the sewer line, the water level in the manhole will be observed to drop in elevation. By measuring the amount of drop in the water surface over a known interval of time, the rate of exfiltration can be computed. Obviously, if the water can leak out of the pipe, it is an indication of a potential infiltration problem during wet-weather and high-groundwater-table conditions.

## EXAMPLE 8.9

An exfiltration test is conducted on a 400-ft-long, 12-in.-diameter sewer reach. The water level in a 4-ft-diameter manhole is observed to drop 2 in. in 1 h. Compute the rate of exfiltration in terms of gallons per day per inch of diameter per mile of pipe.

*Solution* The volume of water that leaked out of the system is equal to the product of the manhole cross-sectional area and the drop in water elevation. This is computed as follows:

$$\text{area} = \pi D^2/4 = \pi \times 4^2/4 = 12.6 \text{ ft}^2$$

$$\text{drop} = 2 \text{ in.} \times 1 \text{ ft}/12 \text{ in.} = 0.167 \text{ ft}$$

$$\text{volume} = 12.6 \text{ ft}^2 \times 0.167 \text{ ft} = 2.1 \text{ ft}^3$$

Converting the volume to gallons yields

$$\text{volume} = 2.1 \text{ ft}^3 \times 7.5 \text{ gal/ft}^3 \approx 16 \text{ gal}$$

$$\text{sewer length} = 400 \text{ ft} \times 1 \text{ mi}/5280 \text{ ft}$$

$$= 0.076 \text{ mi}$$

Since the time interval for the drop was 1 h,

$$\text{leakage rate} = 16 \text{ gal/h} \times 24 \text{ h/d} = 384 \text{ gal/d}$$

and

$$\text{exfiltration} = 384 \text{ gal/d} \div 12 \text{ in.} \div 0.076 \text{ mi}$$

$$\approx 420 \text{ gal/d/in./mi}$$

The tightness of construction of a sewer system can also be evaluated by using a low-pressure air testing procedure; the pressure used is about 28 kPa, or 4 psi. Specifications for allowable air loss rates, in terms of pressure drop, are used in lieu of allowable infiltration or exfiltration rates. In the air pressure test, the sewer pipes are tightly plugged with special *air-lock* balls or other devices that allow the introduction of pressurized air into the sewer reach. The time it takes for the air pressure to drop 7 kPa (1 psi) is determined. This time is compared to a specified allowable time, which is a function of the pipe length and diameter. If the pressure drops in less than this specified allowable time, the sewer line has failed the test and repairs must be made.

Smoke testing can be used effectively to locate sewer leaks or illegal sources of inflow. In the smoke test, a sewer reach is isolated (but flow is not completely blocked), and smoke from a nontoxic smoke bomb is forced through by a blower. The smoke will appear above ground wherever there is a hole or crack in the pipe. It also will appear rising from the roof drain on a home where there is an illegal connection to the sewer line.

To avoid construction and operation problems related to high groundwater tables, pressure sewers or vacuum sewers are sometimes built. An example of a vacuum sewer system is described in the following case study.

## CASE STUDY

### Defying Gravity

Vacuum sewers were once considered viable only for small communities where topography or high groundwater precluded using gravity systems. However, the Englewood Water District (EWD), in southwest Florida, chose to install vacuum sewers to upgrade its system because they are economical, reliable, and easy to install. The project is one of the largest of its kind in the world.

EWD, covering 114 km² (44 mi²) in Sarasota and Charlotte counties, has many small building lots with on-site septic systems. Substandard septic systems and high groundwater tables were polluting local coastal waters and threatening shellfish and recreational tourism. By 1994, the EWD decided to replace the individual septic systems with a municipal system. High groundwater in the district limited maximum line depth to 3.6 m (12 ft), and rights-of-way in highly developed areas also constricted line placement. An estimate for a vacuum system suggested significant cost savings over a gravity system. So the EWD decided to consider vacuum systems. The master plan was redesigned to use nine vacuum areas as the new standard for the collection system, serving about 8500 house connections and a population of about 20,000 people.

### Vacuum Design

The backbone of a vacuum sewer is a centralized collection station and the vacuum station, which contains two premanufactured pallets of tanks and pumping equipment. The first pallet holds a steel collection tank. The second contains two pressure pumps to send collected sewage to a nearby wastewater treatment plant. Three vacuum pumps on the same pallet suction air from the tank to maintain a continuous negative pressure. Sewer mains, also under negative pressure, lead out from the tank to the service area, a network of sewer mains with diameters of 100 to 250 mm (4 to 10 in.). The mains connect to valve pits with 75-mm (3-in.)–diameter PVC pipe. Each fiberglass valve pit, 900 mm (3 ft) in diameter and 1.8 to 2.4 m (6 to 8 ft) deep, serves up to four connections.

Gravity lines carry sewage from each connection point into the valve pit, which releases sewage into the vacuum sewer main when the volume inside the pit reaches a certain level. The actuator controlling the valve between the pit and the sewer main consists of a hollow tube sealed on one end and set upright inside the pit. As the level of sewage rises in the pit, air pressure inside the actuator tube increases. When a certain pressure is reached, the actuator opens the valve and the negative pressure from the main line suctions out the sewage at a rate of 2 L/s (30 gpm). Although the valve shuts after the sewage level drops, it has been designed to stay open just long enough to allow air to enter into the sewer main. The air mixes with the sewage, decreasing sewage density and increasing the rate at which it can be pulled into the central treatment plant. Aeration keeps the sewage from turning septic or degrading under anaerobic conditions, and thus reduces the potential of offensive odors. A biofilter at the vacuum station can eliminate the odors that sometimes develop even with aeration.

Because a vacuum system relies on suction rather than gravity to move sewage, the pipes do not have to follow a continuous slope. Usually installed 0.9 to 1.5 m (3 to 5 ft) below the surface, vacuum pipes can run over or around existing utilities if necessary, offering greater flexibility and shorter construction time than traditional gravity systems. Another advantage of the vacuum system was that it could be sized to accommodate future customers. A gravity system would have required two local lift stations and served only the immediate customers.

### Installation

During the first two phases of construction, which began in 1996, several unexpected advantages surfaced. Excavating and dewatering needs were lower than anticipated. Despite typically high groundwater, no dewatering was necessary because the vacuum sewer trenches were shallow, that is, 0.9 to 1.5 m (3 to 5 ft). Gravity mains, at depths as great as 3.6 m (12 ft), would have required extensive dewatering and some trench shoring.

With more flexible pipe configurations possible, field changes and modifications were easy to make. Because the locations of older septic systems were unknown, the laterals indicated on the plans needed to be field-adjusted to provide a better connection point on each lot. Moving the laterals and valve pits to make the onsite connections was relatively simple, and because of a unit price contract, change orders were not necessary for the field modifications.

With vacuum lines and valve pits installed in the grass swales along the roadways, disruption and restoration of public rights-of-way were minimal. Because only the sewer system laterals crossed the pavement, extensive road restoration and associated repair costs were unnecessary. The project was substantially completed 1 month ahead of schedule.

## Operation

Since the first two phases of the vacuum sewer went on line, in October 1996, there have been no significant operating problems. The entire first construction unit went on line in June 1997. EWD operation personnel have reported the following operating advantages:

- The system is environmentally friendly. The vacuum is monitored continually at the vacuum station with telemetry to EWD headquarters, and system problems are detected and corrected within 45 min. Because the lines are under negative pressure, any line problems that did occur would result in inflow into the vacuum main rather than sewage leaking out, thus minimizing groundwater pollution.
- Because the system is "tight" by nature, infiltration and inflow problems are nonexistent, reducing pumping and treatment costs. An absence of manholes means elimination of the risk of sewer gases. Also, the confined entry problems associated with gravity sewers are obviated.
- The only power required is at the vacuum station, whereas gravity systems would have required lift stations, and pressure sewers would have required a pump and electrical connection at each home. In addition, with all electrical controls in one building, maintenance of the vacuum system is relatively easy.

---

*Source*: Adapted from J. H. Cole and S. F. Torchia, "Defying Gravity," *Civil Engineering*, February 1998. Copyright © 1998 by the American Society of Civil Engineers. Used with permission of the American Society of Civil Engineers.

## 8.5 SEWER REHABILITATION

Sewer systems are part of urban *infrastructure*, the physical facilities and public works that allow communities to function efficiently. Unlike transportation infrastructure, particularly highways and bridges, which are quite visible to the public, sewers receive almost no attention once they are put in the ground in many communities. The old saying, "Out of sight, out of mind" seems to apply.

Often, little thought is given to the need for maintenance of these wastewater collection systems, and only a bare minimum of funds is allocated for this purpose. As a result, many sewer systems in the United States are in a state of disrepair. They are leaky and carry large volumes of infiltration water, they frequently become blocked and surcharge as a result, and occasionally they even collapse.

The cost of excavating and replacing a section of a poorly functioning sewer system is very high. It is usually more economical to apply one of several available methods to repair and rehabilitate the system internally, without having to excavate it. It is even more economical for municipalities and sewer agencies to have continuing sewer maintenance programs designed to prevent unnecessary deterioration of the system in the first place.

## Pipeline Repair

There are several methods of restoring flow capacity, or sealing of leaks in sewer lines, that do not require excavation. To a certain extent, the structural integrity of a pipe weakened by crown corrosion can also be strengthened, as long as a complete collapse has not already occurred. The particular method selected depends on the nature of the problem, the need to maintain flow during the repair, possible traffic disruption, safety, and cost.

### Cleaning and Inspection

Sewer lines must be cleaned before a visual inspection and the selected rehabilitation efforts can be made. Flushing the line, using a fire hose attached to a hydrant and discharging the flow into a manhole, is a common cleaning method. But this must be done with caution to prevent backups into nearby houses connected to the system.

Another cleaning method, for sewers without serious blockages, makes use of a soft rubber ball that is inflated to match the diameter of the pipe and then pulled by rope through a reach between manholes. Sometimes, power rodding machines can be used for mechanical cleaning of a line with blockages. In sewers with accumulations of grit because of relatively low flow velocities, power winches can be used to pull a bucket through the line to scrape up the deposits. Whichever method of cleaning is used, collected sediment and debris should be removed and disposed of properly.

After cleaning, inspections can be made with a flashlight during low flow periods. A much more thorough inspection can be accomplished, however, using closed-circuit television systems. TV inspection allows accurate location of leaks, root intrusion, and structural problems. The camera is pulled through a sewer reach on a special mounting, and a photographic or videotape record of the inspection is made.

## Grouting

Sealing leaks with chemical grout in structurally sound sewer lines is a common method for rehabilitation. The grout is applied internally to joints, holes, or cracks using special tools and techniques, without the need for excavation. The gel-type or foam-type grouts that are used solidify after being forced into the joints or cracks under pressure. In small or medium sewer lines, a *sealing packer* with inflatable rubber sleeves can be pulled through a reach of the system, as illustrated in Figure 8.24. Closed-circuit TV is used to position the packer over the joint or crack to be repaired.

In larger-diameter sewer lines, workers must enter the line to place a sealing ring manually over the defective joint, as illustrated in Figure 8.25. The grout is pumped through a hand-held probe. The air in the sewer must be tested for carbon monoxide, hydrogen sulfide, and explosive gases before the worrkers enter, and appropriate safety precautions must be taken.

## Linings

Larger sewers with structural damage caused by crown corrosion can be reinforced by applying an internal lining of *gunite*, which is a mixture of fine sand, cement, and water. Gunite is applied pneumatically by spraying; it adheres to the vertical and overhead surfaces of the pipe. Long lengths of concrete sewers can be effectively renewed with a gunite lining.

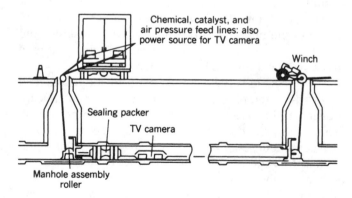

**FIGURE 8.24**

*In small sewers*, a sealing packer *can be pulled through the line for repair*. (From *Existing Sewer Evaluation and Rehabilitation*, with permission of Civil Engineering/American Society of Civil Engineers and the Water Pollution Control Federation)

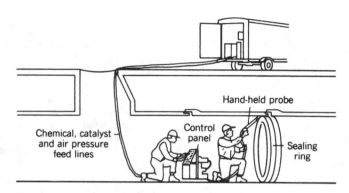

**FIGURE 8.25**

*In large sewers, workers must enter the line for repairs*. (From *Existing Sewer Evaluation and Rehabilitation*, with permission of the American Society of Civil Engineers and the Water Pollution Control Federation)

A method called *sliplining* can be used to rehabilitate extensively cracked pipelines. This involves pulling a flexible plastic liner pipe into the existing pipe and then reconnecting individual service connections to the liner. This is illustrated in Figure 8.26. Sometimes the narrow annular space between the liner and the old pipe is filled with grout to prevent movement. Multiple excavations are required to reconnect each service line to the new liner.

A relatively new method is now available called *inversion lining*. Inversion lining avoids the need for service line excavation. A flexible liner that expands to fit the pipe geometry is thermally hardened and cured. A special cutting device can be used with a closed-circuit TV camera to locate and reopen the service connections. The liner is installed through a tube placed in a manhole, using water or air pressure to push it through the pipe. When the liner is in place, the water or air is heated to begin the curing and hardening process. (See the case study that follows.)

## Manholes and Service Connections

Manholes often need repair to eliminate surface water inflow or groundwater infiltration. Inflow can enter the manhole through holes in the manhole cover, through spaces between the cover and the frame, and under the frame itself if it is poorly sealed. Self-sealing frames and covers are available, but the seals are often damaged by heavy traffic, road work, or snow plowing. Frames may be resealed in place by applying hydraulic cement and a waterproof epoxy coating.

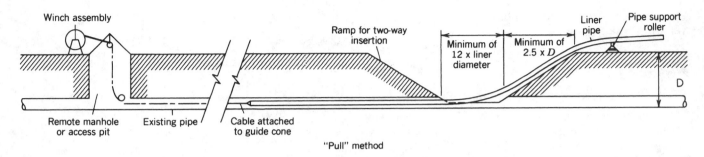

**FIGURE 8.26**

*Sliplining is one method used to repair cracked sewers.* (From *Existing Sewer Evaluation and Rehabilitation*, with permission of the American Society of Civil Engineers and the Water Pollution Control Federation)

Sometimes an effective way to reduce inflow is to raise the frame and cover by adding a manhole adjusting or extension ring and by coating the exposed portion with cement or asphalt. Another method is to install a special insert between the frame and the cover. The insert, illustrated in Figure 8.27, prevents water and grit from entering the manhole, but allows gas to escape through relief vents.

Concrete manholes are also subject to sulfuric acid corrosion and therefore sometimes need structural rehabilitation. Severe structural deterioration is usually solved by excavating and replacing the manhole and applying corrective measures to eliminate the cause of deterioration. Structural repair of less severely damaged manholes first involves removal of the deteriorated

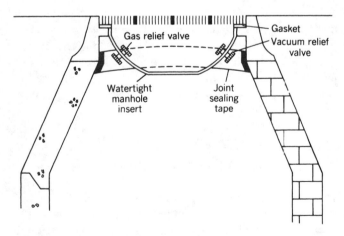

**FIGURE 8.27**

*An insert device installed under the manhole cover prevents surface water from entering the system.* (Courtesy of PRECO Industries, Plainview, New York; illustrator: Deborah Di Lorenzo)

materials, using water or sandblasting or mechanical tools. The remaining material is then stabilized using special chemical preparations, high-strength patching mortar is used to fill in surface irregularities, and a lining or coating is then applied.

One of the most common manhole maintenance problems is infiltration of groundwater through the sidewall and base and around pipe entrances. Chemical grouting can be used effectively to solve this problem. It is less costly than lining or coating methods, and it does not require prior surface restoration. The cracks and openings are sealed by pressure injection of gel or foam grouting materials.

Service or house connections are small-diameter pipelines of 100 mm (4 in.) that connect the lateral sewer line in the street to the buildings it serves. Also called building sewers or service laterals, they can be as long as 30 m (100 ft). The section of the service connection between the sewer line and the property line is installed and maintained by the local public works or sewer department; the section between the property line and the building's drainage system is installed under local plumbing or building code regulations and must be maintained by the property owner.

Defects in service connections, including cracked or open-jointed pipes, can be the cause of a significant portion of the infiltration problem in a sewerage system. This is because the total length of service connections in a community is often greater than the length of sewer mains. In view of this often neglected fact, repair of service connections is an important aspect of a sewer system rehabilitation effort. Several methods are available for applying chemical grout as a repair material, and the inversion lining method can be used. Efforts also must be made to eliminate illegal hookups of roof drains and basement sump pumps to service connections on private property to reduce the inflow problem.

## CASE STUDY

### Custom-Fitted Liner Rehabs Century-Old Sewer

A custom-fitted, cured-in-place liner proved the right solution for fixing a large, 104-year-old combined sewer for Portland, Oregon.

Repairing the antique sewer posed several difficulties. Since the original sewer was built, some 40 ft of fill—in the form of old burned buildings, timbers, and bricks—was placed over the sewer. The fill, much of it unstable material, put more structural stress on the old sewer than it could withstand. A routine inspection revealed numerous structural defects, including a 4-in. crack that ran four city blocks long, and a 3-ft-deep hole in the invert. Officials were concerned that soil could migrate through sewer walls, leave a void, and permit a car or truck to fall from the street above into the hole.

What is more, the sewer's cross section is shaped in an inverted catenary curve. From invert to crown the sewer's height runs between 60 in. and 66 in.; the invert is 6 ft wide. Stretching 3.5 mi in length, the line runs directly beneath a busy two-lane street.

The city's Environmental Services Department considered five alternatives in the design process: liner plate with a gunite coating; a high-density polyethylene (HDPE) liner; shotcrete coating inside the sewer; cured-in-place pipe; and replacement with concrete pipe. A major disadvantage marked the liner plate and shotcrete alternatives: workers would risk having the sewer collapse on them. The HDPE liner proved impractical because of the sewer's depth and the presence of curves in the alignment.

Concrete pipe's price ran too high, because the poor quality of soils above the pipe and the close proximity of structures made the needed 50-ft-deep excavations prohibitively costly. So the only viable alternative was cured-in-place pipe.

Gelco Services Inc., a licensee of Insituform Technologies Inc. based in Salem, Oregon, was the job's low bidder and won the $1.7-million contract.

Crews for subcontractor Benge Construction, Portland, began work by rebuilding a 2-ft-diameter manhole. Benge excavated a 10-ft by 12-ft pit and shored it with steel sheeting. The original 2-ft-diameter manhole was rebuilt with 8-ft-diameter sections of reinforced concrete pipe.

Benge started pushing new pipe downward using an 8-ft starter section of concrete pipe that had a steel cutting ring attached to its bottom. An excavation proceeded and pipe sections were added from the top; the added weight of concrete rings pushed the entire structure downward. Steel straps held the concrete rings together and helped keep the vertical pipe straight.

That procedure was used until the 50-ft manhole reached the old sewer. With jackhammers, workers cut U-shaped sections from the new manhole so that it fit down over the old sewer. Workers next removed the brick in the crown, connected the manhole to the host pipe, and with shotcrete joined the bench and the manhole.

Crews also rebuilt the manhole's bench, placing concrete up to the 60-in. pipe's springline.

### Cooking the Liner

Once the sewer's flow was completely diverted, Insituform's cured-in-place pipe (CIPP) was used to rebuild the sewer. The company manufactured a custom-fitted Insitutube cut to the precise dimensions of the old sewer. The tube is made from an absorbent fabric tube 1.4 in. thick and coated with an elastomeric material.

At the site the Insitutube was saturated with a special polyester resin used for thick laminations. Vacuum pumps drew some 115,000 lb of resin into the absorbent fabric, and rollers assured an even distribution of the liquid. The process was protected by the shelter of low tents that prevented sunlight from accelerating the resin's cure.

Once tributary sewers were plugged in case of rain—which could have ruined the new liner—installation forged ahead with Insituform's inversion process. Hydrostatic water pressure propelled the Insitutube through the sewer, turning it inside out along the length of the old pipe. The job required about 81,000 gal of water to force the tube outward against the sewer walls.

To apply heat for resin curing, water within the tube flowed through aboveground heat exchangers that warmed the water about 180 deg F. By controlling the temperature of the water, Gelco crews regulated the speed of the resin-curing process.

Workers installed a total of 1063 linear feet of Insituform cured-in-place pipe. From the new manhole, one insertion ran 680 ft upstream; the downstream insertion ran 383 linear feet.

With the insertion complete, inspectors entered the rebuilt pipe and found the project successful. To fill voids between the old host pipe and the reconstructed pipe, back-grouting was performed using a foaming liquid grout.

*Source: Engineering News-Record.*

## 8.6 GIS APPLICATIONS

Geographic Information Systems (GIS) are finding widespread use in many areas of environmental technology, including the management of sanitary sewer infrastructure. A GIS is a computerized mapping system capable of storing, manipulating, and displaying geographically referenced information. Locations for the features being mapped can be given by $x$, $y$, $z$, coordinates of longitude, latitude, and elevation, or by ZIP codes, road mile markers, and other such systems. Any variable that can be located spatially can be incorporated into a GIS. In addition to spatial data, attribute data, which describe the various characteristics of all the geographic features, are part of the database.

A GIS makes it possible to quickly integrate and analyze information in a unique way. For example, with GIS technology and household water billing information, it is possible to simulate the discharge of wastewater into the septic systems in a neighborhood close to a wetland area (septic systems are discussed in Section 10.5). The water bills show how much water is used at each address. Since the amount of wastewater is proportional to the amount of water used, areas of excessive septic discharges and possible water pollution problems can be located on a map. The way that a wide variety of data can be stored as different "layers" of information in a GIS makes it possible to depict spatial relationships and perform complex analyses, as in this and in many other environmental applications.

The concept of using layers to depict geographic information is illustrated in Figure 8.28.

### CASE STUDY

**Mapping System Provides Efficient Sewer Management**

To improve management of the municipal sewer system in New Hartford, New York, the town's engineering staff has implemented a GIS system augmented with document and photo images linked to sewer elements that serve 11,000 tax parcels. With this system, day-to-day operations have been streamlined by reducing the time required to retrieve information used to locate and evaluate sewer system problems. What previously required a visit to the records room and a call to the Tax Assessor's Office can now be obtained and printed from the engineer's desk within minutes.

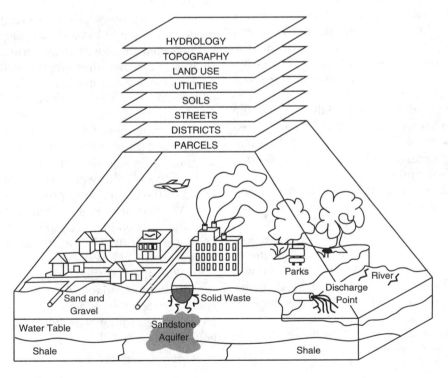

**FIGURE 8.28**
*Schematic illustration of mapping layers used in GIS technology.* (Graphic image supplied courtesy of ESRI. Copyright © 1997 ESRI. All rights reserved)

Typical data stored in the GIS include:

- Property owner information and photos from the Assessor's Office
- As-built drawings for existing sewers
- Elevation and flow data for manholes
- Maintenance records and field notes.

Access to this information is available to other township departments via a local network. This provides simultaneous viewing of the information by multiple users.

GIS layers were derived from existing New Hartford information and maps, including digital tax maps available from the County Planning Department. They were combined to form the base map utilized for creating other GIS layers. The sewer maps were scanned and digitized into seven layers: sewer lines, clean-outs, manholes, force mains, pumping stations, laterals, and county interceptor sewer lines. Lateral attributes are linked to the tax parcels associated with their corresponding digital laterals. Attribute data include addresses, elevations, diameters, materials, slopes, and feature stationing. Map sheets corresponding to features were linked to the document imaging subsystem by entries in the database.

The document imaging software provides virtual folders, which facilitate browsing of documents associated with a geographic feature. The virtual folders can be viewed by clicking on a significant geographic element. For example:

- Clicking on a manhole displays I.D. information and physical attributes of the manhole.
- Clicking on a parcel brings up the tax map number and other information in the folder.
- Clicking on a sewer line displays its corresponding map, plan and profile, and attributes.
- Clicking on a pumping station displays its map sheet and as-built drawing, and attributes.

The GIS application is used by the engineering staff for trouble call responses, landowner notifications, and status review of existing field conditions for new construction. Field technicians can quickly identify the site location, obtain global positioning system (GPS) coordinates, and view or print all necessary documents and other information needed to generate work order requests. The GIS system also enhances analytical planning for future construction needs. An example of some of the elements in a virtual folder is shown in Figure 8.29.

*Source*: Adapted from R. Cleveland and V. La Clair, "Mapping System Provides Efficient Sewer Management," *ArcNews* (ESRI), Spring 2001. Used with permission.

## 8.7  SANITARY SEWER COMPUTER APPLICATIONS

A powerful, easy-to-use sanitary sewer computer-modeling program called SewerCAD (Haestad Methods, Inc., Waterbury, Connecticut) can be used for the design and analysis of wastewater collection systems. This software has a graphical interface and data exchange capabilities that facilitate the development of complex models of combined gravity flow and pressure networks. The gravity portion of the system can be designed and analyzed for either uniform flow or gradually varied flow conditions.

The automatic design capabilities of the program allow design of the sewer system based on user-defined constraints for velocity, cover, and slope. Design can also accommodate partially full pipes, multiple parallel sections, invert/crown matching criteria, and drop manholes. A dry-weather flow library can be customized and supports population, area, discharge, and count-based dry-weather loads. Wet-weather flows, inflow, and infiltration can be added to the model as lump sums or in terms of pipe length, pipe diameter–length, or pipe surface area, or based on a unit count. Peak flow factors can be selected from a user-customizable extreme flow factor library.

SewerCAD can be used to design new systems and analyze the performance of existing systems. The graphical editor and scenario management capabilities facilitate the process of analyzing a large number of design alternatives and finding potential problems in an existing system. The program can be used to track system response to an unlimited range of dry- and wet-weather loading combinations.

 **8.8   RELEVANT WEB SITES**

### AMERICAN CONCRETE PIPE ASSOCIATION

http://www.concrete-pipe.org

This site includes research results and technical data related to the use of concrete pipe for drainage and pollution control.

### MASSACHUSETTS WATER RESOURCE AUTHORITY (MWRA)

http://www.mwra.com/sewer/html/sewhow.htm

This site provides a description of the sanitary sewer system serving the Boston, Massachusetts, area, with a brief history of the system.

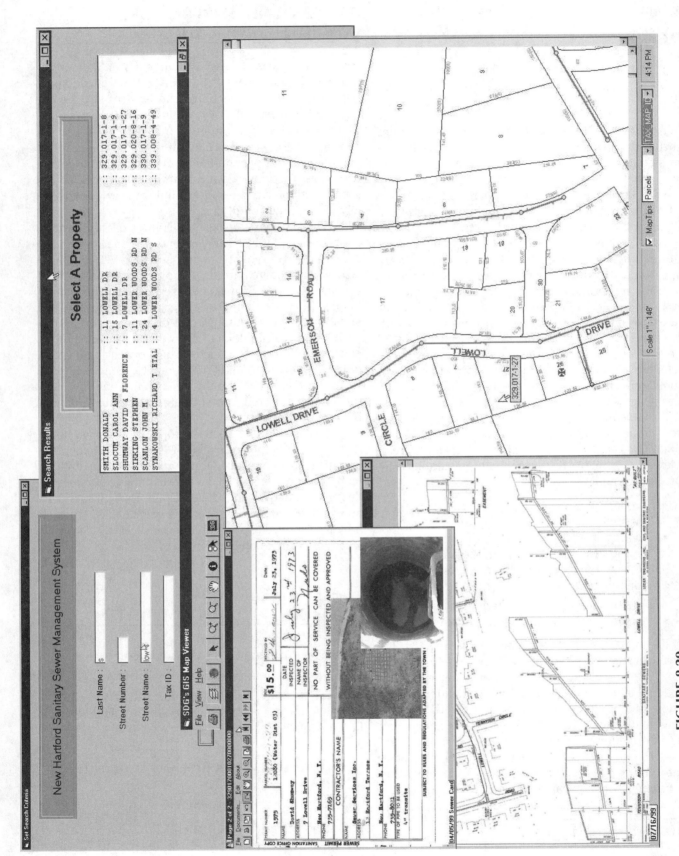

**FIGURE 8.29**
*Example of a GIS "virtual folder."* (Courtesy of Systems Development Group, Inc., Utica, New York)

## Case Studies

The following World Wide Web sites provide information about a variety of sanitary sewer systems throughout the world, including historical background, public issues, and construction methods.

### Parisian Sewer System

http://www.paris.org/kiosque/mar97/egouts.html
http://www.afu.com/sewer/sewer.html

### Hopatcong Sanitary Sewer Project

http://www.hopatcong.org/sewers/sewer.htm

### Infill Sewerage System

http://www.watercorporation.com.au/
infill–sewerage/content–about.asp

### New Orleans Drainage System

http://www.swbnola.org/drain_info.htm

### Story of Sewerage in Leeds

http://www.dsellers.demon.co.uk/sewers/
sew_ch1.htm

### Greater Vancouver Regional District—

http://www.gvrd.bc.ca/services/sewers/
collect/index/html

### Deep Tunnel Sewerage System

http://www.pub.gov.sg/dtss.html

The Government of Singapore is in the process of replacing the existing sewerage system with a *deep tunnel sewerage system*, to provide sewerage infrastructure that will serve Singapore through the 21st century. This Web site provides a description of the project, which comprises 65 km of two large, deep tunnels crisscrossing the island, two centralized water reclamation plants, and a 5-km-long deep-sea effluent outfall pipeline.

# REVIEW QUESTIONS

1. What is the difference between a *combined* sewer system and a *separate* sewer system? Which is preferable? Why?

2. What is the difference between a *lateral* and a *submain* sewer? What is an *interceptor* sewer?

3. List five different materials used in sewer-line construction. Briefly describe the characteristics of three of these.

4. List four purposes of a sewer manhole.

5. List six factors that determine where a manhole is located in a sewer system.

6. What is a *drop manhole?* What is an *inverted siphon?*

7. Describe the general procedure for sanitary sewer design.

8. What key information must be shown on the plan drawings for sanitary sewer construction?

9. Why are sewer pipelines designed to accommodate peak hourly flows? Why is the peak flow factor in a trunk sewer less than that in a lateral?

10. What happens when a sewer system is *surcharged?* Under what circumstances might this occur?

11. Is it good practice to overdesign a sewer pipeline? Why?

12. Describe in general terms how the design flow for a sewer reach is determined.

13. What are typical limitations on the pipeline depth and wastewater flow velocity for a sanitary sewer system?

14. Describe the problem called *crown corrosion* of sewers.

15. Describe two types of sewage lift stations. Under what circumstances might a lift station be needed?

16. List six important factors in sewer construction inspection.

17. What are three key factors that affect the dead load on a buried pipeline?

18. List two key factors that affect the external load-carrying capacity of a pipeline.

19. Describe two different methods for setting line and grade of a sewer pipeline.

20. What is the difference between *infiltration* and *inflow* in a sanitary sewer system?

21. Why is it important to limit the amount of infiltration and inflow into a sewer line? Briefly indicate how each may be controlled.

22. Describe three methods for infiltration testing.

23. Briefly discuss methods used to rehabilitate sanitary sewer systems.

24. Visit the *Parisian Sewer System* Web page and read about the historical development of sewers in Paris. Write a brief report summarizing what you learned.

25. Visit the *Infill Sewerage System* Web page and briefly describe that program; also, from the "Environment" link, read about and describe the Ellenbrook wastewater pipeline.

26. Briefly explain what GIS technology is and how it may be applied in the management of sanitary sewer systems.

27. Visit the *Hopatcong Sanitary Sewer Project* Web page and link to the article titled "The Benefits and Consequences of the Choice Between Septic Systems or Sewers." Read the article and write a brief report to summarize it.

28. Using the EnviroSources search engine (see Section 1.6) or another Internet search directory or search engine, locate at least one additional Web site relevant to one of the topics in this chapter. Add the link(s) to your Environmental Technology Folder. Write a brief description of what the Web site(s) contain.

# PRACTICE PROBLEMS

1. A collector sewer is to be designed to receive flow from 250 ac of a community where the population density is estimated to be 12 persons/ac. The average per capita sewage flow is taken to be 100 gpcd. What is the required design flow for the collector in gallons per minute?

2. A trunk sewer is to be designed to receive flow from a 1-km² area of a community where the population density is 50 persons/ha. The average per capita sewage flow is taken to be 400 L/d. What is the design flow for the trunk sewer in liters per second?

3. A 100-m reach of sewer is to have a minimum capacity of 200 L/s. The street elevation at the upper manhole is 305.55 m and at the lower manhole is 303.05 m. Determine an appropriate pipe diameter and slope for this reach, and establish the pipe invert elevations at the upper and lower manholes. Assume that a minimum earth cover of 2 m above the pipe crown is required.

4. Design the two sewer reaches shown in Figure 8.30. The design flow for reach 1 is 40 L/s and that for reach 2 is 80 L/s. The ground elevation at MH 1 is 350.00 m, at MH 2 is 347.87 m, and at MH 3 is 347.00 m. Use a minimum cover of 2.5 m. Sketch a profile of the street and sewer, using a horizontal scale of 1:1000 and a vertical scale of 1:100.

5. A 12-in.-diameter pipe is placed in a 10-ft-deep, 2.5-ft-wide trench that is backfilled with saturated clay. Compute the dead load on the pipe.

6. A 200-mm-diameter pipe is placed in a 3-m-deep, 0.9 m-wide trench and backfilled with sand. Using a safety factor of 1.5, select an appropriate class of bedding for the pipe if it is standard strength VCP.

7. What would be a suitable type of bedding for the pipe in Problem 6 if it were extra-strength VCP instead of standard strength?

8. What is the total allowable rate of infiltration in a 750-m-long, 600-mm-diameter sewer if 45 L/d/mm/km is allowed?

9. An exfiltration test is conducted on a 350-ft-long, 18-in.-diameter sewer reach. The water level in a 4-ft-diameter manhole is observed to drop 5.5 in. in 1 h. Compute the rate of exfiltration.

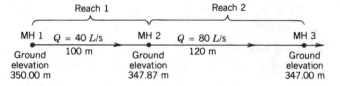

**FIGURE 8.30**
*Illustration for Problem 4.*

# Chapter Outline

C H A P T E R

# 9

# Stormwater Management

Uncontrolled stormwater and surface runoff can cause significant environmental damage. Flooding is one obvious example of a stormwater problem, with the accompanying loss of property and, sometimes, human life. Even when the amount of stormwater runoff is not enough to be characterized as a flood, water pollution problems can be severe, particularly from soil erosion and sedimentation (see Section 5.3). Storm runoff is a major nonpoint source of water pollutants,

including fertilizers, pesticides, oil, organics, and other substances.

Land development and urbanization increase the frequency and severity of these problems. Technical personnel involved in the design and construction of new residential, commercial, or industrial land-use facilities must take steps to reduce the harmful environmental impacts of stormwater runoff. Since 1994, stormwater has been regulated under the National

Pollutant Discharge Elimination System (see Section 10.1). Municipalities, as well as industries, are now required to obtain a *stormwater discharge permit.* New hydraulic control structures and operation and maintenance costs for municipal programs are expected to exceed several billions of dollars per year nationwide.

Until the mid-1970s, the basic approach toward stormwater control was to collect it in underground pipes and to dispose of it as soon as possible at some convenient downstream location. But this is precisely what led to many of the flooding and pollution problems seen today, particularly those in rapidly developing suburban areas. This approach is no longer acceptable in most developing communities.

It is still necessary to provide storm drains to remove excess water from streets and parking lots to prevent inconvenience or flood damage in localized areas, but modern drainage practice recognizes the need to apply *best management practices* to further control or restrain the flow of the collected stormwater and to prevent flooding and pollution problems in the lower part of the drainage basin. In fact, stormwater is increasingly being viewed as a natural resource for use in a beneficial manner, rather than as a waste material to be disposed of quickly.

The use of on-site storage or detention basins is a common method for controlling stormwater. In some cases, the stored stormwater can percolate to the groundwater and help to recharge an aquifer; this is a beneficial use of the water. Even if the stormwater is not used to recharge an aquifer, the fact that it is held temporarily in storage and released slowly from the detention basin protects the environment: Some water pollutants are removed, and downstream runoff flow rates are reduced.

Water pollution control is an important consideration in the design of stormwater management facilities. Computers are frequently used to analyze complex mathematical models of stormwater control systems. The EPA Stormwater Management Model (SWMM), for example, simulates water quality as well as quantity in storm or combined sewer systems; it is capable of analyzing existing systems and designing new systems.

This chapter focuses on some of the ways stormwater is collected and controlled. It builds on some of the topics covered in Chapter 3, particularly those sections dealing with rainfall and surface water. Also, a knowledge of basic hydraulics is necessary to study and understand stormwater control technology. It may be necessary to review the material in Chapter 2, particularly the section on gravity or open channel flow, as well as the pertinent material in Chapter 3.

## 9.1 ESTIMATING STORM RUNOFF

The design or analysis of stormwater control facilities begins with an estimate of the rate and volume of surface runoff to be controlled. Usually, the most important figure to be estimated is the peak flow rate from a storm of a specified frequency and duration. In most cases, it is necessary for the designer to rely on local rainfall data and to use an acceptable formula that relates rainfall intensity and duration to the volume or rate of surface runoff.

### Rational Method

The most popular formula used to correlate rainfall with runoff in relatively small urban and suburban drainage basins is called the *rational formula.* The rational formula expresses the relationship between *peak runoff* and rainfall as follows:

$$Q = C \times i \times A \qquad \text{(9-1)}$$

where $Q$ = peak or maximum rate of runoff
$C$ = a dimensionless runoff coefficient
$i$ = rainfall intensity
$A$ = drainage basin area

It is important to note that $Q$ in the rational formula represents only the maximum discharge caused by a particular storm; it can be visualized as the peak of the storm hydrograph illustrated in Figure 9.1.

In SI metric units, $Q$ is computed in terms of cubic meters per hour ($m^3/h$) and then converted to cubic meters per second or liters per second; in Customary units, $Q$ is computed directly in terms of cubic feet per

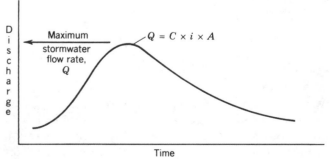

**FIGURE 9.1**
*The* rational formula *is used to estimate the peak or maximum rate of surface runoff due to a particular storm in a specific drainage basin.*

second (ft³/s or cfs). It is important to use appropriate units for rainfall intensity $i$ and area $A$ in each case.

### U.S. Customary Units

In this system of units, rainfall intensity is expressed in terms of inches per hour, and area is expressed in terms of acres. No conversions are necessary. The resulting dimensions for $Q$ become inch-acre/hour, but 1 in.-ac/h is so close to 1 ft³/s that the difference is neglected for practical purposes. (Try the conversion yourself to see how close it is.)

### EXAMPLE 9.1

Estimate the peak rate of runoff on a 5-ac watershed from a storm with rainfall intensity of $i = 3$ in./h. Assume that the dimensionless runoff coefficient $C = 0.4$. (The meaning of this coefficient is explained shortly.)

*Solution* Applying the rational formula (Equation 9-1) yields

$$Q = C \times i \times A = 0.4 \times 3 \times 5 = 6 \text{ ft}^3/\text{s}$$

Remember, there is no need to convert the units for $i$ or $A$; keep them as inches per hour and acres, respectively, and the value for $Q$ is in terms of cubic feet per second.

### SI Metric Units

In this system of units, rainfall intensity is usually expressed in terms of millimeters per hour (mm/h). To obtain a correct value for $Q$, it is necessary to convert millimeters per hour to meters per hour (m/h) and to express the area $A$ in terms of square meters (m²). Drainage areas in SI units are usually expressed in terms of hectares (ha) or square kilometers (km²). The following conversions will be used: 1 ha = 10 000 m² and 1 km² = 1 × 10⁶ m².

### EXAMPLE 9.2

Estimate the peak rate of runoff from a 2-ha drainage basin that has a runoff coefficient of 0.4 for a rainfall intensity of 75 mm/h.

*Solution* Converting the intensity from millimeters per hour to meters per hour is simply a matter of dividing by 1000, since 1 m = 1000 mm; therefore,

$i = 0.075$ m/h.  The area $A = 2$ ha $\times$ 10 000 m²/ha = 20 000 m². Applying the rational formula yields

$$Q = 0.4 \times 0.075 \text{ m}^2/\text{h} \times 20\,000 \text{ m}^2 = 600 \text{ m}^3/\text{h}$$

and

$$Q = 600 \text{ m}^3/\text{h} \times \frac{1 \text{ h}}{3600 \text{ s}} \approx 0.17 \text{ m}^3/\text{s} \quad \text{or} \quad 170 \text{ L/s}$$

### Area

The watershed area $A$ can be easily measured from a topographic map, as described in Section 3.4. Of all the terms in the rational formula, area is the only one that can be determined with some degree of precision. The rational formula has been applied to drainage areas ranging in size from a fraction of an acre to about 5 mi², but for areas exceeding about 200 ac (80 ha), it is considered best to apply some other, more sophisticated hydrologic method to estimate runoff.

### Runoff Coefficient

The coefficient $C$ is a dimensionless number that represents the fraction of rainfall that appears as surface runoff in the drainage basin. If all the rainfall became runoff, as it would on a completely impervious surface, the value of $C$ would be 1.0, its maximum possible value. But as discussed in Section 3.2, some of the rain water is intercepted, evaporates, or infiltrates the ground surface. Therefore, the value of $C$ is always a decimal less than 1.0.

Factors that affect the value of the runoff coefficient include the type of land use in the drainage basin, the ground cover or vegetation, and the ground slope. Typical values of the coefficient $C$ are presented in Table 9.1. Each type of surface or land use has a wide range of possible $C$ values; the selection of a runoff

**TABLE 9.1**
**Typical runoff coefficients**

| Type of surface or land use | Runoff coefficient $C$ |
|---|---|
| Woodland areas | 0.01 to 0.20 |
| Grassland or lawns | 0.10 to 0.25 |
| Pavements and roofs | 0.70 to 0.95 |
| Suburban residential areas | 0.25 to 0.40 |
| Apartment housing areas | 0.50 to 0.70 |
| Industrial areas | 0.60 to 0.90 |
| Business areas | 0.70 to 0.95 |

coefficient for a drainage basin requires good technical judgment and careful evaluation of the physical characteristics of the basin.

### Composite Runoff Coefficient

When several different surface types or land uses make up a watershed, a composite or weighted average value of the runoff coefficient can be computed using the following formula:

$$C = \frac{1}{A_T} \times (C_1 \times A_1 + C_2 \times A_2 + C_3 \times A_3 + \ldots)\ \textbf{(9-2)}$$

where  $C$ = composite runoff coefficient
$A_T$ = total basin area $(A_1 + A_2 + A_3 + \ldots)$
$A_1, A_2$, etc. = areas of different type in the basin
$C_1, C_2$, etc. = respective runoff coefficients for
$A_1, A_2$, etc.

### EXAMPLE 9.3

From an air photo of a 15-ha watershed, it is determined that 6.5 ha is a flat, densely wooded area, 6.0 ha is lawn, and 2.5 ha is paved roadway and parking area. Compute a composite runoff coefficient for the total watershed area.

*Solution*  Refer to Table 9.1. For the wooded area, use $C_1 = 0.01$, since it is flat and densely wooded. For the lawn, use $C_2 = 0.20$, the middle of the range, and for the paved areas, use $C_3 = 0.95$, a conservative judgment. Applying Equation 9-2, we get

$$C = \frac{1}{15} \times (6.5 \times 0.01 + 6.0 \times 0.20 + 2.5 \times 0.95)$$

$$= 0.24$$

If more detailed information about the characteristics of the land had been available, other values of $C$ might have been used. For example, if the lawn area had been very steep, then $C = 0.25$ might have been selected instead of $C = 0.20$. Using the upper and lower limits of each range of $C$ for this problem, the composite $C$ could fall within the range of 0.16 through 0.35. (Try the computation yourself to check it out.)

### Rainfall Intensity

The value of rainfall intensity $i$ used in the rational formula depends on the storm recurrence interval and the storm duration. It is determined from local rainfall intensity–duration–frequency relationships, as discussed and illustrated in Section 3.3.

The first step in establishing the value of rainfall intensity to use in the rational formula is to *select a storm frequency* or *recurrence interval*. More often than not, the recurrence interval is specified by local township, county, or state agencies for various types of projects. Longer recurrence intervals reduce the possibility of the selected storm intensity being equaled or exceeded in any given year.

Accordingly, a local township engineering department may specify a 5-year return period for the design of storm drains in a typical residential neighborhood. This would indicate a willingness to accept some occasional street flooding in order to have a reasonably economical system. On the other hand, a high-value commercial district may have storm sewers designed for a 15- or 25-year storm, reducing the chance of flooding, but increasing the cost of construction.

The spillway structure of a dam must be designed to accommodate stormwater flows from at least the 100-year frequency storm in order to minimize the chance for failure of the dam and loss of life or property downstream. In general, the risks of failure must be balanced with the cost of construction for all stormwater control projects.

The second step in determining the value of rainfall intensity to use is to *select an appropriate storm duration*. The storm duration used in the rational method is called the *time of concentration* of the drainage basin, abbreviated $T_c$. It is defined as the time it takes a drop of water to flow from the hydraulically most distant point of the basin to the outlet of the basin.

If the storm lasts at least as long as the time of concentration, then the entire watershed will be contributing flow to the outlet; only under this condition is there enough time for the maximum peak flow to develop. For shorter duration storms, not all of the basin area contributes flow simultaneously; for storms longer than $T_c$, the rainfall intensity, and therefore the peak flow, will be less than the maximum computed by using the time of concentration of the basin as the storm duration.

### Time of Concentration

There are two parts or components of the time of concentration. These are the *overland flow time* and the *channel flow time*, as illustrated in Figure 9.2. The overland flow time is the time it takes the surface sheetflow to reach the beginning of the stream or storm drain inlet at the upper end of the basin.

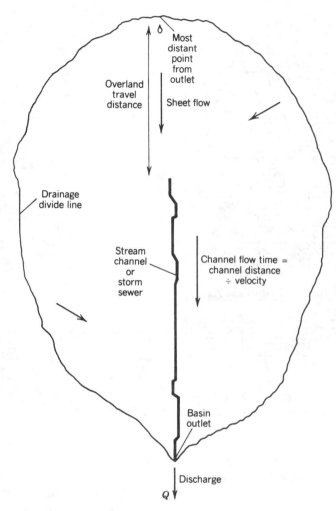

**FIGURE 9.2**

*The* time of concentration *for a drainage basin includes overland flow time and channel flow time.*

Several charts or nomographs are available that correlate overland flow time with ground slope and runoff coefficient or land use. One such chart is illustrated in Figure 9.3. As illustrated, the chart is entered on the left with the known overland travel distance, and a horizontal and then vertical path is traced to the known values of slope and runoff coefficient curves, respectively; finally, the overland travel time is read from the vertical axis on the right side of the chart.

If the velocity of flow in the stream or channel is known or estimated, the channel flow time can be computed from the following simple relationship: *channel flow time = flow distance ÷ flow velocity*. If the channel geometry is known, Manning's formula can be used to estimate the velocity of flow. Otherwise, a nomograph such as the one illustrated in Figure 9.4 may be used. (The minimum value for time of concentration is usually set at 5 min.)

**EXAMPLE 9.4**

A 0.25-km$^2$ watershed has a composite runoff coefficient $C = 0.25$. The overland flow distance to the beginning of the stream that drains the watershed is 150 m, and the slope is 7 percent. The stream is 600 m long and drops 30 m in elevation by the time it reaches the watershed outlet. Estimate the peak rate of runoff for the 5-year storm and the 100-year storm.

*Solution* The drainage area $A = 0.25$ km$^2 \times 10^6$ m$^2$/km$^2$ = 250 000 m$^2$. Enter Figure 9.3 with 150 m or 490 ft for overland travel distance, move horizontally to the right to the slope curves, and estimate where the 7 percent curve would be (about midway between the 5 percent and 10 percent curves); then move down vertically to the runoff coefficient curves and estimate the position for $C = 0.25$. Finally, move horizontally to the right and read 22 min for the time of overland travel.

Use Figure 9.4 to estimate channel flow time; line up the drop in elevation of 30 m (100 ft) with the length of channel, 600 m (2000 ft), and read a channel flow time of about 8 min.

The time of concentration for this watershed is the sum of the overland flow time and the channel flow time, or

$$T_c = 22 \text{ min} + 8 \text{ min} = 30 \text{ min}$$

This value of $T_c$ is used as the storm duration to determine the peak flow at the basin outlet. From Figure 3.6, the rainfall intensity for a 5-year, 30-min-duration storm is 70 mm/h and that for the 100-year, 30-min-duration storm is 150 mm/h. Applying the rational formula yields

5-year storm: $Q = 0.25 \times 0.070 \text{ m/h} \times 250\ 000 \text{ m}^2$
$\approx 4400 \text{ m}^3/\text{h} \approx 1200 \text{ L/s}$

100-year storm: $Q = 0.25 \times 0.150 \text{ m/h} \times 250\ 000 \text{ m}^2$
$\approx 9400 \text{ m}^3/\text{h} \approx 2600 \text{ L/s}$

See Appendix D for a description of a computer program that can be used for stormwater runoff calculations using the rational method described above, as well as SCS methods such as the one described below. A free sample version of the program can be downloaded from the Internet (http://www.hydrocad.net/).

## SCS Method

A method for estimating the volume and rate of storm runoff in rural and suburban areas was developed by

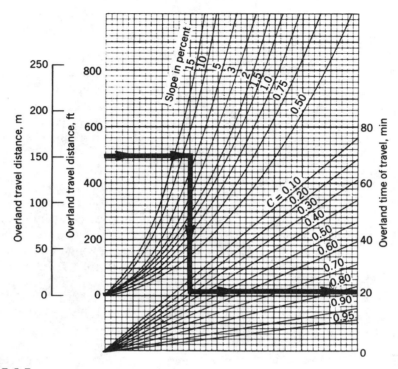

**FIGURE 9.3**

*A chart like this may be used to estimate the overland flow time when the average travel distance, slope, and runoff coefficient are known.* (Courtesy of the Federal Aviation Administration)

the former Soil Conservation Service (SCS) of the U.S. Department of Agriculture (USDA). The Soil Conservation Service is now called the Natural Resource Conservation Service (NRCS); however, the method of estimating runoff described here may still be called the "SCS method." The so-called SCS method is finding widespread application, and in some areas of the United States its use is often required by regulatory agencies for project approval.

A basic distinction between the SCS and rational methods is in the emphasis that the SCS places on the correlation between the type of soil cover in the watershed and the runoff. In this section, a brief overview of the SCS *graphical method* is presented for relatively small watersheds.

Four *hydrologic soil groups* are defined in the SCS method, as follows:

| Soil group | Description |
|---|---|
| A | High infiltration rate/low runoff potential |
| B | Moderate infiltration rate |
| C | Slow infiltration rate |
| D | Very slow infiltration/high runoff potential |

The soil groups in a particular watershed can usually be determined from countywide SCS soil survey maps of the type illustrated in Figure 1.14. Data from field studies of the site and measured infiltration rates can also help to identify the appropriate soil group.

In addition to the soil group, the volume and rate of runoff depend on the type of land use in the watershed. In the SCS method, the effects of both soil group and land use are characterized in a term called the *runoff curve number*, abbreviated CN. Table 9.2

**TABLE 9.2**
**Typical SCS runoff curve numbers**

| Land-use description | CN value for hydrologic soil group | | | |
|---|---|---|---|---|
| | *A* | *B* | *C* | *D* |
| Meadows | 30 | 58 | 71 | 78 |
| Forests | 25 | 55 | 70 | 77 |
| Grass lawns | 39 | 61 | 74 | 80 |
| Commercial–business | 89 | 92 | 94 | 95 |
| Residential | 54 | 70 | 80 | 85 |
| Pavement–roofs | 98 | 98 | 98 | 98 |

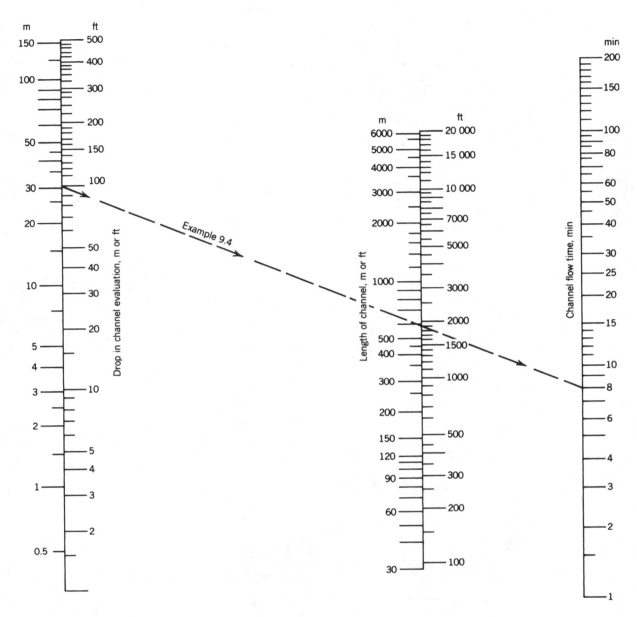

**FIGURE 9.4**

*A rough approximation of channel flow time may be obtained with this nomo-graph using a straightedge. For example, the time of travel in a channel that drops 4 m in elevation over a distance of 500 m is about 15 min. A more accurate estimate can be obtained using Manning's formula.*

presents a summary of typical CN values used in the NRCS method of estimating runoff.

Note that the CN values for pavements and roofs are independent of soil type, as they obviously should be. A completely impervious surface would have a CN = 100, and all the rainfall would become runoff. As the value of CN decreases, the amount of direct runoff will also decrease. Composite or weighted CN values can be computed for watersheds comprising more than

one type of soil or land use in the same way this is done for the composite *C* in the rational method.

A graph showing the relationships among rainfall depth, in inches, the amount of runoff, also expressed in inches, and the CN values is illustrated in Figure 9.5.

The chart in Figure 9.5 is entered on the horizontal axis with the depth of rainfall from an *N*-year storm of 24-h duration. First moving vertically up to the curve matching the CN for the watershed (or the estimated

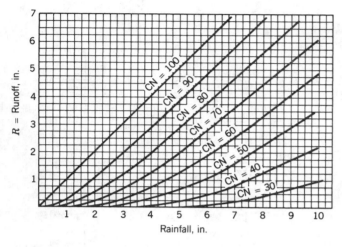

**FIGURE 9.5**

*A selection of SCS rainfall–runoff relationships for several CN values.* (Adapted from "A Method for Estimating Volume and Rate of Runoff in Small Watersheds," with permission of the Soil Conservation Service, U.S. Department of Agriculture)

curve position if it falls between the values shown on the graph) and then moving horizontally to the left, one can read the volume of runoff on the vertical axis. For example, a 24-h rainfall of 6 in. on a watershed with CN = 70 will produce 2.8 in. of direct runoff over the area of the watershed.

To determine the peak rate of runoff, the graph shown in Figure 9.6 is used. The horizontal axis repre-

sents the time of concentration $T_c$ for the watershed; this was defined previously in the discussion for the rational method. The vertical axis has the units of *cubic feet per second per square mile of watershed per inch of runoff (csm/in.).* This value is then applied in the following equation to compute the peak rate of runoff:

$$Q = q \times A \times R \qquad \textbf{(9-3)}$$

where $Q$ = peak rate of runoff, $ft^3/s$
  $q$ = unit peak discharge, csm/in.
    (from Figure 9.6)
  $A$ = drainage age, $mi^2$
  $R$ = direct runoff, in. (from Figure 9.5)

This SCS method for estimating runoff is generally applicable for drainage areas up to about 2000 ac, or 3 $mi^2$. It is considered to be more conservative than the rational method and usually results in higher estimates of runoff. The following example illustrates the SCS graphical procedure for determining peak discharge.

## EXAMPLE 9.5

A 1000-ac watershed has a composite CN = 70 and a time of concentration $T_c$ = 66 min. From rainfall data, it is determined that the 100-year, 24-h storm will cause a total of 7.5 in. of rainfall in the watershed. Compute the peak rate of stormwater runoff from the watershed by the SCS method.

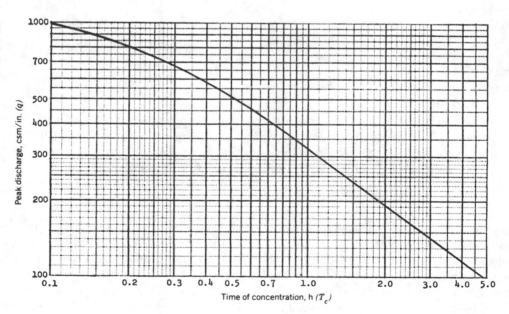

**FIGURE 9.6**

*Unit peak discharge in csm/in. of runoff versus time of concentration for a 24-h storm duration.* (From "Urban Hydrology for Small Watersheds," with permission from the Soil Conservation Service, U.S. Department of Agriculture)

*Solution*   From Figure 9.5, $R = 4.0$ in. The time of concentration $T_c = 66$ min $\times$ 1 h/60 min $= 1.1$ h. From Figure 9.6, $q = 300$ csm/in.

Applying the conversion 1 mi$^2$ = 640 ac yields

$$\text{watershed area } A = 1000 \text{ ac} \times 1 \text{ mi}^2/640 \text{ ac}$$

$$= 1.56 \text{ mi}^2$$

Now, applying Equation 9.3 yields

$$Q = q \times A \times R = 300 \times 1.56 \times 4.0 \approx 1900 \text{ ft}^3/\text{s}$$

## Effects of Land Development

The construction of homes, factories, roads, and other facilities for a growing community changes the runoff patterns of the watershed in which it takes place. Unless appropriate steps are taken, land development can increase the frequency and severity of flooding and water pollution from stormwater runoff.

As land is developed and urbanization takes place, much of what was originally woodland, meadow, or farmland is covered with relatively impervious surfaces—paved roads, driveways, parking lots, and buildings. This has two direct effects on the local hydrology:

1. The amount of infiltration decreases and therefore the volume of direct surface runoff increases.
2. The time of concentration of the watershed decreases, since the runoff flows faster to the outlet over the modified land surfaces and in the drainage channels.

The net result of land development and urbanization is a definite increase in both the peak rate and total volume of runoff from a given storm. This effect is depicted in Figure 9.7.

On a short-term basis, soil erosion and stream siltation problems may be particularly severe during the period of land development and construction. In the long run, the chances for severe flooding increase as a result of land development. In addition, urban runoff can carry oil, lawn fertilizers, organics, and other pollutants into the receiving waters.

Modern land planning practices, however, generally require the implementation of methods to prevent any significant increase in the rate of runoff due to urbanization. Some of the technical solutions to this problem are discussed in Section 9.3. The following example illustrates the effects of land development on runoff.

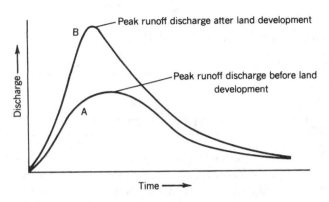

**FIGURE 9.7**
*Curve A shows a storm hydrograph for the original or predevelopment conditions; curve B shows a post-development hydrograph for the same area. Land development or urbanization causes the volume and rate of stormwater runoff to increase.*

## EXAMPLE 9.6

The 0.25-km$^2$ watershed described in Example 9.4 is now to be developed as a residential and business area. The resulting composite runoff coefficient $C$ is estimated to be 0.60. Part of the stream is to be carried in a pipeline, and it is estimated that the new channel flow time after development will be only 5 min. For both the 5-year and the 100-year storms, determine the estimated peak flow rates under developed conditions using the rational method. Compare the increases in peak discharges to those during the predevelopment conditions of Example 9.4.

*Solution*   Entering Figure 9.3 with overland flow distance = 150 m, slope = 7 percent, and $C = 0.60$, read overland travel time = 12 min. Adding the channel flow time of 5 min, a new time of concentration $T_c = 17$ min for the developed watershed is obtained.

Now, from Figure 3.6, read rainfall intensity $i = 90$ mm/h for the 5-year storm and $i = 190$ mm/h for the 100-year storm. Applying the rational formula yields

5-year storm:   $Q = 0.60 \times 0.090$ m/h $\times$ 250 000 m$^2$

$$\approx 14\,000 \text{ m}^3/\text{h}$$

Under predevelopment conditions (Example 9.4), the peak flow was computed to be 4400 m$^3$/h. Therefore, land development is expected to increase the peak stormwater discharge by a factor of 14,000/4400 = 3.2 for the 5-year storm:

100-year storm:   $Q = 0.60 \times 0.190$ m/h $\times$ 250 000 m$^2$

$$= 29\,000 \text{ m}^3/\text{h}$$

Land development will increase the peak discharge by a factor of 29,000/9400 = 3.1 for the 100-year storm.

## 9.2 STORM SEWER SYSTEMS

Storm sewers serve to convey surface runoff to a point of storage or disposal. In separate sewer systems, sanitary or industrial wastewater is excluded from the storm sewers. There are several basic differences in design between storm sewers and sanitary sewers.

First, it can be said that storm sewers are actually designed to fail with a predictable recurrence interval. In other words, a storm drainage system will periodically surcharge and overflow, causing local flooding, with a known frequency. This is because the designer must first select a storm return period in order to estimate peak discharge and then size the pipeline. This limits the capacity of the system to storm intensities equal to or less than the one selected for design.

The probability or frequency of failure can be reduced by selecting a large storm recurrence interval, but the chance for surcharge of a storm sewer can never be completely eliminated. Sanitary sewers, on the other hand, are designed to carry a peak flow rate of sanitary wastewater from a *saturation population* without surcharge or overflow. When sanitary sewers do surcharge, it is usually from excessive infiltration or inflow (that is, poor construction or maintenance).

There are other differences between storm sewers and sanitary sewers. Storm sewers are usually much bigger in diameter than the separate sanitary sewers serving the same area, as illustrated in Figure 9.8. Although they generally carry no flow during dry weather and only partial flows during most rainfalls, storm sewers still must be sized to carry the peak flow from a major storm of specified intensity and duration. Also, unlike sanitary sewers, which are placed in relatively deep trenches for service connections, storm sewers can be placed at shallower depths to minimize excavation.

### Layout and Design

A storm sewage collection system primarily consists of a network of inlets and pipes located in the street right-of-way or in easements along lot lines. A common location within the right-of-way is near the curb or edge of pavement, since inlet boxes can then be connected with fewer manholes and less pipe. Storm drains are located to allow gravity flow to a stream or other body of water, to a storage basin, or in some cases to a spe-

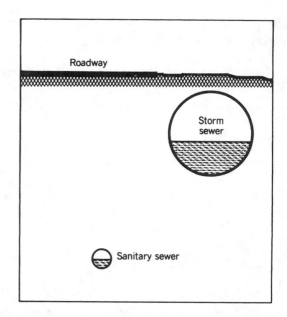

**FIGURE 9.8**
*Sanitary sewers are typically much smaller in diameter and placed in deeper trenches than storm sewers serving the same area.*

cial treatment facility. The installation of stormwater pumping stations is especially avoided because of the very large peak flow capacity that would be required.

Modern design practice makes more use of existing natural streams, open channels, or grass-lined swales to carry runoff before resorting to the use of buried pipes. This is basically to slow down the flow and increase the time of on-site detention (without basin storage), thereby reducing peak flows downstream. Roadway crossings over natural drainage ditches or streams may require the installation of *culverts*. A culvert is a relatively short section of pipe or a cast-in-place concrete structure that carries surface runoff under an obstruction.

Storm sewers are usually built with circular reinforced concrete or plastic pipes. Elliptically shaped pipes are sometimes used when the pipe depth is very shallow, to gain leeway for soil cover over the shorter axis of the pipe. Corrugated metal pipe may be used in some cases for storm drains and culverts.

### Inlets

A stormwater inlet is a structure that intercepts sheet flow and directs the runoff into the underground pipe system. The location and spacing of inlets are important design factors; there must be sufficient inlet capacity to remove the stormwater from the surface and transfer it into the sewers fast enough to prevent backups. Factors such as hydraulic capacity, clogging,

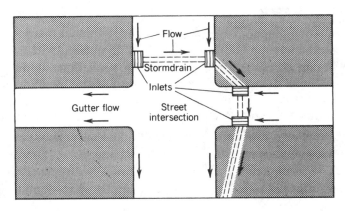

**FIGURE 9.9**

*Proper location of stormwater inlets prevents stormwater runoff from flooding a street intersection.*

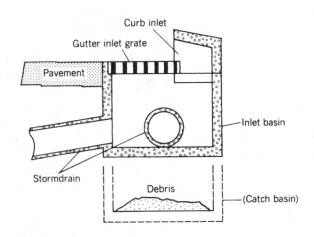

**FIGURE 9.11**

*A section view of a typical stormwater inlet basin. Catch basins that trap grit, leaves, and debris are not commonly used in new storm sewer systems.*

nuisance to traffic, and safety must be evaluated. For example, it is desirable to keep runoff from flooding across street intersections. To accomplish this, inlets are generally located as shown in Figure 9.9.

There are basically three different types of inlet structures: *curb inlets*, *gutter* or *grate inlets*, and *combined inlets*. A curb inlet has a vertical opening along the curb, into which the runoff flows, as shown in Figure 9.10a. A gutter inlet is a horizontal opening in the pavement covered with a cast-iron gate, as shown in Figure 9.10b.

The curb inlet is not an obstruction or nuisance to traffic, but for child safety the opening should be less than 150 mm (6 in.) high; this limits the hydraulic capacity of the inlet. Disadvantages of the gutter-type inlet are that it can obstruct the smooth flow of traffic (including bicycles) and that the iron gate may become plugged with debris.

A combination inlet has both a curb and a gutter opening, as shown in Figure 9.10c. This inlet may be depressed or lowered for additional hydraulic capacity. The combination inlet is the least subject to clogging.

A cross-sectional view of an inlet basin is shown in Figure 9.11. If the pipe invert is above the bottom of the basin, as shown with the dashed lines, the structure is called a *catch basin*. A catch basin serves as a trap for grit, sand, and debris that may be washed into the inlet during a storm. Because of the need for periodic cleaning and mosquito control, and odor problems, catch basins are not always used in new storm drainage systems. Good inlet design and self-cleansing pipe slopes (for adequate flow velocity) can keep the collection system free from debris and blockage problems.

**Design Procedure**

Using a topographic map of the drainage district or area to be sewered, the inlets are first located at street intersections, low points, and appropriate intervals along the streets. Local engineering departments may specify maximum spacing between inlets. The storm drains are located within the public right-of-way or in drainage

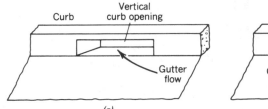
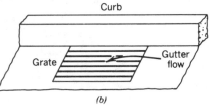
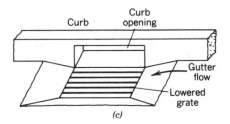

**FIGURE 9.10**

*There are several different types of stormwater inlets, including (a) the curb inlet, (b) the gutter inlet, and (c) the combined inlet. Curb opening inlets are undesirable due to the possibility that small objects or animals can fall into the basin.*

easements, following the natural slope of the land for gravity flow.

A storm sewer reach is a section with constant diameter and slope, usually between two inlets. The boundary of the catchment area that is a tributary to each inlet and downstream reach is sketched on the topographic map (as described in Section 3.4). Each of these areas is measured in terms of acres, hectares, or square meters, using a planimeter if necessary.

The rational method is commonly used to compute peak design flows in urban or suburban drainage districts. An appropriate runoff coefficient $C$ is selected for each of the inlet catchment areas based on type of ground cover and land use. The storm recurrence interval for which the system will be designed is selected by the designer or may be specified in local building and land-use ordinances. Overland flow time, also called *inlet time*, is estimated for each area tributary to an inlet.

Design begins at the uppermost inlet and proceeds downstream; cumulative areas and composite runoff coefficients are computed for each reach of the system. A new time of concentration must be determined for each section of the system to account for the gradually increasing travel time to the reach being designed. At any inlet, the longest time of flow for any water entering it must be selected as the time of concentration for that reach, since there must be enough time for the entire basin to contribute runoff in order for the peak flow to develop. In practice, design computations are usually organized in tabular form.

The pipe locations, diameters, and slopes must be shown on the engineering plan drawings; diameter, slope, and invert elevations are determined in much the same way as for sanitary sewers (see Section 8.1). The self-cleansing velocity for storm sewers is about 0.9 m/s (3 ft/s); it is higher than that for sanitary sewers because stormwater tends to carry heavier and more settleable solids than sanitary sewage does.

The following example illustrates the basic procedure for determining design flows and required pipe diameters. It can be assumed that the given pipe slopes are roughly parallel to the street slope, as determined from a topographic map of the area or street profile. The data for the problem are given in Figure 9.12.

## EXAMPLE 9.7

A storm drainage system has been laid out in a street as shown in Figure 9.12. Individual inlet catchment areas and runoff coefficients have been determined and are shown. Inlet times have been estimated, and the pipe length and slope for each sewer reach are given.

Using the rational method, compute the design flow and the required pipe diameter for each reach for a 10-year storm.

*Solution    Reach 1:* First, determine the rainfall intensity for area 1, using the inlet time as the time of concentration. Enter Figure 3.6 with $T_c = 5$ min, and read $i = 150$ mm/h or 0.15 m/h for the 10-year storm. The catchment area is 1 ha, or 10 000 m². Applying the rational formula yields

$$Q = 0.4 \times 0.15 \text{ m/h} \times 10\ 000 \text{ m}^2 = 600 \text{ m}^3/\text{h}$$

and converting to m³/s gives

$$Q = 600 \text{ m}^3/\text{h} \times 1 \text{ h}/3600 \text{ s} = 0.167 \text{ m}^3/\text{s}$$

To determine the required diameter for reach 1, enter Manning's nomograph with $Q = 0.167$ m³/s and the given slope of 0.0035; read the required diameter $D = 45$ cm and a flow velocity of 1.05 m/s.

*Reach 2:* To compute the peak discharge or design flow for the pipeline between inlet 2 and inlet 3, first compute the composite runoff coefficient for the combined areas 1 and 2, as follows:

$$C = \frac{1}{2.5} \times (1 \times 0.4 + 1.5 \times 0.3)$$

$$= 0.34 \text{ (from Equation 9-2)}$$

Next, compute the time of flow in reach 1, using the relationship

$$\text{time} = \text{distance} \div \text{velocity} = 120 \text{ m} \div 1.05 \text{ m/s}$$

$$= 114 \text{ s} = 1.9 \text{ min}$$

The total time of flow to inlet 2 is the inlet time for area 1 plus the channel flow time in reach 1, or $5 + 1.9 = 6.9$ min. This is larger than the inlet time for area 2; therefore, use a time of concentration $T_c \approx 7$ min for the composite area draining to inlet 2.

From Figure 3.6, read $i = 145$ mm/h for the 10-year storm.

Applying the rational formula yields

$$Q = 0.34 \times 0.145 \text{ m/h} \times 25\ 000 \text{ m}^2$$

$$= 1230 \text{ m}^3/\text{h}$$

and

$$Q = 1230 \text{ m}^3/\text{h} \times 1 \text{ h}/3600 \text{ s}$$

$$= 0.34 \text{ m}^3/\text{s for reach 2}$$

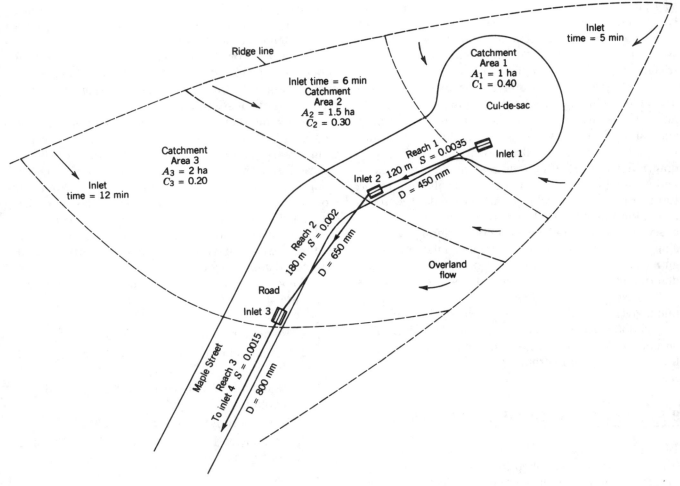

**FIGURE 9.12**

*Illustration for Example 9.7. Plan view of drainage area.*

Now enter Manning's nomograph with $Q = 0.34$ m³/s and the given slope of $S = 0.002$; read the required diameter $D = 65$ cm and flow velocity $V = 1.02$ m/s.

*Reach 3:* The total tributary area to inlet 3 is $1 + 1.5 + 2 = 4.5$ ha, or $45{,}000$ m². The composite runoff coefficient is computed to be

$$C = \frac{1}{4.5} \times (0.4 \times 1 + 0.3 \times 1.5 + 0.2 \times 2)$$

$$= 0.28$$

The time of flow in reach 2 is $180$ m $\div 1.02$ m/s $= 176$ s $= 2.9$ min. The total flow time to inlet 3 is then $5 + 1.9 + 2.9 = 9.9$ min (inlet time for area 1 + travel time in reach 1 + travel time in reach 2). But this is less than the individual inlet time for area 3, which is 12 min. Therefore, the 12-min inlet time dominates and is taken as the time of concentration (or storm duration) for the design of reach 3.

Enter Figure 3.6 with $T_c = 12$ min and read $i = 135$ mm/h or 0.135 m/h for the 10-year storm. Now, applying the rational formula yields

$$Q = 0.28 \times 0.135 \text{ m/h} \times 45\,000 \text{ m}^2 = 1700 \text{ m}^3/\text{h}$$

and

$$Q = 1700 \text{ m}^3/\text{h} \times 1 \text{ h}/3600 \text{ s} = 0.47 \text{ m}^3/\text{s}$$

From Manning's nomograph, with $Q = 0.47$ m³/s and $S = 0.0015$, read $D = 80$ cm and $V = 1$ m/s.

---

A powerful, easy-to-use storm sewer computer-modeling program called StormCAD (Haestad Methods, Inc., Waterbury, Connecticut) can be used for design and analysis of stormwater collection systems. The piping network can easily be drawn on the screen using a tool palette, and the hydraulics of the system can be calculated at the click of a button. Rainfall information is calculated using rainfall tables, equations, or National Weather Service data, and the intensity–duration–frequency (IDF) curves can be plotted.

A choice of stormwater conveyance elements is available, including circular pipes, pipe arches, boxes, and others. Automatic flow calculations handle pressure and varied flow situations, including hydraulic jumps, backwater, and drawdown curves. The program's flexible reporting features allow customization and printing of design and analysis results in a report format or as a graphical plot. An automatic design feature allows design of all or part of a system based on a set of user-defined constraints, including velocity, slope, cover, invert/crown matching, inlet efficiency, gutter spread, gutter depth, and others. The software determines invert elevations and diameters of pipes, as well as the size of a drainage inlet necessary to maintain a given spread or capture efficiency depending on its location in sag or on grade. StormCAD also includes an option to automatically generate storm sewer profiles, including the hydraulic grade line.

## 9.3 BEST MANAGEMENT PRACTICES

Two primary objectives of *best management practices* (BMPs), as they are applied to controlling stormwater runoff from new land development projects, are reproduction of predevelopment flow conditions (runoff attenuation) and pollutant removal. There are two basic types of BMPs: *preventive measures* (nonstructural) and *control measures* (structural).

Preventive measures include source reduction practices—land management techniques to reduce both the amount of runoff and the quantity of pollutants in the runoff. These methods provide cost-effective ways to manage stormwater because they usually require no land or construction. Debris removal from streets, landscaping and lawn management control, vegetated swales, and pervious soil buffers are just some of the preventive measures that can be implemented to achieve the objectives of BMPs. In some drainage basins, however, more sophisticated control measures are necessary to meet the requirements of local water quality control standards.

### Stormwater Flow Attenuation

The most common approach to stormwater flow control generally utilizes temporary storage of the water on site, in the vicinity where it falls, rather than quick discharge

to a nearby body of water. To some extent, stormwater storage occurs naturally in most drainage basins. In some cases, intentional ponding of rain water on rooftops or in parking lots for a short period of time can provide enough storage to reduce peak runoff flow rates. Using open grass–swale drainage channels also can effectively retard the flow of stormwater runoff.

The construction of relatively small reservoirs or basins to hold stormwater after it has been collected from streets, parking lots, and other surfaces is being required in a growing number of communities undergoing urbanization. These basins store or retain the stormwater and allow it to be released slowly under controlled conditions. They can be effective in controlling relatively short, but intense local storms, which tend to cause the most frequent flooding, erosion, and pollution damage in small streams.

Some specific benefits of stormwater storage basins are as follows:

1. Reduction of peak runoff rates
2. Reduction of the severity and frequency of flooding
3. Reduction of soil erosion and stream sedimentation
4. Protection of surface water quality
5. Groundwater aquifer recharge, if soil conditions permit

The basic disadvantage of on-site stormwater storage basins is related to the problem of maintenance. The outlet structures are prone to clogging, and the basins themselves often become the depository for sediment and debris. Weed control is a problem, and mosquitos can breed in pools of water that remain stagnant for a long time. In some cases, safety for children in the area must be considered. Maintenance may be the responsibility of the local municipality or of nearby property owners.

There are basically three different types of stormwater storage basins: *retention basins*, *detention basins*, and *recharge basins*. A retention basin holds some water all the time, forming a permanent pond or small lake. In addition to stormwater control, it may also provide esthetic and recreational benefits on the site. A detention basin, however, only holds the stormwater for a relatively short period of time, during and shortly after periods of rainfall. It is empty and dry most of the time. Sometimes there may be a small stream flowing through the basin, even in dry weather. A section view of a detention basin is shown in Figure 9.13.

The third type, recharge basins, are specifically designed to allow the collected water to percolate into

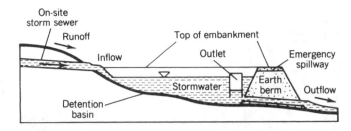

**FIGURE 9.13**

*Section view of a stormwater detention pond. The outlet structure acts as a bottleneck, restraining the rate of discharge from the pond. The pond or basin is usually empty of water during dry weather, except perhaps for a small stream.*

an underlying aquifer. They serve to recharge and replenish groundwater reserves as well as to control storm runoff. For a recharge basin to be effective, the soils underlying the basin must be permeable to allow relatively rapid infiltration, and the seasonal high water table should be at least 0.5 m below the bottom. In some communities, a portion of a recharge basin may be built underground, freeing up valuable land area for other uses; these subterranean basins are sometimes called *dry wells.*

Even with the use of recharge basins and dry wells, it is difficult to recreate the soil recharge rates of predevelopment conditions in any stormwater management project. Basins are fairly good at controlling the peak rate of runoff by extending the discharge over time, but the volume of runoff is increased in most postdevelopment conditions.

**Design Procedure**

Local subdivision regulations and municipal land-use ordinances must be reviewed at the very beginning of the project to determine the specific performance requirements for the basin with regard to stormflow reduction. The computations to determine the *predevelopment peak discharge* (before construction) and the *postdevelopment peak discharge* (after construction) can be made as illustrated in Examples 9.4 and 9.6.

Most communities require that a detention basin provide enough storage volume and outflow control to keep postdevelopment runoff equal to or less than predevelopment runoff. In other words, the developer must build a facility that will effectively maintain the rate of runoff from the site just about as it was in its natural condition, before construction.

This requirement usually pertains to the 100-year storm, but some ordinances also specify that the basin should reduce runoff flows from the 2- and 10-year storms as well. If the basin is designed only for the large 100-year storm, it will have no attenuating effect on the smaller, but more frequent stormflows. Accommodation of more than one storm return period in the basin is accomplished by proper hydraulic design of the basin volume and outlet structure. A concrete structure with several outlets for handling different size storms is illustrated in Figure 9.14.

Using a topographic map of the site, a suitable location for the detention basin should be established. This should be in the lower part of the tributary drainage area. The on-site storm sewer system should be designed, as described in Section 9.2, to direct the runoff into the detention basin. A preliminary estimate of the required basin volume can be made at this point. A simple and quick way of doing this is described later in this section.

After a preliminary basin size is determined, a grading plan should be sketched on the topographic map. The basin can be constructed by balancing excavation and fill in the low-lying area of the site, forming a confining earth embankment that gradually blends in to the natural topography of the land. A thorough hydraulic analysis of flow through the basin and outlet structure should then be conducted.

If the discharge at the outlet is determined to be equal to or less than the maximum allowable discharge, then the basin design is accepted as satisfactory; otherwise, changes are made and the process is repeated until an acceptable design is reached. Finally,

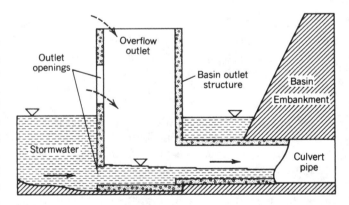

**FIGURE 9.14**

*The outlet structure for a detention basin can be hydraulically designed to reduce peak discharge flows from a range of storm magnitudes or frequencies. Multiple-outlet openings or specially designed proportional weirs can be used for this purpose.*

an emergency spillway is designed for the greatest design storm; freeboard is added to account for the estimations in the calculations.

## Preliminary Design Computations

The basic relationship that expresses the function and operation of a stormwater detention basin is that at any given time the volume of water in storage is equal to the difference between the inflow volume and the outflow volume up to that time, or

$$\text{inflow} - \text{outflow} = \text{storage} \qquad \textbf{(9-4)}$$

The rate of inflow changes with time. It depends on the intensity and duration of rainfall as well as on the physical characteristics of the drainage area. The relationship between inflow and time is shown graphically in an *inflow hydrograph.*

The rate of outflow from the basin also changes with time. It depends on the hydraulic characteristics of the basin outlet structure. The outflow is generally a function of the height or depth of water in the basin; the deeper the water, the faster it flows over or through the outlet.

In mathematical terms, the storage equation is

$$I(\Delta t) - O(\Delta t) = \Delta S \qquad \textbf{(9-5)}$$

where $\Delta t$ = a small time interval, such as 5 or 10 min (pronounced "delta $t$")
$I$ = average inflow rate during $\Delta t$
$O$ = average outflow rate during $\Delta t$
$\Delta S$ = change in storage volume during $\Delta t$ (pronounced "delta $S$")

The solution of this equation leads to the determination of the basin *outflow hydrograph,* which shows the rate of flow out of the basin as a function of time. The procedure for solving the equation and preparing the outflow hydrograph is called *flood routing* or *reservoir routing.*

Typical inflow and outflow hydrographs for a detention basin are shown in Figure 9.15. Initially, the rate of inflow exceeds the rate of outflow and water accumulates in the basin ($\Delta S$ is positive). The outlet structure serves, in effect, as a bottleneck that prevents the water from flowing out of the basin as fast as it flows in. Eventually, the inflowing stormwater subsides, and the basin gradually empties as the water flows through the outlet ($\Delta S$ is negative). A comparison of the inflow and the outflow hydrographs clearly shows the effect of the basin in attenuating or reducing the peak flow rate.

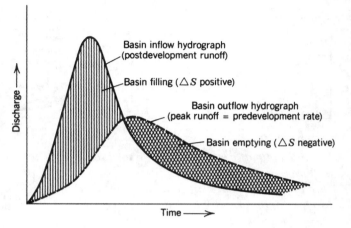

**FIGURE 9.15**
*Typical inflow and outflow hydrographs for a stormwater detention basin. The basin and outlet structure serve to reduce the peak rate of runoff from a developed site.*

The reservoir routing procedure just outlined involves a lot of computation to arrive at a solution of the storage equation and the outflow hydrograph. For this discussion, a simplified procedure is used in order to be able to illustrate the fundamental concept of stormwater detention without getting bogged down in computations. This procedure is sufficient to provide a preliminary or ballpark estimate of the required storage volume or peak outflow rate of a detention basin. It is not an exact method and would not be used for final design or analysis.

In this method, two factors related to the effectiveness of a stormwater detention basin are defined as follows:

$$\text{storage factor (SF)} = \frac{\text{basin storage volume}}{\text{total rainfall volume}} \qquad \textbf{(9-6)}$$

$$\text{flow factor (FF)} = \frac{\text{peak outflow rate}}{\text{peak inflow rate}} \qquad \textbf{(9-7)}$$

The relationship between the storage factor and the flow factor can be approximated by a straight line, as shown in Figure 9.16. The line is a graph of the equation

$$\text{FF} = 1.0 - \text{SF} \qquad \textbf{(9-8)}$$

where FF and SF are the flow factor and storage factor, respectively.

If there is no storage volume at all, then SF = 0, and therefore FF = 1.0. In other words, the outflow rate will equal the inflow rate, and there will be no flow attenuation. On the other hand, if a basin big enough to

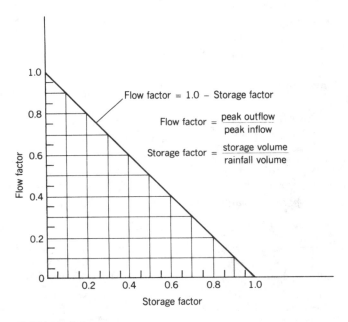

**FIGURE 9.16**

*The relationship between flow factor and storage factor offers a simplified procedure for stormwater detention calculations.* (Adapted from A. Pagan, "Flow Factor Line Used in Storage Calculations," *Irrigation Journal,* 1980, with permission of the American Society of Civil Engineers)

store the total rainfall volume from the storm is provided, then SF = 1.0. Under this circumstance, FF = 0, and there is no outflow at all. The straight-line relationship approximates what happens in between these two extreme conditions when some storage volume is provided.

## Total Rainfall Volume

The total rainfall volume is equal to the area under the inflow (or outflow) hydrograph, since the product of discharge (volume per unit time) and time is equivalent to volume. For practical purposes, it is reasonable to make the simplifying assumption that the inflow hydrograph is triangular in shape.

The peak inflow can be easily computed using the rational method. The time for the rising limb of the hydrograph to reach the peak flow value is taken as the time of concentration $T_c$ of the drainage area. The time for the receding limb to reach the base is conservatively taken as twice the time of concentration, or $2T_c$. This triangular hydrograph is illustrated in Figure 9.17.

The total rainfall volume is the area of the triangular inflow hydrograph. The area of a triangle is equal to the product of one half the base times the height

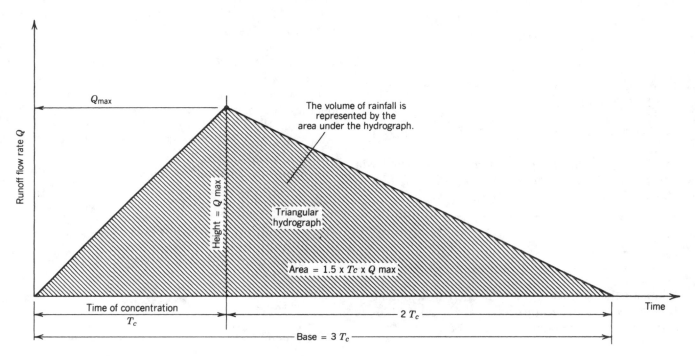

**FIGURE 9.17**

*A triangular hydrograph provides a reasonable estimate of total rainfall volume. The peak flow, or height of the triangle, can be computed using the rational formula.*

($A = bh/2$). Since in this case the base $b = 3T_c$ and the height $h = Q_{max}$, the area is $\frac{1}{2} \times 3T_c \times Q_{max}$ and

$$\text{total rainfall volume} = 1.5 \times T_c \times Q_{max} \quad \textbf{(9-9)}$$

The following examples illustrate the application of this simplified method for preliminary detention basin computations.

## EXAMPLE 9.8

A storm causes a peak runoff rate of 5 m³/s in a drainage basin that has a time of concentration of 30 min. A detention basin with 10,000 m³ of volume can be built onsite for a residential land subdivision. Estimate the peak outflow from the basin for this storm.

*Solution*

$$T_c = 30 \text{ min} \times 60 \text{ s/min} = 1800 \text{ s}$$

Applying Equation 9-9 yields

$$\text{total rainfall volume} = 1.5 \times 1800 \text{ s} \times 5 \text{ m}^3/\text{s}$$
$$= 13\,500 \text{ m}^3$$

Now, applying Equation 9-6 gives

$$SF = \frac{10\,000 \text{ m}^3}{13\,500 \text{ m}^3} = 0.74$$

From Figure 9.16 (or Equation 9-8), we get FF = 0.26, and from Equation 9-7

$$\text{peak outflow rate} = 0.26 \times 5 \text{ m}^3/\text{s} = 1.3 \text{ m}^3/\text{s}$$

In summary, it can be expected that the 10 000-m³ basin will reduce the peak runoff discharge from 5 m³/s to about 1.3 m³/s.

## EXAMPLE 9.9

Estimate the storage volume needed in a detention basin to reduce peak inflow rate of 150 cfs to 100 cfs if the total rainfall volume is 300,000 ft³.

*Solution*   Applying Equation 9-7 yields

$$FF = \frac{100}{150} = 0.67$$

From Figure 9.16,

$$SF = 0.33$$

Now, applying Equation 9-6 gives

$$0.33 = \frac{\text{storage volume}}{300,000 \text{ ft}^3}$$

and

$$\text{basin storage volume} = 0.33 \times 300,000 \text{ ft}^3$$
$$= 100,000 \text{ ft}^3$$

## EXAMPLE 9.10

Referring to the data given in Examples 9.4 and 9.6, assume that the local planning board has required that the land developer provide an on-site stormwater storage basin. The peak runoff after development (Example 9.6) is to be no greater than it was before development (Example 9.4). Assume that the detention basin will have an average water depth of 2.0 m when filled to capacity. How much area of the site, in hectares, will have to be used for the detention basin *(a)* for a 5-year storm and *(b)* for a 100-year storm?

*Solution*   *(a) 5-year storm:* From Examples 9.4 and 9.6, the predevelopment discharge of 4400 m³/h is set equal to the basin outflow rate and the postdevelopment discharge of 14 000 m³/h is set equal to the inflow rate. From Equation 9-7, FF = 4400/14,000 = 0.31, and from Equation 9-8, SF = 0.69. The time of concentration $T_c = 15$ min = 0.25 h.
Applying Equation 9-9 yields

$$\text{total rainfall volume} = 1.5 \times 0.25 \text{ h} \times 14,000 \text{ m}^3/\text{h}$$
$$= 5300 \text{ m}^3$$

Now, from Equation 9-6, 0.69 = storage volume/5300, or

$$\text{storage volume} = 0.69 \times 5300 = 3700 \text{ m}^3$$

Since volume = area × depth or area = volume/depth,

$$\text{basin area} = 3700 \text{ m}^3/2.0 \text{ m} \approx 1900 \text{ m}^2 \text{ or } 0.19 \text{ ha}$$

*(b) 100-year storm:* Following the same procedure as in part *(a)*, using the data for the 100-year storm, the

following results are obtained:

$$FF = \frac{9400}{30{,}000} = 0.31$$

and

$$SF = 1.0 - 0.31 = 0.69$$

$$\text{total rainfall volume} = 1.5 \times 0.25 \text{ h} \times 30\,000 \text{ m}^3/\text{h}$$

$$= 11\,250 \text{ m}^3$$

$$\text{storage volume} = 0.69 \times 11\,250 \text{ m}^3$$

$$= 7800 \text{ m}^3$$

$$\text{basin area} = 7800 \text{ m}^3/2.0 \text{ m}$$

$$= 3900 \text{ m}^2 \quad \text{or} \quad 0.39 \text{ ha}$$

This represents only 1.6 percent of the total site area of 0.25 km$^2$.

---

A computer modeling software program called PondPack (Haestad Methods, Inc., Waterbury, Connecticut) can be used to design and analyze stormwater detention ponds. PondPack can create and analyze urban watershed models, compute maximum pond storage requirements, develop volume rating curves for a variety of storm intensities, handle a variety of pond outlet structures including weirs, orifices, culverts, and perforated riser pipes, and perform reservoir routing calculations for simple ponds or a system of complex interconnected ponds.

## Stormwater Quality Control

There are a number of stormwater quality control techniques that are able to reduce stormwater pollutant load impacts in a watershed. An effective general approach is to create a system with components that resemble natural processes that promote infiltration and flow attenuation as well as biological and physical removal of pollutants. By dividing the watershed into a series of small, linked sub-watersheds, it is often possible to apply and successfully utilize BMPs that are otherwise incapable of managing large volumes of runoff. The following is a brief description of typical BMP components that can be linked in series to provide stormwater quality control.

### Sedimentation Basins

Excavated areas that collect and retain stormwater flows for long enough time periods to trap suspended soil particles are called *sedimentation basins*. They are typically used during the construction phase of a project to eliminate off-site transport of eroded soils and sediment. A sedimentation basin can also be designed to function as a permanent integral part of a stormwater management system.

When used as a fore bay (that is, the first structural component in the system), a sedimentation basin will effectively remove suspended soil particles, leaves, litter, road grit, and other trash and gross particulate pollutants from the incoming runoff. The sedimentation basin also serves as an area where the energy associated with the storm surge can be dissipated, thereby reducing the scour potential and erosive force of the incoming runoff. As is the case with standard detention basins, the pollutant removal efficiencies are highly variable. On the average, the removal efficiency for total suspended solids (TSS) is about 70% and for total phosphorus (TP) about 30%.

### Swales

A swale is a shallow depression or ditch constructed to collect and convey runoff from one point to another. Although swales may have a limited capacity to store and treat stormwater runoff, when combined with other structural stormwater measures, they can substantially improve the quality of stormwater. They do so in two ways. First, the vegetation present in the swale reduces runoff velocity. The extent to which this occurs is dependent on the length, depth, and gradient (or slope) of the swale as well as the density of the vegetation. Second, a portion of the runoff discharged to a swale infiltrates into the soil, thus reducing the quantity or volume of the surface runoff. The amount of infiltration depends on soil moisture conditions, the gradient of the swale, and the velocity of the runoff.

Most swales are constructed for the purpose of collecting runoff and directing it to another BMP. Thus, stormwater runoff typically has a very short contact time in a swale (about 5 to 20 min). This limits the amount of treatment that can occur. The best-designed swale, from a stormwater quality enhancement perspective, is wide and shallow with a slope in the range of 2 to 3 percent. Side slopes should be no greater than 3:1 (horizontal-to-vertical). A water-tolerant, erosion-resistant grass should be established as a dense ground cover on the bottom of the swale. Swale grasses should not be mowed close to the ground since this impedes the filtering and hydraulic functions of the swale.

Also, if the swale is adjacent to a roadway, grass species that are relatively tolerant to road salts should be used. Swales should be designed to generate sheet flow across their point of discharge. This can be accomplished by using rip-rap (large stones), a concrete apron, or another, similar device at the end of the swale. Once sheet flow degrades into concentrated flows, erosion channels are formed, thereby defeating the pollutant removal attributes of the swale.

Groundwater infiltration and recharge will also be limited by the design of the swale. Again, wide, nominally sloped swales constructed over soils with high infiltration rates are most capable of infiltrating runoff. Underlying soils should have a percolation rate more than 0.5 in./h. The intensity and magnitude of a storm, or the length of time between storm events, will affect the opportunity for runoff to infiltrate into the soil. The pollutant removal efficiency grassed swales on average is 70% for TSS, 40% for TP, and 25% for total nitrogen (TN). However, the reported range of removal is highly variable.

## Bioretention Systems

Bioretention systems utilize a combination of settling, filtration, and bioaccumulation processes to treat stormwater runoff. They can be designed with or without a preceding sedimentation basin. In addition, bioretention systems can be built on-line or off-line (operate in series or parallel). In an on-line system, runoff is directed into and through the bioretention area, whereas in an off-line system, runoff is diverted from the main collection system into the treatment area. The general concept of any bioretention system is the slowing and detention of runoff for the purpose of facilitating some form of biologically active treatment. (Biological treatment processes are discussed in Section 10.3.)

In some applications, an off-line *riparian buffer* type of bioretention system can be effective. In addition to attenuating peak flow and filtering particulate pollutants, riparian buffer bioretention systems provide habitat for a wide variety of living organisms. Runoff is diverted, retained, and treated for a period of 18 to 36 h in an area of nominal grade at least 3 to 5 m (10 to 15 ft) wide. Plantings within the created riparian corridor can range from grasses to trees. About 70% TSS, 50% TP, and 80% TN can be removed from the stormwater is riparian buffer zones.

## Created-Treatment Wetlands (CTWs)

A CTW is a constructed shallow wetland area designed specifically to detain and treat stormwater runoff (rather than to create a wildlife habitat.) Most CTWs have a broad, gently graded (1 to 2 percent) bottom, and are designed to accommodate and treat the stormwater runoff volume of the 1-year storm, using specially designed outlet control structures or check dams. Most CTWs are planted with a dense assortment of vegetation capable of existing in saturated conditions. Small standing pools of water are often interspersed within the dense vegetation of a CTW.

CTWs provide very high pollutant removal efficiencies, particularly for nutrients and dissolved pollutants. Pollutant removal is achieved as a result of settling, filtering, and biouptake. In general, expected pollutant removal efficiencies for CTWs are 70 to 80 percent for TSS, 40 to 45 percent for TP, and 25 percent for TN.

## Wet Ponds

Wet ponds (or retention basins) provide substantially higher pollutant removal efficiency rates than conventional dry detention basins. In general wet ponds have relatively high TSS and particulate pollutant removal capabilities. As in any standing water body, incoming particulate pollutants (for example, nutrients and heavy metals) will initially be removed as a result of the settling of the heavier, coarse-grain particulate matter (that is, total suspended solids). Studies have shown that the majority of suspended particulate pollutants will settle out during the first 6 to 12 h. Most wet ponds are designed to store water from storm events of a given magnitude (for example, 1-year, 5-year, 10-year storms).

Unlike most conventional structural stormwater measures, wet ponds can also remove significant amounts of dissolved pollutants, especially soluble nutrients, from the water. This occurs due to bacterial, algal, and aquatic plant uptake of dissolved constituents. Biological assimilation of dissolved pollutants and soluble nutrients represents an important removal pathway, since these types of pollutants are not greatly affected by settling processes. Once assimilated, either the nutrients are trapped in biomass and in the sediments, or microbial activities remove them from the system (for example, denitrification). This combination of the settling of TSS and the assimilation of soluble forms of nutrients, such as phosphorus and nitrogen, makes wet ponds a very effective means of reducing the stormwater-related pollutant load. Pollutant removal efficiencies of wet ponds are approximately 70 percent for TSS, 60 percent for TP, and 40 percent for TN.

## Integration of BMPs

The efficiency of an individual BMP can be increased by creating a routing system that integrates a series of hydrologically linked BMPs, thereby creating a "pollutant removal train." A system of linked BMPs can decrease the load of pollutants discharged from a site to levels no greater than that discharged prior to development. This occurs because each of the interlinked BMPs works in unison with the others to remove pollutants by either different processes or as a result of repeated processes. For example, a vegetated swale linked with a wet pond creates an opportunity for the removal of course particulate materials (via filtration) followed by the removal of dissolved nutrients (via biouptake).

Typically, linked BMPs can be used to augment the pollutant removal capabilities of the individual BMPs (as in the above example) or to pretreat or pre-filter stormwater to increase the efficiency of the primary BMP. An example of such a situation would be the installation of a catch basin or sediment chamber prior to a bioretention system, where the sedimentation chamber's primary function would be to remove coarse sediments. This linking of BMPs is a fundamental feature of the stormwater management project described in the following case study.

## CASE STUDY

### The Pennswood Village Stormwater Management System

Pennswood Village is a retirement community located in Middletown, Bucks County, Pennsylvania. As part of the facility's planned expansion, Pennswood Village entered into a partnership with the Township, which in part involved correction of existing runoff and stormwater quality problems associated with lands both on and off the Pennswood site. The resulting stormwater quality management system was designed to mimic the functional properties of a riparian corridor floodplain. The Pennswood Village design integrated a number of different BMPs, including a sedimentation basin, a vegetated swale, an infiltration basin, a created treatment wetland, and a small wet pond. These BMPs work in series to attenuate peak flows, promote groundwater recharge, and passively remove pollutants through a combination of filtering, settling, and biological treatment mechanisms. The final design exceeded the Township's stormwater management requirements while providing an attractive, passive recreation and learning environment for Pennswood Village and the community at large.

The functional design of the Pennswood Village stormwater management system mimics that of a natural riparian, stream corridor channel (Figure 9.18). The system consists of an integrated series of BMPs, each sized and located to address a specific stormwater management purpose. The alignment and grading of the swales, basins, and wetlands combined with the careful selection of native grasses, shrubs, and trees diminishes the velocity of the runoff, biofilters and settles the pollutants, and creates opportunities for groundwater recharge. The five major elements of the system and their functional attributes are as follows.

At the uppermost end of the system, where runoff is directed by a series of pipes from Route 413 and the contributing watershed, is a sedimentation basin. The sedimentation basin is a stone-lined structure. Its purpose is to slow the storm surge, settle out gross particulate material, and contain the majority of the trash and road debris conveyed along with the runoff. Upon entering the basin, the velocity of the concentrated runoff is reduced by a flow dissipation structure. A weir regulates the volume and time that the collected runoff is detained in the sedimentation basin. The basin is easily accessed for routine maintenance and the periodic removal of accumulated sediments. It is also screened from direct view by a landscape buffer.

Runoff discharged from the sedimentation basin is directed into a grassed swale that conveys the runoff to an infiltration basin. The soils that predominate in this section of the site are highly permeable. The depth to the seasonal high water table is in excess of 2 m (6 ft), as is the depth to bedrock. These conditions of good soil permeability and lack of a constraining horizon are conducive to the infiltration of runoff and the recharge of the shallow aquifer. The infiltration basin is sized to manage the first flush runoff volume of a storm event.

Flows exceeding the infiltration capacity of this basin will be discharged from the basin over a broad crested weir into a long, winding vegetated swale. On either side of the swale is a broad, flat meadow graded and designed to function in a manner similar to a riparian corridor or steam floodplain. That is, it consists of a series of shallow, stepped channels, each of which accommodates and detains the runoff from increasingly larger storm events. Depending on its proximity to the grassed swale and the resulting postconstruction topography, the meadow supports a variety of vegetation, including grasses, shrubs, or trees having different flood and drought tolerances.

At the terminus of the system is a created treatment wetland (CTW) and small wet pond. Outflow from the wet pond is controlled by an outlet structure designed to safely pass the 100-year storm. Initially, during the early part of a storm, runoff that exceeds the capacity of the infiltration basin will flow via the swale into the wet pond. The outlet control structure on the pond will cause water to flood back into the CTW. As water is detained in the CTW and wet pond, it will back up further, eventually overflowing into the broad meadow and created riparian corridor.

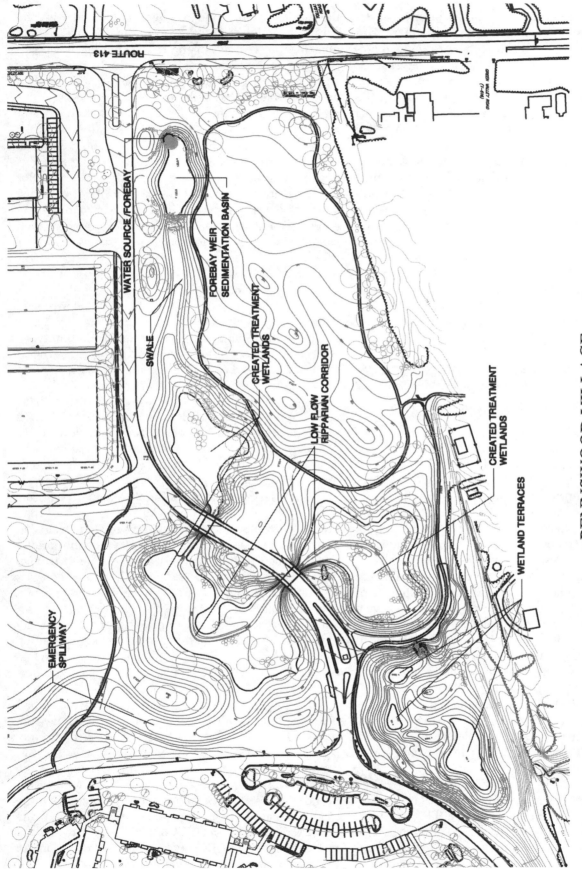

ROUTE 413

WATER SOURCE /FOREBAY

FOREBAY WEIR

SEDIMENTATION BASIN

CREATED TREATMENT
WETLANDS

SWALE

LOW FLOW
RIPPARIAN CORRIDOR

EMERGENCY
SPILLWAY

CREATED TREATMENT
WETLANDS

WETLAND TERRACES

## PENNSWOOD VILLAGE

Phase I- Storm Water Management System

**FIGURE 9.18**

*A plan view of the Pennswood Village stormwater system.* (Courtesy of Princeton Hydro, LCC, and Wells Appel Land Strategies; Lambertville, New Jersey. Used with permission)

The result is a highly functional stormwater management system that exceeded the Township's stormwater management requirements while providing an attractive environment for Pennswood Village and the community at large. As designed, this system will attenuate peak flows, promote groundwater recharge, and passively remove pollutants through a combination of filtering, settling, and biological treatment processes.

*Source*: Adapted from Stephen J. Souza, Mark Gallagher, and Stuart Appel, "The Pennswood Village Stormwater System: The Design and Construction of a Multi-functional Riparian Corridor for the Management of Stormwater Quality," Princeton Hydro, LLC, and Wells Appel Land Strategies, Lambertville, New Jersey. Used with permission.

## 9.4 FLOODPLAINS

Flooding is a natural event that occurs periodically when the water in a stream or river overflows its channel banks and inundates adjacent low-lying land. This land is called the *floodplain*. The portion of the floodplain that is inundated by the 100-year flood is usually called the *flood hazard area*, as illustrated in Figure 9.19. The flood hazard area includes the *floodway*, which carries the major portion of the flood at high velocities, and the *flood fringe*, which is covered with shallower, slowly moving water.

It is important to understand that the 100-year flood is a statistical value; it can occur 2 d in a row, several times in a single year, several times in a century, or not at all in a period longer than 100 years. Larger databases give more accurate values.

Floods are particularly damaging to houses and other structures in the floodway. In addition to economic loss totaling billions of dollars annually in the United States, many lives are lost in severe floods.

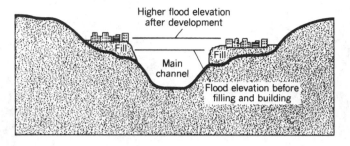

**FIGURE 9.20**
*Urban development in a floodplain causes an increase in flood elevations due to the constricted flow channel.*

Furthermore, significant pollution problems, including contamination of drinking water supplies, occur during flood conditions. Flood damages are the result of poor land use and environmental planning, and they can be avoided with proper floodplain management practices.

Improper floodplain management results in the condition illustrated in Figure 9.20. As the floodway is filled in and built upon, the flow path is restricted. This causes an increase in the flood elevation, making the problem even more severe. The basic objective of regulating land use in the floodplain is to reduce the risk of future flood damages. Floodplain regulation is generally a responsibility of local government, subject to state guidelines. Control is also exercised by the federal government under the requirements of the National Flood Insurance Program (NFIP).

Many states have adopted regulations that restrict or prohibit certain types of construction or activities in the floodplain. These include the following:

1. New buildings
2. Sanitary landfills
3. On-site sewage disposal systems

Recreational facilities such as playgrounds and picnic areas and farming activities are still permitted because they are unlikely to cause obstructions and increase the potential for flooding or environmental damage. Buildings and sanitary facilities constructed in the flood fringe must be floodproofed to a height above the *base flood level* (which has a 100-year recurrence interval) to comply with the requirements of the NFIP. This prevents damage to the structure and contents in the floodplain and reduces pollution problems.

Floodplain management must be preceded by a study that shows the extent of the floodplain on a suitable map. The easiest way to delineate the floodplain is to examine the area topography on available U.S. Geological Survey (USGS) maps. Also, an examination

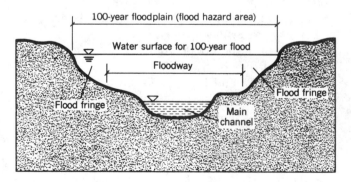

**FIGURE 9.19**
*The 100-year floodplain or* flood hazard area *includes the* floodway *and the* flood fringe *areas.*

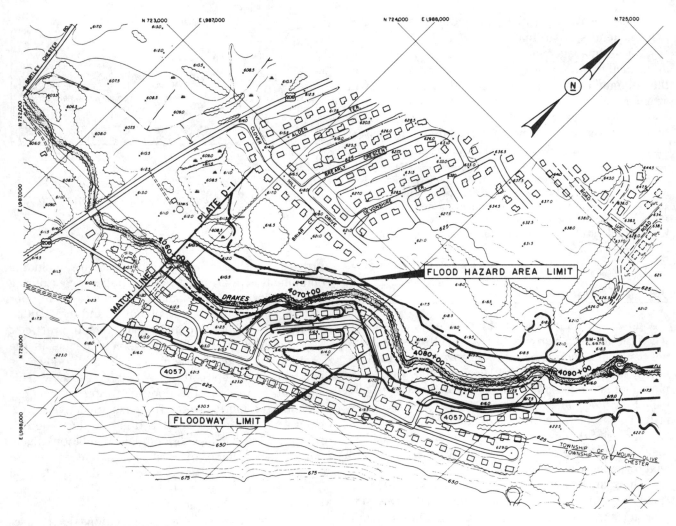

**FIGURE 9.21**
*A typical flood hazard area map, which shows a plan view of the floodway and flood hazard area.* (Courtesy of Division of Water Resources, New Jersey Department of Environmental Protection)

of NRCS soil-type data can provide useful information; floodplain areas are usually associated with deposits of silty soils laid down by the stream over the years.

The most accurate method of delineating the floodplain is to obtain detailed topographic information from field surveys and conduct hydrologic and hydraulic analyses of the flood surface profile. A typical flood hazard area map is shown in Figure 9.21. Hydraulic analyses are typically made using the U.S. Army Corps of Engineers HEC-2 or HEC-RAS computer programs.

Structural methods for flood protection include the use of dams and reservoirs, levees or dikes, and channelization. Reservoirs store the excess runoff in upstream areas. Levees and dikes serve to confine floodwaters to a specific channel or flood zone, and channelization increases existing capacity to carry flow. But all structural methods provide only limited protection up to the storm recurrence interval for which they are designed, and they often spur additional development of the floodplain because of the sense of security that they provide.

Reservoirs and levees take up large land areas, and the expense of construction is usually beyond the means of local municipalities. Overall, the preferred method for preventing flood damage is regulation of land use in the floodplain, rather than use of structural methods. Local zoning and subdivision ordinances can be effective in this regard. The NFIP provides incentive for proper regulation; federal subsidized flood insurance cannot be purchased unless the municipality participates in the NFIP.

## 9.5 CONTROL OF COMBINED SEWER OVERFLOW (CSO)

About 1100 older urban communities in the United States are served by *combined sewer systems* (CSSs) instead of separate sewers; most CSSs were built in the densely populated cities of the Northeast and Great Lakes regions. They were designed to convey sanitary sewage (domestic, commercial, and industrial wastewater) as well as stormwater through a single-pipeline system to a sewage treatment plant. As explained in the beginning of Chapter 8, CSSs are no longer built in the United States, but there are persistent environmental problems that require solutions in many areas due to the CSSs that remain in use.

During dry-weather conditions the sanitary sewage carried in a CSS receives full treatment at the sewage treatment facility, but during periods of rainfall (or snowmelt), when the pipeline also carries stormwater, the capacities of the pipe and treatment plant may be exceeded. Due to excessive flows, *combined sewer overflows* (CSOs) often occur at one or more "relief points" in the pipe system. Relief points are necessary to prevent basement flooding and overloading of the wastewater treatment facilities. Unless modern engineering controls are provided in the CSS and at the relief points, sanitary sewage and stormwater mixtures in CSOs are discharged directly into nearby rivers, lakes, or estuaries without treatment.

CSO contains debris, suspended solids, pathogens, toxic pollutants, floatable solids, nutrients, organic compounds, oil and grease, and other contaminants. Water pollution caused by intermittent CSOs can exceed pollution from treated sewage effluent. Most of the public beach closings imposed after periods of wet weather are due to CSO pollution and its potential for harming the public health. About 16 percent of the shellfish harvest limitations in the Northeast and the Middle Atlantic states and about half of the estuary contamination nationwide are also attributable to CSOs.

In 1994, the EPA established a policy for the control of CSO pollution based on the *National Pollutant Discharge Elimination System* (NPDES) permitting program (see Section 10.1), since each CSO is a point source. This policy recognizes the site-specific nature of CSO problems and provides the necessary flexibility to tailor controls to local situations. It encourages the evaluation of water pollution control needs on a watershed management basis and the coordination of CSO control efforts with other point and nonpoint control activities. Communities served by CSSs must develop and implement long-term plans for CSO control that will attain compliance with water quality standards.

## CSO Control Technology

Complete separation of stormwater and municipal wastewater in older cities is the best long-term solution to the CSO problem, but it can be prohibitively expensive. It requires the construction of either a new sanitary sewer system or a new storm sewer system. Sometimes separation is feasible in the older areas of small towns, as described in the following case study. For most large communities, however, full separation is economically and politically impractical. Other engineering methods must be applied to mitigate the CSO problem.

### CASE STUDY

### Lincoln Sewer Separation Project

The use of a computer hydraulic model saved Lincoln, Illinois, approximately ten percent of the cost of a new storm sewer by allowing designers to optimize the alignment and incorporate existing storm sewer components into the new design. Designers ran multiple hydraulic analyses, evaluating different sewer pipe locations to minimize material and construction expenses. In addition to saving money, the results of the hydraulic analysis provided information that the City Council needed to approve the project.

### The Problem

Thirteen blocks on the west side of Lincoln, a fully developed area in the older part of town, were previously served by a combined sanitary/storm sewer system. During heavy storms, the system's capacity was often overloaded, causing flooding on the streets and backing up into basements.

A study of Lincoln's sewer system conducted in the early 1990s identified this area of the city as problematic and proposed the separation of the sanitary and storm sewer systems as a solution. This meant disconnecting the stormwater inlets from the existing sanitary sewer system and diverting them into a new storm sewer system the city would construct in the same general 13-block area. The size and complexity of the project made it one of the largest sewer construction projects that Lincoln had undertaken in many years and some members of the City Council were reluctant to approve it.

## The Solution

The city hired a firm of consulting engineers to study the situation and design a new storm sewer system. The city provided the engineers with an AutoCAD plot of the existing infrastructure, along with additional topographic data to confirm the invert elevations of the existing pipes. Although the rational method had been used by the engineering firm that performed the earlier study of the system, the city had specified the use of Technical Release 55 (TR-55) methods for this project. First issued by the Soil Conservation Service in 1975, TR-55 incorporates current SCS procedures to calculate runoff volume, peak rate of discharge, hydrographs, and storage volumes required for floodwater retention basins.

The consulting engineers used a computer program called HYDRA to perform the TR-55 analysis and complex routing methods. A digital model of the existing storm sewer system was created in HYDRA by importing the city's AutoCAD file. The model was then loaded with rainfall amounts generated by the 10-year, 24-h storm as determined by the U.S. Geological Survey. The software determined the flows from each subbasin and graphically portrayed the movement of water through the existing system using hydrographs. The hydrographs clearly showed the points where the existing system was overburdened.

Design of the new storm sewer system was complicated by the existence of gas mains, water mains, fiber optic lines, and other underground infrastructure. After creating a preliminary design on paper to get a good idea of the general horizontal configuration, the engineers created a digital model of the new system using HYDRA. The model was loaded with the same 10-year, 24-h storm amounts used previously. The software then calculated the inverts and pipe sizes that would handle this storm based on the given horizontal layout. The results gave all of the parameters for a new storm sewer system that would handle the required rainfall amounts in conjunction with the existing system. Since it was easy to modify the computer model and repeat the analysis, several possible routes were evaluated to see which would cost less for materials and construction. As a check on the results, additional analyses were also conducted using the rational method, providing a conservative check on the final design. This provided the City Council with the quantitative information needed to approve the project, which was completed in 1999, and which, by all indications, is working well.

*Source*: Adapted from "Hydraulic Model Lowers Cost of Storm Sewer Replacement," *Pollution Equipment News*, October 2001. Used with permission. Information on HYDRA can be obtained from Pizer, Inc., 4422 Meridian Avenue North, Seattle, Washington 98103.

Effective CSO control technologies in wide use include:

- In-system controls/In-line storage
- Off-line near-surface storage/sedimentation
- Deep tunnel storage
- Coarse screening
- Swirl/vortex technologies
- Disinfection

These techniques are almost never sufficient separately to meet the needs of a comprehensive CSO control program; two or more are generally combined to produce the required results. Coarse screening, for instance, is usually used to provide pretreatment for any of the other technologies. Where high bacterial levels are the cause of water quality impairment, disinfection may be applied to the CSO along with one or more of the other methods.

A factor that complicates many CSO control efforts is the existence of multiple relief points or overflow locations in any particular CSS; these may often discharge to different receiving bodies of water. They may all be consolidated or combined for treatment, or controlled separately. Sometimes it is best to completely eliminate a particular discharge point at an environmentally sensitive location by rerouting the flows in a modified piping system. Often, a variety of CSO control alternatives are available for a given community.

Each CSS is unique and most are very complex. Determination of an appropriate design approach requires detailed analysis of rainfall records, drainage areas, and piping systems. This usually requires application of computer simulation programs such as EPA's Storm Water Management Model (SWMM). Pollutant levels in CSOs are site-specific and vary widely. They are influenced by the strength of the sanitary sewage flow, the age and condition of the sewer system, and the amount of infiltration/inflow. The highly variable nature of CSOs in quantity and quality and the differences between sites make CSO control a technically challenging design problem.

### In-System Controls/In-Line Storage

This is one of the least costly CSO control methods, but it is not normally sufficient by itself to provide the complete degree of control required. Its objective is to optimize the use of available storage capacity in the existing pipe system and to maximize the conveyance of combined flows to the treatment plant. Since this technique reduces the required design capacity (and cost) of other control methods that must also be applied, it is typically an essential component of all control plans.

There are many "in-system" strategies that can help to optimize in-line storage and maximize the flows to a treatment facility. To mention only a few, these include tidegate repair, the removal of flow obstructions (such as sediment), enlargement of undersized CSS pipe sections to eliminate flow restrictions, and adjustment of *regulator* settings.

Regulators are designed to control the amount of sewage that enters an interceptor from an upstream section of the system and to provide an overflow relief point for flows that exceed the interceptor capacity. Two general categories of regulators are static and mechanical. Static regulators have no moving parts and are not easily adjustable, but they require little maintenance. Mechanical regulators have moving parts, but they are readily adjustable and may be controlled by remote telemetry or designed to respond to variations in local flow conditions. Motor-operated sluice gates, for example, respond to water-level-sensing devices. Controls can be set to fully open or close the gate, or anything in between, as required by flow conditions.

An example of a static type regulator is the *vortex valve*, which allows dry-weather flows to pass without restriction, but controls higher flows by a vortex throttling action (Figure 9.22). Vortex valves can be used to divert flows to CSO treatment facilities and to control flow out of storage facilities. They are typically made of stainless steel and are housed in reinforced concrete structures.

### Off-Line Near-Surface Storage/Sedimentation

This type of CSO control facility comprises tanks that store and treat flow diverted from the combined trunk sewers and interceptors. Coarse screening, sedimentation, and disinfection are commonly provided (these treatment processes are described in more detail in Sections 6.2, 6.5, and 10.2). The tanks are built near the ground surface using open-cut excavation methods, rather than being put in tunnels deep underground. They can be designed to capture the more concentrated flows that occur during the initial stages of a storm and to reduce the number of overflow events, as well as to provide a level of treatment that meets receiving water quality standards for storms of a specified recurrence interval.

The contents of storage/sedimentation tanks are normally returned to the CSS (by gravity, if feasible, or by pumping) and flow to the treatment plant at a rate the plant can accommodate. The tanks may be aerated or mechanically agitated to resuspend settled solids before pump-back to the sewage treatment plant. Some facilities have a separate means to handle grit and heavy solids, such as access for removal by a vacuum truck or a separate set of pumps for solids removal.

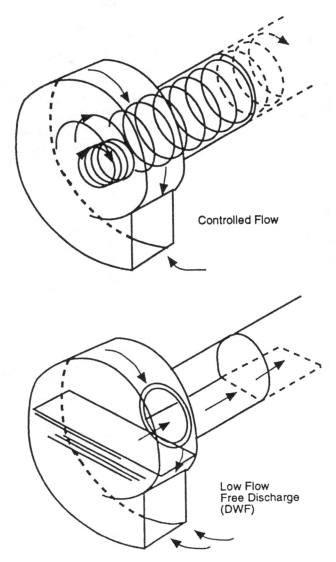

**FIGURE 9.22**
*Example of a vortex valve.* (Courtesy of HRD, Ltd., Clevedon, United Kingdom)

### Deep Tunnel Storage

In congested urban areas, where space constraints or possible construction impacts make near-surface facilities unfeasible, deep tunnel storage can be a viable alternative. Relatively large volumes of combined sewage can be conveyed and stored in deep tunnels, with little disturbance to ground surface features caused by their construction or operation. The feasibility of deep tunneling for CSO control depends on subsurface geological factors to a large degree. The total storage volume needed to accomplish a specified CSO control goal is usually calculated using a computer flow-routing model (such as SWMM).

In addition to the tunnel itself, a typical system includes regulators to divert and control storm flows to the tunnel, coarse screens, vertical dropshafts and air separation chambers, access and vent shafts, and a dewatering system to pump the stored sewage to a treatment plant after the storm event subsides. An example of a deep tunnel project is the Tunnel and Reservoir Project (TARP), which serves Chicago and many nearby communities that have combined sewers. Thirty-one miles of tunnel was put into service in 1985. The "Mainstream tunnel" is 10 m (33 ft) in diameter, bored between 70 and 90 m (about 240 to 300 ft) below the ground, and holds one billion gallons of water. TARP is an ongoing project; more than 120 of the 210 km (75 of the 131 mi) of planned tunnels has been completed and placed into service. CSO is pumped out and conveyed to sewage treatment plants. Completion of the TARP, along with sewer upgrades in the local towns, will eliminate pollution and flooding problems in the Cook County CSS area.

### Swirl/Vortex Technology

Swirl concentrators and vortex separators provide flow regulation and some degree of solids and floatable material removal in combined flows. Typically, sewage entering the device is directed around the perimeter of a cylindrical shell, creating a swirling, vortex flow pattern. The swirling action throttles the flow and causes solids to concentrate at the bottom. Clarified super-

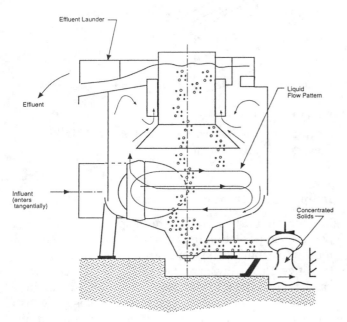

### FIGURE 9.23
*Example of a Fluidsep vortex separator.* (Courtesy of William C. Pisano, "Swirl Concentrators Revisited," American Experience and New German Technology Engineering Foundation Conference, Trout Brook, Missouri, 1988)

natant flows out through the top of the unit, as illustrated in Figure 9.23. The concentrated solids in the underflow are discharged to a treatment plant; floatable material, captured by baffles, is carried out in the underflow when the storm flow subsides and the unit drains. Since swirl/vortex units have no moving parts, operation is governed solely by hydraulic conditions. The devices are used in conjunction with upstream regulators, bar screens, and disinfection systems.

### Other Technologies

Disinfection is a common goal of CSO control strategies. Although liquid sodium hypochlorite is most commonly used, alternatives such as UV light, ozone, or gaseous chlorine are available. (Disinfection methods are discussed in more detail in Section 6.5; coarse screens are discussed in Section 10.2.)

## CASE STUDY

### Tunnel System to Keep Portland's Waterways Clean

Portland, Oregon, is one of hundreds of cities with a combined sewer system that overflows when it rains. Downtown Portland's 32 combined sewer overflow (CSO) outfalls pollute the Willamette River because they are typically 20 percent sewage, according to the city.

To reduce the CSO levels by 550 mil gal (2 million m³) per year (about 94 percent of the total annual CSO discharges) the city is spending $407 million over the next 20 years to control and treat 3 billion gal (11 mil m³) of CSOs that would otherwise affect the Columbia Slough and the Willamette River. An important part of this huge water management project is the 3.7-mi (6.0-km) CSO tunnel for the west side of the Willamette River (Figure 9.24).

The system will allow no more than four overflows a year, compared with a current overflow every time it rains. All told, the west side of the Willamette River will have 10 mi (16 km) of new pipes and tunnels, along with two new pump stations to collect and convey the CSOs to the existing Portsmouth Tunnel and on to a wastewater treatment plant.

This project is one of the largest contracts ever let by the city, with construction for the 14-ft (4.3-m)–diameter west-side CSO tunnel alone costing $84 millions. This tunnel will run through the downtown central business area and a large waterfront park and will have to avoid the footings and abutments of seven bridges.

The tunnel, a central element of the west-side project, will act as a conveyance and storage tunnel. At present it is being designed for a 3-year storm.

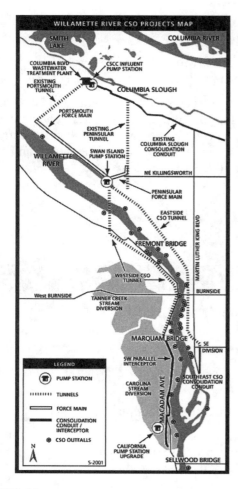

**FIGURE 9.24**

*New combined sewer tunnels and pump stations being constructed along both sides of the Willamette River in Portland, Oregon, are designed to reduce discharges to the river by 550 million gal (2 million m³) a year.* (From *Civil Engineering Magazine*, September 2000. Copyright © 2000, Environmental Services, City of Portland, Oregon. Used with permission of the American Society of Civil Engineers; copyright Environmental Services, City of Portland, Oregon.)

The tunnel will lie 50 to 70 ft (15 to 21 m) below the surface. A one-pass tunneling system will be used because of the high groundwater level. The design was completed in December 2001 and the tunnel should be operational by December 2006. But this is only the first phase of the project. Projects on the east side of the river will have components in place to contain CSOs by 2011.

# 9.6 RELEVANT WEB SITES

## CHICAGO TUNNEL AND RESERVOIR PROJECT (TARP)

http://www.mwrdgc.dst.il.us/plants/tarp.htm

This site provides a description of the vast TARP system for controlling CSO in the Chicago and Cook County, Illinois, area.

## COMBINED SEWERS—A CASE STUDY

http://www.gvrd.bc.ca/services/sewers/collect/CombinedSewers.html

This site presents a discussion of combined sewer systems and how CSO is controlled in the Greater Vancouver Regional District.

## COMBINED SEWER OVERFLOW CONTROL PLAN FOR THE MWRA

http://www.mwra.com/sewer/CSO_Control_Plan.htm

This site provides a report on the CSO control plan for the Massachusetts Water Resource Authority (MWRA), serving the greater Boston, Massachusetts, area.

## CONTROLLING NONPOINT SOURCE POLLUTION

http://www.epa.gov/OWOW/NPS/roads.html

This site presents an EPA report on the best management practices for controlling pollution from roads, highways, and bridges.

## HYDROCAD SOFTWARE SAMPLER

http://www.hydrocad.net

HydroCAD is a computer-aided design system for modeling the hydrology and hydraulics of stormwater runoff. At this site, a free, functional software sampler can be downloaded for trying the program. A tutorial example provides an overview of the applications and capabilities of the software. Rational method or SCS methods can be used to analyze runoff flow rates, inflow and outflow hydrographs, flood routing, and detention ponds.

## SEDIMENT CONTROL AND STORMWATER MANAGEMENT

http://www.dnrec.state.de.us/newpages/stormregs.htm

This site describes Delaware's stormwater management program, which includes the following components: sediment control during construction and postconstruction, stormwater quantity, and water quality control.

## MANAGING URBAN RUNOFF

http://www.epa.gov/OWOW/NPS/facts/point7.htm

This site presents an EPA fact sheet on managing urban runoff, with links to other sources of information.

## URBAN STORMWATER— BEST MANAGEMENT PRACTICES

http://h2osparc.wq.ncsu.edu/descprob/ urbstorm.html#intro

This site discusses "best management practices" (BMPs) for controlling or preventing contamination of stormwater runoff, including the use of preventive measures (largely nonstructural practices) and control measures (structural practices).

## REVIEW QUESTIONS

1. Why is stormwater control an important aspect of environmental technology?

2. Briefly compare the present-day approach toward stormwater control with past practices.

3. Define *time of concentration* of a drainage basin. How is it determined, and what is it used for?

4. What is a *runoff curve number?* What is it used for?

5. What does *csm/in.* stand for?

6. Give two direct effects of urbanization on stormwater runoff. What is the overall effect on a short-term basis and in the long run?

7. Briefly describe three basic differences between storm sewer systems and sanitary sewer systems.

8. List and briefly describe three types of stormwater inlets.

9. Briefly discuss the design procedure for storm sewers.

10. List five specific benefits of on-site stormwater storage.

11. What is a disadvantage of stormwater storage basins?

12. Briefly describe three different types of stormwater storage basins.

13. Briefly discuss the design procedure for a stormwater detention basin.

14. Sketch and label a typical inflow and outflow hydrograph for a stormwater detention basin.

15. What is a *floodplain?* What is a *flood hazard area?*

16. Briefly discuss floodplain management measures.

17. Briefly describe the *CSO* problem in the United States. Name two important CSO control methods and briefly describe them.

18. Visit the *Hydrologic Engineering Center* Web page. Link to the "Mississippi Basin Modeling System Development and Application" report. Read and summarize the Overview section of this report. Peruse the rest of the report. (Though the technical details of this flood-modeling study are beyond the scope of an introductory course, they will give you some insight regarding the sophistication of hydrologic modeling methods and applications.)

19. Visit the *Urban Stormwater—Best Management Practices* Web page and link to the "Control Measures" section. In a brief written report, summarize both the advantages and disadvantages of dry detention basins and wet detention basins.

20. Using the EnviroSources search engine (see Section 1.6) or another Internet search directory or search engine, locate at least one additional Web site relevant to one of the topics in this chapter. Add the link(s) to your Environmental Technology Folder. Write a brief description of what the Web site(s) contain.

# PRACTICE PROBLEMS

1. Estimate the peak rate of runoff on a 15-ac watershed from a storm with rainfall intensity of 2 in./h. Use a runoff coefficient of 0.6.

2. Estimate the peak rate of runoff from a $0.15\text{-km}^2$ drainage basin that has a runoff coefficient of 0.3 if the rainfall intensity is 60 mm/h.

3. From an aerial photo of a 22-ac watershed, it is determined that 4.5 ac is flat grassland, 6.0 ac is lightly wooded, 8.5 ac is a suburban residential area with large lots, and 3.0 ac is impervious pavement and roofs. Compute a composite runoff coefficient for the total watershed area.

4. A drainage basin of 180 ac comprises 40 percent wooded areas, 45 percent grassed areas, and 15 percent paved areas. Estimate a conservative value of the composite runoff coefficient for this drainage basin.

5. A 250-ha watershed comprises 80 ha of woods, 120 ha of suburban residential area, 30 ha of business area, and 20 ha of industrial area. Determine a probable range of values for the composite runoff coefficient in this watershed.

6. A 15-ac catchment area has a composite runoff coefficient of 0.5 and a time of concentration of 20 min. Compute the 25-year peak discharge.

7. A 10-ha drainage basin has a composite runoff coefficient of 0.7 and a time of concentration of 30 min. Compute the peak rate of runoff from a 100-year storm.

8. A $0.45\text{-km}^2$ watershed has a composite runoff coefficient of 0.33. The overland flow distance to the beginning of a stream that drains the watershed is 200 m, and the ground slope is 5 percent. The stream is 750 m long, with an average flow velocity of 0.25 m/s. Estimate the peak rate of runoff for the 100-year storm.

9. The watershed described in Problem 8 is to be developed as an industrial area, with a runoff coefficient of 0.9. Part of the stream is channelized, increasing the average flow velocity to 0.4 m/s. Assume that the overland flow time remains the same. Determine the 100-year peak discharge from the watershed under developed conditions.

10. An 800-ac forested watershed with type C soils has been zoned for the following land uses: 300 ac residential, 200 ac business and commercial, and the remaining 300 ac left as undeveloped forest. Assume that the time of concentration for the watershed decreases from 60 min to 45 min, after development takes place. Using the SCS method, estimate the peak runoff discharge, both before and after development, for a 24-h storm that causes 7 in. of rainfall.

11. A storm drain system is laid out similar to that shown in Figure 9.12, with the following characteristics:

catchment area 1:   $A_1 = 3$ ac, $C_1 = 0.5$, inlet time = 5 min

catchment area 2:   $A_2 = 4.5$ ac, $C_2 = 0.4$, inlet time = 8 min

catchment area 3:   $A_3 = 6$ ac, $C_3 = 0.3$, inlet time = 6 min

reach 1 (inlet 1 to inlet 2): $L = 300$ ft, $S = 0.4$ percent

reach 2 (inlet 2 to inlet 3): $L = 350$ ft, $S = 0.3$ percent

reach 3 (inlet 3 to inlet 4): $S = 0.3$ percent

Using the rational method, compute the design flow and required pipe diameter for each reach of the system for a 25-year storm.

12. A storm drain system is laid out similar to that shown in Figure 9.12, with the following characteristics:

catchment area 1: $A_1 = 1.2$ ha, $C_1 = 0.6$, inlet time = 5 min

catchment area 2: $A_2 = 2.0$ ha, $C_2 = 0.35$, inlet time = 6 min

catchment area 3: $A_3 = 3.0$ ha, $C_3 = 0.2$, inlet time = 7 min

reach 1 (inlet 1 to inlet 2): $L = 100$ m, $S = 0.0015$

reach 2 (inlet 2 to inlet 3): $L = 120$ m, $S = 0.003$

reach 3 (inlet 3 to inlet 4): $S = 0.010$

Using the rational method, compute the design flow and required pipe diameter for each reach of the system for a 5-year storm.

13. A storm causes a peak runoff rate of 8 $\text{m}^3$/s in a drainage basin that has a time of concentration of 20 min. A detention basin with 12 000 $\text{m}^3$ of

storage volume is built on site. Estimate the peak outflow rate from the basin for this storm, using the simplified storage factor method.

14. Estimate the storage volume needed in a detention basin to reduce a peak inflow rate of 10 m³/s to 2 m³/s when the total rainfall volume is 15 000 m³.

15. In a land development project, the peak flow before construction is 40 cfs. The peak flow after construction is estimated to be 120 cfs. The local planning board requires that the peak runoff leaving the site after development be no greater than the 40-cfs predevelopment rate and that on-site storage be provided to accomplish this. If the detention basin will have an average depth of 4 ft at full volume, approximately how many acres of the site will have to be used for the basin? Assume that the time of concentration is 30 min.

## Chapter Outline

CHAPTER

# 10

# Wastewater Treatment and Disposal

R aw or untreated sewage is mostly pure water. In fact, sanitary wastewater comprises about 99.9 percent water and only about 0.1 percent impurities. In other words, if a 1-L (1-kg) sample of wastewater is allowed to evaporate, only about 1 g, or 1000 mg, of solids will remain behind.

In contrast to this, seawater is only about 96.5 percent pure water; it contains about 35,000 mg/L, or 3.5 percent, dissolved impurities. Although seawater con-

tains more impurities than does sanitary sewage, we do not ordinarily consider seawater to be polluted. The important distinction is not the total concentration, but the type of impurities. The impurities in seawater are mostly inorganic salts, but sewage contains biodegradable organic material, and it is very likely to contain pathogenic microorganisms as well.

Actually, sewage can contain so many different substances, both suspended and dissolved, that it is impractical to attempt to identify each specific substance or microorganism. The total amount of organic materials is related to the *strength* of the sewage. This is measured by the *biochemical oxygen demand*, or BOD. Another important measure or parameter related to the strength of the sewage is the total amount of suspended solids, or TSS. On the average, untreated domestic sanitary sewage has a BOD of about 200 mg/L and a TSS of about 240 mg/L. Industrial wastewater may have BOD and TSS values much higher than those for sanitary sewage; its composition is source dependent.

Another group of impurities that is typically of major significance in wastewater is the plant nutrients. Specifically, these are compounds of nitrogen, N, and phosphorus, P. On the average, raw sanitary sewage contains about 35 mg/L of N and 10 mg/L of P.

Finally, the amount of pathogens in the wastewater is expected to be proportional to the concentration of fecal coliform bacteria. The coliform concentration in raw sanitary sewage is roughly 1 billion per liter. Coliform concentration, as well as BOD, TSS, and concentrations of N and P, are parameters of water quality that are discussed in some detail in Chapter 4.

Before discharging wastewater back into the environment and the natural hydrologic cycle, it is necessary to provide some degree of treatment in order to protect public health and environmental quality. The basic purposes of sewage treatment are to destroy pathogenic microorganisms and to remove most suspended and dissolved biodegradable organic materials. Sometimes it is also necessary to remove the plant nutrients—nitrogen and phosphorus. Disinfection, usually with chlorine, serves to destroy most pathogens and helps to prevent the transmission of communicable disease. The removal of organics (BOD) and nutrients helps to protect the quality of aquatic ecosystems.

This chapter describes the most common types of wastewater treatment systems. These treatment methods are grouped into three general categories: *primary treatment*, *secondary* or *biological treatment*, and *tertiary* or *advanced treatment*. First, the topic of sewage effluent standards is discussed to put in perspective the overall goals of the various treatment processes. A later section covers on-site subsurface sewage disposal, and the chapter concludes with the topic of sewage sludge management.

A flow chart showing the typical sequence of wastewater treatment processes is given in Figure 10.1. (Aerial photographs of modern wastewater treatment plants can be seen in Appendix H, Figures 1, 2, and 3.)

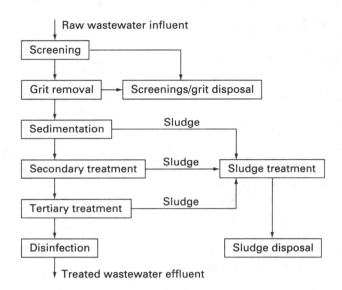

**FIGURE 10.1**
*Schematic overview of a wastewater treatment system. Screening, grit removal, and sedimentation (settling) are primary treatment processes. Secondary treatment usually involves biological processes and additional settling. Not all sewage treatment plants require tertiary (advanced) treatment.*

## 10.1 LEGISLATION AND STANDARDS

In the United States, one early effort at the federal level to guide the nation's clean water strategy was the *Water Quality Act* of 1965 (see Section 5.8). This act was strengthened by the *Federal Water Pollution Control Act Amendments* of 1972. A basic goal of the amendments was to encourage individual states to clean up surface waters to the extent that they would once again be "swimmable and fishable." Reduction of water pollution from point sources was required, funds were made available for construction of sewage treatment plants, and an EPA enforcement program was created.

The strategy of reducing pollution from point sources is implemented by the *National Pollution Discharge Elimination System* (NPDES). Under the NPDES, all municipal or industrial treatment facilities that discharge wastewater effluents must obtain an NPDES Discharge Permit from the EPA or a delegated state agency. The NPDES permits clearly state the allowable amounts of specific pollutants that a particular facility can discharge into the environment.

The *Federal Water Pollution Control Act* was amended again in 1977, at which time the treatment of toxic pollutants was required to include the *best available technology* (BAT) that was economically achievable.

For conventional pollutants (including BOD, suspended solids, fecal coliforms, and pH), the EPA established guidelines reflecting the *best conventional technology* (BCT), which took different cost considerations into account. Since the 1977 amendments, which redefined the focus of EPA priorities and strengthened its enforcement powers, the *Federal Water Pollution Control Act* has been referred to as the *Clean Water Act* (CWA).

A new CWA was passed in 1986; further revision and amendment are likely in the future. The 1986 CWA continued many provisions of the prior laws concerned with water pollution control. It included requirements for joint state–federal programs to clean up streams and rivers still polluted with toxic substances, even after application of BAT to remove them; limits on the ability of polluters to obtain variances from cleanup projects were also established.

### Effluent Standards and NPDES Requirements

Implementation of the NPDES permit program effectively established a system of *effluent standards* for water pollution control. This shifted the focus away from stream standards (see Section 5.8), which regulate the amount of pollutants in the receiving waters, to the amount of pollutants in separate discharges. Effluent standards stipulated in an NPDES permit are easier to enforce than are stream standards. It is quite clear who is responsible when effluent standards are violated, but the responsibility for violation of stream standards is more difficult to determine. For example, if the DO level in a stream drops below the minimum required value, it is difficult to prove that a specific polluter is responsible for the problem. On the other hand, it is easier to prove that a specific polluter is violating the conditions of its NPDES permit with regard to excessive biochemical oxygen demand (BOD) discharges.

NPDES permit requirements and effluent standards may vary for different treatment plants. They depend on the nature of the wastewater, the type of treatment, and the classification of the receiving waters. A typical NPDES permit includes limits as shown in Table 10.1.

**TABLE 10.1**
**Typical NPDES effluent limitations**

| Parameter | Maximum allowable value |
|---|---|
| BOD$_5$ | 30 mg/L |
| TSS | 30 mg/L |
| pH | 6.0 to 9.0 |
| Fecal coliforms | 200 per 100 mL |

Effluent limits are based on the best conventional technology, which includes a combination of primary, secondary, and sometimes tertiary treatment. Most treatment plants in the United States are required to use some form of secondary treatment, which removes at least 85 percent of BOD and suspended solids from the wastewater. When treatment levels are not sufficient for achieving stream standards, requirements can be made more stringent. Some municipal or industrial facilities are required to provide advanced (tertiary) treatment levels, which increase treatment (pollutant removal) efficiencies to 95 percent or more. Removal of phosphorus and nitrogen may also be required to reduce the rate of lake eutrophication and nitrogenous oxygen demand.

### Treatment Efficiency

Treatment efficiency can be defined as the ratio of the amount of pollutants removed to the amount of pollutants in the raw wastewater. In mathematical form, this is

$$\text{efficiency} = \frac{P_{IN} - P_{OUT}}{P_{IN}} \times 100 \qquad \textbf{(10-1)}$$

where $P_{IN}$ = concentration of pollutant flowing into the treatment system

$P_{OUT}$ = concentration of pollutant flowing out of the system

### EXAMPLE 10.1

Raw sewage flowing into a treatment plant (the plant *influent*) has a BOD$_5$ value of 200 mg/L. What is the maximum concentration of BOD$_5$ allowed in the treated sewage discharge (the plant *effluent*) if the required treatment efficiency is 85 percent? If the flow rate is 5 mgd, how many pounds of BOD will be discharged per day?

*Solution*   Applying Equation 10-1 gives

$$85 = \frac{200 - P_{OUT}}{200} \times 100$$

which, after rearranging terms, gives

$$\frac{200 \times 85}{100} = 200 - P_{OUT}$$

from which

$$P_{OUT} = 200 - 170 = 30 \text{ mg/L}$$

Now applying Equation 6-3b (which is not limited to only chlorine in its application), we obtain

$$\text{pounds per day} = 8.34 \times Q \times C$$
$$= 8.34 \times 5 \text{ mgd} \times 30 \text{ ppm}$$
$$\approx 1300 \text{ lb/d}$$

## EXAMPLE 10.2

A sewage treatment plant influent has an average TSS concentration of 250 mg/L. If the average effluent TSS concentration is 20 mg/L, what is the removal efficiency for TSS? If the flow rate is 5 ML/d, how many kilograms of suspended solids are discharged in the plant effluent each day?

*Solution*   First, applying Equation 10-1, we find

$$\text{efficiency} = \frac{250 - 20}{250} \times 100$$
$$= \frac{230}{250} \times 100 = 92 \text{ percent}$$

Now, applying Equation 6-3a, we obtain kilograms per day = $Q \times C$ = 5 ML/d $\times$ 20 mg/L = 100 kg.

### Pretreatment of Industrial Wastewater

Sewage conveyed by municipal sewers into *publicly owned treatment works* (POTWs) comes from several sources, including industrial plants. Wastewater discharged by industry often contains toxic chemicals, such as cyanide from electroplating processes and lead from battery manufacturing plants. Several serious problems can occur when industrial wastewater is discharged into a POTW, such as the following:

*Passthrough:* Nondegradable toxic substances may pass through the treatment plant, causing water pollution; this pollution can pose a threat to aquatic life and, through the food chain, to public health.

*Interference:* Toxic industrial wastes may interfere with the operation of the treatment plant, particularly in those processes that use bacteria to stabilize organic matter in the wastewater.

*Contamination:* Industrial wastes with high levels of toxic metals or organic substances can contaminate sewage sludge, thereby limiting sludge disposal options and raising disposal costs.

*Corrosion:* Industrial wastewater may corrode and damage the pipes and equipment in the sewage collection system and treatment plant.

*Hazards:* Some industrial wastes are highly volatile and can explode. Other wastes may produce toxic gases, posing a threat to persons at the plant and in the local community.

These problems can readily be avoided by *pretreatment* of wastewater at the industrial site before it is discharged into the public sewer system. In some localities, restrictions on the content of industrial discharges into public sewers were imposed many years ago, but these restrictions were limited. In 1972, the EPA created a *National Pretreatment Program;* this uniform nationwide program is comprehensive in scope and must be enforced by all POTWs.

Two sets of rules are now in effect under the National Pretreatment Program. *Categorical pretreatment standards* are industry-specific; they mandate different requirements for each type of industry. For example, there is a categorical standard for the iron and steel industry that limits the ammonia and cyanide discharged by any firm in that industry into a municipal sewerage system. Examples of other major industries subject to categorical pretreatment standards include timber, petroleum, textile, pharmaceutical, and chemical production.

The second category of rules, *prohibited discharge standards*, are substance-specific; they prohibit any discharge to sewer systems of certain types of wastes from all sources. For example, the discharge of any wastewater with pollutants that can create a fire hazard or explosion in the sewage system is not allowed. Also, discharges that have a pH of less than 5.0 or a temperature of more than 40°C (104°F) are prohibited from any industry.

Within the framework of the national standards, POTWs are required to develop local pretreatment programs. These local programs are enforced by the EPA, the state, or the local POTW, depending on the status of program approvals for a given community. As the generator of toxic pollutants, industry must finance, construct, and operate any pollution control facility necessary to comply with federal regulations or local pretreatment rules. Industrial compliance ensures that toxic industrial pollutants will not harm the environment or pose a public health hazard.

According to EPA data, there were more than 16,000 publicly owned treatment plants and 60,000 industrially owned wastewater treatment plants in the United States in the year 2000.

## 10.2 PRELIMINARY AND PRIMARY TREATMENT

Untreated or raw wastewater usually flows by gravity from an interceptor or trunk sewer into the *headworks* of a treatment facility; sometimes wastewater may be pumped to the treatment plant in a force main. The headworks of a treatment plant include a flow measurement device such as a Parshall flume (see page 49) and mechanical systems that provide *preliminary treatment*. Preliminary treatment systems typically include *screens*, *comminutors*, and *grit chambers*.

The first treatment process for raw wastewater is coarse screening. *Bar screens* (or *racks*), as they are called, are made of long, narrow metal bars spaced about 25 mm (1 in.) apart. They retain floating debris, such as wood, rags, or other bulky objects, that could clog pipes or damage mechanical equipment in the rest of the plant. In most large sewage treatment plants, the bar screens are cleaned automatically by a mechanical device, as shown in Figure 10.2. The collected debris or *screenings* are promptly disposed of, usually by burial on the plant grounds.

In some treatment plants, a mechanical cutting or shredding device, called a *comminutor*, is installed just after the coarse screens. A typical comminutor, shown in Figure 10.3, consists of a slotted cylindrical screen with a moving cutter blade. The comminutor shreds and chops solids or rags that passed through the bar screen. The shredded material is removed from the wastewater by sedimentation or flotation later in the treatment plant. In small sewage treatment plants, a manually cleaned bar screen is generally installed in a channel next to the comminutor to serve as an emergency bypass when the comminutor needs repair.

### Grit Removal

A portion of the suspended solids in raw sewage consists of gritty material, such as sand, coffee grounds, eggshells, and other relatively inert material. In cities with combined sewer systems, sand and silt may be carried in the sewage. Suspended grit can cause excessive wear and tear on pumps and other equipment in the plant. Most of it is nonbiodegradable and will accumulate in treatment tanks. For these reasons, a grit removal process is usually used after screening and/or comminuting.

In the sewers, the flow velocity is generally not less than 0.6 m/s (2 ft/s), the self-cleansing velocity.

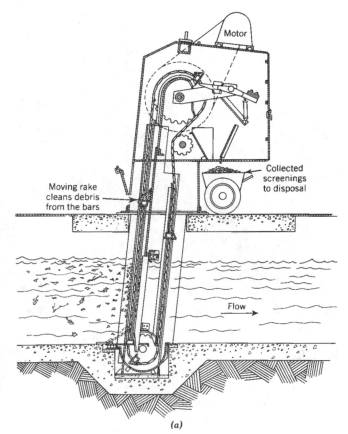

*(a)*

*(b)*

**FIGURE 10.2**
*Typical bar screen installations, which serve to remove sticks, rags, and other debris from the flow of sewage.* (a) (Courtesy of Envirex Inc., a Rexnord Company, Waukesha, Wisconsin), (b) (Courtesy of FMC Corporation, Lansdale, Pennsylvania)

**FIGURE 10.3**
*Typical comminutor installation.* (Courtesy of Worthington Pump Division, Dresser Industries, Inc., Mountainside, New Jersey)

Most gritty material is dense enough to settle out of the flow by gravity if the velocity is reduced to about 0.3 m/s (1 ft/s). Although the grit will settle out at this reduced velocity, the lighter suspended organic solids will still be carried through to the next treatment unit.

The reduction in velocity and the collection of the grit is usually accomplished in long, narrow tanks called *grit chambers.* Even though the sewage flow rate varies throughout the day, the flow velocity can be kept almost constant at 0.3 m/s by the use of a specially shaped outlet weir. Mechanical collectors made of buckets on a continuous chain serve to remove the grit from the tank; the grit is promptly disposed of by burial on the plant grounds. An aerated grit chamber is shown in Figure 10.4. The rising air bubbles help to keep the organic solids in suspension while the grit settles to the bottom.

**Primary Sedimentation (Settling)**

After preliminary treatment by screening, comminuting, and grit removal, the wastewater still contains suspended organic solids that can be removed by plain sedimentation. The basic principles of sedimentation, or plain gravity settling, are discussed in Section 6.2 with regard to

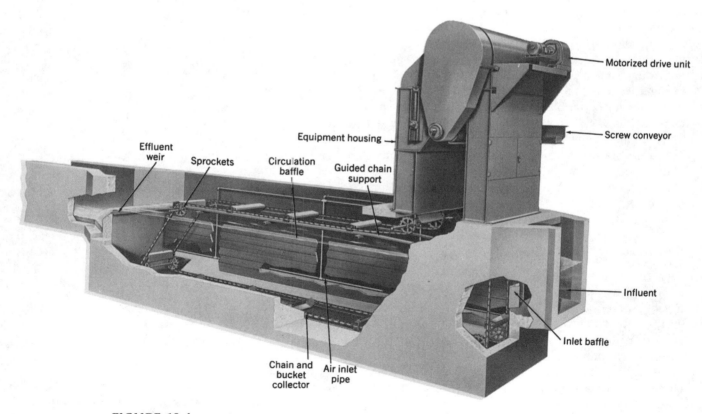

**FIGURE 10.4**
*A mechanically cleaned grit chamber.* (Courtesy of Envirex Inc., a Rexnord Company, Waukesha, Wisconsin)

drinking water purification. For the most part, the same principles apply to sewage treatment. There are, however, differences in the recommended design values for detention time, overflow rate, and tank configuration.

Settling tanks that receive sewage after grit removal are called *primary clarifiers*. They usually provide between 1 and 2 h of detention time; side water depth (SWD) is generally between 2.5 and 5 m (8 and 16 ft). The tanks may be circular or rectangular in shape (see Appendix H, Figures 11 and 15). In addition to mechanical sludge collectors that continually scrape the settled solids along the bottom to a sludge hopper for removal, a surface skimming device is used to remove grease and other floating materials from the liquid surface. Two or more clarifiers should be provided at all but the smallest treatment plants.

The combination of preliminary screening and gravity settling is called *primary treatment*. Chemicals may sometimes be added to the primary clarifiers to promote the removal of very small (or colloidal) particles. Primary treatment usually can remove up to 60 percent of the suspended solids and about 35 percent of the BOD from wastewater, but this relatively low level of treatment is no longer adequate in the United States. In almost all cases, primary treatment must be followed by secondary treatment processes; tertiary treatment may also be required to protect sensitive bodies of water that receive the treated effluent.

## EXAMPLE 10.3

A primary clarifier has an average influent TSS (total suspended solids) concentration of 250 mg/L. If its TSS removal efficiency is expected to be 60 percent, what is the expected average effluent TSS concentration?

*Solution*  Applying Equation 10-1 gives

$$60 = \frac{250 - P_{OUT}}{250} \times 100$$

and rearranging terms yields

$$P_{OUT} = 250 - \frac{250 \times 60}{100} = 100 \text{ mg/L}$$

The removal efficiencies of primary settling tanks can be increased by the addition of chemical coagulants, although this is not common practice in most modern wastewater treatment plants. To achieve BOD and TSS removal efficiencies of at least 85 percent, as

required by the *Clean Water Act*, primary treatment must be followed by an additional treatment process. Generally, this next step is characterized as *secondary treatment*.

## 10.3 SECONDARY (BIOLOGICAL) TREATMENT

Primary treatment processes remove only those pollutants that will either float or settle out by gravity, but about half of the raw pollutant load still remains in the primary effluent. The purpose of *secondary treatment* is to remove the suspended solids that did not settle out in the primary tanks and the dissolved BOD that is unaffected by physical treatment. Secondary treatment is generally considered to mean 85 percent BOD and TSS removal efficiency and represents the minimum degree of treatment required in most cases. In the United States, secondary treatment processes are almost always *biological* systems.

Biological treatment of sewage involves the use of microorganisms. The microbes, including bacteria and protozoa, consume the organic pollutants as food. They metabolize the biodegradable organics, converting them into carbon dioxide, water, and energy for their growth and reproduction. This natural aerobic process requires oxygen and was previously described under the topic of BOD in Section 4.3.

A biological sewage treatment system must provide the microorganisms with a comfortable home. In effect, the treatment plant allows the microbes to stabilize the organic pollutants in a controlled, artificial environment of steel and concrete, rather than in a stream or lake. This helps to protect the dissolved oxygen balance of the natural aquatic environment.

To keep the microbes happy and productive in their task of wastewater treatment, they must be provided with enough oxygen, adequate contact with the organic material in the sewage, suitable temperatures, and other favorable conditions. The design and operation of a secondary treatment plant is accomplished with these factors in mind.

Two of the most common biological treatment systems are the *trickling filter* and the *activated sludge process*. The trickling filter is a type of *fixed-growth* system: The microbes remain fixed or attached to a surface while the wastewater flows over that surface to provide contact with the organics. Activated sludge is characterized as a *suspended-growth* system, because the microbes are thoroughly mixed and suspended in the wastewater rather than attached to a particular surface.

## Trickling Filters

A trickling filter consists basically of a layer or bed of crushed rock about 2 m (6 ft) deep. It is usually circular in shape and may be built as large as 60 m (200 ft) in diameter. Trickling filters are always preceded by primary treatment to remove coarse and settleable solids. The primary effluent is sprayed over the surface of the crushed stone bed and trickles downward through the bed to an *underdrain system*.

A rotary distributor arm with nozzles located along its length is usually used to spray the sewage, although sometimes fixed nozzles are used. The rotary distributor arm is mounted on a center column in the trickling filter; it is driven around by the reaction force or jet action of the wastewater that flows through the nozzles.

The underdrain system serves to collect and carry away the wastewater from the bottom of the bed and to permit air circulation upward through the stones. As long as topography permits, the sewage flows from the primary tank to the trickling filter by the force of gravity, rather than by pumping. A cutaway view of a typical trickling filter unit is shown in Figure 10.5.

As the primary effluent trickles downward through the bed of stones, a biological slime of microbes develops on the surfaces of the rocks. The continuing flow of the wastewater over these fixed biological growths provides the needed contact between the microbes and the organics. The microbes in the thin slime layer absorb the dissolved organics, thus removing oxygen-demanding substances from the wastewater. Air circulating through the void spaces in the bed of stones provides the needed oxygen for stabilization of the organics by the microbes.

Note, however, that the trickling filter is not really a filter at all, in the true sense of the word. The stones are usually about 75 mm (3 in.) in size, much too large to strain or filter out suspended solids. And, by definition, filters have no effect on dissolved solids. The stones in a trickling filter only serve to provide a large amount of surface area for the biological growths, and the large voids allow ample air circulation. Sometimes materials other than rock, such as modules of corrugated plastic or redwood slats, are used to provide the needed surface area and void spaces, but the basic purpose and operation remain the same.

As the microorganisms grow and multiply, the slime layer gets thicker. Eventually, it gets so thick that the flowing wastewater washes it off the surfaces of the stones. This is called *sloughing* (pronounced "sluffing"). Since sloughing does occur periodically, there is a need to provide settling time for the trickling filter effluent, in order to remove the sloughed biological solids. These solids consist basically of billions of microorganisms that have absorbed the dissolved organics into their bodies.

The trickling filter effluent is collected in the underdrain system and then conveyed to a sedimentation tank called a *secondary clarifier*. The secondary clarifier, or *final clarifier* as it is sometimes called, is similar in most respects to the primary clarifier, although there are differences in detention time, overflow rate, weir loading, and other details.

To maintain a relatively uniform flow rate through the trickling filter and to keep the distributor arm

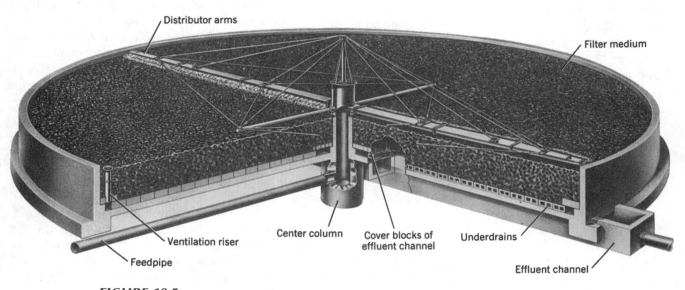

**FIGURE 10.5**
*Cutaway view of a trickling filter. Trickling filters are sometimes enclosed or covered for odor control or for temperature (and process) control.* (Courtesy of Dorr–Oliver Inc., Stamford, Connecticut)

rotating even during periods of low sewage flow, some of the wastewater may be recirculated. In other words, a portion of the effluent is pumped back to the trickling filter inlet so that it will pass through the bed of stones more than once.

Recirculation can also serve to improve the pollutant removal efficiency; it allows the microbes to remove organics that flowed by them during the previous pass through the bed. There are many recirculation patterns and configurations of trickling filter plants. One common pattern, called *direct recirculation*, is shown in a flow diagram in Figure 10.6.

### Recirculation

The amount of recirculation can vary. It is characterized by a *recirculation ratio*, which is the ratio of recycled flow to the raw wastewater flow. In formula form, it is given by

$$R = \frac{Q_R}{Q} \qquad \textbf{(10-2)}$$

where $R$ = recirculation ratio
$Q_R$ = recirculated flow rate
$Q$ = raw sewage flow rate

The recirculation ratio, $R$, is generally in the range of 0.0 to 3.0.

### Hydraulic Load

The rate at which the wastewater flow is applied to the trickling filter surface is called the *hydraulic load*. The hydraulic load includes the recirculated flow $Q_R$; the total flow through the trickling filter is equal to $Q + Q_R$. In formula form, it is given by

$$\text{hydraulic load} = \frac{Q + Q_R}{A_S} \qquad \textbf{(10-3)}$$

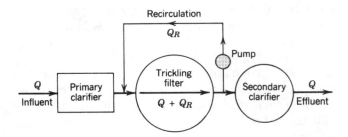

**FIGURE 10.6**
*Schematic diagram showing recirculation of flow through a trickling filter. The rate of sewage flow applied to the filter is the sum of the influent flow rate and the recirculated flow rate.*

where $Q$ = raw sewage flow rate
$Q_R$ = recirculated flow rate
$A_S$ = trickling filter surface area (plan view)

Hydraulic load may be expressed in terms of cubic meters per day per square meter of surface area, or $m^3/m^2 \cdot d$. It also may be expressed in terms of million gallons per acre of surface area per day, or mil gal/ac/d. A typical value for a conventional trickling filter is $20 \ m^3/m^2 \cdot d$ (19 mil gal/ac/d).

### Organic (BOD) Load

The rate at which organic material is applied to the trickling filter is called the *organic* or *BOD load*. It does not include the BOD added by recirculation. Organic load is expressed in terms of kilograms of BOD per cubic meter of bed volume per day, or $kg/m^3 \cdot d$. It is also expressed in terms of pounds of BOD per thousand cubic feet of bed volume per day, or $lb/1000 \ ft^3/d$. A typical value for organic load on a trickling filter is $0.5 \ kg/m^3 \cdot d$, or $30 \ lb/1000 \ ft^3/d$. In formula form, the organic load may be expressed as

$$\text{organic load} = \frac{Q \times \text{BOD}}{V} \qquad \textbf{(10-4a)}$$
$$\text{(SI metric units)}$$

or

$$\text{organic load} = 8340 \times \frac{Q \times \text{BOD}}{V} \qquad \textbf{(10-4b)}$$
$$\text{(U.S. Customary units)}$$

where $Q$ = raw wastewater flow, ML/d (mgd)
BOD = $BOD_5$ in the primary effluent, mg/L (ppm)
$V$ = volume of trickling filter bed, $m^3$ ($ft^3$)

### EXAMPLE 10.4

A 2-m-deep trickling filter with a diameter of 18 m is operated with a recirculation ratio of 1.5. The raw wastewater flow rate is 2.5 ML/d, and the 5-day BOD of the raw sewage is 210 mg/L. Assuming that the primary tank BOD removal efficiency is 30 percent, compute the hydraulic load and the organic load on the trickling filter.

*Solution*  First, compute the surface area of the trickling filter,

$$A_S = \frac{\pi \times D^2}{4} = \frac{\pi \times 18^2}{4} = 254.5 \ m^2$$

Since volume = area × depth,

$$V = 254.5 \text{ m}^2 \times 2 \text{ m} = 509 \text{ m}^3$$

From Equation 10-2, write $Q_R = R \times Q$ and

$$Q_R = 1.5 \times 2.5 = 3.75 \text{ ML/d}$$

Also,

$$Q + Q_R = 2.5 + 3.75 = 6.25 \text{ ML/d} = 6250 \text{ m}^3/\text{d}$$

From Equation 10-3,

$$\text{hydraulic load} = \frac{6250 \text{ m}^3/\text{d}}{254.5 \text{ m}^2} \approx 25 \text{ m}^3/\text{m}^2 \cdot \text{d}$$

The primary effluent BOD can be computed with Equation 10-1:

$$30 = \frac{210 - P_{\text{OUT}}}{210} \times 100$$

and

$$P_{\text{OUT}} = 210 - \frac{210 \times 30}{100} = 147 \text{ ppm}$$

From Equation 10-4a,

$$\text{organic load} = \frac{2.5 \times 147}{509} = 0.72 \text{ kg/m}^3 \cdot \text{d}$$

### Efficiency

BOD removal efficiency of a trickling filter unit depends primarily on the organic load, the recirculation ratio, and the temperature of the wastewater. Generally, the efficiency increases with decreasing organic load, increasing recirculation, and increasing temperature. For example, with no recirculation ($R = 0$) and a temperature of 20°C, a typical trickling filter will have an efficiency of about 60 percent when the organic load is about 2 kg/m$^3$ · d. But if the organic load is 0.5 kg/m$^3$ · d, at the same conditions of recirculation and temperature, the efficiency will be 75 percent.

Furthermore, at the 0.5-kg/m$^3$ · day loading, a recirculation of $R = 1$ instead of $R = 0$ could raise the efficiency to about 80 percent. If the temperature increased to 22°C, the efficiency would be raised to about 85 percent. Because of the marked effect of temperature on treatment efficiency, trickling filters in northern climates are often enclosed under fiberglass domes to provide protection against wind and snow and to reduce the rate of heat loss from the wastewater.

### EXAMPLE 10.5

The BOD removal efficiency of a trickling filter system is 79 percent, and the efficiency of the primary treatment that precedes it is 35 percent. If the raw BOD is 200 mg/L, what is the effluent BOD? Is the treatment plant providing an efficiency that meets the requirement for secondary treatment?

*Solution* Although the biological treatment and secondary clarification provide only 79 percent BOD removal, it should be remembered that 35 percent of the raw BOD was recently removed by primary settling. Thus, 65 percent or $0.65 \times 200 = 130$ mg/L remains as BOD in the primary effluent.

But 79 percent of the BOD entering the trickling filter is removed, leaving 21 percent or $0.21 \times 130 = 27$ mg/L in the secondary effluent. The overall plant efficiency is therefore computed as

$$\text{efficiency} = \frac{200 - 27}{200} \times 100 \approx 87 \text{ percent}$$

This is greater than 85 percent, so the treatment plant is providing secondary treatment.

## Activated Sludge Treatment

The basic components of an activated sludge sewage treatment system include an aeration tank and a secondary settling basin or clarifier. Primary effluent is mixed with settled solids that are recycled from the secondary clarifier and then introduced into the aeration tank. Compressed air is injected continuously into the mixture through porous diffusers located at the bottom of the tank along one side. This is illustrated in Figure 10.7.

In the aeration tank, microorganisms consume the dissolved organic pollutants as food. The microbes absorb and aerobically decompose the organics, using oxygen provided in the compressed air; water, carbon dioxide, and other stable compounds are formed. In addition to providing oxygen, the compressed air thoroughly mixes the microbes and wastewater together as it rapidly bubbles up to the surface from the diffusers. Sometimes, mechanical propellerlike mixers, located at the liquid surface, are used instead of compressed air

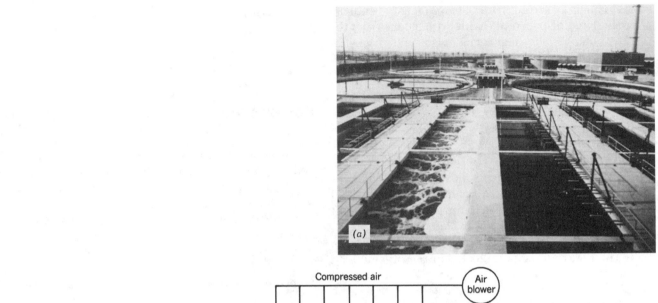

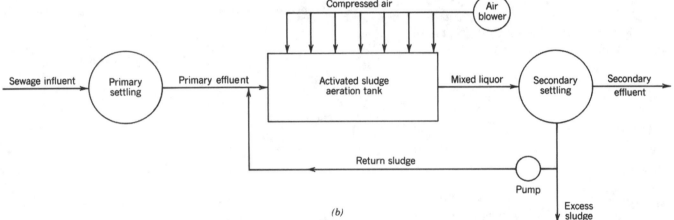

**FIGURE 10.7**

(a) *Typical activated sludge sewage treatment plant. The rectangular tanks in the foreground are the aeration tanks; the air diffusers have been raised out of the tank on the right* (Courtesy of FMC Corporation, Lansdale, Pennsylvania);
(b) *A flow diagram for a conventional activated sludge plant.*

and diffusers. The churning action of the propeller blades mixes air with the wastewater and keeps the contents of the tank in a uniform suspension.

The aerobic microorganisms in the tank grow and multiply, forming an active suspension of biological solids called *activated sludge*. The combination of the activated sludge and wastewater in the aeration tank is called the *mixed liquor*. In the basic or conventional activated sludge treatment system, a tank detention time of about 6 h is required for thorough stabilization of most of the organics in the mixed liquor.

After about 6 h of aeration, the mixed liquor flows to the secondary or final clarifier, in which the activated sludge solids settle out by gravity. The clarified water near the surface, called the *supernatant*, is discharged over an effluent weir; the settled sludge is pumped out

from a sludge hopper at the bottom of the tank. Recycling a portion of the sludge back to the inlet of the aeration tank is an essential characteristic of this treatment process. The settled sludge is in an *active* state. In other words, the microbes are well acclimated to the wastewater and, given the opportunity, will readily absorb and decompose more organics by their metabolism.

By pumping about 30 percent of the wastewater flow from the bottom of the clarifier back to the head of the aeration tank, the activated sludge process can be maintained continuously. When mixed with the primary effluent, the hungry microbes quickly begin to absorb and metabolize the fresh food in the form of BOD-causing organics. Since the microbes multiply and increase greatly in numbers, it is not possible to recycle or return all the sludge to the aeration tank. The excess sludge, called *waste activated sludge*, must eventually be treated and disposed of (along with sludge from the primary tanks). Sludge management is a major aspect of wastewater treatment and is discussed in Section 10.6.

### F/M Ratio

An important factor used in the design and operation of activated sludge systems is known as the *food-to-microorganism (F/M) ratio*. The food is measured in terms of kilograms (pounds) of BOD added to the tank per day. Since the suspended solids in the mixed liquor consist mostly of living microorganisms, the suspended solids concentration is used as a measure of the amount of microorganisms in the tank. This concentration is called the *mixed liquor suspended solids*, or MLSS.

The F/M ratio is an indicator of the organic load on the system with respect to the amount of biological solids in the tank. For conventional aeration tanks, the ratio is in the range of 0.2 to 0.5. It can be computed from the following formula:

$$F/M = \frac{Q \times BOD}{MLSS \times V} \qquad (10\text{-}5)$$

where F/M = food-to-microorganism ratio, in units of kilograms of BOD per kilogram of MLSS per day
$Q$ = raw sewage flow rate, ML/d (mgd)
BOD = applied 5-day BOD, mg/L (ppm)
MLSS = mixed liquor suspended solids, mg/L
$V$ = volume of aeration tank, ML (million gal)

### EXAMPLE 10.6

An activated sludge tank is 30 m long and 10 m wide and has an SWD of 4 m. The wastewater flow is 4.0 ML/day and the raw 5-day BOD is 200 mg/L. The MLSS concentration is 2000 mg/L. Compute the food-to-microorganism ratio for the system.

*Solution*    A conventional activated sludge aeration tank is preceded by primary treatment. Assuming that 35 percent of the raw BOD is removed in the primary clarifier, 65 percent of the BOD will be applied to the aeration tank:

$$0.65 \times 200 = 130 \text{ mg/L}$$

The tank volume is the product of length, width, and depth, or

$$V = 30 \text{ m} \times 10 \text{ m} \times 4 \text{ m} = 1200 \text{ m}^3 \text{ or } 1.2 \text{ ML}$$

Applying Equation 10-5 yields

$$F/M = \frac{4.0 \times 130}{2000 \times 1.2} = 0.22$$

### EXAMPLE 10.7

A conventional aeration tank is to treat a flow of 800,000 gpd of primary effluent with a BOD of 125 ppm. The MLSS concentration is to be maintained at 1800 ppm, and a food-to-microorganism ratio of 0.4 is specified. Compute the required volume of the aeration tank. If the side water depth is to be 15 ft and the tank length is to be three times its width, how long should the tank be?

*Solution*    Rearranging the terms of Equation 10-5 gives

$$V = \frac{Q \times BOD}{F/M \times MLSS} = \frac{0.8 \times 125}{0.4 \times 1800}$$

$$= 0.139 \text{ million gal} \approx 140,000 \text{ gal}$$

and

$$140,000 \text{ gal} \times \frac{1 \text{ ft}^3}{7.5 \text{ gal}} = 18,500 \text{ ft}^3$$

Since volume = length × width × depth, or $V = L \times W \times SWD$, and since the length is to be three times the width, or $L = 3W$, volume is expressed as

$$V = 3W \times W \times SWD = 3 \times W^2 \times SWD$$

Solving for the tank width yields

$$W = \left( \frac{18,500 \text{ ft}^3}{3 \times 15 \text{ ft}} \right)^{1/2} \approx 20 \text{ ft}$$

and solving for lengths yields

$$L = 3 \times 20 = 60 \text{ ft}$$

### Sludge Settling

In the activated sludge process, the organic pollutants are absorbed by the billions of microorganisms in an aeration tank. These microorganisms essentially are the *activated sludge*. But without proper clarification or separation of the sludge from the liquid portion of the mixed liquor, the treatment process will not be effective at all. For this reason, gravity settling in the secondary clarifier is a most important part of the activated sludge treatment system. If the sludge does not settle fast enough, some of it will be carried over the effluent weirs of the clarifier and cause pollution of the receiving body of water.

Under certain conditions in an activated sludge sewage treatment plant, filamentous or stringy bacteria, usually of the species *Sphaerotilus natans*, grow prolifically in the aeration tank, making the sludge very fluffy and light. Sludge with excessive growth of these filamentous organisms settles very slowly, and a clear supernatant is not formed in the secondary clarifier. Much of the sludge flows out with the effluent. This condition is called *sludge bulking*. Bulking of activated sludge may be controlled or limited by appropriate adjustments in the MLSS concentration and F/M ratio; this is accomplished by regulating the rate of sludge return from the clarifier. The amount of aeration may also be a factor; sludge bulking is sometimes associated with too much aeration. Occasionally, adjustments in the mixed liquor pH are made to solve the problem.

A number called the *sludge volume index* (SVI) is used to evaluate the settleability of the activated sludge. It is equal to the volume occupied by 1 g of settled sludge and is expressed in units of milliliters per gram (mL/g).

The determination of SVI involves taking a sample of mixed liquor from the aeration tank and allowing it to settle for 30 min in a 1-L graduated glass cylinder. This is illustrated in Figure 10.8. The volume of settled sludge is read from the markings on the cylinder. The MLSS concentration in the mixed liquor is also measured.

The following formula is used to compute SVI:

$$SVI = \frac{V \times 1000}{MLSS} \qquad (10\text{-}6)$$

where SVI = sludge volume index, mL/g
$V$ = volume of settled sludge, mL/L
MLSS = mixed liquor suspended solids, mg/L

## EXAMPLE 10.8

An aeration tank has an MLSS concentration of 2000 mg/L. After settling for 30 min in a 1-L graduated cylinder, the sludge volume is measured to be 150 mL. Compute the SVI of the sludge.

*Solution*   Applying Equation 10-6 yields

$$SVI = \frac{150 \text{ mL/L} \times 1000}{2000 \text{ mg/L}} = 75 \text{ mL/g}$$

A normal sludge with good settling characteristics generally has an SVI of less than 100. As the SVI increases above 100, sludge settleability decreases and some solids get carried over the effluent weir of the clarifier. Severely bulking sludges have SVI values of over 200. A very high SVI is an indication to a treatment plant operator that the sludge return rate should be increased, the aeration rate should be decreased, or some other process adjustment should be made. Sometimes

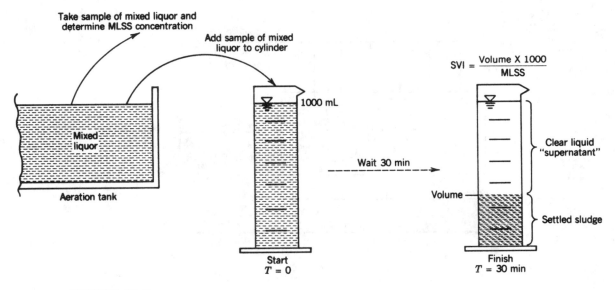

**FIGURE 10.8**
*Schematic illustration of the lab test for the* sludge volume index (SVI), *which is used to evaluate sludge settling characteristics.*

chlorine is added to the aeration tank to destroy the filamentous organisms, but this is a last resort to control the problem.

In general, a well-operated activated sludge treatment system can remove about 90 percent of the raw sewage BOD and TSS; sometimes removal efficiency may be as high as 95 percent. In contrast to the simpler trickling filter sewage treatment system, however, an activated sludge plant requires careful operational control. Energy requirements in an activated sludge plant are also high because of the power consumed for aeration.

## Modifications of the Activated Sludge Process

Several modifications of the conventional activated sludge process have been developed; these serve to increase the treatment plant capacity or to reduce the tank volume requirements.

### Step Aeration

A process called *step aeration* provides multiple feed points of the primary effluent into the aeration tank, as shown in Figure 10.9. By introducing the organics into the tank in increments or steps, rather than only once at the head of the tank, the oxygen demand is spread more uniformly over the length of the tank. In this manner, greater treatment plant capacities can be obtained in a given volume of aeration tank than can be obtained using the conventional process.

### Extended Aeration

For treating small sewage flow rates from suburban residential developments, hotels, schools, and other relatively isolated wastewater sources, a process called *extended aeration* is often used. These small systems are generally in the form of prefabricated steel tanks and are called *package plants*. (Conventional tanks, on the other hand, are usually made with cast-in-place reinforced concrete.) In the extended aeration system, the aeration tank and secondary clarifier are built in a single unit, as illustrated in Figures 10.10 and 10.11.

There are two important distinctions between an extended aeration system and a conventional system. First, screened or comminuted sewage is directed into the extended aeration tank without any primary settling. Second, the detention time or aeration period is about 30 h, whereas the conventional system's detention time is about 6 h.

Another difference is that the extended aeration process operates with F/M ratios as low as 0.05. This means that there is a large population of microorganisms compared to the amount of food (organics). The low F/M ratio and the extended period of aeration allow for the stabilization of most of the organics in the wastewater. But eventually some sludge has to be removed from the aeration tank for disposal.

### Mechanical Aeration

Mechanical aeration systems, which employ oval-shaped basins and horizontal rotor-brush aerators, are efficient and easy to operate. These concrete-lined basins, called *oxidation ditches*, are between 1.2 and

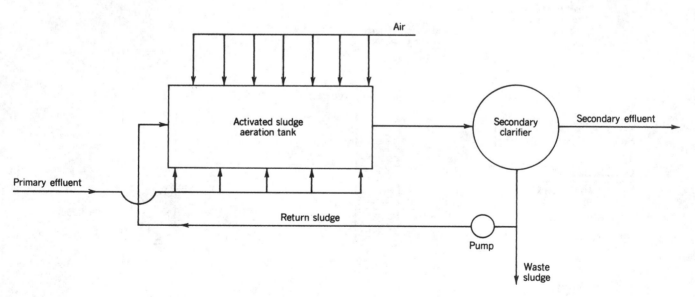

**FIGURE 10.9**
*Flow diagram for the step aeration modification of the activated sludge process.*

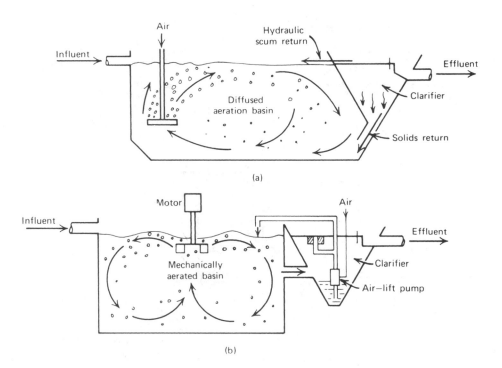

**FIGURE 10.10**

*Two typical small, extended aeration treatment systems using (a) diffused aeration and a slotted bottom clarifier for sludge return and (b) mechanical aeration with an air-lift pump for sludge return.* (From M. J. Hammer, *Water and Wastewater Technology*, 4th ed., Prentice-Hall, Englewood Cliffs, New Jersey, 2001. Copyright 2001 by Prentice-Hall, Inc. Used with permission)

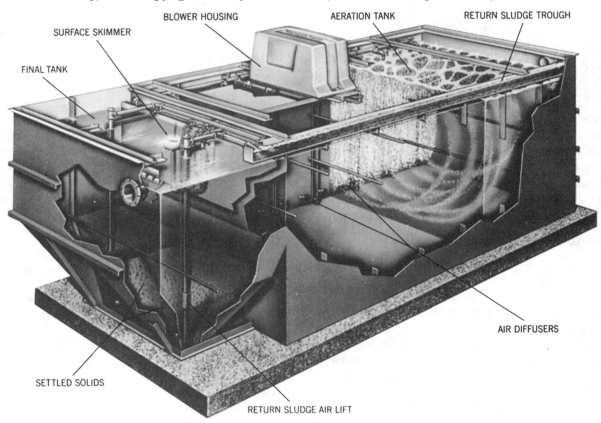

**FIGURE 10.11**

*A typical extended aeration package plant installation.* (Courtesy of FMC Corporation, Lansdale, Pennsylvania)

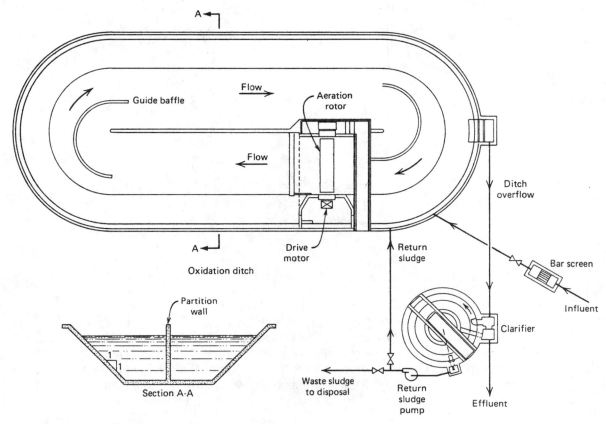

**FIGURE 10.12**

*Flow diagram for an oxidation ditch.* (From M. J. Hammer, *Water and Wastewater Technology*, 2nd ed., Prentice-Hall, Englewood Cliffs, New Jersey, 2001. Copyright © 2001 by Prentice-Hall, Inc. Used with permission)

1.8 m (4 and 6 ft) deep. The horizontal aerator acts like a paddle wheel, propelling the wastewater around in the channel at a velocity sufficient to prevent settling of solids. Atmospheric oxygen is transferred through the free surface of the liquid. Most oxidation ditches are usually operated as extended aeration systems with aeration times greater than 12 h. A flow diagram of a typical oxidation ditch facility is shown in Figure 10.12.

**Contact Stabilization**

In yet another modification of the activated sludge process, the influent sewage is mixed and aerated with return activated sludge for only about 30 min. This process is called *contact stabilization*. The short contact period of 30 min is sufficient for the microorganisms to absorb the organic pollutants, but not to stabilize them.

   After the short contact time, the mixed liquor enters a clarifier and the activated sludge settles out; the clarified sewage flows over effluent weirs. The settled sludge is pumped into another aerated tank, called a *reaeration* or *stabilization tank*. The contents of the stabilization tank are aerated for about 3 h, allowing the microbes to decompose the absorbed organic

material. The total size of a contact stabilization tank is generally less than that of a conventional plant. This is because the volume of activated sludge being stabilized in the reaeration tank is considerably less than the total wastewater flow. A schematic flow diagram of the contact stabilization process is shown in Figure 10.13.

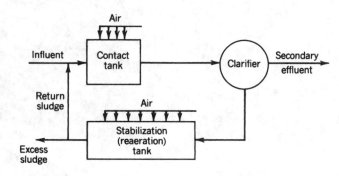

**FIGURE 10.13**

*The contact stabilization modification of the activated sludge process. Organic pollutants are absorbed by the microbes in the contact aeration tank and stabilized in the reaeration tank.*

A typical installation consists of a field-erected circular steel tank, with an inner tank providing a zone for clarification. The annular volume between the inner tank and the outer tank wall provides the room or zones for contact, reaeration, and sludge storage. An air-lift type of pump is usually used to transfer sludge between zones of the tank. With some minor modification of piping and baffling arrangements, a contact stabilization system can also be operated in the step aeration or in the extended aeration mode. A circular sewage treatment unit is shown in Figure 10.14.

## CASE STUDY

### Package System Helps Handle Variable Flows

The village of Dexter, New York, uses a package treatment system that features a submerged, fixed-film aeration system to comply with the state of New York's effluent discharge requirements. Like many eastern cities, the town on the eastern shores of Lake Ontario experiences infiltration and inflow problems as a result of an old sewer system.

The village had faced compliance problems because of wet-weather flows, which can be six times the daily average flow and last up to 2 d. Dexter has spent hundreds of thousands of dollars refurbishing the system, but infiltration and inflow is still a problem throughout the system.

The city and its engineers had the choice of continuing massive rehabilitation of the sewer system or redesigning the treatment plant. Dexter decided to purchase a Smith & Loveless, Inc., factory-built FAST® (Fixed Activated Sludge Treatment) system.

The FAST treatment system combines attached bacterial growth and suspended growth simultaneously, thus ensuring a constant bacterial population even during high inflow and infiltration events. A submerged medium with a high surface area rests within the aeration zone. The aeration system transfers oxygen to the wastewater, and at the same time pumps the wastewater through the submerged medium. Some of the bacterial growth attaches to the surface of the medium while suspended growth occurs within the aeration system.

The biological solids that slough off the medium are transported into a separate clarification system. These solids are generally excess, and the sludge from the clarifier is wasted to sludge holding. Because the system adjusts itself to load conditions, the plant operator does not have to adjust the system in terms of return activated sludge to maintain the bacterial population. The only required operation is occasional air scouring of the submerged medium to dislodge excess bacteria from the medium.

The Dexter plant was designed to provide an average monthly effluent quality of 30 mg/L BOD and 30 mg/L TSS. Established design criteria indicated the total BOD load to the system would not exceed 418 lb BOD/day, while flow to the plant could range from the basic domestic flow of about 250,000 gpd to 1.5 mgd during high infiltration periods. The infiltration flows offer little or no additional BOD contribution above the domestic flow.

The state required duplicity of aeration zones in the system, so the plant was divided into two separate trains, each capable of handling 125,000 gpd under normal conditions (see Figure 10.15). One train was required to hydraulically handle up to 1 mgd in case the other was shut down for maintenance.

The influent is pumped into a mechanically cleaned bar screen, and the flow is then divided equally between the two treatment systems. After initial treatment, the flow is directed into two 30-ft-diameter peripheral feed clarifiers.

The piping arrangement allows flow from the package plants to be diverted into either of the two clarifiers. Return activated sludge from the clarifiers can flow back into the splitter box at the bar screen headworks, or be wasted to an aerobic digester.

The FAST process has minimized washout of bacteria during high flows and lowered the average BOD and TSS levels to 15 mg/L, maintaining consistent effluent quality throughout peak flow periods. Since installing the system, Dexter has not had a wastewater discharge violation.

---

*Source:* Adapted from *WaterWorld*, March 1998. Copyright © 1998 by *WaterWorld*. Used with permission.

### Pure Oxygen Aeration

Air is only 21 percent oxygen. Instead of using air, greater treatment capacities can be achieved by injecting high-purity oxygen into the mixed liquor of an activated sludge sewage treatment plant. A diagram of a *high-purity oxygen aeration system* is shown in Figure 10.16. The oxygen is manufactured at the plant site. Primary effluent, return activated sludge, and oxygen are introduced into the first compartment of a multistaged, covered tank. Mechanical agitators mix the oxygen with the wastewater as it flows through the tank. The total aeration period is only about 2 h, and the F/M ratio is as high as 1.5. Consequently, the aeration tank volume is considerably less than that required for the conventional system.

## Other Secondary Treatment Processes

A biological treatment unit that has been used in Europe and in recent years has become an accepted method of

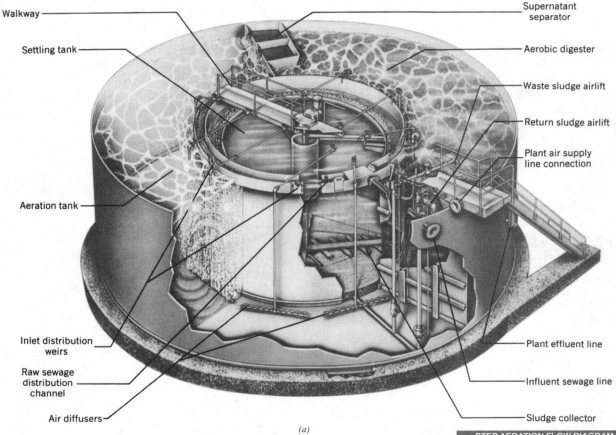

Walkway

Settling tank

Aeration tank

Inlet distribution weirs

Raw sewage distribution channel

Air diffusers

Supernatant separator

Aerobic digester

Waste sludge airlift

Return sludge airlift

Plant air supply line connection

Plant effluent line

Influent sewage line

Sludge collector

*(a)*

## FIGURE 10.14

(a) *Circular prefabricated steel sewage treatment plants are available in diameters up to about 30 m (100 ft).* (b) *They can be used for step aeration, for contact stabilization, or for extended aeration modes of treatment.* (Courtesy of FMC Corporation, Lansdale, Pennsylvania)

## FIGURE 10.15

*Dexter, New York, "package treatment system."*

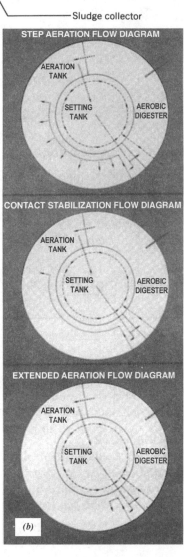

STEP AERATION FLOW DIAGRAM

AERATION TANK

SETTING TANK

AEROBIC DIGESTER

CONTACT STABILIZATION FLOW DIAGRAM

AERATION TANK

SETTING TANK

AEROBIC DIGESTER

EXTENDED AERATION FLOW DIAGRAM

AERATION TANK

SETTING TANK

AEROBIC DIGESTER

*(b)*

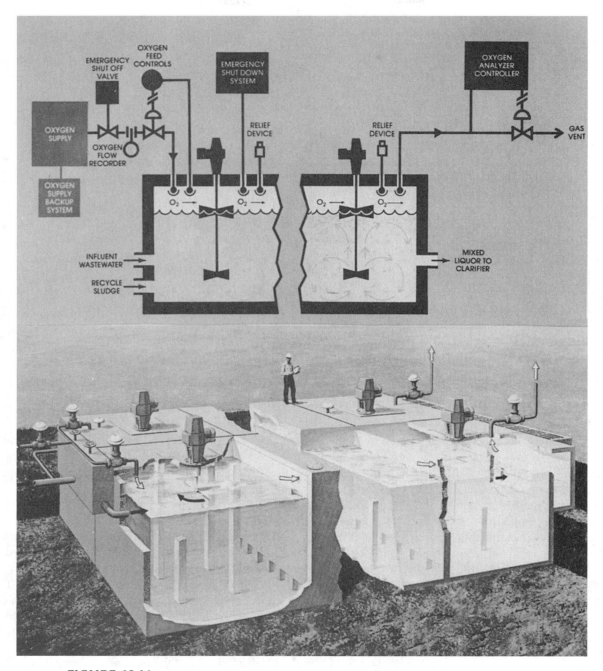

**FIGURE 10.16**
*Pure oxygen can be used in activated sludge wastewater treatment. The tanks are covered to conserve oxygen that is generated on-site.* (Courtesy of Air Products and Chemicals, Inc., Allentown, Pennsylvania)

sewage treatment in the United States is the *biodisc* or *rotating biological contactor.* It consists of a series of large plastic discs mounted on a horizontal shaft. The lightweight discs are about 3 m (10 ft) in diameter and are spaced about 40 mm (1.5 in.) apart on the shaft. A typical biodisc unit is illustrated in Figure 10.17.

The discs are partially submerged in settled sewage (primary effluent). As the shaft rotates, the disc surfaces are alternately in contact with air and with the wastewater. Consequently, a layer of biological slime grows on each disc, and the attached microbes that form the slime absorb the organic material in the wastewater. This process is similar to the trickling filter system, except that the attached microbial growths are passed through the wastewater, instead of the wastewater being sprayed over the microorganisms.

**FIGURE 10.17**
*A series of rotating biological contractors (RBC), or* biodiscs, *for secondary wastewater treatment.* (Courtesy of Bio Systems Division, Autotrol Corp., Milwaukee, Wisconsin)

The speed of rotation and the number of discs can be varied to achieve specific levels of pollutant removal. With several stages of discs, it is possible to remove nitrogenous as well as carbonaceous BOD. This is because growths of nitrifying bacteria predominate in the microbial population on the final disc stages.

In a biodisc system, there is no need to recycle sludge, but a secondary clarifier is needed to settle out the excess biological solids that slough off the discs as the slime layer thickens. As with the trickling filter, the efficiency of the biodisc process is adversely affected by low temperatures. This is because the rate of metabolism of the microbes slows down when the temperature drops.

In suburban or rural areas where land is available at relatively low cost, *sewage lagoons* may be used for secondary treatment. They are also called *stabilization* or *oxidation ponds*. The most common type of lagoon used for treating wastewater is the *facultative pond*. In a facultative pond, which is generally about 2 m (6 ft) deep, both aerobic and anaerobic biochemical reactions take place. This is illustrated in Figure 10.18.

Raw wastewater enters the pond, eliminating the need for primary treatment. Organic solids that settle to the bottom decompose anaerobically, producing such substances as methane, hydrogen sulfide, and organic acids. In the liquid above the sludge zone of the pond, incoming organics and the products of anaerobic decomposition are stabilized by *facultative bacteria* as

well as by aerobic microorganisms. Facultative bacteria can grow in either aerobic or anaerobic environments. The average sewage detention time in a facultative pond may be 60 d or more.

Oxygen is added to the wastewater in the pond by wind action and mixing at the surface and from the daylight metabolism of algae. This oxygen supports the aerobic reactions. The mutually dependent relationship between the algae and bacteria in a stabilization pond is very important. Using energy from sunlight, the algae grow and multiply by consuming the carbon dioxide and other inorganic compounds released by the bacteria. The bacteria use both the oxygen released by the algae and the organics from the wastewater.

Although the algae play an important role in the purification process in a sewage lagoon, they can also cause a problem. When these microscopic plants are carried out of the pond in the effluent flow, the allowable levels of TSS for a secondary effluent are usually exceeded. This problem can be particularly severe during the warm summer months. Sometimes, using two or more ponds in series and with careful control of the effluent flow, algae carryover can be eliminated.

Despite the potential difficulty with regard to TSS removal efficiency, sewage lagoons are being used with increasing frequency in areas where land is readily available. The low construction costs, ease of operation and maintenance, and negligible energy

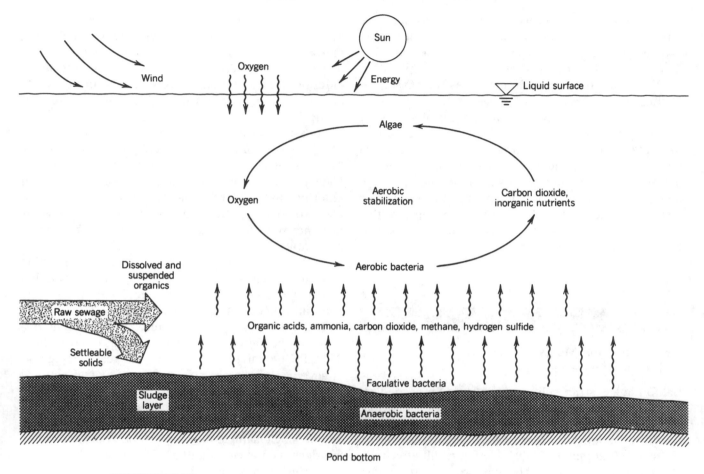

**FIGURE 10.18**
*Schematic diagram of the complex biochemical reactions that take place in a wastewater stabilization pond or lagoon.*

costs offer distinct advantages for this natural purification system. (See Appendix H, Figure 3.)

**Secondary Effluent Disinfection**

The last process in a secondary sewage treatment process is *disinfection*. The purpose of sewage disinfection is to destroy any pathogens in the effluent that may have survived the secondary treatment process, thereby protecting public health. (Removal of BOD and TSS serves primarily to protect the aquatic environment.) Sewage disinfection is particularly important when the secondary effluent is discharged into a body of water used for swimming or water supply by a downstream community.

Like drinking water, sewage is usually disinfected by *chlorination*. The method of application and the chemistry of this process are discussed in Section 6.5. The chlorine demand of wastewater is relatively high when compared to that of drinking water. A chlorine dose of about 10 mg/L is required to leave a combined chlorine residual of 0.5 mg/L in the secondary effluent. A residual of 0.5 mg/L is the minimum required by most environmental regulatory agencies for wastewater effluents.

A separate *chlorine contact tank*, with a series of baffles to eliminate short-circuiting of the flow, is used to ensure at least 15 min of contact time between the sewage and the chlorine. Although the presence of a chlorine residual is a good indication of effective disinfection, more than just residual testing is required; specifically, bacteriological testing may be necessary. NPDES permits typically specify a maximum allowable concentration of 200 fecal coliforms per 100 mL in the plant effluent.

Excessive chlorination of sewage can have an adverse environmental impact. High chlorine concentrations in the vicinity of sewage treatment plant outfall pipes can kill fish and other aquatic life. The treatment plant operator must carefully control the chlorine dose to prevent wasting chlorine as well as to prevent fish kills.

In some cases, it is necessary to dechlorinate the effluent to protect aquatic life. This problem can be avoided if ultraviolet (UV) light rather than chlorine is used for disinfection of effluent. (The application of UV radiation for disinfection is discussed in Section 6.5.)

To test the effectiveness of a particular treatment method on a specific source of water or wastewater, *pilot plant* studies are often performed. This involves building and operating a small-scale treatment facility before making a major investment in the full-scale plant. Prior to construction of a pilot plant, *bench-scale* testing is done in the laboratory. An example of this procedure, as it applies to UV disinfection, is presented in the following case study.

## CASE STUDY

### UV Disinfection Meets Strict California Standards

California has strict coliform standards for secondary, unfiltered wastewater effluent, so strict, in fact, that few engineers thought that UV disinfection could reliably or economically meet these standards. However, the state's first major UV disinfection facility for treating secondary effluent has been approved and is under design at the Central Contra Costa Sanitary District's (CCCSD) 170-ML/d (45-mgd) wastewater treatment plant in Martinez, California. Once completed, the UV facility will have the capacity to treat up to 340 ML/d (90 mgd) at peak wet-weather flow. This will make it one of the largest such facilities in the United States and the world. Full-scale implementation will involve setting up 7500 UV lamps.

UV was proven to be effective through an exhaustive series of bench-scale and pilot-scale tests, completed over a 1-year period. These tests sought to answer several questions: Could UV disinfection consistently meet California's strict criteria of 240 MPN (most probable number) per 100 mL for total coliform density for discharge into Suisun Bay? Could UV be cost-effective for nonfiltered effluent?

The district embarked on this quest for alternative disinfection methods with pilot testing after finding itself in a significant dilemma. The district operates a gaseous chlorine system that provides effluent disinfection as well as odor control and control of activated sludge bulking. Dechlorination is accomplished by injecting sulfur dioxide gas into the chlorinated plant effluent as it is discharged via a 4-mi outfall into Suisun Bay.

The district's existing chlorination/dechlorination system was in need of replacement or a major upgrading. The district saw this as an opportunity to investigate alternative methods and to cease using chlorine as a disinfectant largely because of safety and air and water quality reasons. Chlorination can result in the formation of potential toxic by-products, causing possible air and water problems. In addition, use of gaseous chlorine, which involves storing a highly toxic gas, is a safety concern for treatment plant personnel and neighboring communities. Transporting chlorine gas from the generation site to the point of use also poses a public health risk. With these reasons in mind, the search was on to determine a suitable alternative.

Preliminary screening led to bench-scale testing to evaluate and compare the ability of hypochlorite, ozone, and UV disinfection to meet NPDES discharge and California disinfection requirements. The district's NPDES discharge permit requires that effluent disposed to the bay not exceed a *total* coliform density of 240 MPN/100 mL for a 5-d median—a requirement that is several times more stringent than typical national standards. These typical national standards average 200 MPN/100 mL *fecal* coliform. Also, compliance with the discharge requirement is based on a 5-d moving median and no single sample may have a total coliform bacterial density in excess of 10,000 MPN/ 100 mL.

The bench-scale testing involved sampling the effects of ozone and UV on four different wastewaters: filtered and unfiltered secondary effluent and filtered and unfiltered nitrified secondary effluent. For the purposes of this bench-scale testing, hypochlorite was used to represent chlorine. During cost evaluations, ozone and hypochlorite were eliminated because of their relatively high cost. As the only cost-competitive option of the three, UV disinfection was then pilot tested against chlorination/dechlorination to verify the initial test results and provide data for a more detailed cost comparison.

### Pilot Study Meets Objectives

The objectives of the UV pilot study included the following:

- Confirming the UV dose required for inactivation of total coliform bacteria in unfiltered secondary effluent within discharge limits
- Determining the effects of suspended solids on UV disinfection efficiency
- Evaluating the long-term reliability of UV disinfection efficiency during high suspended solids events

- Evaluating lamp fouling characteristics and determining appropriate lamp cleaning intervals
- Developing preliminary design criteria and refining the total project cost estimate

The pilot plant facilities are shown in Figure 10.19.

To determine the UV dose required for inactivation of total coliform, the standard unit for UV doses—milliwatt seconds/centimeter squared (mW s/cm$^2$) was used. It was determined that, with a clean-lamp dose of 63 mW s/cm$^2$ for dry-weather flows and a 50-mW s/cm$^2$ clean-lamp dose for wet-weather flows, UV can provide inactivation of total coliform bacteria in unfiltered secondary effluent that complies with regulatory limits. In fact, when applied to the plant's nonnitrified secondary effluent, UV disinfection provided more effective virus inactivation than chlorine disinfection. The lower wet-weather UV dose was determined based on historical records, which indicate consistently low suspended solids concentrations during wet-weather periods.

Doses are determined based on the flow rate of the treated wastewater through the UV lamp unit and the transmissibility of the UV to penetrate the wastewater. These doses depend on the clarity of the final effluent.

The strength of the lamp output is controlled by the number of UV lamps used. If a higher dose is necessary, more lamps are used, and if a lower dose is needed, fewer lamps are used. Each lamp's output is constant and cannot be turned up or down. Although this reduces operator control of the system, UV is favorable because of better safety, lower labor costs, and better virus kill.

An evaluation was conducted to determine the rate of lamp fouling and the necessary cleaning frequency, as well as the impact of fouling on the dose rate. The optimum interval was found to be a 2-week rotating cleaning schedule. As the output of the lamps declines over time, the reduced output was taken into consideration in the selection of the design dosage.

**Cost Comparison**

In the past, UV systems entailed high electrical usage and have been costly. However, recent technological advances have lowered overall costs by increasing the efficiency at which UV light is generated, enabling optimization of the UV system's performance. At the same time, costs for using chlorine are also rising, making UV increasingly attractive.

The estimated annualized cost for the new UV system is comparable to chlorination/dechlorination. The comparison is based on using planning level estimates with the capital cost annualized over 20 years at 8 percent and an average wastewater flow of 170 ML/d (45 mgd). The UV system (in the horizontal mode) has a total annual cost of $1.87 million, whereas a new chlorine-based system has a total annual cost of $2.03 million. In the new UV system, chlorine used for purposes other than effluent disinfection (such as controlling odors and activated sludge bulking) will be replaced by hypochlorite. Costs for hypochlorite are factored into the total cost of the UV system.

**Summary**

The pilot testing proved UV disinfection could provide the district with a safe disinfection process in comparison to chemical disinfectants. Additionally, the system does not require the removal of disinfectant residuals before discharge, and there is no risk to the public resulting from accidental release of hazardous gases.

As a result of this pilot study, the CCCSD received concurrence from California Department of Health officials and the Regional Water Quality Control Board that UV disinfection is a safe, reliable, and economically viable alternative to traditional chemical disinfection processes. When the full-scale UV disinfection system is completed, CCCSD, its customers, and the surrounding communities will have a disinfection system that is safe, economical, and environmentally sound.

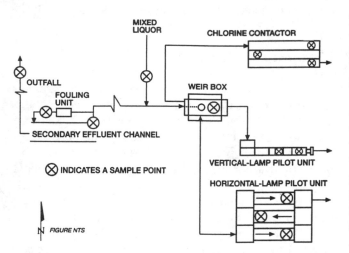

**FIGURE 10.19**

*Equipment layout for pilot study of ultraviolet wastewater disinfection.* (From *Public Works*, October 1994. Copyright © 1994 Public Works Journal Corporation. Used with permission)

## 10.4 TERTIARY (ADVANCED) TREATMENT

Secondary treatment can remove between 85 and 95 percent of the BOD and TSS in raw sanitary sewage. Generally, this leaves 30 mg/L or less of BOD and TSS in the secondary effluent. But sometimes this level of sewage treatment is not sufficient to protect the aquatic environment. For example, periodic low flow rates in a trout stream may not provide the amount of dilution of the effluent that is needed to maintain the necessary DO levels for trout survival.

Another limitation of secondary treatment is that it does not significantly reduce the effluent concentrations of nitrogen and phosphorus in the sewage. Nitrogen and phosphorus are important plant nutrients. If they are discharged into a lake, algal blooms and accelerated lake aging or cultural eutrophication may be the result. Also, the nitrogen in the sewage effluent may be present mostly in the form of ammonia compounds. These compounds are toxic to fish if the concentrations are high enough. Yet another problem with the ammonia is that it exerts a *nitrogenous oxygen demand* in the receiving water as it is converted to nitrates. This process is called nitrification, as discussed in Section 4.3.

When pollutant removal greater than that provided by secondary treatment is required, either to further reduce the BOD or TSS concentrations in the effluent or to remove plant nutrients, additional or *advanced treatment* steps are required. This is also called *tertiary treatment*, because many of the additional processes follow the primary and secondary processes in sequence.

Tertiary treatment of sewage can remove more than 99 percent of the pollutants from raw sewage and can produce an effluent of almost drinking water quality. But the cost of tertiary treatment, for operation and maintenance as well as for construction, is very high, sometimes doubling the cost of secondary treatment. The benefit-to-cost ratio is not always big enough to justify the additional expense. Nevertheless, application of some form of tertiary treatment is not uncommon. Some of the more common tertiary processes are discussed in the following sections.

### Effluent Polishing

The removal of additional BOD and TSS from secondary effluents is sometimes referred to as *effluent polishing*. It is most often accomplished using a granular-media filter, much like the filters used to purify drinking water. Since the suspended solids consist mostly of organic compounds, filtration removes BOD as well as TSS. (See Appendix H, Figure 14.)

Generally, mixed-media filters are used to achieve in-depth filtration of the effluent. Because of the organic and biodegradable nature of the suspended solids in the secondary effluent, tertiary filters must be backwashed frequently. Otherwise, decomposition would cause septic or anaerobic conditions to develop in the filter bed. In addition to the conventional backwash cycle, an auxiliary surface air-wash is used to thoroughly scour and clean the filter bed. Filtration may be done by gravity in an open tank or by pressure in closed pressure vessels.

A schematic diagram of an automatic-backwash tertiary filter is shown in Figure 10.20. The filtered water may be stored in an adjacent tank and used for backwash water when the head loss through the filter reaches a predetermined level. Some filter manufacturers mount the backwash storage tank directly above the filter, forming a single self-contained unit.

Another process, called *microstraining*, also finds application as a tertiary step in wastewater treatment for suspended solids reduction. The microstrainers, also called *microscreens*, are composed of specially woven steel wire cloth mounted around the perimeter of a large revolving drum. The steel wire cloth acts as a fine screen, with openings as small as 20 micrometers ($\mu$m, or millionths of a meter).

The rotating drum is partially submerged in the secondary effluent, which must flow into the drum and then outward through the microscreen. As the drum rotates, captured solids are carried to the top, where a high-velocity water spray flushes them into a hopper mounted on the hollow axle of the drum. A typical microscreen installation is illustrated in Figure 10.21.

### Phosphorus Removal

Phosphorus is one of the plant nutrients that contributes to the eutrophication of lakes. Raw sewage contains about 10 mg/L of phosphorus, from household detergents as well as from sanitary wastes. The phosphorus in wastewater is primarily in the form of organic phosphorus and as phosphate, $PO_4^{3-}$, compounds. Only about 30 percent of this phosphorus is removed by the bacteria in a conventional secondary sewage treatment plant, leaving about 7 mg/L of phosphorus in the effluent.

When stream or effluent standards require lower phosphorus concentrations, a tertiary treatment process must be added to the treatment plant. This usually involves *chemical precipitation* of the phosphate ions and coagulation. The organic phosphorus compounds are entrapped in the coagulant flocs that are formed and settle out in a clarifier.

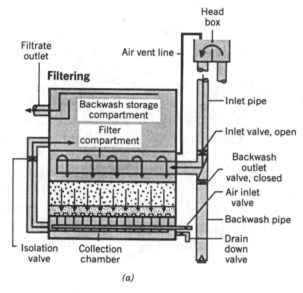

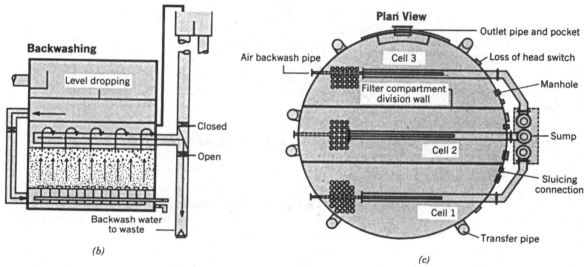

**FIGURE 10.20**

*Auto-backwash rapid filters may be used to polish the effluent in a tertiary or advanced sewage treatment plant. Diagram (a) shows the filtration mode and diagram (b) shows the backwash mode of operation. Three individual filter cells may be constructed in a single prefabricated unit, as shown in diagram (c). (Courtesy of EIMCO Process Equipment Company, Salt Lake City, Utah)*

One chemical frequently used in this process is aluminum sulfate, $Al_2SO_4$. This is called *alum*, the same coagulant chemical used to purify drinking water. The aluminum ions in the alum react with the phosphate ions in the sewage to form the insoluble precipitate called aluminum phosphate. Other coagulant chemicals that may be used to precipitate the phosphorus include ferric chloride, $FeCl_3$, and lime, $CaO$.

Adding the coagulant downstream of the secondary processes provides the greatest overall reliability for phosphorus reduction. It not only removes

about 90 percent of the phosphorus, but it removes additional TSS and serves to polish the effluent as well. But when applied in this manner, as a third or tertiary treatment step, additional flocculation and settling tanks must be built. In some cases, even filters may have to be added to remove the nonsettleable floc.

To avoid the need for construction of additional tanks and filters, in most plants requiring phosphorus removal the coagulant is added to the wastewater at some point in the conventional process. For example, alum may be added just before the primary settling

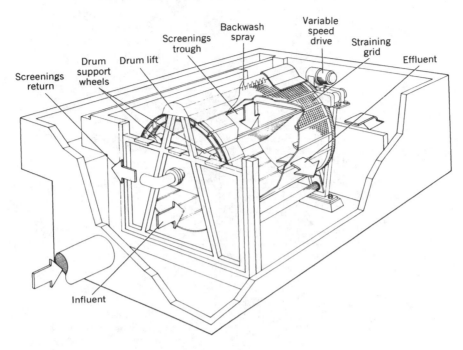

**FIGURE 10.21**
*Perspective view showing the basic components of a microstrainer unit,*
*which may be used for tertiary sewage treatment.* (Courtesy of Permutit Co.,
Inc., Paramus, New Jersey)

tanks. The resulting combination of primary and chemical sludge is removed from the primary clarifiers.

In activated sludge plants, the coagulant may be added directly into the aeration tanks. In this case, the precipitation and flocculation reactions occur along with the biochemical reactions. Sometimes, the coagulant may be added to the wastewater just before the secondary or final clarifiers. Regardless of the point in the process at which coagulant is added, the total volume and weight of sludge requiring disposal increase significantly. The options for phosphorus removal by chemical precipitation are illustrated in Figure 10.22.

## Nitrogen Removal

Nitrogen can exist in wastewater in the form of organic nitrogen, ammonia, or nitrate compounds. The effluent from a conventional activated sludge plant contains mostly the ammonia nitrogen (ammonium ion) form, $NH_4^+$. Effluents from a trickling filter or rotating biodisc may contain more of the nitrate ion form, $NO_3^-$. This is because the nitrifying bacteria, those microbes that convert ammonia to nitrate, have a chance to grow and multiply on some of the surfaces in the trickling filter or biodisc units. They do not survive in a mixed-growth

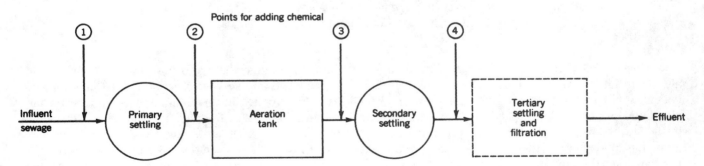

**FIGURE 10.22**
*Phosphorus can be removed from sewage by chemical precipitation; the chemical, usually alum, can be added at one of four different points in the process. Point 2 is the most common point of application.*

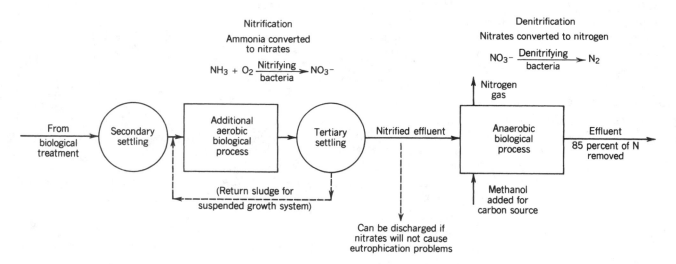

**FIGURE 10.23**

*Nitrogen can be removed from sewage to prevent eutrophication of lakes. The biological processes of* nitrification *and* denitrification *must be carried out after the basic activated sludge process is complete.*

aeration tank, where they are crowded out by faster growing bacteria that consume carbonaceous organics.

Nitrogen in the form of ammonia can be toxic to fish, and it exerts an oxygen demand in receiving waters as it is converted to nitrate. Nitrate nitrogen is one of the major nutrients that causes algal blooms and eutrophication. For these reasons, it is sometimes necessary to remove the nitrogen from the sewage effluent before discharge. This is particularly important if it is discharged directly into a lake.

One of the methods used to remove nitrogen is called *biological nitrification–denitrification.* It consists of two basic steps. First, the *secondary* effluent is introduced into another aeration tank, trickling filter, or biodisc. Since most of the carbonaceous BOD has already been removed, the microorganisms that will now thrive in this tertiary step are the nitrifying bacteria, *Nitrosomonas* and *Nitrobacter.* In this first step, called *nitrification,* the ammonia nitrogen is converted to nitrate nitrogen, producing a *nitrified effluent.* At this point, the nitrogen has not actually been removed, but only converted to a form that is not toxic to fish and that does not cause an additional oxygen demand.

A second biological treatment step is necessary to actually remove the nitrogen from the wastewater. This is called *denitrification.* It is an aerobic process in which the organic chemical methanol is added to the nitrified effluent to serve as a source of carbon. The denitrifying bacteria *Pseudomonas* and other groups use the carbon from the methanol and the oxygen from the nitrates in their metabolic processes. One product of this biochemical reaction is molecular nitrogen, $N_2$,

which escapes into the atmosphere as a gas. A schematic diagram of this process is shown in Figure 10.23.

Another method for nitrogen removal is called *ammonia stripping.* It is a physical–chemical rather than a biological process, consisting of two basic steps. First, the pH of the wastewater is raised in order to convert the ammonium ions, $NH_4^+$, to ammonia gas, $NH_3$. Second, the wastewater is cascaded down through a tall tower; this causes turbulence and contact with air, allowing the ammonia to escape as a gas. Large volumes of air are circulated through the tower to carry the gas out of the system. The combination of ammonia stripping with phosphorus removal using lime as a coagulant is advantageous, since the lime can also serve to raise the pH of the wastewater. Ammonia stripping is less expensive than biological nitrification–denitrification, but it does not work very efficiently under cold-weather conditions.

## Land Treatment of Wastewater

The application of secondary effluent onto the land surface can provide an effective alternative to the expensive and complicated advanced treatment methods previously discussed. A high-quality polished effluent can be obtained by the natural processes that occur as the effluent flows over the vegetated ground surface and percolates through the soil.

Additional benefits of land treatment are that it can provide the moisture and nutrients needed for vegetation growth and can help to recharge groundwater aquifers. In effect, land treatment of wastewater allows a direct recycling of water and nutrients for beneficial use; the sewage becomes a valuable natural resource that is not simply disposed of. But relatively large land areas are needed for this kind of treatment, and soil type as well as climate are critical factors in controlling the feasibility and design of a land treatment process.

There are three basic types or modes of land treatment: *slow rate, rapid infiltration*, and *overland flow*. The conditions under which they can function and the basic objectives of these types of treatment vary.

In the slow rate system, also called *irrigation*, vegetation is the critical component for the wastewater treatment process. Although the basic objective is wastewater treatment and disposal, another goal is to obtain an economic benefit from the use of the water and nutrients to produce marketable crops (that is, corn or grain) for animal feed. Another objective might be to conserve potable water by using secondary effluent to irrigate lawns and other landscaped areas.

A schematic diagram of a slow rate land treatment system is shown in Figure 10.24. Wastewater can be applied onto the land by ridge-and-furrow surface spreading or by sprinkler systems. The most common sprinkler system consists of a long spray boom that rotates on wheel supports around a center pivot. In ridge-and-furrow application, the wastewater flows by gravity in small ditches. In either mode of application, most of the water and nutrients are taken up or absorbed through the roots of the growing vegetation.

Not all the water applied to the land in a slow rate system is absorbed; some of it percolates through the soil to the groundwater zone. Uncontrolled surface runoff of the wastewater is not usually allowed, so the soil must have reasonably good drainage characteristics. The wastewater can be applied to the land at rates of up to 100 mm (4 in.) per week, except during the

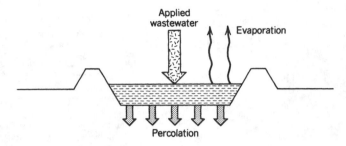

**FIGURE 10.25**
*Rapid infiltration.* (Courtesy of the Environmental Protection Agency)

winter months in northern climates. Of the three basic types of land treatment systems, the slow rate method provides the best results with respect to tertiary treatment levels of pollutant removal. Suspended solids and biochemical oxygen demand are significantly reduced by filtration of the wastewater and biological oxidation of the organics in the top few inches of soil. Nitrogen is removed primarily by crop uptake, and phosphorus is removed by adsorption within the soil.

The rapid infiltration or *infiltration–percolation* mode of land treatment has as basic objectives to recharge groundwater aquifers and to provide advanced treatment of wastewater. A schematic diagram of the rapid infiltration method is shown in Figure 10.25. Most of the secondary effluent percolates to the groundwater; very little of it is absorbed by vegetation. The filtering and adsorption action of the soil removes most of the BOD, TSS, and phosphorus from the effluent, but nitrogen removal is relatively poor. Soils must be highly permeable for the rapid infiltration method to work properly. Usually, the wastewater is applied in large ponds called *recharge basins*.

An overland flow system is illustrated in Figure 10.26. Wastewater is sprayed on a sloped terrace and allowed to flow across the vegetated surface to a runoff

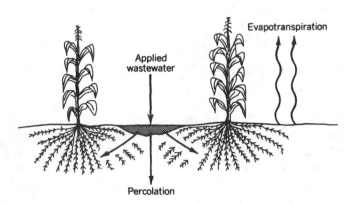

**FIGURE 10.24**
*Slow rate land treatment.* (Courtesy of the Environmental Protection Agency)

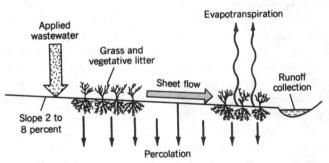

**FIGURE 10.26**
*Overland flow.* (Courtesy of the Environmental Protection Agency)

collection ditch. Purification is accomplished by physical, chemical, and biological processes as the wastewater flows in a thin film down the relatively impermeable surface. Overland flow can be used to achieve removal efficiencies for BOD and nitrogen comparable to other methods of tertiary treatment, but phosphorus removal is somewhat limited. The water collected in the ditch is usually discharged to a nearby body of surface water.

## New Treatment Technologies

Increasingly strict water quality regulations require upgrading many older sewage treatment plants. This trend, along with limited land for expansion at many older plants in urban areas, has led engineers to seek new wastewater treatment technologies. Two new techniques that have already undergone pilot studies are the *membrane bioreactor process* (MBR) and the *ballasted floc reactor* (BFR).

### Membrane Bioreactor Process

In the MBR process, aeration, secondary clarification, and filtration occur within a single bioreactor rather than in three separate basins, providing the required tertiary treatment within a smaller land area compared

to conventional treatment. Hollow-fiber microfiltration membranes are bundled into modules and grouped together in cassettes. Connected by a header pipe to effluent vacuum pumps, the cassettes are submerged in a bioreactor tank. The vacuum pumps pull the effluent through the membranes, but leave the solids behind, eliminating the need for secondary clarification and return sludge pumping. An MBR is shown schematically in Figure 10.27.

Since activated sludge stays in the tank, the mixed liquor suspended solids (MLSS) levels are much higher than in conventional activated sludge systems, thus facilitating treatment within a smaller volume. Air is supplied through coarse bubble diffusers below the membrane cassettes, providing oxygen for biological treatment and agitation to scour and clean the membranes; fine bubble diffusers are also used to supply more air for treatment. Automatic backwash cycles are also used to clean and restore membrane permeability at regular intervals. An MBR system can be built new or retrofitted into an existing activated sludge tank.

### Ballasted Floc Reactor

Ballasted flocculation is a physical–chemical process that increases the settling rate of suspended solids

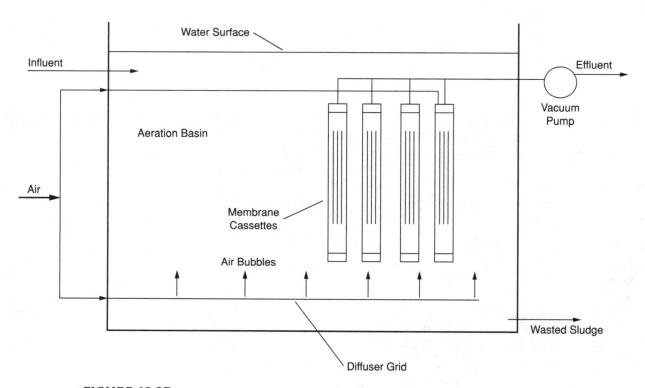

**FIGURE 10.27**
*Schematic diagram of a membrane bioreactor, which combines activated sludge with membrane filtration to accomplish tertiary treatment in a single treatment tank.*

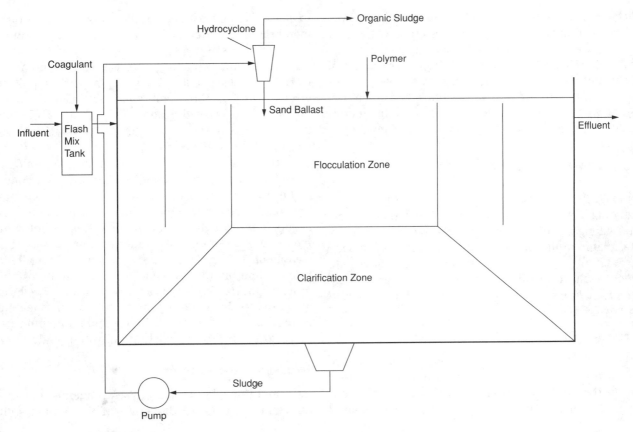

**FIGURE 10.28**

*Schematic diagram of a ballasted flocculation reactor, in which sand is added with coagulant and polymer to form faster settling flocs; the sand is separated from the organic sludge in a hydroclone and recycled.*

over that of conventional primary sedimentation processes. A coagulant is first mixed rapidly with the influent to promote flocculation. Sand and a polymer are then added to form heavier, larger floc particles, which settle rapidly; the polymer binds the sand to the organic floc particles. The clarified effluent is discharged over a weir and settled sludge is pumped to a hydroclone, where the sand is separated by centrifugal force. The sand is recycled back to the BFR and the organic sludge is pumped to an appropriate treatment process. Figure 10.28 depicts a schematic view of a BFR.

MBR and BFR systems can be used in parallel, to allow cost-effective designs. Operation of a successful pilot plant for this type of treatment was completed in 1999.*

---

*For more information see T. P. Giese, "New Wastewater Treatment Technologies," *Public Works Manual*, April 2001, Public Works Journal Corporation, Ridgewood, New Jersey.

## 10.5    ON-SITE WASTEWATER DISPOSAL

Several different sewage treatment methods are described in the preceding sections. They are usually applied at a single and centralized location to wastewater that originated from diverse sources. Usually, these centralized facilities are owned and operated by a local municipality (publicly owned treatment works, POTWs).

In lightly populated suburban or rural areas, however, it is often uneconomical to build a public sewage collection system and centralized treatment plant. Houses may be spread relatively far apart, and the shared cost for each service connection would be excessive. Instead, a system that provides for disposal of wastewater into the ground may be provided for each individual wastewater generator. This is called *on-site subsurface wastewater disposal.*

There is always a strong tendency to construct a centralized POTW as soon as possible in a developing area. Subsurface systems are looked on as a temporary mode of wastewater disposal, and because of frequent failures of such systems, they are often considered unreliable and undesirable. But if properly located, designed, and constructed, an on-site subsurface system can serve effectively for wastewater disposal on a long-term basis. Presently, about 25 percent of the new homes constructed in the United States make use of on-site subsurface disposal.

## Site Evaluation

Surface topography as well as subsurface conditions at the proposed site are important with regard to planning and design of a subsurface wastewater disposal system. One of the most important factors related to the successful operation of a subsurface system is the *texture of the soil* into which the effluent is discharged. Specifically, the permeability, or *hydraulic conductivity*, of the soil must be within an acceptable range. If it is too low, the wastewater will not be able to percolate fast enough for effective disposal. If it is too high, there may not be sufficient time for purification of the effluent before it reaches the water table.

Relatively coarse granular soils, which contain a large percentage of sand, generally have high hydraulic conductivities. They are usually suitable for subsurface disposal of wastewater, as long as the percolation rate is not excessive. Fine-grained soils, such as very fine sands and silts, offer more resistance to the flow of water. Soils containing a large fraction of clay are often unacceptable for subsurface disposal because of their extremely low permeabilities.

In addition to hydraulic conductivity, the *depth to groundwater* and the *depth to bedrock* are important factors with regard to siting and designing a subsurface disposal system. Generally, bedrock and the seasonally high groundwater table must be at least 3 m (10 ft) below the disposal system. These values will vary, depending on local health department and environmental agency regulations.

Before a decision is made regarding the use of subsurface disposal systems, a thorough investigation of site and soil conditions is necessary. Soil survey maps from the local Natural Resource Conservation Service are studied for preliminary data. In the field, a large *test pit* is excavated by a backhoe to a depth of about 4 m (12 ft). Visual observations regarding soil texture, depth to groundwater, and depth to rock are recorded.

### Percolation Test

To evaluate the ability of the soil to transmit the flow of water, a *percolation test*, or "perc test," is conducted. The perc test provides an indirect measure of the soil permeability, or hydraulic conductivity. Usually, several separate perc tests are required in the vicinity of a proposed subsurface disposal system, particularly if soil conditions are highly variable on the site.

A perc test simply measures the rate at which water seeps into the soil in a *test hole*. It is sometimes called a *falling head perc test*, because it is the rate of drop of the water level in the test hole that is measured. In addition to providing data as to whether a conventional subsurface system may be utilized, the perc test provides data for designing the size of the leaching field. Generally, perc tests must be performed by a qualified technician under the supervision of a professional engineer and witnessed by a representative of the local health department.

Specific procedures for conducting perc tests vary somewhat among local environmental agencies and health departments across the country. Although the details vary, including the dimensions of the test hole, most perc test procedures can be summarized as follows:

1. *The test hole.* A test hole about 200 mm (8 in.) in diameter is dug in the soil to the depth of the proposed leaching field, usually about 0.6 m (2 ft). This may be done on a shallow ledge adjacent to the test pit. The sides of the hole may then be scratched with a sharp tool, and the loose material is removed. An inch or two of coarse gravel may be placed at the bottom of the hole to prevent scour later on.

2. *Soaking the test hole.* The soil is soaked and saturated by filling the hole with clean water. For sandy soils, all the water is allowed to drain away; for silty soils or soils with a high clay content, the water is allowed to remain in the hole overnight and the test is conducted on the next day. After thorough soaking, the hole is again filled with clean water to a depth of about 200 mm (8 in.). At a uniform time interval of 5 to 30 min (depending on the rate of fall), the drop in the water level is measured and recorded. This is repeated until a constant rate of drop is observed.

3. *Measurement of the perc rate.* After the rate of drop becomes constant, the hole is again filled to a depth of 200 m (8 in.). The time required for the water level to drop 150 mm (6 in.) is recorded. In tight or slow-draining soils, it may be permissible to measure the drop of water in a 30-min time interval instead of waiting for a 150-mm drop.

4. *Computation of perc rate.* The perc rate for each test hole is computed in terms of minutes per inch of water level drop, or minutes per 25 mm of drop. As long as the perc rates do not vary by more than 20 min, the perc rate for the area is obtained by averaging the rates from each test hole. Otherwise, the variations in soil type are noted.

## EXAMPLE 10.9

During a percolation test, it is observed that the water level in the test hole drops at a steady rate of $\frac{3}{4}$ in. at 10-min intervals. Compute the perc rate in terms of minutes per inch.

*Solution*  The perc rate, in minutes/inch, can be calculated by simply dividing the time interval by the corresponding uniform drop in the water level:

$$\text{perc rate} = \frac{10 \text{ min}}{\frac{3}{4} \text{ in.}} = 14 \text{ min/in.} \quad \text{(rounded up)}$$

In this example, the drop in water level was measured at 10-min intervals; shorter time intervals may be used if the water level drops rapidly, or longer intervals may be used if the level drops slowly.

---

One method for measuring and timing the drop of water in a test hole is illustrated in Figure 10.29. A yardstick or meterstick is simply held against a batter board that is placed across the test hole; the board serves as a fixed reference level for taking the measurement when the bottom of the stick just penetrates the water surface.

Battery-operated devices called *percometers* are available commercially. Their operation is based on the fact that water conducts electricity; when the water level drops below the end of a probe placed in the test hole, a gage indicates that the circuit is broken, and the time interval for a predetermined drop in water level can then be measured.

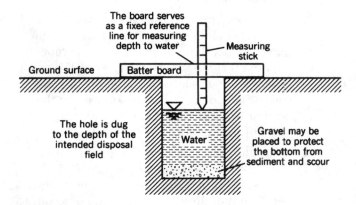

**FIGURE 10.29**
*A perc test can be conducted using a batter board and a meterstick or yardstick.*

Generally, if the perc rate is slower than 60 min/in. (40 min in some states), the soil is considered unsuitable for conventional septic systems. If the rate is faster than 1 min/in., subsurface disposal may also be prohibited in order to protect groundwater quality.

## Septic Systems

The most common type of subsurface disposal system includes a *septic tank* and an *absorption* or *leaching field*, as illustrated in Figure 10.30. Briefly, the tank serves to store settled and floating solids and the leaching field serves to distribute the effluent so that it can percolate through the soil. Decomposition of organics takes place under anaerobic conditions, hence the name *septic system*

### The Septic Tank

Settleable and floating solids must be removed from the raw wastewater before it can be applied to the soil in the leaching field. A buried septic tank is used to provide the necessary primary treatment step, as well as to act as a sludge storage tank. Floating solids and grease are trapped by a baffle at the tank outlet and prevented from entering the leaching field. Settleable solids accumulate in the sludge zone at the bottom of the tank, and primary effluent flows out into the leaching field.

Some of the sludge decomposes under anaerobic conditions, but eventually the undecomposed solids accumulate and the sludge level rises. To ensure proper operation and a long service life, septic tanks should be pumped clean on a routine basis, every few years. If too much sludge is allowed to accumulate in the tank, solids will eventually be carried out in the effluent and will plug up the absorption field. This is one of the most common causes of failure for septic systems.

Septic tanks are manufactured in a variety of shapes and sizes. A typical small rectangular concrete tank is shown in Figure 10.31. The minimum-size tank allowed for an individual household is generally about 3800 L (1000 gal) of liquid capacity. Septic tanks must be watertight and have adequate access for inspection and cleaning. The top of the tank is usually about 300 mm (1 ft) below the ground surface.

### The Leaching Field

The effluent from the septic tank flows into an absorption or leaching field, which distributes the liquid uniformly over a sizable area. From the leaching field, the effluent percolates downward to the water table. As it flows through the soil voids, microorganisms

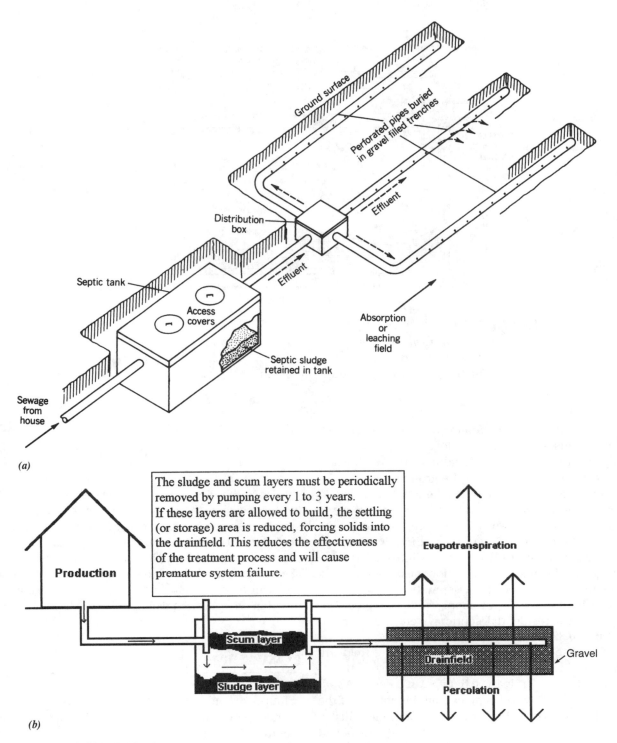

**FIGURE 10.30**

(a) *Perspective view of an on-site subsurface sewage disposal system that uti-lizes a septic tank and leaching or absorption field.* (b) *As the effluent enters the drainfield, most of it percolates through the gravel, in which pockets of oxygen allow aerobic bacteria to live and pathogenic bacteria perish. Phosphorus and nitrogen are metabolized by the vegetation covering the drainfield and a portion of the moisture is returned to the atmosphere through evapotranspiration.* (Adapted from J. vonMeier, *Groundwater Contamination and Septic System Failure,* 1996. Used with permission)

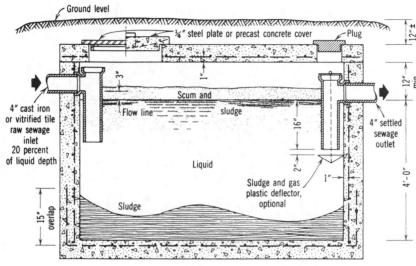

**FIGURE 10.31**
*Cross section of a typical septic tank.* (From *Environmental Engineering and Sanitation*, 4th ed., J. Salvato. Copyright © 1992 John Wiley & Sons, Inc. This material is used by permission of John Wiley & Sons, Inc.)

and other pollutants are removed from the effluent. Filtration, adsorption, and biological decomposition each plays a role in the purification of the wastewater effluent before it is diluted in the groundwater.

A very common type of leaching field consists of two or more separate trenches with pipes that serve to spread the wastewater. Before reaching the trenches, the effluent flows into a *distribution box*, which serves to evenly divide the flow into each trench. Local health department regulations usually require a minimum of 30 m (100 ft) of separation between the leaching field and a water supply well. A typical layout plan for a residential septic system is illustrated in Figure 10.32.

The disposal trenches in the leaching field are shallow excavations that are a minimum of 0.3 m (1 ft) wide and at least 0.6 m (2 ft) deep. About 150 mm (6 in.) of gravel is placed on the bottom of the trench to support the line of perforated effluent distribution pipe. The pipe is covered with more gravel and then with straw or paper to prevent the final layer of backfill soil from penetrating the gravel voids. The slope of the pipe is generally about 0.5 percent or less. On sites with relatively steep slopes, the trenches run roughly parallel to the ground contours. Individual trenches, or *laterals* as they are sometimes called, should be less than 30 m (100 ft) long and should be separated from each other by a distance of at least 1.8 m (6 ft). A typical trench cross section is illustrated in Figure 10.33.

The design of an absorption field involves the determination of the required number and lengths of the trenches. Data from the percolation test are very important in this regard. For residential systems, a chart similar to the one shown in Figure 10.34 may be used by the designer; the requirements differ somewhat for systems serving commercial establishments, and actual design methods vary nationwide.

As seen in the graph of Figure 10.34, the required absorption area is related to the soil perc rate. Absorption area can be converted to linear feet of trench, as illustrated in the following example.

## EXAMPLE 10.10

A four-bedroom home is situated on a site that has an average perc rate of 30 min/in. How many laterals are required if the trench is 2 ft wide? What are the overall dimensions of the leaching field?

*Solution*  From Figure 10.34, determine that 250 ft$^2$ of absorption area is required per bedroom. For a four-bedroom home, a total of $4 \times 250 = 1000$ ft$^2$ of absorption area is needed for the leaching field.

Since the trench width is to be 2 ft, there is a need for 1000 ft$^2 \div 2$ ft $= 500$ ft of trench. Also, since the maximum length of an individual lateral is 100 ft, there is a need for 500 ft $\div$ 100 ft/lateral, or 5 laterals.

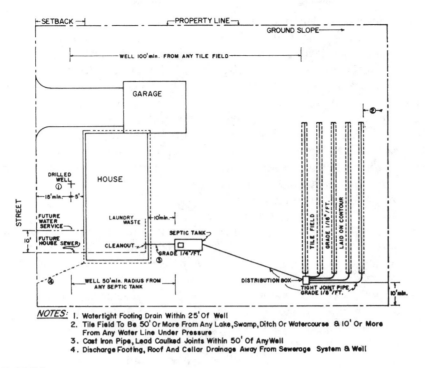

**FIGURE 10.32**

*Plan view of a typical residential on-site subsurface sewage disposal system. (From Environmental Engineering and Sanitation, 4th ed., J. Salvato. Copyright © 1992 John Wiley & Sons. This material is used by permission of John Wiley & Sons, Inc.)*

Each lateral will be separated by a distance of 6 ft. Since there are 4 spaces between the 5 laterals, the width of the absorption field will be $6 \times 4 = 24$ ft. The entire leaching field will occupy $24 \times 100 = 2400$ ft$^2$ of the site (0.022 ha, or 0.0055 ac).

Septic system failures are not uncommon. System failures involve overflow of septic effluent onto the ground and into surface waters, causing public health nuisances and water pollution. Such failures usually can be prevented. All septic tanks require periodic pumping to remove accumulated sludge deposits and prevent

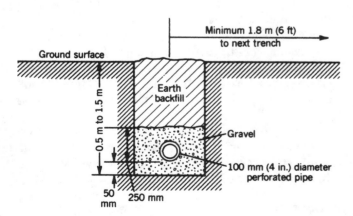

**FIGURE 10.33**

*Cross section through an absorption field trench.*

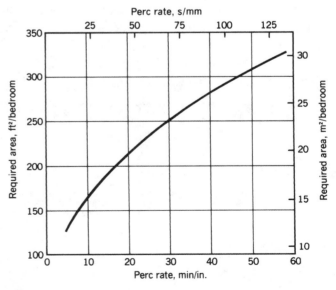

**FIGURE 10.34**

*Typical absorption or leaching field area requirements for private residencies; total required area depends on the perc rate and on the number of bedrooms in the home.*

solids overflow into the drainfield. Neglect of this simple requirement is a frequent cause of drainfield blockage and system failure. Abuse of the septic system, by using it for disposal of harsh chemicals or other inappropriate wastes, and by excessive use of household water, may contribute to the frequency of failures. Washing machines use much water and may be a contributing factor in septic system failures. In addition to the volume of wastewater generated, the amount of lint (material fibers) and chemicals in washing machine effluent discharges can be key factors of system failure. Special filters or "septic protectors" are commercially available for removing fibers from washing machine discharges and prolonging the functional life of most septic systems.

### Seepage Pits

When the site is too small for a conventional leaching field, deeper excavations that take up less area are sometimes used for subsurface disposal. These excavations are called *seepage pits* or *dry wells*. They also may be selected instead of trenches so as to utilize more favorable deeper soil.

Effluent from the septic tank flows into the pit, where it is stored until it seeps out through the side wall and bottom. A cross section of a typical seepage pit is shown in Figure 10.35. Although the use of seepage pits is usually discouraged by local health departments, it is sometimes acceptable for small wastewater flow rates. It is very important that the high water table be at least 1.2 m (4 ft) below the bottom of the pit, to protect groundwater quality.

The required diameter and depth of a seepage pit depend on the perc rates and thicknesses of the different layers of soil encountered in the excavation. Soil layers that have perc rates slower than 30 min/in. are not included in the design computations. A typical seepage pit is about 3 m (10 ft) in diameter and about 4.5 m (15 ft) deep. When more than one pit is required on a given site, a separation of at least three times the diameter is required. A seepage pit can be dug with conventional excavation machinery, but care must be taken not to excavate when the soil is wet; smearing of the soil will reduce its absorptive capacity.

### Mounds

To overcome site restrictions that prohibit the use of either seepage pits or leaching fields, a *mound system* can be used. Typical site restrictions include soils with very slow perc rates and conditions where bedrock or the water table is close to the ground surface. A mound is an effluent absorption system that is raised above the natural ground surface. It provides a suitable fill material for percolation as well as adequate separation from bedrock or the water table. A cross section of a typical mound system is shown in Figure 10.36.

The fill material is usually a medium-texture sand from locally available sources. A bed of gravel and distribution laterals are placed in the upper part of the mound for uniform absorption of the effluent from the septic tank. The effluent must be intermittently applied into the absorption area. The dimensions of a mound system depend on the natural soil conditions, the ground slope, and the depth of fill below the absorption area.

### Other On-Site Systems

The combination of a septic tank and leaching field is the most common type of system for on-site wastewater treatment and disposal. The septic tank provides minimal treatment, but the quality of the effluent need not be high since the assimilative or purifying capacity of the soil is usually enough to protect the groundwater. The leaching field provides a means of uniform effluent distribution into the soil. As previously discussed, certain restrictions on subsurface disposal into the soil can be overcome with the use of seepage pits or mound systems.

In some instances, however, soil or site conditions are such that subsurface disposal is not feasible at all, yet it is still necessary to dispose of the wastewater on site because of the lack of a nearby POTW. Several alternatives to disposal by subsurface absorption are available, but they generally are somewhat more expensive to build, operate, and maintain than subsurface

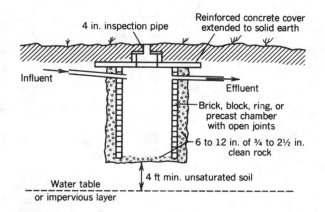

**FIGURE 10.35**
*Cross section of a typical* seepage pit *for on-site disposal of wastewater. A seepage pit is preceded by a septic tank.* (Courtesy of the Environmental Protection Agency)

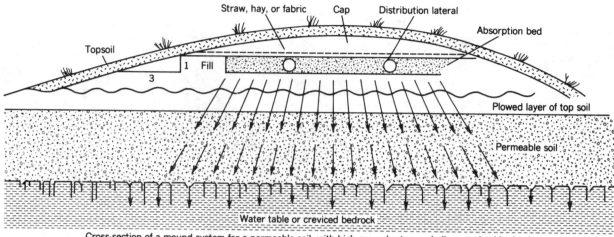

Cross-section of a mound system for a permeable soil, with high groundwater or shallow creviced bedrock

**FIGURE 10.36**

*A* mound system *for on-site sewage disposal.* (Courtesy of the Environmental
Protection Agency)

systems. These include the use of *evapotranspiration
systems* and *intermittent sand filters*. Both of these
systems can be used following a septic tank. If a higher
level of treatment is required, a small aerobic treatment
unit, such as a package extended aeration tank, may be
installed on site.

**Evapotranspiration Systems**

An evapotranspiration (ET) system consists of a sand
bed, a network of perforated distribution pipes, and an
impermeable liner that prevents the treatment effluent
from reaching the water table. In some cases, the liner
may be omitted in order to allow some of the effluent
to seep into the soil. But the basic objective of this type
of system is to dispose of the wastewater into the at-

mosphere and to avoid the need for discharge to either
surface or groundwater. A cross section of a typical ET
system is shown in Figure 10.37.

Effluent from a septic tank is distributed throughout the bed in the perforated pipe network. The effluent
rises through the sand by capillary action and then
evaporates into the air. Grass or other vegetation growing on the top of the bed serves to absorb some of the
wastewater in the root zone and transpire it into the air
through the leaves. Hence the name, *evapotranspiration* system. One of the most critical factors controlling the use and design of an ET system is the local
climate, which affects the rate of evaporation.

Specifically, factors such as average annual rainfall, wind speed, humidity, solar radiation, and temperature are of importance. Evapotranspiration systems

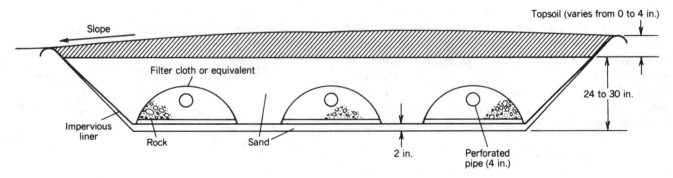

**FIGURE 10.37**

*An* evapotranspiration system *for on-site sewage disposal.* (Courtesy of the
Environmental Protection Agency)

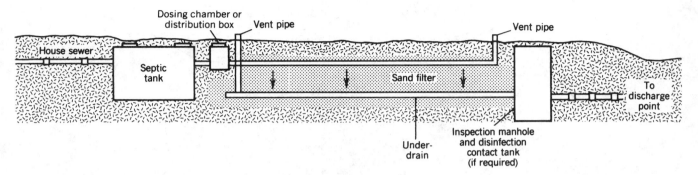

**FIGURE 10.38**
*Cross section through a typical buried sand filter for on-site sewage disposal.*
(Courtesy of the Environmental Protection Agency)

operate effectively in areas where the evaporation rate exceeds the rate of precipitation, such as in the southwestern United States. Transpiration by plants increases the amount of water vapor discharged into the air in soil-covered systems, but only during daylight hours of the growing season.

An important design parameter for an ET system is the hydraulic loading rate, which must be low enough to prevent the bed from filling completely with effluent. A typical loading rate is about $2 \text{ L/m}^2\text{/d}$ ($0.05 \text{ gpd/ft}^2\text{/d}$) in the southwestern part of the country. The hydraulic loading rate is determined from the difference between the rate of evaporation and the rate of precipitation. A typical ET system for a single-family residence in the Southwest would require about $465 \text{ m}^2$ ($5000 \text{ ft}^2$) of land area.

### Intermittent Sand Filters

One of the oldest methods of wastewater treatment involves intermittent application of settled wastewater to a bed of sand. The sand bed is usually about 1 m (3 ft) deep and is underdrained with gravel and collecting pipes. The collected effluent may be disinfected with chlorine before discharge to land or surface waters.

The filters may be built as open units at ground level, or they may be buried in the ground. They provide efficient treatment while requiring a minimum of maintenance. Many of these systems are in use throughout the United States, providing on-site treatment for homes as well as for small commercial establishments. A profile view of a typical buried filter is shown in Figure 10.38.

The upper perforated distribution lines are level and are spaced about 2 m (6 ft) apart. The underdrain pipes are perforated or open joint lines on a slope of about 0.5 percent. They are spaced about 4 m (12 ft) apart. The hydraulic load on a buried filter is about $0.04 \text{ m}^3\text{/m}^2 \cdot \text{d}$ ($1 \text{ gpd/ft}^2$). The finished grade of the topsoil over the filter is mounded to direct runoff away from the bed.

*Dosing*, or application of the wastewater to the sand bed, is an important factor related to the operation and performance of the filter. Large filters must be dosed intermittently, about two times per day, so as to flood the entire unit. It is important to have a sufficient resting period between doses to maintain aerobic conditions in the bed.

Small tanks called dosing chambers store the wastewater until it is applied to the filter by a pump or siphon device. A typical siphon dosing chamber is shown in Figure 10.39. The siphon automatically discharges when the wastewater in the tank reaches a level called the *drawing depth*; no mechanical or electrical controls are needed. It is activated by the pressure head of the wastewater. Cast-iron or fiberglass siphons are commercially available in a range of sizes. Siphon chambers require only minimal routine maintenance.

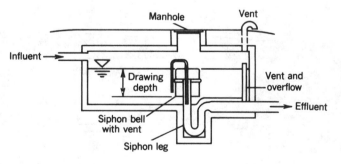

**FIGURE 10.39**
*Section view of a typical dosing chamber and siphon device.* (Courtesy of the Environmental Protection Agency)

## 10.6   SLUDGE MANAGEMENT

Suspended solids removed from wastewater during sedimentation and then concentrated for further treatment and disposal are called *sludge* or *biosolids*. Even in fully aerobic waste treatment processes in which sludge is repeatedly recycled, most of the sludge must eventually be removed from the system. As sewage treatment standards have become more stringent because of increasing environmental regulations, so has the volume of sewage sludge increased. Before it can be disposed of, sludge requires some form of treatment to reduce its volume and to stabilize it. The task of treating and disposing of this material is called *sludge management.*

Sludge forms initially as a 3 to 7 percent suspension of solids, and each person typically generates about 15 L (4 gal) of sludge per week. For a small city of about 500,000 people, that adds up to about 1 mil L (250,000 gal) per day. Because of the volume and nature of the material, sludge management is a major factor in the design and operation of all water pollution control plants. *It can account for more than half of the total costs in a typical secondary treatment plant.*

A discussion of sewage sludge characteristics, common treatment methods, and final disposal options is presented in this section.

### Sludge Characteristics

The composition and characteristics of sewage sludge vary widely. Since no two wastewaters are alike, the sludges produced will differ. Furthermore, sludge characteristics change considerably with time. Wastewater sludge typically contains organics (proteins, carbohydrates, fats, oils), microbes (bacteria, viruses, protozoa), nutrients (phosphates and nitrates), and a variety of household and industrial chemicals. The higher the levels of heavy metals and toxic compounds, the greater is the risk to humans and the environment. A key physical characteristic is the *solids concentration,* because this defines the volume of sludge that must be handled. It also determines whether the sludge behaves as a liquid or a solid. Sludges tend to act like plastic fluids as the solids concentration increases until a relatively solid state is reached.

The amount of sludge solids generated in a wastewater (or drinking water) treatment system depends largely on the degree of treatment provided and on the amount of chemicals added. Sludge initially forms at the bottom of a clarifier or settling tank in the form of a concentrated slurry of solids suspended in water. Its volume depends on the relative amounts of solid

material and water. The concentration of the sludge is expressed as a percentage by weight or mass. For example, a mass of 100 kg of liquid sludge that contains 3 kg of solids and 97 kg of water will have a concentration of 3/100 or 3 percent solids. This is equivalent to a solids concentration of 30,000 mg/L.

The concentration of solids has a very significant effect on the total volume occupied by the liquid sludge. *The total sludge volume is inversely proportional to the solids concentration.* For example, if the percentage of solids is doubled, say from 3 to 6 percent, then the total sludge volume will be decreased to half its original volume. This is a very important relationship. Increasing the solids concentration in the slurry by a few percentage points can significantly reduce the total sludge volume. This reduces the cost of handling, treating, and disposing of the sludge.

For practical purposes, it can be assumed that the weight or mass per unit volume of liquid sludge is the same as that of pure water. For example, the mass of 1 m³ of water is 1000 kg; the mass of the same volume of a relatively thick sludge (such as 10 percent solids) is only about 1020 kg. The difference of 2 percent can be neglected for design or operational purposes. Based on this, the relationship between sludge concentration and sludge volumes can be expressed as follows:

$$S = \frac{M}{V} \times 100 \qquad \textbf{(10-7a)}$$

or

$$S = \frac{W}{8.34 \times V} \times 100 \qquad \textbf{(10-7b)}$$

where  $S$ = sludge solids concentration, in percent
   $M$ = dry sludge solids, kg (Equation 10-7a)
   $W$ = dry sludge solids, lb (Equation 10-7b)
   $V$ = total sludge volume, L (Equation 10-7a) or gal (Equation 10-7b)
   8.34 = number of pounds per gallon (Equation 10-7b)

### EXAMPLE 10.11

A sludge with a 6 percent solids concentration occupies a total volume of 300 m³. *(a)* What is the water content of the sludge? *(b)* What is the mass of the sludge solids? *(c)* If the sludge is further concentrated (or dewatered) to a volume of 200 m³, what will the solids concentration be? What will the water content be?

*Solution*   (a) If 6 percent of the sludge consists of dry solids, then the water content is simply the difference, or $100 - 6 = 94$ percent.

(b) First,

$$V = 300 \text{ m}^3 \times 1000 \text{ L/m}^3 = 300\ 000 \text{ L}$$

Applying Equation 10-7a gives

$$6 = \frac{M}{300,000} \times 100$$

and

$$M = \frac{6 \times 300,000}{100} = 18\ 000 \text{ kg of dry solids}$$

(c) Again applying Equation 10-7a yields

$$S = \frac{18,000}{200,000} \times 10 = 9 \text{ percent solids}$$

The water content will be $100 - 9 = 91$ percent.

---

### EXAMPLE 10.12

---

A volume of 500,000 gal of sludge contains 100,000 lb of dry solids. What is the solids content of the sludge expressed in percent? If the sludge is concentrated to 4 percent solids, what will its volume be?

*Solution*   First, applying Equation 10-7b gives

$$S = \frac{100,000}{8.34 \times 500,000} \times 100 = 2.4 \text{ percent}$$

Again applying Equation 10-7b yields

$$4 = \frac{100,000}{8.34 \times V} \times 100$$

and

$$V = \frac{100,000}{4 \times 8.34} \times 100 = 300,000 \text{ gal}$$

---

Primary sludge typically has a solids concentration of about 7 percent. Secondary sludge has a much lower solids concentration because the suspended solids are mostly biological flocs that settle slowly and do not compact to as high a density as primary sludge. Waste-activated sludge contains only about 2 percent solids or less.

Primary and secondary sludge solids are mostly organic materials, with a volatile fraction of up to 0.8. Primary sludge gives off a strong and offensive odor; it can quickly become septic and difficult to handle. In addition to volume reduction, a basic objective of sewage sludge treatment is to stabilize the biodegradable organic solids and to render the sludge unoffensive and easy to handle.

Water treatment plant sludges are mostly inert chemical precipitates that are relatively stable and unoffensive. Quantities of this type of sludge can vary widely depending on the amount and type of chemicals used and on the composition of the raw water.

The *quantity of primary sludge* produced in a sewage treatment plant depends on the concentration of suspended solids in the raw wastewater and on the TSS removal efficiency. It can be estimated from the following expressions:

$$\text{mass} = E \times \text{TSS} \times Q \qquad \textbf{(10-8a)}$$

$$\text{weight} = E \times \text{TSS} \times Q \times 8.34 \qquad \textbf{(10-8b)}$$

where mass = dry sludge solids, kg (Equation 10-8a)
weight = dry sludge solids, lb (Equation 10-8b)
$E$ = TSS removal efficiency, decimal form
$Q$ = sewage flow rate, ML/d
(Equation 10-8a) or mgd
(Equation 10-8b)

The *quantity of secondary sludge* produced in a sewage treatment plant depends on the BOD concentration and on the fraction of BOD that is converted to biological solids (microbe cells). It can be estimated by the following expressions:

$$\text{mass} = K \times \text{BOD} \times Q \qquad \textbf{(10-9a)}$$

$$\text{weight} = K \times \text{BOD} \times Q \times 8.34 \qquad \textbf{(10-9b)}$$

where $K$ = a coefficient that represents the proportion of BOD converted to biological solids, in decimal form
BOD = applied 5-d BOD, mg/L

The value of $K$ depends on the organic loading or F/M ratio; a typical value for extended aeration or fixed growth systems is $K = 0.25$, and for conventional or step aeration processes, a typical value is $K = 0.35$.

## EXAMPLE 10.13

A flow of 4 ML/d with raw TSS = 240 mg/L and raw BOD = 220 mg/L enters a trickling filter sewage treatment plant. Removal efficiency in the primary settling tank is 50 percent for TSS and 30 percent for BOD. *(a)* Compute the mass of primary sludge solids generated per day. *(b)* Compute the mass of secondary sludge solids generated per day. *(c)* If the primary and secondary sludges are combined and the mixture has a solids concentration of 4 percent, what is the total volume of sludge generated per day?

*Solution* *(a)* Applying Equation 10-8a gives

$$\text{mass} = 0.50 \times 240 \times 4$$
$$= 480 \text{ kg/d of primary sludge}$$

*(b)* First, compute the BOD applied to the secondary system:

$$\text{applied BOD} = (1.0 - 0.3) \times 220 \approx 150 \text{ mg/L}$$

Using Equation 10-9a yields

$$\text{mass} = 0.25 \times 150 \times 4$$
$$= 150 \text{ kg/d of secondary sludge}$$

*(c)* The total mass of combined primary and secondary sludge is 480 + 150 = 630 kg/d. Now, applying Equation 10-7a gives

$$4 = \frac{630}{V} \times 100$$

and

$$V = 63{,}000 \div 4 = 15\,750 \text{ L/d}$$
$$\approx 16 \text{ m}^3/\text{d total volume of sludge}$$

## Sludge Treatment

Sludge is treated prior to ultimate disposal for two basic reasons: *volume reduction* and *stabilization of organics*. Stabilized sludge does not have an offensive odor and can be handled without causing a nuisance or health hazard. A reduced sludge volume minimizes pumping and storage requirements and lowers overall sludge-handling costs.

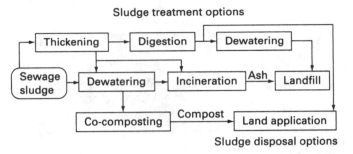

Sludge treatment options

Sludge disposal options

### FIGURE 10.40
*Alternative pathways for sewage sludge treatment and disposal. Sludge disposal in the ocean is no longer allowed in the United States.*

Several processes are available for accomplishing these two basic objectives. They include sludge *thickening*, *digestion*, *dewatering*, and *co-composting*. Typical sludge treatment options are shown in Figure 10.40. Incineration is discussed later as a final disposal option. Co-composting of sludge with garbage and yard waste is discussed in Section 11.4.

### Sludge Thickening

It is usually impractical to treat watery or thin sludges that have solids concentrations of less than about 4 percent. Waste-activated sludge is an example of a thin sludge. Thickening is a physical process that serves to increase the solids concentration of the sludge. Since sludge volume varies inversely with the solids concentration, doubling the solids content from 3 to 6 percent, for example, will cut the total sludge volume in half.

A treatment unit called a *gravity thickener* is usually used to increase the solids concentration. It resembles a circular sewage sedimentation tank that is equipped with vertical slats or pickets attached to the sludge scraper arm. As the scraper arm slowly rotates, the pickets gently stir the sludge. The stirring action serves to release trapped water and gases from the sludge, allowing it to become denser or thicker. The thickened *underflow* of sludge is withdrawn from the bottom of the tank; the effluent or *supernatant* overflows a weir and is pumped back to the inlet of the treatment plant. Sludge can be thickened to more than 10 percent solids in a gravity thickener.

An alternative to gravity thickening is a process called *dissolved-air flotation*, which is particularly effective for very thin sludges. In this process, air is forced into solution under pressure and then mixed with influent sludge, as shown in Figure 10.41. Air bubbles come out of solution in the open flotation tank, carrying sludge flocs to the liquid surface as they rise. A layer of thickened sludge forms and is skimmed from

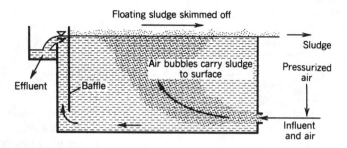

**FIGURE 10.41**
*The dissolved-air sludge flotation process can be used to thicken sewage sludge.*

the surface; the floating sludge layer may be about 400 mm (16 in.) thick. Up to about 6 percent solids concentrations can be obtained in the layer of thickened sludge by dissolved-air flotation.

## Sludge Digestion

Sludge digestion is a process in which biochemical decomposition of the organic solids occurs; in the decomposition process, the organics are converted into simpler and more stable substances. Digestion also reduces the total mass or weight of sludge solids, destroys pathogens, and makes it easier to dry or *dewater* the sludge. Well-digested sludge has the appearance and characteristics of a rich potting soil.

Sludge may be digested under aerobic or anaerobic conditions. Most large municipal sewage treatment plants use anaerobic digestion; aerobic digestion finds application primarily in small, prefabricated (package)

activated sludge treatment systems. Anaerobic digestion offers an energy-saving advantage over aerobic digestion because the anaerobic process produces methane gas. The methane may be burned to provide power for other plant processes and equipment as well as to heat the digester unit. Aerobic systems, on the other hand, are less expensive to build and have fewer operational problems.

In most modern anaerobic systems, digestion takes place in two covered circular tanks. These tanks are typically about 25 m (80 ft) in diameter and about 15 m (50 ft) deep. The sludge in the first tank is heated to a temperature of about 35°C (95°F) and is thoroughly mixed. A diagram of this type of digestion system is shown in Figure 10.42.

The digestion process is essentially completed in the first tank within about 15 d of detention time. The sludge then flows to a second tank, which serves primarily for sludge settling and storage. The digester supernatant, high in BOD and TSS, is pumped back to the inlet of the treatment plant. Digested sludge is removed from the second tank for dewatering and final disposal.

Many complex biochemical reactions occur during anaerobic digestion of sewage sludge. Certain species of bacteria first metabolize the complex organic compounds, breaking them down into simpler molecules called organic acids. Then a different species of bacteria metabolizes the organic acids, forming methane. These methane-forming bacteria grow slower and are much more sensitive to temperature and pH than are the acid-formers. The treatment plant operator must maintain careful control over the process to

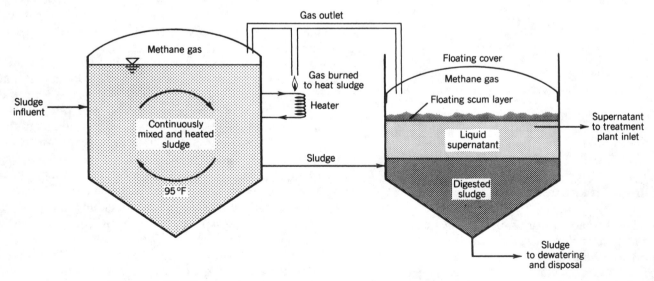

**FIGURE 10.42**
*Schematic diagram of the two-stage anaerobic sludge digestion process.*
(Courtesy of Envirex., a Rexnord Company, Waukesha, Wisconsin)

keep the methane-formers alive and healthy; without them, the digestion process cannot continue.

Methane is an important end product of the anaerobic digestion process. Its energy value can be utilized in the treatment plant. Even if it is not burned for energy, it serves as an indicator to the plant operator of whether the process is working properly. The methane content of the sludge digester gas is monitored; if the methane content starts to decrease, corrective action, such as adjusting the pH, must be taken immediately. Once the methane-forming bacteria die off, the tank must be emptied and the process started up again with fresh sludge.

In aerobic sludge digestion, the alternative to the anaerobic process, the sludge is aerated in an open tank that is similar to an activated sludge tank. Aerobic digestion is usually used in the prefabricated contact stabilization or extended aeration treatment systems. The sludge is aerated for about 30 d, in which time most of the organics are stabilized and the amount of sludge is reduced significantly.

The underlying premise of the aerobic digestion process is that decomposition takes place faster than the rate at which new bacterial cells grow. The BOD loading on the digestion tank is kept very low, so eventually the microorganisms consume their own cellular mass. This is called *endogenous decay*. But not all the sludge decomposes, and some of it eventually has to be removed from the tank for disposal. It is a thin sludge that is difficult to thicken or dewater without some type of additional treatment. Although aerobic digestion is more stable in operation than the easily upset anaerobic process, it has the disadvantage of having high energy requirements for aeration.

### Sludge Dewatering

The process of removing enough water from a liquid sludge in order to change its consistency to that of moist earth is called *sludge dewatering*. Although the process is also called *sludge drying*, the "dry" or dewatered sludge may still contain a significant amount of water, often as much as 70 percent. But at moisture contents of 70 percent or less, the sludge is *liftable*, that is, it no longer behaves as a liquid and can be handled manually or mechanically. Sludge is usually dewatered prior to land burial or incineration.

Several methods are available for dewatering sludge. The simplest method is to spread the digested sludge on an open bed of sand and allow it to remain there until it dries; drying takes place by a combination of evaporation and gravity drainage. A piping system built under the sand bed collects the water that drains from the sludge. The collected water is pumped back to the inlet of the treatment plant. A cross section of a typical *sludge drying bed*, as it is called, is shown in Figure 10.43.

Sludge drying beds usually consist of about 200 mm (8 in.) of sand placed on top of a layer of gravel or crushed stone; the pipes that make up the underdrain system are placed in the gravel or crushed stone layer. Since a relatively large amount of land area may be required to construct the sand beds, this method of sludge drying is more common in rural or suburban communities than in more densely populated urban areas.

Sludge is applied to the sand beds to depths up to 300 mm (12 in.). A typical dewatered *sludge cake*, as it is called, has a solids content of about 40 percent; this level of dewatering can be obtained after about 6 weeks of drying. At this point, the sludge can be removed from the sand manually with a pitchfork or with machinery such as a front-end loader. Sometimes it is necessary to build a glass enclosure, much like a greenhouse, over the sludge beds to protect the sludge from rain and to reduce the drying time in cold weather.

When not enough land is available at the plant site to build sludge drying beds, a mechanical system may be used to dewater the sludge. In addition to requiring less space, a mechanical system offers a greater degree of operational control, particularly when thin secondary

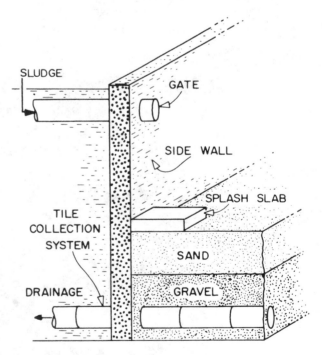

**FIGURE 10.43**

*A section of a sludge drying bed.* (From *Operation of Wastewater Treatment Plants*, MOP/11, 1976, with permission of the Water Pollution Control Federation, Washington, D.C.)

sludges are being handled. The most commonly used mechanical systems for sludge dewatering include the *rotary-drum vacuum filter* and the *centrifuge.*

The rotary-drum vacuum filter consists of a large cylindrical drum covered with a special filtering fabric. The drum rotates partially submerged in a vat that contains the sludge. A vacuum or suction is applied inside the drum, drawing the sludge up against the fabric cover and extracting the water. A thin layer of dewatered sludge, called *filter cake,* adheres to the fabric. Further rotation of the drum carries the filter cake to a stationary blade that scrapes the sludge into a hopper as the drum moves. The fabric is washed by powerful water sprays before reentering the vat.

A typical vacuum filter installation is shown in Figure 10.44. It is often necessary to add certain chemicals to the sludge in order to coagulate the solids and to improve the drainability. This is called *sludge conditioning.* In this process, chemicals such as ferric chloride, lime, or organic polymers are mixed with the sludge before it enters the vat of the vacuum filter.

In the centrifuge dewatering system, the sludge is pumped into a horizontal cylinder, or *bowl,* which rotates at high speed. Sludge conditioning chemicals are also injected into the bowl. The solids are forced to the wall of the rotating bowl by the effect of inertia. The liquid supernatant is pumped back to the head of the plant for treatment. Centrifuges are entirely enclosed, thereby reducing odor problems, but they are sometimes difficult to maintain because of the high speed of the equipment.

## Sludge Disposal

Widely employed methods for final disposal of wastewater sludge have included *ocean dumping, landfilling, incineration, land application,* and *sale as fertilizer.*

Ocean dumping is no longer a viable option in the United States, having been legislated out of existence under the 1988 *Ocean Dumping Act.* Landfilling, or co-disposal with solid waste at sanitary landfills, is severely restricted for a number of reasons. These include the closing of existing landfills and increasing restrictions on new ones opening. Landfills that remain open are raising disposal costs (tipping fees), which is forcing a search for less expensive disposal options. When sludge has to be shipped large distances to a landfill, there is even more of an economic burden. Finally, there may be environmental risks associated with the disposal of sludge in landfills.

### Incineration of Sludge

Sewage sludge can be burned (incinerated), converting the organic solids into mostly water vapor and carbon dioxide and leaving behind an inert ash residue. Properly dewatered sludge can be burned in a *multiple-hearth furnace,* a *fluidized-bed incinerator,* and other types

**FIGURE 10.44**
*A sludge vacuum filter.* (Courtesy of Envirex Inc., a Rexnord Company, Waukesha, Wisconsin)

of incinerators. In the multiple-hearth furnace, dewatered sludge enters at the top (Figure 10.45a). As it passes downward through a series of hearths, it is dried and heated to the ignition point. Gas or oil burners furnish the heat for start-up, and the sludge itself serves as a fuel to sustain the process. Ash is withdrawn from the bottom hearth, and exhaust gases must pass through air-cleaning devices.

In the fluidized-bed incinerator, an upward flow of air mixed with sludge is forced through a bed of hot sand; the air causes the bed to expand or *fluidize*. The sand is preheated to about 800°C (1500°F). The sludge

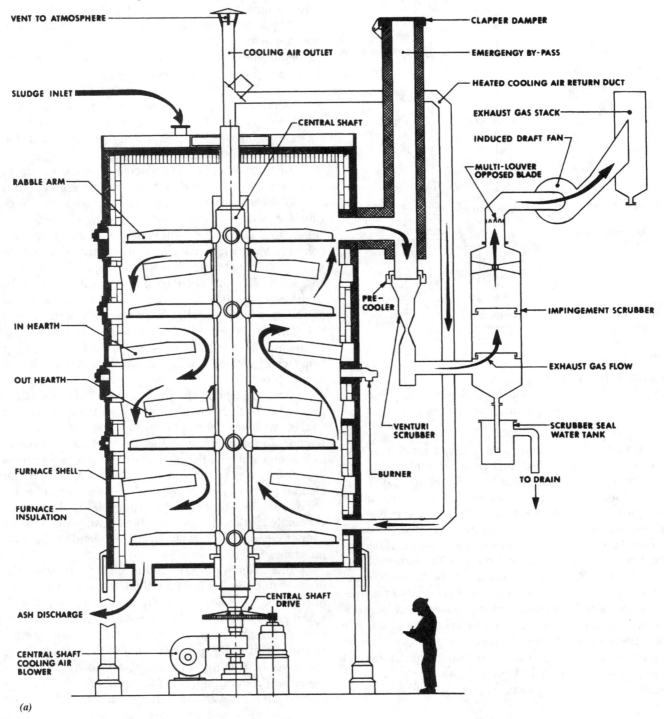

(a)

**FIGURE 10.45**

(a) *Schematic diagram of a multiple-hearth sludge incinerator.* (Courtesy of Zimpro Environmental, Inc., Rothschild, Wisconsin)

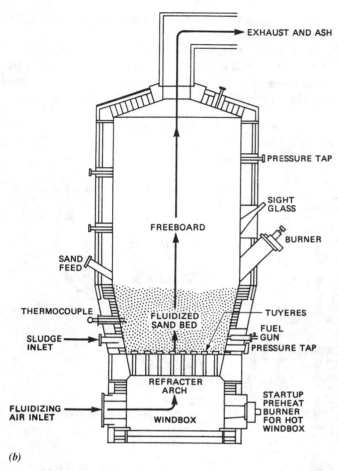

EXHAUST AND ASH

PRESSURE TAP

SIGHT GLASS

BURNER

FREEBOARD

SAND FEED

THERMOCOUPLE

FLUIDIZED SAND BED

TUYERES

FUEL GUN

PRESSURE TAP

SLUDGE INLET

REFRACTER ARCH

STARTUP PREHEAT BURNER FOR HOT WINDBOX

FLUIDIZING AIR INLET

WINDBOX

*(b)*

**FIGURE 10.45** *(continued)*
(b) *Cross section of a fluid bed furnace.* (Courtesy of the Environmental Protection Agency)

is burned as it passes through the hot sand. The ash, carried out with the exhaust gases, is removed in the final air-cleaning equipment (see Figure 10.45b).

Sludge incineration is expensive. As an option for sludge management, it is applicable mostly in congested cities, where land is not available for other disposal methods. Factors of concern in local communities include site availability, odors, truck traffic, esthetics, and ash disposal. Air pollution control is perhaps the most serious environmental and economic factor that requires consideration; incinerator exhaust gases must be stringently treated to meet the provisions of the *Clean Air Act*. Because of all these factors, sludge incineration is under increasing scrutiny by regulators, legislators, and the general public; it is limited in its role as a viable sludge management option.

**Land Application and Marketing**

The ideal management goal for treated sewage sludge (biosolids) is to use it for beneficial purposes. Large-scale *land application* is an option that makes beneficial use of the nitrate and phosphate nutrients contained in the sludge. Biosolids with low levels of heavy metals or toxic compounds can fertilize croplands or trees and can improve the condition of soil on golf courses and large land reclamation sites. The sludge can be applied as ground cover or mixed into the ground during placement.

Liquid digested sludge may be suitable for direct land application. One method of doing this is to haul the liquid sludge in a tank truck equipped with specially designed spreading equipment. Spray irrigation equipment can also be designed to handle liquid sludge. Air-dried sludge is handled by special surface spreading equipment mounted on a truck chassis. This includes conventional rear discharge or *box spreaders* and side discharge or *slinger-type spreaders*.

In 1992, when a ban on ocean disposal of sludge took effect in the United States, coastal communities had to seek alternative disposal methods. New York City, for example, is building processing plants to turn sludge into fertilizer or soil conditioner as a long-term solution. Meanwhile, the city signs contracts with companies that haul the sludge away to disposal in rural areas; about 225 tons per day of sludge (about one fifth of the city's total) is shipped as far away as west Texas by rail. It is placed in slinger-type spreaders that heave it high into the air as they roll across the vast open rangeland. About 3 dry tons are spread per acre per year (which is about equal to a "salt-shaker's worth" per square foot).

Processing, distribution, and *marketing of sludge* for use as fertilizer is a widely considered option for sludges with low levels of heavy metals and toxic compounds. In many communities, dried sludge is ground in hammer mills and bagged for local sale; sometimes it is given away to individuals willing to haul it. It can also be processed in pellet form. Biosolids can be applied to lawns, shrubs, and potting soil. Application to vegetable gardens is questionable because the amount cannot be controlled, as in large-scale land application programs.

Biosolids are distributed in bulk or in bags by a number of cities, including Philadelphia, Houston, and Milwaukee. In some cases, the sludge is mixed with wood chips and allowed to decompose, forming *compost*. Although a municipality may receive some return on the sale of sludge products, it does not usually cover the costs of treating, marketing, and distributing the material. Product liability is a factor of concern, as is the potential for water pollution by leachate generated from the effects of moisture on sludge. If leachate reaches an aquifer used for drinking water, heavy metals and toxic organics are of particular concern because of their possible health effects. Leachate entering surface waters can elevate nutrient levels, causing algal blooms and killing fish.

Standards for final use and disposal of sewage sludge were established in 1992 under the *Clean Water Act*. The rules affect sludge that is applied to land, distributed, and marketed, as well as incinerated. These regulations encourage communities to seek beneficial use options for sludge disposal.

## CASE STUDY

### Baltimore Opts for New Sludge Management Drying Technology

The Wastewater Facilities Division within the City of Baltimore, Maryland Public Works Department has steadily diversified—and privatized—its sludge management program at the 180-mgd Back River Wastewater Treatment Plant. The division previously relied on land application augmented by composting. That will change dramatically with completion of a privatized drying and pelletizing plant that converts the sludge into marketable fertilizer. The technology originated in Belgium, but has never been applied before in a commercial-scale plant.

The department moved to increase its sludge disposal options after an embarrassing accident in 1989 when land application still accounted for 75 percent of the disposed volume. The dewatered sludge was then stored temporarily in two open lagoons with enough combined capacity to hold 40,000 gal prior to transport. A particularly wet August that year blocked access to the farm fields and the by-product sludge from the Back River plant steadily built up. Odors spawned complaints and by autumn the department faced an emergency. A private contractor was finally hired to haul off the sludge—some 63 railcar loads—but for 3 months could not find an out-of-state site to receive it. The train was left wandering through Louisiana, Mississippi, and Alabama before finally returning to Baltimore.

The widely reported incident pointed out one of the vulnerabilities of land application. With the help of a consultant, staffers began formulating a long-range plan. They examined dozens of operating facilities using land application, landfill disposal, composting, heat drying, and incineration. The study team concluded that heat drying and pelletizing the sludge was the most logical method to bring to the 466-ac Back River complex. Adding that capability could shift up to half of the tonnage of digested sludge toward the new process and prevent a recurrence of the former problem. As with the composting plant also in place now, the heat-drying and pelletization plant would be privatized and produce an environmentally acceptable option.

### Setting Up Operations

General design criteria set for the new heat-drying system included provisions for odor control, elimination of outside storage of sludge cake, reduction in volume, and redundancy coupled with ease of expansion, along with a marketable end product. Three companies responded to the city's Request for Proposals to build and operate the venture under a 20-year agreement. Bio-Gro Systems (Annapolis, Maryland) and Enviro-Gro, Inc. (Baltimore, Maryland) proposed plants whose technologies offered comparable capabilities and economics. The contract was therefore split to cover two plants with distinctly different systems for drying and pelletizing the sludge.

Enviro-Gro proposed a plant using a more traditional direct-heat system, whereas Bio-Gro offered a unique process using an indirect-heat process to evaporate the water from the biosolids while producing the pelletized product. Both companies were eventually acquired by and became part of Wheelabrator Clean Water Systems, Inc.

The pelletizing plant utilizing indirect-drying was the first brought on line. The $40-million facility consists of three buildings and storage bins (see Figure 10.46). The 80-ft-high main building houses three dryer/pelletizing process units flanked by two ancillary pumphouses. These structures were designed by the Heavy Structures Group of Butler Manufacturing Company. A 10-ton crane, numerous piping runs, a four-story mezzanine, and conveyors are all structurally supported by the custom metal buildings, which were sold and erected through Nolan-Scott, Inc., the Butler builder serving the Baltimore area. Rust Engineering served as general contractor leading the design/build program.

Located on $1\frac{1}{2}$ ac within the Back River site, the pelletizing facility will find no shortage of raw material for conversion. The largest wastewater treatment plant in the state, Back River generates 220,000 wet tons of sludge annually. Wheelabrator will be paid $100 per wet ton to process the digested sludge. The effluent, initially averaging 2 to 6 percent solids, will be pumped through a 10-in. pipeline to the pelletizing plant's three 37.7-dry tpd centrifuges. As a backup, dewatered sludge cake can also be trucked to a receiving bin at the facility. Using the drying plant's own centrifuges instead of those at the Back River plant will relieve the city of meeting the quality control for the raw material brought to process.

### The Process

The conversion begins with the centrifuges condensing the effluent to a state of 24 percent biosolids. From that stage, the sludge cake feeds into a mixer, where it is combined with some previously dried pellets. This mix

**FIGURE 10.46**
*This custom-fabricated structure was designed to enclose the three dryer/pelletizing process units for the Baltimore, Maryland, Back River Wastewater Treatment Plant.* (Courtesy Public Works Journal Corporation, Ridgewood, New Jersey)

is then conveyed to the drying/pelletizing units, which have vertical arrays of 19 hollow metal trays filled with thermal oil maintained at 500°F by a closed-loop heating system. The sludge mixture spirals steadily downward onto progressively lower trays in a zigzag motion. As they move downward, the pellets dry from the inside out and increase in size as the mix descends from tray to tray. Eventually, the mix is reduced to only 20 percent of the original volume. By the time the pellets reach the bottom of the dryer unit, they are 2 to 4 mm in size and are typically 95 percent solids.

From there, the product is screened and either returned to the recycle bin for additional processing or cooled to 90°F and conveyed pneumatically to a storage silo for finished product. The plant has enough capacity to hold 800 dry tons of finished product.

Air pollution control is provided by a five-stage system consisting of a precooler for exhaust gases, a condenser to remove moisture, a venturi scrubber and packed bed scrubber to remove particulate matter, and a thermal oil heater to destroy volatile organic compounds. The VOC removal system has an 85 percent efficiency factor. Total VOC emissions for the 55-tpd facility will be kept to an estimated 1.4 tons annually, significantly less than the 25 tons per year allowed by Maryland regulations.

Equally important, up to 98 percent of the odor off the wet-process areas is removed. The exhaust air is moved through two packed bed towers, one with a dilute solution of sulfuric acid to purge ammonia, and the second with sodium hydroxide solution to remove hydrogen sulfide. The building's negative air pressure also contributes to emission containment.

Overall, the 210°F heat that the pellets are subjected to during the drying process results in a product that meets the EPA's Class A criterion—the most stringent in terms of pathogen and vector attraction reduction. The pellets have a 4 to 6 percent nitrogen content and their slow-release nature makes them an effective fertilizer for sandy soils in the Southeast.

Owned and operated by the City of Baltimore, the Back River plant was originally constructed in 1907 and services an estimated 1.3 million residents within a 140-mi$^2$ area of Baltimore City and County. Plans are to shift half of the sludge management to heat drying, one fourth to another privatized composting plant, and reserve the balance for continued land application.

*Source:* Adapted from an article that originally appeared in the February 1995 issue of *Public Works*, published by Public Works Journal Corporation, Ridgewood, New Jersey. Copyright © 1995 Public Works Journal Corporation. All rights reserved. Used with permission.

## 10.7 RELEVANT WEB SITES

### BOCA RATON WASTEWATER TREATMENT FACILITY

http://www.ci.boca-raton.fl.us/services/utility/wwater.cfm

This site presents an online case study describing and illustrating the Boca Raton, Florida, activated sludge wastewater treatment plant.

### DEER ISLAND SEWAGE TREATMENT PLANT

http://www.mwra.com/sewer/html/sewditp.htm

A thorough description of the 380-mgd (1400-ML/d) wastewater treatment plant (second largest in the United States) serving the greater Boston, Massachusetts, area.

### EPA's OFFICE OF WASTEWATER MANAGEMENT

http://www.epa.gov/OWM/

This is the Home Page of the U.S. Environmental Protection Agency's Office of Wastewater Management, which is responsible for directing the National Pollutant Discharge Elimination System Permit Program, including stormwater management and control of combined sewer and sanitary sewer overflows, providing oversight of an industrial pretreatment program, and managing the sludge (biosolids) permitting program, including promotion of the beneficial use of biosolids.

### LARGEST WASTEWATER TREATMENT PLANT IN THE WORLD

http://www.mwrdgc.dst.il.us/plants/stickney.htm

This is the Home Page of the Stickney Water Reclamation Plant, the largest wastewater treatment plant in the world. The plant serves 2.38 million people in a 260-mi$^2$ area, including the central part of Chicago and 43 suburban communities. It has a design capacity of 4500 ML/d (1200 mgd).

### LYTTLETON WASTEWATER TREATMENT FACILITY

http://www.w-ww.com/plants/newzealand/lyttelton/index.html

This site presents a case study describing and illustrating the wastewater treatment process serving Lyttleton, New Zealand.

### NATIONAL SMALL FLOWS CLEARINGHOUSE (NSFC)

http://www.estd.wvu.edu/nsfc

This is the Home Page of the NSFC provides information about innovative, low-cost wastewater treatment for small communities, those with populations less than 10,000. Emphasis is placed on finding practical, alternative solutions for "small flows" wastewater problems. A "small flows" system is one that has 1 mil gal or less of wastewater flowing through it each day, ranging from septic systems to small sewage treatment plants.

### ONSITE SEWAGE DISPOSAL

http://www.vdh.state.va.us/onsite/index.htm

This site provides technical information regarding septic systems in Virginia.

### MORE WASTEWATER TREATMENT PLANTS

http://www.sewage.net/plants/

This site provides links to scores of other wastewater treatment facility case studies.

### WATER ENVIRONMENT FEDERATION (WEF)

http://www.wef.org

This is the Home Page of the WEF, a nonprofit technical and educational organization whose mission is the preservation and enhancement of the global water environment. WEF researches and publishes current information on wastewater treatment and water quality protection, provides training, sponsors technical conferences, and reviews and testifies on environmental regulations and legislation. This site provides links to technical discussion groups and chat rooms, provides current water quality news, and features an online bookstore and library.

# REVIEW QUESTIONS

1. Compare untreated sanitary sewage and seawater with respect to the total amount of impurities in each.

2. In raw domestic sewage, what are typical concentrations of BOD, TSS, nitrogen and phosphorus, and coliforms?

3. What are *effluent standards*, and why are they easier to enforce than stream standards? Give an example of typical effluent limitations that would be included on an NPDES permit for a municipal sewage treatment plant.

4. What minimum level of pollutant removal does *secondary treatment* accomplish?

5. What do the terms *influent* and *effluent* refer to?

6. What does *POTW* stand for?

7. Briefly describe preliminary and primary sewage treatment. Approximately what level of BOD and TSS is removed? Is this level of treatment alone adequate by today's standards?

8. What is the difference between a *bar screen* and a *comminutor?*

9. What is the purpose of *grit removal?* How is it accomplished?

10. Give a brief description of a *primary clarifier.*

11. Sketch a flow diagram that shows the overall preliminary and primary sewage treatment processes.

12. What is the purpose of *secondary treatment?* Why is it called *biological treatment?*

13. What is the difference between a *fixed-growth* secondary treatment system and a *suspended-growth* system? Name one common type of treatment process in each of those two categories.

14. Briefly describe the configuration and operation of a *trickling filter* unit. Why is a *final clarifier* needed?

15. What is the purpose of *recirculation* in a trickling filter?

16. What is meant by *hydraulic load* and *organic load?*

17. How does temperature affect a trickling filter operation?

18. Give a brief description of the configuration and operation of a conventional activated sludge sewage treatment system.

19. What is meant by the term *mixed liquor?*

20. What is *waste-activated sludge?*

21. What is the significance of the *F/M ratio?*

22. What does *MLSS* stand for?

23. What is meant by *sludge bulking?* How is the *SVI* related to bulking? How may bulking be controlled?

24. Sketch a flow diagram of an activated sludge treatment process.

25. Briefly describe four different modifications of the conventional activated sludge process. What is a *package plant?*

26. Briefly describe the configuration and operation of a *biodisc* sewage treatment system. In what way is it similar to a trickling filter system?

27. What is a *stabilization pond?* How does it function? In what ways do algae help and hinder the process?

28. What is the purpose of a *chlorine contact tank?* Are there any problems associated with chlorination of secondary effluent?

29. What is the basic purpose of *tertiary treatment* of sewage?

30. What is meant by *effluent polishing?* Briefly describe two different systems used to polish a secondary effluent.

31. Describe a method for removing phosphorus from sewage.

32. Describe two methods for removing nitrogen from sewage.

33. What is a *nitrified effluent?* What is *denitrification?*

34. Describe three types of *land treatment* for sewage.

35. Under what circumstances is on-site sewage disposal warranted? What is a secondary benefit of on-site subsurface disposal?

36. What factors are of importance in planning and designing a subsurface wastewater disposal system?

37. Describe the purpose and procedure of the *perc test.*

38. What are the functions of a *septic tank* in a subsurface disposal system? Is it ever necessary to clean out the tank? Why?

39. What is the function of a *leaching field* in a sub-surface disposal system? Briefly describe the configuration of a common type of leaching field.

40. What is a *seepage pit?* When is it used?

41. Under what circumstances is a *mound system* used for on-site sewage disposal? Briefly describe this system.

42. Briefly describe the configuration and operation of an *evapotranspiration system.* Under what circumstances is it used for on-site sewage disposal?

43. Briefly describe the configuration and operation of an *intermittent sand filter system.* Under what conditions is it used?

44. Briefly discuss some general options for *sludge disposal.*

45. Does sludge volume vary with solids concentration? Describe the relationship in one sentence. Is there a significant difference in volume between a 2 percent sludge and a 4 percent sludge, each of which contains the same mass of solids?

46. Approximately what is the solids content in primary sludge? In secondary sludge? Why is there a difference?

47. What are two basic reasons for treating sewage sludge?

48. Name four different sludge treatment processes.

49. Briefly describe the operation of a *gravity thickener.*

50. Briefly describe the *dissolved-air flotation* process.

51. What is the purpose of *sludge digestion?* Describe the *anaerobic digestion* process. Describe *aerobic digestion.*

52. Describe three methods for drying sewage sludge.

53. What is *filter cake?* Would it be advisable for a treatment plant operator to have filter cake and mixed liquor for lunch?

54. Briefly describe two types of sludge incinerators.

55. Visit and explore the *Deer Island Sewage Treatment Plant* Web page. Write a brief report that describes the treatment methods used at the Deer Island Sewage Treatment Facility and the effluent outfall system.

56. Visit and explore the EPA *Office of Wastewater Management* Web page. Search for and link to the "Industrial Pretreatment" site. Write a brief report that summarizes the main elements of the National Pretreatment Program.

57. Using the EnviroSources search engine (see Section 1.6) or another Internet search directory or search engine, locate at least one additional Web site relevant to one of the topics in this chapter. Add the link(s) to your Environmental Technology Folder. Write a brief description of what the Web site(s) contain.

58. Find out where the sewage from your college or town is treated and discharged (by contacting the city or local municipal utilities authority, public works department, or health department). Make an appointment to visit the facility, if possible, and tour the plant. Write a brief report on what you learned. Include data about the average and maximum plant capacities, the total population served, and the type of treatment processes used. Draw a flow diagram showing the sequence of all unit processes and treatment steps.

## PRACTICE PROBLEMS

1. The influent of a sewage treatment plant has a TSS concentration of 180 mg/L. What is the concentration of suspended solids in the plant effluent if the plant is achieving a 90 percent TSS removal efficiency? If the flow rate is 10 mgd, how many pounds of solids are discharged per day into the receiving stream?

2. What is the efficiency of a sewage treatment plant that has an influent BOD of 240 mg/L and an effluent BOD of 10 mg/L? If the flow rate is 15 ML/d, how many kilograms of BOD is the plant discharging into the receiving stream?

3. The influent to a primary clarifier has a TSS concentration of 200 mg/L. If the effluent from the tank has a TSS concentration of 60 mg/L, what is the TSS removal efficiency of the clarifier?

4. The influent to a primary settling tank has a BOD of 210 ppm. It is estimated that 33 percent of the BOD will be removed in the tank. What is the expected BOD concentration in the tank effluent?

5. A trickling filter is used to treat a sewage flow of 2 mgd. A direct recirculation ratio of 2 is being utilized. What is the rate at which sewage is applied to the surface of the trickling filter?

6. A trickling filter has a diameter of 20 m and a depth of 2.5 m. It is operated with a direct recirculation ratio of 1.0, and the influent sewage flow rate is 3 ML/d. Influent BOD to the primary tank is 200 mg/L, and the BOD removal efficiency in that tank is 35 percent. Compute both the hydraulic load and the organic load on the trickling filter.

7. The BOD removal efficiency of a trickling filter system is 80 percent and the efficiency of the primary treatment that precedes it is 30 percent. If the raw sewage BOD is 220 mg/L, what is the secondary effluent BOD? What is the overall BOD removal efficiency?

8. An activated sludge tank is 100 ft long and 30 ft wide, and has a SWD of 15 ft. The wastewater flow rate is 1 mgd and the primary effluent BOD is 130 mg/L. The MLSS concentration in the aeration tank is 1800 mg/L. Compute the food-to-microorganism ratio for the system.

9. Sewage is to be aerated in an activated sludge tank; the flow rate is 3 ML/d and the primary effluent has a BOD of 120 mg/L. The MLSS is to be kept at 2000 mg/L and the F/M ratio is to be 0.3. If the SWD is 5.0 m and the tank is to be 20 m long, what should be its width?

10. An aeration tank is operating with a MLSS concentration of 1800 mg/L. After settling for 30 min in a 1-L cylinder, the sludge occupies a volume of 450 mL. What is the SVI of the sludge? Would you expect this sludge to settle satisfactorily in the secondary clarifier?

11. In a time interval of 30 min, the water level in a perc test hole is observed to drop 15 mm. What is the perc rate?

12. A time interval of 2 h is recorded for a 6-in. drop of water level in a perc test hole. Compute the perc rate.

13. A three-bedroom home is situated on a lot that has a perc rate of 50 min/25 mm. How many laterals are required in the leaching field if the trench width is to be 0.6 m and the maximum trench length is 20 m?

14. A five-bedroom home is situated on a lot that has a perc rate of 17 min/in. How many laterals are required in the leaching field if the trench width is to be 2 ft and the trench length is to be 100 ft? What will be the width of the leaching field?

15. A sludge with a 3 percent solids concentration has a volume of 600,000 gal. How many tons of dry solids are in the sludge?

16. A wastewater treatment plant generates 300,000 L/d of a 2 percent sludge. What is the mass of sludge solids removed from the wastewater each day? If the sludge was thickened to 6 percent solids, what would be the sludge volume and the mass of sludge solids?

17. A flow of 3 ML/d of raw sewage, with TSS = 220 ppm and BOD = 200 ppm, enters a conventional activated sludge treatment plant. Removal efficiencies in the primary clarifier for TSS and BOD are 60 percent and 35 percent, respectively. The primary and secondary sludges are combined and thickened to 6 percent solids. What is the total volume of thickened sludge generated in the treatment plant each day?

## Chapter Outline

# CHAPTER

# 11

# Municipal Solid Waste

Any material that is thrown away or discarded as useless and unwanted is considered *solid waste*. At first glance, the disposal of solid waste may appear to be a very simple and mundane problem. In this age of lasers, microcomputers, and space flight, it hardly seems possible that garbage disposal should present any great challenge. But many factors make solid waste disposal a complex problem of huge proportions for a modern (and somewhat wasteful) industrial society.

First, there is no question that *improper* disposal of solid waste can cause serious environmental or ecological damage. Air pollution can result from inadequate solid waste incineration, and soil contamination, as well as surface water and groundwater pollution, can be caused by the disposal of solid waste in improperly built landfills. These kinds of pollution can lead to a variety of diseases in humans, thereby threatening public health.

There is another connection between improper solid waste disposal and public health. Solid waste can harbor rodents and insects, which may act as vectors of infectious diseases such as typhoid, plague, and dysentery. In addition, waste deposited in open dumps causes a variety of nuisances, including odors, fire hazards, and windblown debris. Solid waste disposal in open dumps is no longer an acceptable practice in the United States, where modern waste disposal methods are designed to prevent the spread of disease and protect environmental quality. But transmission of diseases due to improper waste disposal remains a problem in some areas, particularly in developing countries.

Further complicating the solid waste disposal problem are the very large and ever-increasing amounts of solid wastes generated by society and the difficulty in finding sites for new processing and disposal facilities. In addition to the technical challenges that must be met, there are significant social and economic problems. For example, the NIMBY ("not in my backyard") syndrome, whereby some citizens protest and resist the siting of waste processing or disposal facilities in their communities, is a major problem for engineers and public officials. It is best solved by adequately informing the public and involving them in the decision-making process before final plans and designs are completed. Also of concern is the need to ensure that waste disposal facilities do not inflict unfair harm on people by being sited in low-income communities without full consideration of all possible alternative solutions.

Municipal solid waste is not generally considered hazardous, but certain types of commercial or industrial wastes are poisonous, explosive, or otherwise very dangerous and can cause immediate and direct harm to people or the environment if disposed of improperly. This *hazardous waste* typically requires transport, processing, and disposal methods that are different from those required for nonhazardous municipal solid waste. Some household products (paints, cleaners, pesticides, and the like) may contain hazardous components; it is important that these materials be disposed of safely and kept out of the municipal waste stream. Many states, local governments, and civil organizations are developing programs (for example, used motor oil recycling) to minimize the amount of this type of waste.

Clearly, problems related to solid waste go beyond merely their proper disposal. In addition to many technical and environmental difficulties, administrative, economic, and political problems must be solved. The effort to address all these problems is usually referred to as the practice of *solid waste management*. In this context, management encompasses the planning, design, financing, construction, and operation of facilities for the collecting, transporting, processing,

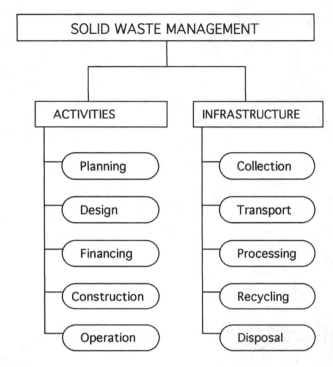

**FIGURE 11.1**
*An overview of solid waste management activities and facilities. Infrastructure refers to all the types of public and private works that are built to handle the solid wastes.*

recycling, and final disposal of the residual solid waste material (see Figure 11.1). Overall, solid waste management activities constitute a multibillion dollar per year industry in the United States.

Modern municipal solid waste collection, processing, recycling, and final disposal methods (for example, landfilling) are discussed in the following sections. (The topic of hazardous waste management is discussed in Chapter 12.)

## 11.1 HISTORICAL BACKGROUND

In ancient cities, food scraps and other wastes were simply thrown into the unpaved streets, where they accumulated. Around 320 B.C., in Athens, the first known law forbidding this practice was established, and a system of waste removal began to evolve in several eastern Mediterranean cities. In ancient Rome, property owners were responsible for cleaning streets fronting their property; disposal methods were very crude, including open pits just outside the city walls. As populations increased, efforts were made to transport the wastes farther out, and the city dump was thus created.

In the United States, near the end of the 1700s, municipal collection of garbage was started in Boston, New York, and Philadelphia. Disposal methods were still very crude. In fact, garbage collected in Philadelphia was simply dumped into the Delaware River downstream from the city. This kind of practice is unthinkable today in the United States; the dumping of solid waste into any body of water, including ocean dumping, is not allowed.

The first municipal refuse incinerator was built in England around 1875. In the United States, several cities started incinerating solid wastes at the beginning of the 20th century. Most of the largest cities, though, were still dumping solid waste on land or in water at that time.

Early incinerators caused noticeable air pollution. The *sanitary landfill* was developed as a relatively inexpensive alternative to refuse incineration, especially for cities or towns with ample land areas. It was an improvement over the city dump, but soon it became clear that unlined landfills were not environmentally safe in the long run. Today, extensive air pollution control devices are required on all incinerators, and municipal sanitary landfills must be lined and have other environmental safeguards.

In the United States, the first attempt at the federal government level to improve solid waste disposal practices was passage of the *Solid Waste Disposal Act* of 1965. This law provided state funding for solid waste management programs and established disposal regulations. It has been amended and greatly enhanced by the *Resource Conservation and Recovery Act* (RCRA) of 1976, which also addresses the management of hazardous waste. RCRA began a significant policy change; rather than focusing only on disposal, the emphasis is now on reducing waste volumes at the source, recycling materials, and recovering thermal energy.

Today, the disposal of solid waste should be part of an *integrated waste management plan*. This means that the methods of collection, processing, resource recovery, and final disposal should mesh with one another to achieve a common objective. Co-disposal of sewage sludge with refuse, for example, could be one part of an integrated waste management plan. Recycling (resource recovery and reuse) is playing an ever-increasing role in these plans, and recycling technology has evolved considerably since the 1970s; now, almost every municipal solid waste management plan has at least some recycling component. Despite these advances, there is still (and probably always will be) a need to dispose of some solid waste in the ground; in the early 1990s, federal standards (under RCRA) for landfills were strengthened to further protect groundwater quality.

## 11.2  SOLID WASTE CHARACTERISTICS

Solid wastes are grouped or classified in several different ways. These different classifications are necessary in order to be able to address, in an effective manner, the complex challenges of solid waste management. Factors related to waste generation rates, sources, and quantities and the terms used to describe waste characteristics are discussed here.

### Types of Solid Waste

The term *municipal solid waste* (MSW) is generally used to describe most of the nonhazardous solid waste from a city, town, or village that requires routine or aperiodic collection and transport to a processing or disposal site. Sources of MSW include private homes, commercial establishments, and institutions (for example, schools), as well as industrial facilities. However, MSW does not include industrial process wastes, construction and demolition debris, sewage sludge, mining wastes, or agricultural wastes.

Municipal solid waste comprises two types of materials: *refuse* and *trash*. Refuse includes *garbage* and *rubbish*. Garbage contains putrescible or highly decomposable food waste, such as vegetable and meat scraps. Rubbish contains mostly dry, nonputrescible material, such as glass, rubber, metal cans, and slowly decomposable or combustible material, such as paper, textiles, or wood objects. Actually, only about 10 percent of refuse is garbage; most of it is rubbish. Trash includes bulky waste materials that generally require special handling and is therefore not collected on a routine basis. An old couch, mattress, television, or refrigerator and even a large uprooted tree stump are examples of trash items. The classification of different types of nonhazardous municipal solid wastes is illustrated in Figure 11.2.

### Quantities and Components

Information regarding the weight, volume, and composition of municipal solid waste is necessary for the proper planning, design, and operation of collection and disposal facilities. Although average data are available, it is usually necessary to make measurements and evaluate a community's MSW in detail before preparing specific plans or final facility designs. (See Appendix H, Figure 5.)

#### Generation of Solid Waste

Studies by the EPA indicate that, on average, about 20 N (4.5 lb) per person per day of MSW is generated

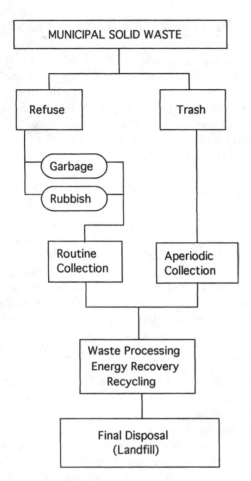

**FIGURE 11.2**

*A general classification of municipal solid waste (MSW).*

in the United States. The actual number will vary for each community depending on the time of year, the location, the amount of commercial and industrial activity, and other factors. Typically, solid waste generation rates vary between 16 and 32 N (3.5 and 7 lb) per person per day. (Keep in mind that these figures include nonhazardous solid waste from commercial, industrial, and institutional facilities, as well as residential or household waste.)

Nationwide, about 220 million tons of MSW are generated every year. To help visualize the vast magnitude of this quantity of material, consider that one year's worth of loose (or uncompacted) MSW would cover the entire area of New York City to a depth of approximately 2.5 m (8 ft) assuming that the uncompacted MSW weighs 1 kN/m$^3$ (170 lb/yd$^3$). Also consider that on a daily basis, the MSW generated would fill roughly 30 sports stadiums the size of the Astrodome. (For practice in solving problems involving unit weights and volumes, look up the area of New York City on the

Internet and verify the estimated depth of MSW given above. Do the same for the estimate of the number of sports stadiums using rounded approximations of the stadium's dimensions. A review of unit weight and volume calculations is given in Appendix C.)

According to EPA estimates, between 55 percent and 65 percent of MSW originates from residential sources and between 35 percent and 45 percent originates from commercial establishments and schools. Local and regional factors, such as climate and level of commercial activity, contribute to the variations.

More solid waste is generated in the United States, both in terms of a per-person basis and the total amount, than in any other country in the world. Waste generation rates are not necessarily correlated with gross national product. Japan and Western European nations, for instance, generate less than half as much solid waste as does the United States, although they have strong economies. In Calcutta and other cities in India, municipal solid waste generation rates can be less than 5 N/person/day (about 1 lb/person/day). Definitions of what materials make up municipal solid waste vary among nations, so precise comparisons of generation rates are difficult to make.

### Composition of Solid Waste

Municipal solid waste contains a wide variety of materials. Some can be burned, others cannot; some can be recovered and recycled, others cannot. If the only waste management option is disposal in a sanitary landfill, it matters very little what the relative proportions of glass, paper, garbage, and other materials are. But federal regulations under RCRA require that communities consider all possible options in order to be able to develop comprehensive waste management plans. Because of this, identifying the specific components of the waste stream is an important step for setting waste management goals. (See Figure 5 in Appendix H).

In the United States, the largest component (more than one third) of MSW is paper and paper products. Only a small fraction (10 percent) is actually garbage. The relative amounts, by weight, of the materials most commonly found in American MSW are shown graphically in Figure 11.3. Yard wastes include tree trimmings, grass clippings, and leaves. "Other" types of waste may include rubber, leather, textiles, and additional miscellaneous materials.

### Construction and Demolition (C&D) Waste

Although C&D waste (or debris) is not considered to be a component of MSW, it is a significant component of the total solid waste stream generated in the United States. C&D debris is produced when new structures

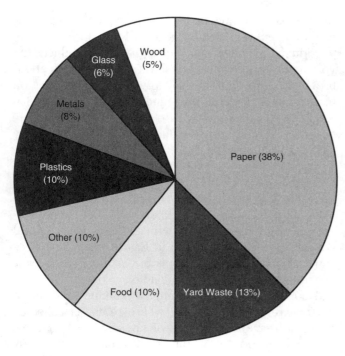

**FIGURE 11.3**

*MSW components in the United States (percent by weight), before recycling. In 1998, U.S. residents, businesses, and institutions produced a total of approximately 220 million tons of MSW. This is equivalent to a waste generation rate of about 4.5 lb per person per day, up from 2.7 lb per person per day in 1960. About 28% of the 220 million tons is recovered for recycling annually.* (Source: Environmental Protection Agency)

are built and when existing structures are renovated or demolished. About 27% of C&D waste is wood, 23% is asphalt, concrete, and brick, 13% is drywall (often called wallboard), 12% is roofing material, 9% is metal, and the rest is plastic, paper, and other miscellaneous materials.

According to a recent EPA estimate, about 136 million tons of building-related C&D debris was generated in the United States in 1996, from both residential and nonresidential sources. Construction and demolition debris is considered a nonhazardous waste because it comprises mostly inert and benign materials.

Construction and demolition debris is most commonly discarded in landfills (see Section 11.6 for a description of landfill technology). Since the waste material is inert, these landfills are typically not required to provide the same level of environmental protection as are landfills licensed to receive MSW. A significant amount of C&D debris is discarded in MSW landfills, especially in regions where landfill tipping (disposal) fees are relatively low.

There is an increasing trend to recover and recycle uncontaminated materials from C&D debris, most frequently including metals, concrete, asphalt, and wood. Most of the C&D recycling facilities are concrete or asphalt crushing plants and wood waste processing plants. In 1996, roughly 25 percent of building-related C&D debris was reprocessed and recycled. (Recycling of C&D debris is discussed more thoroughly in Section 11.5.)

**MSW Management Strategy**

EPA's *integrated waste management strategy* includes three main components, listed in order of preference:

1. Source reduction
2. Recycling
3. Disposal

*Source reduction* (waste prevention) includes reuse of products on-site, designing products or packaging to reduce their quantity or make them easy to reuse, lengthening the useful lives of products, as well as on-site composting of yard trimmings (composting is discussed in Section 11.4). *Recycling*, the off-site recovery and processing for reuse of various MSW components, including off-site composting, is discussed in Section 11.5. Final *disposal* includes incineration (preferably with energy recovery) as well as land disposal (in engineered landfills); incineration is discussed in Section 11.4 and landfills are discussed in Section 11.6.

In the United States in 1998, about 55% of MSW was disposed of in landfills, about 28% was recovered and recycled, and about 17% was incinerated. The EPA goal for the nation is to recycle at least 35% of MSW by the year 2005, while reducing the generation of MSW to 4.3 lb per person per day. Because economic growth results in more products and materials being generated, there will be an increased need to further develop recycling and composting infrastructure, buy more recycled products, and invest in source reduction activities.

## 11.3  SOLID WASTE COLLECTION

Of the billions of dollars expended each year for municipal solid waste management, about two thirds is needed to cover the cost of waste collection. *Collection* includes temporary storage or containerization, transfer to a collection vehicle, and transport to a site where the waste undergoes processing and ultimate disposal. Processing and final disposal are challenging problems, but waste collection is the most expensive phase, largely because it is labor-intensive. In addition, proper collection techniques are important to protect public health, safety, and environmental quality.

Solid waste collection may be a local municipal responsibility, whereby public employees and equipment are assigned to the task. Sometimes it is more economical to have private collection companies do the work under contract to the municipality. In some communities private collectors are paid for the service by the individual homeowners. Whatever the actual administrative arrangement, proper planning, operation, and regulation of the collection activity are necessary. The EPA has developed recommended procedures for the storage and collection of solid waste; these activities must be done in a way that will not cause fire, health, or safety hazards or provide food and shelter for vectors of disease (for example, rats and flies). Enforcement of these rules is left up to the states and local communities.

Proper on-site storage is of particular importance for municipal refuse that contains a significant amount of putrescible garbage. Watertight, rust-resistant containers with suitable covers reduce the incidence of rodent or insect infestation, and offensive odors and unsightly conditions may be kept to a minimum if the containers and storage areas are washed periodically. The EPA recommends that refuse be collected at least once per week, and trash should be collected at least once every 3 months. The EPA also suggests a limit of 333 N (75 lb) for the weight of manually emptied waste containers; for most residences, 115-L (30-gal) galvanized metal or plastic containers are effective. Larger containers can be used along with mechanical collection trucks. Bulk containers or dumpsters should be used where large volumes of refuse are generated, such as at shopping centers, restaurants, apartment buildings, and hotels.

The collection truck and crew make up the most important element of a collection system. Collection trucks most commonly used in the United States are of the enclosed, compacting type. The collection capacity can vary, but is typically about 24 m³ (32 yd³). Compaction in the collection truck significantly reduces the volume of loose refuse. Loose refuse may weigh about 0.75 to 1.5 kN/m³ (130 to 260 lb/yd³). After compaction in a truck, it may weigh as much as 2 to 4 kN/m³ (350 to 700 lb/yd³). In other words, compaction in a collection vehicle temporarily reduces the refuse volume by as much as 80 percent. (The volume will increase again to some degree when the refuse is unloaded at a processing or disposal facility.)

The majority of refuse collection trucks are front-, side-, or rear-loading compactors, although some noncompaction closed-body trucks are also used. Typical collection trucks are shown in Figure 11.4. Crew sizes can vary from one to four people. A one-person crew generally uses a special side-loading vehicle or mechanical collection truck to minimize walking time. Four-person crews may be used where refuse is collected from the rear or side yards of houses. Two- or three-person collection crews are most typical, particularly with curbside collection routes.

Technical decisions must be made with regard to frequency of collection or pickup from each waste generation site and the point of pickup (curb, alley, backyard, or other). These decisions depend on the type of community, the type of waste (mixed or separated), the population density, and the land use in the collection area. Cost is also a major consideration; to lower costs, there is a tendency toward reduced collection frequency, an increased use of curbside (rather than walk-in) collection, and an increase in mechanical collection from standardized waste storage containers.

Mechanical collection systems are becoming popular in many communities because of improved esthetics of curbside container placement as well as lower costs. These systems consist of standardized containers and truck-mounted lifting mechanisms. In fully automatic systems, an articulated arm mechanism on the vehicle engages, lifts, empties, and replaces the

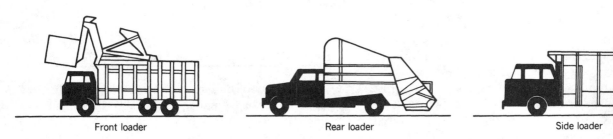

Front loader          Rear loader          Side loader

**FIGURE 11.4**

*Enclosed compaction-type refuse collection vehicles reduce the volume of collected waste material by at least 50 percent.* (Courtesy of the New York Department of Environmental Conservation)

container without manual assistance. Semiautomatic systems require a truck crew member to place the container in position to be automatically hoisted and emptied into the collection truck and then manually returned to its setout position. Standardized MSW collection containers for single-family homes may vary from 225 to 360 L (60 to 95 gal) in size. Larger containers, up to 1200 L (315 gal) or greater, may be used at multifamily dwellings. All containers are wheel-mounted for ease of movement.

Combined collection of garbage and rubbish is generally more economical than separate collection of these types of refuse. In many communities, however, certain materials are recycled. Homeowners practice *source separation;* that is, they separate glass, metal, paper, and plastic from the remainder of their refuse. The recyclable materials are then picked up in a separate collection truck to prevent the refuse from contaminating the recyclable component and lowering its resale value. Recycling collection trucks often have side-loading compartments for the various recyclables (as discussed in Section 11.5 and shown in Figure 11.14).

One of the most effective ways to minimize waste collection costs is to optimize the collection route; an optimum route is one that results in the most efficient use of labor and equipment. Selected characteristics of an optimum route include the following:

1. Collection vehicles should not travel twice down the same street; that is, collection paths should not overlap.
2. Refuse collection on crowded streets and roads should not occur during morning or afternoon rush hours.
3. Collection should occur in the downhill direction as much as possible to conserve fuel.
4. The starting point should be close to the collection vehicle garage, and the last collection point should be as close as possible to the destination of a filled vehicle (that is, transfer station, incinerator, processing plant, or sanitary landfill).

These characteristics may seem to be simplistic examples of what is only common sense. Actually, they place significant constraints or limits on the collection routes, especially for large and densely populated urban areas. In fact, a sophisticated branch of mathematics, called *systems analysis* or *operations research*, is needed to solve this complex problem. Computers are routinely used to conduct the analysis, providing engineers and managers with collection routes that can be modified periodically to accommodate changes in community growth or development.

## Transfer Stations

It is not always feasible for individual collection trucks to haul refuse to a waste processing plant or final disposal site, especially if the ultimate destination is not in the immediate vicinity of the community in which the waste is collected. To solve this waste transport problem efficiently, one or more *transfer stations* may be used.

A transfer station is a facility at which solid wastes from individual collection trucks are consolidated into larger vehicles, such as tractor-trailer units. It is more economical for a few of these larger vehicles to transport the consolidated solid waste over the long-haul distance to the processing or disposal location, rather than have each collection truck make the trip. A one-way haul distance of about 20 km (12 mi) may be a typical upper limit for an individual waste collection truck, but thorough engineering and cost–benefit comparison studies are generally conducted to determine the need for and advantages of a transfer station.

Individual transfer station capacities may vary from somewhat less than 100 tons to more than 500 tons of waste per day, depending on the size of the community. There are two basic modes of operation: *direct discharge* or *storage discharge*. In a storage discharge transfer station, the refuse is first emptied from the collection trucks into a storage pit or onto a large platform. Grapples or front-end loaders are then used to load or push the waste, respectively, into large trailer units.

In a direct discharge station, each refuse truck empties directly into the larger transport vehicles. The trailers typically have a capacity of about 75 m$^3$ (100 yd$^3$) and hold the solid waste from four collection vehicles if it is not compacted and from up to eight collection vehicles if it is compacted. A direct discharge transfer station requires a two-level arrangement, as depicted in Figure 11.5. A backhoe equipped with a tamping device may be used to compact refuse dumped directly into an open trailer. Mechanical top-closing panels are used to produce a closed vehicle during transport.

In addition to open-top trailers, two types of closed compactor trailers are available. In one type, the compactor is built into the trailer and compacts the waste for later ejection at the process plant or disposal site. The second type of trailer is anchored to a separate compactor unit during the loading process; the trailers must be equipped with conveyors or ejection devices for unloading the compacted refuse. Another

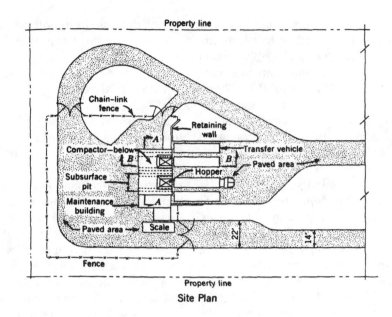

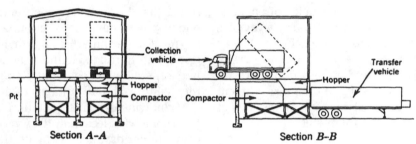

**FIGURE 11.5**

*At a direct-discharge transfer station, several collection trucks deposit refuse into a larger vehicle for hauling to a more distant disposal site.* (From J. A. Salvato, *Environmental Engineering and Sanitation*, 4th ed., Wiley, New York. Copyright © 1992. Reprinted with permission of John Wiley & Sons, Inc., and by courtesy of Malcolm Pirnie Engrs., New York)

type of transfer station design, called the *push pit station*, includes a storage pit with a ram at one end and a hopper at the other end. After the pit is filled by the collection trucks, the ram pushes the refuse into the hopper, which loads the trailers.

## Other Collection Methods

Before leaving the topic of municipal solid waste collection, a few other waste collection and transport methods should be mentioned. Some homes are equipped with garbage grinders, for example. These devices reduce the amount of food wastes in refuse. Since biweekly collection frequencies are usually only necessary because of the rapid decomposition of garbage, if all homes in a community had grinders, the collection frequency could be cut in half. Although the ground garbage winds up in sewage and flows to

a wastewater treatment plant, most sewer systems and treatment plants can handle the extra load; an engineering study would have to confirm this.

Innovative collection systems involving pneumatic pipeline transport have been tried. In pneumatic systems, refuse is pulled by suction or vacuum through underground pipes to a central processing plant. Waste collection at the Disney World amusement park in Florida, for example, is done by a system of this type. It eliminates the need for noisy and unsightly refuse collection trucks. But complex controls, valves, and high-speed turbines are required for operation of the system, and installation costs are high. Pneumatic collection systems have also been installed in some small communities in Sweden and Japan. Sweden was the first country to use pneumatic transport of refuse in a large pipeline (300 mm), from an apartment house to an incinerator that also provided energy for space heating. Despite the high-tech appeal of pneumatic waste collection and transport systems, they are feasible only in specialized local situations and are unlikely to replace conventional methods in the foreseeable future.

## 11.4  SOLID WASTE PROCESSING

Municipal solid waste may be treated or processed prior to final disposal. Solid waste processing provides several advantages. First, it can serve to *reduce the total volume and weight* of waste material that requires final disposal. Volume reduction helps to conserve land resources, since the land is the ultimate *sink* or repository for most waste material. It also reduces the total cost of transporting the waste to its final disposal site.

In addition to volume and weight reduction, waste processing *changes its form* and *improves its handling characteristics*. Processing can also serve to recover natural resources and energy in the waste material for reuse, or *recycling*. The most widely used municipal waste treatment processes, including *incineration, shredding, pulverizing, baling,* and *composting*, are discussed in this section. Although incineration (burning) greatly reduces the waste volume, it is a processing rather than a disposal operation; land burial is still required for final disposal of the ashes and other unburned residue that remains behind. (Recycling is discussed in Section 11.5, and final disposal in *sanitary landfills* is discussed in Section 11.6.)

### Incineration

One of the most effective methods of reducing the volume and weight of municipal solid waste is to burn it in a properly designed furnace under suitable temperature and operating conditions. This process is called *incineration*. It is expensive, primarily because extensive air pollution control equipment is required. An incinerator also requires high-level technical supervision and skilled employees for proper operation and maintenance. The advantages of incineration, however, often outweigh these disadvantages. According to EPA estimates, roughly 16 percent of the total MSW stream in the United States was incinerated in 1993. (By contrast, Japan burns about 75 percent of its MSW.)

Incineration is a *chemical process* in which the combustible portion of the waste is combined with oxygen, forming mostly carbon dioxide and water. This chemical reaction is called *oxidation*, and it results in the release of thermal energy, that is, heat. The carbon dioxide gas and water vapor are released into the atmosphere. For complete oxidation, the waste must be mixed with appropriate volumes of air, and a proper temperature must be maintained for a suitable length of time. Typically, furnace temperatures are about 815°C (1500°F), and the waste must remain in the furnace for about 1 hr. For some types of solid waste, temperatures up to 1400°C (2550°F) may be reached during combustion.

Incineration can reduce municipal refuse by about 90 percent in volume and 75 percent in weight. In densely populated urban areas, where large sites suitable for sanitary landfilling are not available within reasonable hauling distances, incineration may be the most economical option for refuse processing. In many cases it is feasible to design and operate the incinerator so that the heat from combustion can be recovered and used to produce electricity or steam; this type of system is often called a *resource recovery* or waste-to-energy facility.

### Incinerator Residues and Emissions

Incineration does not completely destroy all the solid waste. *Bottom ash*, the solid residue remaining in the furnace after combustion, includes glass, metal, fine mineral particles, and other unburned substances. The volume of MSW bottom ash is about 5 percent of the original solid waste volume. Another type of incinerator ash, called *fly ash*, is carried along in the combustion airstream (or *flue gas.*) Fly ash consists of finely divided particulate matter, including cinders, mineral dust, and soot. Most of MSW incinerator ash (about 80 percent by weight) is bottom ash; the remainder is fly ash.

The possible presence of heavy metals in incinerator ash is a factor to be considered. By destroying the burnable portion of the waste, incineration concentrates the metals, which remain in the ash. Metals such as lead and cadmium can be harmful if they are present in high enough concentrations. Some organic substances, such as dioxins, may also be present in the ash. One way to minimize this potential problem is to keep toxic products and materials containing heavy metals (for example, batteries and plastics) out of the municipal waste stream.

Fly ash generally has a higher concentration of toxic substances than does bottom ash. Testing of fly and bottom ash, which are often combined, is needed to determine whether they must be managed as hazardous waste. In any case, it is important that all incinerator ash be disposed of in lined landfills, preferably separate from other solid wastes. The ash can be chemically or physically treated prior to disposal or reuse. For example, if mixed with lime and water, the ash forms a cementlike material that immobilizes the metals. In this form, it can be used for road base construction and other beneficial purposes.

Modern air pollution control devices are installed on all municipal solid waste incinerators to remove the fly ash and potentially harmful gaseous contaminants. This equipment is located after the furnace, but before the tall chimney or stack. (Air pollution control devices, such as fabric filters, acid-gas scrubbers, and electrostatic precipitators, are discussed in Chapter 13.) Incinerator stacks may be between 60 and 180 m (200 and 600 ft) high; the discharge of cleaned flue gases at these heights increases the rates of dilution and dispersion, further reducing potential air pollution. The height of the incinerator stack or chimney depends on several factors, including local topography, land use, climate, and average wind conditions, as well as Federal Aviation Agency (FAA) regulations.

Efficient operation of the incinerator is essential for proper air pollution control. High enough combustion temperature, adequate burning time, and sufficient air supply are needed to destroy harmful organic compounds in the incoming waste. Frequent emissions (stack) testing is important to ensure that all systems are working properly to prevent air pollution. The EPA has set maximum emission levels that are used for guidance by state and county regulatory agencies. States can set their own emission standards as long as they are equal to or stricter than federal standards. One type of substance regulated in emissions from MSW incinerators in New Jersey, for example, is mercury, a toxic heavy metal; between 90 and 95 percent of the mercury must be removed from the flue gas. (Mercury is found in a wide range of consumer products, such as batteries and thermostats, and usually winds up in the MSW stream.)

## Design and Operation

Most large, modern municipal incinerators are designed for *continuous-feed operation*, as opposed to the less desirable intermittent or *batch-feed* mode of operation. A continuous feed of refuse allows for uniform furnace temperature; this provides more efficient combustion and reduces potential thermal shock damage to the incinerator components.

A typical incinerator plant includes a below-grade refuse storage pit or *tipping area*, which provides volume for at least 1 d of refuse storage. Sufficient storage volume is necessary to allow for continuous operation of the facility. Refuse is lifted from the pit by a crane with a grab bucket and deposited into a charging hopper and chute. Then it is released from the chute onto a charging grate or stoker. Various types of mechanical traveling or rocking grates are available to agitate and move the burning material through the furnace in a manner that allows for a proper draft or flow of air.

Municipal incinerators are built in a variety of configurations, including the *rectangular furnace* and the *rotary kiln furnace*. In the rectangular type, two or more moving grates are arranged in tiers; in a rotary kiln furnace, a drying grate precedes the rotating drum (kiln) where burning is completed. A rotary kiln furnace is shown schematically in Figure 11.6.

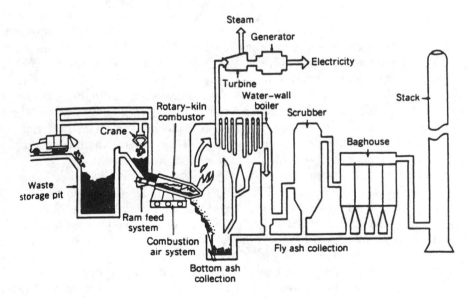

**FIGURE 11.6**
*Schematic of a rotary kiln waste-to-energy furnace.* (From J. A. Salvato, *Environmental Engineering and Sanitation*, 4th ed., Wiley, New York, 1992. Copyright © 1992 John Wiley & Sons, Inc. Used with permission)

Inside a furnace, combustion occurs in two phases: *primary combustion* and *secondary combustion*. In primary combustion, moisture is driven off; then the burnable waste is volatized and ignited. In the secondary stage of combustion, the remaining unburned gases and particulates, which are entrained in the airstream after primary combustion, are oxidized. The secondary stage of combustion helps to eliminate odors and reduces the amount of unburned particulates in the exhaust gases. Auxiliary gas or fuel oil is sometimes used for furnace warmup and to initiate primary combustion when the refuse is very wet. Auxiliary fuel also facilitates complete secondary combustion and provides additional smoke and odor control in the exhaust gases.

Sufficient quantities of air must be thoroughly mixed with the refuse so that oxygen is available for both primary and secondary combustion. In a rectangular-type furnace, air can be supplied from openings beneath the grates (underfire air) and admitted to the area above the grates (overfire air). The relative amounts of underfire and overfire air are determined by the facility operator for efficient operation of the incinerator. Airflow is obtained from the natural draft of a chimney or from forced-draft fans.

The furnaces of incinerators that are not used for energy recovery are typically built with *refractory* materials, which resist the effects of very high combustion temperatures. Refractory bricks are made of alumina, magnesia, silica, and a clay mineral called kaolin; the refractory walls are only about 225 mm (9 in.) thick.

### Energy Recovery

Recovery of the heat given off by burning refuse in an incinerator can be accomplished using a refractory-lined furnace followed by a boiler. The boiler converts the heat from combustion into steam or hot water. In this way, the energy content of the refuse can be recycled and put to beneficial use. Municipal solid waste in the United States contains a large amount of paper, wood, and other combustible material, giving it an energy content roughly one third that of coal.

Another type of energy recovery system makes use of a *water-tube wall* furnace. A water-tube wall furnace is lined with closely spaced welded steel tubes that are arranged vertically to form continuous walls. Insulation on the outside of the walls reduces heat loss. Heat is absorbed by the water that circulates through the tubes, and the heated water is used to produce steam. An advantage of this type of system is that the water also serves to control furnace temperature, eliminating the need for excess air. Smaller volumes of airflow result in lower air pollution control costs compared to the costs for a refractory-lined furnace.

When raw or unprocessed solid waste is fed as a fuel directly into a heat recovery type of facility, the process is referred to as *mass burning*. In some incineration systems, though, refuse may be treated or processed by shredding and by separation of the noncombustible waste material before being fed into the furnace. In this case, the solid waste is called *refuse-derived fuel*, or RDF. Roughly 2 kg of steam can be produced per kilogram of incinerated RDF, but variation in the characteristics of the waste leads to variation in the rate of steam production. Provision must be made to burn auxiliary fuels when the volume or recoverable heat content of the waste temporarily decreases.

Heat recovery and reuse from MSW incineration is a very attractive waste management option from an environmental and ecological perspective. But the high costs for equipment and controls, the need to have highly skilled technical personnel, and the need for auxiliary fuel systems are some of the factors that must be taken into consideration when developing a solid waste management plan. Almost all new incinerators, however, make use of heat recovery systems. And incineration is becoming a more attractive waste management option when compared to landfilling, largely because suitable landfill sites are becoming increasingly difficult to find.

Due to their complexity, large incinerators (with capacities exceeding 100 tons per day) are often custom-designed, constructed, and operated as part of a turnkey operation. Following a competitive bidding process, the incinerator vendor arranges to finance (privately or through public bonding) and operate the facility under contract to the municipality or regional solid waste authority.

For relatively small facilities, a *modular incinerator* with capacities of about 1 to 3 tons per hour can be supplied as a completely prefabricated package by the manufacturer. Some modular units are *batch-fed* and operate using a *starved air furnace* to reduce air pollution. Most state regulatory agencies have set emission standards specifically for modular incinerators. Modular incinerators can also be equipped with boilers to recover waste heat energy.

A law enacted by Congress, the *Public Utilities Regulatory Policies Act* (PURPA), requires large power companies to purchase energy from smaller generators of electricity who offer it for sale. Consequently, MSW heat recovery plants that generate more steam than can be used or sold can convert their excess steam into electricity for sale to the local power company. A turbine is required at the incinerator facility to convert the steam into electricity. A facility at which both electrical and thermal energy is produced from the same primary fuel source is called a *cogeneration* facility.

Although the process of converting municipal solid waste into electricity is relatively inefficient (as compared to producing electricity from fossil fuel), co-generation offers some significant advantages to energy recovery from MSW. It is a reliable technology and is an important component in comprehensive solid waste management in the United States.

### Pyrolysis

*Pyrolysis* is a high-temperature thermal process that can provide an alternative to incineration; it takes place in a low-oxygen or oxygen-free environment and produces by-products that can be used as fuels. Natural gas is burned to start the process. Instead of oxidation, a complex series of decomposition and other chemical reactions takes place. Air pollution with pyrolysis is less of a problem than with incineration due to the reduced volume of waste gases. Pyrolysis can be used for the processing of discarded rubber tires; rubber is reduced to oil and methane gas, which can be sold. (Rubber tires can also be shredded and added to asphalt paving material for road construction.)

## Shredding and Pulverizing

Size reduction of municipal solid waste is accomplished by the physical processes of *shredding* or *pulverizing.* Shredding refers to the actions of cutting and tearing, whereas pulverizing refers to the actions of crushing and grinding. These two terms are frequently used synonymously with regard to solid waste management. Note that the size reduction obtained by shredding or pulverizing refers to the size of individual components or pieces of the solid waste material. However, shredding and pulverizing also reduce the overall volume of the original or raw waste material, sometimes by as much as 40 percent.

There are many reasons for size reduction of municipal solid waste. The production of refuse-derived fuel, or RDF, requires processing of the raw solid waste; this typically includes shredding and pulverizing. Composting, which will be discussed in the next section, also frequently requires some type of size reduction process. Shredding and pulverizing may first be applied where the basic objective is to recover material from the waste that can be recycled and marketed. The size reduction and homogenizing processes improve the performance of the mechanical separation machinery. Finally, shredding of refuse prior to land burial can increase the capacity of the landfill; it also reduces the potential of rodent infestation, since the animals have difficulty finding food scraps or voids for a habitat in the homogeneous material.

### Hammer Mills

One of the most common types of equipment used for processing MSW into a uniform or homogeneous mass is the *hammer mill.* A hammer mill is a mechanical impact device in which the raw solid waste material is hit with a force sufficient to crush or tear individual pieces of the waste. Impact is provided by several hammers that rotate at high speeds (up to 1500 rev/min) around a center horizontal or vertical shaft. A vertical hammer mill is shown in Figure 11.7.

In horizontal-shaft hammer mills, cutting bars or a breaker plate attached around the periphery of the mill chamber also help to reduce the size of the waste. When size reduction is complete, the processed waste simply falls through the grate at the bottom of the chamber. Not unexpectedly, repair and replacement of hammer mill components are part of a frequent maintenance routine due to the high speeds and impact action. In addition to the cost of maintenance, electric power requirements are high.

A hammer mill is a very versatile size reduction device because it will accept almost any type of waste material (except of course very bulky or dense items such as tree stumps or engine blocks). It is possible to reduce the size of solid waste material components to uniform fragments between 25 and 50 mm (1 and 2 in.) with proper operation. A typical size for a hammer mill is a 150-hp unit capable of processing about 100 kN

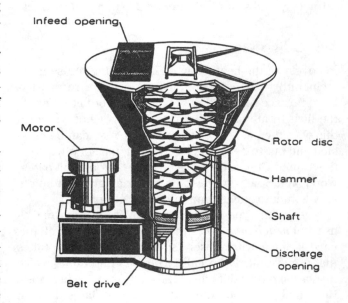

**FIGURE 11.7**

*Vertical hammer mill.* (Reprinted with permission from L. P. Diaz, *Composting and Recycling*, Lewis Publishers, Boca Raton, Florida, 1993. Copyright © 1993, Lewis Publishers, an imprint of CRC Press, Boca Raton, Florida)

(11 tons) of solid waste per hour. In addition to relatively high costs for operation and maintenance, the disadvantages of size reduction by hammer mills include noise and dust generation.

A modern innovation in the equipment used for MSW size reduction is the *rotary shear shredder*. This is a high-torque, relatively low-speed (up to 60 rev/min) machine, consisting of two or more parallel horizontal shafts that rotate in opposite directions. Each shaft has cutters that shear and tear the waste material. The high torque and shearing action allow this type of machine to shred difficult materials, such as tires. (More than 200 million used tires are discarded each year in the United States.)

## Baling

Compacting solid waste into the form of rectangular blocks or bales is called *baling*. MSW bales are typically about 1.5 m³ (2 yd³) in size and weigh roughly 1 kN (or 1 ton). Solid waste can be compacted under high pressures (about 700 kPa or 100 psi) in either vertical or horizontal presses; the bales are frequently wrapped

with steel wire to help retain their rectangular shape during handling. (They also may be enclosed in hot asphalt, plastic, or Portland cement, or tied with metal bands, depending on the intended use or disposal method. If moisture content and compaction pressures are high enough, they may retain their shape without being wire-wrapped or encased.) Semiautomatic horizontal presses can bale up to 36 kN (or 4 tons) per hour of MSW (see Figure 11.8). Volume reduction can be as much as 90 percent of the original waste volume.

Solid waste volume reduction may be expressed in terms of a compaction ratio, as well as in percent. An understanding of the relationship between percent volume reduction and compaction ratio is important, particularly when reviewing and interpreting the manufacturers' data and selecting or specifying suitable compaction equipment. Appropriate formulas are

$$\text{percent volume reduction} \qquad (11\text{-}1)$$
$$= \frac{\text{initial volume} - \text{final volume}}{\text{initial volume}} \times 100$$

$$\text{compaction ratio} = \frac{\text{initial volume}}{\text{final volume}} \qquad (11\text{-}2)$$

**FIGURE 11.8**
*High-pressure compaction units can be used for making rectangular bales or blocks of solid waste.* (Courtesy of American Solid Waste Systems—Division of American Hoist & Derrick Company, St. Paul, Minnesota)

## EXAMPLE 11.1

The initial volume of a mass of solid waste is 15 m³. After compaction, the volume is reduced to 3 m³. *(a)* Compute the percent volume reduction and the compaction ratio. *(b)* If it is desired to obtain a volume reduction of 90 percent, what will the compaction ratio have to be?

*Solution*

*(a)* Applying Equation 11-1 gives

$$\text{percent volume reduction} = \frac{15 - 3}{15} \times 100$$

$$= 80 \text{ percent}$$

Applying Equation 11-2 yields

$$\text{compaction ratio} = \frac{15}{3} = 5$$

The compaction ratio is typically expressed as 5:1, or "five to one."

*(b)* Using Eq. 11–1 to obtain the compaction ratio that would correspond to a 90 percent volume reduction yields

$$90 = \frac{15 - \text{final volume}}{15} \times 100$$

and rearranging terms yields

$$\text{final volume} = 1.5 \text{ m}^3$$

Using a final volume of 1.5 m³ and applying Equation 11-2 gives

$$\text{compaction ratio} = \frac{15}{1.5} = 10 \quad \text{or} \quad 10{:}1$$

From this example, it can be seen that to improve the solid waste volume reduction by 10 percent, from 80 percent to 90 percent, it is necessary to double the compaction ratio from 5:1 to 10:1.

---

The basic advantages of an MSW baling process include the significant decrease in waste volume, the ease of handling the compacted refuse, and the reduction of litter and nuisance potential. Additionally, the compacted waste can be hauled to a landfill by conventional vehicles, and the service life of the landfill can be greatly increased (by as much as 60 percent) because of the smaller volume of waste requiring burial. At the landfill, the bales can be neatly stacked in place without a problem of windblown debris, the likelihood of animal or insect infestation is decreased, soil cover requirements are reduced, and the need for on-site compaction is eliminated. Sometimes the bales may be left uncovered temporarily, as shown in Figure 11.9. A case study of a baling project follows.

## CASE STUDY

### Baler Is Heart of New Transfer Station

The city of Canadian and Hemphill County, Texas, have pooled their resources to install a transfer station to serve the disposal needs of the 4000-population west Texas community. The new facility replaces the city's municipal landfill, which had reached capacity and was closed. Solid waste and recyclables are collected at the transfer station and transported 45 mi to Pampa, Texas, for landfill disposal or recycling. City crews are responsible for collection and processing of materials, while Hemphill County personnel handle transportation and disposal of the city's refuse.

The 12,000-ft² facility consists of the central operations building, the main transfer station, the hazardous waste disposal area, and the composting center plus on-site parking, aprons, and approaches. The materials processing center includes the main tipping floor, baling equipment with conveyors, the recyclable sorting area, the glass crusher, the bale storage, and the loading dock. The facility is situated on 7 ac of land south of Canadian's corporate limits.

The heart of the transfer station's operation is a two-ram "Bulldog" baler manufactured by Mosley Machinery Company. The baler is used to compress materials ranging from food scraps to plastic containers to construction materials into like-material, wire-wrapped bales weighing up to 2500 lb each. Compacted bales are loaded by fork lift into a trailer for transport to the landfill. Recyclables such as newsprint, corrugated cardboard, plastics, aluminum, and steel cans are compacted in the same manner for transport to the processors.

The baler has enabled the city to reduce a truckload of trash into four small bales, or about one fifth of its original volume. Instead of 12 loads per week, the city now hauls the same amount of trash to Pampa in about three loads every 2 weeks. The fully automated baler can bale up to 24 tons of solid waste per hour.

---

*Source:* Adapted from an article that originally appeared in the April 1995 issue of *Public Works Manual*, © 1995, Public Works Journal Corporation, 200 South Broad Street, Ridgewood, New Jersey. All rights reserved. Used with permission.

**FIGURE 11.9**
*Uncovered bales of solid waste at a disposal site.* (American Solid Waste Systems—Division of American Hoist & Derrick Company, St. Paul, Minnesota)

## Composting

*Composting* is a process in which the organic portion of MSW is allowed to decompose under carefully controlled conditions. It is a biological rather than a chemical or mechanical process; decomposition and transformation of the waste material are accomplished by the action of bacteria, fungi, and other microorganisms. Composting, where it is applied, must be part of a comprehensive MSW management system that handles other components of the solid waste stream as well as the organic portion. In some applications, sewage sludge and agricultural wastes may be combined with the organic portion of the municipal solid waste.

With proper control of moisture, temperature, and aeration, a composting plant can reduce the volume of the raw organic material by as much as 50 percent. In addition, composting can stabilize the waste and produce an end product that may be recycled for beneficial use. The end product is called *compost* or *humus*. It resembles potting soil in texture and earthy odor, and it may be used as a soil conditioner or mulch.

A complete municipal solid waste composting operation includes *sorting* and *separating*, *shredding*

and *pulverizing*, *digestion*, *product upgrading*, and, finally, *marketing*. Sorting and separation operations are required to isolate organic, decomposable waste materials from the plastic, glass, metal, and other nonbiodegradable substances. Solid waste sorting and separation methods are a key part of MSW recycling operations and are discussed in detail in Section 11.5. Shredding and pulverizing (discussed previously) serve to reduce the size of the individual pieces of the organic waste, resulting in a relatively uniform mass of material. This facilitates handling, moisture control, and aeration of the decomposing waste. Size reduction also helps optimize bacterial activity and increases the rate of decomposition. After size reduction, the wastes are ready for the actual composting or *digestion* step. Digestion may take place in open *windrows* or in an enclosed mechanical facility.

A windrow is a long, low pile of the prepared organic waste, usually about 3 m (10 ft) wide at the base and about 2 m (6 ft) high. Most windrows are conical in cross section and about 50 m (150 ft) in length. The composting waste is aerated by periodically *turning* each windrow. This can be done manually with a pitchfork, but at most large facilities it is accomplished mechanically by specially designed machinery. Some of

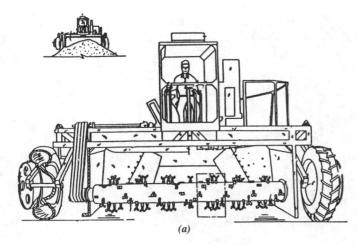

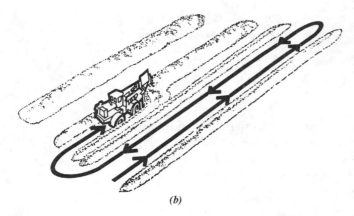

**FIGURE 11.10**

(a) *Windrow turning machine;* (b) *windrow turning arrangement.* (Reprinted with permission from L. F. Diaz, *Composting and Recycling,* Lewis Publishers, Boca Raton, Florida, 1993. Copyright © 1993, Lewis Publishers, an imprint of CRC Press, Boca Raton, Florida)

these machines turn and rebuild the windrow directly behind the machine; others rebuild the turned windrow adjacent to its original position (see Figure 11.10). Turning frequency varies with moisture content and other factors. When moisture content is maintained at about 50 percent, windrows are turned two or three times a week, and in some cases daily.

Generally, open-field windrow composting takes about 5 weeks for digestion or stabilization of the waste material. An additional 3 weeks may sometimes be required to ensure complete stabilization. Temperatures in an aerobic compost windrow may reach 65°C (150°F) because of the natural metabolic action of *thermophylic* microbes that thrive at such elevated temperatures. The relatively high temperatures destroy most of the pathogenic or disease-causing organisms that may be present in the waste.

Open-field windrow composting requires relatively large land areas. A community population of 250,000 for example, requires about 24 ha (60 ac) of land for an MSW composting facility. To reduce land requirements, various types of enclosed mechanical systems can be used in lieu of the open-field method. A typical enclosed system will reduce the land area needed to serve 250,000 people by a factor of 6, from 24 to 4 ha (60 to 10 ac). A variety of mechanical-type compost systems are available. Oxygen is supplied to the waste material by forced aeration, stirring, or tumbling. An example of a system that utilizes a rotating drum to tumble the waste is shown schematically in Figure 11.11.

In addition to reducing land requirements, enclosed mechanical compost facilities can reduce the time required for stabilization from about 5 weeks to about 1 week. Some mechanical systems use several enclosed vessels or tanks for digestion of the waste. They may be equipped with rotating plows, vanes, or augers to mix the shredded waste material and to facilitate aeration. In some systems, compressed air is blown into the vessel for aeration.

Before the stabilized compost or humus can be sold for use as a mulch or soil conditioner, it must be processed further to upgrade or improve its quality and appearance. This includes drying, screening, and granulating or pelletizing. Sometimes, the compost is placed in bags, although bulk sale is more efficient and economical. The basic purpose of compost upgrading is to make the final product more readily marketable. Marketability is the most serious bottleneck or obstacle to the usefulness of composting as a major MSW management option. Only a small portion of MSW is composted in the United States (and other countries), primarily because of the low demand for the end product. The largest *potential* market for compost is agriculture. Land reclamation and landscaping rank next.

Compost can increase the organic and nutrient content of soil and improve its texture and ability to retain moisture. But the demand for compost by farmers is low, largely because of its relatively poor quality as fertilizer, the cost of transporting it, and the availability and ease of applying inorganic chemical fertilizers. MSW compost does find some use for landscaping, mulching, and erosion control at parks and golf courses, and it is used as a soil conditioner in residential gardens and nurseries. Home or backyard composting is practiced by many people; if properly done, home composting can divert up to 30 percent of household waste from a municipal landfill or incinerator.

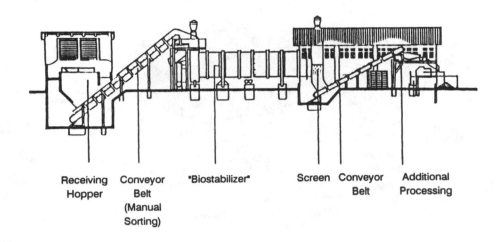

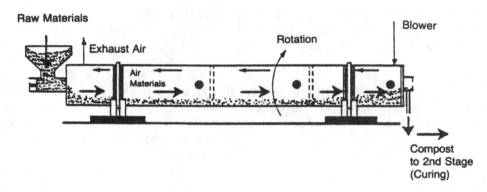

**FIGURE 11.11**

*Enclosed mechanical-type composting system.* (Reprinted with permission from
L. F. Diaz, *Composting and Recycling*, Lewis Publishers, Boca Raton, Florida, 1993.
Copyright © 1993, Lewis Publishers, an imprint of CRC Press, Boca Raton, Florida)

Negative impacts of composting are possible if inadequate technology is used or if the facility is poorly managed and operated. The potential for water pollution may exist if the moisture content of the compost is too high (more than about 65 percent); provisions should be made to properly channel and divert runoff around an open windrow site. Suitable measures should be taken to minimize rodent and insect problems.

Odor control is a big problem at many compost facilities, particularly for open systems. They can become particularly foul and intense if anaerobic conditions occur due to poor operation. Proper windrow turning and aeration of the waste can minimize the odors.

### Co-composting

An interesting example of *integrated waste management* is co-composting of municipal solid waste and sewage sludge. Sewage sludge adds nitrogen, phos-

phorus, and other elements that enrich the solid waste and help the composting process. The sludge is first dewatered so that it does not add too much moisture to the compost pile. The dewatered sludge and organic portion of MSW must be thoroughly mixed. At a time when ocean disposal of sludge has been banned and sludge incinerators meet with much public opposition, co-composting may offer an increasingly viable technique for processing both sludge and MSW organics prior to final disposal.

When sludge is composted by itself or co-composted with MSW organics, the *static windrow* method or mechanical agitators may be used to aerate the waste (see Appendix H, Figure 8). In the static aeration method, compressed air is either forced up through the composting waste or pulled down through it, as shown in Figure 11.12. A length of perforated pipe about 150 mm (6 in.) in diameter is located under the windrow, which is roughly the same size and shape as the windrow aerated by the turning method. The

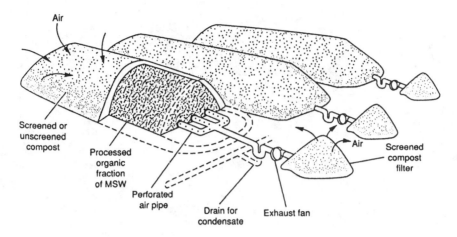

**FIGURE 11.12**

*Schematic of aerated static pile composting system.* (From G. Tchobanglous *et al.*, *Integrated Solid Waste Management*, McGraw-Hill, New York, 1993. Copyright © 1993, McGraw-Hill. Used with permission)

windrow may be covered with a layer of finished compost to absorb any objectionable odors. Air is forced in (or out) by a blower for about 5 min at about 15-min intervals.

## 11.5 RECYCLING

Residential, commercial, industrial, and institutional activity will always result in the generation of solid waste. But the public view of what constitutes waste, that is, useless and unwanted material, is changing. As people become more aware of ecological or environmental imperatives and as the space available to landfill waste in certain regions of the country shrinks, the need for *recovering and reusing* much of what was previously thrown away or dumped is becoming more evident. Also, the cost of disposing of solid waste material in an environmentally sound manner makes it more necessary to consider alternative waste management techniques.

The ideal approach to solid waste management is to first reduce waste at the source and then to recover reusable materials from the waste stream prior to disposal. This is accomplished by *recycling*, that is, by separating out and reusing these components of the waste stream that may have some economic value. According to EPA estimates, about 28 percent of the MSW waste stream in the United States was recycled in 1998, up from about 6 percent in 1960.

Recycling returns various materials to the production cycle and saves natural resources along the way. Some materials, such as aluminum and steel, can be recycled many times. Energy, as well as material, can be recycled. A resource recovery incinerator is an example of a facility where thermal energy is recycled by extracting the heat content of MSW and converting it to steam and even to electricity. Composting is an example of a process that allows recycling of the organic component of municipal solid waste. Usually, however, when people think of recycling, they think of the collection, processing, and resale of materials commonly found in homes and businesses.

### Recyclable Materials

Many materials can be recycled. Some materials are the result of commercial and industrial production and never enter the solid waste stream. The major components of municipal solid waste that have some economic value and are *recyclable* include metals, paper, glass, and plastics.

#### Metals

Metals are classified as being either *ferrous* or *nonferrous*. Ferrous metal contains iron and is magnetic; nonferrous metal does not contain iron and is nonmagnetic. Steel, for example, is a ferrous metal, and aluminum is a nonferrous metal.

Although aluminum is one of the smallest components of MSW (less than 1.5 percent), it has one of the highest values as a recyclable material and is generally considered the "cash cow" of municipal recycling programs (see Figure 11.18). It is one of the most abundant metals found in Earth's crust, but the process of extracting aluminum from mined ore is expensive. Since new aluminum is relatively costly, recycled aluminum has a strong market and plays a significant role in the aluminum industry. Aluminum scrap can be magnetically separated from ferrous metals, shredded in a hammer mill, and transported to a smelting plant in car-load lots. About 2.7 million tons of aluminum was generated in the year 1990, and roughly 38 percent of it was recycled. Most recycled aluminum comes from used beverage cans and is used to make new cans. Recycled aluminum cans may be either shredded (see Figure 11.13) or crushed and baled.

Ferrous metals make up roughly 6 percent of MSW, and one of the major components of this ferrous waste is the steel can. The "tin can" is made of steel and then coated with tin to reduce corrosion and to stabilize the flavors of foods and beverages. Tin coatings can be removed from used cans, although current steel–making technology can handle up to 10 percent

**FIGURE 11.13**

(a) *Aluminum cans are placed in a hopper and carried along a moving belt; then they move through a magnetic separator that removes any steel cans.*
(b) *The cans may be crushed and baled or shredded as shown here, prior to reprocessing.* (Courtesy of Aluminum Association, Inc., Washington, D.C.)

of tin in the recycled steel. There is an unlimited market for steel scrap. Roughly 12 million tons of steel waste is generated each year, and about 15 percent of it is recycled (mostly in the form of food and beverage containers). However, much larger quantities of scrap steel are recycled at the point of origin or obtained from scrapped steel girders and junked cars. This source of scrap steel is not part of the MSW stream.

**Paper**

Paper and paperboard products constitute the largest fraction of MSW (37 percent in 1993). Recycling rates vary, depending on the type of paper. Old newspaper (ONP), for example, has about a 30 percent recovery rate, while corrugated paper has almost a 50 percent recovery rate. Some other types of used paper that can be recycled include office paper, tissue, cardboard, bags, magazines, and cartons.

Recycled paper can be used for several purposes, but it is never "as good as new" after reprocessing. The fibers are weakened and it is difficult to control the color of the recycled product. Federal law also prohibits the use of recycled paper products for many types of food containers in order to prevent the possibility of contamination. It often costs less to transport raw paper pulp than scrap paper; collection, sorting, and transport account for about 90 percent of the cost of paper recycling, and the processes of pulping, deinking, and screening wastepaper are generally more expensive than making paper from virgin wood or cellulose fibers. The market for recycled paper is quite volatile. As more mills and deinking plants come on line, demand for recycled paper should stabilize and price fluctuations may settle down.

It is often noted that paper recycling will help to preserve forests, since it takes about 17 trees to make 1 ton of paper. But this may be an unrealistic notion. The fact that selective harvesting of mature trees is a necessary forest management technique must be considered. Thinning out timberland by the harvesting or cropping of trees increases the health and productivity of the forest. Therefore, from an ecological perspective, paper recycling does not necessarily lead to the preservation of forests.

**Glass**

Glass makes up almost 7 percent of MSW and has a recovery rate of about 20 percent; most glass waste is from food and beverage containers. Glass is the least troublesome material in MSW. It is an inert, nonpolluting substance made primarily from silica sand, an abundant natural resource. Even though the raw material from which it is made is so plentiful, there is still a market demand for waste glass. But recycled glass has economic value only when it can be separated by color and then crushed to make new glass.

Crushed glass, or *cullet*, can be remelted to produce a new batch of glass; it is estimated that cullet could make up as much as 30 percent of the material needed to produce new glass containers. If glasses of different colors (clear, brown, and green) are mixed together, however, the cullet has little value for the remanufacture of new glass containers. At many municipal recycling drop-off centers, the glass is kept separate in large containers to prevent the mixing of different colors.

## Plastic

Plastic, a nonbiodegradable, petroleum-derived substance, made up about 9 percent of MSW in 1993. Plastics are difficult to recycle because of the many different types of polymer resins used in their production. To be of most value, plastics have to be separated by type; if mixed, the materials can only be used to make lower quality products, such as *plastic lumber*.

Because many people have difficulty in identifying different plastic types, a recycling symbol and code have been developed; it is usually stamped on all recyclable plastic products. For example, the code for PET-type plastic (polyethylene teraphthalate) is *1*, and for HDPE-type plastic (high-density polyethylene) the code number is *2*. Soft drink containers are typically made of PET; milk and detergent containers are typically made from HDPE.

Plastic's share of the waste stream is growing in both weight and volume as more companies rely on plastic products and packaging. Collection of plastic containers still remains a problem, though, because of their large volume but very low density. In effect, most of a plastic container is air. This creates difficulties in collection vehicles, which have limited volume. Unfortunately, some municipalities have abandoned plastic recycling due to its low percentage by weight in MSW and its very high collection cost.

## Construction and Demolition (C&D) Debris

A significant type of community solid waste that is not considered to be part of the MSW stream (according to EPA definitions) is *construction and demolition waste* (or *debris*). This material comprises mostly wood, concrete, brick, and metals that are the result of building construction, renovation, or demolition. As landfill space becomes more expensive, more of these materials are being recycled. This recycling usually occurs at the source by placing the wood, brick, and other materials in separate containers for transfer to market. In some regions, both public and private *bulky waste recycling facilities* have been developed to enclose and automate portions of the sorting operation. Even though the market value of the materials is relatively low, the hauler can save money by diverting the waste from a landfill facility.

Metals, asphalt concrete, Portland cement concrete, and wood are the materials most frequently recovered from C&D debris; gypsum wallboard and roofing shingles are recovered and recycled to a much lesser degree. The major obstacles to recycling C&D debris include the high cost of sorting and processing the material and the relatively low cost of disposal in landfills in many areas. The number of C&D recycling facilities is growing, however, because of the increasing efforts being made to develop good markets for the recovered materials. Metals recovered from C&D debris have the highest recycling rates.

Portland cement concrete and asphalt concrete consist mostly of aggregates (sand and gravel). Concrete debris can be crushed and recycled as a replacement for road subbase gravel, especially in regions where aggregates are in short supply. Many wood processing facilities accept clean wood from C&D debris; processed (chipped) wood can be used as mulch, animal bedding, and fuel. Asphalt recovered from asphalt roofing shingles can be reused in cold mix asphalt paving products and new roofing; however, it is often difficult to meet the quality specifications for those products. Gypsum wallboard (drywall) can be recycled by separating the paper backing from the gypsum; the paper and the gypsum can be processed and reused in the manufacture of new wallboard. Recovered wallboard can also be used as cat liter.

According to EPA estimates, demolition debris accounts for roughly half of the total building-related C&D waste stream. This can be significantly reduced if new buildings are designed more for disassembly or *deconstruction* (that is, for the selective dismantling or removal of materials from a building before or instead of demolition) than they have been in the past. This could be as simple as using screws instead of nails.

## Other Recyclables

Reclaimed rubber must be shredded in special shredding machines and broken down chemically before it can be rebonded and remolded. Rubber produced by this process, called *revulcanization*, is not usually as strong as the original product. As an alternative to revulcanization, the use of shredded rubber in asphalt pavements is being investigated by engineers. Another interesting use for discarded tires is the *tire playground* for children, which is becoming a more familiar sight in many communities.

Hundreds of millions of rubber tires are discarded each year in the United States. Some wind up in landfills, but most are now burned as fuel for power plants and cement kilns. Rubber tires provide more energy per pound than coal and burn more cleanly. Temperatures up to 1900°C (3500°F) destroy most organic pollutants.

Used motor oil is widely recycled through drop-off centers at automobile service stations or through municipal recycling programs. Grass clippings are being eliminated from the MSW stream as more homeowners practice "cut it and leave it" mowing or backyard composting. Textiles are usually recycled through relief agencies such as the Salvation Army or processed into rags; roughly 75 percent of the textiles collected in the United States are exported and sold as used clothing.

## Source Separation

Before a recyclable material can be used by a market, it must be separated from the bulk of the waste and sorted with other materials of the same type or classification. Separation of recyclables can be accomplished at their source or point of waste generation by the home or business owner, or they can be separated at a centralized waste processing plant.

In the 1970s, large centralized waste processing plants were constructed. It was thought that recyclable materials could be removed from the mixed municipal waste stream and then sold to secondary materials vendors. But the quality of the recycled material fell short of market expectations. For example, newspaper and corrugated paper became wet and contaminated from the food waste in the refuse. Glass bottles broke and mixed with other waste, making separation virtually impossible. In general, cross-contamination was so great that, although some materials were successfully removed from the waste stream, they had little, if any, economic value.

Current thinking is that recovery of recyclables from mixed MSW does not produce viable market commodities. The message learned from recycling in the 1980s and early 1990s was that the *end market* controlled the process. Many recycling programs failed because they collected materials of such poor quality that no market wanted them, and they ended up being diverted to a landfill anyway.

Meanwhile, during this time, residential and commercial *source-separated programs* were developed in many communities. Source-separated means that the homeowner or business keeps the recyclable materials separate from the garbage in the home or business (source), thereby dramatically increasing the cleanliness and marketability of the material.

Source-separated programs initially consisted of local drop-off centers at public works garages, where homeowners could bring their materials and place them in separate containers. The homeowner was responsible for keeping all the materials separate in their home and in transport. But not all citizens could be relied on to voluntarily separate their refuse into homogeneous groups, and many were not willing to transport the materials to the appropriate recycling center. In addition, as pressure was placed on public officials to increase their recycling amounts, more recyclables were added to the list. This caused most people to complain about the number of different containers needed and the inconvenience. One solution to the inconvenience of homeowners having to transport the materials was to institute *curbside collection* programs, where residents placed recyclables at the curb for collection separate from their garbage. Collection was done by the town or by private contractors.

Specialized compartmentalized collection trucks were developed; they allow many materials to be collected in one truck, but kept separate with baffle walls. Glass was separated by color, and metals and paper were placed in separate compartments. Paper was always kept separate because broken glass will embed itself in the paper and ruin its value. As more materials were added to the recyclable stream, such as plastic with eight different resin types, the practicality of using separate curbside containers for each commodity and separate compartments on the truck was lessened. In effect, the collection activity itself became a limiting factor in advancing the recycling effort.

In answer to this dilemma, many communities began to mix or *commingle* the nonpaper materials (glass, metals, and plastics) in 7- to 32-gal containers at the home. Commingled material was then collected in one compartment of the truck; this greatly simplified the recycling process for the homeowner. To continue to preserve marketability, waste paper and commingled materials are never mixed. Although commingling and curbside recycling programs make recycling much more convenient, they have dramatically increased the cost of solid waste management because a separate collection was now necessary for the recyclable materials. A modern recycling truck is shown in Figure 11.14.

To encourage a higher level of citizen participation, some states mandated that municipalities reach a recycling percentage goal. In New Jersey, for example, a 25 percent recycling goal was set in 1980. This was later increased to 60 percent of all MSW by 1995. Nonrecyclable waste would then be incinerated and the residual ash would be disposed of in modern lined landfills. The effect of the legislation was to dramatically increase citizen participation rates. Other states soon passed similar laws. Ironically, the success of the recycling industry began to adversely affect the marketability of the material.

The amount of high-quality material available to the market in the late 1980s and early 1990s increased dramatically, causing a glut as more communities tried to sell their materials to the markets. Since so much

**FIGURE 11.14**

*A medium-duty truck for applications in residential refuse and recycling curbside collection.* (Courtesy of Freightliner Corp., Portland, Oregon)

material was available, the markets tightened up their specifications and paid for only the highest quality material. This caused some marginal programs with poor-quality or contaminated material to fail for lack of an end market. To address this new problem, the industry turned to *material processing* or *beneficiation* to improve the quality of the materials prior to marketing.

## Processing Recyclables

Where communities have commingled collection, the materials must be separated by type before they can be sold. The most common system for processing or *beneficiating* the source-separated but commingled material is called a *materials recycling facility* or MRF (often pronounced "murf" in waste management jargon).

Separation and sorting at a MRF offers an efficient and reliable method for processing recyclable materials; it also allows for marketing of larger amounts of quality material. A large, centrally located MRF serving many communities can accomplish separation and sorting of recyclables on a relatively economical basis through economies of scale.

A typical MRF is an industrial-type metal or concrete building similar to a warehouse. It has high doors to allow trucks to dump inside. An average MRF building has a floor area of about 4000 m² (0.4 ha or 1 ac) and contains separate areas for depositing paper and commingled materials. In many cases, air pollution control equipment is installed to collect and remove dust from the air. Sometimes they are heated, and special sorting rooms are constructed to provide workers with proper lighting and ventilation.

In general, the following types of materials are received at a MRF in a collection truck separated into two or three compartments:

**Paper Compartment**

- Old newspaper (ONP), generally bundled
- Old corrugated cardboard (OCC)
- Mixed paper, including envelopes, magazines, and junk mail

**Commingled Compartment**

- Clear, brown, and green glass
- Ferrous metal ("tin") food and beverage containers
- Aluminum food and beverage containers
- HDPE and PET plastics (milk, detergent, and soft drink containers)

A MRF utilizes a combination of labor and equipment to sort the source-separated, commingled materials from the collection trucks into various categories for marketing. A variety of technologies are used to improve the quality of the material by carefully separating glass (by color), steel, aluminum, and plastic (by resin type) into uniform commodities. Paper is also processed. A schematic diagram of a typical 100- to 300-ton-per-day MRF is shown in Figure 11.15. MRFs may be custom-designed by consulting engineers and built by contractors or constructed as turnkey facilities by companies that design, build, and operate the systems.

### Typical MRF Operation

Incoming trucks are weighed and the waste is then dumped on a concrete *tipping floor.* Any unsuitable materials seen on the floor are removed by hand and placed in containers for haul to the landfill. A large front-end loader then pushes the material onto inclined rubber-belt conveyors, which may pass by one or more workers who remove deleterious materials, such as large metal pots, bricks, or garbage. The conveyor then passes under an electromagnetic separator that removes the tin cans and other ferrous metals from the commingled waste stream. This metal is then conveyed to a baler for compression and baling prior to shipment to steel mills.

The remaining commingled material is then screened on a shaker table to remove the dirt and broken glass. This broken glass has little or no resale value and frequently will be used in the manufacture of paving material such as "glassphalt." The rest of the material is then classified with large blowers to remove the plastic and aluminum containers from the remaining unbroken glass bottles. These glass bottles are then conveyed into a sorting room, where workers separate the bottles by color; *if the glass is not separated by color, it has no*

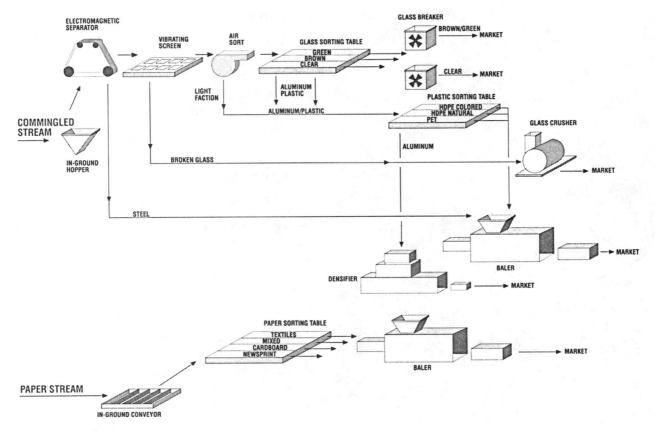

**FIGURE 11.15**
*Schematic diagram of material flow at a typical MRF.* (Courtesy of Killam
Associates, Consulting Engineers, Millburn, New Jersey)

*resale value.* In most MRFs, the glass bottles are further
processed by crushing them into roughly 12-mm ($\frac{1}{2}$-in.)
particles and then removing bottle caps and other such
material in a *rotating drum trommel screen classifier* (see Figure 11.16). The resultant glass product is
marketable as furnace-ready cullet.

The plastic and aluminum containers are also
conveyed to a sorting area. The separation of plastic is
typically accomplished by hand with skilled sorters
who can identify the plastic types through experience.
Recently, equipment has been developed that can identify plastic by its chemical properties and can allow
the complete automation of plastic sorting. The equipment is expensive, but it can eliminate three or four
sorters who normally have to separate the materials
manually. After the plastic is separated, it is either baled
or chipped for shipment to market. Other types of plastic, such as polystyrene and styrofoam, are not generally recovered because of low tonnages and high
processing costs.

The aluminum is usually separated from the plastic and other remaining nonrecyclable materials on
the conveyor belt through the use of a device called an

*eddy current separator.* This device repels aluminum
up into the air and off of the belt, allowing it to be captured and baled or densified for shipment to market (see
Figure 11.17). While aluminum typically accounts for
about 5 percent of the commingled waste stream, it produces almost 80 percent of the commingled revenue.

Quality control inspectors watch the mechanized
separation processes carefully to ensure that the
various materials do not mix in the plant and that
the individual commodities remain clean and pure. By
mechanizing the process and maintaining high quality
control, a large MRF can easily process up to 300 tons
of commingled material per day.

Paper is processed in a MRF on a separate conveyor line. Old newspaper is generally bundled at the
curb and not allowed to mix with the other papers in
the collection truck. At the MRF, the ONP is dumped
in a separate area prior to being conveyed past quality
control workers, who remove materials that are considered to be contaminants. It is baled and loose-loaded
into tractor trailers and shipped to paper mills for use
in making newspaper. Much of the remaining paper is
suitable for sale to tissue mills.

**FIGURE 11.16**
*Trommel screen classifier.* (Courtesy of Triple/S Dynamics, 1031 S. Haskell, Dallas, Texas)

## CASE STUDY

### Somerset County Materials Recycling Facility

#### Environmental Impact

The Somerset County Materials Recycling Facility (MRF) was developed to allow the County to comply with one of New Jersey's toughest environmental orders—the recycling of 60% of the state's solid waste by 1995. Governor Florio enacted this standard in 1990 to ease the solid waste crisis by significantly reducing the amount of waste which requires disposal. In order to meet this 60% goal, counties must recycle and market more types of materials. The Somerset County MRF plays an important role in achieving this goal, by allowing residents to more easily and conveniently put additional materials at the curb for recycling pickup and by processing those materials to meet tough market specifications. This is especially important because increasing quantities of recyclable materials are competing for limited market capacity.

The MRF also reduces the potential adverse environmental impact of siting and operating solid waste disposal facilities by reducing the amount of waste requiring conventional disposal. The MRF has no environmental impact of its own.

#### Social Significance

The MRF has had a positive impact on the County, in that it has, through the level of processing it offers, allowed the County to recycle additional types of materials at a reasonable cost. For example, Somerset was the first county in the United States to include textiles in its curbside program. The County will enjoy significant economic benefit not only from the revenues from the sale of recyclables, but also from the "avoided cost" of not having to haul the waste to out-of-state landfills.

#### Originality and Innovativeness

The Somerset County MRF is unique not in the types of equipment used in the facility, but in their application to the site and building. The County began recycling in the early 1980s by collecting materials and dumping them in a warehouse, where up to 85 people hand-processed the materials to marginal specifications. The County originally conceived the new MRF as an upgrade of the existing manual operation. The building was designed in 1988 and constructed in 1990 for that purpose. However, in the late 1980s, the recycling markets became saturated with materials, and only those materials with the highest quality were marketable. Accordingly, the County commissioned Killam Associates to design automated lines to process paper and commingled materials. The small building size necessitated shoe-horning $2.8 million worth of equipment to process 150 tons of material per day into 18,500 ft$^2$ of floor space. The remaining 8000 ft$^2$ of floor space was needed for material tipping. The conveyors and processing equipment rise 25 ft from the slab and utilize multiple levels for smooth and uninterrupted material flow.

#### Complexity

The design of the MRF was complicated by the need to place bulky, but sophisticated equipment in the building while still maintaining substantial clear floor space for material storage. The goal was accomplished by using the building's height and constructing much of the equipment at a second story level and allowing the material to move downward as it progressed through its various processing stages. All equipment is computer-interconnected for startup and operation of the materials processing line.

#### Technical Value

The MRF uses the latest state-of-the-art equipment to achieve the goal of processing and beneficiating the recyclables to the highest market specification. For example, the aluminum is separated from the plastic through the use of an eddy current separator. This device generates an eddy current field under the aluminum and plastic passing over it and creates a repulsion force in the aluminum. The aluminum is then "thrown" from the belt into a conveyor, thereby effectuating the separation.

**FIGURE 11.17**
*County of Somerset MRF; "biscuit" of aluminum being extruded.* (Courtesy of Killam Associates, Millburn, New Jersey)

Other equipment used includes a magnetic separator to recover steel cans, glass breakers, crushers, trommels, and air cleaners to beneficiate glass, balers for compacting steel, plastic, and paper, and air handling equipment to collect glass and paper dust.

**Cost-Effectiveness**

The MRF greatly improves the economic viability of the County's recycling program. For example, the new MRF operates with approximately one half of the labor force needed for the original facility. In addition, the materials marketed from the MRF are of significantly higher market value and are generating higher revenue. In fact, the revenue derived from the sale of the recyclables will more than offset the MRF operating cost.

The percentages of commingled materials recycled, and the percentage contribution of revenue for each material, are shown in the pie chart in Figure 11.18. It is interesting to note that the mixed broken glass had the highest tonnage processed, yet contributed no revenue. On the other hand, the aluminum had the lowest tonnage, but clearly contributed the highest revenue. In 1994, Somerset County processed 37,400 tons of material and received $1.4 million in revenue.

*Source:* Used with permission of Killam Associates, Consulting Engineers, Millburn, New Jersey.

# COMMINGLED SHIPPED YTD 1994
Source: Somerset County, NJ

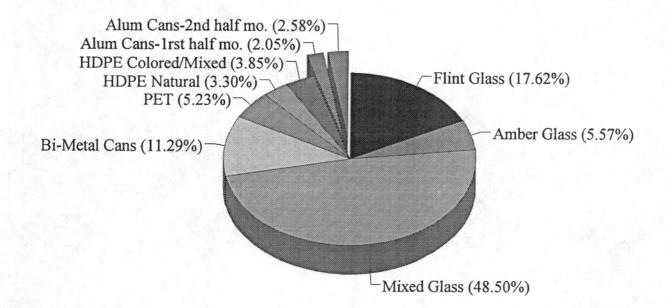

Alum Cans-2nd half mo. (2.58%)
Alum Cans-1rst half mo. (2.05%)
HDPE Colored/Mixed (3.85%)
HDPE Natural (3.30%)
PET (5.23%)
Bi-Metal Cans (11.29%)
Flint Glass (17.62%)
Amber Glass (5.57%)
Mixed Glass (48.50%)

# COMMINGLED REVENUE YTD 1994
Source: Somerset County, NJ

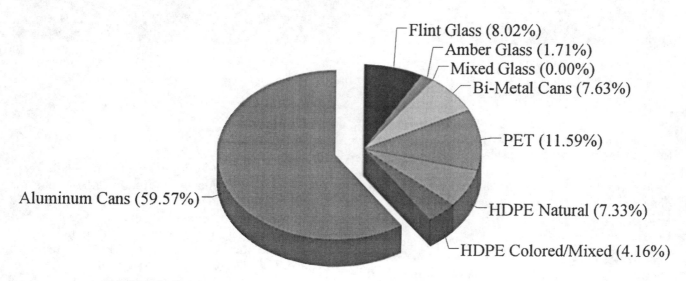

Flint Glass (8.02%)
Amber Glass (1.71%)
Mixed Glass (0.00%)
Bi-Metal Cans (7.63%)
PET (11.59%)
Aluminum Cans (59.57%)
HDPE Natural (7.33%)
HDPE Colored/Mixed (4.16%)

**FIGURE 11.18**
*Summary of Somerset County Recycling Program. Aluminum had the lowest tonnage shipped, but contributed the highest revenue.* (Courtesy of Killam Associates, Millburn, New Jersey)

## 11.6  SANITARY LANDFILLS

The oldest and most widely used method for ultimately disposing of solid waste is *land disposal.* Until the middle of the 20th century, waste was simply placed in a heap on top of the ground, and the disposal site was called a *dump.* These uncontrolled, open dumps quickly became breeding grounds for many vectors of disease, including rats, mosquitoes, and flies. In addition to posing a direct threat to public health, open dumps were smelly, unsightly nuisances. They also polluted surface water and groundwater, and they regularly caught fire. Open dumping of solid waste material is no longer an acceptable (or legal) disposal method in the United States and in many other countries.

Nevertheless, most municipal waste is still disposed of on land. But the waste is now buried in a *sanitary landfill,* not simply deposited in a pile on the ground. *A sanitary landfill is not a dump.* It is a carefully planned and engineered facility for solid waste disposal. This means that it is designed, constructed, and operated in an environmentally sound manner that does not threaten public health or safety, and that also minimizes public nuisances (such as windblown litter and unpleasant odors).

Three key characteristics of a municipal sanitary landfill distinguish it from an open dump:

1. Solid waste is placed in a suitably selected and prepared (for example, lined) landfill site in a carefully prescribed manner.
2. The waste material is spread out and compacted with appropriate heavy machinery.
3. The waste is covered each day with a layer of compacted soil.

Perhaps the most salient feature of modern sanitary landfill design is the technology used to prevent groundwater pollution. In the recent past, it was believed that a suitable depth or thickness of naturally occurring soil between the bottom of the landfill site and the groundwater table or bedrock would suffice to prevent pollution. It was thought that pollutants seeping out from the bottom of a MSW landfill (in a liquid called *leachate*) would be filtered and absorbed (in a process called *natural attenuation*) as they percolated down into the groundwater aquifer.

But, by the 1980s, it was discovered that soil does not necessarily absorb or attenuate all the contaminants seeping out from a sanitary landfill, no mater how thick the underlying soil layer. Construction of *natural attenuation landfills* for MSW is now banned in most locations, and suitable bottom liners and leachate collection systems are required at new landfill sites.

Current EPA rules for the operation of municipal solid waste landfills (MSWLFs) also require installation of monitoring systems to detect groundwater pollution; if contamination is detected, it must be reduced below the federal limits for safe drinking water.

In addition to providing an option for waste management, an MSWLF also serves to improve or reclaim poor-quality land. The landfill gradually raises the ground elevation or surface grade of the site. Some completed sanitary landfills have been successfully converted into municipal parks, playgrounds, golf courses, and other community land-use projects. The ultimate use of the site should be decided at the outset of the project so that the landfill can be constructed and operated with that goal in mind.

Landfilling is generally the most economical alternative for solid waste disposal, which accounts for its frequent application. In recent years, however, the practice of landfilling has declined in popularity as an option for waste disposal. For many urban communities, it has become increasingly difficult to find suitable landfill sites that are within economical hauling distances (less than 25 km). Public opposition (the NIMBY syndrome) has also become a major consideration. The suitability of a landfill site is determined by its size or volume, as well as a host of other technical and environmental factors. A few big cities in the United States are presently embarking on plans to build MSW incinerators (with energy recovery) due to the rapidly declining space for landfill within city limits.

Another reason for the declining popularity of sanitary landfilling is the realization that, in the long run, some environmental damage may occur no matter how well-engineered are the design, construction, and operation of the site. It is believed by many people that it is not possible to build a completely safe and secure solid waste landfill (particularly for hazardous waste). Some of the confined pollutants may eventually escape into the environment (in the form of leachate), with the potential to harm public health and damage the local ecosystem.

Despite these concerns, however, landfilling of solid waste will continue to be practiced worldwide to a significant extent for many years to come simply because of economic and technical necessity. *It is not possible to reclaim and recycle all solid waste materials.* There will always be a solid residue from incineration (or any other waste processing method), which will require ultimate disposal on land or underground, and there is no guarantee that other disposal methods are entirely safe. Incineration, for example, will always cause some degree of air pollution, even with use of the most sophisticated air-cleaning devices.

According to EPA estimates, about 55 percent of the total MSW stream in the United States was deposited in landfills in 1996, down from 62 percent in 1993. The number of landfills is also decreasing—from about 8000 in 1988 to 2400 in 1996. The total capacity, though, has remained constant because new landfills are much larger than those provided in the past.

## Site Selection

Many technical factors are involved in selecting a location for a new sanitary landfill. Perhaps the most important include the site's *volume capacity, accessibility*, and *hydrogeologic conditions*. Other factors include the climate as well as local socioeconomic conditions. One of the most difficult and frustrating problems with regard to siting a new landfill, however, is more political than technical in nature.

The general public equates *landfill* with *dump* and takes little note of the fact that a landfill is designed, built, and operated according to up-to-date engineering principles. Citizens, including local politicians, are reluctant to allow construction of a new landfill in their communities, and long, drawn-out legal battles over proposed sites are not uncommon. A possible way to overcome this problem may be to take the siting approval authority out of the hands of county officials and politicians and leave the final decision up to an appropriate nonpartisan statewide commission.

The total capacity and *design life* of a new landfill depend on the size and topography of the site, the rate of refuse generation, and the degree of refuse compaction. The amount of daily soil cover material adds roughly 20 percent to the overall fill volume and must be considered when evaluating the capacity of a landfill. There should be enough volume capacity within the working area of the site so that the landfill will have a design life of about 25 years. The longer the useful life of the site, the more economical is the overall solid waste management operation for the communities involved.

A generally accepted rule of thumb regarding municipal landfill capacity is that roughly 1 ha–m (8 ac–ft) of volume is needed each year to serve a population of 10,000 persons. (This can be visualized as 1 ha of land covered to a depth of 1 m, or 1 ac of land covered to a depth of 8 ft, with compacted refuse.) The following examples illustrate typical computations related to landfill capacity.

### EXAMPLE 11.2

A rural community of 15,000 persons generates refuse at an average rate of 5 lb per person per day. A 25-ac landfill site is available, with an average depth of compacted refuse limited to 20 ft by local topography. It is estimated that the compacted refuse will have a unit weight of 1000 lb/yd$^3$ and that an additional 25 percent of volume will be taken by the cover material. What is the anticipated useful life of the landfill?

*Solution*  The total weight of refuse generated per year is

$$\frac{5 \text{ lb}}{\text{person--d}} \times \frac{365 \text{ d}}{1 \text{ yr}} \times 15,000 \text{ persons}$$
$$= 27.4 \times 10^6 \text{ lb/yr}$$

The total yearly volume of refuse is

$$27.4 \times 10^6 \frac{\text{lb}}{\text{yr}} \times \frac{1 \text{ yd}^3}{1000 \text{ lb}} = 27,400 \text{ yd}^3/\text{yr}$$

The additional volume for cover material is

$$0.25 \times 27,400 = 6850 \text{ yd}^3/\text{yr}$$

Therefore, the total landfill volume required is

$$27,400 + 6850 = 34,250 \text{ yd}^3/\text{yr}$$

The available volume of the landfill is

$$25 \text{ ac} \times \frac{43,560 \text{ ft}^2}{1 \text{ ac}} \times 20 \text{ ft} \times \frac{1 \text{ yd}^3}{27 \text{ ft}^3} = 8.07 \times 10^5 \text{ yd}^3$$

The useful life of the site is estimated to be

$$8.07 \times 10^5 \text{ yd}^3 \div 34,250 \text{ yd}^3/\text{yr} \approx 24 \text{ years}$$

### EXAMPLE 11.3

Estimate how many hectares of land would be required for a sanitary landfill, under the following conditions:

| | |
|---|---|
| Design life of the site | 30 years |
| MSW generation rate | 25 N/person/d |
| MSW compacted unit weight | 5.0 kN/m$^3$ |
| Average fill depth | 10 m |
| Community population | 50,000 |
| MSW-to-cover ratio | 4:1 (20 percent of volume for cover) |

*Solution*  The quantity of MSW generated each year is

$$25 \, \frac{N}{person\text{–}d} \times \frac{365 \, d}{1 \, yr} \times 50{,}000 \, persons$$
$$= 4.56 \times 10^5 \, \frac{kN}{yr}$$

The volume of compacted refuse is

$$4.56 \times 10^5 \, kN/yr \div 5.0 \, kN/m^3 = 91 \, 250 \, m^3/yr$$

The additional volume required for soil cover is

$$91 \, 250/4 = 22 \, 813 \, m^3/yr$$

The total required volume is $91 \, 250 + 22 \, 813 = 114 \, 063$ $m^3/yr$.

The area required is computed as

$$area = \frac{volume}{depth} = \frac{114 \, 063 \, m^3/yr}{10 \, m} = 11 \, 406 \, m^2/yr$$

Since 1 ha = 10 000 $m^2$ and the design life is 30 years, the total required landfill area can be estimated as

$$11 \, 406 \, \frac{m^2}{yr} \times \frac{1 \, ha}{10 \, 000 \, m^2} \times 30 \, years \approx 34 \, ha$$

(Additional land would be needed for access roads, buildings, and the like.)

---

In addition to capacity and useful life, *site accessibility* is another important factor related to the location of a sanitary landfill. Accessibility refers to the ease with which refuse collection or transport vehicles can reach the disposal area on streets and highways without causing a public nuisance or traffic hazard.

Sites that require trucks to travel relatively long distances, particularly through residential neighborhoods, are not desirable. It would be best to locate the landfill in a nonresidential area, where the constant truck traffic will not be so much of a nuisance (litter, noise, and odor) or hazard. Since suitable landfill sites are increasingly difficult to find in most urban communities, the landfills are often located well outside city limits. In such cases, centralized *transfer stations* (see page 341) are used to improve the waste hauling operation. The collection trucks first deliver the refuse from the individual neighborhoods to the transfer station, where it is accumulated in large tractor–trailer vehicles for final transport to the landfill.

Capacity and accessibility have always been important factors related to the location of a sanitary landfill, but public concern for environmental health protection is now a more significant force with respect to landfill siting decisions. It is extremely important to minimize contamination from the landfill of the surrounding environment, particularly of surface water or groundwater. Local hydrology and geology (or *hydrogeology*) of the site has a direct influence on the possibility of water pollution. Soil borings and a thorough study of subsurface geologic conditions are of importance for landfill siting and design. Data regarding underlying rock types, soil gradation, permeability, and other factors that affect groundwater flow must be obtained.

Perhaps the key imperative for landfill location and design is that at *no time should the waste be in contact with surface water or groundwater.* Accordingly, landfills should not be located in low-lying wetland areas such as swamps or marshes, nor should they be located in the floodplains of streams or rivers; EPA rules (under RCRA) place strong restrictions on site requirements in this regard. Also, a minimum distance of 60 m (200 ft) from any lakes or ponds should generally be maintained, and a vertical separation of at least 1.5 m (5 ft) between the base of the landfill and the seasonally high groundwater table elevation is generally required.

It is important that new landfills not be located in unstable areas where landslides or sinkholes may occur. If a sanitary landfill must be located in a *seismic impact zone* where it may be damaged by mild earthquake activity, it must be designed to resist the forces caused by the seismic activity. New MSWLFs or extensions of existing facilities should not be located within 60 m (200 ft) of active geologic fault lines. Finally, the proximity to airports is of concern in siting new sanitary landfills. A landfill closer than about 1.5 km (1 mile) to an airport runway must be operated in a manner that prevents hazards to aircraft from birds that may be attracted to the site.

## Leachate Containment

Sanitary landfills generate significant amounts of a highly contaminated liquid called *leachate*, and past practice in landfill design relied on the natural ability of soil to filter and adsorb the pollutants. This concept of "natural attenuation" has been abandoned, and the new design concept is based on the *control, containment,* and *treatment* of the leachate; modern MSWLFs are also called *containment landfills.*

Some leachate results directly from the moisture and decomposition of garbage and other putrescible material in the waste material, but much of it may come from runoff or surface water that first infiltrates the fill and percolates downward through the waste material. Direct contact with the waste results in severe contamination of this water. If leachate then reaches and mixes with groundwater or seeps out of the fill into a nearby stream or lake, significant environmental damage can occur. Generally, as more water infiltrates and flows through the landfill, more pollutants are leached. However, if a sanitary landfill is located in an arid region with very little rainfall, the production of leachate is much less of a problem. It is important for engineers to be able to estimate the rate of leachate generation in a landfill to be able to adequately design the facility. A hydrogeologic technique called the *water balance method* may be used to calculate leachate amounts; computer programs are utilized to speed up these and other computations related to the design of a sanitary landfill.

MSWLF leachate quality is highly variable, but it generally contains more pollutants than raw sewage or many industrial wastes. The strength of leachate typically depends on the type and depth of the waste, age of the landfill, rate of water infiltration, landfill operation, and other factors. In some cases, when sludge from municipal sewage treatment plants is disposed of in an MSWLF (a practice called *co-disposal*), the generation and quality of leachate may be significantly affected. Many parameters are used to measure the quality of leachate, including BOD, dissolved solids, pH, heavy metals, and others. (Leachate BODs of up to 20,000 mg/L have been reported.) Hazardous wastes inadvertently disposed of in an MSWLF can lead to more serious environmental threats from a wide variety of toxic substances carried in the leachate.

There are basically two ways to prevent leachate problems at a sanitary landfill. One is to intercept and channel surface runoff to prevent it from entering the landfill. Another is to provide a suitable type of impermeable barrier or liner between the waste and the underlying aquifer. This is now required for all new municipal solid waste landfills and expansions of old landfills in the United States under RCRA Subtitle D regulations.

### Control of Surface Water

Surface water can be controlled and diverted away from the landfill by proper grading; generally, a 6 to 12 percent slope of the daily cover will allow water to drain freely from the landfill. Concrete-lined channels or drainage ditches can be constructed around the perimeter of the site to divert upland rainfall and to collect surface runoff from the site.

The final cover or *cap* at the top of a completed landfill is constructed so that it is impermeable and

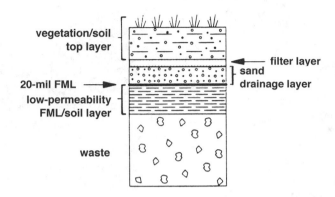

**FIGURE 11.19**
*Final cover or cap of an MSW sanitary landfill.*
(Courtesy of the Environmental Protection Agency)

prevents infiltration of rain water directly on the landfill areas; it is designed and graded to keep water away from the buried waste. The cap may typically consist of a soil layer at least 600 mm (24 in.) thick, with a 3 to 5 percent slope; it should be underlain by a sandy-soil drainage layer and a flexible *membrane liner* (FML) with a low-permeability soil layer (see Figure 11.19).

### Bottom Liners

Most new MSWLFs are to be constructed with composite liners made of a synthetic material over a 600-mm (24-in.) layer of clay; this forms an impermeable barrier that contains the leachate and prevents it from mixing with groundwater. (In some cases, state regulatory agencies may allow some variations in the design and thickness of the bottom liner, depending on local conditions.)

The synthetic material, usually made of HDPE plastic, is called a flexible membrane liner (FML) or *geomembrane*. It is placed over a very carefully prepared subbase to prevent its puncture or ripping. A typical cross section of a composite liner is shown in Figure 11.20. The clay must have a hydraulic conductivity of less than $1 \times 10^{-6}$ mm/s. That is, any water or leachate that happens to enter the clay will flow at a rate less than one millionth of a millimeter per second (equivalent to about 30 mm or a little more than 1 in. per year).

### Leachate Collection and Treatment

The leachate that is intercepted and contained by the bottom liner system cannot be allowed to simply accumulate indefinitely. It must be collected and treated at a centralized location prior to disposal. Options for final disposal primarily include discharge into a municipal

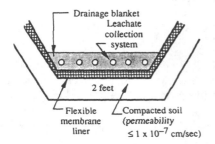

**FIGURE 11.20**
*Cross section of an MSWLF composite bottom line.*
(Courtesy of the Environmental Protection Agency)

sewage treatment plant, land application, or on-site treatment and discharge into a stream. A network of perforated plastic pipes located in a drainage layer located just above the impermeable liner is generally used to collect the leachate. The perforated pipes, called *laterals* (100 mm or 4 in. in diameter), direct the leachate to a larger header system and finally to a sump pit or pump station. Provisions are made for inspection and maintenance to ensure free flow of leachate.

MSWLF leachate is generally a high-strength wastewater characterized by high BOD, low pH, and the possible presence of toxic substances. It is sometimes, but not always, amenable to conventional sewage treatment processes. In addition to technical problems related to treatment and disposal, nontechnical factors must often be considered, such as legal and regulatory constraints as well as public participation.

If a nearby sewer system is available and the municipal treatment plant has adequate capacity, it is most convenient to treat the leachate at the existing public facility; the leachate often requires some degree of onsite pretreatment, depending on its characteristics. If it is treated on-site only and then discharged to a nearby body of water, it is subject to point source permit requirements. Many of the physical, biological, and chemical treatment processes described in Chapter 10 can be applied, in a variety of combinations, to adequately treat the leachate.

## Construction and Operation

A construction feature common to all MSW landfills is the *refuse cell.* It is the basic building block of the landfill and must be formed in a proper manner. The *ramp method* is one that may be used to construct refuse cells, as illustrated in Figure 11.21.

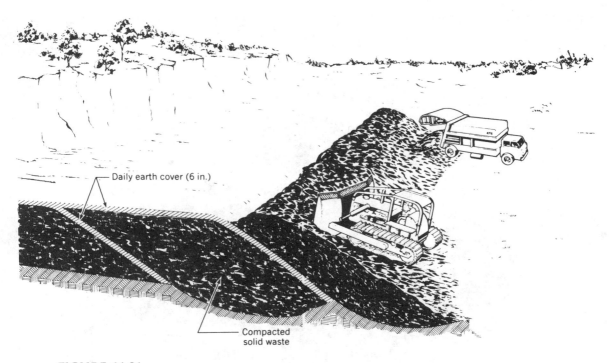

**FIGURE 11.21**
*In the ramp method of landfill construction, refuse is placed and compacted on a slope. Bottom liners are not shown.* (Courtesy of the Environmental Protection Agency)

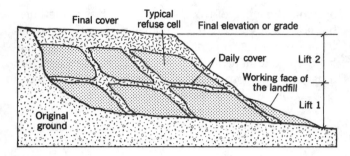

**FIGURE 11.22**
*The basic building block of a sanitary landfill is a compacted cell of solid waste, which is separated from other cells by a layer of compacted soil. (Bottom liner and cap details are not shown.)*

Incoming refuse is spread out and promptly compacted in thin layers at a confined portion of the site. To ensure proper compaction, these layers are generally no more than 1 m (3 ft) thick; several such layers are placed and compacted on top of each other to a maximum height of about 3 m (10 ft). The unit weight of the compacted refuse is about 6 kN/m$^3$ (1000 lb/yd$^3$), more than four times its initial loose density when it was collected.

At the end of each day's operation, the compacted refuse is covered with a layer of soil, which may also be compacted to a thickness of at least 150 mm (6 in.). Both the compacted refuse and the soil cover make up a single cell of the landfill. Several adjoining cells, all of the same

height, make up a *lift*, and the completed landfill may consist of several lifts, as illustrated in Figure 11.22. Cover material is preferably obtained directly from the landfill site, but it also can be hauled in from offsite locations.

The daily soil cover serves several useful purposes, although it uses up a significant fraction of the landfill volume. It reduces odors, windblown debris, and the risk of fires; it minimizes the risk of infectious disease transmission (for example, by rats or flies), and it improves site esthetics and access. When the landfill is finally completed, the cap placed on top of the uppermost lift must be impermeable to prevent water from entering; the cap may also be covered with about 150 mm (6 in.) of soil (see Figure 11.19).

Various types of heavy machinery are needed at MSW sanitary landfills to spread and compact the refuse, as well as to haul, spread, and compact the soil cover material. Traditional earthwork construction equipment can be used. The bulldozer, for example, is one of the most versatile machines at a sanitary landfill. It can be used for a variety of operations, including spreading and compacting.

Large landfills may require several additional machines, including large earthmovers called scrapers or pans. These machines excavate, haul, and spread cover material when the haul distance exceeds the range of dozers (about 100 m or 300 ft). Special landfill equipment, such as the steel-wheeled compactor shown in Figure 11.23, may be used to achieve higher compaction densities of the refuse. The large-toothed wheels serve as load

**FIGURE 11.23**
*A steel-wheeled compactor, one of several types of heavy machinery needed for daily operation of a municipal sanitary landfill.* (Courtesy of BOMAG: A Unit of AMCA International Corporation, Springfield, Ohio)

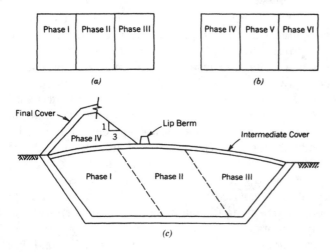

**FIGURE 11.24**

*Phasing plan for a multiphase landfill:* (a) *lower phase;* (b) *upper phase;* (c) *cross section.* (From *Design, Construction, and Monitoring of Landfills*, A. Bagchi. Copyright © 1994 John Wiley & Sons, Inc. This material is used by permission of John Wiley & Sons, Inc.)

concentrators, providing more compaction pressure than conventional rubber-tired or crawler-tractor equipment.

A *phasing plan* is important for operating large sanitary landfills. This simply means that selected areas of the landfill site will be scheduled to receive final cover or *capping* in a particular sequence, after final grade is reached. When the total depth of the landfill exceeds 9 m (30 ft) from the base, it is not unusual for an intermediate cover to be used at mid-depth, as illustrated in Figure 11.24. The intermediate cover is typically 600 mm (2 ft) of clay soil and 150 mm (6 in.) of topsoil over the area; new phases are started on top of the lower phases. The sequence and direction of filling should be indicated in the phasing plan; it must be prepared so as to ensure that all the waste is deposited in its final resting place.

In flat-lying areas of the country, or where the ramp method of landfill construction may not be appropriate, an excavation or trench is prepared (and lined) to receive the waste. Some communities, however, have piled the waste into enormous heaps ("Mt. Trashmores"), some of which have been converted into ski slopes.

### Landfill Gas

The organic material in buried solid waste will decompose from the action of microorganisms. The rate of decomposition depends primarily on the amount of moisture present in the waste. Decomposition proceeds at a relatively rapid rate when moisture content exceeds 50 percent, but complete breakdown of the waste and landfill stabilization typically takes about 20 years.

At first, the refuse decomposes aerobically until the oxygen that was present in the freshly placed fill is used up by the aerobic microorganisms. Then the anaerobes take over, producing *methane*, $CH_4$, and other gases as they metabolize the organics. Methane mixed with air in concentrations between 5 and 15 percent is highly explosive. It is also poisonous and can cause death by asphyxiation.

The methane will follow the path of least resistance; it can travel considerable distances horizontally through porous layers of sand and gravel. Hazardous conditions can occur if the gas rises to the surface and accumulates in basements, buildings or other enclosed areas. The explosions and tragic loss of life that have occurred in the past because of landfill gas could have been avoided by proper design and construction of the landfill. This is why the possible flow of the gas through the refuse cells and soil must be taken into account during the design stage of an MSW sanitary landfill.

Gas movement can be controlled by providing impermeable barriers in the fill. A 1-m (3-ft)–thick layer of moist clay soil, for example, will act as an excellent barrier to the flow of landfill gas at the bottom and periphery of the fill. A venting system can be constructed to collect the blocked gas and vent it to the surface, where it will be safely diluted and dispersed into the atmosphere, as illustrated in Figure 11.25.

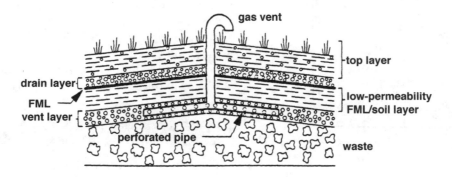

**FIGURE 11.25**

*Methane gas can be vented through the cap of a completed landfill to dilute it below dangerous levels.* (Courtesy of the Environmental Protection Agency)

Instead of venting the gas to the atmosphere, the methane can be collected and recovered for use as a fuel. Actually, methane is one of the most important recoverable resources from MSW landfilling operations. At several landfills in the United States methane recovery is now practiced, and more such gas recovery projects are being planned. Recovery systems that are designed to process the collected landfill gas are commercially available. They use a principle called *membrane permeation* to separate traces of carbon dioxide, nitrogen, and hydrogen sulfide from the gas, leaving a purer methane product for sale as a fuel. An MSWLF gas extraction system is shown in Figure 11.26.

### Landfill Maintenance and Monitoring

In addition to setting minimum standards for siting, design, construction, and operation of MSWLFs, federal regulations in the United States address the manner in which the facilities are completed or *closed*. The owner or operator of the landfill is required by law to prepare a written 30-year plan for final cover maintenance, leachate collection system operation and maintenance, groundwater monitoring, and methane gas monitoring. It is also necessary that a description of planned uses of the site be prepared. Some of these requirements can be waived by state regulatory agencies if it can be demonstrated that there is no potential for hazardous wastes to pollute the uppermost aquifer.

Groundwater monitoring requires the installation of monitoring wells at appropriate locations around the site boundary. Semiannual sampling and testing for a wide variety of inorganic compounds and volatile organics is required. Remedial action must be taken if it is found that pollution has occurred and exceeds *trigger levels* established by state public health or environmental agencies.

## A Landfill Landmark

Operation of the first sanitary landfill in the United States began in 1937, in Fresno, California. This was a landmark for public health protection. In fact, in the summer of 2001, the U.S. Department of the Interior was preparing to include the Fresno landfill on the National Register of Historic Places. This idea was sponsored by the National Park Service, as part of a program to document the history of civil engineering. But an ironic complication put the effort on hold. It turns out that the landfill was already on a special federal register—the Superfund list, which involved the landfill site in hazardous waste cleanup activities (see Section 12.4).

The Fresno landfill used the trench method for solid waste disposal, and although it was an engineered facility, the site was not lined (since that was not a requirement at the time.) In the early 1980s, it was discovered that methane gas and leachate were causing pollution problems. Consequently, in 1989, the landfill was placed on the Superfund list. Since then, roughly $38 million has been spent on cleaning up the site, which contains about 60 million $m^3$ (79 million $yd^3$) of MSW. Methane is now vented, leachate is now treated, and the capped site is being transformed into a public park and sports complex. The possibility of listing the completed landfill on the national Register of Historic Places in the future has not been ruled out completely.

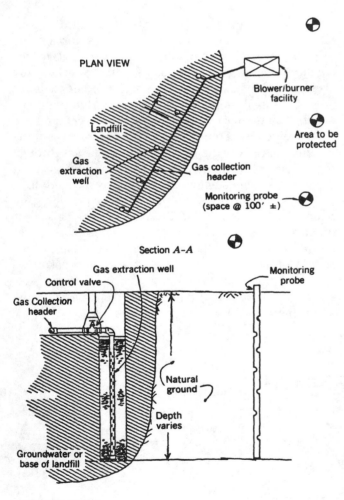

**FIGURE 11.26**
*Landfill gas extraction.* (Courtesy of the Environmental Protection Agency)

## CASE STUDY

### Closure of a Sanitary Landfill

Yarmouth, Massachusetts, located in the heart of Cape Cod, is one of the East Coast's most desirable spots, serving as both a year-round community and a vacation retreat. In 1990, federal and state regulations required that Yarmouth close its 23-ha (57-ac) landfill site, which had been operating since the early 1950s as a municipal landfill. Capping was required to prevent groundwater contamination and protect the town's water supply.

The town determined that reusing the site was the most cost-effective plan, and in 1993 retained a consulting engineering firm to collaborate on a comprehensive landfill closure and reuse plan. The plan included a 9-hole expansion to the town's existing 18-hole golf course located adjacent to the landfill site, a regional bikeway connection, a park, and improved solid waste disposal facilities.

Due to lack of good-quality topsoil on Cape Cod, it was decided to manufacture topsoil to cover the capped landfill. A topsoil blend was developed that included a mix of biosolids (sewage sludge), composted yard waste (leaves and grass clippings), and naturally available sand. In addition, it was determined that the sewage treatment facility located adjacent to the landfill produced a high-quality effluent that could supply the golf course spray irrigation system; this would help supplement the town's limited potable water supply. To accommodate town waste disposal needs once the landfill was closed, a new combined residential waste drop and recycling facility, a yard waste composting area, and a construction/demolition waste transfer station were designed.

Throughout the project development, the consultant and town officials held public meetings to receive citizen input and address concerns. Informational newsletters also updated residents on project progress and costs. The public participation process, which included working with several boards and committees, holding open public meetings, and airing cable television presentations, was integral to the project's success. Town support of the project became clear when, at a town meeting, the vote was unanimous to spend $12 million for the project.

The consulting firm helped Yarmouth to quickly obtain permits from the Department of Environmental Protection, ensuring that closure was completed on schedule. Capping of the landfill area was completed in 1997; the golf course and the park recreational areas were planned for public use in 1999. Yarmouth's unique approach to landfill closure and reuse optimized the town's resources and demonstrated that a potential liability can be transformed into a long-term asset.

---

*Source:* Adapted from an article that originally appeared in the June 1998 issue of *Public Works*, published by Public Works Journal Corporation, Ridgewood, New Jersey. All rights reserved. Used with permission.

## CASE STUDY

### Largest Landfill Gas-to-Energy Plant in U.S.

Landfills can provide an easy-to-tap source of methane gas to fuel power-generation gas turbines. But landfills may also produce hydrogen sulfide, $H_2S$, which causes odors and, when burned in the gas turbines, produces acidic sulfur dioxide, $SO_2$, which will corrode plant equipment.

The Central Sanitary Landfill and Recycling Center in Broward County, Florida, operates the nation's largest landfill gas-to-energy plant. About 9 million $ft^3/d$ of landfill gas, collected from 300 wells on the site, feeds gas turbine generators to produce electricity. Originally the gas generated by the landfill contained only small amounts of $H_2S$, well below levels that would cause corrosion or odor problems. But as debris from Hurricane Andrew (August 1992), as well as waste from the large construction projects it spawned, entered the landfill, the nature of the collected gas changed considerably. The $H_2S$ levels rose to as much as 5000 ppm, requiring its removal before it entered the gas turbines.

There are many processes available for removing $H_2S$ from combustible gas streams produced in a landfill. For example, $H_2S$ may be removed using adsorbent chemicals such as metal oxides. But the cost of the chemicals plus the disposal cost of the spent material can be quite high, especially at the high $H_2S$ levels

**FIGURE 11.27**
*View of the Yarmouth, Massachusetts sanitary landfill site.* (From *Public Works*, June 1998. Copyright © 1998 Public Works Journal Corporation. Used with permission)

collected at the Broward County landfill. To keep costs low, a hydrogen sulfide oxidation system, which uses ferric, $Fe^{3+}$, ions to oxidize the $H_2S$ to elemental sulfur, was selected; the ferric ions are reduced to ferrous, $Fe^{2+}$, ions in the process. Atmospheric oxygen continuously regenerates the ferrous ions back to ferric ions.

In this process, gas from the landfill first enters a countercurrent-flow, "mobile-bed" absorber column. As the $H_2S$-contaminated gas flows up through the column, it contacts a downward flow of water solution containing ferric ion. The $H_2S$ is absorbed in the solution, where it reacts with the ferric ion to produce ferrous ion and sulfur. The mobile-bed packing comprises hollow plastic balls that are in a constant state of agitation. This constant movement and the smooth surface of the balls prevents the sulfur from sticking to the packing.

Solution from the scrubber, containing reduced ferrous ion and elemental sulfur, flows to an atmospheric oxidizing tank. The oxidation tank is mixed with air and equipped with baffles to provide successive stages of oxidation. Atmospheric oxygen regenerates the iron back to ferric ion for return to the absorber. A side stream from the oxidizer is pumped to a settling tank, where it is concentrated before filtering and dewatering. The dewatered sulfur is removed and the filtrate returned to the oxidation tank. Since elemental sulfur is not hazardous, the dewatered sulfur (60% solids) produced by this process is sent right back to the landfill. (A photograph of this facility is included in Appendix H, Figure 9.)

---

*Source:* Adapted from an article that originally appeared in the May 1997 issue of *Public Works*, published by Public Works Journal Corporation, Ridgewood, New Jersey. All rights reserved. Used with permission.

 ## 11.7 RELEVANT WEB SITES

### ENVIRONMENTAL INDUSTRY INTERACTIVE

http://www.envasns.org/

This Environmental Industry Associations Web site provides information about solid waste recycling rates and market trends, landfill capacity, standards, and operations, interstate movement, and waste management equipment.

### EPA's OFFICE OF SOLID WASTE

http://www.epa.gov/epaoswer/osw/index.htm

This Home Page of EPA's Office of Solid Waste leads to general information and technical articles for students,

teachers, and concerned citizens on the topics of solid waste transport, treatment, disposal, laws and regulations, waste source reduction, waste cleanup programs, and more.

### INTERNATIONAL SOLID WASTE ASSOCIATION (ISWA)

http://www.iswa.org/

This is the Home Page of ISWA, an organization that promotes and develops solid waste management worldwide to protect public health, natural resources, and the environment. ISWA provides information about research and development, education, and training. This site also provides access to the ISWA Times, bookstore, and member discussion board.

### MUNICIPAL SOLID WASTE PUBLICATIONS

http://www.epa.gov/epaoswer/non–hw/
muncpl/pub97/msw-pub.htm

This Web site lists more than 100 publications available online from EPA, covering municipal solid waste source reduction, recycling, incineration, landfills, used oil, educational materials, and newsletters.

### SOLID WASTE ASSOCIATION OF NORTH AMERICA (SWANA)

http://www.swana.org/

This is the Home Page of SWANA, an organization that serves to exchange ideas and foster professionalism and education in the field of solid waste.

### SOLID WASTE DISPOSAL FACILITIES CRITERIA

http://www.epa.gov/epaoswer/non–hw/muncpl/criteria.htm

This EPA guide provides basic information on municipal solid waste landfill regulations including location, operation, design, groundwater monitoring, corrective action, closure, and financial factors.

### SOLID WASTE ONLINE

http://www.solidwaste.com/

This Web site provides news, technical articles, an online resource library, and other information pertaining to the solid waste management industry.

### SOLID WASTE REPORT

http://www.bpinews.com/enviro/pages/swr.htm

This site provides a weekly online newsletter focusing on solid waste topics, such as resource recovery, recycling, collection, disposal, and other topics.

# REVIEW QUESTIONS

1. Investigate how MSW is managed in the town or city in which you live or attend school. How much waste is generated per person per day? Who is responsible for collection? How much recycling is done? Where is the waste processed and disposed of? You probably can obtain most of this information at the local department of public works. Write a brief report describing the results of your investigation.

2. Why is solid waste disposal a significant problem? Briefly discuss the historical development of solid waste management in the United States.

3. Give a brief definition of *MSW*, including the meaning of the terms *refuse, garbage, rubbish,* and *trash.* Roughly how much MSW is generated in the United States each year? About how much of it is *garbage?*

4. What does the practice of *solid waste management* encompass? What is generally the most expensive function or activity?

5. Briefly discuss the basics of MSW collection and transport. What are some of the characteristics of an optimum collection route? What is a *transfer station* and when is it needed? Briefly discuss an example of a nonconventional method for solid waste collection.

6. List three purposes or advantages of solid waste processing.

7. Discuss the process of incineration, including design and operation factors. What is a resource recovery incinerator?

8. Discuss the two kinds of solid residue remaining after incineration.

9. Briefly discuss the purpose of *shredding* or *pulverizing* MSW. How is it accomplished?

10. What is the purpose or advantage of *baling* solid waste? How is it accomplished?

11. What is *composting?* Briefly describe two methods by which it is accomplished, and discuss any advantages or disadvantages. What is *co-composting?*

12. What materials in the MSW stream are recyclable? Briefly discuss some of the factors related to the recovery and reuse of each type of material. Which recyclable material provides the largest revenues?

13. Briefly discuss the collection of MSW recyclables. What is *source separation?* What is meant by *comingling* and *beneficiation?*

14. What is a *murf?* Make a sketch or flow diagram showing the major steps in its operation, and briefly describe each step.

15. What is a *cullet?* Is clear cullet less valuable than mixed color cullet? Explain.

16. What are three key characteristics of an MSW *sanitary landfill* that distinguish it from an open dump? What are some potential disadvantages of landfilling as a waste disposal technique? Why will it continue to be used for many years to come despite any disadvantages?

17. Briefly discuss the technical and environmental factors involved in selecting a site for a new sanitary landfill. What are some of the nontechnical factors involved? If a new sanitary landfill was being proposed for construction in your town, what are some of the questions you would ask at a town meeting? What is the NIMBY syndrome? Would you have it?

18. What does site *hydrogeology* refer to? Why is it of importance with regard to sanitary landfill design? What is a *natural attenuation landfill?*

19. What is the basic building block of a sanitary landfill? How is it constructed? Make a sketch to illustrate your answer. What kind of equipment or machinery is used to construct and operate the landfill?

20. What is the purpose of daily cover at a sanitary landfill? How does it differ from the final cover or *cap?* What is a landfill *phasing plan?*

21. What is *leachate?* How is it controlled? Make a sketch of the cross section of a *containment landfill.* Briefly describe how a MSW landfill is monitored after it is completed.

22. Discuss the generation of gas at a landfill and its control and possible dangers. Are there any potential advantages of landfill gas generation?

23. What is a landfill phasing plan?

24. Visit the EPA *Office of Solid Waste* Web page. Link to the "Waste Cleanup" page. Write a brief report explaining the difference between RCRA's Superfund Program and the RCRA Corrective Action Program. Also, research and describe the purpose of the "*Battery Act.*"

25. Visit the *Solid Waste Association of North America* Home Page (SWANA). Link to the "SWANA Press Release" page; read the latest news story and write a brief summary of it.

26. Using the EnviroSources search engine (see Section 1.6) or another Internet search directory or search engine, locate at least one additional Web site relevant to one of the topics in this chapter. Add the link(s) to your Environmental Technology Folder. Write a brief description of what the Web site(s) contain.

# PRACTICE PROBLEMS

1. If the initial loose volume of a mass of MSW is 20 yd$^3$ and after compaction the volume is reduced to 4 yd$^3$, what is the percent volume reduction and the compaction ratio? If it is desired to increase the volume reduction to 95 percent, what would the compaction ratio be?

2. The minimum waste compaction ratio reported by the manufacturer of a high-pressure compaction machine is 8:1. What is the corresponding percent volume reduction of the waste? What was the uncompacted waste volume of a 2.5 yd$^3$ bale?

3. The waste compaction ratio reported by the manufacturer of a high-pressure compaction machine is 10:1. What is the corresponding percent volume reduction of the waste? What was the uncompacted waste volume of a 2.0 m$^3$ bale?

4. If the initial loose volume of a mass of MSW is 18 m$^3$ and after compaction the volume is reduced to 2 m$^3$, what are the percent volume reduction and the compaction ratio? If it is desired to increase the volume reduction to 95 percent, what would the compaction ratio be?

5. A community of 50,000 people generates MSW at a rate of 8 lb per person per day. It is compacted in a sanitary landfill to a unit weight of 1200 lb/yd$^3$. After 1 year of operation, to what depth will a 25-ac landfill be covered? Assume an MSW-to-cover ratio of 4:1.

6. A community of 240,000 people generates MSW at a rate of 25 N per person per day. It is compacted in a sanitary landfill to a unit weight of 10 kN/m$^3$. After 1 year of operation, to what depth will a 10-ha landfill be covered? Assume an MSW-to-cover ratio of 4:1.

7. A community of 20,000 people generates MSW at a rate of 5 lb per person per day. The compacted unit weight of the MSW in the collection truck is 500 lb/yd$^3$. If the capacity of a collection truck is 15 yd$^3$, how many truckloads of MSW, on average, will be unloaded at the landfill each day?

8. A community of 30,000 people generates MSW at a rate of 30 N per person per day. The compacted weight of the MSW in the collection trucks is 3 kN/m$^3$. If the capacity of a collection truck is 12 m$^3$, how many truckloads of MSW, on average, will be unloaded at the landfill each day?

9. A community of 20,000 people uses a 10-ac landfill site that can be filled to an average depth of 50 ft. If MSW is generated at a rate of 6 lb per person per day, its compacted unit weight in the fill is 1000 lb/yd$^3$, and the MSW-to-cover ratio is 4:1, what is the useful life of the site?

10. A community of 50,000 people uses a 12-ha landfill site that can be filled to an average depth of 20 m. If MSW is generated at the rate of 25 N per person per day, its compacted unit weight in the fill is 8 kN/m$^3$, and the MSW-to-cover ratio is 5:1, what is the useful life of the site?

11. A community of 35,000 people generates MSW at the rate of 30 N per person per day. It is compacted in a sanitary landfill to a unit weight of 6 kN/m$^3$. To what depth would a 2-ha area of a landfill site be covered after 1 year of operation? Assume that soil cover adds 25 percent to the required volume.

12. A community of 70,000 people generates MSW at the rate of 4 lb per person per day. It is compacted in a landfill to a unit weight of 1200 lb/yd$^3$. To what depth would a 3-ac area of a landfill site be covered after 1 year? Assume that soil cover adds 20 percent to the required volume.

13. How many acres of land are required for a sanitary landfill under the following conditions: design life = 25 years, waste generation = 5 lb per person per day, compacted weight = 1000 lb/yd$^3$ in the fill, average fill depth = 35 ft, population = 30,000, and MSW-to-cover ratio = 4:1?

14. How many hectares of land would be required for a sanitary landfill under the following conditions: design life = 40 years, waste generation = 20 N per person per day, compacted weight = 8 kN/m$^3$ in the landfill, average fill depth = 10 m, population = 50,000, and MSW-to-cover ratio = 4:1?

## Chapter Outline

# C H A P T E R

# 12

# Hazardous Waste Management

In a modern society, large quantities of dangerous wastes are generated by chemical manufacturing companies, petroleum refineries, paper mills, smelters, and other industries. Even commercial establishments, such as dry cleaners, machine shops, and automobile repair shops, generate some dangerous waste. These *hazardous wastes*, as they are called, can result in serious illness, injury, or even death of the individuals and populations exposed to them; they can also pose an immediate and significant threat to environmental quality when improperly stored, transported, or disposed of. Because of these dangers, current waste management practice does not allow the intermingling or mixing of municipal refuse and hazardous waste. Chapter 11 focused on nonhazardous municipal refuse (garbage, rubbish, and trash); this chapter covers the management of hazardous waste from industrial and commercial facilities as well as from smaller household sources.

Hazardous wastes differ from other wastes in form as well as behavior. They often are generated as liquids, but they can occur as solids, sludges, or gases. The wastes are often contained and confined in metal drums or cylinders. There are hundreds of incidents on record in which illegal or inadequate handling and disposal of such hazardous wastes caused harm to the public and the environment. Many cases involved surface and groundwater contamination, including public water supplies. Other incidents involved the emission of harmful substances into the atmosphere. The illegal, but frequent practice of "midnight dumping," as it was called, may have saved the waste generators money, but it took a toll on public and environmental health as well as public tax dollars (for site cleanups and proper disposal).

The storage, collection, transport, treatment, and disposal of hazardous waste requires special attention. Years ago, most people did not fully understand or appreciate how these types of wastes might adversely and seriously affect public health and natural ecosystems. Many hazardous wastes were simply left out in the open on the ground, placed in pits or lagoons, or dumped in rivers. As a result of this practice, thousands of uncontrolled or abandoned hazardous waste sites were created, including old warehouses, manufacturing facilities, and landfills. One of the most significant environmental problems facing society today is the challenge of identifying, containing, and remediating (cleaning up) these old sites, to protect public health and environmental quality. In addition to the need to remediate uncontrolled hazardous waste sites (see Figure 12.1), modern design requirements for proper treatment, storage, and disposal of newly generated hazardous wastes are extremely complex and costly. Synthetic bottom liner systems, for example, are just one of the many construction features now used at modern waste landfill sites (see Figure 12.2).

During the 1970s, the seriousness of the hazardous waste problem came into focus, and in 1976 in the United States the first law to deal with it on the national level was enacted by Congress. This was the *Resource Conservation and Recovery Act*, or RCRA (pronounced "rekra" in waste management jargon). This act is administered and enforced by the EPA or by individual states that have established approved hazardous waste programs no less stringent than the federal program. The federal regulations that control hazardous waste management (under Subtitle C of RCRA), particularly disposal of the waste on land, are more stringent than those for nonhazardous municipal solid waste. In 1980, another federal law was passed to provide funds for the cleanup of the many uncontrolled and abandoned hazardous waste sites that were created before RCRA took effect. This law, the *Comprehensive Environmental Response, Compensation, and Liability Act* (CERCLA), is commonly referred to as the "Superfund" program.

Under RCRA, a "cradle-to-grave" system for managing hazardous waste from its point of origin to its final disposal was established by the EPA. Design requirements for hazardous waste landfills were also established. By 1984, when RCRA was strengthened by the *Hazardous and Solid Waste Amendments*, it had become clear that even well-regulated land disposal could result in environmental damage and that hazardous waste problems could not be solved by land disposal alone. Greater emphasis was placed on effective treatment prior to land disposal; treatment can neutralize or stabilize the waste by changing its chemical or physical characteristics.

Today, it is widely recognized that even strict controls on land disposal coupled with modern treatment technology cannot fully solve the nation's hazardous waste problem. It is imperative that the amounts of hazardous wastes generated in the first place be reduced and, wherever feasible, completely eliminated. *Waste minimization* (applying source reduction and recycling) is now the main strategy for managing hazardous (as well as municipal) waste in the United States and many other countries. The EPA promotes voluntary waste minimization by providing technical guidance and assistance to waste generators; individual states play a central role in these efforts.

Several options are available for hazardous waste management and disposal. In order of preference, these can be summarized as follows:

1. *Eliminate or reduce waste quantities* at their source by modifying industrial processes and other techniques.

**FIGURE 12.1**
*Improper storage of hazardous waste material.*
(Courtesy of Division of Waste Management, New Jersey Department of Environmental Protection)

**FIGURE 12.2**

*A synthetic bottom liner installed at a hazardous waste landfill site; a layer of soil covers the liner to protect it from damage by construction equipment. See Figures 12.9 and 12.10 for details.* (Courtesy of Gundie Lining Systems, Inc., 1340 E. Richey Rd., Houston, Texas, 77093)

2. *Reclaim and recycle the waste*, using it as a resource for some other industrial or manufacturing process.

3. *Stabilize the waste*, rendering it nonhazardous, by using appropriate chemical, biological, or physical processes.

4. *Incinerate the waste* at temperatures high enough to destroy or detoxify it.

5. *Apply modern land disposal methods*, preferably after providing some form of containerization or appropriate treatment.

Note that there is a difference between the terms *hazardous wastes* and *hazardous materials*, although many people use these terms interchangeably. A hazardous waste is a *used* substance or by-product of some process or activity, whereas a hazardous material is new or unused. This may seem like a trivial distinction, but it is significant with regard to the myriad laws now on the books to control the management of these substances. One federal law intended to regulate hazardous materials before they become hazardous wastes is the *Toxic Substances Control Act* (TSCA), passed by Congress in 1976. In this chapter, the focus is on the management of *hazardous waste* rather than hazardous materials.

## 12.1 CHARACTERISTICS AND QUANTITIES

Effective management of hazardous waste requires a straightforward and unambiguous means of identifying the type of waste that is to be managed and regulated. A definition of hazardous waste should be workable, that is, reasonably easy to apply. Also, the use of technical terms must be precise. For example, many people frequently use the terms *toxic* and *hazardous* interchangeably. But this is not accurate, as is explained shortly. In addition to a workable definition for a waste management plan to be effective, data are also needed on waste sources and quantities.

## Definition of Hazardous Waste

Published definitions and classifications of hazardous waste vary. In the United States, wastes are defined under RCRA as being hazardous if they "cause or significantly contribute to an increase in mortality or an increase in serious irreversible, or incapacitating reversible illness; or pose a substantial present or potential hazard to human health or the environment when improperly treated, stored, transported, or disposed of, or otherwise managed." This may give a broad view of hazardous waste, but it is definitely not a workable definition; more specific criteria are needed to facilitate an accurate identification of a particular waste material as being hazardous or not.

The necessary criteria for identifying hazardous wastes have been set up by the EPA. First, a material may be defined as hazardous if it is specifically named or *listed* as such in the federal regulations. This is a very direct and unambiguous method of definition and makes it easy for waste generators to know if they must manage their waste as hazardous; if it is on the list, it is hazardous, period. The list adopted by the EPA includes *nonspecific source wastes*, *specific source wastes*, and certain *commercial chemical products*.

Nonspecific source wastes commonly generated by industrial processes include materials such as degreasing solvents, dioxin wastes, and other very dangerous materials that come from a wide variety of manufacturing plants. Source-specific wastes, on the other hand, come from identifiable industries, such as petroleum-refining or wood-preserving facilities, and include wastewaters, sludges, and other residues. Commercial chemical products include discarded acids, chloroform, creosote, and pesticides such as DDT. Also, any waste material that contains a listed waste, regardless of the percentage, is considered a hazardous waste. (This prevents generators from simply diluting their listed wastes with nonhazardous wastes to evade the RCRA regulations.)

If a material is not on the EPA list of hazardous substances, that does not imply that it is nonhazardous. It still may be defined as a hazardous waste if it exhibits any of the measurable *characteristics* of a hazardous waste. The four primary characteristics, under the RCRA, are based on the physical or chemical properties of *toxicity*, *reactivity*, *ignitability*, and *corrosivity*. Two additional types of hazardous materials include waste products that are either *infectious* or *radioactive*.

(Note that toxicity is only one of several possible characteristics of a hazardous waste material, and so the term *toxic* should not be used synonymously with the term *hazardous*.)

## Toxicity

Toxic wastes are poisons, even in very small or *trace* amounts. Some may have an *acute* or immediate effect on humans or animals, causing death or violent illness. Others may have a *chronic* or long-term effect, slowly causing irreparable harm to exposed persons. Certain toxic wastes are known to be *carcinogenic*, causing cancer (sometimes many years after initial exposure). Others may be *mutagenic*, causing biological changes in the children or offspring of exposed people and animals.

Most toxic wastes are generated by industrial activities, including the manufacture of chemicals, pesticides, paints, petroleum products, metals, textiles, and many other products. The toxicity of any particular waste is determined by an EPA-specified test called the *toxicity characteristics leaching procedure* (TCLP). The TCLP is used to determine the mobility of organic and inorganic compounds present in the waste. This procedure attempts to mimic conditions a waste may be exposed to in a landfill, thus projecting the potential mobility of those compounds. In general, the extract from a representative sample of the waste is analyzed to see if it contains more than the allowable concentrations of one or more of the specific toxic substances listed by the EPA. Currently 39 specific organic and inorganic chemicals are listed. An abbreviated list showing some selected toxic chemicals is provided in Table 12.1.

**TABLE 12.1**
**Maximum concentration of contaminants for the TCLP[a]**

| Contaminant | Maximum level (mg/L) |
| --- | --- |
| Arsenic | 5.0 |
| Benzene | 0.5 |
| Carbon tetrachloride | 0.5 |
| Chlordane | 0.03 |
| Chloroform | 6.0 |
| Chromium | 5.0 |
| Endrin | 0.02 |
| Lead | 5.0 |
| Mercury | 0.2 |
| Pentachlorophenol | 100.0 |
| Silver | 5.0 |
| Trichloroethylene | 0.5 |
| Vinyl chloride | 0.2 |

[a]This is an abbreviated list, for illustration purposes only.

## Other Characteristics of Hazardous Wastes

*Reactive wastes* are unstable and tend to react vigorously with air, water, or other substances. The reactions cause explosions or form very harmful vapors and fumes. *Ignitable wastes* are those that burn at relatively low temperatures (less than 60°C, or 140°F) and are capable of spontaneous combustion (that is, they present an immediate fire hazard) during storage, transport, or disposal. Many waste oils and solvents are ignitable. *Corrosive wastes*, including strong alkaline or acidic substances, destroy materials and living tissue by chemical reaction. The pH value is used as an indicator of this characteristic; typically, liquids with pH less than 2 or greater than 12.5 are considered to be corrosive. Such wastes can rust or corrode unprotected steel at a rate more than 6 mm (0.25 in.) per year at a temperature of about 55°C (130°F).

*Infectious* or *medical waste* includes human tissue from surgery, used bandages and hypodermic needles, microbiological material, and other substances generated by hospitals and biological research centers. This type of material must be handled and disposed of properly, following EPA guidelines, to avoid infection and the spread of communicable disease.

*Radioactive waste*, particularly high-level radioactive waste from nuclear power plants, is also of special concern as a hazardous waste. Excessive exposure to ionizing radiation can harm living organisms. Radioactive material may persist in the environment for thousands of years before it decays appreciably. Because of the scope and technical complexity of this problem, radioactive waste disposal is always considered separately from other forms of hazardous waste.

Even though a waste exhibits a characteristic of hazardous waste, it may be exempt from regulation. These wastes include listed exempted wastes such as drilling muds or wastes that are recyclable, such as lead acid batteries. States have the right to list and regulate certain wastes as hazardous in addition to those listed or determined to be hazardous by RCRA. New Jersey and South Carolina, for example, regulate waste oil.

The EPA estimates that, until the mid-1980s, only about 10 percent of hazardous waste was actually disposed of in an environmentally sound manner. Much of it was disposed of in unlined landfills, waste piles, or lagoons and poses a potential threat to public health and environmental quality. The cost of cleanup is expected to reach billions of dollars and is currently being paid for through industrial contributions and federal funds allocated through the *Superfund* program (as covered later in this chapter).

## Waste Sources and Amounts

Under RCRA, the *generator* or creator of hazardous waste is responsible for identifying it as such. The generator is any person or company that produces material that is listed by the EPA as hazardous or has any of the defined characteristics of a hazardous waste. It is the generator's task to analyze all solid wastes to determine if they meet the RCRA definitions or are on the list. Once a waste is identified as being hazardous, it becomes subject to RCRA rules, and the generator assumes legal responsibilities for its proper management and disposal.

Three different categories of generators are recognized by RCRA, including *large-quantity, small-quantity*, and *conditionally exempt small-quantity generators*. Large-quantity generators create more than 1000 kg (2200 lb) of hazardous waste per month or more than 1 kg (2.2 lb) of *acutely hazardous* waste per month. Acutely hazardous wastes are those considered to be so dangerous that even small amounts are regulated in the same way as are larger amounts of other hazardous wastes; they are specifically identified as being acutely hazardous on the EPA list.

Small-quantity generators (SQGs) create between 100 and 1000 kg (220 and 2200 lb) of hazardous waste per month or less than 1 kg (2.2 lb) of acutely hazardous waste per month. Typical SQGs include dry cleaning facilities and automobile service stations. Conditionally exempt small-quantity generators create less than 100 kg of hazardous waste per month.

Large-quantity waste generators and SQGs must comply with the RCRA regulations, including obtaining an EPA identification (EPA ID) number, properly handling the waste before transport, manifesting the waste (discussed in the next section), and properly keeping records and reporting. Conditionally exempt SQGs do not require EPA ID numbers. Appropriate pretransport handling requires suitable packaging to prevent leakage and labeling of the packaged waste to identify its characteristics and dangers.

Large-quantity generators may accumulate waste on site for a maximum of 90 d if it is properly stored and labeled and if a written emergency plan is prepared. SQGs may accumulate waste on site for up to 180 d. The complete details of these regulations are extensive and complex and require full and careful attention by the responsible parties. There are close to 70,000 generators of RCRA-regulated hazardous waste in the United States.

### Hazardous Waste Quantities

Databases regarding the total amount of hazardous waste generated each year in the United States differ with respect to scope, content, and time frame. It can be said, however, that of the several hundred millions of tons of hazardous waste generated each year, most of it is corrosive liquid industrial waste that is discharged into sewerage systems after appropriate pretreatment. Roughly 35 million tons per year of additional types of hazardous waste, though, must be stored, incinerated, treated, or disposed of on or in the land. Much of the total production of hazardous waste in the United States comes from the northeastern and midwestern regions of the country.

Hazardous wastes are generated by a wide variety of manufacturing and nonmanufacturing facilities and processes, much too numerous to give a comprehensive list here. About half of the hazardous waste comes from the chemical products industry; the rest of it is generated by the petroleum, electronics, and metal-related industries, as well as numerous others.

There are about 750 *listed* hazardous wastes and countless more of the *characteristic* hazardous wastes. Examples of the types of hazardous wastes produced by the chemical industry include spent solvents and still bottoms (acetone, benzene, toluene, trichloroethylene, and others), strong acids (nitric, sulfuric, and hydrochloric acid, and others), strong alkaline wastes (ammonium and potassium hydroxide, and others), and reactive wastes (sodium permanganate, potassium sulfide, and others).

Metal manufacturing produces spent plating wastes, heavy metal sludges, cyanide wastes, chromic acid, and many others. The paper industry creates carbon tetrachloride, acids, ammonium hydroxide, and petroleum distillates. The list of examples can go on and on, and the few given here only serve to indicate the scope and wide variety of hazardous wastes that must be managed. Adding to the vast scope of the problem is the introduction of new products by industry and therefore new wastes; the tremendous pace and volume of hazardous waste generation point to the need to intensify waste minimization and recycling efforts.

### Household Hazardous Wastes

About 1.6 million tons of household hazardous wastes is generated each year in the United States. This is a small fraction (less than 1 percent) of the total amount of MSW generated, but when it is improperly disposed of it can create potential risks to people and the environment. Household hazardous wastes should not be intermingled with municipal refuse.

As much as 45 kg (100 lb) of hazardous substances can accumulate in the basements, garages, and storage closets of the average American home. These substances include leftover paint, stains, and varnishes, batteries, motor oil, and pesticides. They are sometimes disposed of improperly in sewers, on the ground, or by intermingling them with ordinary household refuse. Many of these wastes can injure sanitation workers, contaminate wastewater treatment systems, and pollute the air, surface water, and groundwater. Individuals can minimize these risks by reducing the use of hazardous products and by taking their hazardous wastes to local waste collection facilities, if available. Many states are drafting legislation to encourage this practice, and many civic organizations and private firms have implemented successful recycling programs (for example, for used motor oil, paint, and solvents).

## 12.2 TRANSPORTATION OF HAZARDOUS WASTE

One of the most important of the laws that regulate the handling of hazardous waste in the United States is RCRA; under Subtitle C of this law, the EPA is given the responsibility for establishing and enforcing a national hazardous waste management system. In addition to identifying the hazardous wastes that are subject to the regulations, the EPA sets and administers standards related to hazardous waste transport, treatment, storage, and disposal. This section focuses on the *transport* of hazardous waste.

Hazardous waste generated at a particular source requires movement to a *licensed* facility for proper treatment, storage, or disposal. Because these three activities are so closely related, the RCRA legislation links them all together and refers to their locations as *treatment, storage, and disposal facilities* (or TSDFs). Moving or transporting the waste from its source to a TSDF requires special care and attention due to the potential threats to environmental quality and public health and safety. Not only is there a possibility of an accidental spill, there has been (before RCRA) a tendency for unscrupulous waste transporters to abandon the wastes at random locations or to open a valve and dump them "on the run." This illegal practice (often called "midnight dumping") has been significantly curtailed by the enactment of RCRA and other legislation.

In the United States, the transport of hazardous materials and hazardous waste is regulated by the EPA, as well as by the Department of Transportation (DOT). A law that specifically focuses on transport activities is the *Hazardous Materials Transportation Act* (HMTA), passed by Congress in 1975. The HMTA requires proper labeling and transport of all hazardous materials, including hazardous waste.

Hazardous waste that is moved from its source is transported mostly by truck over public highway routes that are less than 160 km (100 miles) long. Only a very small amount of hazardous waste is transported by rail and almost none by air or inland waterways. Highway shipment is used the most because the trucks can access most industrial sites and TSDFs; trains require expensive siding facilities and are suitable only for very large waste shipments.

Hazardous wastes can be shipped in cargo tank trucks made of steel or aluminum alloy, with capacities between 7600 and 34000 L (2000 and 9000 gal). They also can be containerized and shipped in 210-L (55-gal) drums. Specifications and standards for cargo tank trucks and shipping containers are included in RCRA and HMTA and are enforced by the EPA and the DOT. The DOT regulations deal mostly with the "hardware," that is, with the containers, trucks, and shipping descriptions; the EPA regulations pertain primarily to a comprehensive tracking or *manifest system*, which follows the waste from "cradle to grave." All transporters of hazardous waste are required to have an EPA ID number.

## Manifest System

One of the key features of RCRA with regard to the transport of hazardous waste is the "cradle-to-grave" *manifest system* for monitoring the journey of the waste from its point of origin to point of final disposal. The manifest is a record-keeping document that *must be prepared by the generator* of the hazardous waste, such as a chemical manufacturing plant or other industry. The generator has primary responsibility for managing the waste and gives the manifest to a licensed waste transporter, along with the waste itself. The transporter must comply with the regulations of the DOT and the EPA with regard to transport vehicle standards, operation, and response to accidental spills.

The manifest must be delivered by the waste transporter or hauler to the recipient of the waste at an authorized off-site storage, processing, or disposal facility. *Each time the waste changes hands, the manifest form must be signed.* Copies of the manifest are kept by each party involved, and copies are also sent to the appropriate state environmental agency. Figure 12.3 illustrates a typical manifest cycle followed by the state of New Jersey.

In addition to curtailing the practice of improper disposal (for example, midnight dumping), the manifest system serves several other purposes. It provides data with regard to sources, types, and quantities of hazardous wastes. These data are valuable for future

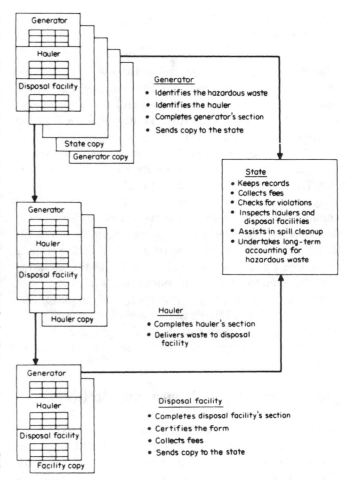

**FIGURE 12.3**

*Hazardous waste "cradle-to-grave" manifest system.* (From G. F. Bennett, *et al.*, *Hazardous Materials Spills Handbook*, McGraw-Hill, New York, 1983. Reproduced with permission of The McGraw-Hill Book Companies)

planning and design of hazardous waste management systems. The manifest ensures that the nature of the waste is described to haulers and operators of processing or disposal facilities; this prevents accidents due to improper handling or disposal of the waste. It also provides information regarding recommended emergency response procedures in the event of accidental spills or leaks.

In the event of a leak or accidental spillage of hazardous waste during its transport, the EPA and the DOT require the truck driver to take immediate and appropriate actions, including notifying local authorities of the discharge. An area may have to be diked to contain the wastes and cleanups may have to be started to remove it and reduce environmental or public health hazards.

## 12.3 TREATMENT, STORAGE, AND DISPOSAL

The RCRA links together the activities of hazardous waste *treatment*, *storage*, and *disposal*. A site to which hazardous wastes are transported and at which any one or a combination of these three activities occur is called a TSDF (treatment, storage, and disposal facility). TSDFs are subject to the RCRA rules and regulations and must obtain an EPA ID number and operating permit. Treatment, storage, and disposal include a wide variety of methods and technologies, and, as a result, TSD regulations are even more extensive than those for hazardous waste generators and transporters. Personnel at TSDFs must be well trained to ensure that the wastes are correctly identified and handled, and the facilities must have controlled entry systems of 24-h security surveillance to prevent the possibility of unauthorized entry. In addition, TSDFs must have contingency plans and emergency procedures, equipment and facility inspections, closure and postclosure plans, and other requirements.

### Hazardous Waste Treatment Methods

Some types of hazardous waste can be detoxified or made less dangerous by chemical, biological, or physical treatment methods. Treatment of hazardous waste may be costly, but it can serve to prepare the material for recycling or ultimate disposal in a manner that is safer than disposal without treatment. It also can reduce the volume needing final disposal. Many effective treatment methods are available. The hazardous waste treatment industry is in a phase of rapid development and innovation because of the need for even more economical treatment techniques. An overview of some of the prevalent treatment methods is given here.

### Chemical Treatment Processes

Chemical processes often used for treatment of hazardous waste include *incineration, ion exchange, neutralization, precipitation,* and *oxidation–reduction*.

*Incineration.* Incineration is a thermal–chemical process that not only can detoxify certain organic wastes, but can essentially destroy them as well. It is preferred by some people in the waste management industry over most other hazardous waste treatment processes, particularly because of the economic and public pressures to reduce or eliminate land disposal.

The burning of organic wastes at very high temperatures converts them to an ash residue and gaseous emissions. Combustion detoxifies hazardous waste material by altering its molecular structure and breaking it down into simpler chemical substances. Although the ash itself may have to be treated as a hazardous waste (if so indicated after a TCLP test), a much smaller volume of waste is left for ultimate disposal.

Stack emissions from a properly designed and operated incinerator burning organics, such as chlorinated hydrocarbons, for example, include $CO_2$, $H_2O$, $N_2$, and HCl (hydrochloric acid). Only the HCl is hazardous, but it is readily reacted with lime to produce nonhazardous salts, which can be landfilled.

Not all hazardous waste can be incinerated. Heavy metals, for example, are not destroyed, but enter the atmosphere in vapor form. However, incineration has been successfully applied to potent hazardous wastes such as chlorinated hydrocarbon pesticides, PCBs, and many other organic substances.

Special types of thermal processing equipment, such as the *rotary kiln*, the *fluidized bed incinerator*, the *multiple-hearth furnace*, and the *liquid injection incinerator*, are available for burning hazardous waste in either solid, sludge, or liquid form. Figure 12.4 is a diagram of a liquid injection incinerator. It is of interest to note that incineration of hazardous wastes on ocean-going ships (to reduce air pollution concerns) has been tried, but the future of this practice is doubtful due to technical difficulties.

Like any other TSDFs, hazardous waste incinerators must be licensed and must follow federal and state environmental rules and regulations. They generally must conduct a trial burn to determine the optimum operating methods for the incinerator. As much as 99.99 percent of organic waste constituents must be destroyed or removed (and 99.999 percent of dioxin wastes). Particulate and HCl emissions in the stack gas are also limited.

*Other Chemical Processes.* In the *ion-exchange* process, industrial wastewater is passed through a bed of resin that selectively adsorbs charged metal ions. An example of its use in the metal finishing industry is for removal of waste chromic acid from production rinse water.

*Neutralization* refers to pH adjustment for reducing the strength and reactivity of acidic or alkaline wastes. Limestone, for example, can be used to neutralize acids, and compressed carbon dioxide can be used to neutralize strong bases.

*Precipitation* refers to a type of reaction in which certain chemicals are made to settle out of solution as a solid material. An example of its application is in the battery industry, where the addition of lime and sodium hydroxide to acidic battery waste causes lead and nickel (both are toxic heavy metals) to precipitate out of solution.

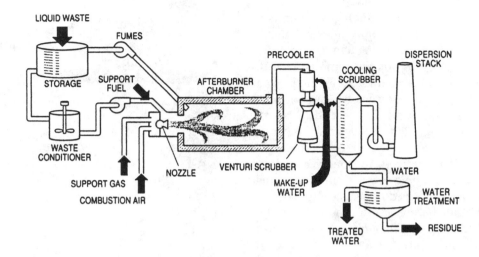

**FIGURE 12.4**
*Schematic of a liquid injection incinerator.* (Reprinted with permission from
W. C. Blackman, *Basic Hazardous Waste Management*, 1992. Copyright © 1992,
Lewis Publishers, an imprint of CRC Press, Boca Raton, Florida)

*Oxidation and reduction* are complementary chemical reactions involving the transfer of electrons among ions. Oxidation of waste cyanide by chlorine, for example, renders it less hazardous.

A schematic diagram showing a chemical treatment facility is shown in Figure 12.5.

### Biological Treatment Processes

Biological treatment involves the action of living microorganisms. The microbes utilize the waste material as food and convert it, by natural metabolic processes, into simpler substances. It is most commonly used for stabilizing the organic waste in municipal sewage (see Section 10.3), but certain types of hazardous industrial waste can also be treated by this method. Organic waste from the petroleum industry, for example, can be treated biologically. It is necessary, though, to inoculate the waste with bacteria that are readily acclimated to it and can use it for food. In some cases, *genetically engineered* species of bacteria may be used.

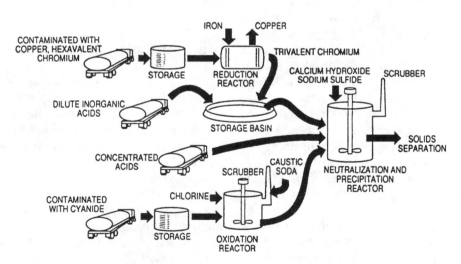

**FIGURE 12.5**
*Schematic chemical treatment: neutralization, precipitation, and chemical oxidation–reduction.* (Reprinted with permission from W. C. Blackman, *Basic Hazardous Waste Management*, 1992. Copyright © 1992, Lewis Publishers, an imprint of CRC Press, Boca Raton, Florida)

In addition to the traditional biological treatment systems, including the activated sludge and trickling filter processes, a treatment method called *landfarming* or *land treatment* (which is not the same as landfilling) may be used. The waste is carefully applied to and mixed with surface soil; microorganisms and nutrients may also be added to the mixture, as needed. The toxic organic material is degraded biologically, whereas inorganics are adsorbed and retained in the soil.

Landfarming is a relatively inexpensive method for treatment, as well as being a way to ultimately dispose of certain types of hazardous waste. But food or forage crops must not be grown on the same site because they could take up toxic material. In this regard, a disadvantage of land treatment for hazardous waste is that relatively large tracts of land may have to be withdrawn from potentially productive agricultural use. The surface topography and subsurface geological conditions of the site must be suitable so that surface or groundwater contamination will not occur.

Certain organic hazardous wastes can be treated in slurry form in an open lagoon or in a closed vessel called a *bioreactor*. A bioreactor may have fine bubble diffusers to provide oxygen and a mixing device to keep the slurry solids in suspension, as illustrated in Figure 12.6.

### Physical Treatment Processes

Physical treatment can be used to concentrate, solidify, or minimize the volume of hazardous waste material. Solidification can be accomplished by encapsulating the waste in concrete, asphalt, or plastic. This produces a solid mass of material that is resistant to leaching. Hazardous wastes can also be solidified by mixing them with lime, fly ash, and water to form a solid, cement-like product. Another process that is used to solidify and stabilize contaminated soil is *vitrification*. This involves the melting and fusion of the materials at high temperatures (about 1600°C, or 2900°F), thereby reducing the potential for leaching of contaminants. It is possible to remediate old waste dumps using in situ (on-site) vitrification; the required high temperatures can be achieved by inserting electrodes in the ground and allowing heat to build up as current flows through the soil.

The simplest physical process that can concentrate and reduce wastewater volume is *evaporation*, which may be facilitated by using mechanical sprayers. Other physical processes utilized to separate hazardous waste from a liquid include *sedimentation*, *flotation*, and *filtration*.

Two examples of physical processes used to remove specific hazardous components from a liquid

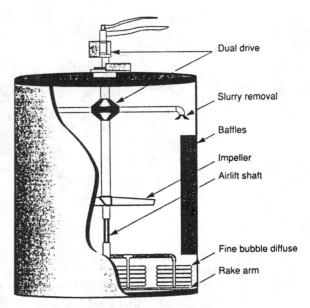

**FIGURE 12.6**
*Bioreactor for treatment of hazardous waste.* (From M. D. LaGrega, *et al.*, *Hazardous Waste Management*, 1994, McGraw-Hill, New York. Reproduced with permission of The McGraw-Hill Companies)

waste include *activated carbon adsorption* and *air/gas stripping*. Hazardous substances can be adsorbed onto a porous granular or powdered carbon matrix. Used or spent carbon is regenerated or *activated* for reuse. Air/gas stripping in cascade or countercurrent towers has been used to remove volatile organics from wastewater and contaminated groundwater.

## Storage Tanks and Impoundments

Proper storage of hazardous waste is imperative because of the potential for serious harm to public health and environmental damage in the event of an accidental discharge. Many generators of hazardous waste store the material on site for varying periods of time. Relatively large quantities may be stored in aboveground basins or lagoons. Aboveground basins may be constructed of steel or concrete, but they are subject to corrosion or cracking and are not suitable for storing reactive or ignitable waste.

Relatively small amounts of hazardous waste that are generated on an intermittent basis may be placed in 210-L (55-gal) fiberglass, plastic, or steel drums for ease of handling, temporary storage, and transportation. Corrosive material is stored in fiberglass or glass-lined containers to reduce deterioration and leakage. Toxic chemical liquids may be stored in metal drums.

Containers or drums of hazardous waste must be labeled properly before transport to a processing or disposal facility. The label must identify the contents as an explosive, flammable, corrosive, or toxic material. Appropriate signs or placards must be placed on the transport vehicle to warn the public of potential danger and to assist emergency response workers if there is an accidental spill along the transport route.

## Underground Storage Tanks

The design and construction of tanks used to store hazardous materials as well as hazardous wastes is a major environmental concern. Thousands of cases of environmental damage (particularly groundwater contamination) caused by leaking tanks are known to have occurred in the past. The bulk of these cases involved leaks from old underground gasoline or oil storage tanks, but many old underground tanks holding hazardous wastes have also been known to corrode and leak. Since more than half of the population in the United States depends on groundwater for public and household use, leaking waste storage tanks are a threat to public health as well as to environmental quality. Underground storage tanks (USTs) pose even more of a threat than aboveground tanks. USTs are unseen, and their existence is often unknown to people living in their vicinity until a problem occurs, damage is done, and expensive cleanup or remediation actions are required.

Under RCRA and its 1984 amendments, standards have been set by the EPA to control the use of UST systems. These standards apply to tanks with more than 10 percent of their volume below ground, containing hazardous wastes, petroleum products, or other hazardous substances; underground piping and pumping systems associated with the tanks are also regulated. UST regulations are intended to prevent leaks and spills as well as to detect them if they do occur.

Regulations governing the design, installation, and operation of aboveground as well as underground storage tanks are numerous and complex. New UST systems must be made of fiberglass or cathodically protected steel for corrosion control. A secondary containment system, including either double-wall tanks or cutoff walls with impervious underlayment, must be provided.

Automatic leak-detection systems and alarms must be installed as well as devices that prevent overfills (see Figure 12.7). Owners or operators of the tanks are responsible for monthly inspections and periodic removal of sludge from the tanks. Existing storage tanks that are not protected from corrosion or leakage must be taken permanently out of service; they must be emptied of all liquids, vapors, and sludge and then either removed or filled with sand or concrete.

## Surface Impoundments

Before implementation of the RCRA regulations, large volumes of liquid hazardous wastes were deposited in surface excavations such as pits, ponds, and lagoons (PPLs, in waste management jargon). Most PPLs were unlined impoundments or holding facilities, which

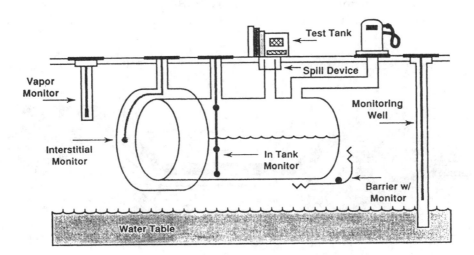

**FIGURE 12.7**
*Leak-detection alternatives include automatic monitors to detect loss of volume, vapors from leaked petroleum products, and leaked liquids. Surrounding soil, as well as the interstitial space between the tank and secondary liner or barrier, is monitored.* (Courtesy of the Environmental Protection Agency)

provided no protection against leakage and ground-water contamination. Except for sedimentation, evaporation of volatile organics, and possibly some surface aeration, they provided no treatment of the waste. A large number of these old PPLs are today's hazardous waste remediation sites.

Surface impoundments of liquid hazardous waste are allowed under the RCRA if they meet stringent design criteria, and they are still widely used by generators of liquid waste. All existing surface impoundments must have at least one liner and be located over an impermeable base if they are to continue in use. New impoundments must have at least two liners and a leachate collection system; groundwater monitoring is also required for all surface impoundments. A schematic diagram of a surface impoundment for liquid hazardous waste is shown in Figure 12.8. Accumulated sludge must be periodically removed and provided further handling as a hazardous waste.

### Waste Piles

As with unlined lagoons used for liquid wastes, many old piles of solid hazardous waste are now in need of cleanup or remediation. But with proper precautions, generators of certain hazardous wastes are allowed to use a waste pile for temporary accumulation of hazardous waste; the material must be landfilled when the pile size becomes unmanageable. Only noncontainerized solid, nonflowing material can be stored in a waste pile; examples of such materials include the nonmagnetic materials from automobile shredding operations, which contain fabric, rubber, plastic, insulation, lead, and cadmium. Waste from aluminum salvage operations may also be stored temporarily in waste piles.

Waste piles must be carefully constructed over an impervious base and must comply with requirement for landfills. The pile must be protected from wind dispersion and erosion; if hazardous leachate or runoff is generated, monitoring and control systems must be provided.

## Land Disposal of Hazardous Waste

Certain hazardous wastes may be disposed of in the ground. Nonliquid or containerized hazardous waste can be buried in a *secure landfill;* liquid waste can be disposed of using *deep-well injection* systems. Disposal in underground mines, caves, or concrete vaults may sometimes be allowed, but only after proper treatment and containment.

Land disposal of hazardous waste is not an attractive option because of the inherent environmental dangers involved in its practice and because of future liability. In the past, it did not provide sufficient environmental protection, and many uncontrolled and abandoned waste sites now need to be cleaned up. But with proper site selection, engineering design, and operational safeguards, land disposal is the least expensive alternative for many types of hazardous waste. It will continue to be necessary in the future. Even with waste minimization and incineration, it is not possible to completely eliminate the generation of hazardous waste.

To discourage land disposal activities, particularly when other treatment or destruction methods exist, the EPA and many states have placed strict limitations on their use. Certain hazardous wastes, such as dioxins, PCBs, cyanides, halogenated organic compounds, and acids with pH less than 2, are banned from land

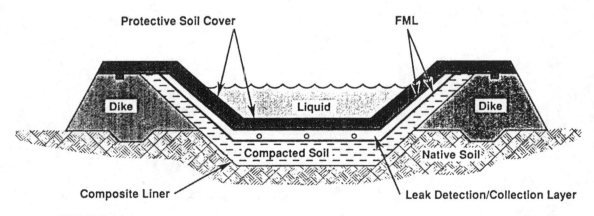

**FIGURE 12.8**
*Schematic of a cross section of a liquid waste impoundment double liner system.* (Reprinted with permission from W. C. Blackman, *Basic Hazardous Waste Management,* 1992. Copyright © 1992, Lewis publishers, an imprint of CRC Press, Boca Raton, Florida)

disposal. These wastes must be treated or stabilized before land disposal or meet certain concentration limits on the hazardous constituents.

### Secure Landfills

A secure landfill must have a minimum of 3 m (10 ft) of height separating the base of the landfill from underlying bedrock or a groundwater aquifer; this is twice the minimum separation needed for a municipal solid waste landfill. *All secure landfills also must have a double-liner and leachate collection system for increased safety, a network of monitoring wells to sample groundwater, and an impermeable cap or cover when completed.* A cross-sectional view of a typical secure landfill is shown in Figure 12.9.

An impermeable liner on the bottom and sides of a secure hazardous waste landfill serves as a barrier, preventing groundwater from entering or any leachate from leaving the fill. But no liner can be considered 100 percent effective; some leakage of liquid can always be expected. The double-liner and leachate collection system at a secure landfill provide redundancy to ensure protection of environmental quality and public health.

A synthetic flexible filter called a *geotextile* may be used to separate soil and waste material from the upper or primary leachate piping system. Under that is the primary liner, which must be a synthetic material with a permeability coefficient equivalent to $10^{-6}$ mm/s or less. It is called a *geomembrane* or *flexible membrane liner* (FML). Geotextiles and FMLs are generally made from plastic polymer resins. The secondary or lowermost leachate collection piping network serves as a leak detection and backup system. The lower or secondary liner is a composite layer of compacted clay and an FML, as illustrated in Figure 12.10. Collected leachate can be pumped to a treatment facility for processing.

A secure landfill is considered to have four phases in its total operation. During the first or *active phase*, the hazardous wastes are deposited in the prepared fill area. Incompatible chemical wastes are placed at separate locations in the fill to avoid explosions or other dangerous reactions. The waste material should first be solidified or containerized in drums, and care must be taken to avoid rupturing the individual containers as they are placed in the fill.

During the second or *closure phase*, the impermeable cap or cover is constructed over the landfill site.

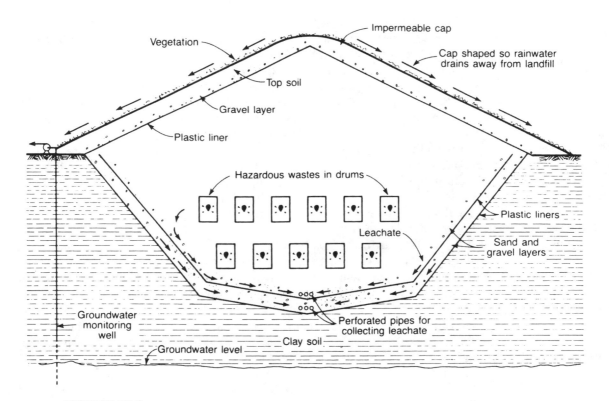

**FIGURE 12.9**
*Cross section of a typical secure landfill.* (From P. Montague, "Hazardous Waste Landfills: Some Lessons from New Jersey," *Civil Engineering*, September 1982, with permission from the American Society of Civil Engineers)

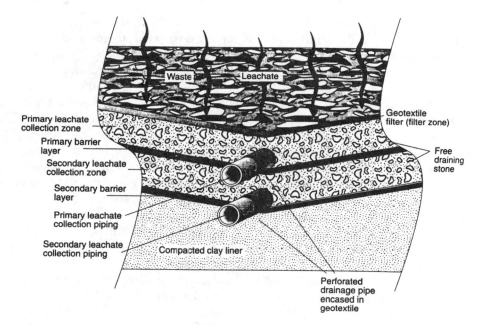

**FIGURE 12.10**

*Section view of a double liner and leachate collection system for a secure landfill.* (From M. D. LaGrega, *et al., Hazardous Waste Management,* McGraw-Hill, New York, 1994. Reproduced with permission of The McGraw-Hill Companies)

The composite bottom liner, the impermeable cap, and the double leachate collection systems serve as a first line of defense against leakage during both the active and the closure phases of the landfill operation. A third or *postclosure* phase, defined as the 30-year period after closure of the site, involves continuous operation of the monitoring well system (see Figure 12.11). This is a second line of defense. A routine program of sampling and testing must be implemented to detect any plumes of chemical leakage or contaminated groundwater.

It is believed that no landfill can be completely secure forever; the natural degradation of the protective liner or natural geological forces, including the possibility of earthquake, will eventually destroy the structural integrity of the landfill. A fourth and last phase for the landfill, called the *eternity phase,* is expected to involve some leakage of the waste material into the environment. When the leakage is detected, pumps can be installed in the monitoring wells to intercept the polluted water and bring it to the surface for treatment. This last line of defense is intended to protect the underlying aquifer from any significant damage from the landfill operation.

### Underground Injection

An option for land disposal of liquid hazardous waste is called *deep-well* or *underground injection.* This involves pumping the liquid down through a drilled well into a porous layer of rock. The liquid is then injected under high pressure into the pores and fissures of the rock. The layer or stratum of rock in which the waste is stored, usually limestone or sandstone, must lie between impervious layers of clay or rock; this injection zone can be from a few hundred meters to a few kilometers below the surface. The capacity of the geological strata to accept an injected waste depends on its porosity (void space), permeability (ability to transmit a liquid), and other factors.

The injection well must be at least 0.4 km (0.25 mi) from an underground source of drinking water, and the waste must be injected into a separate geological formation free of faults or fractures. The well must be cased and cemented for added protection against contamination of any drinking water supplies. The casing usually includes three concentric pipes. The outermost or surface casing should extend below the deepest usable water aquifer, as illustrated in Figure 12.12. A long string casing pipe and inner injection tubing extend into the injection layer or strata. Well injection pressures and flow rates must be carefully monitored; injection pressure must not be so great as to fracture the geological formation receiving the waste.

The chemical and petroleum refining industries are the largest users of injection wells, together accounting for about 90 percent of their use as a disposal

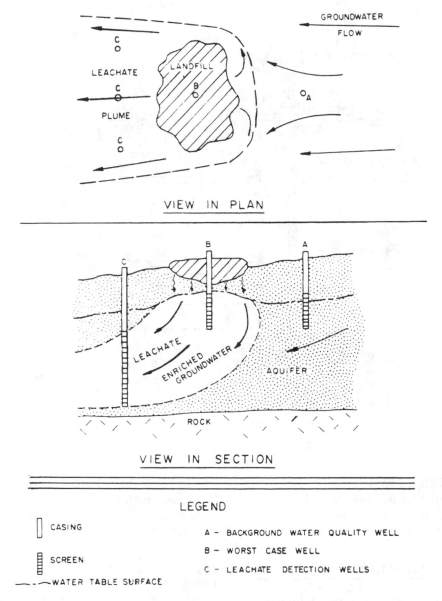

**FIGURE 12.11**

*A groundwater monitoring system consists of a series of wells strategically located around the waste disposal site.* (From C. W. Francis and S. I. Auerbach, *Environment and Solid Wastes.* Copyright © 1983, Butterworth Publishers. Used with permission of the publishers)

option. Most active waste injection wells are located in heavily industrialized states or in oil- and gas-producing states. Siting criteria for new wells focus on the geological requirements. There must be a water-bearing stratum of nonbeneficial use that has sufficient volume, porosity, and permeability to accept the injected waste. Confining strata below and above the injection zone must be sufficiently thick and impermeable so as to confine the waste.

Comprehensive engineering and geological studies must be conducted to show that underground drinking water supplies will not be degraded by operation of the injection well. In addition to the requirements of the Underground Injection Control program (under the SDWA), each state may have regulatory requirements for injection wells. Underground injection takes up little land area and requires little or no waste pretreatment, but it is unlikely to become a predominant method for disposal of hazardous waste because of the inherent risk of eventual leakage into drinking water aquifers, despite the most thorough engineering and subsurface geological investigations.

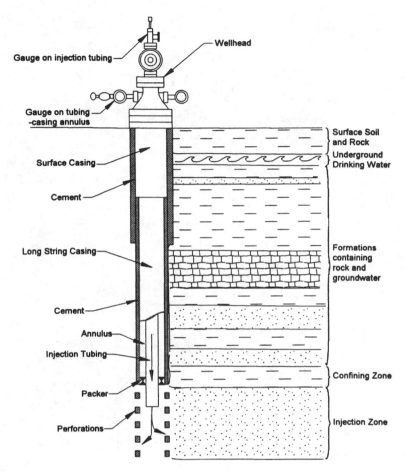

**FIGURE 12.12**
*Typical hazardous waste injection well.* (Reprinted with permission from S. M. Testa, *Geological Aspects of Hazardous Waste Management*, 1993. Copyright © 1993, Lewis Publishers, an imprint of CRC Press, Boca Raton, Florida)

## 12.4  SITE REMEDIATION

Before the enactment of legislation prohibiting uncontrolled disposal of hazardous wastes, many industrial and manufacturing facilities had long been storing or disposing of their hazardous waste materials in unlined waste piles, lagoons, and landfills. Many thousands of these old abandoned and unlined waste sites exist in the United States, posing a serious threat to public health and environmental quality. Efforts to remediate or clean them up will continue for years to come. The cradle-to-grave hazardous waste tracking or manifest provision of the RCRA was implemented primarily to eliminate this problem in the future.

### Superfund Program

Although the RCRA of 1976 greatly expanded the role of the government in regulating solid and hazardous

waste management, it did not address the serious problems at existing or abandoned hazardous waste dump sites. As a result, a federal law, the *Comprehensive Environmental Response, Compensation, and Liability Act* (CERCLA), was enacted by Congress in 1980. This act is commonly referred to as the *Superfund* legislation because it authorized expenditures of billions of dollars of federal funds. The money is intended primarily to pay for cleanup action *(site remediation)* at the most dangerous hazardous waste dumps in the United States; these sites are identified on the *Superfund List* or *National Priority List* (NPL). In 1986, the *Superfund Amendments and Reauthorization Act* (SARA) allocated $8.5 billion for the cleanup of these sites over a 5-year period, and $5 billion in additional funds was made available after 1991.

More than 1230 hazardous waste sites are listed on the National Priority List. Sites are reported by local and state agencies, businesses, and citizens as well as by the EPA. To evaluate the extent of danger at a site, the EPA uses a scoring system called the *Hazard Ranking System* (HRS); sites that score high enough on the HRS are eligible for the NPL. The site is also evaluated to determine if it requires *early action, long-term action,* or both. The evaluation (called a *Remedial Investigation/Feasibility Study,* or RI/FS) includes a description of the existing extent of pollution and an evaluation of alternative solutions to the problem.

Early actions are taken at sites that may pose threats to people or the environment in the very near future. Long-term actions are planned for sites that require extensive cleanup efforts. (An early action site should not be confused with a *hazardous material incident,* such as an emergency involving a spill of dangerous chemicals. Those "hazmat emergencies" require immediate attention by trained hazardous material technicians, usually called an *environmental response team,* or ERT.)

Remedial action at an abandoned hazardous waste site is typically quite complicated and cannot be accomplished overnight. There is a significant distinction between an immediate emergency response to an accidental spill of hazardous material and an engineered solution to a long-abandoned festering waste dump. The total size or area of the dump site and the degree of damage already done to the local environment cannot be known until thorough studies have been conducted. Engineers, geologists, environmental scientists, and technicians play a significant role in the site studies and in the design of remedial action plans, particularly with regard to the hydrogeological and groundwater contamination aspects of the project. A simplified critical path network diagram that may be used for the planning and scheduling of the many tasks that make up a site remediation project is shown in Figure 12.13.

One problem involved in financing the cleanup of old and abandoned hazardous waste sites is identification of the company responsible for the dumping. This can be difficult and sometimes impossible. A basic purpose of the Superfund is to raise money for remedial action when the responsible party cannot be identified and government has to oversee the cleanup activity. Potentially responsible parties must share in the cleanup costs, if they can be identified, but most of the Superfund is financed by a tax on oil and chemical companies.

The work of the Superfund got off to a slow start in the 1980s, largely because of political difficulties and legal problems; unfortunately, the Superfund is still struggling to achieve its goal. In the coming years, considerable progress will have to be made if environmental health and public safety are to be protected.

# Field Sampling Methods

Site remediation must be preceded by a comprehensive study of surface and subsurface site conditions. Environmental sampling is a key part of any field investigation, and it is important that an appropriate sampling method be chosen if the investigation is to be successful. Field studies of a site can be characterized as being either *nonintrusive* or *intrusive.*

Nonintrusive investigative methods include *aerial photography,* surface *geophysical surveys,* and surface *radiological surveys.* One or more of these methods may be employed prior to on-site sampling for laboratory analysis. Current and historical aerial photographs can be used to identify the boundaries of fill areas, drainage pathways, previous storage and process areas, and pollutant source locations. Geophysical techniques include electromagnetic, ground-penetrating radar, gravity, and seismic surveys. They can provide additional site data, such as the location of buried drums and tanks and, in some cases, the boundaries of contamination plumes. Radiological surveying techniques are used to locate areas where radioactive wastes are concentrated. A common and effective radiological method is the *gamma walk-over survey.* This is done by carrying a gamma radiation detector over the surface, following a marked survey grid pattern.

## Intrusive Sampling Techniques

Intrusive sampling techniques involve penetration of the ground or water surface. These include soil-gas surveys as well as the sampling of soil sediment, surface water, and groundwater. To prevent damage and for safety reasons, local utilities are notified before intrusive sampling procedures are conducted.

*Soil-Gas Surveying.* This is a relatively inexpensive technique used where volatile organic compounds (VOCs) may be among the site contaminants. Soil gas (methane plus other hydrocarbons) is formed naturally in soils as organic material decomposes. It is necessary to collect background soil-gas samples from undisturbed areas near the site to distinguish between natural deposits of organics and contamination due to waste deposits or plumes of migrating contaminants.

Among the soil-gas surveying methods, *field screening* provides a quick and inexpensive way to get results. It employs a portable photoionization detector with a hand-driven steel sampling rod and is primarily useful where contamination levels exceed 1 ppm. The sampling rod can be driven to a maximum depth of

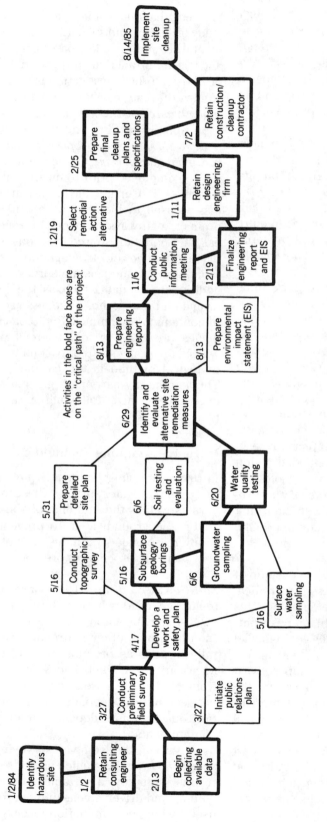

**FIGURE 12.13**

*A network diagram and schedule may be used to help manage the many activities for a hazardous site remediation project. More than 1 year is typically needed for investigation and planning before cleanup and confinement work can actually begin. Additional borings, sampling, and testing activities are frequently required throughout the project, to accurately define the limits of soil, sediment, and groundwater contamination. (Activities on the "critical path" control the duration of the project.)*

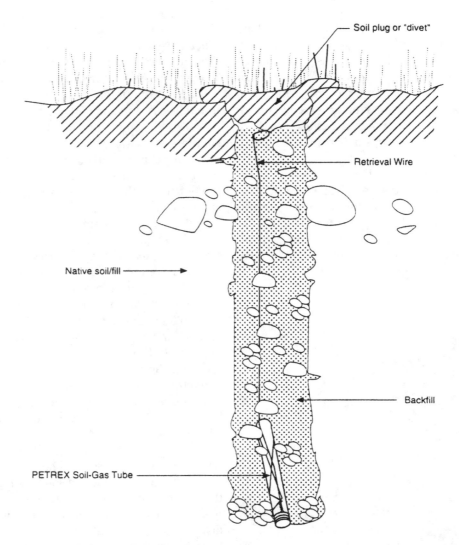

**FIGURE 12.14**
*Shallow installation of a PETREX soil-gas sample tube.* (Reprinted with permission from M. E. Byrnes, *Field Sampling Methods for Remedial Investigations*, 1994. Copyright © 1994, Lewis Publishers, an imprint of CRC Press, Boca Raton, Florida)

about 1.5 m (5 ft) into the ground. Another soil-gas survey technique is the *sample tube method* shown in Figure 12.14. Glass sampling tubes containing ferromagnetic filaments and an activated carbon adsorbent are installed about 1 m below the ground surface and left for a period of time. After they are retrieved, the tubes are shipped to a laboratory, where the sample is analyzed using a mass spectrometer, which ensures accurate identification of most organic compounds. This method is very effective at sites where the contaminant levels are very low, because the sampling tubes can be left in the ground as long as needed to obtain results. The results, though, may take several weeks to obtain.

*Soil Sampling.*  One objective of a site field investigation is to determine the extent of contamination present in the soil. Soil can be contaminated by chemical spills, leaking underground storage tanks (USTs), buried drums, and improper waste disposal. Soil-gas sampling techniques (discussed earlier) are useful for volatile organics; other sampling methods are used when the pollutants are nonvolatile.

Soil sampling tools include the scoop, the hand auger, the slide hammer, and a variety of tube samplers. A hand-held scoop can be used to collect grab samples of surface soil; the sample is placed directly into a sample jar. A hand-held auger is an effective shallow soil sampling tool; although extensions for the shaft are available, hand-held augers are most useful for collecting samples less than 1.5 m (5 ft) deep. Since rotation of the auger mixes the soil, this tool is useful primarily for composite samples. Shallow soil samples can also be obtained using a slide hammer coring tool, comprising a stainless steel core barrel, an extension rod, and a slide hammer. The slide hammer is used to drive the core barrel into the ground; soil is transferred from the barrel into a sample jar.

Soil samples collected at depths exceeding 1.5 m are usually obtained by driving a split-tube or solid-tube sampler; a drill rig is used to drive the sampler deep into the ground (see Figure 12.15). In some cases, a thin-walled sample tube can be hydraulically pushed into the ground instead of being driven by a hammer; in this way the soil is not compacted in the sampling process.

**FIGURE 12.15**
*Drill rig used to collect soil samples at a hazardous waste site. Level B protective equipment is worn by the technicians.* (Courtesy of Killam Associates, Consulting Engineers, Millburn, New Jersey)

*Sediment and Surface Water Sampling.* Sometimes it is necessary to determine if pollutants are present in the sediment of streams or rivers on or near the site. This is important because of the possibility for rapid spread of contaminants in the flowing water. In addition to the scoop and slide hammer methods, sediment may also be sampled using the *box sampler* method. A stainless steel sample box with spring-loaded sample jaws and a pole with a spring release mechanism is pushed into the sediment. When the jaws are released, the bottom of the box is closed and the sample can be retrieved from the water.

In addition to soil and sediment contamination, it is often necessary to determine if pollutants are present in surface waters on or near the site. The depth at which samples are collected depends on the suspected level and nature of contamination, but usually samples are collected from the water surface. For samples collected less than 1.5 m below the water surface, an *extendible bottle sampler* can be used. After lowering the bottle to the desired depth, a seal ring is lifted, allowing the bottle to fill with water. Samples can be collected at depths well beyond 1.5 m using the *Kemmerer bottle method;* the sample bottle is lowered by means of a flexible cable and opened by use of a trip weight.

*Groundwater Sampling.* There is always a possibility that contaminants found in soil will migrate or leach into the groundwater underlying a polluted site. Once they reach the groundwater, the contaminants will eventually disperse or spread out into the aquifer. Application of a technique called the *direct push method*, in combination with a mobile laboratory, allows samples to be collected and analyzed quickly and inexpensively. A truck-mounted hydraulic press and slide hammer are used to force a sample probe up to 60 m (200 ft) through the soil into the groundwater. A small-diameter bailer or weighted sample vial is then lowered inside the probe. After a preliminary assessment of groundwater conditions is made in this manner, permanent *monitoring wells* can be accurately positioned for long-term sampling as well as possible site remediation.

Monitoring wells allow groundwater samples to be repeatedly collected; they also allow observation of changes in groundwater elevations. The main components of a groundwater monitoring well are shown in Figure 12.16. The stainless steel or PVC well screen has slots large enough to allow groundwater and pollutants to flow freely into the well, but small enough to keep most soil particles out. A sump is used to catch any fine-grained soils that do penetrate the screen; it may be cleaned out regularly to keep the well screen clear.

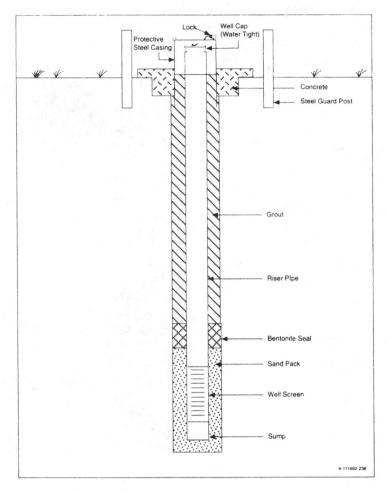

**FIGURE 12.16**
*Primary components of a groundwater monitoring well.* (Reprinted with permission from M. E. Byrnes, *Field Sampling Methods for Remedial Investigations*, 1994. Copyright © 1994 Lewis Publishers, an imprint of CRC Press, Boca Raton, Florida)

Before samples are collected from a monitoring well, it must first be developed and purged. This restores the aquifer's natural hydraulic and geochemical characteristics, which are disturbed during the well drilling operation. Development involves surging and pumping long enough to clean out the well and for the water to stabilize in turbidity, pH, temperature, and conductivity. Purging refers to the removal of stagnant water from the well so that a representative sample can be collected. Sample bottles for volatile organics should be filled first, since VOCs are continuously being lost to the atmosphere during the sampling procedure. A water-level reading is commonly taken before a sample is collected from the well.

Groundwater samples can be removed from the well by one of several methods. A bailer, the least complicated

of sampling tools, can be lowered into a well on a fishing line; after the bailer is raised, the collected water is transferred into a set of sample bottles. A device called a bomb sampler can also be lowered into the well; this allows water to be collected at specific depths in the water column. A piston pump can be installed in a well that is to be sampled regularly for an extended time, and submersible pumps can be used in deep wells.

*Sampling Containerized Wastes.*   Abandoned drums or tanks that may contain hazardous wastes pose special safety problems. They must be opened and sampled in a way that prevents sudden releases of vapors or liquids, fires, or explosions. Representative samples of the waste must be obtained throughout the depth of the tank or drum. (See Appendix H, Figure 6.)

## Field Investigation and Safety

Many risks and hazards may be encountered during the field investigation of an old, abandoned hazardous waste site. Technicians who perform the fieldwork must have a thorough knowledge of the potential dangers and must conduct the site investigation with extreme caution. Ideally, a field investigation team should consist of at least three people, for safety purposes. In addition to knowing all the chemical, physical, and biological characteristics of hazardous waste materials, they should be familiar with first aid techniques. They should also be familiar with the proper use and operation of personal protective clothing and equipment as well as with sampling and gas detection devices.

One primary danger for personnel at a waste disposal site is the inhalation of toxic gases or vapors. Self-contained breathing apparatus (SCBA) and cartridge respirators or adsorbent gas masks are frequently used. The equipment must be well maintained, and the technician must be thoroughly familiar with its use.

It is also necessary that devices for detecting and measuring combustible or toxic gases and oxygen-deficient atmospheres be used during the initial investigation of the site. Such devices must be calibrated before use. In addition to respiratory protection, suitable clothing and equipment for eye and body protection are used. Eye protection includes safety glasses, protective goggles, or face shields attached to a hard hat, depending on the risk involved. In some cases, fully enclosed suits are used for protection. A technician wearing protective equipment while sampling groundwater in the vicinity of a landfill site is shown in Figure 12.17. The types and levels of protection needed are predetermined before a worker is allowed to enter a site.

**FIGURE 12.17**

*An example of good safety procedure—a technician wearing protective clothing (Level C) while collecting samples of groundwater at a waste landfill site.* (Courtesy of Chris Venezia)

## Levels of Protection for Site Entry

Equipment for protecting the body against contact with known or anticipated chemical hazards is divided into four categories according to the degree of protection afforded.

1. *Level A.* Selected when the highest level of respiratory, skin, and eye protection is needed.
2. *Level B.* Selected when the highest level of respiratory protection is needed, but a lesser level of skin protection. This is the minimum level recommended on initial site entries until the hazards have been further defined by on-site studies.
3. *Level C.* Selected when the types of airborne substances are known, the concentrations are measured, and the criteria for using air-purifying respirators are met.
4. *Level D.* A work uniform providing minimal protection; not worn on any site with respiratory or skin hazards.

Minimum requirements for Level A protection include a pressure-demand, self-contained breathing apparatus, a fully encapsulating chemical-resistant suit, inner and outer chemical-resistant gloves, chemical-resistant steel toe and shank boots, and two-way radio communications.

Minimum requirements for Level B protection include a pressure-demand, self-contained breathing apparatus, chemical-resistant clothing, inner and outer chemical-resistant gloves, chemical-resistant steel toe and shank boots, a hard hat, and two-way radio communications.

Minimum requirements for Level C protection include a full-face, air-purifying, canister-equipped respirator, chemical-resistant clothing, chemical-resistant gloves and boots (steel toe and shank), a hard hat, and two-way radio communications.

Minimum requirements for Level D protection include coveralls and leather or chemical-resistant shoes or boots (with steel toe and shank).

## Site Work Zone Controls

To prevent the inadvertent transfer of hazardous substances from the waste site to unaffected areas, work at hazardous waste sites must be controlled. Control can be effected in a number of ways, including:

1. Setting up security and physical barriers to exclude unnecessary personnel from the area
2. Establishing work zones on the site and regulating access to them
3. Conducting operations in a way that reduces the exposure of workers and equipment and minimizes the potential for airborne dispersion of contaminants
4. Implementing appropriate decontamination procedures

*Work Zones.* One method of reducing the migration of contaminants is to delineate zones on the site where prescribed operations occur. Movement of personnel and equipment between zones is limited by access control points. Three common zones are the *exclusion zone*, the *contamination reduction zone*, and the *support zone* (Figure 12.18).

The exclusion zone, the innermost of three concentric areas, is the zone where contamination does or could occur. An access control point is set up at the periphery of the zone (called the *hotline*) to regulate movement of equipment and personnel in and out; all people entering the zone must wear prescribed levels of protection. The hotline is fenced or defined by landmarks and may be adjusted as more information becomes available.

The support zone, the outermost part of the site, is a noncontaminated or clean area. An equipment trailer and command post are located in this area, and traffic is restricted to authorized personnel. The contamination reduction zone provides a buffer between the contaminated (exclusion) and the clean (support) zones. Decontamination stations are established in this area, one for personnel and one for equipment. Exit from the exclusion zone is through a decontamination reduction corridor. The boundary line between the contamination reduction zone and the support

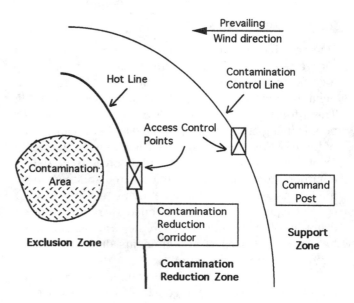

**FIGURE 12.18**
*Diagram of site work zones.* (Courtesy of Killam Associates, Consulting Engineers, Millburn, New Jersey)

zone is called the *contamination control line*. Access to the contamination reduction zone from the support zone is through a control point. Personnel entering must wear the prescribed protective equipment; entering the support zone requires removal of any protective equipment worn in the contamination reduction zone.

## Remediation Techniques

The basic objective of site remediation is to eliminate any immediate danger caused by the spread or migration of waste material as well as to reduce any long-term threat to public health and environmental quality (especially to groundwater). The required course of action depends on the type of waste material, the extent of existing contamination, the location of the site, and other factors. Since each site is unique, particularly with regard to hydrogeological conditions, no two hazardous waste remediation projects are identical; sound engineering judgment must be used in each case.

A site-specific goal regarding the appropriate degree of remediation is necessary, but there is no general method for determining "how clean is clean." Some states simply require that site cleanups achieve background levels of the waste contaminants. For known carcinogens, however, where risk assessment data are available, the EPA may set specific exposure risk levels.

## Removal of the Waste

One possible course of action is to physically remove the waste material from the site by excavation or dredging and transport it to some other location for treatment, incineration, or final disposal in a secure landfill. This *off-site solution* may be the most desirable for people living in the vicinity of the site, but it can be one of the most expensive options. Also, moving the waste from one location to another still involves some risk of environmental pollution. Contaminated soil may be removed using standard earth-moving equipment, but special equipment is often needed to remove buried drums or other containers. Extreme care must be taken to prevent releases of contaminants during the excavation or removal operation.

Dredging contaminated sediments from hazardous waste ponds or lagoons also requires special precautions to prevent further pollution. Clamshell, dragline, or backhoe machines are used to dredge consolidated sediments; in some cases, stream diversion or diking may first be necessary. When sediments are unconsolidated or have a high water content, the material may be pumped in slurry form; this is called *hydraulic dredging*. Dredged material can be dewatered, and both the solid and liquid fractions can be subjected to appropriate treatment and disposal operations.

## On-Site Remediation

On-site remediation, in which the waste is not removed to another location, generally focuses on the need to minimize the production of leachate and to eliminate groundwater pollution. Another primary goal is to contain or prevent the further migration of any groundwater pollution that may have already occurred. This could involve the temporary removal of the waste, construction of a lined, secure landfill on the same site, and replacement of the waste in the new landfill. It could involve the extraction of soil or groundwater, treatment or destruction of the pollutants, and replacement or reinjection of the cleaned soil or groundwater. Finally, on-site actions could involve the isolation and containment of the waste, without moving it, by the construction of impermeable barriers to block the flow of water or other liquids.

*Extraction, Treatment, and Replacement.* Groundwater at an old waste site may be pumped or extracted to lower the water table below the waste material. In many cases, the groundwater is extracted from a contaminated area or plume and treated to remove the pollutants. The treated water may be discharged on the ground surface or reinjected around the perimeter of the plume. This *pump-and-treat* method is illustrated in Figure 12.19.

Toxic or flammable soil gas generated from anaerobic decomposition or volatilization of the buried waste can be removed using an induced draft extraction fan and treated using granular activated carbon adsorption; in some cases, the gases can be destroyed by flares or combustion devices. Contaminated soils can also be excavated from a hazardous waste site, treated, and then replaced on site, as illustrated in the following case study.

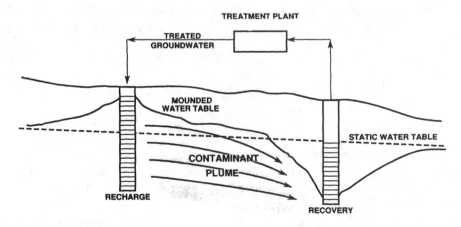

**FIGURE 12.19**
*The "pump-and-treat" method for remediating contaminated groundwater.*
(Reprinted with permission from W. C. Blackman, *Basic Hazardous Waste Management*, 1992. Copyright © 1992, Lewis Publishers, an imprint of CRC Press, Boca Raton, Florida)

### Cleanup of a "Town Gas" Hazardous Waste Site

The EPA estimates that there are about 1800 hazardous waste sites in the United States created from former manufactured-gas plants. These facilities, also known as "town gas" plants, were built beginning in the early 19th century to produce methane gas for street lighting, domestic hot water heating, and cooking. Hazardous by-products and wastes from the gas manufacturing process, including coal tar and other substances, became contaminated with volatile organic compounds (VOCs), such as benzene and toluene. Most of these plants were shut down after the mid-1900s, but the hazardous wastes remained where they were originally dumped and the sites now require remediation.

One of these gas plants, located in Paterson, New Jersey, was left with a legacy of over 100,000 tons of highly contaminated soil when it was finally closed. This much soil could cover a football field to a depth of about 3.5 m (12 ft). The former Paterson Gas Plant was constructed and began producing manufactured gas in the mid- to late 1860s. Various gas manufacturing processes were utilized during its operating history.

During the period of manufactured-gas plant operation, substantial quantities of raw material feed stock (such as coal, coke, oil, natural gas, propane, butane, and kerosene) were stored, handled, or used at the site. As a result of the manufacturing processes, various by-products and wastes were generated, stored, and handled, including tar, light oils, ammonia liquors,

coke, clinkers (clumps of ash), spent lime, spent oxide, and carbon. Manufacturing operations continued until the plant was retired in the late 1970s. Dismantling of the facility occurred into the early 1980s.

In the late 1980s a Remedial Investigation (RI) was initiated by the owner of the site. A series of RIs was conducted through the early 1990s to delineate the vertical and horizontal extent of contamination. All these investigative efforts were conducted upon approval by the New Jersey Department of Environmental Protection (NJDEP). After completing the RIs a series of pilot studies evaluating different treatment technologies were conducted. The site owner retained a firm to construct and operate an on-site thermal desorption system. To control fugitive air emissions during excavation of the contaminated soil, it was decided to place the excavation area and stock pile/feed areas within tents (see Figure 12.20).

The Paterson plant had been in operation since the 1860s, and over the years, organic contaminants seeped through about 8 m (26 ft) of soil down to bedrock and to the Passaic River. To safely excavate the contaminated soil, a "freeze wall" was constructed to surround the work area and block the flow of water. The "wall" consists of 216 drilled shafts, spaced about 1 m (3 ft) apart, and penetrating about 3 m (10 ft) into bedrock. A brine solution, chilled to about $-18°C$ ($0°F$) and pumped through the shafts for up to 2 months, established a wall roughly 1.5 m (5 ft) thick.

Excavation of contaminated soil with on-site thermal treatment commenced in July 1993. In February 1995, the processing of material was discontinued

**FIGURE 12.20**
*Enclosed remediation site.* (Courtesy of Seaview Thermal Systems, Blue Bell, Pennsylvania)

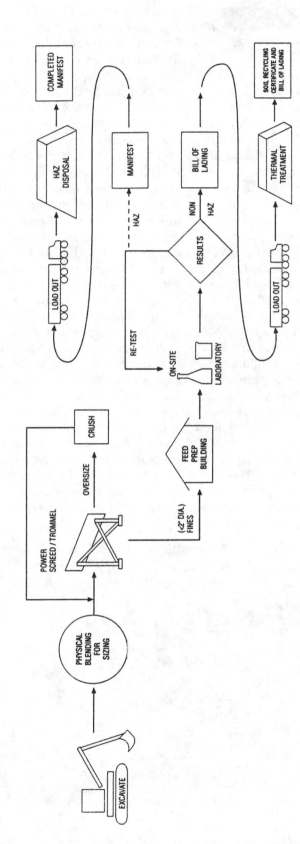

**FIGURE 12.21**
*Soil management flow chart for remediation of a former "town gas" plant.*
(Courtesy of Killam Associates, Consulting Engineers, Millburn, New Jersey)

because of various mechanical, processing, and site failures, including collapse of the tent (which occurred during a snow storm). In late 1995, a new approach to remediating the site was developed. The primary difference in this new approach is that excavated soils will be analyzed for waste characteristics (see Section 12.1) and removed from the site. Nonhazardous soils will be delivered to a soil remediation facility, thermally treated, and then recycled for beneficial reuse (e.g., landfill cover). The hazardous soil will be properly disposed of at other approved facilities.

In general, soils will be excavated with a backhoe and transferred to a containment building. Within the containment building, the material will be moved from a storage pile to a trommel by a front-end loader. The trommel will separate materials less than and greater than 1.5 in. in diameter. This material larger than 1.5 in. in diameter will be further managed through a portable crusher, to further reduce its size. Crushed material will then be returned to the trommel for sizing. This cycle will continue until all material passes the size criterion.

The managing of the soils on-site is required both to create the proper size and to reduce the benzene concentrations to levels that are not RCRA hazardous. The managing is allowed under EPA, Section 40 of the *Code of Federal Regulations*, paragraph 262.34 (40 CFR 262.34), as described in *Manufactured Gas Plant Site Remediation Strategy Memo*, Edison Electric Institute (Washington, D.C., 1993). The managing is allowed as long as it occurs in a containment building that meets 40 CFR 265 Subpart DD.

The nonhazardous soil removed from the site will be treated at off-site soil remediation facilities. At the off-site facilities, the soil will be off-loaded into metal storage buildings which have lined concrete floors and concrete retainment walls. The soils will be treated in thermal soil remediation units. After thermal treatment, the soil will be certified as clean through analysis of soil samples. Treated soil will have beneficial reuse, such as cover materials at local landfills.

The managed soils deemed RCRA hazardous by waste characterization testing will be manifested, transported, and disposed of at an approved hazardous waste disposal facility. A flow chart of soil management is illustrated in Figure 12.21.

Coal tar, which cannot be treated at the off-site thermal facilities, will be segregated, placed in closed containers such as drums, and recycled through incorporation into either asphalt paving products (such as asphalt cement) or a fuel oil in accordance with applicable rules and regulations. In the event that the mixing is not feasible, the coal tar will be shipped off-site as RCRA hazardous waste.

---

*Permanent Containment Methods.* Vertical walls can be constructed of steel sheet piling or slurry trenches to block the horizontal movement of liquid. But these *subsurface cutoff walls* are feasible only if there is an *aquiclude*, or naturally occurring impermeable layer, below the waste site. The cutoff wall must penetrate or be keyed into the aquiclude, which serves as a natural bottom liner. A wall made of steel sheet piling can be driven to depths of about 30 m (100 ft) if the soil is not too rocky, but steel piling is not suitable if the waste contains corrosive liquids. A *slurry trench cutoff wall* can be built to contain most hazardous wastes, as shown schematically in Figure 12.22.

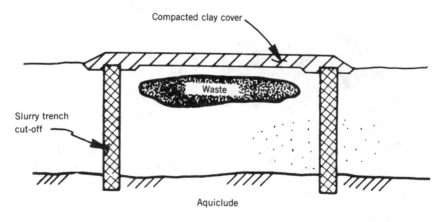

**FIGURE 12.22**
*Slurry trench cutoff walls can be used to prevent the spread of polluted groundwater at an abandoned hazardous waste dump site.* (From T. L. Sweeney, *et al.*, *Hazardous Waste Management for the 80's.* Copyright © 1982, Ann Arbor Science Publishers. Used with permission of the publisher)

A slurry trench cutoff wall can be excavated from the ground surface with a clamshell or backhoe excavator, as illustrated in Figure 12.23, without moving or otherwise disturbing the waste material. The trenches, roughly 1 m (3 ft) wide and up to 30 m (100 ft) deep, can be dug without collapse by filling them temporarily with a bentonite-clay slurry. The dense slurry (a mixture of clay and water) maintains the stability of the trench during excavation until it is backfilled with a material that forms the vertical barrier. The backfill material may be a mixture of soil and cement or soil and clay; it must be blended properly to have a permeability of $10^{-6}$ mm/s or less to block the flow of water. In some cases, a double trench wall, or a trench wall with an installed synthetic liner sheet, may be built for extra ensurance that there will be no liquid or water movement into or out of the site.

**FIGURE 12.23**
*Excavation of a deep slurry trench cutoff wall.*
(Courtesy of Geo-Con., Inc., Pittsburgh, Pennsylvania)

*In Situ Bioremediation.* Bioremediation is a technique used for cleaning up contaminated soil or groundwater at hazardous waste sites. It relies on the biological action of microorganisms to convert the contaminants into harmless substances. The use of microorganisms for environmental purposes is not new; it has existed for decades at secondary wastewater treatment plants. What is unique about bioremediation is the way in which biological processes are applied *in situ*, that is, on the site of old hazardous waste dumps.

In soil above the water table, contamination may exist in four phases. It may be in the *vapor phase*, within the pore spaces; the *adsorbed phase*, attached to soil particles; the *aqueous phase*, dissolved in water; and the *liquid phase*, in the form of *nonaqueous-phase liquids* (NAPLs). Below the water table, contaminants may exist in all but the vapor phase. Certain contaminants, such as *petroleum hydrocarbons* (for example, benzene, gasoline, and oil), are less dense than water and are called *light nonaqueous-phase liquids* (LNAPLs). These tend to float on the surface of the water table and spread laterally, as illustrated in Figure 12.24.

Other contaminants, including *chlorinated hydrocarbons* (for example, carbon tetrachloride and trichloroethane), are denser than water and are called *dense nonaqueous-phase liquids* (DNAPLs). These substances tend to migrate downward, sometimes penetrating deep into the groundwater. Chlorinated hydrocarbons are used widely as industrial degreasing solvents. They have often been disposed of improperly in refuse sites, lagoons, and storage tanks; as a group, chlorinated solvents are the most prevalent type of groundwater contaminants found in the United States. They are not as readily biodegradable as petroleum products.

Bioremediation of sites contaminated with petroleum hydrocarbons often relies on indigenous (native) bacteria found in soil. The addition of suitable bacterial mixtures to "jump start" the process can significantly reduce the treatment time. This is called *bioaugmentation*. If oxygen and nutrients (that is, nitrogen and phosphorus) are not present in sufficient quantities, one or more of these elements may also be added. Systems for treating groundwater typically comprise recovery wells, monitoring wells, and injection wells. Recovered groundwater is often treated prior to the addition of nutrients or oxygen and reinjection. Treatment may involve the use of an air stripper tower or activated carbon unit an oil–water separator, a biological unit or a combination of these. Wells are located so that nutrient- or oxygen-enriched groundwater flows toward the recovery wells, and only a portion of the recovered groundwater may be reinjected; to control flow rate and flow path, a portion may be discharged into surface waters.

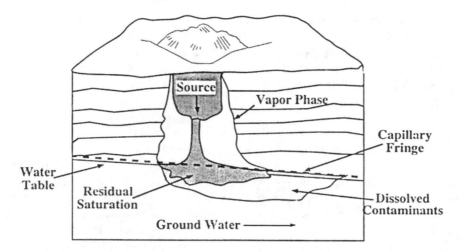

**FIGURE 12.24**

*Underground distribution of LNAPLs.* (Reprinted with permission from R. D. Norris, *et al.*, *Handbook of Bioremediation*, 1993. Copyright © 1993, Lewis Publishers, an imprint of CRC Press, Boca Raton, Florida)

*Bioventing and Air Sparging.* To stimulate aerobic biodegradation of petroleum hydrocarbon contaminants, air may be injected below the ground surface. When air is injected through boreholes above the water table, the process is known as *bioventing*. When it is injected below the water table, the technique is termed *air sparging*, as illustrated in Figure 12.25.

Air sparging is an effective means of treating petroleum hydrocarbons because it also enhances physical removal by direct extraction of VOCs. Airflow must

be carefully controlled during a sparging operation to avoid the possible increase in the spread of contaminants. Too low a flow can significantly reduce treatment efficiency, and too high a flow can result in a loss of control. Because of this, installation of an air sparging system must always be preceded by a pilot test.

*Remediation Methods for Chlorinated Solvents.* Chlorinated solvents can be transformed biologically by microorganisms, thereby remediating contaminated

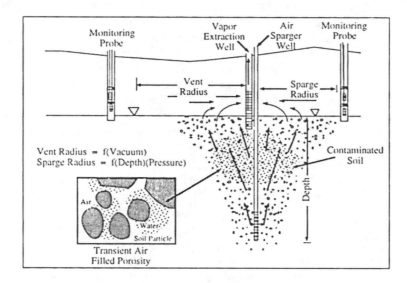

**FIGURE 12.25**

*Air sparging.* (Reprinted with permission from R. D. Norris, *et al.*, *Handbook of Bioremediation*, 1993. Copyright © 1993, Lewis Publishers, an imprint of CRC Press, Boca Raton, Florida)

soils. In addition, soils contaminated with chlorinated solvents may be treated effectively by soil vacuum extraction to remove volatile vapors; groundwater can be treated by air stripping. These are physical treatment processes; the vapors are discharged into the atmosphere or treated with activated carbon, catalytic combustion, or incineration.

*Natural Bioremediation.* Naturally occurring microorganisms can degrade contaminants in the subsurface environment. Studies have shown that plumes of dissolved hydrocarbons will eventually degrade without human intervention. The natural assimilative capacity of an aquifer depends on the local hydrogeology, the geochemistry, and the metabolic characteristics of indigenous microorganisms. In some cases, removal of leaking tanks or contaminated soil may be all that is necessary for natural bioremediation to complete the cleanup. In most cases, though, natural bioremediation is used only to supplement the methods discussed previously.

Hydrocarbons that do not completely degrade in the unsaturated zone will be transported within the water table aquifer, where the extent of further biodegradation is likely to be limited by the available oxygen supply; oxygen is only slightly soluble in water. As the plume migrates, though, contaminated water will disperse and mix with clean, oxygenated water.

Depending on site conditions, problems related to risk analysis, property rights, third-party liability, and the possible need for variances from existing regulations will have to be considered if natural bioremediation is used. A comprehensive monitoring system is needed, including interior wells to monitor the plume and guardian wells at the outside edge of the contaminated area; guardian wells are needed to monitor potential off-site migration of contaminants and to determine if additional remedial steps are necessary. Computer models may also be used to predict the extent of plume migration and rate of biodegradation in an aquifer. In some cases, particularly for chlorinated contaminants, the inoculation or addition of nonindigenous microorganisms into the ground can speed up the rate of biological action and remediation.

## Brownfields

The previous section focused on the remediation of sites that are definitely contaminated with potentially harmful substances. There are many industrial and commercial properties, though, which, while they may not be on the Superfund list, are abandoned sites where redevelopment is hindered by perceived or low-level environmental contamination. These properties or sites are called *brownfields*. For example, old warehouses or abandoned factories in cities may become brownfields, as may abandoned mines or fields in rural areas where illegal dumping may have occurred.

Although brownfields do not pose severe or immediate environmental threats, it is important that they be cleaned up and renovated to promote redevelopment and economic growth. The EPA, beginning in 1995, has provided legal and technical guidance to prospective purchasers, current landowners, and land developers on issues related to liability and site cleanup. In 1997, a federal tax incentive was begun to promote cleanup and redevelopment of brownfields in economically distressed urban and rural areas and to bring thousands of abandoned and underused industrial sites back into productive use.

## 12.5 HAZARDOUS WASTE MINIMIZATION

The most desirable solution to the hazardous waste problem is to reduce or minimize the quantity of waste at its source. *Waste minimization* is, in fact, a primary waste management goal in the United States. In the 1984 *Hazardous and Solid Waste Amendments*, it is stated that "The Congress hereby declares it to be the national policy of the United States that, wherever feasible, the generation of hazardous waste is to be reduced or eliminated as expeditiously as possible." Waste minimization policies are also followed by Japan and many European countries.

Waste minimization techniques focus on source reduction and recycling activities that reduce either waste volume or hazards. Recycling is an increasingly attractive option, particularly for waste that includes high concentrations of metals, oils, acids, and other substances that may have economic value. A diagram outlining waste minimization techniques is shown in Figure 12.26.

Waste minimization is now a top priority or goal for industry, not because of mandatory requirements (there are none), but because of the very high costs of the other options, including disposal on or in the ground. Disposal sites are scarce and prices keep rising. It has become clear that simply storing or burying hazardous waste in the land is not the ultimate solution or management option for industry. Even treatment and stabilization of the waste prior to its land disposal will not remedy the problem. Because of restrictions on land disposal, many untreated wastes that were sent to landfills in the past must now be incinerated, at costs much higher than those for landfilling. Waste minimization also can reduce a generator's financial and legal liability; the less waste generated, the lower are the costs and environmental risks.

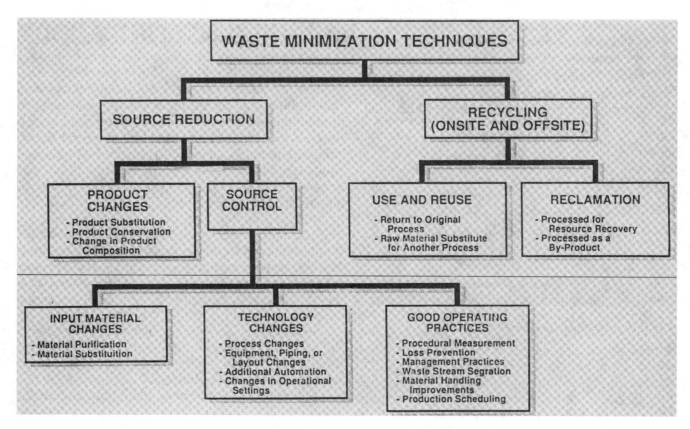

**FIGURE 12.26**

*Methods to reduce waste.* (Reprinted with permission of W. C. Blackman, *Basic Hazardous Waste Management*, 1992. Copyright © 1992, Lewis Publishers, an imprint of CRC Press, Boca Raton, Florida)

Waste minimization is not regulated as a mandatory program; the United States and other countries that promote minimization and recycling as a management option rely on cooperative, voluntary efforts by the industrial sector. Mandatory performance standards and other regulatory approaches have been rejected because they would be counterproductive. They would second-guess industry's production decisions and would be very difficult to administer. It is considered best to rely on existing strong economic incentives for industry to reduce their wastes. The most constructive role for government in this effort is to provide technical information and other assistance to waste generators. Since the states deal first-hand with the generators, they also should play a central role in assisting industry with waste minimization programs. A few states provide incentives and encourage waste reduction through tax preferences.

## Waste Minimization Audits

The first step in establishing a waste minimization program is to conduct a *waste minimization assessment* or audit. This is a careful review of an industry's potential opportunities to reduce or recycle its waste; it may be done by in-house staff or an independent consultant. An effective way to begin such an audit is to select a few waste streams or processes for intensive review, rather than to attempt to cover all waste streams at once. Waste minimization assessments identify and characterize waste streams, the production processes that are generating the waste, and the quantity of waste generated.

The EPA suggests the following steps for a waste minimization assessment:

- Prepare background material for the assessment.
- Conduct a preassessment visit to identify candidate waste streams.
- Select waste streams for detailed analysis.
- Conduct a detailed site visit to collect data on selected waste streams and controls and related process data.
- Develop a series of potential waste minimization options.

- Undertake preliminary option evaluations (including development of preliminary cost estimates).
- Rank options according to waste reduction effectiveness, extent of current use in the industry, and potential for future application at the facility.
- Present preliminary results to plant personnel along with a ranking of options.
- Prepare a final report, including recommendations to plant management.
- Develop an implementation plan and schedule.
- Conduct periodic reviews and updates of assessments.

## CASE STUDY

### A Waste Minimization Assessment

In 1986, the EPA sponsored a waste minimization assessment at an electric arc furnace steel-making facility. The assessment team examined waste minimization options, including source reduction and resource recovery, for the company's corrosive and heavy metal wastes. The assessment revealed that calcium fluoride (fluorspar) in the sludge generated during neutralization of the pickling line wastewater could be economically recovered.

Previously, the company had disposed of the sludge and purchased 1000 tons of fluorspar per year as flux material for the steel-making process. The waste minimization option identified by the assessment team saved the company $100,000 per year in costs avoided to purchase fluorspar and an additional $70,000 per year because of a 30 percent reduction in the volume of sludge to be disposed of.

*Source:* Environmental Protection Agency.

## Waste Reduction Methods

Methods of reducing hazardous waste quantities can be grouped into several categories, including *product changes, input material changes, technology changes, good operating practices,* and *recycling* (see Figure 12.26). These methods can be applied in a wide range of industries and manufacturing operations. Most involve source reduction, the EPA's preferred option. Others involve on- and off-site recycling or reuse. The most suitable method for any particular industry can be determined after a waste minimization audit.

Improved operations may include the purchase of fewer toxic production materials, improved material

storage and handling practices, employee training, and better housekeeping practices. Hazardous waste can be separated from nonhazardous waste to save money for disposal and find new opportunities for reuse. Production equipment can be modified to produce less waste and to enhance material recovery operations; an improved preventive maintenance program can also be helpful. End products can be redesigned or reformulated to be less hazardous, and the sources of leaks and spills can be eliminated. Closed-loop production systems can be designed and implemented to increase on-site recycling.

Metal parts cleaning and paint application processes are essential for many industries and businesses. Hazards from metal parts cleaning can be minimized by reducing the volume or the toxicity of the cleaning agents used. Low-toxicity paints (for example, water-based) that do not contain heavy metals can be used to reduce paint-related hazardous wastes.

Sometimes, one company's waste can be used as the feedstock or raw material for another company's production process. Organizations that serve as *hazardous waste clearinghouses* or *material exchanges* are able to facilitate exchange and recycling efforts. A clearinghouse can help to make arrangements between waste generators and potential users of the waste material. It can serve as a matchmaker or directly as a transfer agent, purchasing the waste from the generator, reprocessing it if needed, and selling it for reuse by some other industry. Waste clearinghouses and exchanges have found some success in several European and Scandinavian countries; private or government-funded waste exchange organizations are likely to play an increasingly important role in managing hazardous waste in the United States in the coming years.

 ## 12.6 RELEVANT WEB SITES

### EPA's Brownfields Web Page

http://www.epa.gov/swerosps/bf/

This site offers information about the EPA's activities to help states and communities prevent, assess, safely clean up, and reuse brownfields.

### EPA Reach It

http://www.epareachit.org

This Homepage of the **EPA Re**mediation and **Char**acterization **I**nnovative **T**echnologies (EPA REACH IT) site provides access to comprehensive and up-to-date information about site remediation technologies and their applications.

**GROUND WATER PROTECTION COUNCIL**

http://gwpc.site.net/default.htm

This site provides information about groundwater regulations and the most effective methods regarding comprehensive groundwater protection and underground injection techniques.

**GROUNDWATER REMEDIATION TECHNOLOGY ANALYSIS CENTER (GWRTAC)**

http://www.gwrtac.org/

This site provides information that the GWRTAC compiles, analyzes, and disseminates on innovative groundwater remediation technologies.

**HAZARDOUS WASTE CLEANUP INFORMATION**

http://www.clu-in.com/

This site provides information about innovative hazardous waste treatment technologies and programs.

**HAZARDOUS WASTE DATA OVERVIEW**

http://www.epa.gov/enviro/html/rcris/rcris_overview.html

This site presents information about generators, transporters, treaters, storers, and disposers of hazardous wastes, as reported by the states to the EPA, under the RCRA.

**OFFICE OF SOLID WASTE—HAZARDOUS WASTE**

http://www.epa.gov/epaoswer/osw/hazwaste.htm

This EPA site provides information on all forms of hazardous wastes, storage, reuse, recycling, and cleanup, as well as laws, regulations, policies, and frequently asked questions about hazardous waste.

**SOIL AND GROUNDWATER CLEANUP**

http://www.sgcleanup.com/

This is the site of an informative online magazine that focuses on hazardous waste, site remediation, groundwater treatment, site assessment, risk assessment, and many other relevant topics.

**SUPERFUND BASIC RESEARCH PROGRAM**

http://www.niehs.nih.gov/sbrp/home.htm

This National Institute of Environmental Health Sciences (NIEHS) site provides information about research efforts that are focused on understanding the risks hazardous waste exposures pose to humans, and developing new technology which will help remediate contaminated sites.

**SUPERFUND SITE**

http://www.epa.gov/superfund/

This is the EPA's gateway to a wealth of useful information about hazardous waste management and the Superfund program.

**UNDERGROUND INJECTION CONTROL (UIC)**

http://www.epa.gov/ogwdw/uic.html

This site provides information about five classes of injection wells, defined by EPA according to the type of waste they inject and where the waste is injected.

## REVIEW QUESTIONS

1. There have been many episodes or cases publicized since the 1950s involving hazardous wastes and the damage they caused to public health and the environment. The Love Canal case is perhaps one of the most famous, but there are many others, involving such substances as DDT, PCBs, mercury, and dioxin. Visit your public or college library or the World Wide Web to research the Love Canal episode in New York State, the Bhopal, India, incident (1984), and at least one other case. Write a brief report on your findings.

2. List five options for the management and control of hazardous waste. Which is the least desirable? Which is the most desirable? Explain why.

3. Give a workable definition of hazardous waste, including the basic properties that are characteristics of such wastes. Briefly describe these properties. Is there a distinction between hazardous wastes and hazardous materials? Should the term *toxic waste* be used interchangeably with the term *hazardous waste?* Why? What is the *TCLP?*

4. Briefly discuss the primary mode of transport for hazardous waste. What is an *SQG*? What is a *TSDF*?

5. Briefly discuss *RCRA*, *CERCLA*, *HSWA*, *SARA*, *TOSCA*, and *HMTA*.

6. What is the *manifest system*? Briefly discuss its basic purposes and its role in hazardous waste management.

7. Briefly discuss hazardous waste storage methods.

8. Briefly discuss the role of incineration in hazardous waste treatment. What are other common chemical treatment processes that may be applied to hazardous waste? What are the basic purposes of those processes?

9. Briefly discuss some of the common physical treatment processes that may be applied to hazardous waste. What are the basic purposes of these processes?

10. Briefly discuss biological treatment methods for hazardous waste. What is the difference between *land farming* and *landfilling* of hazardous waste?

11. What are the basic differences in the design requirements for an MSWLF and a secure hazardous waste landfill? Describe four different phases of the operating life of a secure landfill. Is it really secure?

12. Make a sketch showing a cross section of a secure landfill. Briefly describe the bottom liner and leachate collection systems.

13. Briefly describe the application of *deep-well injection* as a land disposal option for liquid hazardous waste. Are there other options that you think may be preferable? Why?

14. Briefly describe restrictions on land disposal of hazardous waste.

15. Briefly describe the role and application of the *Superfund* in hazardous waste management. What is the *NPL?*

16. What is meant by *site remediation?* Briefly describe the difference between off-site and on-site remediation options.

17. Briefly discuss two basic alternatives for on-site remediation.

18. What is *bioremediation?* Briefly discuss how it is accomplished for VOCs and for chlorinated hydrocarbons. What is *natural bioremediation?*

19. Briefly describe three different nonintrusive field sampling methods used to investigate hazardous waste disposal sites.

20. Briefly describe soil and soil-gas sampling techniques used at waste remediation sites.

21. Discuss some methods used to sample sediment, surface water, and groundwater at hazardous waste remediation sites.

22. Briefly discuss some of the safety factors related to remediation site field investigation. What is one of the primary dangers for personnel at the site? Why is it important for field technicians to be knowledgeable about the properties and characteristics of hazardous wastes?

23. Why are there no mandatory requirements for *waste minimization?*

24. Briefly describe the purpose and major steps of a waste minimization audit or assessment.

25. Briefly describe four different strategies for waste minimization. Explain the role of a hazardous waste clearinghouse or material exchange.

26. Visit and explore the *Ground Water Protection Council* Web page. Briefly describe the five different types of waste injection wells (Class I through Class V) that have been categorized by the Underground Injection Control Program.

27. Visit and explore the EPA *Office of Solid Waste–Hazardous Waste* Web page. Link to the "Waste Cleanup" page. Write a brief report explaining the difference between the RCRA's Superfund Program and the RCRA Corrective Action Program.

28. Visit and explore the EPA *Superfund Site.* Find your way to the *Site Assessment Homepage* after first linking to "Researchers and Scientists." Read about hazardous waste site assessment and the Hazard Ranking System (HRS). Write a brief report summarizing your findings.

29. Using the EnviroSources search engine (see Section 1.6) or another Internet search directory or search engine, locate at least one additional Web site relevant to one of the topics in this chapter. Add the link(s) to your Environmental Technology Folder. Write a brief description of what the Web site(s) contain.

## Chapter Outline

CHAPTER

# 13

# Air Pollution and Control

Air is necessary for the survival of all higher forms of life on Earth. On the average, a person needs at least 30 lb of air every day to live, but only about 3 lb of water and 1.5 lb of food. A person can live about 5 weeks without food and about 5 days without water, but only 5 minutes without air. Naturally, everyone likes to breathe fresh, clean air. But the *atmosphere*, that invisible yet essential ocean of different gases called air, is as susceptible to pollution from human activities as are water and land environments.

The atmosphere appears to be so vast in volume that at first glance it may be difficult to believe that human activities could have any lasting impact on it. Smoke and gases discharged into the air are soon diluted to very low concentrations by mixing and dispersion. It seems improbable that human activity could pollute the entire atmosphere and very unlikely that such small amounts of contaminants could harm public health or even change Earth's climate. So it may seem, but most people know better.

The atmosphere is extensive, but not infinite. Recall the fact that if Earth is imagined to be about the size of an apple, the depth of the atmosphere would be equivalent to only the thickness of the apple's skin. Personal experience demonstrates the discomfort of being enveloped in the exhaust fumes of motor vehicles on a congested highway or when passing by certain manufacturing plants. And the unpleasant but all too common sight of smog is well known to most people who live in cities.

Public health studies tend to demonstrate that over the long term the standard of health for people living and working in highly industralized urban areas is lower than that for populations in rural areas, where air pollution is much less severe. There is much scientific evidence of a distinct relationship between generally dirty air and a higher incidence of respiratory diseases, including lung cancer. Atmospheric pollution and air quality control are clearly issues of major significance in the fields of public health engineering as well as in civil and environmental engineering technology.

The objective of this chapter is to present some of the basic facts and concepts regarding air pollution and its control. After a brief review of the history of air pollution, atmospheric and meterological factors are discussed. Types and effects of ambient air contaminants are presented, and the topics of indoor air pollution as well as global climate change are covered. Air sampling and measurement techniques, air quality rules and regulations, and, finally, air pollution control strategies and equipment for stationary as well as mobile sources are discussed.

## 13.1 HISTORICAL BACKGROUND

Air pollution from human activities can be said to have originated with the discovery of fire; even today, most air contaminants originate in combustion processes from mobile as well as stationary sources. As early as 4000 B.C., when copper and gold were beginning to be forged and clay was glazed, there must have been significant amounts of air pollution in the local areas nearby those activities. And there is little doubt that the use of coal for fuel, which began around the year A.D. 1000, resulted in considerable amounts of air pollution. Early in the 14th century, for instance, King Edward II prohibited the burning of coal in London while Parliament was in session because of the smoke it caused. At the start of the 17th century, when coal was converted to coke and used for iron smelting, air pollution increased substantially. By the middle of the 17th century, the city of London had such severe smoke problems that a brochure containing ideas for reducing air pollution was written by a special commission and submitted to King Charles II and to Parliament.

Later in the 17th century, the development of the steam engine and the beginning of the Industrial Revolution heralded a new era of air pollution. Early in the 20th century, yet another significant air pollution era began when the gasoline-powered automobile became a major contributing source. But it was not until the mid-1900s that any lasting attempts were made to protect air quality. Meaningful efforts to control air pollution were initiated about that time, largely as the result of deadly *air pollution episodes*, particularly those in Donora, Pennsylvania; London, England; and the Meuse Valley, Belgium.

Donora, a small industrial town in western Pennsylvania, experienced a *temperature inversion* (a weather effect explained in the next section) in the fall of 1948. The inversion trapped air pollutants, mostly from the local steel mill, wire mill, and zinc plating plant, in the valley in which the town is situated. An intense smog hung in the air and persisted for about 1 week. By the time the smog dissipated, the death of more than 20 people and the illness of about 600 others were attributed to its effects. Since the population of Donora at the time (about 14,000 people) was relatively small, the per capita death rate was actually the highest ever recorded during an air pollution episode. A similar episode occurred in London in 1952, also due to a weather inversion. An extremely severe ground-level smog lasted for more than a week, reducing visibility to only a few meters. The average death rate in London more than tripled during that week (about 4000 people died), but it dropped back to normal when the smog abated. The Donora episode, London's "Killer Smog," and other events helped focus public attention on air pollution and prompted development of new strategies, laws, and technologies needed to solve the problem.

## 13.2 ATMOSPHERIC FACTORS

To understand topics related to the effects and control of air pollution, it is first necessary to know something about the composition and physical behavior of the atmosphere itself. What does "pure" air consist of, and how do meteorological or weather conditions affect the mixing and dispersion of pollutants? *Meteorology*, the science of the atmosphere and weather forecasting, involves the study of both large-scale and small-scale atmospheric circulation patterns. During some types of adverse weather conditions, small-scale circulation patterns are such that emitted pollutants are confined to a restricted volume of air. It is necessary to know how these weather patterns develop and to understand their impact on air pollution control requirements. Large-scale weather patterns are also of concern with regard to global air pollution problems.

## Composition of the Atmosphere

The atmosphere is a mixture of many different gases, but mostly it consists of molecular *nitrogen* and *oxygen* ($N_2$ and $O_2$). About 78 percent of dry air is nitrogen and about 21 percent is oxygen. These percentages are expressed on a volume basis. For example, a container holding 1 m$^3$ or 1000 L of air (at standard pressure) includes about 780 L of nitrogen and 210 L of oxygen.

The nitrogen and oxygen combined make up approximately 99 percent of the atmosphere. The remaining 1 percent of clean air is a mixture of several other gases. Most of that remaining 1 percent (about 0.9 percent) is the inert gas *argon*. The rest of it includes *carbon dioxide, methane, hydrogen, helium, neon, ozone,* and numerous other gases in trace (very small) amounts. Figure 13.1 shows the relative amounts of atmospheric gases. It is seen that the "pure" atmosphere is normally a mixture of many different substances. Later in this chapter, several specific substances called air pollutants or contaminants are delineated.

The relative amounts or concentrations of gases in air can be expressed in terms of *parts per million* (ppm) as well as in terms of percentage. For example, since 1 percent = 10,000 ppm (see Section 4.1), an oxygen level of 21 percent in air can also be expressed as 21,000 ppm. Obviously, it is more convenient to express

that concentration in percent. On the other hand, the concentration of carbon dioxide in the atmosphere, about 0.034 percent, may be more conveniently expressed as 340 ppm. A very wide range of concentrations is related to air quality; for instance, natural ozone concentrations can be as low as 0.02 ppm. Concentrations of air pollutants are also expressed in terms of mass per unit volume; conversion calculations are explained in Section 13.6.

Water vapor is also a normal component of the atmosphere, but the amount may vary significantly over time and location. Local climate is a major factor that affects the amount of water vapor or moisture in the air. In very humid regions, for example, moisture content may be as high as 5 percent. Moisture affects air quality in several ways. For instance, as it condenses or evaporates, water releases or absorbs heat, which affects atmospheric stability and air circulation patterns. Also, when atmospheric moisture condenses, fogs are formed; fogs tend to occur more frequently in urban areas due to higher levels of particulates, which serve as nuclei for the formation of fog droplets. Fogs are typically involved in serious pollution episodes (for example, at Donora) because the droplets help the conversion of sulfur oxides into sulfuric acid. Fogs also block heat energy from the sun and prolong weather circulation patterns that tend to trap the pollutants.

## Atmospheric Layers

The full atmosphere extends upward roughly 160 km (100 mi) above the surface of Earth. But the mixture of gases just discussed refers only to the *troposphere*, the lowermost surface layer of the atmosphere. The troposphere, which is roughly 12 km (8 mi) in depth, contains about 95 percent of the total air mass. It is in this relatively thin layer of air that oxygen-dependent life is sustained, clouds are formed, weather patterns develop, and most air pollution problems occur. The density of air increases significantly with a decrease in altitude or distance above Earth's surface, and it is for this reason that most of the total air mass is in the bottom layer. The "skin of the apple" mentioned previously refers to this life-supporting layer. Above the troposphere, there is not enough oxygen to support life.

The layer of air above the troposphere, called the *stratosphere*, is a stable layer (in terms of air circulation patterns) that extends upward from Earth's surface to an altitude of about 50 km (30 mi). Even though it is deeper than the troposphere, the stratosphere contains only a small part of the total air mass because of its lower air density. It does, however, contain much more naturally occurring *ozone* ($O_3$) than the troposphere.

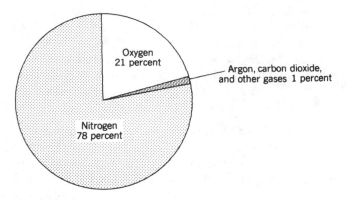

**FIGURE 13.1**
*Molecular nitrogen and oxygen are the main constituents of the atmosphere, but "clean" air also contains argon, carbon dioxide, and trace amounts of several other gases.*

As will be discussed later, this stratospheric ozone plays a crucial role as a barrier to harmful ultraviolet (UV) radiation from the sun. Layers of the atmosphere existing above the stratosphere include the *mesosphere*, the *ionosphere*, and the *thermosphere*. These portions of the atmosphere are essentially unaffected by air pollution.

A basic physical characteristic that distinguishes one atmospheric layer from the next is the *temperature gradient*, that is, the gradual change of air temperature with altitude. For example, air temperature normally decreases with increasing altitude in the troposphere, but increases with altitude in the stratosphere. This is illustrated in Figure 13.2. Naturally occurring variations in the temperature gradient of the troposphere cause certain weather patterns that directly affect air quality.

## Effects of Weather

Air quality at any given location can vary greatly over time, even though the rate of emission of pollutants remains relatively constant. This is because the air pollutants are mixed, dispersed, and diluted within the troposphere by movement of air masses, both horizontally and vertically. Air movements and therefore air quality are very dependent on local as well as regional weather conditions. Knowledge of horizontal and vertical circulation patterns is of importance with regard to implementation of air pollution surveys, site selection for new industrial plants, establishment of maximum allowable air pollutant emission rates, and design of tall stacks or chimneys.

### Horizontal Dispersion of Pollutants

Horizontal dispersion or spreading of air pollutants depends on wind speed and direction. The concentration of air pollutants decreases with increasing wind speed because, as the pollutants are discharged from the source, they are more rapidly separated and dispersed by the swiftly moving air. Knowledge of prevailing wind speed and direction in a given locality makes it possible to select sites for new industrial facilities or power plants so as to minimize local air pollution effects. Locating such sites downwind of residential areas is preferable, naturally, to upwind location. Wind velocity data, plotted in a graph called a *wind rose*, give a picture of the speed and direction from which the wind tends to come; the graph provides information regarding the *prevailing winds*. A typical wind rose is shown in Figure 13.3.

Winds develop because of the combined effects of temperature gradients and the rotation of Earth. Warm air near the equator rises and moves toward the poles while friction and the forces resulting from Earth's rotation deflect the air movement, eastward in the northern hemisphere and westward in the southern hemisphere. Important factors that affect circulation patterns include topography, daily and seasonal variations in surface heating, proximity to large bodies of water and to mountains, and cloud cover. Since soil and rock warm up and cool faster than water, winds near shorelines are directed toward the water at night and inland during the day. In an urban area, where steel, concrete, and masonry absorb and hold heat, a *heat island* encompasses the city at night, with a self-contained circulation pattern from which pollutants cannot readily escape.

### Vertical Dispersion of Pollutants

In addition to wind patterns and horizontal dispersion, the vertical motion of air is very important with regard to air quality. Vertical mixing of air and dispersion of pollutants depends on the kind of *atmospheric stability* prevailing at any given time. The atmosphere is considered to be *stable* when there is little or no vertical movement of air masses and therefore little or no mixing and dispersion of pollutants in the vertical direction. Air pollutants tend to accumulate near the ground under stable conditions, and severe pollution episodes

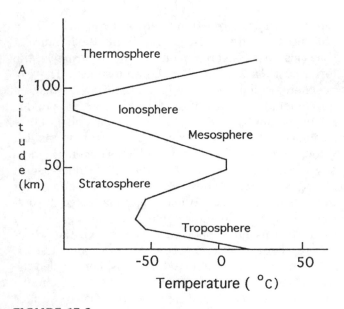

**FIGURE 13.2**
*In the troposphere, temperature normally drops with increasing distance from Earth's surface; in the stratosphere, temperature increases with increasing distance or altitude.*

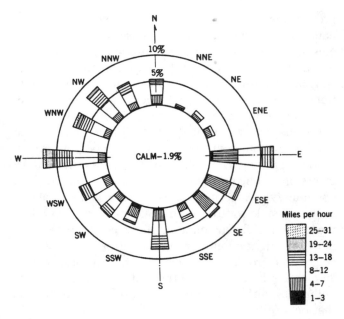

**FIGURE 13.3**

*Example of a wind rose. Positions of spokes show wind direction; total length shows percentage of time, for the reporting period, that the wind was blowing from that direction. Shaded segments show the percentage of time the wind was blowing at the indicated speed.* (From *Environmental Engineering and Sanitation*, J. A. Salvato. Copyright © 1992 John Wiley & Sons, Inc. This material is used by permission of John Wiley & Sons, Inc.)

may occur. An *unstable* atmosphere, on the other hand, is one in which the air moves naturally in a vertical direction, increasing mixing and dispersion of the pollutants. With regard to local or regional air quality, *a condition of atmospheric unstability is preferable to a stable condition.*

Atmospheric stability depends on the rate of change of air temperature with altitude, that is, on the temperature gradient (or profile) that may prevail at a particular time and location. Normally, air temperatures in the troposphere decrease with increasing altitudes (see Figure 13.2). But the *rate at which the air temperature drops*, called the *environmental lapse rate*, is of crucial importance with regard to atmospheric stability. (Environmental lapse rate is also called *prevailing* or *ambient* lapse rate.) To understand the relationships among lapse rates, stability, and vertical mixing, a brief discussion of basic atmospheric physics is necessary.

Atmospheric pressure decreases with increasing height above the ground. As a volume or parcel of air

rises, it naturally expands and cools, provided that heat energy is not added or withdrawn. A physical process in which there is no heat transfer is called an *adiabatic process*. Since a parcel of air that is not in immediate contact with the ground is well insulated by its surroundings, it can be assumed that naturally occurring expansion (or compression) of air parcels, as they move vertically, is adiabatic.

Using the first law of thermodynamics, one can calculate a theoretical *dry adiabatic lapse rate* to be 1°C per 100 m (5.4°F per 1000 ft). This lapse rate is independent of the prevailing atmospheric temperature gradient at any given time. In other words, regardless of what the actual air temperature profile may be, a moving parcel of air will always cool down 1°C for each 100 m it rises in the atmosphere and will warm up 1°C for every 100 m it sinks. [The dry adiabatic lapse rate is calculated on the assumption that there is no moisture or water vapor in the air. Water, usually present in the atmosphere to some degree, emits or absorbs heat as it condenses or evaporates. As air rises, condensation occurs, producing a slower cooling rate (lower lapse rate) than for dry air. The average wet adiabatic lapse rate is about 3.5°F per 1000 ft.] The adiabatic lapse rate usually differs from the environmental lapse rate because of factors such as geographic features, wind, and sunlight.

When the environmental lapse rate exceeds the adiabatic lapse rate, the actual air temperature drops more than 1°C per 100 m. The atmosphere is unstable when this happens, and vertical mixing of air masses will occur. This is because an upward-moving parcel of air will cool adiabatically at a slower rate than the surrounding air. Since the parcel of air then becomes warmer (and lighter) than its surroundings, it continues to be accelerated upward by buoyant forces. Conversely, a downward-moving parcel of air will become cooler (and denser) than its surroundings and will continue its descent under the influence of gravity.

Consider, for example, a parcel of air at an altitude of 1000 m. It will have the same pressure and temperature as the air surrounding it, say 20°C. Suppose that the environmental lapse rate at the time is −2°C per 100 m. If the parcel rises to an altitude of 1100 m, its temperature will drop to 19°C, while the surrounding air will have a temperature of 18°C. Since the parcel is warmer than the surrounding air, it will continue to rise, as if it were a hot-air balloon. This is illustrated in Figure 13.4. If the parcel were to move to a lower altitude, say 800 m, its temperature would become 22°C, while the surrounding air temperature would be 24°C; the colder (and denser) air parcel would continue to descend.

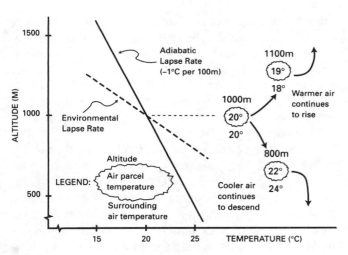

**FIGURE 13.4**
*Illustration of unstable atmospheric conditions, when the environmental lapse rate (for example, −2° per 100 m) exceeds the adiabatic lapse rate. In this example, buoyant forces keep the air parcels moving in a vertical direction.*

Environmental lapse rates are classified as being either *strong* or *weak* (*superadiabatic* or *subadiabatic*), as illustrated in Figure 13.5. Strong lapse rates are associated with an unstable atmosphere, while weak lapse rates are associated with a stable atmosphere. A lapse rate characterized by an increase in actual air temperature with increasing altitude is called a *temperature inversion* and results in an extremely stable atmosphere. Temperature inversions, which effectively prevent the upward mixing and dispersion of contaminants, are usually the major causes of air pollution episodes, such as those in Donora and London. In urban areas, air quality will decrease rapidly due to the stagnation caused by an inversion until weather conditions change and a superadiabatic lapse rate is restored.

Temperature inversions can be caused by a variety of local meteorological conditions, and they can occur just about anywhere, but certain geographic conditions can increase their frequency and duration. An inversion can be particularly severe, for example, for a community located in a valley, which acts as a holding basin or sink for the cold, denser air masses near the ground. Hills surrounding the valley tend to block much of the horizontal air motion, thus adding to the stagnation problem. The city of Los Angeles, for example, lies in a mountain-rimmed bowl that traps air pollutants during the frequent temperature inversions that occur there.

Three types of temperature inversions, each associated with a specific weather pattern, are *frontal inversions, subsidence inversions,* and *radiation inversions*. Frontal inversions generally develop at relatively high altitudes when a warm air mass overruns a cold air mass; they are not particularly important with regard to air quality control. Subsidence inversions, however, are of major importance. Although they also develop at relatively high altitudes, subsidence inversions may persist for several days or even weeks, especially in the summer months. This inversion is caused, as its name implies, when a large warm mass of air subsides or descends over a community (for example, Los Angeles). The actual inversion may occur at an altitude of 300 m, for example, while the lapse rate below is sub- or superadiabatic. This kind of inversion forms a lid or cap that literally traps pollutants and prevents further vertical mixing, as shown in Figure 13.6. The plume from a smokestack will abruptly

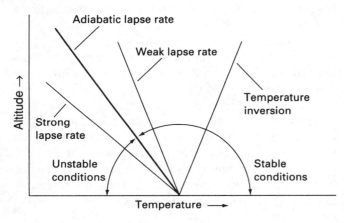

**FIGURE 13.5**
*The temperature profile that separates stable atmospheric conditions from unstable conditions is the adiabatic lapse rate. The air is very stable during an inversion, when temperatures increase with height above the ground or altitude.*

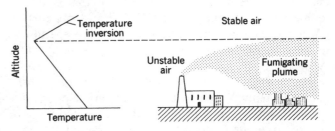

**FIGURE 13.6**
*When a temperature inversion begins above the ground because of local weather conditions, it acts as a lid or ceiling that prevents further vertical mixing and traps pollutants below it.*

stop rising when it reaches the inversion altitude. If the ambient lapse rate below the inversion is superadiabatic, a *fumigating* condition may develop near ground level due to the strong mixing effects. Fumigation results in very high ground-level pollutant concentrations.

Radiation inversions develop at low altitudes and are caused by the rapid cooling of the ground due to radiation, mostly on clear winter nights. The temperature of the air in contact with the ground also drops, causing the inversion. Air pollutants emitted during the night are trapped and do not disperse until later the next day, when the ground warms sufficiently to break the inversion. Fumigating conditions are common during radiation inversions; fortunately, these inversions are not long-lasting, although they are frequent.

Sophisticated mathematical models have been developed to calculate and predict ground-level concentrations of pollutants downwind from sources such as smokestacks. Most computer programs that relate emissions to air quality use the *Gaussian dispersion equation*. This model is based on a variety of simplifying assumptions, including a uniform steady wind, a constant emission rate, flat terrain, a conservative pollutant (one that is not lost by decay, chemical reaction, or deposition), and others. It is assumed to be accurate to within only ±50 percent. Although the results of the calculations are uncertain, the Gaussian model provides engineers and scientists with an analytical tool useful for comparing various pollution control strategies. A demonstration of its application is beyond the scope of this book.

## 13.3 TYPES, SOURCES, AND EFFECTS

Air pollution may be simply defined as *the presence of certain substances in the air in high enough concentrations and for long enough durations to cause undesirable effects.* "Certain substances" may be any gas, liquid, or solid, although certain specific substances are considered significant pollutants because of very large emission rates or harmful and unwanted effects. "Long enough durations" can be anywhere from a few hours to several days or weeks; on a global scale, durations of months and years are of concern. Before considering those substances considered to be major pollutants, it is useful to first consider some of the general distinctions among the various types of air contaminants.

Implicit in this definition is the assumption that the air pollution is *anthropogenic*, that is, caused by human activities. But air pollution may also result from natural causes. In fact, at certain times pollution from natural sources can be far more severe and longer lasting than air pollution from human activities. For example, the 1991 volanic eruption of Mount Pinatubo in the Philippines spewed vast quantities of dust and gases into the atmosphere in a relatively short period of time. Pollutants from this eruption reached the upper atmosphere and acted as sunlight reflectors, causing global temperatures to fall slightly for a few years. In addition to volcanic discharges such as the one from Mount Pinatubo, other natural air pollutants include smoke and gases from forest fires, windblown dust from deserts, salt seaspray, pollen grains, and other naturally occurring substances. Obviously, there is not much that can be done to control or regulate these natural occurrences, although there may be certain actions that can be taken to mitigate their harmful effects.

Another general distinction among air pollutants involves the difference between *primary pollutants* and *secondary pollutants*. Primary pollutants are emitted directly into the air from a specific source, such as a power plant stack. Secondary pollutants, however, are not emitted directly from a source, but are formed in the atmosphere by complex chemical reactions involving the primary pollutants and sunlight. The sources of primary air pollutants are either *mobile* (for example, automobiles) or *stationary* (for example, coal-fired electric power generating stations). The distinction between mobile and stationary sources of pollutants is important because of the different pollution control technology applied to each type, as well as the different kinds of contaminants that they emit.

Air pollution occurs indoors as well as outdoors. Until recently, outdoor or *ambient* air quality problems received most of the attention of scientists, engineers, and regulatory agencies. Two kinds of ambient pollutants are regulated under the *Clean Air Act: criteria pollutants* and *hazardous air pollutants*. The National Ambient Air Quality Standards (NAAQS) promulgated under the *Clean Air Act* characterize five primary pollutants and one secondary pollutant as criteria air pollutants. These six pollutants are emitted in relatively large quantities by various sources and tend to threaten human health or welfare. Although the discharge of hazardous air pollutants is not as voluminous as that of the criteria pollutants, they are considered to be immediately harmful to human health and are, for the most part, associated with certain specific sources. (The *Clean Air Act* and the NAAQS are discussed further in Section 13.7.)

### Criteria Air Pollutants

The five primary criteria pollutants include the gases *sulfur dioxide* ($SO_2$), *nitrogen oxides* ($NO_x$), and *carbon monoxide* (CO), solid or liquid *particulates* (smaller than 10 $\mu m$), and *particulate lead.* Except perhaps for lead, the primary pollutants are emitted in industrialized countries at very high rates, usually measured in millions of tons per year. *Ozone* ($O_3$) is the secondary criteria pollutant regulated under the NAAQS. Although ozone near the ground is a harmful pollutant, in the stratosphere it helps block harmful ultraviolet radiation. This is discussed in more detail later.

## Sulfur Dioxide

Certain fossil fuels, particularly coal, may contain the element sulfur. When these fuels are burned for power or heat, the sulfur is also burned, or *oxidized.* This chemical reaction can be described by the following equation:

$$S + O_2 \Rightarrow SO_2$$

Sulfur dioxide is a colorless gas with a sharp, choking odor. It is a primary pollutant because it is emitted directly in the form of $SO_2$. Ordinarily, most people do not think of a gas as having much weight, but when the volumes are large, the total weight can be substantial. Roughly 25 million tons of $SO_2$ is discharged into the atmosphere in the United States each year. More than 80 percent of the emissions result from fossil fuel combustion, and most of that is from electric utility power plants. Only a very small amount comes from mobile sources. Other sources of $SO_2$ emissions are petroleum refining, copper smelting, and the making of cement. Since the early 1970s, total sulfur oxide emissions have been reduced significantly as a result of fuel desulfurization, flue gas scrubbing, and other controls.

In the presence of oxygen, water vapor, and sunlight, $SO_2$ can be involved in additional chemical reactions. It reacts with oxygen to form *sulfur trioxide,* which then reacts with water vapor to form a mist of *sulfuric acid.* The chemical reactions are shown in the following equations:

$$2\,SO_2 + O_2 \Rightarrow 2\,SO_3$$
$$SO_3 + H_2O \Rightarrow H_2SO_4$$

The sulfuric acid ($H_2SO_4$) mist is a secondary pollutant because it is not emitted directly, but is formed subsequently in the atmosphere. It is a constituent of *acid rain,* an important regional air pollution problem (as discussed in Section 13.4). Sulfuric acid molecules may also condense on existing particles in the air, and sulfate ($SO_4^{2-}$) aerosols often compose a significant fraction of particulate pollution in the atmosphere. The sulfur pollution eventually reaches the ground by *wet deposition* (that is, as acid rain) or *dry deposition* (without precipitation).

## Nitrogen Oxides

There are many forms of nitrogen oxides (characterized collectively as $NO_x$), but the one that is of greatest importance is *nitrogen dioxide* ($NO_2$). Most emissions are initially in the form of nitric oxide (NO), which by itself is not harmful at concentrations usually found in the atmosphere. But NO is readily oxidized to $NO_2$, which in the presence of sunlight can further react with hydrocarbons to form *photochemical smog.* Smog is, of course, harmful. $NO_2$ also reacts with the hydroxyl radical ($OH^-$) to form nitric acid ($HNO_3$), which contributes to the problem of acid rain. Although NO is colorless, $NO_2$ is a pungent, irritating gas that tends to give smog a reddish-brown color.

More than 20 million tons of nitrogen oxides is discharged each year in the United States. The largest source is from the oxidation of nitrogen compounds during the combustion of certain fossil fuels, such as coal or gasoline. Nitrogen oxides are also formed when temperatures are high enough to oxidize molecular nitrogen in the combustion air. Stationary sources are the major contributions of nitrogen oxides, although mobile sources are also important. (Ironically, modifications in the operation of internal combustion engines meant to control carbon monoxide emissions, such as increasing the air supply and raising the combustion temperature, tend to make the $NO_x$ problem worse.)

## Carbon Monoxide

During complete combustion of fossil fuels, carbon atoms in the fuel combine with oxygen molecules to form carbon dioxide ($CO_2$). But the process of combustion is rarely complete. *Incomplete combustion* of the fuel may occur when the oxygen supply is insufficient, when the combustion temperatures are too low, or when residence time in the combustion chamber is too short. Carbon monoxide (CO), a product of incomplete combustion, is the most abundant of the criteria pollutants. About 100 million tons of it is emitted each year in the United States. The chemical equation that describes its formation is written simply as

$$2\,C + O_2 \Rightarrow 2\,CO$$

Carbon monoxide is completely invisible; it is colorless, odorless, and tasteless. (The harmful effects of this invisible gas are discussed in the next section.) Almost 70 percent of the total carbon monoxide emissions come from highway vehicles, and atmospheric concentrations are very much a function of urban traffic patterns. CO levels, which typically range from 5 to 50 ppm in city air, may often reach 100 ppm on congested highways. (Cigarette smoke contains more than

400 ppm of carbon monoxide.) A significant fraction of total CO emissions comes from residential heating systems and from certain industrial processes. Power-generating plants, on the other hand, are typically designed and operated in a manner that reduces incomplete combustion, so they do not emit very much CO at all.

**Particulate Matter**

Extremely small fragments of solids or liquid droplets suspended in air are called *particulates*. Except for lead, particulates are distinguished on the basis of particle size and source, rather than by chemical composition. Since they typically have irregular shapes, *size* is defined in terms of an equivalent *aerodynamic diameter*, which refers to a perfect sphere with the density of water that would have the same settling velocity as the particle.

Most particulates range in size from 0.1 to 100 $\mu$m (one micrometer, or 1 $\mu$m, is one millionth of a meter; it may also be called a micron). The particulate materials of most concern with regard to adverse effects on human health are generally less than 10 $\mu$m in size and are referred to as $PM_{10}$.

Even though individual particles are so small, the total mass of particles discharged into the atmosphere is very large. Total suspended particulate (TSP) emissions are approximately 7 million tons per year in the United States. Particulate emissions come from materials handling processes, combustion processes, or gas conversion reactions in the atmosphere. Major sources include industrial processes, coal- or oil-burning power plants, residential heating systems, and highway vehicles.

Particles smaller than 1 $\mu$m tend to remain suspended in the atmosphere indefinitely, whereas those larger than 1 $\mu$m tend to settle out under the force of gravity. Particulates are generally classified by name according to their size and phase (solid or liquid).

Suspended solids roughly 1 to 100 $\mu$m in size are called *dust* particles, while smaller suspended solids (less than 1 $\mu$m) may be called either *smoke* or *fumes*. Dust is formed from materials handling activities or mechanical operations, including grinding, woodworking, and sandblasting. Smoke is a common product of incomplete combustion; smoke particles consist mostly of carbonaceous material. (*Soot* is a term that refers to visible clusters of carbon particles.) Fumes, usually consisting of very small metallic oxide particles, are typically formed during certain high-temperature chemical reactions and vapor condensation.

A suspension of liquid particles between 0.1 and 10 $\mu$m in size is called a *mist*, whereas a *spray* consists of liquid particles greater than 10 $\mu$m in size. Finally, the term *aerosol* refers to a quantity of any small particles, liquid or solid, suspended in air. To help to put this array of terms in proper perspective, Figure 13.7 depicts the relative size ranges of various solid and liquid particulates.

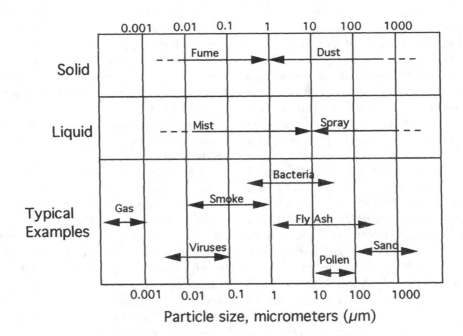

**FIGURE 13.7**
*Characteristic sizes of liquid and solid particulates.*

## Lead Particulates

This toxic metal, in the form of a fume (less than $0.5 \mu m$ in size), is one of the criteria pollutants. In the past, major sources of lead (Pb) fumes were motor vehicles that burned gasoline containing a lead-based antiknock additive. The EPA now requires the use of unleaded gasolines, but lead is still emitted from petroleum refining and smelting operations, battery recovery, and other industrial activities. Young children are particularly at risk from lead poisoning because even slightly elevated levels of lead in the blood cause learning disabilities, seizures, permanent brain damage, and even death. Ambient lead levels have decreased by more than 90 percent since the mid-1970s, when restrictions on leaded gasoline took effect. Under the 1990 *Clean Air Act Amendments*, a complete ban on the use of lead in motor vehicle fuel took effect in 1996.

## Ozone and Photochemical Smog

Ozone ($O_3$), a secondary air pollutant in the troposphere, is formed by a set of exceedingly complex chemical reactions between nitrogen dioxide ($NO_2$) and volatile organic compounds (VOCs). VOCs are hydrocarbons that quickly evaporate under normal atmospheric conditions. The reactions are initiated by the ultraviolet energy in sunlight. Actually, a number of secondary pollutants (collectively termed *photochemical oxidants*) are formed in the reactions. Ozone, the most abundant of the oxidants, is the key component of *photochemical smog.*

The word *smog,* a contraction of "smoke and fog," was initially coined in England to describe the visible air pollution caused by elevated levels of particulates and sulfur oxides. While this kind of smog differs in origin and composition from photochemical smog, a common characteristic of both types is the low-lying, irritating brownish haze that prevails in the atmosphere when smog occurs.

Since nitrogen oxides and VOCs are both emitted in large quantities by motor vehicles, photochemical smog is especially common in urban areas with lots of highway traffic and ample sunshine. Los Angeles, California, for example, is generally noted for its chronic smog problem. Geographic features and weather patterns contribute to the situation. Mountains up to 2500 m (8000 ft) high surround the Los Angeles basin, and a steady easterly flow of cool upper air often subsides over the city, forming a temperature inversion. The inversion traps the pollutants over the city, and the clear skies (about 200 d per year) ensure enough sunlight to produce smog.

## Hazardous Air Pollutants

Air pollutants associated with certain specific sources and that pose an immediate threat to human health are called *air toxics* or *hazardous air pollutants* (HAPs). The risks are greatest for people living in heavily polluted and industrialized urban areas. Hazardous air pollution increases the incidence of cancer and other adverse health effects and causes widespread environmental harm. A large fraction of toxics found in lake water, for example, is deposited from the air rather than from surface runoff (for example, mercury in the Great Lakes). While emission of the criteria pollutants is, for the most part, steady and gradual, there is the possibility for hazardous air pollutants to be released in sudden and often catastrophic accidents. One of the most notorious of these accidents is the 1985 event at a pesticide factory in Bhopal, India, where a release of toxic methyl isocyanate into the air killed approximately 3000 people and injured hundreds of thousands more. Although not as severe as the incident in Bhopal, there are hundreds of instances each year in the United States in which extremely hazardous substances are released into the air accidentally.

The air toxics program established by the *Clean Air Act Amendments* of 1990 includes strong incentives for innovation and pollution prevention in industry. A list of 189 toxic air pollutants has been created, along with a 10-year schedule for setting standards for all major sources of hazardous air pollutants. A major source is one that discharges more than 10 tons per year of any single HAP or 25 tons per year of any HAP combination. Industries will be required to install the best control technologies that are achievable to meet the new standards. It is expected that this 10-year regulatory effort will reduce emissions of hazardous air pollutants by more than 1 million tons per year in the United States.

Standards have already been set for six toxic air pollutants: asbestos, benzene, beryllium, mercury, vinyl chloride, and radionuclides (radioactive air pollutants). Maximum levels for arsenic and many others are now being developed. Asbestos, which can become airborne during the demolition or renovation of old buildings containing asbestos fireproofing, can cause a variety of lung diseases, including cancer. Benzene, also a potent carcinogen (cancer-causing substance), is emitted into the air mostly by gasoline-powered vehicles. Beryllium comes from foundries, ceramic factories, incinerators, and other sources. Mercury is a metal found in trace amounts in coal and is released into the air when coal is burned for power generation. It also can be released during incineration of garbage and from the weathering of latex-based paints, which contain mercury to prevent mildew. Mercury can cause brain and kidney damage. Vinyl chloride is emitted by various plastics processing facilities.

## Effects of Ambient Air Pollution

Air pollution is known to have many adverse effects, including those on human health, building facades and other exposed materials, vegetation, agricultural crops, animals, aquatic and terrestrial ecosystems, and the climate of Earth as a whole. Some of the health effects have already been alluded to, but are considered here in greater detail. Effects on materials, vegetation, and so on—largely *economic impacts*—will also be considered here. The long-term effects on global climate are considered later in this chapter (as are the specific impacts of indoor air pollutants).

### Health Effects

Perhaps the most important effect of air pollution is the harm it causes human health. Generally, air pollution is most harmful to the very old and the very young. Many elderly people may already suffer from some form of heart or lung disease, and their weakened condition can make them very susceptible to additional harm from air pollution. The sensitive lungs of newborn infants are also susceptible to harm from dirty air. But it is not just the elderly or the very young who suffer; healthy people of all ages can be adversely affected by high levels of air pollutants.

Major health effects are categorized as being either *acute*, *chronic*, or *temporary*. An acute effect is short-lasting, but severe, and may even result in death. Chronic (or long-term) effects usually include respiratory illnesses such as bronchitis, emphysema, asthma, and perhaps lung cancer. Temporary effects include intermittent periods of eye or throat irritation, coughing, chest pain, malaise, and general discomfort. Intermittent *air pollution episodes* are temporary, but can be devastating. The Donora and London episodes are important examples, but many others have occurred in the past in many different countries, and the possibility of their recurrence cannot yet be ruled out. An episode typically lasts about 2 to 7 d, during which time illnesses and excess deaths occur in all age groups; but mostly the very old, the very young, and previously ill persons are affected.

It is difficult for public health experts to correlate or match up specific air pollutants with specific diseases with absolute certainty, but some general conclusions can be drawn from available data, particularly data obtained during air pollution episodes. Some of the factors that must be considered include *threshold*

*dosage*, *total body dose*, *exposure time*, and *synergistic effects*. A threshold level for a given pollutant is a minimum level below which there will be no health effects. A substance with no definitive threshold will have some detectable effect or response at any level of the pollutant. The *dose–response relationships* for most air pollutants, unfortunately, do not reveal clear-cut threshold values. The fact that people are exposed to some of the same pollutants in water and food, as well as the air, further complicates public health studies of air pollution effects because the total body dose is not just a function of airborne pollutants. Also, the health effects of many pollutants (for example, CO) are strongly related to the time of exposure, and, further, some pollutants work synergistically, meaning that the total health effect of two or more substances together is worse than any of them acting alone.

There is much evidence linking lung cancer to air pollution, although the actual cause-and-effect relationship is still unknown. Some of the facts that support the linkage are that lung cancer deaths are more frequent in urban areas compared to rural areas, carcinogens are typically found in polluted urban air, and the carcinogens cause cancer in animals tested in the laboratory.

Air pollutants enter the body by the respiratory system, through the throat, nasal cavities, and trachea, into the bronchial tubes and alveoli of the lungs. Gas transfer takes place in the small alveolar sacs, where the pollutants can then be absorbed into the blood. The lungs, a major target of air pollution, can be damaged by gaseous and particulate air pollutants.

Typical effects of sulfur dioxide, oxides of nitrogen, and ozone include eye and throat irritation, coughing, and chest pain. These pungent gases can harm lung tissue when inhaled and are associated with bronchitis, emphysema, and other lung diseases. Sulfur dioxide can constrict the bronchial tubes and adversely affect the cilia, the very small hairs that are part of the defense mechanism of the respiratory tract. Nitrogen dioxide is known to cause pulmonary edema, an accumulation of excessive fluids in the lungs. Ozone, a highly irritating gas, produces pulmonary congestion; symptoms of ozone exposure may include dry throat, headache, disorientation, and altered breathing patterns. Chronic bronchitis, which is manifested by a severe cough, is caused by excessive mucous secretions in the bronchial tubes; deaths from bronchitis have been correlated with high levels of air pollution. The same is true for emphysema, which is characterized by the breakdown of the alveoli, causing great difficulty in breathing. This disease is common in urban areas and is a leading cause of death, especially when coupled with the synergistic effects of tobacco smoke.

Carbon monoxide, a colorless and odorless gas virtually unnoticeable to the senses, is especially dangerous because it is inhaled without causing any irritation or discomfort. Carbon monoxide is acutely toxic because it readily combines with hemoglobin in blood, taking up the place normally occupied by oxygen (which the body needs continuously). The formation of carboxyhemoglobin reduces the ability of the blood to transfer oxygen to body cells, leading to asphyxiation or suffocation.

A carbon monoxide level of about 1000 ppm can cause unconsciousness in a healthy person after 1 h of exposure; death by asphyxiation will occur in about 4 h at that concentration. Even much lower levels can cause illness or reduced mental awareness. As a result, a maximum allowable 8-h exposure limit for workers in the United States has been set at 50 ppm. Under certain circumstances, particularly in the immediate vicinity of heavily congested highways, atmospheric carbon monoxide levels may reach 1-h peaks as high as 400 ppm.

The adverse health effects of inhaled particulate matter depend mostly on the size of the particles. Particles that penetrate deep into the lungs and reach the alveoli are especially harmful because they remain there for relatively long periods of time. Certain particulates are dangerous due to their toxic or carcinogenic properties; lead fumes and asbestos fibers are two such examples. Many carbonaceous particles are also suspected to be carcinogenic.

While particles smaller than 0.1 $\mu$m in size will readily reach the alveoli, particles larger than 1 $\mu$m in size are usually trapped by the protective mucous lining and cilia hairs in the nose and throat. A factor that affects the deposition of particles between 0.1 and 1 $\mu$m in size in the lungs is the number of breaths per minute; another factor is the total volume of air moved in and out of the lungs with each breath. Particles larger than 10 $\mu$m tend to settle quickly in the air; they are not considered to cause many health problems because those that are inhaled are usually removed in the upper respiratory tract and do not penetrate deep into the lungs. High levels of particulates, working synergistically with high levels of sulfur dioxide, have been linked to increased rates of respiratory infections and cardiac disorders. Sulfuric acid mist is especially irritating to mucous membranes; the tiny droplets of sulfuric acid are more than four times as harmful as dry sulfur dioxide gas. Lead particulates eventually interfere with the action of red blood cells, and beryllium dust causes pulmonary fibrosis, with symptoms of shortness of breath, coughing, and weight loss.

## Other Effects of Air Pollution

Air pollution causes significant damage to material objects, particularly in heavily polluted urban areas. This includes soiling and deterioration of building surfaces (or facades) and public monuments, corrosion of metals, and the weakening of textiles, leather, rubber, nylon, and other synthetic products.

Deposition or settling of particulates on exposed materials is one of the causes of soiling, and the frequent cleaning of these soiled surfaces may contribute to rapid deterioration. Abrasion, caused by particles carried in the wind at high speeds, can eventually erode and wear away hard, solid surfaces, even stone.

Sulfur dioxide, one of the pollutants that corrodes metals and weakens synthetic fibers (like nylon hose), also can chemically discolor painted surfaces. Sulfuric acid mist can cause serious damage to exposed marble, limestone, and mortar. The acid reacts with calcium carbonate in these building materials, forming water-soluble calcium sulfate, which is easily washed away by rain, leaving an eroded surface. Leather becoming brittle when exposed to sulfur dioxide pollution, as well as the cracking of rubber that is exposed to ozone, are additional examples of the direct and irreversible chemical attack of materials by air pollution.

The damage of materials by air pollution is more than just an esthetic problem; it is an *economic problem* of major proportions. Although this is not immediately apparent to the casual observer, the total cost of cleaning and repairing damage caused by air pollution is estimated to exceed $1 billion per year in the United States.

Air pollutants can damage trees, flowers, fruits, and vegetables in various ways. Some pollutants cause collapse of the leaf tissue; others may bleach or discolor the leaves. Ozone, in particular, causes damage to tree foliage and reduces the growth rate of sensitive tree species. Ozone alone accounts for approximately 90 percent of pollution damage to agricultural crops (for example, corn and wheat), amounting to between $2 and $3 billion per year in the United States. Certain air pollutants also harm cattle and other livestock, but this is usually a local problem on farms near specific industrial sites that cause the pollution.

To the general public, the most noticeable effect of air pollution is on the atmosphere itself. Specifically, this is the unsightly haze and reduction of visibility caused by the scattering of light. In the eastern region of the United States, most of the haze is due to suspended sulfate particles, but in the West, it is usually caused by dust, nitrogen oxides, or photochemical smog. Suspended particulates and haze can affect weather conditions by increasing the frequency of fog formation as well as rainfall. Perhaps not as noticeable, but of great significance in the long run, is the fact that an accumulation of suspended particles in the atmosphere can appreciably reduce the amount of solar energy that reaches Earth's surface. This, in effect, will increase Earth's reflectivity (albedo) and could lead to a decline in average global temperatures. At the present time, however, it seems that any increase in Earth's reflectivity is counterbalanced by a global phenomenon called the *greenhouse effect*.

## 13.4 GLOBAL AIR POLLUTION

Air pollution problems are not necessarily confined to a local or regional scale. Atmospheric circulation can transport certain pollutants far away from their point of origin, expanding air pollution to continental or global scales; it can truly be said that air quality problems know no international boundaries. Some air pollutants are known to be associated with changes in Earth's climate, requiring consideration of governmental actions to limit their impacts. Describing the pollution as *global* does not mean that the pollutant in question is necessarily dispersed over the entire world. It means that the pollution occurs at various places around the globe and is not unique to individual locations.

Scales of pollutant transport in the atmosphere can be described as follows:

- *Local:*       Up to a few kilometers from the source and mainly associated with plumes
- *Regional:*   Up to 1000 km from the source and associated with the merging of individual pollutant plumes
- *Continental:* Up to a few thousand kilometers from the source; an interchange of pollutants between the troposphere and stratosphere is possible at this scale
- *Global:*     More than a few thousand kilometers from the source and potentially throughout the entire atmosphere

At the global scale, identifiable pathways of pollutant transport are not necessarily discernible because of extensive atmospheric mixing. Two important air pollution problems that are generally considered worldwide in scope are *global warming* and *depletion of stratospheric ozone*. The environmental problems associated with *acid rain*, of major concern primarily on a regional and continental scale, are also considered here.

## Global Warming

Average land surface temperatures are increasing worldwide. In fact, the decade of the 1990s was the warmest ever recorded, and the trend of gradually rising average temperatures seems to be continuing. By some estimates, global mean temperature has risen roughly 0.5°C (1°F) since the end of the 19th century. This may seem to be an insignificant rise, given the wide variation in temperatures that occur on a daily and annual basis at any given location, as well as the obvious difficulty in measuring, collecting, and interpreting worldwide temperature records dating as far back as a century or more ago. But most atmospheric scientists think that even a small increase in average global temperature can have a noticeable impact on Earth's climate.

The fact that Earth is getting warmer is not really in dispute among scientists. There has been some controversy, though, about the causes of global warming, about how warm it will get, and about the effects of warming on climate patterns. Relatively few scientists believe that the amount of warming so far is still too small to be distinguished from natural causes. The prevalent view among most atmospherics researchers attributes the warming trend to an increase in concentrations of carbon dioxide and other trace gases that trap and retain heat from the sun. In fact, in June 2001, a report from the U.S. National Academy of Sciences reaffirmed the view that Earth's atmosphere is getting warmer, that human activity is largely responsible for the warming trend, and that the situation is worsening.

Carbon dioxide and other gases trap heat from the sun in a process that seems similar to what happens in a greenhouse. Consequently, the process of global warming is commonly called the *greenhouse effect*, although a direct comparison between the atmosphere and a glass-enclosed greenhouse is not strictly accurate. The reason for this is explained later in this section.

### Greenhouse Effect

To understand this theory of global warming, it is necessary first to distinguish between a *natural greenhouse effect* and an *anthropogenic greenhouse effect*. The natural greenhouse effect is a normal result of the presence of a blanket of air around Earth, while the anthropogenic greenhouse effect is considered to be a direct result of the *accumulation of trace gases in the air from human activities*. It is important to note that without the natural greenhouse effect, life in its present form would not be possible on Earth.

Since liquid water is necessary for life, the average temperature on Earth must be in the range where liquid water can exist, that is, between 0°C and 100°C. Earth is warm enough for liquid water (and life) to exist because of the blanket of air we call the atmosphere. If Earth had no atmosphere, the average temperature at the surface would be about −18°C, as it is on the airless moon. In fact, the temperature on the moon falls to about −150°C in the dark. But the average temperature of Earth, in the layer of air just above the surface, is about 15°C. In other words, the atmosphere and its natural greenhouse effect can be thought of as keeping Earth about 33°C warmer than it would otherwise be. The reason for this is that the atmosphere affects the amount of radiant heat energy from the sun that reaches the ground, as well as the amount of heat that is radiated by Earth back out into space. Here is how that happens.

All objects radiate heat energy, and the hotter the object, the more heat it radiates. The type of energy radiated, which is characterized by its wavelength, depends on the temperature of the object. Radiation from the surface of the sun, which has a temperature of about 1600°C, is mostly in the part of the spectrum we call *visible light*. Light passes through Earth's atmosphere without being completely absorbed. Overall, about 45 percent of incoming solar radiation reaches Earth's surface, about 15 percent is absorbed by the atmosphere, and about 40 percent is reflected back into space by clouds. Some light is also reflected at the surface, but most of it is absorbed by soil, rocks, water, and vegetation, warming the land and sea.

Part of the sun's radiant energy, with wavelengths shorter than visible light, is in the range of the spectrum called *ultraviolet light*. An even smaller part of the total energy from the sun is emitted at the other end of the spectrum in a band known as *infrared radiation*, with wavelengths longer than those of visible

light. Ultraviolet radiation is absorbed in the stratosphere (by oxygen and ozone molecules) and warms that layer of the air. But this is not related to the greenhouse effect. Infrared radiation, on the other hand, plays a key role in the greenhouse effect. However, it is the infrared energy reradiated by Earth, and not the infrared energy from the sun, that is involved.

The lower the temperature of an object, the longer the wavelengths of its radiant energy. Earth's surface, at a much lower temperature than the sun, radiates energy that is mostly in the infrared range of the spectrum. Unlike visible light, this infrared radiation is absorbed by the atmosphere, primarily by water vapor and carbon dioxide molecules. Because of this, the lower layer of the atmosphere, the troposphere, is warmed.

The warm air itself also radiates infrared heat, some of which goes back down toward the ground and keeps it warmer than it would otherwise be. This is essentially what is meant by the greenhouse effect. The rest of Earth's radiation goes upward through the atmosphere, where it is repeatedly absorbed and reradiated and eventually escapes into space, as shown in Figure 13.8.

The lowest part of the troposphere is the warmest, and temperature drops with increasing altitude, as shown in Figure 13.2. This cooling trend is reversed in the stratosphere because incoming ultraviolet radiation from the sun is absorbed by molecules of oxygen and ozone. Even though the surface of Earth is about 33°C

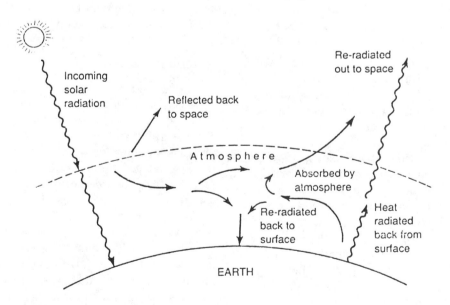

**FIGURE 13.8**
*The greenhouse effect. About 65 percent of the radiation reaching the ground is infrared energy reradiated from the atmosphere.* (From J. Gribbin, *Hothouse Earth,* Grove Widenfeld, New York, 1990. Used with permission)

warmer than it would be without the greenhouse effect, the average surface temperature will not keep rising indefinitely as long as the chemical composition of the atmosphere stays the same.

Overall, there must be an equilibrium between the amount of heat reaching Earth from the sun and the amount of heat that Earth reradiates back into space. Earth would eventually warm up if it radiated less heat than it received from the sun. But then a warmer Earth would radiate more heat, and the warming trend would stop when the outgoing heat was again in balance with the incoming heat.

Earth's radiation balance is expressed in terms of watts per square meter. A watt (W) is a unit for power, that is, energy per second. A typical household light bulb, for example, may have a power output of 150 W, mostly in the form of light. Worldwide, the annual average emission of long-wave (infrared) radiation from Earth's surface is about 390 $W/m^2$. But the radiation emitted into space from the top of the atmosphere is about 240 $W/m^2$, as measured by satellites. The difference, about 150 $W/m^2$, is the radiation energy absorbed by *greenhouse gases* in the atmosphere and reradiated back to Earth's surface. The greenhouse effect can be considered, then, to be equivalent to the energy emitted by a typical household light bulb shining over every square meter of Earth. This, along with the direct energy from the sun, is enough to keep the average surface temperature at 15°C (59°F), rather than the −18°C (−0.4°F) it would be with no atmosphere.

For a given composition of the atmosphere, there is a balance point at which the average temperature of Earth's surface remains constant (as long as the sun's radiation does not change). But a change in the composition of the atmosphere (that is, more greenhouse gases) would cause Earth's surface to get warmer until a new balance point was reached.

The natural equilibrium that existed a couple of hundred years ago, prior to the Industrial Revolution, has noticeably changed. This is largely due to the burning of fossil fuels and an increase in the amount of carbon dioxide in the atmosphere, an *anthropogenic greenhouse effect*.

It was mentioned earlier that the use of the term *greenhouse effect* is not strictly correct in describing the phenomenon of global warming. This is because a glass-pane enclosure prevents *convective heat loss* as well as infrared radiation loss, raising temperatures inside a real greenhouse much higher than is possible by the effect of Earth's atmosphere alone. (Convection refers to transfer of heat by the motion of a mass of fluid from one region of space to another.) The atmosphere also allows certain wavelengths of infrared radiation to escape into space through what is called the *atmospheric window*. Nevertheless, the term greenhouse effect is widely used by the media and recognized by the public as being the cause of global warming.

## Greenhouse Gases

Nitrogen and oxygen, the main constituents of the atmosphere, play no part in the greenhouse effect. But there are approximately 35 *trace gases* that scientists believe contribute to global warming. Carbon dioxide ($CO_2$) is considered to be one of the most important of these *greenhouse gases*, absorbing most of the heat trapped by the atmosphere.

Carbon dioxide is not considered to be an air pollutant in the usual sense of the word. In fact, it is a normal trace (very small) component of the atmosphere. Now at an average concentration of about 350 ppm (0.035 percent), it does not cause any adverse effects on human health. But since the Industrial Revolution in the mid-19th century, an ever-increasing amount of carbon dioxide has been emitted into the air from the burning of fossil fuels (oil, gas, and coal) for energy. About 100 years ago, carbon dioxide levels were roughly 270 ppm, as determined by analysis of air bubbles in ice cores taken from Greenland and Antarctica. And average carbon dioxide concentrations appear to have been increasing at a rate of almost 2 percent per year since the late 1950s.

About 90 percent of anthropogenic carbon dioxide emissions occur in the northern hemisphere. The main source of carbon dioxide is the burning of fossil fuels, which provides more than 80 percent of human energy needs. Oil accounts for about one half of the fossil fuels used for energy, coal for about one third, and the remainder is from natural gas. It is also thought that the destruction of forests contributes to increasing carbon dioxide levels, but the effect of deforestation is much smaller than that from the burning of fossil fuels.

Other gases of special importance in global warming are chlorofluorocarbons (CFCs), methane, nitrous oxide, and ozone. Although the average concentrations of these gases are much lower than that of carbon dioxide, they are much more efficient than carbon dioxide at soaking up long-wave radiation. Overall, carbon dioxide is estimated to cause almost 60 percent of the warming effect and CFCs about 25 percent, and the remainder is caused by methane, nitrous oxide, ozone, and other trace gases.

CFCs are synthetic chemicals that enter the atmosphere solely as a result of human activities. They are also of concern because of their impact on stratospheric ozone depletion and are discussed in more detail later. Methane and nitrous oxide, naturally occurring gases, are increasing to some degree due to human actions.

Each of these trace gases has a very long residence time in the atmosphere, so their effects are long-lasting.

Natural sources of nitrous oxide include nitrogen-based fertilizer and manure from farm animals. Ironically, recent studies show nitrous oxide is also emitted into the air from catalytic converters, the devices installed on automobile tailpipes to reduce smog pollution. Nitrous oxide accounts for slightly more than 7 percent of greenhouse gases, and nearly half of that comes from catalytic converters. Although ambient nitrous oxide levels are much lower than carbon dioxide levels, nitrous oxide is about 300 times more potent than carbon dioxide with regard to its ability to absorb heat energy and warm up the atmosphere. So its role as a greenhouse gas is significant.

Nitrous oxide ($N_2O$) has not been regulated as a criteria pollutant because the *Clean Air Act* was originally implemented to control smog and other types of air pollution, not global warming. Although it technically may be possible to reduce $N_2O$ emissions by redesigning the catalytic converters, gradually moving from gasoline-powered cars to electric cars could provide a more comprehensive solution to the problem of pollution from auto emissions.

## Potential Impacts of Global Warming

One of the methods that scientists used to estimate the impacts of global warming involves computer analysis of mathematical equations that model Earth's atmosphere. Typically, these sophisticated computer programs are called *general circulation models* (GCMs). As a basis for predicting future global impacts, most models assume that the concentration of greenhouse gases will effectively double. On this basis, the GCMs generally predict an average global warming of up to 4.2°C (7.5°F) and an overall increase in precipitation of about 10 percent by the year 2050. It is also expected that global warming will create a more active hydrologic cycle, increasing cloudiness as well as precipitation.

Not all scientists agree that human activities have been affecting global temperatures. Despite controversy about the anthropogenic greenhouse gas theory and the accuracy of GCMs, atmospheric scientists are able to establish scenarios of general effects that might occur if average global temperatures continue to rise. Two important potential impacts of global warming include the effects on sea levels and on ecosystems.

Recent estimates suggest that global sea level has risen by about 0.15 m during the 20th century, with most of the rise occurring since 1930. Some scientists believe that, because of greenhouse warming, average sea levels may rise by at least 0.3 m and as much as 1.4 m by the year 2030. This is likely to cause extensive economic and social hardship in coastal areas all over the world.

Sea levels may rise for several reasons, including thermal expansion of seawater, melting of glaciers and polar ice sheets, changes in patterns of precipitation and runoff, and changes in evaporation rates into a warmer atmosphere. Rising sea levels will increase the danger of flooding for low-lying coastal areas, which include many cities. Higher sea levels may also increase beach erosion problems and change the shape of the coastline. Saltwater intrusion into coastal wetlands or groundwater aquifers will also be likely to occur more often.

Potential impacts of global warming on ecosystems mainly include the effects on agriculture and forest growth. Plant growth and development will be influenced by an increase in carbon dioxide levels, which stimulates photosynthesis and decreases water losses from transpiration. But any benefit from increasing carbon dioxide levels may be either enhanced or offset by the changing patterns of climate associated with the effect of global warming. Changes in precipitation and climate may present a series of new stresses to agricultural and forest ecosystems, such as pests, disease, drought, or flood.

In addition to affecting agriculture and forests, greenhouse warming is expected to have other impacts. For example, higher temperatures and humidity may increase the chances of disease in humans and animals in some parts of the world. The physical effects of warming may require adjustments in flood plain designations, public water supplies, building codes, engineering and architectural projects, and other important areas that affect human lifestyles.

## International Control Efforts

In 1990, atmospheric scientists meeting in Geneva recommended that all countries take immediate steps to reduce emissions of greenhouse gases. They reported that many industrialized nations could cut carbon dioxide emissions by 20 percent by the year 2005, using existing technology, without causing significant economic hardship. At the meeting, all 12 of the European nations agreed to set targets for the reduction of carbon dioxide emissions, but the United States (responsible for more than 20 percent of the world's carbon dioxide output), Russia, and oil-producing countries such as Saudi Arabia felt that more research about global warming was needed before major policy decisions could be made.

In 1992, at the Earth Summit in Rio de Janeiro, a treaty was signed that obligated industrialized countries, including the former Soviet bloc, to stabilize emissions of greenhouse gases at 1990 levels by the year

2000. At a 1995 global warming conference held in Berlin, 120 countries agreed to negotiate binding timetables for actual reductions in emissions after the year 2000. This accord, called the *Berlin Mandate*, also calls for explicit measures to reduce greenhouse gases from developing countries.

At an international conference held in 1997 in Kyoto, Japan, agreement was reached on more specifics for reducing greenhouse gases. A key feature of the agreement, called the *Kyoto Protocol*, is the set of binding emissions targets and timetables for developed nations. Specific targets vary from country to country, though those for the major industrial powers of Europe, Japan, and the United States are very similar. The target for the United States, for example, is to cut emissions to a level 7 percent below 1990's levels. Emissions targets for the major greenhouse gases are to be reached over 5-year budget periods, helping to smooth out short-term fluctuations in economic performance or weather, either of which could spike pollutant emissions in a particular year.

The first budget period will be 2008 through 2012. Having a full decade before the start of the binding period will allow more time for companies to make the transition to greater energy efficiency and improved emission control technologies. Activities that absorb carbon, such as planting trees, will be taken into consideration in meeting the targeted emission reductions. Accounting for the role of forests in absorbing carbon dioxide is considered critical to a comprehensive and environmentally responsible approach to global warming and climate change. It also provides the private sector with low-cost opportunities to reduce emissions. International emissions trading is another important feature of the Kyoto Protocol. With emissions trading, countries can purchase emissions permits from those that have more permits than they need (because they have met their targets with room to spare).

Implementation of the Kyoto Protocol requires formal ratification by all the nations involved. Since implementation will have economic consequences, some nations are more reluctant to participate fully than others. In fact, in the spring of 2001, the United States suddenly rejected the Kyoto accord, largely because the plan set no standards for two other major emitters of greenhouse gases (China and India). Securing meaningful participation in any plan for emission reduction from developing countries (like the aforementioned) remains a top priority for the United States before ratification will occur. One thing is clear, though: All nations now recognize the importance of acting to reduce emissions and to slow the rate of global warming as soon as possible. It may be expected that a revised version of the Kyoto accord will be agreed upon in the near future.

### Innovative Technology for Controlling Global Warming

Presently, there are only two approaches to controlling anthropogenic global warming. One is to develop alternative energy sources (like nuclear and solar power), which do not produce $CO_2$ the major greenhouse gas. The other is to cut back on energy use through conservation efforts and use of more efficient automobiles and appliances. In the United States, scientists have recently begun to explore a third strategy called *carbon sequestration*, in which $CO_2$ would be removed from power plant emissions by scrubbers and then stored in compressed liquid form under land or under ocean waters. This new technology may be useful in the future, when the cost of capturing $CO_2$ in scrubbers decreases significantly, and when the environmental challenges related to its storage are resolved.

## Depletion of Stratospheric Ozone

Ozone ($O_3$) plays an important role with regard to atmospheric chemistry in both the troposphere and the stratosphere. At ground level, it is a pollutant, but stratospheric ozone (at altitudes between 12 and 30 km) is crucial for life on Earth. The stratospheric ozone layer blocks most of the harmful ultraviolet (UV) rays coming from the sun, thus protecting plants and animals.

Since 1985, when atmospheric scientists made a dramatic announcement about the discovery of a large hole in the ozone layer over Antarctica, public awareness and concern about the situation have grown significantly throughout the world. The *ozone hole*, which is actually an ozone-depleted region that seems to occur each year from about August to November, encompasses an area that is roughly the same size as the entire Antarctic continent. Although the hole later closes, the brief, but severe depletion of *ozone* has an effect that is eventually distributed around the world. Even if the major ozone losses over Antarctica are not included, however, measurements have shown a slow, but steady decrease (roughly 1 percent per year) in total global ozone levels since the late 1970s. Atmospheric ozone levels are routinely measured by spectrographic instruments on the ground and by satellites.

Ultraviolet energy from the sun is absorbed by ozone molecules before it can reach Earth's surface. When ozone concentrations decline, more UV radiation reaches the ground. In effect, ozone serves as a protective shield from UV radiation. This is important because increased exposure to UV radiation is linked to human skin cancer, eye cataracts, and suppression of immune system response. Many species of plant and aquatic organisms can also be harmed by increased

UV exposure. Furthermore, UV radiation reacts with formaldehyde in the troposphere, adding to the formation of photochemical smog.

Ozone is basically an unstable molecule, and there is a delicate balance between its formation and removal in the stratosphere. It is continuously being formed as short-wavelength UV energy is absorbed, and at the same time it is being converted back to molecular oxygen by a wide variety of photochemical reactions. This equilibrium is affected by the presence of certain substances, particularly chlorine, that speed up the rate of ozone removal.

One of the most significant causes of ozone depletion is attributed to the presence in the atmosphere of organic chemicals called *chlorofluorocarbons* (CFCs). Sources of CFC gases include aerosol spray cans, refrigerants, industrial solvents, and foam insulation. CFCs are not biodegradable. They remain in the troposphere for a very long time before eventually drifting up into the stratosphere, where they are broken down by UV radiation. This reaction releases chlorine, which is then available to act as a catalyst in speeding up the destruction of ozone molecules. In a complex series of photochemical reactions, one chlorine molecule can destroy several hundred thousand ozone molecules before it eventually leaves the stratosphere.

Several theories have been suggested by scientists as to why there is such a sharp seasonal depletion of ozone over Antarctica. Although the exact nature of this phenomenon is still being investigated, scientists generally believe that CFCs from sources in the northern hemisphere are transported to Antarctica by stratospheric circulation. Microscopic ice crystals are also thought to provide reaction sites at which chlorine from the CFCs can interact with ozone. A technique using laser radar (called *lidar*) is being used by scientists to study how stratospheric ice crystals change in size and shape as the ozone hole is created. As a lidar beam hits air molecules, trace chemicals, and ice particles, a part of the beam is scattered back to an array of detectors. Lidar beams reach up to 100 km (60 mi) and can create three-dimensional maps depicting the chemical composition and physical characteristics of the atmosphere throughout the year.

In the mid-1970s, when CFCs were first recognized as a serious threat to the ozone layer, more than 200,000 tons of CFC were being used annually in the United States alone as a propellant in aerosol cans. In 1979, under the *Toxic Substances and Control Act*, the EPA banned the use of CFCs for most aerosol propellant uses, and several other countries adopted similar CFC restrictions. The subsequent drop in aerosol applications has been offset by increases in other uses, however, including the manufacture of plastic foam insulating materials.

In 1986, at an international meeting in Montreal, Canada, representatives from more than 35 nations signed the *Montreal Protocol*, which called for a 50 percent reduction in CFC use by the year 1998. In the United States, production of CFCs ended in 1995, but the chemicals could still be used in existing equipment. Some people believe that a complete elimination of all ozone-depleting substances should be set as a goal. But a total ban of CFC use would interfere with lifestyles in many developed countries. Refrigeration has become almost a necessity, and luxuries such as air conditioning in buildings and cars are considered the norm. A total ban on CFCs can also interfere with the attempts of developing countries to modernize industrial facilities and raise living standards. For this reason, the governments of many developing countries are resisting full implementation of international controls unless there is financial assistance to compensate for the lack of CFCs.

Environmentally safe alternatives to CFCs for refrigeration and thermal insulation are available and should be used, or major changes in lifestyles may be necessary. A delay in limiting CFC use now can potentially have serious consequences in the future, and many atmospheric scientists believe that a significant reduction in CFC consumption is needed to prevent continued destruction of the ozone layer.

## Acid Deposition

Since the early 1970s, problems associated with acidic precipitation have gained worldwide attention. *Acid rain*, as it is also called, is believed to have damaged or destroyed fish and plant life in thousands of lakes throughout central and northern Europe (especially in Scandinavia), the northeast United States, southeast Canada, and parts of China. Many species of trees in forests throughout these regions have been in decline, largely due to soil acidification. Acid rain also causes pitting and corrosion of metals and the deterioration of painted surfaces, concrete, limestone, and marble in buildings, monuments, works of art, and other exposed objects.

The acidity of a liquid is measured on a logarithmic *pH scale* that ranges from 0 to 14. A pH of 7 is neutral, higher pHs are basic or alkaline, and lower pHs are acidic (see Section 4.3). Rainwater is naturally acidic, even in regions far removed from human activity, because atmospheric carbon dioxide ($CO_2$) reacts with water vapor ($H_2O$) to form carbonic acid ($H_2CO_3$). As a result, a pH of about 5.6 is usually considered to represent background rainwater acidity in a pristine environment. Acid rain, then, can be defined as rainwater with pH less than 5.6 that results from air pollution caused by human activities.

Recent scientific studies show that in certain urban and industrial areas of the United States and in other countries the average pH of rainwater is less than 4.5. In parts of Southern California, acid fogs sometimes have a pH of less than 3.0, and a pH of 2.2 (as acidic as vinegar) was reported during a rainfall in Scotland in 1974. Since the pH scale is logarithmic, a drop of one pH unit represents a tenfold increase in acidity, a drop of two pH units represents a hundredfold increase, and so on. Rainwater with a pH of 3.6, for example, is 100 times more acidic than normal or "clean" rainwater.

Acid rain is caused by the emission of sulfur and nitrogen oxides into the atmosphere, mostly from the burning of fossil fuels for electric power. Other sources from human activities include certain industrial processes and the gasoline-powered automobile. Sulfur dioxide reacts with water vapor in the air to form sulfuric acid; nitrogen dioxide reacts with water vapor to form nitric acid. It has been found, though, that the contribution of sulfur dioxide to acid rainfall is more than twice that from nitrogen oxides. Contributions of these gases from natural sources, such as from swamps and volcanoes, are small in comparison to human sources.

Atmospheric mists of sulfuric acid and nitric acid eventually reach the surface of Earth in the form of rainfall, snow, fog, or dew. Some sulfur oxides also can exist as tiny particulates and settle out of air in a dry form. The term *acid deposition* is used to describe the overall effect of both wet and dry precipitation at ground level when acidic materials have reached the surface environment. It is the long-term accumulation of acid deposition, rather than individual rainfall events, that has a significant effect on the environment. Acid precipitation in the form of snow (where pH is measured from melted snow water) is also a concern in the higher elevations or latitudes, particularly when the spring melt creates a sudden low-pH impulse to surface water supplies.

A major environmental impact of acid deposition is the lowering of pH in lakes and rivers. In New York State, for example, the pH of lakes in the Adirondacks averaged about 6.7 in the 1930s, but has since decreased to about 5.1. Most aquatic life is disrupted as the pH drops. Phytoplankton populations are reduced, and many common water-dwelling invertebrates, such as mayflies and stoneflies, cannot survive when the pH falls below about 6.0. Some sensitive species of fish, including trout and salmon, are harmed when pH levels fall below 5.5. Acidity has a deleterious effect on the reproductive cycle of fish; when the pH is less than 4.0, reproduction of most fish species is unlikely. *Acid dead* lakes have pHs below about 3.5.

The pH of lakes receiving acid deposition is greatly influenced by the presence of *buffers* in the water. Buffers are capable of neutralizing acids and maintaining a stable pH level, despite the addition of acidic materials to the liquid. Calcium carbonate, for example, is a natural buffer released from limestone. Lakes, streams, and ponds underlain by limestone rocks or soil deposits usually have a large buffering capacity for acid rain due to the dissolved calcium carbonate in the water. But the buffering action in many lakes can eventually be exhausted due to the reaction of the acids with the calcium carbonate. For some lakes with little natural buffering capacity, the effects of acid deposition can be lessened by adding limestone to the water.

There is no direct link between acid deposition and public health, but groundwater with pH less than 4.5, which can result from acid deposition, may corrode pipes and cause leaching of heavy metals into drinking water. Accumulation of toxic metals (for example, mercury and cadmium) in the food chain may also be caused by acid deposition.

A factor that complicates the acid rain problem and makes finding a solution difficult is its regional and continental scale. Most oxides of sulfur and nitrogen are emitted from tall smokestacks at power plants in order to increase the dispersion and dilution of the stack gases. This may protect nearby communities from the immediate effects of air pollution, but discharge from tall chimneys allows the pollutants to be carried for long distances in the atmosphere. The pollution is, in effect, "air-mailed" to other regions and even other continents.

Most acid deposition in the northeastern region of the United States is believed to be the result of fossil fuel combustion by power plants and industries located in the midwestern section of the nation. In fact, in the 1980s about 16 million tons of sulfur emissions each year came from the Midwest. It has also been estimated that 50 percent of the acid rain in eastern Canada comes from the United States, and about 25 percent of the acid rain in the New England region of the United States originates from Canadian sources. In addition, acid rain in Norway is believed to come mostly from industrial areas in Great Britain and continental Europe.

The transport of pollutants across political boundaries is a significant issue. In 1984 in the United States, several northeastern states petitioned the EPA to order the reduction of emissions from coal-burning power plants in the Midwest. The request was denied by the EPA at that time on the basis that the existing requirements of the *Clean Air Act* were not being violated. In 1988, Canada decided to cut acid rain pollutants in half and requested that the United States do the same.

The *Clean Air Act* of 1990, however, addressed this issue by setting an annual limit on sulfur dioxide emissions from power plants at 8.9 million tons by the year 2000 and by requiring significant reductions of nitrogen oxide emissions as well. Research projects and economic incentives for industry to reduce pollution were also initiated. Economic incentives include allowing the sale of *emission credits*. When a polluter reduces its emissions below a specified level, it can sell the difference as a credit to some other company or power plant that is planning to expand. The credit can help to offset some of that company's cost of installing new pollution control equipment. Research on mitigating the acid rain problem includes a $5 billion program (begun in 1986 as a joint Canadian and United States effort) to develop cleaner coal-burning technology.

## 13.5 INDOOR AIR QUALITY

Up until the mid-1980s, the focus of air pollution control technology was on the outdoor environment. At about that time, health risks related to indoor air pollution became a concern. After all, most people spend the majority of their time (up to 90 percent) indoors, at home and at work. This fact, along with increased energy conservation measures (following in the wake of the oil shortages of the 1970s), brought indoor air pollution to the attention of environmental scientists and public health officials. Energy conservation measures tend to lower air exchange rates, allowing air contaminants to accumulate. Consequently, indoor air can be more polluted than outdoor air.

Although there are no federal or state regulations specifically controlling indoor air quality, the EPA has developed a comprehensive program that focuses on the indoor air quality (IAQ) problem. The EPA is working to implement this program using nonregulatory as well as regulatory tools available under several existing federal laws. A high priority has been set on improving the way buildings are designed and operated; exposure to air pollutants can be significantly reduced by applying sound construction and building maintenance practices. In addition, much attention is being given to the identification of specific indoor pollutants and the means to reduce their levels indoors. Adequate ventilation is essential to achieving and maintaining good IAQ levels, but some architects and engineers think that limiting pollutant sources is far more effective.

### Air Exchange

Outdoor air leaves or enters a house in three ways: *infiltration, natural ventilation,* and *forced ventilation.*

Infiltration refers to air exchange that occurs when windows and doors are closed; air leaks through cracks, joints, and holes in exterior walls and the roof of a building. Typically, infiltration rates in U.S. houses range from 0.5 ach (*air changes per hour*) to 4 ach; newer buildings have rates closer to the lower end of the range, while older buildings have rates closer to the upper end. A rate of 1 ach means that, on average, the air in the entire house is exchanged with outdoor air in 1 h.

Large amounts of energy are lost due to infiltration exchange of heated or cooled indoor air with outside air. Infiltration rates can be minimized or tightened by use of proper construction and insulation techniques, and this has become a means of energy savings in new home construction. Lower infiltration rates, however, tend to worsen problems related to IAQ.

In natural ventilation, air circulates through opened doors and windows. Both natural ventilation and infiltration are caused by indoor and outdoor air temperature differences, as well as by wind. Natural ventilation rates are higher than infiltration rates under the same temperature and wind conditions. Forced ventilation, using fans or blowers, can provide the highest rates of air exchange. Forced (mechanical) ventilation can be either intermittent, removing air from a single room, such as a kitchen, or continuous, removing indoor air and distributing filtered and conditioned outdoor air throughout the house or building.

Infiltration and ventilation dilute and remove indoor air pollutants. When there is not enough infiltration, natural ventilation, or forced ventilation, air exchange rates are low and pollutant levels tend to increase. Overall, the most effective strategy for maintaining good, energy-efficient IAQ is to reduce sources of pollutants as much as is feasible and use ventilation to limit pollutant concentrations to acceptable levels. In some cases, the use of air cleaners can be effective for the removal of particulate pollutants.

### Indoor Pollutants

There are many sources of indoor air contaminants. Combustion products, especially tobacco smoke, are particularly important. Radon, continuously produced from the radioactive decay of naturally occurring radium found in soil and rock, is also a significant indoor air contaminant. Other indoor air pollutants are asbestos, organic chemicals from household products, formaldehyde, lead, and biological substances.

#### Combustion Products

Environmental tobacco smoke (ETS) is a self-evident air pollutant from combustion that many people produce knowingly in the home, despite the warnings of adverse health effects. ETS is a mixture of more than 4000 compounds, at least 40 of which are carcinogenic and many of which are strong irritants. Indoors, tobacco smoke can harm occupants as well as the smoker; ETS is also referred to as *secondhand* or *sidestream* smoke, and exposure to it is often called *passive smoking*. The EPA in 1992 concluded that exposure to ETS is responsible for about 3000 lung cancer deaths each year in nonsmoking adults. Long-term exposure to secondhand smoke combined with radon is considered to be especially dangerous.

Short-term health effects of exposure to ETS include eye, nose, and throat irritation. Infants and young children, particularly asthmatic children, are especially at risk of experiencing the harmful effects of ETS. Clearly, the most effective way to avoid exposure is to eliminate smoking in the home. If smoking indoors cannot be avoided, increased ventilation rates by open windows or exhaust fans will reduce, but not eliminate exposure to ETS.

In addition to secondhand smoke, other sources of indoor combustion products include stoves, space heaters, fireplaces, and chimneys. Pollutants from these sources include carbon monoxide, nitrogen dioxide, and particulates; these are also major outdoor air contaminants (see Section 13.3). Kerosene space heaters may also emit sulfur dioxide and acid aerosols. In enclosed environments where there is no tobacco smoke, the level of suspended particulates (smaller than 15 $\mu$m) may be as much as 40 $\mu$g/m$^3$. With ETS, particulate levels may approach 700 $\mu$g/m$^3$, which greatly exceeds EPA standards. Steps to reduce exposure to indoor combustion products include using exhaust fans over gas cooking stoves, inspecting furnaces, flues, and chimneys annually, and using properly sized wood stoves that are certified as meeting EPA emission standards.

### Radon

Radon is a colorless, odorless, radioactive gas that is part of a natural decay process beginning with uranium and radium. It is found in varying concentrations in soils and rocks that make up Earth's crust. Because it is a gas, radon flows easily through porous soil and fissures in rock. When it reaches the ground surface, the gas disperses and is diluted to very low concentrations in the outdoor environment. But when radon enters a building through openings, cracks, or joints in the basement floor or walls, the gas can accumulate. Indoor radon concentrations may build up to very high levels unless appropriate steps are taken to prevent its entry or accumulation. Radon may also be found in

groundwater in some regions; it can be released into household air as water is used for routine domestic purposes. Major radon entry routes into a house are shown in Figure 13.9.

Radon is chemically inert, but it breaks down into chemically active decay products (*radon progeny*) that easily become attached to particles in the air. The health risk associated with radon, lung cancer, is considered to be caused mostly by inhalation of particulate matter with attached radon progeny. Further decay of radon progeny releases radiation that damages lung tissue. The degree of risk depends on total exposure time and average radon concentrations in the home. Inhalation of a mixture of both radon and tobacco smoke is thought to increase health risks significantly. The EPA estimates that between 5000 and 20,000 lung cancer deaths per year in the United States can be attributed to household radon exposure.

Two common devices used to measure radon concentrations are the *charcoal canister* and the *alpha-track detector*. The charcoal canister provides quicker results (3 to 7 d) than the alpha-track detector (30 to 90 d) and is more widely used. Radon concentrations are expressed in units of pCi/L (picocuries per liter). A *curie* (Ci) is a basic radiation unit corresponding

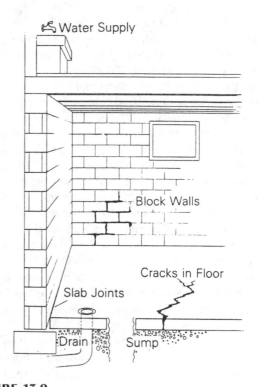

**FIGURE 13.9**
*Common radon entry points.* (Courtesy of the Environmental Protection Agency)

roughly to the decay rate of 1 g of radium. One *pico-curie*, a much smaller unit, corresponds to the disintegration of about two atoms per minute (pico $= 10^{-12}$). Radon concentrations may also be expressed in terms of *working level* (WL), where 1 WL $= 100$ pCi/L. Working levels are associated with historical lung cancer data of underground miners, from which much of the current information regarding household radon effects is extrapolated.

Remediation methods are available to reduce indoor radon exposure. The EPA has set a recommended level for remedial action at 4 pCi/L; some other countries, such as Canada and Finland, do not recommend remedial action unless radon levels exceed 20 pCi/L. All scientists agree that prolonged exposure to high levels of radon is hazardous, but uncertainty remains about the risks of low-level exposure. No relationship between lung cancer and radon in homes has yet been definitively established in a scientific study.

With regard to radon in groundwater, the problem is not in drinking the water, but in inhaling the released gas. A level of 10,000 pCi/L of radon in water will produce about 1 pCi/L of radon in indoor air. Under the *Safe Drinking Water Act*, the maximum contaminant level for radon in potable water is 300 pCi/L. Radon can readily be removed from water by aeration or by activated charcoal filtration.

Two general approaches for reducing indoor air radon levels are *reducing radon entry* into the house and *removing radon* after it has entered the house. Radon entry can be reduced by several methods. One is to cover exposed earth and seal all pores, cracks, joints, and other openings in the crawl space or basement of the structure. Another method is to install a subslab and perimeter soil ventilation system. This system reverses the predominant direction of airflow, which normally brings the radon into the house. A subslab suction system is shown schematically in Figure 13.10.

Other methods of reducing radon entry include avoiding the use of building materials that may contain radium and removing radon from potable water supplies with the use of aeration or activated carbon filtration. An effective method for removing radon after it has entered the house is ventilation of the affected living space. Ventilation with outside air can reduce indoor radon levels substantially. If radon levels are not too high, natural ventilation can be sufficient to eliminate radon pollution.

Forced ventilation relies on the use of fans to increase house air exchange rates by blowing in outside air. Skillful design, installation, and testing of a forced ventilation system are necessary to assure that air entry and exhaust points are properly balanced. Otherwise, additional radon can be drawn in. Application of ventila-

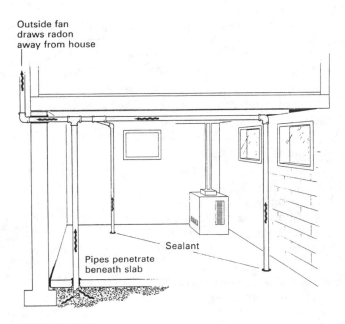

Outside fan
draws radon
away from house

Sealant

Pipes penetrate
beneath slab

**FIGURE 13.10**
*Subslab ventilation.* (Courtesy of the Environmental Protection Agency)

tion, whether natural or forced, is limited by the increased energy costs for heating or cooling needed to maintain comfortable conditions in the home. This problem can be mitigated by use of *heat-recovery ventilation*, as shown in Figure 13.11. This kind of system, also called an air-to-air heat exchanger, uses heat in the exhaust air to warm the incoming air; the process can be reversed in an air-conditioned house in warm weather. Heat-recovery ventilation can reduce energy loss significantly.

## Asbestos

Asbestos is a mineral fiber that has been used as insulation and as a fire-retardant in buildings. Many asbestos products have been banned, and its use is now limited. In older buildings asbestos is still found in pipe and furnace insulation, asbestos shingles, floor tiles, textured paints, and other construction materials. If these materials are disturbed by cutting, sanding, or other activity, excessive airborne asbestos levels can occur. Improper attempts to remove these materials can also release asbestos fibers into indoor air. As a guideline, average asbestos levels should not exceed 0.1 fibers/mL for fibers longer than 5 $\mu$m.

If very small asbestos fibers are inhaled, they accumulate in the lungs. Adverse health effects, which can take many years to develop, include asbestosis (lung scarring), mesothelioma (cancer of the chest and abdominal linings), and lung cancer. Most asbestos-related illness is associated with occupational exposure

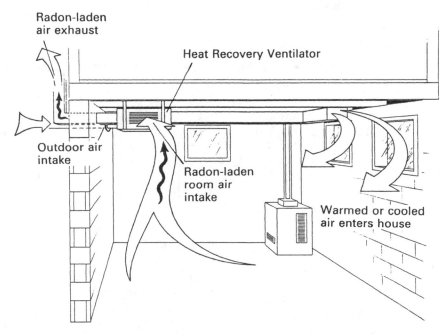

**FIGURE 13.11**
*Heat-recovery ventilation.* (Courtesy of the Environmental Protection Agency)

to high concentrations of fibers, but elevated levels can also occur in homes.

Since there is no danger from asbestos unless the fibers are airborne, it is advisable to leave undamaged asbestos materials alone if they are not likely to be disturbed. If it is necessary to remove or clean up damaged asbestos materials, the work should be done by qualified contractors. Sealing off the materials in lieu of removing them is sometimes an option.

### Organics from Household Products

Organic chemicals are widely used as ingredients in paints, varnishes, waxes, cleaning agents, cosmetics, pesticides, hobby materials, and other products used in the home. They all can release organic compounds into the air when they are used and, to some degree, when they are stored. Tetrachloroethylene, methylene chloride, and paradichlorobenzene are just a few of the multitude of potentially harmful organic chemicals found in household products. Health effects vary greatly, depending on the level of exposure and length of time exposed. Some of these chemicals are highly toxic or carcinogenic; others may have no known health effect. Eye and respiratory tract irritation, headaches, dizziness, visual problems, and memory impairment are some of the immediate symptoms experienced after exposure to some organics.

Household use of pesticides is a notable source of organics in indoor air. Pesticides commonly used in and around the home include products to control insects, termites, rodents, and fungi. They are sold as sprays, liquids, powders, and crystals. These products are all dangerous if not used properly. The EPA registers pesticides and requires manufacturers to put information on the label regarding proper use. The sale of chlordane, aldrin, dieldrin, and heptachlor for home or commercial use is no longer permitted.

### Formaldehyde

Formaldehyde is an organic chemical widely used in the manufacture of many building materials and household products. It is also a by-product of combustion and may be present in significant amounts in indoor air. It is used, for example, to add permanent-press qualities to drapes and other textiles, as a component of glues and adhesives, and as a preservative in some paints and coatings. Among the most significant household sources of formaldehyde are pressed wood products that are made using adhesives containing urea formaldehyde (UF) resins. These include particleboard and plywood paneling.

Formaldehyde is a colorless, pungent-smelling gas that can cause eye and throat irritation, nausea, and respiratory distress in some people exposed to high concentrations. It has also been shown to cause cancer in animals and may cause cancer in humans. Average concentrations in older homes (emissions generally decrease over time) are generally below 0.1 mg/m$^3$; in homes with significant amounts of new pressed wood products, levels can exceed $0.3 \, mg/m^3$. A suggested guideline is for levels not to exceed $0.12 \, mg/m^3$.

Formaldehyde emissions increase with temperature and humidity. Thus, the use of dehumidifiers and air-conditioning can help reduce indoor concentrations. Increased ventilation rates also reduce formaldehyde levels in the home. Use of exterior-grade pressed wood products in the home helps to minimize formaldehyde emissions because they are made with phenol-formaldehyde resins rather than UF resins.

### Lead

Lead is a harmful environmental pollutant. People can be exposed to lead in drinking water and food, as well to lead dust in the air. The most significant source of lead dust, old lead-based paint, is a particular threat to the health of children. Lead can become airborne in the home when lead-based paint is improperly removed from surfaces by scraping, sanding, or open-flame burning. In 1978, residential use of lead-based paint was banned by the Consumer Product Safety Commission. Old lead-based paint in good condition is generally not a problem and can be left undisturbed. Lead dust from contaminated soil tracked into the house can also cause high levels of airborne lead.

At high levels, lead can cause convulsions, coma, and even death. Lower levels can adversely affect the brain, central nervous system, blood cells, and kidneys. Infants and children are especially vulnerable because lead is more easily absorbed into their bodies. Harmful effects of lead in infants and children include delays in physical and mental development.

### Biological Contaminants

Airborne contaminants of a biological nature include bacteria, fungi, viruses, animal dander, house dust mites, pollen, and other tiny forms or products of life. There are many sources of these biological pollutants. Contaminated central heating or cooling systems can become breeding grounds and then distribute these contaminants throughout the home. Standing water, water-damaged materials, or wet surfaces can also serve as breeding grounds.

Allergic reactions, such as asthma and rhinitis, can be triggered by many biological contaminants in sensitive people. Infectious diseases, such as influenza, measles, and chicken pox, can be transmitted through the air, and some fungi release disease-causing toxins that can become airborne. The deaths from Legionnaire's disease of 29 visitors at a convention in a Philadelphia hotel in 1976 is a notable example of the potential hazards. The hotel's ventilation system had nurtured the bacterium *Legionella*.

By controlling the relative humidity level in a home, growth of some types of biological contaminants can be minimized; a relative humidity between 30 and 50 percent is generally recommended. Kitchen and bathroom exhaust fans vented to the outdoors will help to eliminate moisture that builds up from everyday activities. Naturally, keeping a house clean will also help reduce airborne biological contaminants, particularly house dust mites, animal dander, pollen, and other allergy-causing agents.

## Sick Building Syndrome

Indoor air pollution is not limited to individual homes; many multistory commercial and office buildings have significant air quality problems. A number of well-identified illnesses (for example, Legionnaire's disease) have been directly traced to specific building problems. These are called *building-related illnesses*.

When the occupants of a building have symptoms that do not fit the pattern of any particular illness and are difficult to trace to a specific source, the phenomenon is referred to as *sick building syndrome*. This term is applied to a building when more than 20 percent of its occupants complain of health problems for 2 weeks or more and the symptoms are relieved when the occupants leave the building. It is estimated that as many as 30 percent of new or remodeled commercial buildings induce sick building syndrome. Health complaints typically include sneezing, fatigue, headache, dizziness, nausea, dry throat, and irritability.

Air pollution sources in commercial buildings are similar to those found in the home. One major cause of sick building syndrome is poor design, operation, or maintenance of the complex mechanical ventilation system needed to heat, cool, and circulate air throughout a large building.

Ventilation problems arise when, in an effort to save energy, inadequate amounts of outdoor air are exchanged with the indoor air. Problems also may occur if air supply and return vents within each room are blocked or placed improperly. In the wrong locations, outdoor air intake vents can bring in automobile exhaust, boiler emissions, or even air vented from restrooms. The ventilation system itself can be a source of air pollution, spreading microbes that thrive on the inside surfaces of duct work, as well as in humidifiers, dehumidifiers, and air conditioners. Finally, air pollutants can be circulated into offices from restaurants, print shops, and dry cleaning stores located in the same building, as well as from underground parking garages.

Sick building syndrome usually cannot be effectively remedied without a comprehensive air quality survey and investigation. These investigations may begin with questionnaires and telephone interviews to assess the nature and extent of occupant symptoms. The ventilation system is often the most important factor to investigate; inadequate ventilation accounts for about half of sick building syndrome cases. Air quality testing may help to identify contaminants, but air sampling and analysis are not always effective in solving the problem due to the very low levels of pollutants.

## 13.6  AIR SAMPLING AND MEASUREMENT

To evaluate air quality and to design efficient air pollution control systems, it is necessary to determine emission rates from the sources of pollution and to analyze the type and amounts of pollutants in the gas and surrounding air. Before such measurements can be made, appropriate samples must be collected. The quantification and evaluation of air quality involves accurate determination of pollutant concentrations, which are typically expressed as a *ratio* of the *mass* of the pollutant to the *volume* of air or gas in which it is found.

Measurement of mass is primarily carried out in an analytical laboratory, while determination of volume is usually done in the field at the time of sampling. There are hundreds of methods and types of instruments that can measure pollutant mass. Some instruments require a few milligrams of pollutant for accurate analysis, while others can detect and accurately measure extremely small amounts, even in the range of a few molecules. There are also many methods and devices for measuring volume, velocity, and flow rate. The choice of a technique for sampling and measuring the mass and volume depends on the properties of the sampled gas or air and the specific pollutants to be analyzed. The purpose of this section is to highlight some of these techniques and to present an overview of the technology involved in air sampling and measurement.

### Air Sampling Methods

There are three distinct kinds of air sampling: *source sampling*, *ambient sampling*, and *indoor sampling*.

#### Source Sampling

Source (or emission) sampling is performed at the location of a pollutant discharge, such as the exhaust gas from a chimney, ventilation system, the tailpipe of an automobile. Source sampling is also termed *stack sampling* at power plants, solid waste incinerators, or factories where the discharge comes from a chimney or smokestack.

A basic purpose of source sampling is to evaluate the pollution discharged from a specific generator of pollution and to use the results to determine whether *emission standards* are being complied with. Emission standards may be set by the EPA, state or local environmental agencies, manufacturers, and some professional organizations. Other purposes of source sampling include the collection of data for designing and operating air cleaning equipment and for measuring the effectiveness or efficiency of that equipment.

For accurate and meaningful results in determining the components of a flowing gas stream in a stack or duct, sampling devices and procedures must follow formal requirements known as *EPA Reference Methods for Stationary Source Air Emissions Testing*. When large smokestacks or gas ducts are involved, it is necessary to obtain gas samples from several positions in the gas passage. EPA Method 1 gives specific guidance for selecting the proper locations in the stack to measure velocity and to sample for pollutants.

Method 2 deals with the use of pitot tubes for velocity measurements, and Method 3 describes several ways to measure gas properties and compute the average dry gas mixture molecular weight (as is needed for velocity calculations). Stack exhaust gas contains mostly nitrogen, carbon dioxide, oxygen, carbon monoxide, and water vapor; this mixture is called the *carrier gas*. The pollutants being measured are dispersed in the carrier gas in much smaller quantities. A device called the *Orsat analyzer* is one of the instruments referred to in EPA Method 3. It can measure the volume percentages of the major exhaust gas constituents, allowing computation of the average molecular weight. Since the chemical analysis of the Orsat analyzer is very time-consuming, EPA Method 3 allows the use of alternative devices, including the *Fryite flue gas analyzer* and electric or battery-powered gas analyzers that use selective absorption of infrared light to measure the gas components. EPA Method 4 focuses on measurement of gas moisture content and allows calculation of total molecular weight.

EPA Method 5, one of the most fundamental of the EPA source testing methods, is concerned with the determination of particulate emissions. An important requirement of Method 5 is that the gas sample be obtained *isokinetically* to assure that the sample is truly representative of the stack gas. An *isokinetic sample* is one in which the velocity of the gas stream entering the sampling probe is the same as the velocity of the carrier gas. The temperature, the pressure, and the total volume flow rate of the gas stream being sampled must be known, as well as those of the main gas stream. For the emission test to be valid, the sampling velocity must be within 10 percent of the average stack gas velocity.

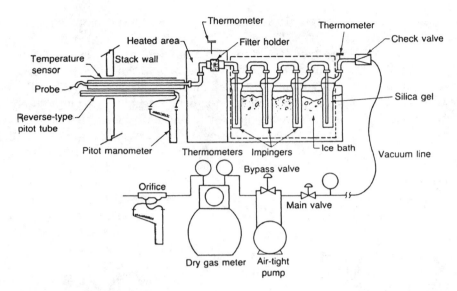

**FIGURE 13.12**
*EPA Method 5 sampling train for particulates.* (Courtesy of the Environmental Protection Agency)

A stack *sampling train* for EPA Method 5 is shown schematically in Figure 13.12.

## Ambient Sampling

Ambient (or atmospheric) sampling pertains to the measurement of outdoor air pollution levels. Samples are collected from the air after pollutants from various sources have been thoroughly dispersed and mixed together under natural meteorological conditions. Ambient sampling provides broad area or background air quality data in urban or rural areas and serves as a basis for assessing health effects, determining compliance with federal or state ambient air quality standards, and predicting the effects of proposed new sources of air pollution. Ambient sampling methods are discussed later in this section.

## Indoor Sampling

Indoor air sampling includes industrial hygiene sampling and residential sampling. Industrial (occupational) hygiene air sampling is done in factories or other workplaces to protect the health of people who may be exposed to pollutants throughout the workday. Residential air sampling, on the other hand, provides data regarding the quality of indoor air in private homes to protect the health of residents.

*Monitoring.* A distinction should be made between air *sampling* and air *monitoring*. Air sampling refers to the acquisition of a representative quantity of polluted air or gas for subsequent laboratory analysis and evaluation of the types and amounts of contaminants present. Air monitoring, on the other hand, refers to the collection of pollutant concentration data on an effectively continuous basis. Continuous (or intermittent) instruments are available for use as stack-gas monitors as well as ambient or indoor air monitoring devices. But continuous monitoring devices generally are not sensitive enough to measure extremely low concentrations of pollutants. Due to the dangers of even small amounts of organic or other toxic air pollutants and the recent development of toxic air pollution regulations, collection of air samples for later laboratory analysis has increased in importance. This is particularly so for emission sampling and indoor sampling.

## Units of Measurement

Concentrations of air pollutants are commonly expressed as the *mass of pollutant per unit volume of air mixture*, that is, as milligrams per cubic meter ($mg/m^3$) or, for very small concentrations, as micrograms per cubic meter ($\mu g/m^3$) and nanograms per cubic meter ($ng/m^3$). Analytical devices that can detect trace substances in the nanogram range are available; a nanogram is one billionth ($10^{-9}$) of a gram.

Concentrations of gaseous pollutants may also be expressed as *volume of pollutant per million volumes of the air plus pollutant mixture* (parts per million, ppm) or in percent, where 1 ppm = 0.0001 percent by volume. It is sometimes necessary to convert from volumetric units to mass per unit volume and vice versa. The relationship between ppm and $mg/m^3$ depends on the gas density, which in turn depends on the temperature, the pressure, and the molecular weight of the pollutant. Concentration values expressed in volumetric units (ppm), however, do not vary as temperature or pressure vary. The EPA ambient air quality standards (discussed in Section 13.7) are based on a temperature of 25°C (77°F) and a barometric pressure of 1 atm (where 1 atm = 760 mm Hg).

The following expression can be used to convert between ppm and mg/m³ at any temperature or pressure:

$$mg/m^3 = \frac{273 \times ppm \times \text{molecular weight} \times \text{pressure}}{22.4 \times \text{temperature}}$$

$$(13-1)$$

Simply multiply the calculated value of mg/m³ by 1000 to obtain $\mu g/m^3$.

The constant 22.4 is the volume in liters occupied by 1 *mole* of an ideal gas at standard conditions (0°C and 1 atm). One mole of any substance is a quantity of that substance whose mass in grams numerically equals its molecular weight. In Equation 13-1, temperature must be expressed in kelvins (K, units of absolute temperature), where K = °C + 273. Pressure must be expressed in atmospheres.

## EXAMPLE 13.1

Federal standards limit hourly ozone ($O_3$) levels to 0.12 ppm. Express this concentration in terms of $\mu g/m^3$ at 25°C and 1 atm pressure.

*Solution*  The molecular weight of ozone is the sum of three atomic weights of oxygen, or $16 \times 3 = 48$ (see Table 4.1 for atomic weights). The temperature, in kelvins, is 25 + 273 = 298 K. Applying Equation 13-1 gives

$$mg/m^3 \ \text{(ozone)} = \frac{273 \times 0.12 \times 48 \times 1}{22.4 \times 298}$$

$$= 0.235 \ mg/m^3$$

and

$$0.235 \times 1000 = 235 \ \mu g/m^3 \quad \text{(ozone)}$$

## EXAMPLE 13.2

Federal standards limit daily sulfur dioxide ($SO_2$) levels to 365 $\mu g/m^3$ at 25°C and 1 atm pressure. Express this concentration in terms of ppm as well as a percentage by volume.

*Solution*  The molecular weight of sulfur dioxide is the sum of the atomic weights of one sulfur atom and two oxygen atoms, or 32 + 16 + 16 = 64. Applying Equation 13-1 gives

$$0.365 = \frac{273 \times ppm \times 64 \times 1}{22.4 \times 298} = 2.617 \times ppm$$

Solving for the concentration in ppm yields

$$ppm \ (SO_2) = \frac{0.365}{2.617} = 0.14 \ ppm \quad \text{(rounded off)}$$

A volumetric concentration of 0.14 ppm is equivalent to 0.14 volume of the gas in $1 \times 10^6$ volumes of mixture, or $0.14/10^6 = 14 \times 10^{-8}$. To convert this into a percentage, multiply by 100, obtaining 0.000014 percent.

## EXAMPLE 13.3

The exhaust of an automobile contains 2.2 percent carbon monoxide and has a temperature of 82°C. Express the concentration in ppm and mg/m³.

*Solution*  The molecular weight of carbon monoxide (CO) is 12 + 16 = 28. Convert 2.2 percent to ppm, as follows:

$$2.2 \times \frac{1 \ ppm}{0.0001 \ \%} = 22,000 \ ppm$$

The absolute temperature in kelvins is 82 + 273 = 355 K, and the exhaust gas can be assumed to be at atmospheric pressure.

Applying Equation 13-1 gives

$$mg/m^3 = \frac{273 \times 22,000 \times 28 \times 1}{22.4 \times 355} = 21\ 148 \ mg/m^3$$

This concentration is equivalent to 21.1 g/m³ (rounded off) because 1 g = 1000 mg. It will be greatly reduced when the exhaust gas is dispersed in the atmosphere. (The EPA ambient standard is only 10 mg/m³ averaged over an 8-h period.)

## Particulates

Collection and measurement of suspended particulates in air depends on principles very different from those used in the analysis of gases. The mass (inertia) and size of the particles are key physical properties that affect the sampling process. Three general methods for collecting and measuring particulate air pollutants, which rely on physical properties, are the *gravimetric technique*, the *filtration technique*, and the *inertial technique*.

The gravity technique is the simplest method, but it can only measure the amount of *settleable* particulates (dust and fly ash) in the air. A basic and inexpensive device, called a *dustfall bucket*, is an example of a first-generation method used to determine how much particulate material settles to Earth. In this technique, an open bucket containing water to trap and hold the particles is exposed in a suitable location, such as a building rooftop. After a collection time of usually 30 d, the water is evaporated and the particulates weighed. The results may be reported in terms of *grams per square meter per month* (g/m²/month), *kilograms per hectare per month* (kg/ha/month), or *tons per square mile per month* (tons/mile²/month). The total amount of dust that will settle out of the atmosphere in a typical urban area can be quite high; as much as 50 tons/mile²/month of dustfall has been observed in some cities.

## EXAMPLE 13.4

The mass of a 6-in.-diameter dustfall bucket is 120.00 g when empty. After 30 d of exposure, the bucket and collected particulates have a combined mass of 120.30 g. Compute the dustfall.

*Solution*

$$\text{mass of particulates} = 120.30 - 120.00 = 0.30 \text{ g}$$

$$\text{area of the bucket} = \frac{\pi D^2}{4} = \pi \times \frac{0.5^2}{4} = 0.196 \text{ ft}^2$$

Since 1 mi = 5280 ft, 1 lb = 454 g, and 1 ton = 2000 lb,

$$\text{dustfall} = \frac{0.3 \text{ g/month} \times 1 \text{ lb}/454 \text{ g} \times 1 \text{ ton}/2000 \text{ lb}}{0.196 \text{ ft}^2 \times (1 \text{ mi})^2/(5280 \text{ ft})^2}$$

$$= 47 \text{ tons/mi}^2/\text{month}$$

Also, since 1 ft = 0.305 m and 1 ha = 10 000 m²,

dustfall =

$$\frac{0.3 \text{ g/month} \times 1 \text{ kg}/1000 \text{ g}}{0.196 \text{ ft}^2 \times (0.305 \text{ m})^2/(1 \text{ ft})^2 \times 1 \text{ ha}/10 \text{ 000 m}^2}$$

$$= 165 \text{ kg/ha/month}$$

(As an exercise, convert this dustfall rate to g/m³/month.)

Dustfall buckets have several limitations, including the long sampling period needed to get results. They have largely been replaced by more accurate second-generation particulate sampling devices known as *high-volume (hi-vol) samplers*. These devices, which reduce sampling time to 24 h, use the filtration technique rather than gravity for capturing the particulates. A hi-vol sampler draws a large volume of air through a glass-fiber or membrane filter. The filter is weighed before and after sampling, and the airflow rate, which gradually decreases as particulates accumulate on the filter, is accurately metered and recorded. Results are expressed as μg/m³. In cities, *total suspended particulate* (TSP) concentrations are typically around 100 μg/m³, although peak hourly values may be several times that amount; in rural areas, the levels are generally about 30 μg/m³. Expressing results in terms of μg/m³ may give an erroneous impression that the quantities of material are exceedingly small or negligible. However, a value of 200 μg/m³ is roughly equivalent to 1 ton of particulates per cubic mile of air.

## EXAMPLE 13.5

The airflow through a high-volume sampler was recorded at 55 ft³/min at the beginning of sample collection and 35 ft³/min after 24 h of continuous sampling. The filter weighed 10.000 g before and 10.200 g after sample collection. What was the TSP level measured in the sample?

*Solution* The average rate of airflow through the filter is

$$\frac{55 + 35}{2} = 45 \text{ ft}^3/\text{min}$$

The total volume of air passing through the filter in 24 h is

$$45 \text{ ft}^3/\text{min} \times 60 \text{ min/h} \times 24 \text{ h} = 64,800 \text{ ft}^3$$

Converting to cubic meters, we get

$$64,800 \text{ ft}^3 \times 0.028 \text{ 32 m}^3/\text{ft}^3 = 1.835 \times 10^3 \text{ m}^3$$

The weight of particulates is

$$0.200 \text{ g} \times 10^6 \text{ μg/g} = 200 \times 10^3 \text{ μg}$$

The total suspended particulate concentration is

$$\frac{200 \times 10^3 \ \mu g}{1835 \ m^3} = 110 \ \mu g/m^3$$

## PM$_{10}$ Samplers

Hi-vol samplers are now used primarily to obtain samples of particulates for the analysis of metals, organics, sulfates, and nitrate compounds. They are also used to continue long-term historical records of particulate levels at certain locations. Since the development of the hi-vol sampler in the late 1940s, there have been several improvements, including better filter materials and air inlet design configurations. Measurement of particulate levels in ambient air is now done primarily by *PM$_{10}$ samplers*, which are modified hi-vol devices. PM$_{10}$ refers to particulate matter equal to or less than 10 $\mu m$ in size; it has been found that particles in this size range are of primary concern with respect to harmful respiratory health effects in humans. PM$_{10}$ sampler inlets are designed to remove the larger particles before the sample reaches the filter, and flow rates are controlled and metered by electronic sensors. A PM$_{10}$ sampler is depicted in Figure 13.13.

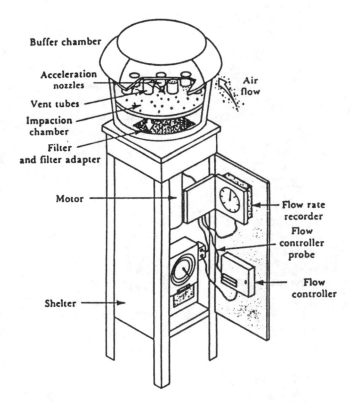

**FIGURE 13.13**
*High-volume sampler with PM$_{10}$ inlet.* (Courtesy of the Environmental Protection Agency)

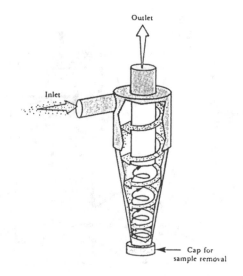

**FIGURE 13.14**
*Cyclone collector for particulates.* (Reprinted with permission from G. D. Wight, *Fundamentals of Air Sampling*, 1994. Copyright © 1994, Lewis Publishers, an imprint of CRC Press, Boca Raton, Florida)

## Inertial Samplers

These operate on the principle that when the airflow direction is changed suddenly, the inertia of the particulates will cause them to hit an impaction surface, on which they can be trapped. The impaction surface may be a glass-fiber mat or a solid surface coated with oil and is weighed before and after sampling to measure the amount of material collected. The impaction surface may simply be a hard surface nearly parallel to the gas flow, as in a *cyclone collector* (shown in Figure 13.14). Here the impacted particulates fall to the bottom of the collector for removal and weighing.

Another type of inertial sampling device used to collect and analyze specific particulates, such as pollen grains or bacteria, is the *cascade impactor*, shown schematically in Figure 13.15. This device captures particles on a series of slides that is placed in the air stream. The small orifices through which the air flows are progressively decreased in size, thereby increasing flow velocity. Particles of different size ranges are captured on each slide because of the particles' inertia, the sudden change in direction of flow, and the different flow velocities. The particulates can be observed and measured on the slides with a microscope.

## Other Types of Particulate Collectors

These include the *polyurethane foam* (PUF) *sampler* and the *dichotomous sampler*, which allows separation of the particles into two size ranges (typically between

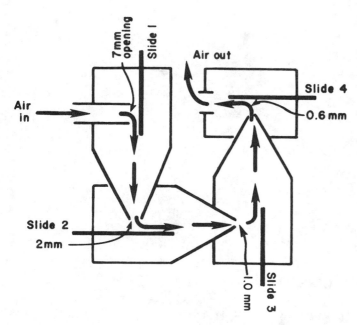

**FIGURE 13.15**

*A cascade impactor for collecting and analyzing particulate air pollutants.* (From P. A., Vesilind and J. J. Pierce, *Environmental Pollution and Control,* 2nd ed. Copyright © 1983, Ann Arbor Science Publishers. Used with permission of the publisher)

10 and 2.5 $\mu$m and less than 2.5 $\mu$m in diameter). The dichotomous sampler, shown schematically in Figure 13.16, uses both filtration and impaction techniques to operate. All particulate sampling devices require careful flow control and accurate measurement of flow rate, time, and sample volume.

### Smoke Readings

A *Ringlemann smoke chart* may be used to determine if plumes of smoke are within allowable standards (but it cannot be used for dusts, mists, or fumes). Dense plumes are generally viewed as emissions that are out of compliance. The density of a plume is compared visually to five standard shades of gray, between white and black; smoke readings range from 0 (all white) to 5 (all black or opaque). A moderately dense smoke, for example, may be said to be Ringlemann 3. The ability to obtain accurate and consistent smoke readings requires special training and skill. The chart is located at a distance of about 15 m from the observer, and 30 observations are made within 15 min; an average is computed from these readings. Ringlemann smoke chart readings are being replaced by a measure of the *percent opacity;* a Ringlemann reading of 1, for example, would correspond to an opacity of about 20 percent.

## Gases

The physical properties and behavior of gases differ markedly from those of particulates. One important difference is that gas molecules are small enough to pass through the finest filter, and they are certainly too small to settle out by gravity. Techniques for sampling and measuring gaseous pollutants generally involve either the process of *absorption* or the process of *adsorption.* (Although these two words look similar, students should carefully note the difference in spelling; the actual processes are very different.)

*Absorption* is a process involving the uniform distribution of the gaseous compound in a liquid or solid, called the *absorbent.* For air sampling, the absorbent is usually a liquid; absorption in a liquid for later analysis in a laboratory is the EPA-approved method for sulfur dioxide and other low-molecular-weight compounds.

Absorption can be either a physical or chemical process; dissolving a gas in liquid is a physical process. Chemical absorption, though, uses a liquid containing a substance that reacts with the gas to form an easily detected by-product. A typical absorption sampler, called

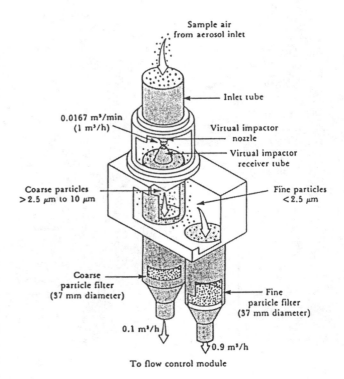

**FIGURE 13.16**

*Dichotomous sampler flow diagram.* (Reprinted with permission from G. D. Wight, *Fundamentals of Air Sampling,* 1994. Copyright © 1994, Lewis Publishers, an imprint of CRC Press, Boca Raton, Florida)

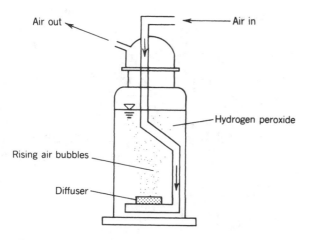

**FIGURE 13.17**

*A glass bubbler or absorber may be used for sampling specific gaseous pollutants. For example, hydrogen peroxide will absorb sulfur dioxide from the air, forming sulfuric acid. The level of sulfur dioxide in the air can be computed after measuring the amount of sulfuric acid in the bubbler.*

a *bubbler* or *impinger tube*, is shown in Figure 13.17. The air is pumped through a small glass diffuser (with 50-$\mu$m or smaller pores) and bubbled up through the

liquid, which will either dissolve the gas or react with it chemically. If a known volume of air containing sulfur dioxide, for example, is bubbled up through hydrogen peroxide ($H_2O_2$), sulfuric acid is quickly formed, as described by the following chemical equation:

$$H_2O_2 + SO_2 \rightarrow H_2SO_4$$

The amount of sulfuric acid that is formed in the reaction can be measured by standard chemical techniques; from that, the amount and concentration of sulfur dioxide in the air sample can be computed. Bubblers are not 100 percent efficient because not all the air sample is absorbed by the liquid. They must be tested in advance to determine their efficiencies in order to be able to adjust the final quantitative results.

An absorption instrument called a *24-h bubbler*, shown in Figure 13.18, can be used to test for three different gases at the same time. Separate sampling trains with suitable collecting liquids in the bubbler are connected in parallel to a vacuum pump. The rate of airflow can be controlled and measured. A similar device, called a *sequential sampler*, can be used to collect up to 12 samples in sequence for fixed periods of time, typically 2 h. The sequential sampler allows peak concentrations of a specific pollutant to be determined on a daily basis.

(a)                                                                    (b)

**FIGURE 13.18**

(a) *A three-gas sampler and* (b) *the sampler in an all-weather shelter.*
(Courtesy of RAC Division, Andersen Samplers, Inc., Atlanta, Georgia)

*Adsorption* is a physical process in which the gas molecules are attracted to the surface of a solid and held there by molecular bonding forces. The adsorbent is usually a granular, porous material (for example, charcoal or activated carbon) with a very high surface area–to–volume ratio. A reverse process by which collected molecules may be removed from the surfaces is called *desorption*. Desorption can be achieved by use of a solvent or heat to drive off the captured molecules.

Industrial hygiene surveys to detect organic compounds in the workplace often use charcoal adsorption tubes and solvent desorption. After the pollutants are desorbed, gas chromatography–flame ionization analysis techniques are used in a laboratory to measure the level of contamination.

In *colorimetric tubes*, the adsorbent may be precoated with a chemical that reacts with and changes color in proportion to the amount of gas adsorbed. For example, the adsorbent in a carbon monoxide detector tube will change from yellow to blue-green as an air sample containing CO passes though the tube. The CO concentration is measured by comparing the color to a calibrated color chart. Other materials, such as silica gels, alumina, and organic polymers, may also be used as adsorbents. Organic polymer adsorbents are particularly useful in air sampling for organic compounds.

## Whole-Air Samplers

Whole-air samples (or *grab samples*) are collected in a number of ways, including use of a simple evacuated canister or flask. When the flask is opened at the sampling location, the air sample is drawn into it by the vacuum; the sample is analyzed later in a laboratory.

Another grab sampling device that is effective if the gas under study is insoluble is a *displacement bottle*, shown schematically in Figure 13.19. As water or some other liquid is drained or siphoned out, air is drawn in by the lowered pressure to take the place of the liquid removed. The volume of the air sample drawn

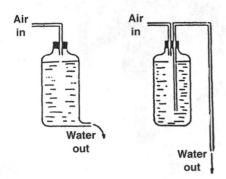

**FIGURE 13.19**
*Displacement bottle.* (Courtesy of the Environmental Protection Agency)

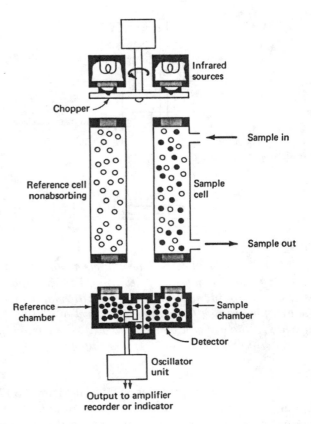

**FIGURE 13.20**
*Nondispersive infrared analyzer.* (From K. Wark and C. Warner, *Air Pollution: Its Origin and Control,* 2nd ed., Harper & Row, New York, 1981. Courtesy of Dr. K. Wark.)

in is equal to the volume of the displaced liquid. In general, grab or whole-air samples are not very useful when extremely small quantities of pollutants are present in the air, since the collected volumes may not be large enough for accurate analysis.

## Gas Analysis

Hundreds of modern instruments are available from manufacturers for analyzing gases. Colorimetric and bubbler techniques are what may be called first- and second-generation air analysis devices. An example of a widely used third-generation device is the *nondispersive infrared analyzer*, illustrated in Figure 13.20. This is used to measure gases that absorb infrared radiation, such as CO, $SO_2$, $NO_x$, and HC. The measurement is based on the principle of selective absorption; different gases transmit and absorb different wavelengths of infrared radiation.

The detector at the bottom of Figure 13.20 has two chambers containing equal volumes of the gas to be studied. The chambers are separated by a flexible metallic diaphragm and a stationary metallic button, which form a capacitor. Infrared radiation is directed through two identical cells, one of which is a reference cell and is filled with nitrogen. The other is a sample cell, through which the gas to be analyzed flows. When a pollutant of interest is present in the sample, it absorbs an amount of radiation proportional to the pollutant concentration, but no absorption occurs in the reference cell.

After passing through either the reference cell or the sample cell, the radiation is then absorbed by the gas in the detector cells. Because some energy was already absorbed in the sample cell, the gas in the refer-

ence detector cell gets warmer than the gas in the sampler detector cell. The pressure then increases in the reference cell; this deflects the diaphragm, changes the capacitance, and sends an electrical signal to a recorder. The signal is proportional to the concentration of pollutant.

Additional modern analysis techniques include gas chromatography, mass spectrometry, flame emission spectrometry, and electrochemistry, among others. Sophisticated *continuous monitoring* (CM) instruments, as shown in Figure 13.21, are available. These

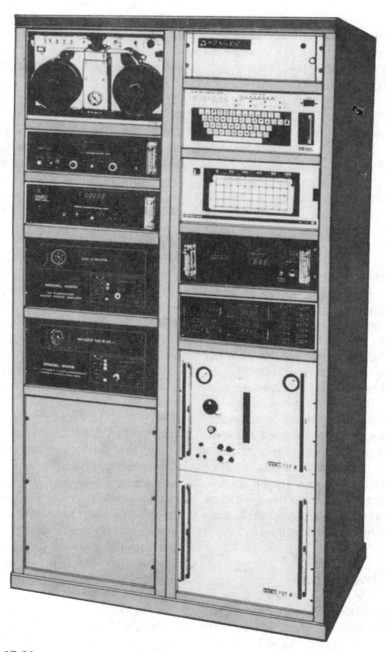

**FIGURE 13.21**

*A continuous ambient monitoring system.* (Courtesy of Dasibi Environmental Corp., Glendale, California)

expensive instruments combine collection and automatic analysis for many different air pollutants. Electronic detectors, meters, and recording devices are part of the sampling train for this kind of equipment. Continuous graphs that show the hourly change in pollutant levels or concentrations can be obtained. This is particularly useful in urban areas when used as part of a pollution episode warning system.

## 13.7 AIR POLLUTION CONTROL

The need to minimize or control air pollution has been evident for several hundred years. As mentioned in Section 13.1, one of the earliest attempts to legislate air pollution control was in England in the 14th century, when King Edward II prohibited the burning of coal. In fact, he decreed that anyone found guilty of burning coal while Parliament was in session would be executed. Apparently, the smoke bothered him somewhat.

Today, it is known that smoke can be more than just a temporary annoyance; smoke and other air pollutants affect health and well-being, esthetic sensibilities, and even Earth's climate on a global scale. It is also known, certainly, that threats of medieval-type punishments cannot solve the problem. In a modern industrial society, it is difficult to avoid generating at least some waste products that will enter the atmosphere, in one way or another. Realistic pollution control strategies, laws, and air quality standards are necessary, and logical principles of engineering and technology must be directed toward the development and use of pollution control equipment.

This section begins with an overview of air pollution control strategies. Also included here is a discussion of air quality legislation and standards. Air quality standards serve as a framework within which the success of pollution control strategies can be measured. They are based on the best available data regarding the health and environmental impacts of various air pollutants. Finally, this section includes a discussion of the different types of add-on devices and equipment that are used to reduce pollutant emissions from both stationary and mobile sources and to keep air quality within the legally established standard limits.

### Pollution Control Strategies

There are several approaches or strategies for air pollution control. The most effective control would be to prevent the pollution from occurring in the first place. Complete *source shutdown* would accomplish this, but shutdown is only practical under emergency conditions, and even then it causes economic loss. Nevertheless, state public health officials can force industries to stop operations and can curtail highway traffic if an air pollution episode is imminent or occurring. A source shutdown can only offer, at best, a very temporary solution to local problems of air pollution.

Another option for air pollution control is *source location* in order to minimize the adverse impacts in a particular locality. Community *air zoning* may be included in municipal master plans, requiring power plants or industrial facilities to be located where fewer people will be affected by the pollutants. The location of these zones can be established on the basis of prevailing wind patterns and weather conditions. This option has a limitation; although local air quality may be somewhat protected, the pollutants can still be "air-mailed" to neighboring communities by the dispersion of the plumes emitted from tall chimneys or smoke stacks.

Tall smoke stacks take pollutants high into the atmosphere, allowing the processes of mixing and dispersion to dilute the contaminants, reducing pollution levels. However, "what goes up must come down," and when it comes down, the problem of air pollution returns. Acid deposition (or acid rain) is a notable example of an air pollution problem that transcends local boundaries.

An important approach for air pollution control is to encourage or require industries to make *fuel substitutions* or *process changes*. For example, making more use of solar, hydroelectric, and geothermal energy would eliminate much of the pollution caused by fossil fuel combustion at power-generating plants. Nuclear power would do the same, but other problems related to high-level radioactive waste disposal and safety remain to be solved. Also, using natural low-sulfur coal and oil would reduce $SO_2$ emissions from existing power-generating stations. Technology is available for treating and *desulfurizing* dirty fossil fuels prior to their combustion, but it is expensive. A complete change of some industrial manufacturing processes can also reduce air pollution; one example is the use of electric furnaces instead of open-hearth furnaces in the steel industry.

Fuel substitutions are also effective in reducing pollution from mobile sources. For example, the use of reformulated gasolines or alternative fuels such as liquefied petroleum gas, compressed natural gas, or methanol for highway vehicles would help to clear the air. (This is discussed in more detail later in this section.) Ultimately, complete replacement of gasoline-powered vehicles with electric-powered vehicles may eliminate one of the major sources of ambient air pollution. California was the first state to require that, by the year 1998, 2 percent of new car sales be *zero-emission vehicles* (that is, electric cars). The requirement increases to 10 percent by the year 2003. Other states, including Massachusetts and New York, have adopted similar quotas and timetables.

The use of *correct operation and maintenance practices* is important for minimizing air pollution and should not be overlooked as an effective control strategy. For example, if a power plant operator allows too much air into the boiler furnace, fly ash emissions will increase. Adding too much sulfur at a sulfuric acid manufacturing plant, without providing enough air, can cause excessive $SO_2$ emissions. Even a failure to properly lubricate a fan motor at an incinerator can lead to unnecessary pollution. Finally, one of the most important strategies for controling emissions from mobile sources is an effective motor vehicle inspection and maintenance program.

## Air Quality Legislation and Standards

Up until the mid-1900s, air pollution was viewed primarily as a state or local problem. Today, however, it is recognized to be not only nationwide in scope, but of global scale and international significance. In the United States, federal laws controlling air pollution began with the *Clean Air Act* of 1963. Although governmental responsibility for air quality control was thereby raised to the national level, state and local efforts are still very important for the success of the federal air pollution control program.

A few years after the *Clean Air Act*, the *Air Quality Act of 1967* extended the authority of the federal government in ambient (outdoor) air pollution control and introduced the idea that regulatory standards could precede existing technology, a concept known as *technology-forcing legislation*. Another federal law (passed in 1970), the *Occupational Health and Safety Act* (OSHA), also pertains, in part, to air pollution; it establishes limits of exposure to certain air pollutants for workers in various industries.

Air quality laws and guidelines are under constant review and are often modified. The set maximum allowable levels, discharge rates, or exposure times for specific substances. Three basic types of guidelines are *threshold limit values, source performance standards,* and *ambient air quality standards.*

Threshold limit values (TLVs) focus on specific air contaminants that have well-recognized cause-and-effect health relationships. They serve primarily as occupational guidelines, limiting the exposure of workers in industrial environments to specific vapors or chemical dusts. TLVs are typically established on the basis of exposure for 8 h/d, 5 d/week.

Source performance standards (also called *emission standards*) focus on the major municipal and industrial generators of air pollutants for both stationary and mobile sources. These include power plants, solid waste incinerators, manufacturing plants, oil refineries, highway vehicles, and many more. They are usually expressed in terms of the mass of pollutant emitted per unit of time, production volume, or distance.

Ambient air quality standards focus on the allowable levels of out-of-doors atmospheric pollutants. They are intended to set limits that will minimize the overall adverse effects on health, comfort, and property. Exposure time is assumed to be for 24 h/d, 7 d/week (in contrast to the more limited exposure times inherent in TLV standards). Emission standards are established at levels intended to satisfy or meet the ambient air quality standards in an *air quality control region* (AQCR). These regions are identified on the basis of meteorological and social or political factors.

### Clean Air Act

There have been several amendments to the *Clean Air Act* of 1963 and the *Air Quality Act* of 1967; together, the original acts and their amendments may be referred to simply as the *Clean Air Act* (CAA). To understand the current status of federal legislation in the national effort to control air pollution, it is of value to have a perspective of the evolution of the CAA over the years since its inception.

The *Clean Air Act Amendments of 1970* began to strengthen the federal air pollution control effort. Under these amendments, *National Ambient Air Quality Standards* (NAAQS) were officially established to protect the public health and welfare. These amendments also required the states to prepare *State Implementation Plans* (SIPs) for achieving and maintaining the national standards. For certain industries, *New Source Performance Standards* (NSPS) were established to limit emissions. In addition, strict automobile emission standards were set; these standards were an example of technology-forcing legislation. The automobile industry was unable, at that time, to reach the stated goals with existing technology, and this part of the legislation has been amended several times.

The primary NAAQS (as of 1990) for six *criteria pollutants* are presented in Table 13.1. *Primary standards* were established to protect public health; *secondary standards* (not shown) were established to protect against nonhealth effects, such as crop damage or visibility. (These terms should not be confused with primary and secondary *pollutants*.) States can set their own standards, but they must meet or exceed the national primary ambient standards. Since ambient air

**TABLE 13.1**
**Primary national ambient air quality standards (NAAQS)**

| Pollutant | Averaging time | Allowable concentration |
|---|---|---|
| $PM_{10}$ | Annual arithmetic mean | 50 $\mu g/m^3$ |
| | 24 h | 150 $\mu g/m^3$ |
| $SO_2$ | Annual arithmetic mean | 80 $\mu g/m^3$ |
| | 24 h | 365 $\mu g/m^3$ |
| CO | 8 h | 10 $mg/m^3$ |
| | 1 h | 40 $mg/m^3$ |
| $NO_2$ | Annual arithmetic mean | 100 $\mu g/m^3$ |
| $O_3$ | 1 h | 235 $mg/m^3$ |
| Pb | 3 mo | 1.5 $\mu g/m^3$ |

quality varies significantly with time, the NAAQS specify one or more averaging or measuring time periods for each pollutant. The allowable concentrations may be expressed as $\mu g/m^3$ or as ppm.

Emission standards pertain to specific sources and are too numerous to list here. For example, a coal-fired electric power plant is allowed to discharge no more than 1.2 lb of $SO_2$ per million Btu (British thermal units) of heat input. At a sulfuric acid manufacturing plant, the NSPS for $SO_2$ is 2 kg per metric ton of acid produced. For a sludge incinerator, the particulate emission standard is a maximum of 0.65 g per kg of sludge input. Tailpipe (exhaust) emission standards for cars include maximum allowable values of 0.4 g per mile (gpm) of nitrogen oxides, 3.4 gpm of carbon monoxide, and 0.25 gpm of nonmethane hydrocarbons.

By the mid-1970s, only about one third of the air quality control regions in the United States were meeting national standards, prompting Congress to pass the *CAA Amendments of 1977*. The regions not meeting standards were termed *nonattainment areas*. One of the innovations of the 1977 Act was a policy of *emission offsets*, which created a way to allow industrial expansion in the nonattainment areas. If a particular company in a nonattainment area plans to build a new factory, for example, one of the ways offsets can be obtained is for the company to reduce pollutant emissions from its existing sources. This can be done by installing new and more efficient control equipment on an older factory. The company may also buy, sell, or trade emission offsets with other companies in the area, and the offsets can be "banked" for future use.

The 1977 Amendments also established the concept called *prevention of significant deterioration* (PSD), which applies to attainment areas where the air is even cleaner than the standards require. Three dif-

ferent kinds of attainment areas were defined: class I, in which no deterioration of air quality is allowed; class II, in which moderate deterioration is allowed; and class III, which allows somewhat more deterioration than does class II, but not more than the NAAQS would permit. The National Parks and Wilderness Areas are class I (pristine) areas. Industrialized PSD areas fall into the class III category.

The *Clean Air Act Amendments of 1990* introduced significant changes and added greatly to the potential for improving nationwide air quality. They also addressed, for the first time, global climate effects. The EPA is now allowed to impose sanctions (for example, loss of highway construction funds) against states and cities that do not comply with federal standards. For highway vehicles, stricter tailpipe emission standards were set for CO, $NO_x$, and hydrocarbons; cleaner burning fuels (for example, reformulated gasoline, or RFG) are also required in certain cities with severe air pollution. And under this act, lead is completely banned from use in motor vehicle fuel as of the year 1996.

Additional 1990 amendments include regulation of toxic emissions from major sources such as oil refineries, as well as much smaller sources such as dry cleaners and gasoline stations; a phase-out schedule for chemicals that deplete stratospheric ozone; a 10-million-ton-per-year reduction in $SO_2$ emissions to reduce acid rain; a new operating permit program to be administered by the states; and many others. The amendments of 1990 are an extensive and complex piece of legislation, one that will take decades to fully implement and will cost a great deal of money. Maximum allowable concentrations for some hazardous air pollutants are shown in Table 13.2. Additional air toxics are now being identified and regulated.

## Air Quality Index

To be able to provide the public with timely, easy-to-understand information about outdoor air quality and to evaluate national air quality trends, the EPA publishers a daily *Air Quality Index* or AQI (prior to July

**TABLE 13.2**
**Hazardous air pollution standards**

| Pollutant | Averaging time (h) | Allowable concentration |
|---|---|---|
| Asbestos | 8 | 1 fiber/$cm^3$ |
| Beryllium | 8 | 2 $\mu g/m^3$ |
| Mercury | 10 | 0.01 $mg/m^3$ |
| Vinyl chloride | 8 | 1 ppm |

1999, this was called the *Pollutant Standards Index* or PSI). In addition to its use as a public information tool, EPA and local officials use the AQI to help determine which precautionary steps need to be taken if air pollution levels rise into the unhealthful range.

An AQI is determined for each of the five criteria air pollutants—particulates, sulfur dioxide, carbon monoxide, nitrogen dioxide, and ozone. This is done by converting the measurement of a pollutant's concentration in a community's air to a number on a scale of 0 to 500. There are five intervals on the AQI scale, each of which is related to the potential health effects for each of the five criteria air pollutants. These intervals are:

| AQI | Health Effect |
| --- | --- |
| 0 to 50 | Good |
| 51 to 100 | Moderate |
| 101 to 200 | Unhealthful |
| 201 to 300 | Very unhealthful |
| 301+ | Hazardous |

The most significant number on the AQI scale is 100, since this number corresponds to the standards established under the *Clean Air Act* for each pollutant. A $365$-$\mu g/m^3$ reading for sulfur dioxide, for example, would translate to an AQI level of 100. Similarly, a $235$-$mg/m^3$ measured concentration for ozone would also convert to an AQI of 100. In most communities in the United States, AQI levels generally fall between 0 and 100; readings in excess of 100 are likely to occur only a few times a year, if at all. However, several metropolitan areas in the United States do have more severe air pollution problems, and may often experience AQI levels in excess of 100. Even in such areas, though, AQI readings in excess of 200 are quite rare.

EPA reports the highest of the five AQI figures for each major metropolitan area and identifies which pollutant corresponds to the figure that is reported. For example, if EPA reports an AQI level of 110 for ozone for a given metropolitan area, residents of that area would know that the ozone level for the region is at the low end of the unhealthful range; they would also know that ozone is the pollutant with the highest AQI reading for the day, and that all other pollutants are in the good or moderate range. On days when two or more pollutants exceed the standard (that is, have AQI values greater than 100), the air pollutant with the highest index level is reported, but information on any other pollutants above 100 may also be reported.

*Episodes Criteria.* AQI levels above 100 may trigger preventive actions by state or local officials. These could include health advisories for susceptible people to limit certain outdoor activities and potential restrictions on industrial activities. The 200 level is likely to trigger an *Alert* stage. Activities that might be restricted by local governments include incinerator use and open burning of leaves or refuse. A level of 300 will probably trigger a *Warning*, which is likely to prohibit the use of incinerators, severely curtail power plant operations, cut back operations at some manufacturing facilities, and require the public to limit driving by using car pools and public transportation. An AQI level of 400 or above would constitute an *Emergency*, and would require stopping most industrial and commercial activity, plus a prohibition of almost all private use of motor vehicles.

## Emission Controls for Stationary Sources

The optimum strategy for air quality protection is to reduce the amount of pollution at its source, primarily by fuel substitutions or process changes. When this is not possible or is simply insufficient to accomplish the goal, some type of air cleaning equipment must be installed at the source.

Several types of air cleaning devices can collect or trap air pollutants before they are emitted into the atmosphere. Some of these devices serve to control only suspended particulates and others control only gaseous pollutants. The design or selection of a particular type of air cleaning apparatus depends on the physical and chemical properties of the pollutant to be removed as well as on temperature, corrosivity, and other characteristics of the pollutant and the carrier gas.

### Control of Particulates

Most ambient particulate air pollutants come from stationary sources, particularly from power plants, industrial processes, mining activities, and refuse incinerators. The primary particulate characteristics that determine which type of air cleaning equipment is best for a specific pollution control application are average particle size, size distribution, density, and reactivity. Properties of the carrier gas, such as flow rate, moisture content, temperature, and flammability, are also of importance.

Particulates come in a variety of shapes and sizes, and they can be liquid droplets or solid fume and dust particles. Most particulates are not round, but are quite irregularly shaped. It is usually necessary to refer to an effective diameter of an irregularly shaped particle when one or two of its dimensions are much larger than the other. The behavior of a particle that is suspended in a gas stream is a function of its density and its *aerodynamic diameter*. The aerodynamic diameter of an irregularly shaped particle is defined as the diameter of a sphere with the density of water that will settle in still air at the same rate as the actual particle itself. This is the diameter referred to when discussing particulates and particulate control equipment. A device called a *cascade impactor*, which separates suspended particles on the basis of aerodynamic diameters, is used to determine the size distribution of particles in an exhaust stream.

There are many different kinds of particulate control equipment, including *gravity settlers*, *cyclones*, *electrostatic precipitators*, *fabric filters*, and *wet scrubbers*. Some are much less expensive than others, but are also less efficient in removing fine particulates from the carrier gas. Some are limited to cleaning dry, low-temperature gases. Since air pollution control problems are generally unique, engineering principles must be used to select the best air cleaning device(s) for specific control applications.

*Gravity Settlers.* The simplest kind of air cleaning device is the *settling chamber*, which is shown schematically in Figure 13.22. It is basically an enlarged section or compartment in a flue in which the velocity of the carrier gas is reduced. When the gas flow velocity is slowed sufficiently, coarse particles (those more than about 40 $\mu$m in diameter) can settle out by gravity. A settling chamber of this type usually serves as a precleaner, to prevent clogging of the more efficient small-particle collectors that must follow it. Baffles are usually added inside the chamber to increase settling and particulate removal efficiency.

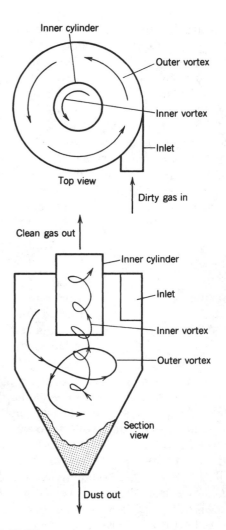

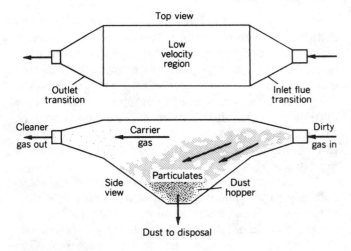

**FIGURE 13.22**
*A settling chamber or enlarged flue section provides a simple way to remove settleable particulates at the source.*

**FIGURE 13.23**
*In a cyclone collector, particulates are spun out toward the outer wall by inertia. They are slowed down by friction and settle to the bottom; clean air flows upward and out the top.*

*Cyclones.* Another air cleaning device, called a *cyclone*, removes many of the particulates by causing the carrier gas to flow in a spiral path inside a cylindrical chamber. Instead of using velocity reduction and quiescent gravity settling, a cyclone subjects the particulates to inertial forces and friction, which separate them from the carrier gas. As shown in Figure 13.23, carrier gas enters the cyclone chamber from a tangential direction at the outer wall of the chamber and forms a vortex as it swirls around inside the cylindrical and conical shell. The larger particulates move outward and are forced against the wall (as they obey Newton's first law of motion, or the law of inertia). Particles are slowed down by friction and slide down the wall into a dust hopper at the bottom, where they are removed. The cleaned gas stream then swirls upward in a narrower spiral, through an inner cylinder and an outlet at the top.

Cyclones are most efficient for removing relatively coarse particulates, but they are also somewhat efficient for removing smaller particles. They can routinely achieve efficiencies of 90 percent for particles larger than about 20 $\mu$m. The typical *cut diameter* for a cyclone is about 15 $\mu$m; the cut diameter is that for which 50 percent of the particles are collected and 50 percent are not. In other words, most of the particles larger than 15 $\mu$m are removed, as well as about half of those as small as 15 $\mu$m. Unlike the gravity settling chamber, removal efficiency for small particles in a cyclone increases as the flow velocity increases. In some advanced designs, cyclones are capable of more than 98 percent removal efficiencies for particles larger than 5 $\mu$m in diameter.

Cyclones, which are low in cost compared to other air cleaning devices, are particularly useful for the removal of industrial dusts from process gases. They are one of the most widely used of all industrial gas cleaning devices. But cyclones by themselves are generally not sufficient to meet stringent air quality regulations. Low first cost and almost maintenance-free operation make them most suitable for use as precleaners for more expensive final air cleaning devices, such as electrostatic precipitators or "baghouse" fabric filters.

*Wet Scrubbers.* Wet collection devices, called *scrubbers*, trap suspended particles by direct contact with an aerosol spray of water or some other liquid. In effect, a scrubber washes the particulates out of the carrier gas as they collide with and are intercepted by the countless number of tiny droplets in the aerosol. There are several types of wet scrubbers on the market, including *spray-chamber scrubbers*, *cyclone spray scrubbers*, *orifice scrubbers*, *wet-impingement scrubbers*, and *venturi scrubbers*.

In a spray-chamber (or spray-tower) scrubber, the upward-flowing carrier gas is washed by water sprayed downward from a series of nozzles, as shown schematically in Figure 13.24. The water is recirculated after it is sufficiently clarified to prevent clogging the nozzles. Spray chambers can remove particles larger than 8 $\mu$m from the gas with efficiencies of about 90 percent. They are widely used to control dust in fiberglass and paper towel production and in other industrial applications. Dust removal efficiencies can be increased by introducing the carrier gas tangentially into the chamber, as in a cyclone chamber scrubber. The inertial effects and higher velocities caused by the spiral motion in the cyclone enhance aerosol droplet formation and removal of particulates.

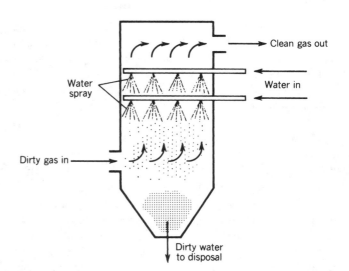

**FIGURE 13.24**
*The spray tower is a type of wet scrubber that removes suspended particulates from the carrier gas. Most of the dirty water is clarified and recycled.*

In orifice and wet-impingement scrubbers, the gas and droplet mixture collides with a solid surface. These scrubbers have the advantage of lower water recirculation rates and treatment efficiencies of about 90 percent for particles larger than 2 $\mu$m. Venturi scrubbers are the most efficient of the wet collectors, achieving removal efficiencies of more than 98 percent for particles larger than 0.5 $\mu$m in diameter. Scrubber efficiency depends on the relative velocity between the droplets and the particulates. Venturi scrubbers achieve the high relative velocities by injecting water into the throat of a venturi (a construction in the flow path) through which air is passing at about 75 m/s (250 ft/s). Venturi-type scrubbers are easy to install and maintain, but typically require relatively large pressure drops to operate.

*Electrostatic Precipitators.* Electrostatic precipitation is a commonly used method for removing fine particulates from gas streams, particularly at power-generating stations. In an electrostatic precipitator, particles suspended in the carrier gas stream are given an electric charge as they enter the unit and are then removed from the gas stream by the influence of an electrical field. The precipitation chamber or box includes gas flow distribution baffles, discharge and collection electrodes, a dust cleaning system, and collection hoppers. A high direct-current voltage is applied to the discharge electrode system in order to charge the particles, which then are attracted to the collecting surfaces where they become trapped.

In a typical electrostatic precipitator, illustrated in Figure 13.25, the collecting surfaces comprise a group of large, parallel, rectangular metal plates suspended vertically inside the boxlike structure. There are often hundreds of plates with a combined total surface area in the range of tens of thousands of square meters. A series of negatively charged discharge electrode wires hang between all the grounded collector plates, and voltages up to 100,000 V are applied. The suspended particles acquire a negative charge as they flow into and through the precipitation chamber.

Particles that stick to the positively charged collector plates are removed periodically when the plates are mechanically vibrated or *rapped*. Rapping is a mechanical technique for separating the trapped particles from the electrode plates, which typically become covered with a 6-mm (0.25-in.) layer of dust. Rappers are either of the impulse (single blow) or vibrating type. The dislodged particles are collected in a hopper bin at the bottom of the unit and removed for disposal (usually in a landfill).

An electrostatic precipitator can remove particles as small as 1 $\mu$m with an efficiency exceeding 99 percent. For larger particles, the removal efficiencies may exceed 99.9 percent. When the concentration of particles in the carrier gas is very high, a gravity settling chamber or a cyclone is generally installed in front of the precipitator for precleaning of the gas. The removal efficiency of an electrostatic precipitator is very sensitive to the speed and distribution of the gas stream flowing across the plates. The flow must be slow and uniform, that is, the same across each plate, from top to bottom. Removal efficiency is also a function of the total collection surface area and may be expressed by the following formula:

$$E = 100 \times (1 - e^{-wA/Q}) \qquad \text{(13-2)}$$

where $E$ is the percent removal efficiency, $e$ is the natural logarithm base, $w$ is the effective drift velocity, $A$ is the total area, and $Q$ is the gas flow rate through the unit. Effective drift velocity, defined as the speed with which a particle approaches a plate in the electric field, is obtained from pilot studies.

## EXAMPLE 13.6

What is the expected efficiency of an electrostatic precipitator that has a total collector plate area of 5000 m², a flow rate of 150 m³/s, and a drift velocity of 0.1 m/s? What is the efficiency if the plate area is increased to 7500 m² or to 10 000 m²?

*Solution* For an area of 5000 m²,

$$E = 100 \times (1 - e^{-0.1 \times 5000/150}) = 96.4\%$$

For an area of 7500 m²,

$$E = 100 \times (1 - e^{-0.1 \times 7500/150}) = 99.3\%$$

For an area of 10,000 m²,

$$E = 100 \times (1 - e^{-0.1 \times 10,000/150}) = 99.9\%$$

It can be seen that above 99% efficiencies, relatively large increases in collector surface area achieve minimal increases in removal efficiency.

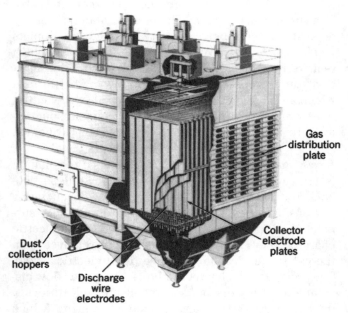

**FIGURE 13.25**
*Cutaway view of an electrostatic precipitator.*
(Courtesy of Research-Cottrell/Air Pollution Control Division)

The installation of precipitator units at a power-generating facility or at an industrial plant represents a very large financial investment for the owner. Removal efficiencies usually are guaranteed by the precipitator manufacturer so as to assure the owner that federal and state air quality or source emission standards will not be violated. Because of this, the precipitator manufacturers build and test exact scale models of each unit in a laboratory before they are actually constructed in the field. In this way, the required configuration or shape of ductwork and baffles can be predetermined so as to provide uniform flow velocities across the plates.

Although installation costs and space requirements are high, the operating costs of electrostatic precipitators are relatively low. Gas flows as high as 5 million cfm (cubic feet per minute) at temperatures of up to 650°C (1200°F) can be handled effectively. Particles in the range of 0.05 to 200 $\mu$m can be removed by electrostatic precipitation with very low pressure losses.

*Fabric Filters.*   One of the most efficient devices for removing suspended particulates from a gas stream is the fabric filter, commonly referred to as a *baghouse.* A typical baghouse comprises a series of long and narrow filter bags, which are suspended upside down in a large enclosure (shown schematically in Figure 13.26) The bags are usually between 125 and 300 mm (5 and 12 in.) in diameter. Dust-laden carrier gas is blown through the bottom of the enclosure by fans; particulates are trapped inside the filter bags, whereas the gas passes through the filter fabric and exits at the top of the baghouse. Filter baghouses are one of the oldest and most commonly used air cleaning devices, and it is estimated that there are more than 200,000 of them currently in operation in the United States.

The removal efficiency of a fabric-filter dust collector can approach 100 percent for particles as small as 1 $\mu$m in diameter, and particles as small as 0.01 $\mu$m can be removed to a significant extent. But the fabric filters cause relatively high pressure losses, and they are expensive to operate and maintain, and the carrier gas generally must be cooled before passing through

the unit; cooling coils needed for this purpose add to the expense. Baghouse filter installations are classified as being either low capacity (handling up to a few thousand cfm), medium capacity (handling up to 100,000 cfm), or high capacity (handling up to 1 million cfm).

Baghouses are generally preceded by gravity settling chambers or cyclones to reduce the particulate load on the filter bags as well as to reduce the required cleaning frequency for the bags. Several compartments of filter bags are often used at a single baghouse installation so that an individual compartment can be cleaned while the others remain in service. The bags are cleaned by one of several methods, and the loosened particulates are collected and removed for disposal.

Three important filter cleaning methods include *shakers, reverse-flow cleaning,* and *pulse-jet cleaning.* A shaker cleaning system separates the collected dust from the fabric by mechanical shaking, using a motor-driven mechanism at the top of the bag. The flow of dirty carrier gas is stopped during the intermittent cleaning process; one isolated filter compartment is cleaned at a time. After a minute or so of shaking, the dust is allowed to settle, and the filtration process resumes.

In reverse-flow cleaning, the dust is separated from the fabric by a low-pressure backflow of air; in high-temperature applications, though, just-cleaned hot gas, rather than ambient-temperature air, is used. The reverse flow of air or gas is driven by a separate fan, which is much smaller than the main system fan. The filter bags generally contain rings that keep them from collapsing during flow reversal. Reverse-flow cleaning is used primarily in high-capacity baghouse installations.

In the pulse-jet cleaning method, pulses of high-pressure compressed air create shock waves that knock the dust away from the filter bag surface. The pulses last only a fraction of a second, and the baghouse does not have to be subdivided for cleaning isolated units; each bag is pulsed every few minutes, allowing a continuous flow of dusty air through the baghouse. A cage inside each bag prevents it from collapsing.

## Control of Gases

Gaseous air pollutants can be controlled using the methods of absorption or adsorption, similar to the techniques used for air sampling, but on a much larger scale. A third method for controlling gaseous emissions involves combustion or incineration. The specific control method selected depends on the nature and properties of the gas or vapor to be controlled.

The terms *gas* and *vapor* are often used synonymously, but there is a basic distinction between the two. The difference is that a gas is not readily condensed into a liquid, whereas a vapor, although existing in a

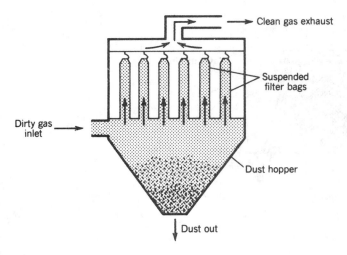

**FIGURE 13.26**
*Section view of a baghouse filter. The filters may be cleaned by mechanical vibrations or by blowing clean air back through the unit.*

gaseous state, is not far from being a liquid. A vapor can exist as dispersed and freely moving molecules at a temperature not far above the dew point, or temperature at which a pure vapor at atmospheric pressure will condense into liquid form. Oxygen and nitrogen in the atmosphere, for example, are considered to be gases, but water or moisture in the air is considered a vapor. Most VOCs (volatile organic compounds) are vapors (with a few exceptions, such as methane and other VOCs with low boiling points). The distinction between gases and vapors is important because different control techniques are applied. Vapors, for example, are more readily adsorbed onto surfaces or condensed into a liquid.

*Absorption.*   The transfer of a gaseous pollutant into a contacting liquid, such as water, is an absorption process; either the gas must be soluble in the liquid or a reactive liquid absorbent is used to capture the pollutant by chemical reaction. Gas removal efficiency depends on the solubility or reactivity of the gas, gas and liquid flow rates, and contact time between the gas and the liquid. Applications of gas absorption in air pollution control include removal of sulfur dioxide from combustion sources, recovery of ammonia in fertilizer manufacture, and control of odors from rendering plants.

Wet scrubbers or washers, similar to those used for suspended particulate control, may be used for gas absorption. Gas absorption can also be carried out in *packed scrubbers* or *towers*. A common type is the *countercurrent* tower: After entering the bottom of the

tower, the gas flows upward through a wetted column of a light, chemically inactive packing material. The liquid flows downward and is uniformly spread throughout the column packing, which increases the area of gas and liquid contact. Thermoplastic materials are the most widely used for gas absorption; certain metals and ceramic materials are sometimes used. Countercurrent scrubbers usually have efficiencies in the range of 90 to 95 percent.

*Cocurrent* and *cross-flow* packed scrubber designs are also available. In the cocurrent design, both the gas and liquid flow in the same direction—vertically downward through the scrubber. Although not as efficient as a countercurrent design, it can work at higher liquid flow rates to prevent plugging of the packing when high particulate loadings are present in the carrier gas. Pressure drops can also be kept low, and tower area can be reduced. The cross-flow design, in which the gas flows horizontally through the packing and the liquid flows vertically downward, also can operate with lower pressure drops when high particulate loadings are present. A cross-flow packed scrubber is illustrated in Figure 13.27.

*Flue Gas Desulfurization.*   Oxides of sulfur, one of the major air pollutants emitted by coal-fired power plants, can be controlled by gas absorption in a scrubber, as well as by several other methods. Options other than scrubbing include changing to a low-sulfur fuel (that is, natural gas or oil) or removing the sulfur from the coal.

An option that is based on absorption is called *flue gas desulfurization* (FGD). FGD systems may

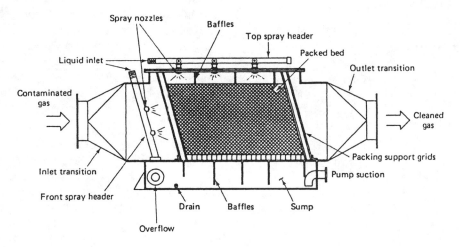

**FIGURE 13.27**
*A cross-flow scrubber.* (From J. Macdonald, "Absorbers," in *Air Pollution Control Equipment*, L. Theodore and A. J. Buonicone, eds., Prentice-Hall, Englewood Cliffs, New Jersey, 1982, p. 48. Copyright © 1982 by Prentice-Hall. Reprinted with permission)

involve wet scrubbing or dry scrubbing. In wet FGD systems, flue gases are contacted with a liquid absorbent, with or into which the sulfur dioxide reacts or dissolves. This forms a slurry or liquid that contains the sulfur compounds. A dry scrubbing system offers cost and energy savings as well as relatively easier operation compared to wet FGD systems, but it requires higher chemical consumption and is limited to low-sulfur coal applications. FGD systems are also classified as either *regenerable* or *nonregenerable*. Most systems are nonregenerable because of their lower capital and operating costs. They produce a sludge that requires appropriate disposal. Regenerable FGD systems are not as common; they require additional steps to convert the sulfur dioxide into useful by-products such as sulfuric acid.

One of several methods for flue gas desulfurization involves contact of the sulfur oxides with lime (CaO) in a wet scrubber. The lime is first reacted with water to produce a slurry of calcium hydroxide. In this *lime scrubbing process*, sulfur dioxide reacts with calcium from the lime to form calcium sulfite and carbon dioxide. If limestone ($CaCO_3$) is used instead of lime, calcium sulfate (gypsum) is formed. The lime process is more efficient (up to 95 percent) than the limestone process, but lime is more expensive than limestone. Both calcium sulfite and calcium sulfate are not very soluble in water and can be precipitated out as a slurry by gravity settling. The thick slurry, called *FGD sludge*, creates a significant sludge disposal problem. Flue gas desulfurization helps to reduce ambient $SO_2$ levels and mitigate the acid rain problem. However, in addition to the expense (which is passed on directly to the consumer as higher rates for electricity), there remains a major disposal problem for the millions of tons of FGD sludge generated each year. A schematic flow diagram for a limestone scrubbing system is shown in Figure 13.28.

*Adsorption.* Gas *adsorption*, as contrasted with *absorption*, is a surface phenomenon: The gas molecules are *sorbed* (attracted to and held) on the surface of a solid. Activated carbon is one of the most common adsorbent materials; it is very porous and has an extremely high surface area-to-volume ratio. Activated carbon (heated charcoal) is particularly useful as an adsorbent for purifying gases containing organic vapors as well as for solvent recovery and odor control. A well-designed carbon adsorption unit can have an efficiency exceeding 95 percent. Figure 13.29 is a schematic diagram of an activated carbon adsorption system.

*Incineration.* The process of combustion or *incineration* (rapid oxidation) can be used to convert VOCs

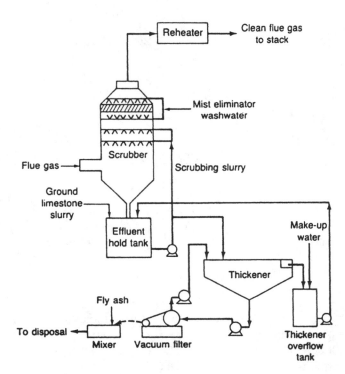

**FIGURE 13.28**
*Limestone FGD flow diagram.* (Reprinted by permission of Waveland Press, Inc., from C. D. Cooper and F. C. Alley, *Air Pollution Control: A Design Approach*, 2nd ed., Waveland Press, Inc., Prospect Heights, Illinois, 1994. All rights reserved)

and other gaseous hydrocarbon pollutants to carbon dioxide and water. Incineration of VOCs and hydrocarbon fumes can be accomplished in a *thermal incinerator* or *afterburner*, as illustrated schematically in

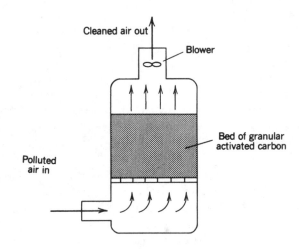

**FIGURE 13.29**
*Activated carbon can be used to adsorb certain gaseous air pollutants.*

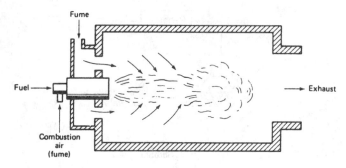

**FIGURE 13.30**

*Schematic of a thermal incinerator.* (From *Air Pollution Control Equipment*, L. Theodore and A. J. Buonicore, eds., Prentice-Hall, Upper Saddle River, New Jersey, 1982, p. 135. Copyright © 1982 by Prentice-Hall. Used with permission)

Figure 13.30. To achieve complete combustion, it must be designed to provide the proper amount of turbulence, burning time, and sufficiently high temperatures. Maintenance of good turbulence (mixing) is a key factor in combustion because it reduces the required burning time and temperature. A process called *direct flame incineration* can be used when the waste gas is a combustible mixture by itself, not needing the addition of extra air or fuel.

Thermal incinerators are typically constructed of a steel shell lined with refractory material (for example, fireclay bricks) 100 to 200 mm (4 to 8 in.) thick. The refractory lining protects the shell and serves as a thermal insulator; temperatures may be as high as 1100°C (2000°F) inside the combustion chamber, but less than 100°C (212°F) on the outer surface of the steel shell.

Incineration can be very efficient (up to 99.95 percent efficient); with enough time and high enough temperatures, the organics can be almost completely oxidized. Overall reduction in VOC emissions from a source depends not only on the efficiency of the thermal oxidizer, but also on the *capture efficiency* of the system. Capture efficiency is the fraction of pollutants emitted that are actually sent through the incinerator. VOCs that escape capture are called *fugitive emissions*. The total efficiency of the system is equal to the product of the afterburner efficiency and the capture efficiency. For instance, if combustion efficiency is 98 percent and the capture efficiency is only 50 percent, then the total pollution control efficiency is only 0.5 × 0.99, or 49 percent. A 100 percent capture efficiency can be achieved in a *total enclosure* in which all airflow is *into* the enclosure, except for the exhaust points, which are directed to the incinerator.

Certain substances, such as platinum, may act in a manner that will assist the combustion or oxidation reaction. Such substances, called *catalysts*, allow complete oxidation of the combustible gases at low temperatures. A *catalytic incinerator* allows the combustion reactions to occur rapidly at temperatures as low as 400°C (750°F). Incineration devices called *catalytic converters* are installed in exhaust systems of automobiles to reduce the carbon monoxide and hydrocarbon emissions by oxidizing them to $CO_2$ and $H_2O$.

*Direct flare systems* provide yet another incineration method for safely disposing of waste gases (like methane) at refineries, chemical process plants, and other facilities. The gas is continually discharged from an elevated stack, with combustion occurring in a chamber at the top of the stack. A blower may be used to supply the air and turbulence needed for smokeless burning of the exhaust gases.

## Emission Controls for Mobile Sources

Highway vehicles—cars, trucks, buses, and motorcycles—are the principal *moving sources* of air pollution, although boats, trains, and airplanes also contribute to the problem. Pollutant emissions from these mobile sources are significant. In 1990, for example, about two thirds of CO emissions in the United States came from mobile sources, mostly from automobiles; in some urban areas, as much as 95 percent of CO emissions came from these sources. It is the mobility of huge numbers of these diverse, decentralized sources that makes their regulation and control even more of a challenge than control of stationary sources. Also, mobile sources such as highway vehicles are typically very close to the *receptors*—people living in cities.

In addition to carbon monoxide, the mobile emissions of most concern in the United States include nitrogen oxides and volatile organic compounds. Sulfur dioxide and particulate emissions are less of a problem, largely because of fuel desulfurization, fewer diesel vehicles, and the generally good condition of motor vehicles. In many other countries, though, sulfur dioxide and particulate emissions are of concern, as are emissions of lead. Vehicular emissions of lead have effectively been eliminated in the United States because of the ban on the use of leaded gasoline.

### Internal Combustion Engines

The four-stroke, gasoline-burning internal combustion engine is the prime mover of most highway vehicles, although motorcycles and small three-wheeled vehicles are also used, particularly in developing countries. There are four major emission points from internal combustion engines, including the exhaust or tailpipe, the engine crankcase vent, the carburetor, and the fuel tank.

Hydrocarbons, carbon monoxide, and nitrogen oxides come from the exhaust, unburned gasoline and hydrocarbons come from the crankcase, and hydrocarbons evaporate from both the carburetor and fuel tank.

Evaporative losses from the fuel tank and carburetor are controlled by the use of an activated carbon canister that stores the vapors emitted when the engine is turned off. When the motor is on, the vapors are purged from the canister and burned in the engine. Crankcase vent emissions are controlled by *positive crankcase ventilation* (PCV) *systems*, which serve to recycle gases that slip by the piston rings back to the engine intake manifold; a PCV check valve is used to prevent the buildup of excessive pressure in the engine crankcase.

Emissions from the tailpipe of an automobile, which account for roughly 60 percent of the hydrocarbons and almost all the carbon monoxide and nitrogen oxide emissions, depend on the mode of operation of the vehicle. Acceleration results in less carbon monoxide and hydrocarbon emissions, but more nitrogen oxides; deceleration results in less nitrogen oxides, but much higher hydrocarbon levels from the partially burned fuel. Because of these variations, exhaust emissions testing procedures are standardized so that test results will be meaningful.

Exhaust emissions from an automobile tailpipe are controlled by use of a *catalytic converter*, a device installed in the tailpipe ahead of the muffler. It allows almost complete oxidation of combustible gases in the exhaust to occur at relatively low temperatures; in effect, a catalytic converter allows "flameless combustion" to take place. The carbon monoxide and unburned hydrocarbons in the exhaust are converted to carbon dioxide and water. But any sulfur in the gasoline is oxidized to particulate sulfur trioxide, thereby increasing sulfur levels in the air.

To further reduce pollutant emissions from gasoline-powered motor vehicles, certain engine operating or design features can be controlled or modified. For example, increasing the air-to-fuel ratio helps combustion, thereby reducing carbon monoxide and hydrocarbon emissions. Carburetors are not very effective in accurately controlling the air-to-fuel ratio; new cars are equipped with computer-controlled fuel injectors that can optimize the air and fuel mixture for different driving conditions.

Other key engine design and operation features that affect emissions include the compression ratio and timing of the spark plugs. High compression ratios produce greater power, but the accompanying higher temperatures produce more nitrogen oxide emissions. Hydrocarbon and nitrogen oxide emissions can be reduced by retarding the spark, but this reduces the performance of the engine. Clearly, control of air pollutant emissions from internal combustion engines is a complex problem.

As a result of emission controls required for gasoline-powered internal combustion engines, today's cars typically emit about 80 percent less pollution than did cars in the 1970s. On the other hand, total vehicle miles traveled have more than doubled since then, and many challenges remain in the effort to further reduce vehicle-related air pollution.

*Changes in Fuels.*   Fuel composition is yet another factor in mobile source emissions, and changing or modifying the fuel can play an important role in controlling air pollution from these sources. Removing impurities in the fuel is an obvious way to reduce pollution. Lead, a substance that was blended with gasoline to increase octane ratings and engine performance, is one substance no longer allowed in gasoline. Airborne lead is not only a health hazard, it also damages catalytic converters. The elimination of leaded gasoline in the United States has significantly reduced lead levels in urban air.

Another example of an undesirable substance in gasoline is sulfur, which not only leads to increased sulfur dioxide emissions, but interferes with the operation of catalytic converters, thereby contributing to increased carbon monoxide and hydrocarbon emissions. Because of more strictly controlled refining processes compared to other countries, gasolines produced in the United States contain less sulfur than fuels produced elsewhere and therefore cause less sulfur dioxide emissions.

Alternative fuels that can replace gasoline include compressed natural gas, liquefied petroleum gas, methanol, ethanol, propane, and hydrogen. Each of these possible replacements has a variety of advantages as well as disadvantages compared to gasoline. For example, compressed natural gas produces about 50 percent less carbon monoxide and volatile organic emissions and no air toxics such as benzene, but it has significant problems related to on-vehicle fuel handling, storage, and refueling. Hydrogen gas produces no carbon oxides or hydrocarbons and has the highest energy content of any combustible fuel, but there are numerous technological and safety problems to solve before hydrogen becomes a practical alternative fuel for automobiles.

*Reformulated Gasoline* (RFG).   As required by the 1990 *Clean Air Act Amendments*, a Reformulated Gasoline Program went into effect in 1995 in areas of the United States designated as having severe ozone pollution and smog, as well as in other areas of the country that opt to participate in the program. RFG is an *oxygenated fuel*, containing at least 2 percent of oxygenates by weight. The oxygenates replace or dilute other, less desirable compounds and enhances cleaner burning, especially in cold climates. This blend of gasoline also has limits on the amounts of benzene and other toxics, as well as hydrocarbons, it may contain.

RFG is expected to reduce emissions of hydrocarbons and toxics by at least 15 percent as compared to 1990 levels; emission of other criteria pollutants will also be reduced. Although the price of RFG is expected to raise fuel costs by 2 to 4 cents per gallon, no additional investments are necessary to retrofit highway vehicles or gasoline filling stations to accommodate the new fuel.

*Diesel Standards.* In 2001, the EPA implemented standards to control emissions from diesel engines, a significant source of air pollution. Under these new standards, diesel producers will be required to virtually eliminate sulfur from the fuel. Sulfur produces soot and clogs up a vehicle's catalytic converter. With the removal of sulfur, manufacturers of diesel engines will be required to incorporate the sophisticated pollution control devices that are now standard equipment in all gasoline-burning cars. These changes will be phased in over time—80 percent of all diesel fuel must be virtually sulfur-free by year 2006, and the rest by 2010. Diesel engine manufacturers will have until 2010 to complete the required engine modifications. These new standards will cut air pollution from trucks and buses and other commercial vehicles by 95 percent, the equivalent of removing 13 million trucks from the road.

### Zero-Emission Vehicles

Alternatives to internal combustion engines include solar-powered cars and electric vehicles. Solar-powered cars are still very much in the experimental stage, but electric cars are of particular interest at the present time. Some states (for example, New York and California) require that, by the year 2003, 10 percent of all motor vehicles sold in the state must be *zero-emission vehicles.* The only zero-emission vehicle presently capable of mass production is the electric-powered car, which uses lead–acid batteries to store the electricity.

Most electric-powered cars and vans have a range of only about 130 km (80 mi) before the batteries need recharging, and they are considerably more expensive than conventional gasoline-powered vehicles ($10,000 to $25,000 more by some estimates). Some researchers think that emissions from the mining, smelting, and recycling of lead needed to operate a large fleet of electric cars and vans could pose serious threats to public health. In addition, the extra electricity needed to recharge car batteries will come from power plants that also generate pollution.

Clearly, there are trade-offs that must be considered by environmental engineers and policy makers in regard to zero-emission vehicles. Although electric-powered vehicles operate adequately for certain light-duty applications, more research and improvement in battery technology is needed before they are likely to become a significant alternative to internal combustion engines.

### Hybrid Vehicles

A *hybrid* vehicle is one that can use two sources of power to turn the wheels. Some automobile makers, for example, have begun to make prototypes of hybrids that use a gasoline-powered internal combustion engine with an electric motor. The electric motor powers the car up to a speed of about 20 km/h (12 mph), below which gasoline engine emissions are highest. Above 20 km/h, the vehicle's relatively small (1.5-Li) gasoline engine starts up and becomes the main power source. The electric motor can also operate when peak power is needed, as when climbing a hill. When the gasoline engine operates, it also recharges the batteries for the electric motor, so the vehicle does not require any external recharger. It is very possible that hybrid vehicles such as this will be used until a technical breakthrough makes zero-emission vehicles fully competitive with other types of vehicles.

 ## 13.8 RELEVANT WEB SITES

### AIRNow

http://www.epa.gov/airnow/index.html

This is the Home Page of the EPA's AIRNow Web site, which provides easy access to national air quality information for more than 100 cities in the United States. This site provides daily air quality forecasts as well as real-time air quality (AQI) data in a visual and easy-to-understand format. Four main parts of the site include "Ozone Maps," "Air Quality Forecasts," "Where I Live," and "Publications."

### AIR RELEASES OVERVIEW

http://www.epa.gov/enviro/html/airs/airsover.html

This site contains emissions and compliance data on air pollution point sources regulated by the U.S. EPA and state and local air regulatory agencies.

## CLEAN AIR MARKET PROGRAMS

http://www.epa.gov/airmarkets/

This is the Home Page of the EPA Clean Air Market Programs, which focus on emissions allowance trading to reduce air pollution. The site offers much information about programs, regulations, progress, and results of efforts to control acid rain, smog, and ozone pollution.

## GLOBAL CLIMATE CHANGE

http://www.ec.gc.ca/climate/index.html

This is Canada's Home Page for global warming, and provides much information on climate change, causes, effects, solutions, and what is being done in Canada to address the issue.

## GLOBAL WARMING SITE

http://www.epa.gov/globalwarming/

This EPA site provides the latest news about global climate change, information about the causes and impacts of global warming, the U.S. inventory of greenhouse gases and sinks, and a wealth of other information on the greenhouse effect.

## INDOOR AIR QUALITY

http://www.epa.gov/iaq

This is EPA's award-winning Home Page for indoor air quality, and provides much useful information and many links to publications on the sources and effects of indoor air pollutants in homes, schools, and offices.

## OFFICE OF AIR AND RADIATION

http://www.epa.gov/oar/

This Home Page of the EPA's Office of Air and Radiation deals with technical policies and regulations for air pollution control. Topics include indoor and outdoor air quality, stationary and mobile sources of air pollution, radon, acid rain, stratospheric ozone depletion, and pollution prevention.

## OFFICE OF AIR QUALITY PLANNING AND STANDARDS

http://www.epa.gov/oar/oaqps/

This is the Home Page of EPA's Office of Air Quality Planning and Standards (OAQPS), which directs national efforts to meet air quality goals, particularly for smog, air toxics, carbon monoxide, lead, particulate matter (soot and dust), sulfur dioxide, and nitrogen dioxide. The OAQPS is responsible for more than half of the guidance documents, regulations, and regulatory activities required by the *Clean Air Act Amendments of 1990.*

## OZONE DEPLETION

http://www.epa.gov/ozone

This EPA Web site contains information about the science of ozone depletion, regulations in the United States designed to protect the ozone layer, information on methyl bromide, flyers about the UV index, information for the general public, and other topics.

## PLAIN ENGLISH GUIDE TO THE CLEAN AIR ACT

http://www.epa.gov/oar/oaqps/peg_caa/pegcaain.html

This site provides an easy to understand summary of the 1990 version of the *Clean Air Act.*

## UNITED NATIONS CONVENTION ON CLIMATE CHANGE

http://www.unfccc.de/

This site provides information about the Kyoto Protocol, the UN Framework Convention, emissions, and other data.

# REVIEW QUESTIONS

1. Briefly discuss the historical background of air pollution. What is an *air pollution episode?* Give two examples.

2. What are the major gaseous constituents of fresh (clean) air?

3. Give a brief definition of air pollution. What is meant by the term *anthropogenic air pollution?*

4. List three sources and types of natural air pollutants.

5. What is the *troposphere?* How does it differ from the *stratosphere?*

6. What is meant by *atmospheric stability?* How does it affect air quality?

7. Define the terms *environmental lapse rate, adiabatic lapse rate,* and *temperature inversion.* How does a weak prevailing lapse rate differ from an inversion? Does a superadiabatic lapse rate cause poor air quality? Briefly explain your answer.

8. Describe three types of temperature inversions. How do they affect air quality? Which type of inversion causes *fumigation?*

9. Consider a parcel of air at an altitude of 1000 m. It will have the same pressure and temperature as the air surrounding it, say 18°C. Suppose that the prevailing lapse rate at the time is +2°C per 100 m. If the air parcel rises to an altitude of 1100 m, how will its temperature compare to that of the surrounding air? Will the parcel of air continue to rise, or will it tend to descend? Why? What would happen if the parcel of air moved to a lower altitude, say 800 m? Would you say that the atmosphere was stable under these conditions? Make a sketch similar to Figure 13.4 to illustrate your answer.

10. What is the difference between primary air pollutants and secondary air pollutants? What is the difference between criteria air pollutants and hazardous air pollutants?

11. What are the six criteria air pollutants regulated under the *Clean Air Act?* What are their major sources?

12. What is the difference between a *dust* and a *fume?* Between a *mist* and a *spray?* Is *fly ash* synonymous with *smoke?*

13. What size range of particulates is of most significance with respect to human health? What size range of particulates will readily settle out of the air?

14. Briefly discuss *photochemical smog.* Is it a primary pollutant? Why?

15. Briefly discuss typical types and sources of hazardous air pollutants.

16. Briefly discuss the adverse health effects of the six criteria air pollutants. What are some other harmful effects of air pollution in addition to those on human health?

17. Briefly describe four different scales of air pollutant transport in the atmosphere.

18. Briefly explain the difference between the *anthropogenic greenhouse effect* and the *natural greenhouse effect.* Why is the term *greenhouse effect* not completely accurate as a synonym for global warming?

19. Briefly discuss the role of the wavelength of light (that is, short versus long or visible versus infrared light) as a factor in the natural greenhouse effect. How does the troposphere serve as a blanket to keep Earth warm?

20. Identify and briefly discuss the major *greenhouse gases.*

21. Briefly discuss the potential impacts of global warming.

22. Why is stratospheric ozone not considered an air pollutant? What is the *ozone hole,* and what harm is done by global ozone depletion?

23. What is the major cause of stratospheric ozone depletion?

24. What is *acid deposition?* Briefly discuss its causes and effects.

25. List the major sources and effects of indoor air pollution.

26. Describe methods for controlling and reducing indoor air pollution for each of the major sources.

27. Briefly discuss *sick building syndrome.*

28. Briefly explain the differences between *source sampling* and *ambient sampling.* What is meant by *isokinetic sampling?* Is there a difference between air quality sampling and air quality monitoring?

29. What are the typical units of measurement of gaseous and particulate air pollutants? Do they all vary with air temperature and pressure?

30. Discuss the major techniques and devices used to sample and measure particulate air pollutants.

31. Discuss the major techniques (that is, absorption and adsorption) and the devices used to sample and measure gaseous air pollutants.

32. Discuss five strategies for air pollution control.

33. With regard to air pollution regulations, briefly discuss the meanings of *threshold limit values, emission standards,* and *ambient standards.*

**34.** Briefly discuss the history and scope of the *Clean Air Act*. What is the Pollution Standards Index?

**35.** Describe the basic operating principles for five different particulate emission control devices for stationary sources. Which of these are used primarily as precleaning devices?

**36.** Describe the key devices used to control gaseous air pollution? What is an FGD system used for?

**37.** Combustion is a major source of air pollution. Can it also be used for air pollution control? Explain.

**38.** Describe the basic emission control techniques for highway vehicles. What is RFG?

**39.** Visit and explore the EPA *Office of Air and Radiation* Web site. Link to the "Indoor Air" page and check out "What's New at IAQ." Read one of the documents and write a brief summary of what you learned from the article.

**40.** Visit and explore the EPA *Global Warming Site.* Link to the "What's in the News" page. Read one of the recent documents and write a brief summary of what you learned from the article.

**41.** Using the EnviroSources search engine (see Section 1.6) or another Internet search directory or search engine, locate at least one additional Web site relevant to one of the topics in this chapter. Add the link(s) to your Environmental Technology Folder. Write a brief description of what the Web site(s) contain.

## PRACTICE PROBLEMS

**1.** Plot a graph of altitude versus prevailing air temperature, using the following data:

| Altitude, m: | 0 | 50 | 100 | 150 | 200 | 250 | 300 |
|---|---|---|---|---|---|---|---|
| Temperature, °C : | 21 | 20 | 19 | 22 | 21 | 20 | 19 |

  *(a)* What kind of lapse rate exists between 0 and 100 m?

  *(b)* What kind of lapse rate exists between 200 and 300 m?

  *(c)* Do these data show a temperature inversion? If so, where?

  *(d)* Would fumigation occur if smoke at 20°C was discharged from a chimney at a height of 50 m? At a height of 250 m?

**2.** Plot a graph of altitude versus prevailing air temperature, using the following data:

| Altitude, m: | 0 | 50 | 100 | 150 | 200 |
|---|---|---|---|---|---|
| Temperature, °C : | 19 | 19.5 | 20 | 20.5 | 19 |

  *(a)* What kind of lapse rate exists between 0 and 150 m?

  *(b)* What kind of lapse rate exists between 150 and 200 m?

  *(c)* Do these data show a temperature inversion? If so, where?

  *(d)* Would fumigation occur if smoke at 20°C was discharged from a chimney at a height of 100 m? At a height of 200 m?

**3.** Federal standards limit annual mean sulfur dioxide ($SO_2$) levels to 80 $\mu g/m^3$ at 25°C and 1 atm pressure. Express this concentration in terms of ppm.

**4.** Federal standards limit annual mean nitrogen dioxide ($NO_2$) levels to 100 $\mu g/m^3$ at 25°C and 1 atm pressure. Express this concentration in terms of ppm, as well as a percentage by volume.

**5.** Federal standards limit 8-h average carbon monoxide (CO) levels to 9 ppm. Express this concentration in terms of $mg/m^3$ at 25°C and 1 atm pressure.

**6.** Federal standards limit 1-h average carbon monoxide (CO) levels to 40 $mg/m^3$ at 25°C and 1 atm pressure. Express this concentration in ppm, as well as percent by volume.

**7.** The exhaust of an automobile contains 1.8 percent by volume of carbon monoxide (CO) at a temperature of 85°C. Express this concentration as $g/m^3$.

**8.** The exhaust of an automobile contains 0.1 percent of nitrogen dioxide ($NO_2$) at a temperature of 90°C. Express the concentration in $g/m^3$.

**9.** The mass of a 6-in.-diameter dustfall bucket is 125.00 g when empty. After 30 d of exposure, the bucket and collected particulates have a combined mass of 125.25 g. Compute the dustfall in terms of tons per square mile per month, as well as kilograms per hectare per month.

10. The airflow through a hi-vol sampler was recorded as 54 ft³/min at the beginning and 48 ft³/min at the end of a 24-h sampling period. The filter weighed 9.80 g before and 10.05 g after sample collection. What was the suspended particulate concentration?

11. What is the expected efficiency of an electrostatic precipitator that has a total collector plate area of 6000 m², a flow rate of 200 m³/s, and a drift velocity of 0.15 m/s? What is the efficiency if the plate area is increased to 7500 m² or to 15 000 m²? Plot a graph that shows the relationship between efficiency and total plate area, assuming that drift velocity and flow rate remain constant.

12. What is the expected efficiency of an electrostatic precipitator that has a total collector plate area of 4000 m², a flow rate of 100 m³/s, and a drift velocity of 0.12 m/s? What is the efficiency if the flow rate is increased to 150 m³/s or decreased to 50 m³/s? Plot a graph that shows the relationship between efficiency and flow rate.

## Chapter Outline

CHAPTER

# 14

# Noise Pollution and Control

Noise is perhaps one of the most undesirable by-products of a modern mechanized lifestyle. It may not seem as insidious or harmful as the contamination of, for example, drinking water supplies from hazardous chemicals, but it is a pollution problem that affects human health and well-being and that can contribute to a general deterioration of environmental quality. It can affect people at home, in their community, or at their place of work.

Simply defined, *noise is undesirable and unwanted sound.* It takes energy to produce all sound, so, in a manner of speaking, noise is a form of *waste energy.* It is not a substance that can accumulate in the environment, like most other pollutants, but it can be diluted with distance from a source. All sounds come from a *sound source*, whether it be a radio, a machine, a human voice, an airplane, or a musical instrument. Not all sound is noise. What may be considered music to one person may be nothing but noise to another. To an extent, noise pollution is a matter of opinion.

Noise from highway traffic, construction activities, and other sources in the community is of special concern to civil and environmental engineers and technicians. Architects, builders, urban planners, and public health officials also get involved in problems related to noise, as do industrial hygienists, mechanical engineers, and equipment manufacturers. Noise is a ubiquitous problem and can have impacts ranging from temporary public or personal nuisance to permanent hearing loss in individuals.

Laws and regulations regarding noise pollution exist at the federal, state, and local levels. More and more communities are imposing noise controls on construction activities, and the environmental impact reports for land development or transportation projects usually must include a section on noise. To understand noise regulations and to plan for effective community and occupational noise control, it is necessary to have a basic understanding of sound and its measurement and the effects of noise.

## 14.1 BASIC PHYSICS OF SOUND

Sound energy is *produced by mechanical vibrations of a sound source.* The vibrations are transmitted or carried away from the source in the form of *sound waves.* Sound waves can be transmitted through solids, liquids, or gases, but they cannot be transmitted in a vacuum, where there is no medium or material to transmit the vibrations.

In air, sound waves can be visualized as a series of tiny, quick pulses or oscillations of air pressure (or air density), slightly above and slightly below the ambient atmospheric pressure. This is shown schematically in Figure 14.1. The pulses of pressure impinge on the ear, creating what the mind perceives as *sound.*

Molecules in the high-density regions are relatively close together; molecules in the low-density regions are spread further apart. The regions of high and low air pressures may be represented graphically as the peaks and valleys of a trigonometric sine curve. This looks much like a wave traveling along a string that is being moved side to side or up and down, but, of course, this is only a schematic diagram. Air molecules carrying a sound wave are actually vibrated or displaced rapidly back and forth in the direction of the wave; they do not move or travel along a sine curve.

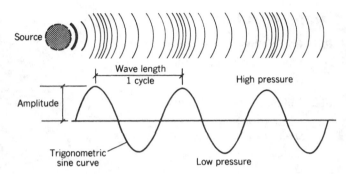

**FIGURE 14.1**
*The alternating high- and low-pressure regions in a sound wave can be represented by a trigonometric sine curve. The peaks represent high-pressure regions, and the valleys represent low-pressure regions, above and below average atmospheric pressure.*

## Sound Wave Characteristics

The distance between the pressure peaks (or valleys) is called the *wavelength.* As sound is transmitted, the wave appears to move outward from the source of vibration. The number of wavelengths that appear to pass a fixed point in 1 s is called the *frequency* of the wave. The height of the peaks, called the *amplitude* of the wave, represents the pressure intensity and is related to the volume or loudness of the perceived sound.

A single wavelength is also called a *cycle.* Frequency can be expressed in units of *cycles per second* (cps), but the term *hertz* (Hz), named after the 19th-century physicist Heinrich Hertz, is now more commonly used. One hertz is equivalent to 1 cycle per second (1 Hz = 1 cps). A sound wave with a frequency of 1000 Hz, for example, is one in which the pressure waves pass by a given point at a rate of 1000 cps. Another way to view the frequency of a sound wave is to consider it as the number of times the air pressure reaches the peak value (or wave amplitude) in 1 s at any particular point in the air. If the sound pressure (above and below the barometric pressure) at any fixed point is plotted on a graph as a function of time, the sine curve will be formed.

The frequency should not be confused with the *speed of sound,* which is a constant in a given transmission medium. For example, a sound wave travels at a speed of about 1500 m/s in water and about 5000 m/s in steel. In air, at standard temperature and pressure, sound travels at a constant speed of about 340 m/s (1100 ft/s). If a bolt of lightning is seen, for example, and it then takes about 3 s to hear the thunder, it is safe to say that the lightning occurred (3 s) × (340 m/s) = 1020 m, or about 1 km, away from the observer's location.

There is an important relationship among the three key characteristics of a sound wave: wavelength, frequency, and speed. The speed of a wave is equal to the product of its wavelength and frequency. This is expressed simply as

$$v = \lambda \times f \qquad \textbf{(14-1)}$$

where $v$ = speed of sound, m/s or ft/s
 $\lambda$ = wavelength, m or ft per cycle
 ($\lambda$ = Greek "lambda")
 f = frequency, Hz or cps

The dimensional validity of this expression is easily verified as follows:

$$\frac{\text{meters}}{\text{second}} = \frac{\text{meters}}{\text{cycle}} \times \frac{\text{cycle}}{\text{second}}$$

Since the speed of sound is a constant, there is an inverse relationship between the frequency and the wavelength. In other words, the higher the frequency, the shorter is the wavelength and vice versa.

## EXAMPLE 14.1

Sound waves in air at 21°C travel at a speed of 344 m/s. How long would it take to hear thunder from a bolt of lightning that occurred 5.0 km from an observer?

*Solution* From physics, it is known that distance equals the product of speed and time. Expressing this algebraically, after first converting km to m, yields

$$5000 \text{ m} = 344 \text{ m/s} \times \text{time}$$

and

$$\text{time} = \frac{5000 \text{ m}}{344 \text{ m/s}} = 15 \text{ s} \qquad \text{(rounded off)}$$

## EXAMPLE 14.2

What is the wavelength of a sound traveling through the steel rails of a railroad track if the frequency of the sound caused by a moving train is 500 Hz? (Assume that sound travels at a speed of 5000 m/s in steel.)

*Solution* Applying Equation 14-1 gives $\lambda = v/f = 5000/500 = 10$ m.

## Loudness and Pitch

Two terms used to describe human perception of sound are *loudness* and *pitch*. Loudness is related to the amplitude of the wave as well as other factors, and is discussed in more detail in the next section. Pitch is a function of the frequency of the wave that produces it. A high-pitched sound (for example, a shrill whistle) has a relatively high frequency, compared to a low-pitched sound (for example, a fog horn), which has a lower frequency. (It also follows that a high-pitched sound must have a shorter wavelength than a low-pitched sound.)

Sound waves cause eardrums to vibrate, activating middle and inner ear organs and sending bioelectrical signals to the brain. The human ear can detect sounds in the frequency range of about 20 to 20,000 Hz, but for most people hearing is best in the range of 200 to 10,000 Hz. A sound of 50-Hz frequency, for example, is perceived to be very low-pitched, and a 15,000-Hz sound is very high-pitched. The middle C note on a piano has a frequency of 262 Hz. In normal conversation, the human voice covers a range of about 250 to 2000 Hz. The audibility of a sound depends on both frequency and amplitude. As people age, hearing often become less acute. This natural loss of hearing ability, called *presbycusis*, is not related to noise pollution exposure.

The sine curves or sound waves illustrated in Figure 14.2*a* and Figure 14.2*b* are *pure tones*. Sound, and especially noise, is typically much more complex than a pure tone; it comprises a combination of many pressure waves, each with a different frequency and amplitude. A sound wave formed from such a combination does not exhibit the simple pattern of a sine curve. Two pure tones, such as these shown in Figure 14.2*a* and Figure 14.2*b*, can combine to form a wave like the one shown in Figure 14.2*c*. A random pattern of pressure intensity is typical of most unwanted sounds or noises, as shown in Figure 14.2*d*.

The pressure of a cyclical function, such as a sound wave, is characterized by a mathematical term called the root-mean-square value, or *rms*. This is necessary because the arithmetic average of the sound pressure peaks and low points is zero; half are above and half are below the ambient air pressure. The rms value is equal to the *square root of the mean of the squares* of the sound pressure fluctuations. In the following discussion on noise measurement, reference to sound pressure refers to the rms value.

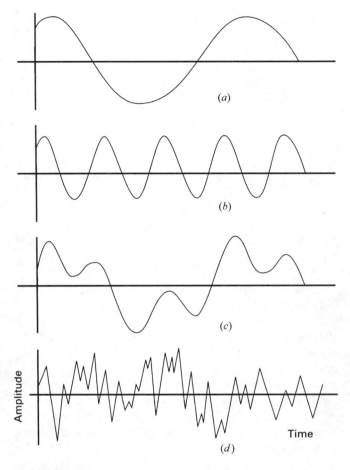

**FIGURE 14.2**
*Pure tones (a and b) can combine to form a wave pattern like (c); random noise is depicted in (d).*

## 14.2 MEASUREMENT OF NOISE

Sound waves are characterized by cyclical changes of air pressure above and below average atmospheric pressure. Compared to average atmospheric or barometric pressure, sound pressure fluctuations are exceedingly small. Average atmospheric pressure at sea level is about 100 kPa (14.7 psi); it may also be expressed as 1 *bar*. Sound pressures are so small that they are more conveniently expressed as *microbars*, or $\mu$bar. A $\mu$bar is one millionth ($1 \times 10^{-6}$) of average atmospheric pressure at sea level. One $\mu$bar is equivalent to about 0.1 Pa (0.0000147 psi).

The human ear is a remarkably sensitive organ. The average person with normal hearing can detect a sound with a pressure amplitude as small as 0.0002 $\mu$bar. This value is, by convention, taken to be the lowest audible sound for humans and is used as a base or reference level in most noise measurements. This *threshold of hearing* may also be expressed in terms of micropascals ($\mu$Pa), where 0.0002 $\mu$bar = 20 $\mu$Pa.

The highest sound pressure that can be perceived by the human ear without causing pain is about 1000 $\mu$bar. The range of pressure, from 0.0002 $\mu$bar up to 1000 $\mu$bar, covers a very wide spectrum; it is equivalent to a range of 5 million to 1 (since 1000/0.0002 = 5,000,000). The human ear can actually perceive sound pressures as high as 10,000 $\mu$bar (or 1 kPa) before immediate physical damage occurs to the eardrum or middle ear.

### The Decibel Scale

Measuring sound levels with pressure units that can vary over such a wide range of values is impractical and inconvenient. Another disadvantage of measuring sound in terms of microbars or micropascals is the fact that the ear responds nonlinearly with respect to pressure. This means that a person's perception of loudness is not simply a direct function of sound pressure. For example, a doubling of sound pressure levels is not necessarily heard or perceived as a doubling of the loudness of the sound. Other factors such as frequency are involved in this phenomenon, as are discussed shortly.

To avoid the disadvantages of using pressure directly for sound or noise measurement, a logarithmic relationship called the *decibel scale* is used. The measurement units, called *decibels* (dB), do not represent actual physical qualities, such as pressures. A decibel is essentially a *ratio* of two pressures; logarithms are used to convert the range of the ratios into more manageable and convenient numbers. The magnitude of *volume* or a sound expressed in decibels is called a *sound pressure level* (SPL). An SPL is defined mathematically as

$$\text{SPL} = 20 \times \log\left(\frac{P}{P_0}\right) \qquad \textbf{(14-2)}$$

where SPL = sound pressure level, dB
  $P$ = rms sound pressure, $\mu$bar
  $P_0$ = reference pressure, $\mu$bar

The reference pressure generally used for $P_0$ is the *hearing threshold* or lowest audible sound pressure of 0.0002 $\mu$bar. The base of the logarithm or log function is 10; recall that the logs are essentially exponents or powers of 10. For example, $\log(100) = \log(10^2) = 2$; $\log(1000) = 3$, and so on. Also, since $10^0$ is defined mathematically as being equivalent to unity or 1, $\log(1) = \log(10^0) = 0$.

Based on the preceding definition of sound pressure level, an SPL of 0 dB does not represent the complete absence of sound. It represents the hearing threshold or lowest audible sound for most people. This can be seen by applying Equation 14-2, using 0.0002 $\mu$bar for both $P_0$ and $P$; SPL = 20 × log(0.0002/0.0002) = 20 × log(1) = 20 × 0 = 0. In fact, some people with particularly acute powers of hearing can detect sounds with negative SPL decibel values.

## EXAMPLE 14.3

An ambulance siren causes a sound pressure of 200 $\mu$bar. What is the SPL of the siren?

*Solution*  Applying Equation 14-2, we obtain

$$\text{SPL} = 20 \times \log\left(\frac{200}{0.0002}\right) = 20 \times \log(10^6)$$
$$= 20 \times 6 = 120 \text{ dB}$$

The threshold of pain for humans, about 1000 $\mu$bar, is equivalent to about 134 dB (try the computation using Equation 14-2). It can be seen, then, that the decibel scale serves to reduce the scope or spread of sound measurement to a reasonably convenient range of 0 dB to about 140 dB. To put the decibel scale in perspective with regard to people's perceptions of the loudness of common sounds and levels of noise, a list of typical SPL values is shown in Figure 14.3.

### Combined Noises

In many instances, it is necessary to predict what the combined sound pressure level will be when two or more nearby noise sources act at the same time. For example, it may be required to estimate the noise level expected at the boundary of a construction site caused by the combined operation of trucks, dozers, pavers, and other machinery. First, it is important to realize that it would be incorrect to add together the SPLs of each individual noise source by using simple arithmetic. This is because of the logarithmic nature of the decibel scale.

For practical purposes, it is convenient to know that the combination of two sounds with equal SPL values always results in only a 3-dB increase over the SPL of one source alone. For example, if a construction drill has an SPL of 90 dB and two identical drills are operating simultaneously next to each other, the resulting noise or SPL from both together will be 93 dB (but not 180 dB!). Likewise, if one loud ambulance siren causes 120 dB of noise, then two identical sirens will result in

| Sound level in dB | Environmental conditions |
|---|---|
| 140 | Threshold of pain |
| 130 | Pneumatic chipper |
| 120 | Loud automobile horn (distance 1 m) |
| 110 | Overhead jet plane |
| 100 | Inside subway train (New York) |
| 90 | Inside motor bus |
| 80 | Average traffic on street corner |
| 70 | Conversational speech |
| 60 | Typical business office |
| 50 | Living room, suburban areas |
| 40 | Library |
| 30 | Bedroom at night |
| 20 | Broadcasting studio |
| 10 | Threshold of hearing |
| 0 | |

**FIGURE 14.3**
*The decibel (dB) scale is used to measure noise levels.* (Courtesy of Bruel & Kjaer Instruments, Inc., Marlborough, Massachusetts)

an SPL of 123 dB. It does not matter what the original SPL value is; the combined SPL will only be 3 dB more than the SPL of the single source.

On the other hand, when two sounds that differ by more than 15 dB in SPL are combined, the contribution of the weaker sound will not be noticeable to even the most sensitive ear or measuring device. For example, if a 95-dB rock drill is operating simultaneously with and next to an 80-dB loader at a construction site, the combined SPL will be measured and perceived as 95 dB. The weaker sound is, in effect, drowned out by the louder one. This phenomenon is called *masking* of one sound by another. Most people have experienced this at one time or another. For example, industrial noise in a factory at a level of about 80 dB will mask most human voices, making conversations difficult, if not impossible, under those conditions.

Sometimes the SPL values of noise sources that differ between 0 and 15 dB must be combined. To avoid complicated mathematics, it is convenient to use a chart like the one illustrated in Figure 14.4. The chart is entered first with the numerical difference between the two noise levels to be added. Moving up to the curve

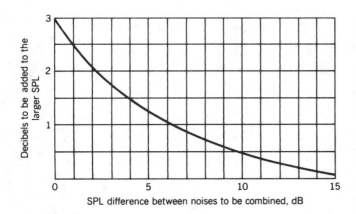

**FIGURE 14.4**

*A chart like this may be used to simplify the addition of decibel values when combining two or more sound levels. The chart is entered on the horizontal axis with the numerical difference between the sound levels to be combined, and the corresponding number of dB to be added to the larger of the two sound levels is read on the vertical axis.*

and then across to the left axis, the corresponding number of decibels to be added to the higher SPL value is read from the chart (see Example 14.4).

**EXAMPLE 14.4**

Four identical dozers are available for excavation of soil at a construction site. Each dozer has an SPL of 90 dB when operating alone. What is the SPL when the four dozers are operating at the same time? (For simplification, the effect of distance from the noise source has been ignored in the discussion of SPL. For this problem, it is assumed that the dozers are operating together in a confined area.)

*Solution*   First, consider what happens when only two of the machines are operating. The numerical difference between the two SPL values is $90 - 90 = 0$. Entering the curve in Figure 14.4 with a difference of 0, read a corresponding 3-dB value on the vertical axis. Therefore, the two dozers operating simultaneously will generate an SPL of $90 + 3 = 93$ dB. This is an expected from the previously discussed rule for combining identical SPL values.

With a third dozer operating, add another 90 dB to the previous level of 93 dB for two dozers. The difference between the two SPL values is now $93 - 90 = 3$ dB. Entering the chart with a 3-dB difference, find that it is necessary to add an additional 1.7 dB to the 93 dB to obtain the combined SPL. Thus, $93 + 1.7 = 94.7$ dB from

three dozers operating simultaneously. Remember that the SPL increment is always added to the larger of the two SPL values being combined.

Finally, with a fourth dozer, combine the 94.7 dB with 90 dB. The difference is 4.7 dB, and the required increment from the chart is 1.3 dB. This results in $94.7 + 1.3 = 96$ dB being generated by the four machines. This can be checked by considering that combining two pairs of dozers, at 93 dB a pair, also yields $93 + 3 = 96$ dB.

### Sound Intensity

It has been noted that all sounds are produced by mechanical vibrations of some physical medium and that sound waves transmit energy. The rate at which this energy is transmitted through a unit area perpendicular to the direction of the sound wave is called *sound intensity* and is expressed in terms of watts per square meter (W/m$^2$). Sound intensity is proportional to the square of the rms value of sound pressure or SPL.

For every 10-dB increase in the SPL, there is a tenfold increase in sound intensity. For instance, an SPL of 10 dB is ten times more intense than an SPL of 1 dB, 20 dB is 100 times ($10^2$) more intense, 30 dB is 1000 times ($10^3$) more intense, and so on. Also, for every doubling of sound intensity (for example, doubling of the number of equal sound sources), the SPL increases by 3 dB (as illustrated in Example 14.4).

For a single-frequency sound or pure tone, there is a direct correspondence between loudness and intensity. However, most sounds comprise several frequencies, and the correspondence is affected by interference effects of the sound waves. Most people *perceive* a 6- to 10-dB increase in the SPL to be equivalent to a doubling of loudness. It is important to remember that there is a difference between the scientific description and measurement of sound and an individual's personal or subjective opinion about it.

### Sound-Level Measurements

The human ear responds to sounds in a complex way. There is really no simple relationship between the physical measurement of sound pressure levels and an individual's perception regarding the loudness of sounds. To a certain extent, the perception of relative loudness is subjective and depends on the individual's opinion. But one physical characteristic of sound (other than amplitude) that is known to have a definitive effect on the perception of loudness is the pitch, or frequency, of the sound wave.

Experiments have shown that the average person with normal hearing will perceive a high-pitched sound to be louder than a low-pitched sound, even though both sounds have exactly the same intensity or SPL. For example, a sound with SPL = 70 dB at a frequency of 1000 Hz is usually perceived as being louder than a 70-dB sound at a frequency of 100 Hz. In fact, a sound with a frequency of 100 Hz must have an SPL of about 76 dB for it to be judged as loud as a 1000-Hz sound with an SPL of 70 dB. A higher pressure level (that is, more energy) is needed at the lower frequency for the average person to perceive the same loudness because the human ear is somewhat inefficient in detecting low-pitched sounds.

Because sounds with the same SPL intensity but varying frequencies are not perceived as being of equal loudness or volume, a method that allows meaningful and consistent sound- or noise-level measurements is needed. One way to accomplish this is to use a chart showing *equal loudness contours*, as depicted in Figure 14.5.

The contour curves represent loudness or *sound levels* called *phons*. At a frequency of 1000 Hz, the reference pitch, sound pressure levels are the same as sound levels; both are expressed in terms of decibels. Consider, for example, a listener with normal hearing who hears a 100-Hz tone with an SPL of 70 dB. What loudness does the listener perceive? Enter the chart on the bottom axis at 100 Hz and follow the vertical line

upward to a point at the 70-dB value for SPL; the closest contour curve is labeled 60. In other words, the sound will be judged to have a loudness of 60 phons. Using the same contour, it can be seen that a person hearing a 65-dB SPL sound at 200 Hz would perceive the same loudness, that of 60 phons.

Many other methods are used to measure apparent loudness. For example, another unit of loudness that is used is the *sone*. A loudness of 40 phons corresponds to 1 sone; each doubling of the sones increases the phons by 10. For example, a sound of 2 sones is equivalent to 50 phons, 4 sones is equivalent to 60 phons, and so on. Other units are used as well, but a full discussion is beyond the scope of this text. The important point to note is the complexity involved in measuring both noise levels and effects.

**Sound-Level Meters**

Many electronic instruments are available for measuring noise. The basic components of a typical noise meter include a microphone, an amplifier, a frequency filter, and a readout device. The readings or measurements are called *sound levels* and take into account the variation of perceived noise or loudness with frequency. Some noise surveys can be conducted with a battery-operated, hand-held sound-level meter, as shown in

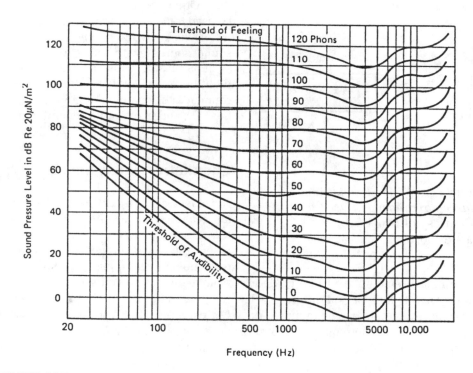

**FIGURE 14.5**

*Equal loudness contours, in phons.* (From D. M. Lipscomp and A. C. Taylor, *Noise Control: Handbook of Principles and Practices*, Van Nostrand Reinhold, New York, 1978. Copyright © 1978 Van Nostrand Reinhold Company. Used with permission)

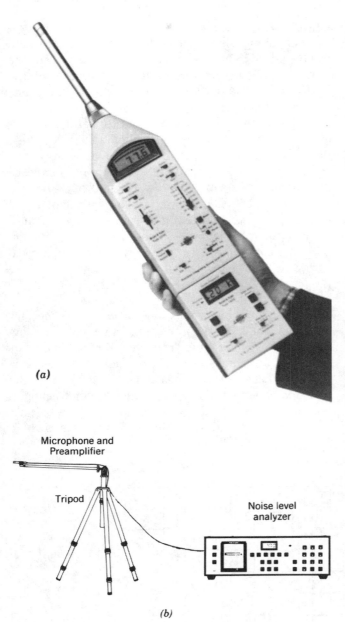

**FIGURE 14.6**
(a) *Typical hand-held sound level meter and*
(b) *typical measuring arrangement used for traffic noise surveys.* (Courtesy of Bruel & Kjaer Instruments, Inc., Marlborough, Massachusetts)

Figure 14.6. If noise measurements are to be made at one location over a relatively long period of time, such as for a traffic noise survey, meters can be mounted on a tripod and a recording device and a frequency analyzer can be added to the system. Noise meters generally cover a range of 20 to 180 dB. The selection of noise-measuring instruments depends on the type of noise, the environmental conditions, and the general purpose of the survey.

### Frequency-Weighting Networks

When measuring total noise levels of any source, it is necessary to break down the noise into different ranges of frequencies, or *bands* (for example, low-frequency middle-frequency, and high-frequency bands). One reason for this is the fact that engineering solutions to many noise problems differ for different frequency ranges; low-frequency noises, for instance, are generally more difficult to control. Another reason is that the human ear responds differently to different frequencies. It is not very sensitive to low-frequency noise, so mid- and high-frequency noises are perceived as being more annoying. Also, high-frequency noises are more capable of causing a hearing loss.

Sound-level meters are typically supplied with built-in frequency filters or weighting networks. The A filter is most commonly used. Sound measurements made with this filter are referred to as A-weighted decibels, or *dBA*. The A-weighted network filters out low-frequency and very high frequency sounds, where the human ear is less efficient. This helps to match the meter readings with the sensitivity of the ear and with the average person's judgment of the relative loudness of various sounds. Two sound levels of equal dB level but of different frequencies have different dBA levels; the lower frequency sound has the lower dBA level.

Other frequency filters or weighting networks may be used for specialized noise-level surveys. For example, the C-weighted filter may be preferable for analyzing sound sources with many low-frequency components, such as blasting noise or artillery fire. The A-weighted filter, however, is taken to be the best measure of environmental or community noise; it is usually required by the EPA and many other federal and state agencies.

## Cumulative Noise Levels

Noise levels almost always vary with time, and the overall measurement results must be reported in statistical terms. Several methods and parameters are used to represent the time-averaged values of individual or short-term noise events, as well as overall noise exposure from many noise events. One technique is to determine the percentage of time a sound level (SL) is exceeded. For example, after a series of SL measurements, it may be reported that the $L_{90} = 75$ dBA; this means that during the time period in which readings were taken, sound levels were equal to or higher than 75 dBA 90 percent of the time. Other noise parameters discussed here include the *sound exposure level*, the *equivalent sound level*, and the *day–night sound level*.

## Sound Exposure Level

Many types of noise pollution problems involve a series of intermittent noise events that occur over a period of time; each noise event may last for a short time interval, from a few seconds to a few minutes. For example, noise from a single airplane flying over a community may last for 1 min or so; in addition, many planes may fly over the town in the course of a day. The first step in analyzing the noise pollution effects on the community from these sources is to compute a *sound exposure level* (SEL) for each noise event (or flyover, in this case). SEL values provide a basis for computing noise events of variable durations in a manner that can match the average person's impression of the noise.

An SEL expresses the equivalent accumulation of sound energy over the duration of a noise event; duration is defined as extending from the time when the SPL first exceeds the normal ambient noise level until the time when the SPL drops back down to that level. This is illustrated in Figure 14.7, which shows the variation in SPL over 1 min. The shaded area covers the duration of the noise event, where the SPL values exceed (in this case) 60 dBA.

In Figure 14.7, it is seen that the maximum SPL is 85 dBA ($L_{max} = 85$), but $L_{max}$ values do not accurately represent an average person's perception of the noise, due to the variety of durations that occur among different noise events. To account for this, SEL values are standardized to a 1-s duration. In other words, the total sound energy represented by the lightly shaded area

is compressed into an equivalent amount of energy for only 1 s of time; this *noise dose* can be viewed as the solid dark line. Because the SEL is standardized to 1 s, it almost always is larger in magnitude than the $L_{max}$ of the actual noise event. In this illustration, the SEL value is about 95 dBA.

At first, it may seem strange to describe the magnitude of a noise event using a decibel level that actually exceeds the maximum measured level. Keep in mind, though, that SEL values are intended for use in noise surveys, where it is necessary to compare many noise events of varying duration in a way that matches human perceptions. No matter what the duration of a noise event, the higher the SEL, the more annoying or harmful the noise is likely to be.

## Equivalent Sound Level

SELs measure the noise associated with individual events such as the flyover of a single airplane. It is important to be able to objectively describe the cumulative effects of a series of many similar noise events. One method used to do this is to determine the *equivalent sound level* (or $L_{eq}$). The $L_{eq}$ may be conceptualized as an average or constant SPL over the period of interest. It contains as much sound energy as all the actual noise events combined during that time frame. The period of interest may be 1 h, an 8-h workday, or a full 24-h day. It is important to identify the time period used for any noise survey; for example, an 8-h workday $L_{eq}$ would be identified as $L_{eq(8)}$.

The average sound level represented by an $L_{eq}$ is not an arithmetic value, but a logarithmic or *energy-averaged* value. Because of this, higher SPL values receive greater emphasis than lower values. For example, if an SPL is constant at 50 dBA for 30 min and then is constant at 100 dBA for the next 30 min, the $L_{eq}$ for the full 60-min time frame will be 97 dBA, not 75 dBA (as would be expected using an arithmetic average). Loud events typically dominate the noise environment described by an $L_{eq}$.

## Day–Night Sound Level

The *day–night sound level*, or DNL, is the $L_{eq(24)}$ with a 10-dBA penalty added to the sound pressure levels that occur during the night (from 10 P.M. to 7 A.M.). Another symbol used for the DNL is $L_{dn}$. The 10-dBA penalty for nighttime noises represents an adjustment for the additional annoyances caused by those noises. Most people are more sensitive to noise during the night, particularly because ambient or background noise levels are lower at that time. DNL values can be measured with standard noise-metering equipment; they can also be predicted using computer models.

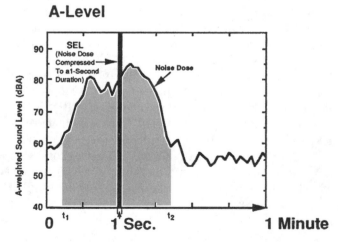

**FIGURE 14.7**

*Sound exposure levels are used to account for the time interval of a noise event.* (Courtesy of Harris Miller Miller & Hanson Inc., Sacramento, California)

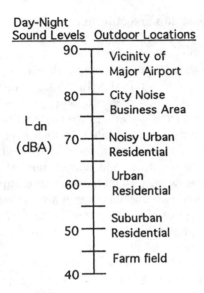

**FIGURE 14.8**
*Typical ranges of day–night environmental noise levels.*

Typical DNL values are shown in Figure 14.8. They range from a low of 40 dB in extremely quiet, remote locations to a high of almost 90 dBA in urban neighborhoods close to major airports. In some studies of noise pollution, computed DNL values in a town or neighborhood can be shown on a map as noise contours depicting lines of equal noise exposure in the same way that elevation contours depict topography. DNL measurements are of particular importance for analyzing and describing the exposure of communities to aircraft noise; in the United States, $L_{dn}$ has been selected by the EPA and the Federal Aviation Administration (FAA) as the optimum parameter for evaluating aircraft noise.

## EXAMPLE 14.5

Transient noise levels due to traffic along a highway were measured at 10-min intervals for 1.5 h; nine individual SL readings (in dBA) were taken in sequence: 72, 76, 79, 81, 84, 76, 75, 75, 74. Determine the sound level that was exceeded 50 percent of the time over the total measurement period. Also determine the $L_{10}$ and $L_{90}$ values.

*Solution* First, rearrange and rank the sound-level readings in descending order, as shown in the first two rows:

| dBA: | 84 | 81 | 79 | 76 | 76 | 75 | 75 | 74 | 72 |
|---|---|---|---|---|---|---|---|---|---|
| rank: | 1 | 2 | 3 | 4 | 5 | 6 | 7 | 8 | 9 |
| % Time exceeded: | 10 | 20 | 30 | 40 | 50 | 60 | 70 | 80 | 90 |

The *percent time exceeded* is easily computed by dividing the rank number by the number of readings plus 1 and then multiplying by 100. For example, for the sound level with rank 3 (79 dBA), $[3/(9 + 1)] \times 100 = 30$ percent. In other words, during the time period that sound measurements were made, they equaled or exceeded 79 dBA about 30 percent of the time ($L_{30} = 79$ dBA).

From the table, it can be seen that $L_{50} = 76$ dBA, which is, in effect, the *median* of the readings. It can also be seen that $L_{10} = 84$ dBA. In other words, it can be assumed that 10% of the time the sound levels equaled or exceeded 84 dBA (even though there was no measurement of an SL above that value). Although 72 dBA appears, from the given data, to have been exceeded most of the time, the computations show that level to be the $L_{90}$ value. (This is a simplified example of a statistical method that may be used to describe transient or intermittent noise levels; the results become more accurate as the number of SL readings increases.)

## 14.3 EFFECTS OF NOISE

The difference between sound and noise is often subjective, a matter of personal opinion. Despite differences of opinion, some very definite harmful effects are caused by exposure to high sound levels, whether or not they are called noise. These effects may be physical or emotional, and they can range in severity from being merely annoying to being extremely painful and hazardous. Excessive noise levels can cause environmental problems and workplace hazards.

The most direct harmful effect of excessive noise is *physical damage to the ear* and the temporary or permanent hearing loss that results from the damage. Temporary hearing loss, often called *temporary threshold shift* (TTS), refers to a reduced ability to detect weak sounds; hearing ability is usually recovered within 1 month of exposure. Permanent loss, usually called *noise-induced permanent threshold shift* (NIPTS), represents a loss of hearing ability from which there is no recovery.

Below a sound level of 80 dBA, hearing loss does not usually occur at all. But temporary (or TTS) effects are noticed at sound levels between 80 and 130 dBA. About 50 percent of people exposed to 95-dBA sound levels at work will develop NIPTS, and most people exposed to more than 105 dBA will experience permanent hearing loss to some degree. A sound level of 150 dBA or more can physically rupture the human eardrum.

In general, high- and middle-frequency sounds are more harmful than low-frequency sounds at the same level, and the degree of hearing loss depends on the duration as well as on the intensity of the noise. For example, 1 h of exposure to a 100-dBA sound level can produce a TTS that may last for about 1 d. But in some cases, particularly in factories with noisy machinery, workers are subjected to high sound levels for several hours a day. Exposure to 95 dBA for 8 h/d over a period of 10 years may cause about 15 dBA of NIPTS.

In addition to hearing losses, excessive sound levels can cause harmful effects on the circulatory system by raising blood pressure and altering pulse rates. Noise can also cause emotional or psychological effects, such as irritability, anxiety, and stress. Even lack of concentration and mental fatigue are significant health effects of noise. In industry, it results in lowered worker efficiency and productivity, increased employee absences from work, and higher accident rates on the job. In the community it interferes with sleep, recreation, and personal conversations.

Outdoor sound levels ($L_{dn}$) of less than 55 dBA and indoor sound levels of less than 45 dBA are considered by the EPA as best to prevent annoyance and not interfere with conversation or sleeping. DNLs greater than 65 dBA are not usually compatible with residential land-use or public-use facilities such as schools or libraries; noise attenuation is needed above this level.

In general, then, noise is more than a mere nuisance or annoyance. Noise can have a direct effect on public health as well as esthetic sensibilities. There can be no question that noise affects the *quality of life.* Mitigation or control of noise in the community and in the workplace is an important aspect of environmental technology.

## 14.4 NOISE MITIGATION

Noise can be controlled to mitigate or reduce its intensity, and steps can be taken to reduce durations of exposure to it. All noise control problems involve three essential elements: a noise source, direct and indirect paths of noise transmission, and one or more recipients of the noise (see Figure 14.9). Noise levels and exposure time are generally limited by community ordinances, federal and state regulations, and occupational safety laws.

## Regulations and Standards

Noise pollution problems are significant in today's industrial society, but they are not only a modern phenomenon. Urban noise levels were high even in ancient

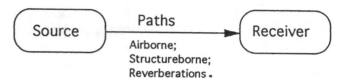

**FIGURE 14.9**
*Basic elements in noise control. Air provides a direct path for noise transmission; solid materials and reverberations (reflections) provide indirect paths.*

cities; it has been said, for example, that Julius Caesar banned the use of chariots on the cobblestone streets of Rome during the night so that he could sleep. Even before widespread use of personal automobiles began in the 20th century, the clamor of iron horseshoes on the stone pavements of major cities was a big problem. In addition to automobiles, increasing aircraft traffic added to community noise levels. Increased use of machinery in the workplace after the Industrial Revolution added significantly to occupational noise hazards.

### State and Local Regulations

As with other forms of environmental pollution, laws and regulations are necessary to limit and control problems caused by excessive noise levels. Many municipal governments have enacted local noise control ordinances; local enforcement of regulations consistent with federal and state laws is considered the most practical approach for mitigating noise pollution.

Modern community noise control ordinances are usually based on objective *performance standards,* which specify maximum noise emission levels for operation of certain types of equipment and the conduct of specific local activities. Some ordinances set maximum sound levels for new products or machines sold in the community, such as motorcycles and power tools. Most have curfews that limit nighttime operation of certain types of equipment; zoning restrictions may also apply, limiting noise-producing activities in residential areas or in the vicinity of hospitals and schools. Depending on a community's requirements for noise control, other limits may be included, such as land-use planning standards and noise exposure criteria and building code provisions specifying noise controls in buildings and residential structures.

Laws established at the state level can be effective in mitigating noise pollution problems and encouraging towns and cities to set equal or even more stringent requirements. New York State, for example, enacted a highway noise law in 1965, as did California in 1967; a comprehensive antinoise law was enacted in

New Jersey in 1972. The weakest link in noise control legislation at all levels of government is enforcement of the regulations.

### Federal Noise Regulations

Under the *Noise Control Act* (1972), the EPA was authorized to set limits on community or environmental noise, and manufacturers were required to label and warrant product performance with regard to noise levels. The 1972 law was amended in 1978 by the *Quiet Communities Act* in order to encourage noise control problems at state or local government levels. But in 1982, EPA's Office of Noise Abatement and Control was eliminated due to federal budget cuts, and its responsibilities were shifted to state and local governments. Other federal agencies, such as OSHA (Occupational Safety and Health Administration), continue to take an active part in noise pollution control.

Maximum permissible workplace noise levels in the United States were first established in 1970 under the *Occupational Health and Safety Act*. OSHA does not allow worker exposure to sudden impact or impulse noise levels above 140 dBA. Limits have been set on the duration of exposure to a range of noise levels. For example, a maximum sound level of 90 dBA is allowed for an 8-h exposure. Higher sound levels are permitted, but for shorter durations (for example, 100 dBA is allowed for up to 2 h and 115 dBA for 15 min). No exposure to continuous noise levels above 115 dBA is allowed. The National Institute for Occupational Safety and Health (NIOSH) recommends that the OSHA standards be reduced by 5 dBA; for example, it is recommended that the maximum 8-h exposure level should be reduced from 90 to 85 dBA to more adequately protect workers from hearing loss.

In addition to the OSHA standards, noise regulations have also been set by other federal agencies. The Department of Housing and Urban Development (HUD), for example, has provided guidelines for residential buildings that are to qualify for HUD mortgage insurance. Unfortunately, enforcement of the noise standards for new buildings leaves much room for improvement.

The Department of Transportation (DOT) has central authority for highway noise abatement and for directing activities at the state and the local levels. DOT regulations establish maximum noise levels for interstate trucks and buses; for example, new trucks over 5 tons must achieve sound levels less than 83 dBA in speed zones of 35 mph or less at a distance of 50 ft from the road centerline. The Federal Highway Administration has adopted the noise-level and land-use relationships summarized in Table 14.1. The $L_{10}$ values refer to sound levels (SLs) not to be exceeded more than 10 percent of the time.

**TABLE 14.1**
**Land-use and outdoor noise levels**

| Land-use category | Land-use description | $L_{10}$ noise level (dBA) |
|---|---|---|
| A | Land areas where serenity and quiet serve an important public need, such as amphitheaters or specially designated portions of certain public parks | 60 |
| B | Residential areas, motels, hotels, schools, libraries, hospitals, public parks, recreation areas, and meeting rooms | 70 |
| C | Developed land not included in category A or B | 75 |

Responsibility for aircraft and airport noise control is assumed primarily by the Federal Aviation Administration (FAA). The FAA has set limits on commercial aircraft noise in areas surrounding airports; these limits vary for different classes of aircraft. Efforts to decrease flyover noise levels include designing quieter aircraft, improving airport operating procedures, and changing flight patterns. Environmental noise due to aircraft flyovers, however, remains one of the most intractable noise pollution problems for many communities.

## Noise Control Techniques

There are four fundamental ways in which noise can be controlled: protect the recipient, increase the path length, block the path, or reduce the noise at the source. In general, the best control method is to reduce noise levels at the source. Sometimes it is necessary to use two or more techniques, though, because no single method would be sufficiently effective.

### Protection of the Recipient

One way to protect individuals from excessive noise levels is to require the use of ear plugs or ear muffs. Specially designed ear muffs can reduce the sound level reaching the eardrum by as much as 40 dBA; they are very useful for protecting industrial or construction workers who are exposed to noise for long periods of time. But it is obviously unrealistic to consider the use of personal ear protectors as a practical solution to community noise for the average citizen. Even in occupational settings, ear muffs cannot be relied on as the only solution. Workers tend not to wear them on a regular basis, despite company requirements for their use.

**Increasing Path Distance**

Sound levels drop significantly with increasing distance from the noise source. Increasing the path length between source and recipient offers a passive means of control; it requires no effort on the part of the recipient. Municipal land-use ordinances pertaining to the location of airports and minimum distances between houses and roads or highways make use of the attenuating effect of distance on sound levels. Poor planning efforts, on the other hand, can result in situations where houses are built too close to major highways; in California, for example, some houses have been built as close as 6 m (20 ft) from interstate highways. At that distance, sound levels inside a house with closed windows can be as much as 70 dBA from a passing truck. Noise is more than just an annoyance at that level; normal speech communication is almost impossible.

The decrease or attenuation of sound levels with increasing distance from the source occurs because the fixed amount of sound energy is spread and diluted over an increasing area. As a sound wave spreads out from a source, its intensity decreases; from a *point source*, the sound waves tend to form concentric spheres and the intensity decreases with the square of the distance from the source. For example, if the path length is doubled, the intensity reaching the recipient of the sound is one fourth of its original intensity. This is known as the *inverse square law*.

Most noise sources are not points, and reflections of the sound waves from the ground or walls and ceilings prevent the formation of spherical waves. A highway, for example, is considered to be a *line source* of noise, from which sound is propagated in the form of a half-cylinder (see Figure 14.10). The relationship between the sound level and distance from a line source can be written as

$$SL_B = S_A - 10 \times \log\left(\frac{D_B}{D_A}\right) \qquad \textbf{(14-3)}$$

where $SL_A$ = sound level at distance $D_A$ from the source

$SL_B$ = sound level at distance $D_B$ from the source

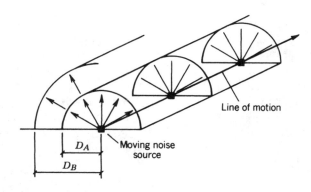

**FIGURE 14.10**

*As a sound wave spreads out from a line source of noise, its sound level decreases at about the rate of 3 dBA for each doubling of the distance from the source.*

**EXAMPLE 14.6**

The sound level measured at a 4-m distance from the centerline of a busy highway is 85 dBA. *(a)* What is the sound level at a distance of 12 m from the road centerline? *(b)* At what distance from the road centerline is the sound level reduced to 79 dBA?

*Solution*   *(a)* Applying Equation 14-3 gives

$$SL_{12} = 85 - 10 \times \log\left(\frac{12}{4}\right)$$

$$= 85 - 10 \times 0.477 = 80 \text{ dBA}$$

*(b)* Again applying Equation 14-3 yields

$$79 = 85 - 10 \times \log\left(\frac{D_B}{4}\right)$$

and

$$\log\left(\frac{D_B}{4}\right) = \frac{79 - 85}{-10} = 0.6$$

From the definition of the log function,

$$\frac{D_B}{4} = 10^{0.6} = 3.98$$

and

$$D_B = 4 \times 3.98 = 16 \text{ m}$$

Sound levels from a line source decrease 3 dBA for each doubling of the distance from the source (use Equation 14-3 to confirm this). However, sound levels from a point source decrease at twice that rate, or 6 dBA, for each doubling of distance. Use of a chart as shown in Figure 14.4 for adding sound levels, along with an appropriate distance formula, allows the prediction of noise levels from transportation, construction, or other noise sources in a community or neighborhood.

### Noise Barriers

A barrier placed in the path of a sound wave will absorb and reflect some of the sound energy, reducing sound levels on the side opposite the source. Different materials absorb (soak up) or reflect (bounce back) sound energy in different amounts. Absorptive materials include heavy drapes, carpets, and special ceiling and wall acoustic materials. Absorptive materials can control interior noise, sound reflection, and reverberation (reflected sound that persists in an enclosed space after the source has stopped). Most hard, smooth, impervious materials reflect sound. For any sound barrier to work effectively, all cracks or other openings must be sealed.

Highly absorptive interior finish materials for walls, ceilings, and floors can decrease indoor noise levels significantly. In concert halls, meeting rooms, classrooms, and theaters, reflective and absorptive surfaces must be carefully proportioned to create optimum hearing conditions. Acoustic privacy between rooms can be easily created by partitions that are both heavy and airtight. Most building materials and components are tested for acoustic properties and assigned *sound transmission class* (STC) numbers. The higher the STC rating for a partition, the better is the acoustical privacy between adjacent rooms. For example, $1/2$-in. gypsum boards nailed on wood studs provide an STC rating of about 30 dB, and a 4-in.-thick brick wall has an STC of about 45 dB.

Ceilings made from fibrous materials, often called *acoustic ceilings*, are highly absorptive of sound (unlike plaster or gypsum board, which are highly reflective). Acoustic properties of these ceiling materials are described in terms of *noise reduction coefficients* (NRCs). An NRC rating of 0.80, for example, indicates that the material absorbs 80 percent of the sound that reaches it and reflects only 20 percent of it. Most acoustic ceilings have NRCs between 0.5 and 0.9, while plaster and gypsum board ceilings have NRC ratings below 0.1. Acoustic ceilings are most useful for reducing interior noise levels in offices, restaurants, retail stores, and noisy factories. However, they do not block the transmission of noise.

Lightweight, porous materials that produce high NRC ratings allow most sound to pass through. In fact, a ceiling with a high NRC generally has a low STC number. To reduce interior noise and sound transmission at the same time, composite ceiling panels can be used; these panels are made with a highly absorbent material laminated to a denser substance that provides acoustic privacy.

### Source Reduction

Perhaps the most direct approach for noise control is to reduce the sound produced by the source itself. Vehicles and machinery can be effectively muffled to reduce the noise. In industrial facilities, noise reduction can be obtained by using rigid, sealed enclosures around machinery; acoustic absorbing material can line the inside of the enclosure. Machines or their enclosures can be isolated from the floor using special spring mounts or absorbent mounts and pads; flexible couplings can also be used for interior pipelines to reduce noise transmission.

One of the best, but often overlooked, methods of noise source reduction is regular and thorough maintenance of operating machinery. Even machines of the best design periodically need lubrication and realignment to keep vibrations to a minimum. Proper balancing of rotating components and replacement of worn bearings and gears is necessary; fasteners must be checked periodically to assure that they have not loosened. All machinery should be operated with its original design limits to reduce vibrations.

### Construction and Industrial Noise

The construction industry has long been a focus of complaints related to excessive noise. Construction activities typically require the use of heavy machines, each of which can be a significant source of noise (see Figure 14.11). Noise levels at construction sites can be controlled using proper construction planning and scheduling techniques. For example, locating noisy air compressors and other equipment away from the site boundary will help to mitigate noise levels outside the site. Temporary barriers may also be used to physically block the noise. Construction managers must be aware of any local municipal noise ordinances that may restrict the hours of construction activity, as well as the allowable noise levels emitted from various pieces of equipment.

Construction machinery such as portable air compressors, rock drills, and paving breakers must be sound-tested by the equipment manufacturers and must meet EPA as well as any local noise emission standards;

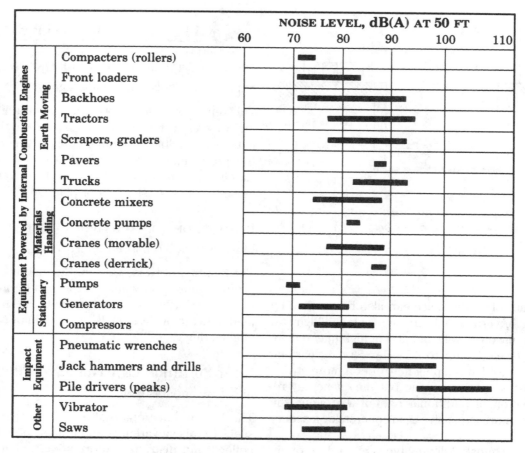

**FIGURE 14.11**

*Relative ranges of noise levels for some common types of heavy construction machinery.* (Courtesy of the Environmental Protection Agency)

A-weighted sound levels are generally measured at a distance of both 1 and 7 m. The International Organization for Standardization (ISO) has also proposed noise emission standards for construction machines, but they are based on sound power levels rather than sound pressure levels.

Portable air compressors that provide power to certain construction machines have been identified by the EPA as one of the major sources of overall noise at a construction site. Large compressors require complete enclosures, including vibration isolators and sound-absorption panels; air must be brought in through acoustically treated ducts, and an adequate exhaust silencer must be used on the engine. These enclosures can reduce the noise from about 110 to 85 dBA at a distance of 1 m. But there are some types of construction machinery for which it is extremely difficult to reduce the noise emissions. Special exhaust mufflers, for example, can reduce the noise levels emitted by paving breakers and rock drills only from about 108 dBA to about 100 dBA at 1 m.

**Transportation Noise**

Most automobile traffic noise comes from the movement of the vehicle tires on the pavement and wind resistance. Noise from the engine and the exhaust of a properly maintained automobile adds little to the noise level. Traffic volume and speed have significant effects on the overall sound levels: doubling the speed increases sound levels by about 9 dBA; doubling the traffic volume (number of vehicles per hour) increases sound levels by about 3 dBA. Also, a smooth flow of traffic causes less noise than does a stop-and-go traffic pattern.

Proper highway planning and design are essential for controlling traffic noise. Establishing lower speed limits for highway segments that pass through residential areas, limiting traffic volume, and providing alternative routes for truck traffic can all be effective noise control measures. The use of a *cut* or depressed roadway passing through urban neighborhoods is also very effective for reducing traffic noise in the adjacent buildings. This is illustrated in Figure 14.12.

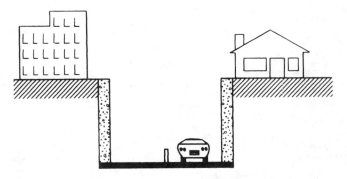

**FIGURE 14.12**
*One way to reduce traffic noise in a city is to build sections of the roadway in a cut below street level.*

The path of traffic noise can also be blocked by construction of vertical barriers alongside the highway. Concrete or masonry walls about 4 m high may reduce peak noise levels by up to 15 dBA. Sloping earthen berms with dense landscaping may offer a more attractive solution as a noise barrier, but they generally require acquisition of additional right-of-way along the route and are more expensive than concrete or masonry barriers.

One of the most annoying types of outdoor or environmental noise comes from aircraft flyovers in the vicinity of airports. Most of the noise is from the aircraft propulsion system: propellers, rotors, turbine blades, and jet-engine exhaust streams. Additional noise comes from airframe vibrations and aerodynamic turbulence. Noise mitigation can be achieved by improvements in engine and airframe design as well as by appropriate control of flight patterns. Many parameters are used to measure and describe aircraft noise, and there is much room for improvement in its control; a full discussion of the topic is beyond the scope of this textbook.

 ## 14.5 RELEVANT WEB SITES

### HIGHWAY TRAFFIC NOISE IN THE UNITED STATES

http://www.nonoise.org/library/highway/probresp.htm

This report provides information about the problem of highway traffic noise and the response to that problem in the United States, summarizing the general nature of the problem, the response of the Federal Highway Administration (FHWA) to the problem, and highway noise barriers constructed or planned.

### INSTITUTE OF NOISE CONTROL ENGINEERING

http://users.aol.com/inceusa/ince.html

The Institute of Noise Control Engineering is a not-for-profit organization whose purpose is to promote engineering solutions to environmental noise problems. Members are professionals in the field of noise control engineering.

### NOISE POLLUTION CLEARINGHOUSE (NPC)

http://www.nonoise.org/

The NPC is a national not-for-profit organization that seeks to raise awareness about noise pollution, create, collect, and distribute information and resources regarding noise pollution, strengthen laws and governmental efforts to control noise pollution, and establish networks among environmental, professional, medical, governmental, and activist groups working on noise pollution issues.

### QUIET COMMUNITIES ACT OF 1997

http://pages.prodigy.net/rockaway/s951.htm

This site gives the full text of the legislation intended to reestablish the Office of Noise Abatement and Control in the Environmental Protection Agency.

## REVIEW QUESTIONS

1. Give a brief definition of noise pollution. Is the distinction between loud sounds and noise a matter of opinion or an objective and scientific fact?

2. Briefly discuss the physical characteristics of sound and sound waves.

3. How is a plot or graph of the sine function related to a sound wave?

4. What characteristic of a sound wave relates the speed to the frequency of the wave? Describe the relationship.

**5.** What does the term *hertz* refer to? How is it related to *pitch?*

**6.** In what range of frequencies can individuals with normal hearing perceive sounds? What is the frequency range of normal conversation? What is the average *threshold of hearing?*

**7.** What are the characteristics of sound that affect a person's perception of the volume or loudness of the sound?

**8.** What is a *decibel?* Briefly describe why sound is measured in units of decibels instead of sound pressures in units of microbars.

**9.** What is the difference between *sound pressure level* (SPL) and *sound intensity?* Is it possible to hear a sound with a negative SPL? Why?

**10.** How much does sound intensity increase for each 10-dB increase in SPL? When sound intensity is doubled, what is the increase in the SPL?

**11.** Is the sound pressure level from two identical noisy machines twice the level from one machine operating alone? Briefly explain your answer. What is meant by *masking* of sound? Under what conditions does it occur?

**12.** Briefly explain the difference between *sound pressure levels* measured in decibels and *sound levels* measured in A-weighted decibels. Which is the more common unit of sound measurement and why?

**13.** What is a *phon?* Is there any relationship between a phon and an SPL?

**14.** What is a *sound exposure level* (SEL)? Briefly explain its application.

**15.** Briefly describe the meanings and applications of $L_{eq}$ and $L_{dn}$ values as measures of noise pollution.

**16.** Discuss the effects of excessive noise. At what SL, approximately, does physical harm occur? At what level does it interfere with indoor sleep or conversations? Is the duration of time of exposure to noise a factor with regard to its effect?

**17.** Briefly discuss how noise pollution is regulated at the local, state, and federal levels. What is considered the most effective way to regulate noise pollution?

**18.** Investigate the noise pollution ordinance in your town or city, and write a brief summary or report of your findings.

**19.** What are the four basic ways in which noise can be controlled or reduced? Which is considered to be the best or most effective way?

**20.** What are the effects of distance on sound levels from a *point source* and a *line source?* Why do they differ?

**21.** Briefly discuss the application of barriers in noise control. What is the difference between an absorptive material and a reflective material in a noise barrier? Are acoustic ceiling tiles helpful in providing acoustic privacy; that is, do they block the transmission of sound?

**22.** Briefly discuss STCs and NRCs as related to noise pollution control.

**23.** How can noise be reduced at the source? What is one of the best, but often overlooked, methods of reducing noise from machinery?

**24.** Briefly discuss sources of construction noise and control methods.

**25.** What is the source of most of the noise from automobile traffic? What is the effect on noise levels (in dB) of doubling speed or traffic volume?

**26.** Describe some ways in which automobile traffic noise can be controlled. Briefly discuss problems related to aircraft noise.

**27.** Visit the *Highway Traffic Noise in the United States* Web page and read the section on "Noise Abatement Procedures." Write a brief report to summarize what you learned.

**28.** Visit the *Quiet Communities Act of 1997* Web page. Summarize the Act in a brief written report.

**29.** Using the EnviroSources search engine (see Section 1.6) or another Internet search directory or search engine, locate at least one additional Web site relevant to one of the topics in this chapter. Add the link(s) to your Environmental Technology Folder. Write a brief description of what the Web site(s) contain.

## PRACTICE PROBLEMS

1. About how long would it take to hear thunder from a bolt of lightning that occurred 2.5 mi from an observer's location?

2. If a bolt of lightning is seen and after 5 s the thunder is heard, about how far away is the storm?

3. What is the wavelength of a 15,000-Hz sound in water, assuming that the speed of the wave is 1500 m/s?

4. The speed of sound in concrete is 3400 m/s. What is the frequency of a sound transmitted through concrete if the sound wavelength is 500 mm?

5. A machine causes an rms sound pressure of 160 $\mu$bar. What is the SPL?

6. What is the sound pressure in microbars caused by a noise with an SPL of 95 dB? By what factor is that sound pressure higher than the pressure corresponding to the threshold of human hearing?

7. A bulldozer is being used to push-load a scraper at a construction site. Each machine has an SPL of 95 dBA operating alone. What is the SPL of the two machines operating together?

8. What is the combined SPL from three vibratory rollers operating at the same time on a construction site if each one alone has an SPL of 75 dBA? What will be the SPL if four identical rollers operate simultaneously?

9. What is the combined SPL from a 75-dB roller and a 90-dB scraper at a construction site?

10. What is the combined SPL from an 80-dB vibratory roller and a 90-dB bulldozer at a construction site?

11. A listener with normal hearing is exposed to a 300-Hz tone that has an SPL of 40 dB. How loud is the sound in phons? What is the SPL for a tone at 100 Hz that seems to be equally loud to the listener?

12. A listener with normal hearing is exposed to a 10,000-Hz tone that has an SPL of 62 dB. How loud is the sound in phons? What is the SPL for a tone at 50 Hz that seems to be equally loud to the listener?

13. The sound level measured at a distance of 25 ft from the center of a road is 76 dBA. What is the sound level at 100 ft from the road under the same traffic conditions? What would you expect the level to be at 100 ft if the traffic volume doubled, but average speed remained the same?

14. The sound level at a distance of 60 m from the center of a roadway is 65 dBA. What is the sound level at a distance of 10 m from the center of the road under the same traffic conditions? What would you expect the level to be if the average speed doubled, but traffic volume was the same?

15. Transient noise levels due to traffic along a highway were measured at 10-min intervals, and the following individual SL readings (in dBA) were taken in sequence: 73, 78, 82, 84, 84, 76, 75, 75, 73, 72, 76, 78, 81, 85, 81, 80, 78, 78, 74. Compute the $L_{10}$, $L_{50}$, and $L_{90}$ sound levels.

16. Transient noise levels due to traffic along a roadway were measured at 10-min intervals, and the following individual SL readings (in dBA) were taken in sequence: 70, 75, 79, 81, 81, 73, 72, 71, 69, 68, 72, 75, 79, 82, 77, 76, 75, 74, 73. Compute the $L_{10}$, $L_{50}$, and $L_{90}$ sound levels.

## Appendix Outline

# A P P E N D I X

# A

# Environmental Impact Studies and Audits

Traditionally, the planning process for civil engineering and construction projects has always included consideration of the economic as well as the technical problems involved. It was not until 1970, when the *National Environmental Policy Act* (NEPA) took effect, that *environmental impacts* were also given due consideration in the project planning process.

The NEPA regulations focused on major federal projects that could damage the environment. Shortly after NEPA, many states also adopted similar laws of their own. Eventually, local municipalities began to incorporate thorough environmental planning requirements into land-use or zoning ordinances. Now, even small and privately funded construction projects usually must include an environmental impact study before approval for the project will be granted. Many existing industrial facilities are also *audited* for compliance with environmental regulations and other purposes. It is important that all civil design and construction professionals have a basic understanding of what environmental impact studies and environmental audits are and how they are used.

## A.1 THE ENVIRONMENTAL IMPACT STATEMENT (EIS)

An *environmental impact statement* (EIS) for a proposed project is a written report that summarizes the findings of a detailed environmental review process. The writing of an EIS is preceded by two steps. First, an *environmental inventory* must be conducted for the site and vicinity of the proposed project. This inventory includes a thorough description of the existing physical environment and serves as the basis for evaluating the possible impacts of the project. The second step involves a systematic and comprehensive *environmental assessment*. This assessment, a crucial part of the EIS process, identifies and analyzes the potential adverse environmental consequences of the project. This analysis includes a prediction of all possible environmental changes as well as a consideration of the magnitude and overall importance of those changes. In many assessments, an attempt is made to measure and describe qualitative environmental impacts in quantitative or numerical terms.

The format of the EIS document or report may vary to some degree, depending on the requirements of the municipality or regulatory agency that will review and approve it. Generally, the following topics or sections are included in a final draft of the EIS:

1. Description of the existing environment
2. Description of the proposed project
3. Environmental assessment
4. Unavoidable adverse environmental impacts
5. Secondary or indirect impacts
6. Methods for reducing adverse impacts
7. Alternatives to the proposed project
8. Irreversible commitments of energy and resources
9. Consideration of public input and review

The EIS is meant to be used as a planning and decision-making tool. It is supposed to be objective and unbiased, and it is not meant to either promote or block the implementation of a proposed project. The greatest benefit of the EIS process is that environmental concerns must be examined thoroughly, and the chances for severe or unexpected damage due to a construction project are significantly minimized.

Unfortunately, EIS reports are sometimes manipulated by developers to promote a construction project, or they are misused by special interest groups to stop a project completely. A criticism often directed toward the EIS is that it may be imposed on small projects that might not warrant so much concern. But the role of the EIS as an environmental planning tool is still evolving. We can expect that it will eventually be fine-tuned to the point where environmental protection will be achieved in a cost-effective manner. Because it is a permanent and important aspect of civil construction technology, some of the key features of an EIS are discussed in the following sections.

## A.2 DESCRIPTION OF THE EXISTING ENVIRONMENT

A basic objective of an environmental study is to anticipate any potential impacts of a proposed construction project on the environment. It is first necessary to have an accurate and thorough picture of what the predevelopment (existing) environmental conditions are at and near the proposed site. Sometimes an environmental inventory report is already available for an entire city or township. Usually, though, the interdisciplinary team that is preparing the EIS must conduct a more detailed and site-specific environmental survey.

The inventory of existing natural resources and urban facilities in the vicinity of the project site typically includes the following categories of data:

- *Geology, soils, and topography.* This includes a description of the types of bedrock that underlie the site, soil types and characteristics, and existing ground slopes or topography. Soil erosion potential is a particularly important factor, as are percolation rates, depth to groundwater, and the location of aquifer recharge areas.
- *Water resources.* Streams and lakes on or near the project site are studied and described. Data on existing surface and groundwater quality are discussed, as are drainage patterns, flood hazards, and streamflow rates. Existing or predevelopment runoff rates are evaluated so that appropriate measures can be taken to ensure that they are not increased later on.
- *Vegetation and wildlife.* The extent and type of woodlands or plant growth on the site are described, and any rare or unique species are inventoried. Species of animals that use the site as a habitat are also discussed, and

the presence of any endangered species is determined. Usually, environmental resource data are presented graphically for clarity. An example of a site-specific map showing existing vegetation is illustrated in Figure A.1.

- *Air quality and noise.* Existing air quality data from the nearest state or federal air sampling station are obtained and discussed. Local meteorological conditions, including average wind speed and direction and the frequency of temperature inversions, are studied and summarized. Noise levels, frequency, and duration in the vicinity of the site are evaluated.

- *Transportation.* Existing modes of transportation are described, including automobile, bus, rail, and aircraft. Local traffic volumes, patterns, and existing roadway capacities are evaluated. An example of a site-specific traffic survey is shown in Figure A.2.

- *Public utilities.* The location and service capacities of nearby water supply mains and sewerage systems are described and shown on a site plan. Gas, telephone, electric, and refuse collection service in the area are evaluated.

- *Population, land use, and socioeconomics.* Existing population densities and land-use patterns are studied and described, including residential, commercial, industrial, and agricultural areas. Local incomes and economic levels, the local tax base, and the capacities of school, fire, and police services in the area are evaluated.

- *Historical or unique cultural features.* The possibility of an archaeological site existing within the project boundary is investigated. The locations of historical landmarks, museums, or libraries are described. Any unique esthetic features, such as a beautiful view or the last remaining open space in the community, are also noted.

## A.3 DESCRIPTION OF THE PROPOSED PROJECT

In addition to a complete environmental inventory, it is also necessary to have a clear picture of the nature and extent of the proposed project. Although detailed engineering plans are not generally needed for this, a preliminary plan must be made available by the project owner. This plan must be comprehensive enough to allow for a meaningful assessment of environmental impacts.

To illustrate, consider again the fictitious land development project discussed in Section 1.1. An EIS would have to be prepared for this project. The developer's consulting engineer or architect would provide information regarding the total area of the project, the number of building lots, the relative distribution of residential, commercial, and industrial facilities, and other data. A preliminary plan showing the proposed alignment and grading of roadways would be submitted. First-floor elevations of proposed structures and any changes or regrading of topography would be indicated.

A proposed stormwater drainage system would have to be shown, including underground pipelines and any stormwater detention basins. The points of stormwater discharge would be shown. Plans for a proposed water supply and wastewater collection system would be submitted, showing the location and capacities of pipelines and other utilities. In some cases, information regarding the type of construction, the landscaping, and the expected market value of the constructed facilities would be required.

## A.4 ASSESSMENT OF ENVIRONMENTAL IMPACTS

The primary goal of the EIS procedure is to predict any adverse (or beneficial) effects of a proposed project on the natural and urban environment. This is done so that measures can be taken to minimize or eliminate the harmful impacts when the project is implemented. The prediction or assessment of environmental impacts is not an easy task. It must be conducted by an interdisciplinary team including civil engineers and technicians, geologists, urban planners, and biologists or ecologists. For large and complex projects and particularly for sensitive environmental settings, the team may also include archaeologists, architects, and social scientists.

Certain environmental impacts can be evaluated directly and objectively. They are not really subject to conflicting subjective or personal opinions. For example, the expected increase in stormwater runoff due to the project can be computed and compared to the existing runoff rates and volumes. The effect of the increase on the site and on downstream properties may then be predicted. As discussed in previous chapters, these effects might include flooding, soil erosion, and water pollution.

Air quality impacts can also be assessed using sophisticated mathematical models. Usually, emission of carbon monoxide from cars is of particular significance in land development projects; the increase in automobile traffic would contribute directly to this effect. Basic traffic engineering principles can be applied to

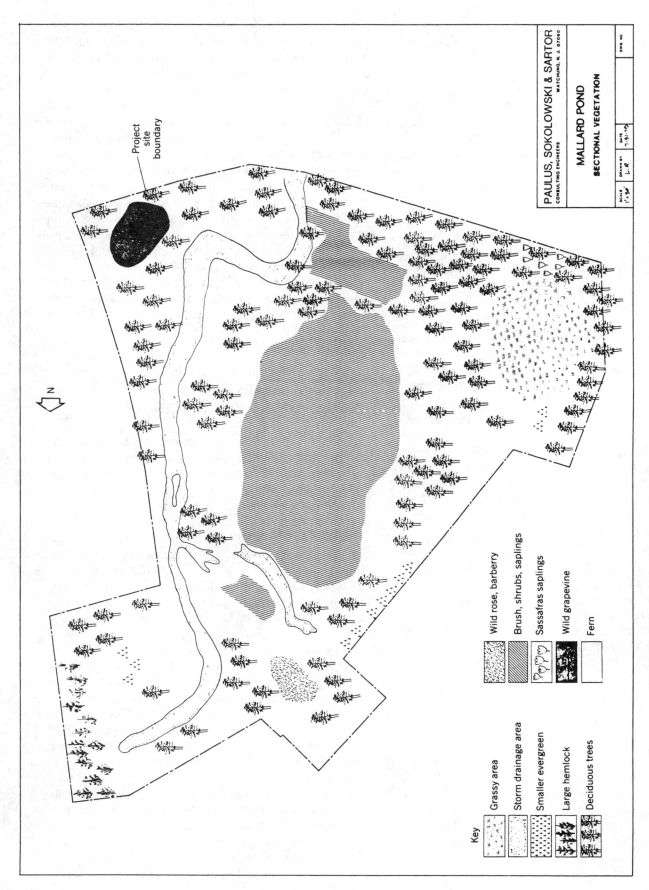

**FIGURE A.1**

*Typical project site map showing an inventory of existing environmental conditions—in this case, vegetation.* (Courtesy of Paulus, Sokolowski & Sartor, Inc., Warren, New Jersey)

Key

Grassy area

Storm drainage area

Smaller evergreen

Large hemlock

Deciduous trees

Wild rose, barberry

Brush, shrubs, saplings

Sassafras saplings

Wild grapevine

Fern

Project site boundary

N

PAULUS, SOKOLOWSKI & SARTOR
CONSULTING ENGINEERS
WATCHUNG, N. J. 07060

MALLARD POND
SECTIONAL VEGETATION

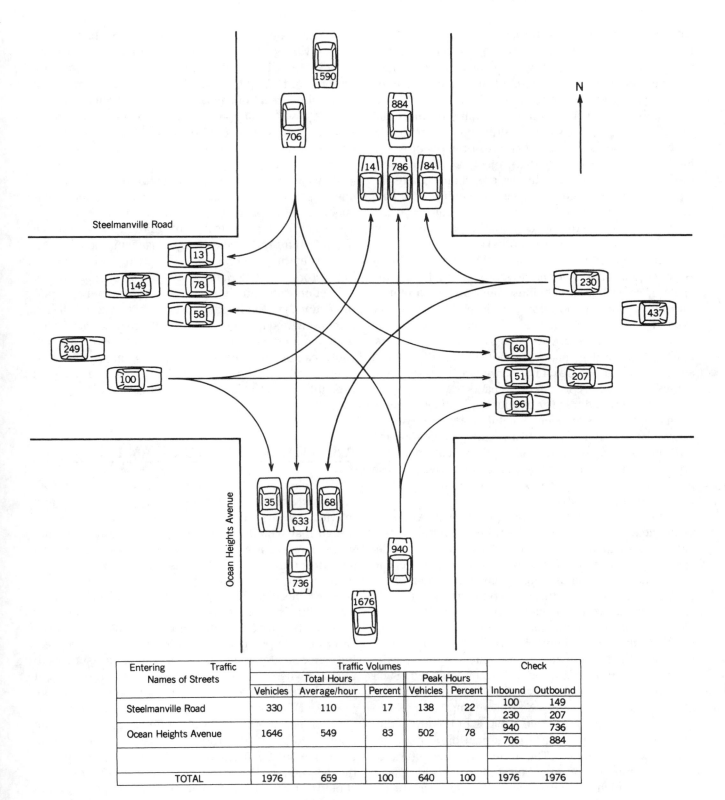

| Entering            Traffic | Traffic Volumes | | | | | Check | |
| Names of Streets | Total Hours | | | Peak Hours | | | |
| | Vehicles | Average/hour | Percent | Vehicles | Percent | Inbound | Outbound |
| Steelmanville Road | 330 | 110 | 17 | 138 | 22 | 100 | 149 |
| | | | | | | 230 | 207 |
| Ocean Heights Avenue | 1646 | 549 | 83 | 502 | 78 | 940 | 736 |
| | | | | | | 706 | 884 |
| | | | | | | | |
| TOTAL | 1976 | 659 | 100 | 640 | 100 | 1976 | 1976 |

## FIGURE A.2

*A traffic survey is usually part of the EIS section on the existing environmental conditions.* (Courtesy of Paulus, Sokolowski & Sartor, Inc., Warren, New Jersey)

estimate the increase in traffic as a function of population density and land use. Using this information, along with data on existing air quality and prevailing weather conditions, the impact of the project on ambient air quality can be anticipated.

Impacts on vegetation and wildlife are more difficult to evaluate objectively. Although it is relatively easy to estimate how many hectares or acres of woodland will be destroyed by the project, it is much more difficult to agree on the value or importance of this impact. If the project site is the last remaining woodland in an urban community or if some of the trees are among an endangered species, the impacts would be considered more severe than otherwise.

It is important to distinguish between *short-term impacts* and *long-term impacts*. For example, the impacts of construction activities might include a temporary increase in neighborhood noise levels from the heavy machinery. Once the project is completed, these impacts cease; they are therefore considered short term. But the effect of the project on, say, runoff patterns and local aquifer recharge rates will not cease when construction is finished; these are long-term impacts.

Many procedures for conducting an environmental assessment have been developed over the years. They share the basic goal of providing a comprehensive and systematic environmental evaluation of the project, with the greatest degree of objectivity. These procedures range in complexity from simple checklists to more complex matrix methods.

In the *checklist method*, all potential environmental impacts for the various project alternatives are listed, and the anticipated magnitude of each impact is described qualitatively. For example, negative impacts can be indicated with minus signs. A small or moderate impact could be shown with, say, two minus signs (− −), whereas a relatively more severe impact could be shown with three or four minus signs (− − − −). Beneficial or positive impacts can be shown with plus (+) signs. If the environmental impact is not applicable for a particular project alternative, a zero (0) would be shown. Such a list presents a visual overview of the assessment.

In the *matrix methods*, an attempt is made to quantify or grade the relative impacts of the project alternatives and to provide a numerical basis for comparison and evaluation. The anticipated magnitude of each potential impact may be rated on a scale of, say, 0 to 10; the higher numbers may represent severe adverse impacts, whereas the lower numbers represent minor or negligible effects. Zero (0) would indicate no expected impact for a particular activity or environmental component.

Numerical weighting factors are also used in the matrix method to indicate the relative importance of a particular impact. These weighting factors are agreed on by the assessment team and are site- and project-specific. For example, the impacts on groundwater quality may be considered more important in a particular area than impacts on air quality, particularly if the groundwater is a sole source of potable water. Groundwater quality could be assigned a relative importance or weight of, say, 0.5, compared to 0.2 for air quality.

Weighting factors can be multiplied by the respective impact magnitudes to put each impact in perspective. For example, consider that the impact on groundwater quality has a magnitude of 4 and the impact on air quality has a higher magnitude of 6. But after weighting the impacts (multiplying by the weighting factors), we see that the overall significance of the impact on water quality, $0.5 \times 4 = 2$, is more important or severe than the impact on air quality, $0.2 \times 6 = 1.2$. If the weighted impacts for all the listed items are added together, a composite score or *environmental quality index* can be obtained for each project alternative. The alternative with the lowest index is the one that will probably cause the least harmful environmental impact overall.

## A.5 OTHER ASPECTS OF AN EIS

An EIS should include a section in which *mitigating measures* are discussed. These mitigating measures are, in effect, suggested changes of details regarding project design that would tend to reduce or eliminate the adverse impacts. For example, one of the most serious short-term impacts due to construction activities is an increase in soil erosion and sedimentation in local streams; this leads to degradation of surface water quality. Specific measures for controlling erosion and preventing stream sedimentation can be described in the EIS (for example, using hay bales and temporary seeding). Another example of a mitigating measure is to relocate where facilities are constructed on the site, if possible, to preserve valuable trees or other vegetation.

An EIS report must also focus on *unavoidable adverse impacts*—those harmful effects that simply cannot be avoided if the proposed project is implemented. For example, if construction of the project must involve the destruction of a mature stand of beautiful trees, this should be identified as an unavoidable environmental impact. Short of not building the project at all (the *no-action alternative*) and therefore not having to cut down the trees, nothing can be done to mitigate this impact.

All reasonable *project alternatives* should be evaluated and discussed in the EIS. These may include changes in scope or location as well as the *no-project* or *no-action alternative.* The no-action alternative causes no environmental disruption to the proposed site and environs, but it generally has some adverse socioeconomic side effects. For example, suppose that the project involves a residential subdivision; the no-project option preserves the site in its natural state, but a shortage of available housing in the community could be an unwelcome result.

Many EIS reports must include an evaluation of possible *secondary* or *indirect impacts* that would be caused by project implementation. Secondary impacts are those that are not immediately apparent and that are not directly caused by the project itself, but probably would not occur if the project were not built.

For example, consider what could happen if a new water main and sewer line are built along a rural road to connect a new subdivision of homes to existing municipal water and sewerage facilities. This is illustrated in Figure A.3. Before long, new homes will be built along that road, causing *strip development*, because water and sewerage utilities are readily available. In effect, construction of the original planned subdivision may indirectly lead to the less desirable future development.

Most EIS documents contain a section pertaining to the *irreversible* or *irretrievable* commitments of resources that will result if the proposed project is built. Supplementing the section on unavoidable adverse impacts, this serves to review and focus attention on energy and material consumption, loss of wildlife habitat, loss of rare or endangered species, and permanent changes in land topography and use.

Finally, a complete EIS will contain a section that responds to public opinion and input. The EIS report is initially prepared in a draft form, which is distributed to the appropriate agency for review and to interested citizens and public interest groups. In most cases, a public hearing is announced and held so that the environmental issues can be discussed openly.

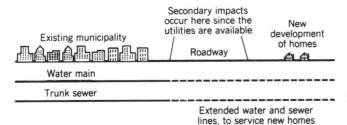

**FIGURE A.3**
*Undesirable strip development may be a secondary or indirect environmental impact of utility construction.*

Public input is considered important in the EIS process because it can provide a perspective or identify an alternative that can otherwise be overlooked by the professionals who prepared the report. Public involvement also serves to resolve disputes early in the planning process. The final copy of the EIS document will reflect this public feedback and input.

## A.6 ENVIRONMENTAL AUDITS

An *environmental audit* is an evaluation of an industrial facility's waste generation and waste management practices, as well as an assessment of the facility's compliance with environmental laws at the local, state, and federal levels. Environmental auditing is a management tool that enhances the overall environmental performance of manufacturing facilities and is now generally a requirement for property transfers and reduction of legal liabilities due to improper or inadequate waste management operations. The *Comprehensive Environmental Response, Compensation, and Liability Act* (CERCLA) and the *Superfund Amendments and Reauthorization Act* (SARA) assign liability for hazardous waste conditions to property owners who knowingly purchase contaminated industrial sites. From the findings of an environmental audit, all parties to a transaction will have accurate knowledge of the conditions of the property and its true value.

There are several different types and purposes of environmental audits. *Waste minimization audits*, for example, are discussed in Section 12.5. Another type is the *transactional audit*, performed prior to the sale or refinancing of manufacturing or commercial facilities; it is typically required by lending institutions, insurance companies, buyers, and state regulatory agencies before the transaction is finalized.

The audit team must be impartial in their review of a facility's status. Company personnel are often part of the team, since they are the most knowledgeable about plant operations; independent consultants and specialists may also serve as audit team members or may conduct the audit in its entirety for a completely objective assessment.

An audit conducted to assess liability typically consists of three phases. In the first phase, a determination of whether contamination exists is made by means of a site survey, a historical property evaluation, and a regulatory file check. The second phase characterizes types, sources, and extent of any contamination. The third phase usually includes a detailed assessment of environmental (and related financial) risks or liabilities. Key steps involved in an auditing process include preaudit planning and development of protocols, checklists, and questionnaires; field surveys and sampling activities; record keeping; and final evaluation of the findings.

A report is prepared at the completion of the audit, candidly addressing deficiencies. Recommendations in the audit report may include corrective actions, such as physical plant upgrades, revisions to standard operating procedures, improved sampling, and obtaining appropriate permits. Action plans and follow-up procedures are also essential to the effectiveness of the audit.

 **A.7   RELEVANT WEB SITES**

### EPA's Office of Federal Activities

http://es.epa.gov/oeca/ofa/

EPA's Office of Federal Activities maintains the national system for the filing of all federal Environmental Impact Statements (EISs), publishes a weekly notice of availability of the EISs, and works with federal agencies sponsoring projects to improve the environmental impact assessment process and to modify those projects determined to be environmentally unsatisfactory.

### International Association for Impact Assessment (IAIA)

http://www.iaia.org

The IAIA is an organization of researchers, practitioners, and users of various types of impact assessment from all parts of the world, whose purpose is to advance the state of the art and science of impact assessment in applications ranging from local to global. IAIA members include corporate planners and managers, public interest advocates, government planners and administrators, private consultants and policy analysts, university and college teachers, and students.

### NEPA Purpose and Implementation

http://water.usgs.gov/public/eap/
NEPA.PURPOSE.html

This Web site presents implementation and management procedures for NEPA, the *National Environmental Policy Act.*

# APPENDIX

# B

# Role of the Technician and the Technologist

## Appendix Outline

The engineering team includes *technicians* and *technologists* as well as *engineers*. It is important for students to have a clear understanding of their future role on this team and to be aware of the educational requirements necessary to begin a career in the field of civil–environmental engineering or technology. It is also helpful for students to be aware of the wide variety of employment opportunities and job responsibilities that exist as they relate to the different levels of education and training.

## B.1 EDUCATION

There are no less than six different levels of education at which a person can begin a career in the field of civil–environmental engineering or technology. As would be expected, a higher level of education requires a greater investment of time and stronger academic abilities than does a lower level of education. These educational levels include the following:

| Engineering | Technology | Certification |
|---|---|---|
| Doctorate degree | Bachelor's degree | Various levels |
| Master's degree | Associate degree | |
| Bachelor's degree | | |

The basic difference between bachelor's degree programs in *engineering* and in *technology* is in the sequence and level of technical courses in the curriculum. Engineering programs place much more emphasis on math, science, and general analytical abilities than do the technology programs. Specific engineering courses are taken by the student in the junior and senior years of college, after a solid foundation in theoretical principles has been established in the freshman and sophomore years. Most engineering courses rely on a thorough knowledge of calculus.

Engineering is often defined as the *application of science and math* to solving problems for the benefit of people. Technology, on the other hand, can be defined as the *application of engineering principles* for the benefit of people. There is less emphasis on math and theory in the technology programs. Instead, practical applications and hands-on skills are stressed. Technology courses usually require knowledge of algebra and trigonometry, but do not rely on calculus, particularly in the freshman and sophomore years. And specific technical subjects may be studied in the freshman year of a technology curriculum.

Generally, a minimum of 7 years of full-time university study is required for the doctorate degree (Ph.D.), 5 years is needed for the master's degree (M.S.), and 4 years is needed for the bachelor's degree in engineering (B.S. or B.E.). A minimum of 4 years is required for the bachelor's degree in technology (B.E.T. or B.S.E.T.), and 2 years is needed for the associate degree in technology (A.A.S.). Some schools offer a master's degree in technology, but this is not very common.

Certification as an operator of a public water supply or sewerage system requires a high school diploma and the passing of a written exam; in many states, several years of operating experience are also required. The levels of certification depend on the type and size of the water or sewerage facility being operated.

Graduates of the bachelor's degree program in engineering technology are called *technologists*, whereas graduates of the associate degree program are called *technicians*. Many employers, however, do not make a distinction between the technologist and the bachelor's-degree-level engineer; some technologists are hired with a job title that includes the word *engineer*. Most states allow technologists to take the professional engineering (P.E.) licensing exam, but the requirements for years of experience vary. In general, the role of the technician and technologist is that of a liaison between the engineer and builder.

## B.2    EMPLOYMENT

For the purpose of discussion, employment opportunities can be categorized into eight different types of activities:

1. *Research and development:* Conducting laboratory and theoretical investigations to further the understanding of environmental processes and to develop new applications and environmental control equipment.
2. *Teaching:* Instructing and guiding engineering and technology students, developing educational curricula and new courses, writing textbooks, and preparing other instructional material.
3. *Project planning and management:* Conducting technical, economic, and environmental feasibility and impact studies evaluating project alternatives, overseeing the progress of engineering studies and design projects.
4. *Project design:* Conducting design computations and preparing detailed plan drawings and specifications to guide the construction of the project.
5. *Construction management:* Estimating construction costs, scheduling equipment, material delivery, and labor, and supervising and coordinating field activities, construction inspection, material testing, and quality and safety control.
6. *Facility operation and maintenance:* Conducting daily process evaluations and control, water and wastewater testing, and supervising maintenance and repair activities.
7. *Regulation and enforcement:* Monitoring environmental quality, enforcing environmental rules and regulations, reviewing and approving plans for new water supply and waste disposal facilities, and inspecting existing facilities.
8. *Marketing and sales:* Providing technical support and liaison between manufacturers and users of environmental control products and equipment.

Of course, it is possible to have a career that includes more than one of these eight activities. But the likelihood of working in a specific activity depends somewhat on educational level and training. The likelihood of having a supervisory role in any of these activities is even more dependent on education. This relationship is shown in Figure B.1.

An engineer with a master's degree, for example, would have opportunities in all eight activities, but engineers at that level are most frequently employed in responsible positions related to project planning and project design. Technicians with associate degrees have significant opportunities to hold responsible positions in construction and facility operation. Technicians may also be employed in planning, design, enforcement, and perhaps research activities, but this is usually under direct supervision of more highly trained and experienced professional engineers. It can be seen from Figure B.1 that the range of employment opportunities and the likelihood of having a position of responsibility in any activity increase as education and training increase.

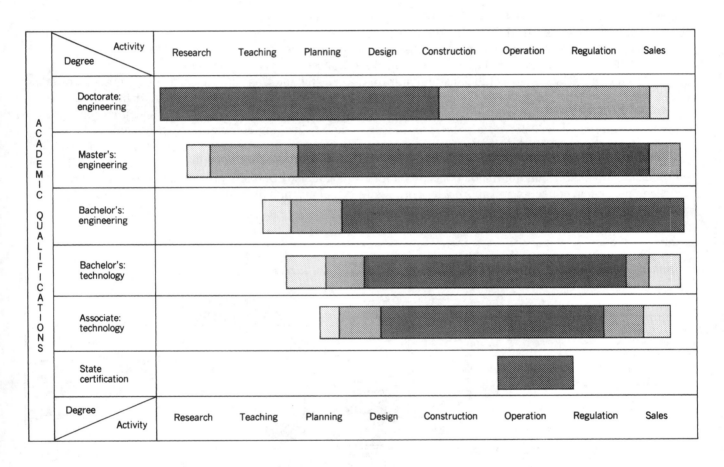

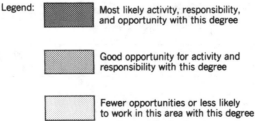

### FIGURE B.1

*Employment opportunities in civil–environmental technology are related to the level of education. In general, greater opportunities for responsibility and advancement are available with higher levels of education.*

## Types of Employers

Several different types of organizations employ civil–environmental engineers, technologists, and technicians. These include colleges or universities, consulting engineering firms, municipal engineering or public works departments, construction contractors, industries, water and sewerage utilities, and regulatory government agencies. The activities that these different employers are most likely to be engaged in are shown in Figure B.2.

## Job Activities

Some activities, such as project design, overlap or span the educational spectrum from Ph.D. to the associate degree. How do job duties and responsibilities vary with education in that particular activity? What role do engineers, technologists, and technicians play in project design?

Generally, an engineer with a high level of education or experience will serve as *project manager*. This involves meeting with the client or project owner, preparing the budget, scheduling and coordinating specific design tasks, and making major decisions about the technical design concepts and approach to problem solving. A *project engineer* (with a master's or bachelor's degree in engineering) will work under the direction of the project manager; he or she assumes overall responsibility for the daily design activities involved in preparing plans and specs for that particular project.

If it is a large and complex project—such as the design of a modern water treatment plant—several other engineers, technologists, and technicians will work under the direct supervision of the project engineer. Engineers with bachelor's and master's degrees will be involved in the detailed design of specific project components, such as the coagulation or filtration processes. This includes design computations using electronic calculators and desktop computers, preparation of sketches and plans, as well as equipment selection and specification writing. Technologists may also be involved in these detailed design activities.

Technicians with associate degrees are involved in assisting the design engineers and technologists. Under direct supervision, they perform specific routine tasks, such as making computations, preparing and inputting data for computer analysis, preparing detailed plan drawings, plotting data, and other activities. The technician also conducts project site surveys, soil or water sampling and testing, and other field investigations.

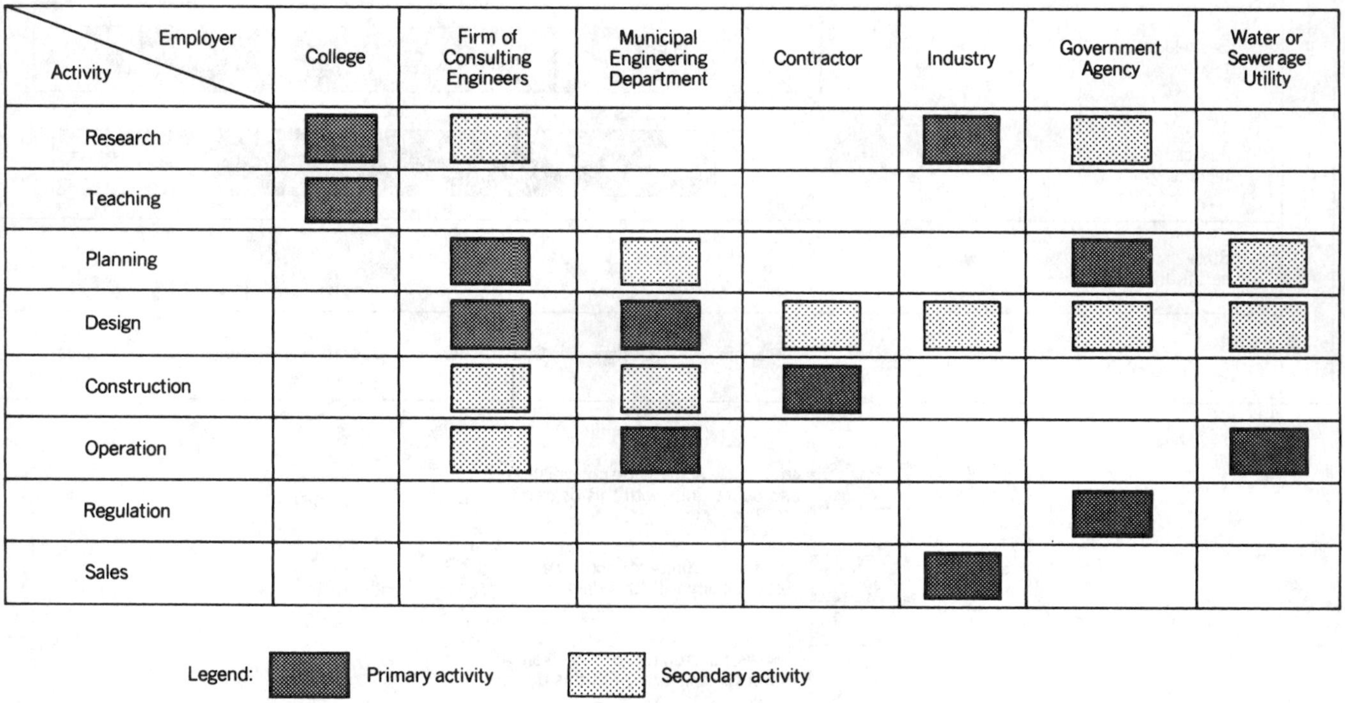

| Activity \ Employer | College | Firm of Consulting Engineers | Municipal Engineering Department | Contractor | Industry | Government Agency | Water or Sewerage Utility |
|---|---|---|---|---|---|---|---|
| Research | ■ Primary | ▢ Secondary | | | ■ Primary | ▢ Secondary | |
| Teaching | ■ Primary | | | | | | |
| Planning | | ■ Primary | ▢ Secondary | | | ■ Primary | ▢ Secondary |
| Design | | ■ Primary | ■ Primary | ▢ Secondary | ▢ Secondary | ▢ Secondary | ▢ Secondary |
| Construction | | ▢ Secondary | ▢ Secondary | ■ Primary | | | |
| Operation | | ▢ Secondary | ■ Primary | | | | ■ Primary |
| Regulation | | | | | | ■ Primary | |
| Sales | | | | | ■ Primary | | |

Legend: ■ Primary activity   ▢ Secondary activity

### FIGURE B.2

*There are many different types of employers in the field of environmental technology. Most employers focus on one or two principal tasks or activities, such as design or construction.*

Senior technicians with good technical abilities and several years of experience can assume more design responsibility and supervise other, less experienced technicians.

## Conclusion

It is not possible to discuss here all the employment opportunities and activities related to environmental engineering and technology. However, this discussion should help the student to appreciate the wide range of job activities and types of employers and to understand the general relationship between level of education and opportunities for responsibility and advancement. In particular, it is important that the student understand the distinction between *engineering* and *engineering technology.*

In the coming years, there will be a need in the field of environmental technology for technical personnel at all levels of education and training. Protection of public health and environmental quality is a top priority goal for most citizens, including politicians and legislators. As researchers and engineers develop new techniques for waste management and pollution control, more and more opportunities will be available for technologists and technicians to apply and implement the principles of modern environmental technology.

## B.3   RELEVANT WEB SITES

### AMERICAN SOCIETY FOR ENGINEERING EDUCATION

http://www.asee.org/precollege/
This site is a guide for students interested in engineering and technology careers. It includes information about the different engineering and technology fields, links to colleges, and other useful information for students.

### AMERICAN SOCIETY OF CIVIL ENGINEERS

http://www.asce.org

This is the Home Page of the American Society of Civil Engineers, which gives information on environmental engineering as well as other subdisciplines that make up the civil engineering profession.

The following sites offer information pertaining to environmental jobs and careers:

### ENVIRONMENTAL CAREER OPPORTUNITIES

http://www.gwu.edu/~greenu/jobs.html

### ENVIRONMENTAL CAREERS ORGANIZATION

http://www.eco.org

### ENVIRONMENTAL JOBS AND CAREERS

http://www.ejobs.org

### U.S. ENVIRONMENTAL PROTECTION AGENCY JOBS

http://www.epa.gov/ezhire

### WATER RESOURCES JOBS

http://www.uwin.siu.edu/~awra/hydata/jobs.html

## Appendix Outline

# APPENDIX

# C

# Review of Basic Mathematics, Units, and Unit Conversions

Environmental technology involves the application of mathematics for problem solving. Most of the mathematics used in this introductory text does not go very much beyond the level of elementary algebra and geometry. It is assumed that most engineering technology students have already studied these topics and are prepared to apply them. Many students, though, can benefit from a brief review of some fundamentals before reading the example problems in the text and working out the practice problems. This Appendix can serve as a refresher or primer for relevant computational skills, and is intended primarily for review by self-study.

Computation of *area*, whether it is of the cross section of a pipe or a watershed, is a very common task in many environmental technology applications. So is the computation of *volume*, whether it is of a water supply reservoir or a municipal solid waste landfill. Area and volume computations require skill in *algebraic substitution*, also called *evaluation of literal expressions* (since the variable parts of a formula, or equation, are called *literal* parts). The ability to *solve simple equations* is also of importance in basic environmental technology applications, including some skills in using *exponents* and understanding *logarithms* (as used in the determination of biochemical oxygen demand and dissolved oxygen profiles, for example). In practical applications, knowledge of *significant figures* and *rounding* is important, as is knowledge of *units of measurement* and the *conversion of units* from one type or system to another.

## C.1  REVIEW OF BASIC MATHEMATICS

### Algebraic Substitution

An algebraic expression is made up of some combination of additions, subtractions, multiplications, and divisions of *constants* and *variables*. A constant is simply a number (or symbol, like $\pi$) with a value that does not change, whereas a variable is a symbol that stands for an unknown number. It is convenient to use letters to represent variables, such as $x$, $y$, $A$, $Q$, etc. For instance, $2L + 2W$ is an algebraic expression which means "two times $L$ plus two times $W$" (when a constant is adjacent to a variable, without any arithmetic symbols, the multiplication is implied). Remember that multiplication (or division) is always done before addition (or subtraction). If the variable $L$ represents the length of a rectangle and $W$ represents its width, this expression would represent the so-called "perimeter" of the rectangle (that is, the total length or distance around its four sides.)

### EXAMPLE C.1

Evaluate the expression $2L + 2W$ for $L = 9$ and $W = 5$ (for now, we will ignore units).

*Solution*  To evaluate an expression for specific values of the variables, simply substitute the values of the variable into the expression and do the arithmetic. In this example, for $L = 9$ and $W = 5$, the following is obtained after making the substitutions:

$$2 \times 9 + 2 \times 5 = 18 + 10 = 28$$

The symbol "$\times$" as used here stands for multiplication, not the variable "$x$." Remember, do multiplication or division *before* addition or subtraction.

## Exponents

Use of an *exponent* or *power* is a simple way of expressing *repetitive multiplication*. For example, to "multiply 3 times itself five times," we can write $3 \times 3 \times 3 \times 3 \times 3$, or, in a much simpler form, we could write $3^5$, where the small number 5 in the upper right corner is called the *exponent* or *power*. In this case, the number 3 is called the *base*. Here is another example: For "multiply 4 times itself three times," we can write $4^3$ instead of $4 \times 4 \times 4$. The number 4 is the base, and the number 3 is the power. For the product of "10 times itself six times," we can write $10^6$ instead of $10 \times 10 \times 10 \times 10 \times 10 \times 10$. The use of "exponent notation" with 10 as the base is most convenient for writing very large (and very small) numbers (as will be seen later in the discussion of scientific notation.)

### EXAMPLE C.2

Evaluate each of the following:
*(a)* $3^3$  *(b)* $2^4$  *(c)* $x^5$ for $x = 4$  *(d)* $10^2$  *(e)* $10^5$

*Solution*

*(a)* $3^3$ means $3 \times 3 \times 3$, or 27 (**not** $3 \times 3$, or 9):

$$3^3 = 27$$

*(b)* $2^4$ means $2 \times 2 \times 2 \times 2$, or 16 (**not** $2 \times 4$, or 8):

$$2^4 = 16$$

*(c)* $x^5$ for $x = 4$ means $4 \times 4 \times 4 \times 4 \times 4$, or 1024 (**not** $4 \times 5$, or 20):

$$4^5 = 1024$$

*(d)* $10^2 = 10 \times 10 = 100$ (Note: When 10 is the base, the result is simply the numeral 1 followed by the same number of zeros as the power or exponent.)

*(e)* $10^5 = 10 \times 10 \times 10 \times 10 \times 10 = 100{,}000$ (Note: $10^5 = 1$ followed by 5 zeros.)

The exponents 2 and 3 have special names. When a base is raised to the second power it is said to be *squared*. A base raised to the third power is said to be *cubed*. For example, "5 squared" is $5^2$ or $5 \times 5 = 25$, and "4 cubed" is $4^3$ or $4 \times 4 \times 4 = 64$. These special names for the powers of 2 and 3 come from the calculation of areas and volumes. Area and volume calculations are very common in environmental technology, and typical problems are reviewed later in this Appendix.

## Square Roots and Cube Roots

Sometimes it is necessary to determine the *square root* or the *cube root* of a number. The square root of a number $n$, written as $\sqrt{n}$, is the number that must be squared to get $n$. In symbols, if $x = \sqrt{n}$, then $x^2 = n$. (The symbol used to denote square root, $\sqrt{\phantom{n}}$, is also called a *radical.*) For example, the square root of 25, written $\sqrt{25}$, is 5, because $5^2 = 25$. In a similar fashion, the cube root of a number $n$ is the number that must be cubed to get $n$. For example, the cube root of 27 is 3, because $3^3 = 27$. The symbol for cube root is $\sqrt[3]{\phantom{n}}$.

### EXAMPLE C.3

Find the following roots:
(a) $\sqrt{81}$
(b) $\sqrt{225}$
(c) $\sqrt[3]{64}$
(d) $\sqrt[3]{343}$
(e) $\sqrt[3]{1000}$

*Solution*
(a) $\sqrt{81}$ is 9, because $9^2 = 81$
(b) $\sqrt{225}$ is 15, because $15^2 = 225$
(c) $\sqrt[3]{64}$ is 4, because $4^3 = 64$
(d) $\sqrt[3]{343}$ is 7, because $7^3 = 343$
(e) $\sqrt[3]{1000}$ is 10, because $10^3 = 1000$

The examples above involve only the roots of *perfect squares* and *perfect cubes*. A perfect square has an integer (whole number) square root, and a perfect cube has an integer cube root.

In many applications, the roots are not integers or whole numbers, and they can only be expressed as decimal approximations. For instance, $\sqrt{7} = 2.65$, as determined with a calculator and rounded to the nearest hundredth. (Rounding off is discussed later in this Appendix.) Note that $2.65^2 = 2.65 \times 2.65 = 7.0225$, so 2.65 is only an approximation of the square root of 7. Although we could be more accurate by expressing the root as 2.646, since $2.646^2 = 7.001316$, we can never reach a number in decimal form whose square is exactly 7, no matter how many places to the right of the decimal we use. At some point, we always have to round off.

## Negative and Fractional Exponents

Mathematical expressions sometimes have terms with negative exponents or powers. For any integer $n$, $a^{-n} = 1/a^n$. In words, a number taken to a negative power is equal to the reciprocal of the same number to the positive power (remember that the reciprocal of $x$ is $1/x$). This definition of what a negative exponent means is necessary for the general rules of exponents to work in all cases.

It is also possible to have fractional (and decimal) exponents. By definition, $a^{1/2}$ is the square root of $a$, and $a^{1/3}$ is the cube root of $a$. For example, $9^{1/2} = 3$. In words, 9 to the one-half power is the square root of 9, which equals 3. In general, $a^{1/n} =$ the $n$th root of $a$. Another example is $27^{1/3} = 3$. In words, 27 to the one-third power is the cube root of 27, which equals 3.

### EXAMPLE C.4

Evaluate the following expressions:
(a) $2^{-3}$
(b) $3^{-2}$
(c) $16^{1/2}$
(d) $27^{1/3}$
(e) $25^{-1/2}$
(f) $36^{0.5}$

*Solution*
(a) $2^{-3} = 1/2^3 = 1/8$
(b) $3^{-2} = 1/3^2 = 1/9$
(c) $16^{1/2} = 4$
(d) $27^{1/3} = 3$
(e) $25^{-1/2} = 1/25^{1/2} = 1/5$
(f) $36^{0.5} = 6$

## Scientific Notation

Very large or very small numbers are often written in *scientific notation*, which has the form $a \times 10^n$, where $a$ is a number between 1 and 10 and $n$ is an integer. To convert a number from standard notation to scientific notation and vice versa, it is convenient to remember that multiplying a number by 10 (that is, $10^1$) moves the decimal point one place to the right, multiplying by 100 (that is, $10^2$) moves the decimal point two places to the right, and so on. Also, dividing a number by 10 (or multiplying by $1/10 = 10^{-1}$) moves the decimal point one place to the left, dividing a number by 100 (or multiplying by $1/100 = 10^{-2}$) moves the decimal point two places to the left, and so on.

## EXAMPLE C.5

| Standard Notation | Scientific Notation |
|---|---|
| 85 | $8.5 \times 10^1$ |
| 924 | $9.24 \times 10^2$ |
| 3576 | $3.576 \times 10^3$ |
| 0.123 | $1.23 \times 10^{-1}$ |
| 0.0345 | $3.45 \times 10^{-2}$ |
| 0.00678 | $6.78 \times 10^{-3}$ |

## Logarithmic Notation

A *logarithm* (log) is just an exponent. For instance, given that $100 = 10^2$, then we say that the logarithm of 100 is 2 to the base 10, or simply $\log_{10} 100 = 2$. Also, since $1000 = 10^3$, then $\log_{10} 1000 = 3$. (Logs can be expressed with any number for a base. For logs to base 10, the subscript is often omitted; for example, $\log 1000 = 3$.) Logarithmic notation is just the opposite of exponential notation. For example, given $x = b^y$, then $y = \log_b x$. In words, if $x = b$ to the $y$ power, then $y =$ the log of $x$ to the base $b$. Logarithms were very useful for doing precise computations before the advent of electronic calculators and digital computers. It is still important to understand the use of logarithms for many applications in engineering technology and higher mathematics.

## EXAMPLE C.6

Write each of the following in logarithmic form:
*(a)* $2^3 = 8$   *(b)* $10^4 = 10,000$   *(c)* $100^{1/2} = 10$

*Solution*
*(a)* $2^3 = 8$ means $\log_2 8 = 3$
*(b)* $10^4 = 10,000$ means $\log 10,000 = 4$
*(c)* $100^{1/2} = 10$ means $\log_{100} 10 = \frac{1}{2}$

## EXAMPLE C.7

Write each of the following in exponential form:
*(a)* $\log 100 = 2$   *(b)* $\log_2 32 = 5$   *(c)* $\log 100,000 = 5$

*Solution*
*(a)* $\log 100 = 2$ means $10^2 = 100$

*(b)* $\log_2 32 = 5$ means $2^5 = 32$
*(c)* $\log 100,000 = 5$ means $10^5 = 100,000$

## Combined Operations

It is frequently necessary to evaluate mathematical expressions that involve a series of arithmetic operations. The order in which the operations are performed is important in order for the result to be correct. There are three basic steps to follow:

1. Evaluate expressions within pairs of parentheses or brackets first, starting with the innermost parentheses.
2. Perform multiplications and divisions, working from left to right.
3. Perform additions and subtractions, working from left to right.

## EXAMPLE C.8

Compute
*(a)* $9 - 4(7 - 5)$
*(b)* $17 - 2(3) + 4$
*(c)* $18 - 3(4) - (6/2 - 1)$

*Solution*
*(a)* $9 - 4(7 - 5) = 9 - 4(2) = 9 - 8 = 1$
*(b)* $17 - 2(3) + 4 = 17 - 6 + 4 = 15$
*(c)* $18 - 3(4) - (6/2 - 1) = 18 - 12 - (3 - 1)$
$= 18 - 12 - (2) = 4$

## EXAMPLE C.9

Evaluate Equation 5-2 using the data for Example 5.3 (on page 132), using a scientific hand-held calculator:

*Solution* The method for evaluating this expression, which deals with stream pollution and dissolved oxygen profiles, depends on the type of calculator being used. Many modern calculators allow the entire expression to be entered just as it appears on page 132, and it is then automatically evaluated with the click of one button. Other calculators allow for storage of intermediate results in memory, so the expression can be evaluated without the need to write down intermediate results. In the example on page 132, intermediate results are written out, to illustrate the proper *order of operations.* For further clarification, it can be noted that the first step involves the difference between 0.5 and 0.2, or 0.3, divided by the product of 0.2 and 14.8, or 2.96. The quotient is then multiplied by 3.7, yielding 0.375, shown in the example on page 132.

## Solving Equations

Applications of mathematics in most cases involve finding the *solution* (or solutions) to an equation. There are many types of equations. The most basic type is the *first-degree equation in one variable* (or one "unknown"). For example, the equation $x - 7 = 2$ is a first-degree equation because the variable, $x$, has an exponent or power of 1 (although a power of 1 is not shown). The solution to this simple equation is $x = 9$, which is evident by inspection. A solution to a first-degree equation is a number that, when substituted for the variable term in the equation, makes the left side of the equation equal to the right side. In other words, the solution makes the equation a true statement. In this simple example, it is clear that $9 - 7 = 2$. We also say that *the solution satisfies the equation.*

Not all equations can be easily solved by simple inspection. A method for systematically determining the solution to an equation involves application of the *properties of equality*, which can be summed up as "what you do to one side of an equation, you must do to the other side." In other words, a number can be added to or subtracted from both sides of an equation without changing the validity of the equation. Both sides of an equation can also be multiplied (or divided) by a number without changing the validity of the equation. The objective of applying these properties of equality is to *isolate the unknown* on one side of the equals sign. For example, the equation $x - 7 = 2$ can be solved by adding 7 to both sides of the equation. By adding 7 here, the variable $x$ is isolated on the left, and the sum of 2 and 7, or 9, remains on the right, thus $x = 9$.

### EXAMPLE C.10

Solve the following equations:
(*a*) $15 - y = 6$
(*b*) $3x + 7 = 19$
(*c*) $3.7 - 7.4y = 0$

*Solution*
(*a*) First subtract 15 from both sides to get $-y = -9$, and after multiplying both sides by $-1$, $y = 9$ is the solution

(*b*) First subtract 7 from both sides to get $3x = 12$, and after dividing both sides by 3, $x = 4$ is the solution
(*c*) First subtract 3.7 from both sides to get $-7.4y = -3.7$, and after dividing both sides by $-7.4$, $y = 0.5$ is the solution

### EXAMPLE C.11

The distance $d$ traveled in time $t$ by a car traveling at average speed $v$ is the product of its speed and time of travel, or $d = vt$, as all physics students soon learn. If a car travels a distance of 200 mi in 4 h, what is its average speed?

*Solution* Substituting the given values of distance (200) and time (4) in the equation gives $200 = (v)(4)$. Since this is a linear equation in one unknown ($v$ in this case), it can be solved by dividing both sides of the equation by 4 to give a solution of $v = 200/4 = 50$ miles per hour, or 50 mph.

## Significant Figures and Rounding

Most applications of mathematics in environmental technology involve using numbers that come from field or laboratory measurements, such as pollutant concentrations and flow rates. But *no measurement can be perfect or exact* because of the physical limits in human perception and limitations of the measuring instruments. There just are no perfect measuring instruments of any kind. Because of this, it is important to know how to use and display all computational results with an appropriate number of *significant figures*.

A digit in a number is a *significant figure* when it is known with some reliability. For example, if the thickness of a book was measured with a metric ruler graduated in *mm* (millimeters), it could be reported as, perhaps, 21.5 *mm*, where all three of the digits are significant. The 5 in that number (which would be estimated by eye as half a millimeter) is the least accurate digit, but it is still significant. The number 21.5 can be described as having three significant figures. *The number of significant figures in a measured quantity is the number of sure or certain digits, plus one estimated digit,* which is a function of the least count or graduation of the measuring instrument.

The number 21.55 would be interpreted as having four significant figures, and it would be assumed that the second five was reliable. It is very important not to display a number with more significant figures than are justified by the actual measurement or calculation. If a number is written as 21.55079, it will be assumed that it has seven significant figures. If this is not the case, the number should be *rounded off* to display an appropriate number of significant figures. Before looking at the topic of rounding off numbers, here are a few more facts and rules regarding significant figures.

In decimal numbers less than one, zeros just to the left and right of the decimal are not significant, since they serve only to locate the place values of the digits. For example, the numbers 0.543, 0.0543, and 0.000543, all have three significant figures. Intermediate zeros, however, are significant. For example, 6078, 607.8, 60.78, 6.078, and 0.6078 all have four significant figures. Simply moving the decimal point does not change the number of significant figures. An intermediate zero does not serve to locate the decimal point.

Numbers with trailing zeros can be ambiguous with regard to the number of significant figures. For example, the number 140,000 could be interpreted as having six significant figures, or as having only two significant figures (the 14) if the zeros are used only to place or locate the decimal point. The best way to avoid ambiguity is to use *scientific notation* to display the number. In this case, if the number has six significant figures, it would be written as $1.40000 \times 10^5$. On the other hand, if it has only two significant figures, it would be written as $1.4 \times 10^5$. When using scientific notation, the factor $a$ preceding $10^n$ should be written as a number between 1 and 10 with the desired number of significant figures.

### Rounding of Numbers

When doing computations with measured quantities, it is often necessary to round off the numbers to display the appropriate number of significant figures. The steps used to round off are:

1. Look at the digit just to the right of the place to which you will round off.
2. If that digit is less than 5, replace it and all digits to its right with zeros.
3. If that digit is 5 or more, replace it and all digits to its right with zeros, and add 1 to the digit to its left. (Note: Trailing zeros after a decimal point can be omitted, except in the case of measured quantities.)

For example, the number 1234.5678 rounded to the nearest thousand is 1000, to the nearest hundred is 1200, to the nearest ten is 1230, to the nearest whole number is 1235, to the nearest tenth is 1234.6, to the nearest hundredth is 1234.57, and to the nearest thousandth is 1234.568. Rounding to one significant figure would be $1 \times 10^3$, to two significant figures would be $1.2 \times 10^3$, to three significant figures would be $1.23 \times 10^3$, to four significant figures would be $1.235 \times 10^3$, to five significant figures would be $1.2346 \times 10^3$, and so on.

### Computations with Measured Quantities

A computed number can be no more precise than the least precise number in the original data. Consequently, *the computed result of multiplication (or division) should be rounded off so that it has as many significant figures as the least precise quantity used in the calculation.* For example, the product of $5.1 \times 9.52 = 48.552$, but it should be rounded off to 49 since 5.1 has two significant figures. The quotient of 48.552 and $5.12 = 9.4828125$ should be rounded to 9.48 since 5.12 has three significant figures.

*The computed results of an addition (or subtraction) should be rounded off so that it has the same number of decimal places (to the right of the decimal point) as the number in the calculation that has the least number of decimal places.* For example, the sum of 1.2 and 3.456 must be written as 4.7, since the result should not have more than one decimal place to the right of the decimal point (as in 1.2). And the result of $3.456 - 1.2$ should be written as 2.3.

## Area and Volume Computations

Environmental technology often involves computation of areas and volumes, ranging from the area of a watershed to the volume of a water storage tank or sanitary landfill.

*Area* is a measure of the planar surface enclosed within a boundary. It has physical dimensions of length squared, or $L^2$. If length is measured in units of meters, then area is expressed in units of square meters, or $m^2$. If length is measured in units of feet, then area is expressed in square feet, or $ft^2$.

*Volume* is a measure of the three-dimensional space enclosed within a continuous surface or solid figure. It has physical dimensions of length cubed, or $L^3$. If length is measured in units of meters, then volume is expressed in units of cubic meters, or $m^3$. If length is measured in units of feet, then volume is expressed in units of cubic feet, or $ft^3$.

*Surface area* is measure of the total surface encompassing a solid figure. For example, for a box, the total surface area is the sum of the flat areas of the top, bottom, front, back, and two ends of the box. (This is true for a prismatic rectangular solid figure, in which the opposite sides have the same dimensions.) For a cylinder, the total surface area is the sum of the circular areas of the top and bottom plus the area of the curved surface. Like areas of flat, two-dimensional shapes, surface area has the units of length squared.

The formulas for areas of common shapes and volumes of common solids are shown in Figure C.1. Notice that the surface area of the curved surface of a cylinder ($2\pi Rh$) is formed by the product of a line of height $h$ and a line equal in length to the circumference of the base of the cylinder ($2\pi R$). Also, the volume of a cylinder is the product of the area of the base and the height. The volume of a lake or reservoir of irregular shape with an area $A$ is the product of the area and the average depth $h$, as shown in the bottom right corner of Figure C.1.

## EXAMPLE C.12

(a) Compute the total surface area of a rectangular box 1.2 m long, 3.4 m wide, and 5.67 m high ($l = 1.2$ m, $w = 3.4$ m, $h = 5.67$ m).

(b) Compute the volume of the box described in part (a) above.

*Solution*

(a) The area of the front of the box is length times height, or 1.2 m $\times$ 5.67 m = 6.804 m$^2$. But this area must be rounded off to 6.8 m$^2$ because of the rules for significant figures and rounding. The area of the top of the box is length times width, or 1.2 m $\times$ 3.4 m = 4.08 m$^2$, which must be

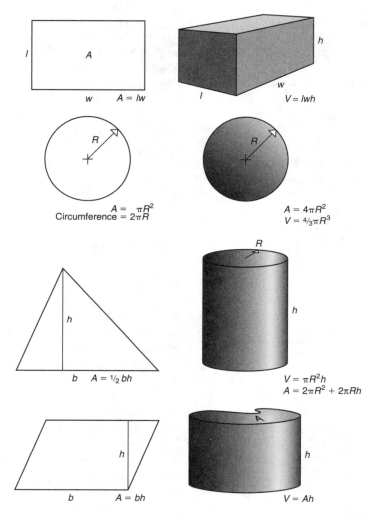

**FIGURE C.1**
*Areas and volumes.*

rounded to 4.1 m². The area of the end of the box is width times height, or 3.4 × 5.67 = 19.278 m², which must be rounded to 19. The total surface area of the box is the sum of the two ends, the top and the bottom, and the front and the back, resulting in 2 × 6.8 + 2 × 4.1 + 2 × 19 = 13.6 + 8.2 + 38 = 59.8, which must be rounded to 60 m². [Notice that when multiplying by 2 (for two ends, etc.), the 2 is an exact number and the products are not rounded off. Also, when adding the three products, the sum is rounded so that there are no digits to the right of the decimal point (as in the 38).] The total surface area of the box should be written as 60 m², with the appropriate number of significant figures.

(b) The volume of the box is the product of length, width, and height ($V = lwh$), which gives $V = 1.2 \times 3.4 \times 5.67 = 23.1336 = 23$ m³ (rounded to two significant figures.)

---

### EXAMPLE C.13

A cylindrical tank is filled with 925 m³ of water. The tank has an inner diameter of 3.0 × 10¹ m (or 30 m with two significant figures). Compute the depth of the water in the tank.

*Solution* From Figure C.1, $V = \pi R^2 h$. First substitute 925 for volume $V$ and 15 m for the radius $R$ (which is half the diameter), to get the equation $925 = \pi(15)^2(h)$. For practical purposes the value of $\pi$ can be taken as 3.14, giving $925 = 706.5h$. Solving for $h$ by dividing both sides of the equation by 706.5, and rounding off gives $h = 1.3$ m.

---

## C.2 UNITS OF MEASUREMENT

Two systems of measurement units are used in this book—the *International System of Units* and the *U.S. Customary System of Units*. The International System, also called *SI metric units*, is used in most countries of the world (SI stands for Systèm International). U.S. Customary units are also called *conventional, English, Imperial*, or *inch–pound* units. They are still used in the United States.

Although all federally funded projects in the United States are now designed using metric units, inch–pound units are still often used in the private sector. It is important for engineering and technology professionals in the United States to be capable of using both systems. In this book inch–pound units are noted parenthetically after metric units for help in "thinking metric." (For Americans accustomed to inch–pound units, "thinking metric" means developing a feeling for the equivalent values of the units rather than memorizing exact conversions.) In engineering practice it is generally best to utilize only one system of units on design drawings and written specifications, rather than using dual metric and inch–pound units.

### Base Units

Base SI metric units include the *meter* (m) for length, the *kilogram* (kg) for mass, and the *second* (s) for time. In the conventional system, base units include the *foot* (ft) for length, the *pound* (lb) for force, and the *second* (s) for time. Note that in the SI metric system, mass is a base unit, whereas in the U.S. conventional system, force is a base unit.

In the metric system, force is a "derived" physical quantity, with units of *kilogram-meters per second per second* (kg · m/s²). This somewhat strange-looking combination of units for mass, length, and time stems from Newton's second law of motion, which states that force equals the product of the mass and its acceleration ($F = ma$). For convenience and in honor of Sir Issac Newton, units of Kg · m/s² are called *newtons* (abbreviated N). One newton is defined as the force that will cause a mass of 1 kg to accelerate at a rate of 1 m/s².

In the U.S. Customary or inch–pound system, mass is a "derived" quantity, with units of *slugs*. A mass of 1 slug will accelerate at the rate of 1 ft/s² when acted upon by a force of 1 lb. (Units of slugs are rarely used in practical applications of environmental technology.)

### Weight

Mass and weight are different physical quantities. The mass of a given quantity of matter is constant anywhere, whereas the weight of that matter depends on the strength of the gravitational field. *Weight is the force due to gravity.* From Newton's second law of motion, it follows that weight equals mass times the acceleration due to gravity, $W = mg$, where weight $W$ is expressed either in terms of newtons or pounds, mass $M$ is expressed in either kilograms or slugs, and $g$ is the *acceleration due to gravity*. At mean sea level on Earth, the average value for $g$ is 9.81 meters per second per second (m/s²), or 32.2 feet per second per second (ft/s²).

Since $W = mg$, a mass of 1.00 kg weighs $W = (1 \text{ kg})(9.81 \text{ m/s}^2) = 9.81$ N. An object with a mass of 50.0 kg would weigh 490 N, and so on. It is technically incorrect to say that an object "weighs 50 kg," although many laymen use such expressions. It is very important to understand that *mass and weight are distinctly different* physical quantities.

It is often convenient to remember that a force (or weight) of 1 N is roughly equal to the weight of an apple (about $\frac{1}{4}$ lb), and a mass of 1 kg weighs about 2.2 lb (on Earth). More discussion and examples of conversions of units from one system to another follows in Section C.3.

## Temperature

In addition to mass, force, length, and time, *temperature* is also a fundamental physical quantity used in environmental technology. In the S.I. system, the *kelvin* (abbreviated K) is the standard unit for temperature, but for practical purposes the *Celsius scale* is used (formerly called the *centigrade scale*). The Celsius scale sets the temperature of boiling water at 100° and the temperature of freezing water at 0°, and 0°C is equal to 273.15 K. (Note that °C is the abbreviation for *degrees Celsius*.) In the United States, temperature is expressed in *degrees Fahrenheit* (°F). The temperature of freezing water is 32°F and the temperature of boiling water is 212°F.

Temperatures in Celsius and Fahrenheit units are related by the following formulas: $T_F = 32° + (9/5)T_C$ or $T_C = (5/9)(T_F - 32°)$, where $T_F$ and $T_C$ are corresponding degrees Fahrenheit or degrees Celsius. For example, a temperature of 20°C is equal to $32 + (9/5)(20) = 68°F$ and a temperature of 50°F is equal to $(5/9)(50 - 32) = 10°C$.

## Other Derived Units

Other physical quantities of importance in environmental technology, such as *area, volume, pressure,* and *flow rate,* are expressed in units *derived from the base units.* For example, area is expressed as square meters ($m^2$) or *square feet* ($ft^2$) and volume is expressed as *cubic meters* ($m^3$) or *cubic feet* ($ft^3$). Other common units for area are *acres* (ac) and *hectares* (ha), and common units for volume include *liters* (L) and *gallons* (gal). (These are further discussed in relevant sections of the text.)

Pressure, defined as force per unit area, is expressed in derived units of $N/m^2$ (newtons per square meter) or as $lb/in.^2$ (pounds per square inch). For convenience, a pressure of $1 \ N/m^2$ is called a *pascal,* and $1 \ lb/in.^2$ is abbreviated *psi.* The symbol for the pascal is Pa (with a capital P). Other key derived measurement units include *joule* (J) for *energy* and *watt* (W) and *horsepower* (hp) for *power.* These and other derived units are discussed in the relevant chapters of this book.

Unit weight (denoted by the Greek letter $\gamma$, pronounced "gamma"), defined as weight per unit volume, is expressed as newtons per cubic meter ($N/m^3$), pounds per cubic yard ($lb/yd^3$), or some other ratio of weight and volume. In algebraic form $\gamma = w/v$, where $\gamma$ = unit weight, w = weight, and $v$ = volume. Rearranging terms yields $w = \gamma v$ as well as $v = w/\gamma$. Uncompacted municipal solid waste (MSW), for example, has a unit weight of $\gamma = 1000 \ N/m^3$. An MSW volume of $v = 5 \ m^3$ would therefore weigh $w = 1000 \times 5 = 5000 \ N$, and 2500 N of MSW would occupy a volume of $v = 2500/1000 = 2.5 \ m^3$.

## Metric Prefixes and Multipliers

Metric units are a *decimal-based* (or base-10) measurement system. They avoid units like inches, feet, miles, gallons, pounds, and tons. SI metric units are based on multiples of 1000, or $10^3$. It is not necessary to use fractions, like 1/16 in., or unique and esoteric conversion factors (like 1 mi = 5280 ft) when using the metric system. This helps to prevent mistakes in mathematical analysis, in engineering design, and in construction.

Metric units use *prefixes* to express multiples of 1000 for convenience in expressing very large or very small numbers without using scientific notation. For example, a pressure of 1 Pa (one pascal) is a very low pressure; pressures in many engineering applications are often much higher. In a typical water main, for instance, the water pressure may be about 40 000 Pa (60 psi). (In the SI unit system, a space is used instead of a comma to separate groups of three zeros.) It is better to write the metric pressure value as 40 *kilopascals* (or kPa) rather than 40 000 Pa, to avoid using all the zeros. The prefix *kilo* stands for 1000 or $10^3$, so 40 kPa means $40 \times 10^3$ Pa or 40 000 Pa.

The metric prefixes most commonly used in environmental engineering applications are:

| Prefix | Symbol | Multiplier |
|--------|--------|------------|
| giga | G | $10^9$ |
| mega | M | $10^6$ |
| kilo | k | $10^3$ |
| milli | m | $10^{-3}$ |
| micro | $\mu$ | $10^{-6}$ |
| nano | n | $10^{-9}$ |
| pico | p | $10^{-12}$ |

For example, a volume of five *milliliters* (5 mL) is equivalent to $5 \times 10^{-3}$ or 0.005 L. A mass of 0.000 003 grams ($3 \times 10^{-6}$ g) is equivalent to 3 $\mu$g (3 *micrograms*). And seven million ($10^6$) liters of water is equivalent to 7 ML (7 *megaliters*) in volume.

Other important metric prefixes, such as centi (c), with a multiplier of $10^{-2}$, and deci (d), with a multiplier of $10^{-1}$, are often used in scientific applications. One *centimeter* (1 cm), for example, is a length equal to 0.01 m, and one *deciliter* (1 dL) is a volume of 0.1 L.

## C.3 CONVERSION OF UNITS

It is often necessary to convert the units that express a measured or calculated physical quantity from one type to another. This can be within a given system or between systems.

In both cases, the best way to do the conversion is to use a ratio equal to unity (one) as a multiplier, with the desired unit on the top and the unit to be replaced on the bottom. The units can be "canceled" in the same manner as common factors are canceled in division. For instance, since $1\,m = 10^3\,mm = 1000\,mm$, the ratio of 1 m/1000 mm = 1, as does the ratio of 1000 mm/1 m. To convert a length of 0.5 m to an equivalent length expressed in mm, simply multiply as follows:

$$0.5\ m = 0.5\ \cancel{m} \times \frac{1000\ mm}{1\ \cancel{m}} = 500\ mm$$

Notice how the units of *m* cancel, leaving *mm*.

To convert a mass of 125 g to an equivalent mass in kg, multiply as follows:

$$125\ g = 125\ \cancel{g} \times \frac{1\ kg}{1000\ \cancel{g}} = 0.125\ kg$$

In this example, the units of *g* cancel, leaving the unit *kg*.

Using this method of "unit cancellation" helps to avoid errors when making conversions. It is necessary to have access to a list of appropriate unit equivalencies in order to convert units. Selected lists of unit equivalencies are given in Table C.1 and Table C.2. Additional unit equivalencies can be found in most physics and engineering textbooks.

## EXAMPLE C.14

Convert a water main diameter of 60 in. to an equivalent pipe diameter expressed in *(a)* feet and *(b)* millimeters.

*Solution*

*(a)* The appropriate unit conversion, of course, is 1 ft = 12 in., and simply *dividing* 60 *by* 12 results

in a diameter of 5 ft. But it is possible to make the inadvertent error of multiplying by the 12 rather than dividing by 12. Using unit *cancellation* helps to avoid such an error. In this case, since feet (ft) is the desired unit, it should be in the top of the conversion ratio, and the inches should cancel out, as follows:

$$60\ in = 60\ \cancel{in} \times \frac{1\ ft}{12\ \cancel{in}} = 5\ ft$$

(If we assume the 60-in. length is accurate to two significant figures, we could express the diameter of the pipe as 5.0 ft.)

*(b)* From Table C.1, 1 mm = 0.03937 in., and in this case the desired unit is mm, which must be in the top of the ratio, as follows:

$$60\ in. = 60\ \cancel{in} \times \frac{1\ mm}{0.03937\ \cancel{in}} = 1524\ mm = 1500\ mm$$
$$\text{(to two significant figures.)}$$

## EXAMPLE C.15

Convert a volume of 15 ft$^3$ to an equivalent volume in m$^3$.

*Solution*   Since 1 ft$^3$ = 7.48 gal and 1 m$^3$ = 264 gal we can write

$$15.0\ ft^3 = 15.0\ \cancel{ft^3} \times \frac{7.48\ \cancel{gal}}{1\ \cancel{ft^3}} \times \frac{1\ m^3}{264\ \cancel{gal}}$$
$$= 0.425\ m^3 = 0.43\ m^3$$

*Alternate Solution*   Using the unit equivalency of 1 m = 3.281 ft, the conversion can be done as follows:

$$15.0\ ft^3 = 15.0\ \cancel{ft^3} \times \frac{(1\ m^3)}{(3.281\ \cancel{ft})^3}$$
$$= 15.0\ \cancel{ft^3} \times \frac{1\ m^3}{35.32\ \cancel{ft^3}} = 0.42\ m^3$$

(Note that 35.32 equals 3.281 cubed. Also, the difference between 0.43 and 0.42 is "rounding-off error" and is of little consequence.)

**TABLE C.1**
**Unit conversions and equivalencies**

| *SI metric to U.S. Customary equivalencies* | *Unit abbreviations* |
|---|---|
| *Length* | ac = acre |
| 1 mm = 0.03937 in. | atm = atmosphere |
| 1 m = 3.281 ft | cfs = cubic feet per second |
| 1 km = 0.6214 mi | ft = feet |
| | $ft^2$ = square feet |
| *Area* | $ft^3$ = cubic feet |
| $1 m^2 = 10.76 ft^2$ | gal = gallon |
| $1 ha = 10\ 000\ m^2 = 2.471\ ac$ | gpg = grains per gallon |
| $1 km^2 = 0.3861 mi^2$ | gpm = gallons per minute |
| | ha = hectare |
| *Volume* | hp = horsepower |
| $1 L = 0.2642\ gal = 0.03531\ ft^3$ | in. = inch |
| $1 m^3 = 264.2\ gal = 35.31\ ft^3$ | kg = kilogram |
| | km = kilometer |
| *Volume flow rate* | $km^2$ = square kilometer |
| 1 L/s = 15.85 gpm = 0.02282 mgd = 0.03531 cfs | kN = kilonewton |
| $1 m^3/s = 15{,}850\ gpm = 22.82\ mgd = 35.31\ cfs$ | kPa = kilopascal |
| $1 ML/d = 1000\ m^3/d = 0.264\ mgd$ | kW = kilowatt |
| | L = liter |
| *Mass and weight (force)* | L/s = liters per second |
| 1 kg = 2.205 lb | lb = pound |
| 1 N = 0.2248 lb | m = meter |
| 1 ton (metric) = 1000 kg = 2205 lb | $m^2$ = square meter |
| 1 kg/L = 8.345 lb/gal | $m^3$ = cubic meter |
| $1 kN/m^3 = 172\ lb/yd^3$ | $m^3/s$ = cubic meters per second |
| | mg/L = milligrams per liter |
| *Pressure* | mgd = million gallons per day |
| 1 kPa = 0.147 psi | mi = mile |
| 1 atm = 100 kPa = 14.7 psi | $mi^2$ = square mile |
| | ML/d = megaliters per day |
| *Chemical concentrations* | mm = millimeter |
| 1 mg/L = 1 ppm = 0.0584 gpg = 8.345 lb/million gal | N = newton |
| 1 $\mu$g/L = 1 ppb | ppb = parts per billion |
| 1% = 10 000 ppm | ppm = parts per million |
| | % = percent |
| *Power* | psi = pound per square inch |
| 1 kW = 1.341 hp | $yd^3$ = cubic yard |
| 1 hp = 550 ft · lb/s | $\mu$g/L = micrograms per liter |

**EXAMPLE C.16**

Convert an area of 350 ha to square kilometers ($km^2$).

*Solution*  Since 1 ha = 10 000 $m^2$ and 1 $km^2$ = $10^6\ m^2$,

$$350\ ha = 350\ ha \times \frac{10\ 000\ m^2}{1\ ha} \times \frac{1\ km^2}{10^6\ m^2} = 3.5\ km^2$$

**EXAMPLE C.17**

(a) Convert a flow rate of 100 gpm (gallons per minute) into an equivalent flow rate expressed in units of cubic feet per second (cfs, or $ft^3/s$).

(b) Convert a flow rate of 5.0 cubic meters per second ($m^3/s$) into an equivalent flow rate expressed in terms of megaliters per day (ML/d).

**TABLE C.2**
**SI metric and U.S. Customary equivalencies**

| *Selected SI metric equivalencies* | *Selected U.S. Customary equivalencies* |
|---|---|
| *Length* | *Length* |
| 1 km = 1000 m | 1 ft = 12 in. |
| 1 cm = 0.01 m = 10 mm | 1 mi = 5280 ft |
| 1 m = 100 cm = 1000 mm | 1 yd = 3 ft |
| *Area* | *Area* |
| 1 ha = 10 000 m$^2$ | 1 ac = 43,560 ft$^2$ |
| *Volume* | *Volume* |
| 1 m$^3$ = 1000 L | 1 ft$^3$ = 7.48 gal |
| 1 L = 1000 cm$^3$ | 1 gal = 4 qt |
| *Mass and weight* | *Mass and weight* |
| 1 kg = 9.81 N | 1 slug = 32.2 lb |
| *Pressure* | *Pressure* |
| 1 atm = 100 kPa | 1 atm = 14.7 psi |
| 1 kPa = 1000 Pa | 1 atm = 29.92 in Hg |
| 1 Pa = 1 N/m$^2$ | 1 atm = 33.9 ft H$_2$O |

*Solution*

(a) 100 gpm = 100 gal/min × 1 ft$^3$/7.48 gal × 1 min/60 s = 0.223 ft$^3$/s, or 0.223 cfs

(b) 5.0 m$^3$/s = 5.0 m$^3$/s × 10$^3$ L/1 m$^3$ × 1 ML/10$^6$ L × 3600 s/1 h × 24 h/1 d = 432 ML/d = 430 ML/d (to two significant figures)

---

 ## C.4   RELEVANT WEB SITES

### CONSTRUCTION METRICATION NEWSLETTER

http://www.nibs.org/cmcnews.html

This site provides current and back issues of the *Construction Metrication Newsletter*, a quarterly periodical that focuses on technical issues and news about metrication progress in the U.S. construction industry.

### DIGITAL GENERATION UNIT CONVERSIONS

http://www.webcom.com/~legacysy/convert2/convert2.html

This site provides an easy-to-use Java-based software program that computes unit conversions.

### INTERNATIONAL SYSTEM OF UNITS

http://physics.nist.gov/cuu/Units/index.html

This site provides an easy-to-read introduction to the SI metric system, including prefixes, rules and style conventions, historical background, and a bibliography of other online references pertaining to the metric system.

### TOWARD A METRIC AMERICA

http://ts.nist.gov/ts/htdocs/200/202/mpo_home.htm

This Web site provides information about the U.S. Department of Commerce – National Institute of Standards and Technology *Metric Program*, including the history of metric conversion in the United States and a table of conversion factors.

### UNITED STATES METRIC ASSOCIATION (USMA)

http://lamar.colostate.edu/~hillger/

Home Page of the USMA, this site provides a gateway to a wide variety of metrication links and information about the metric system, standards, and conversion factors.

# D

# HydroCAD™ Software for Stormwater Computations

A free version of HydroCAD™ is available for downloading (http://www.hydrocad.net/). It provides most of the capabilities of the full HydroCAD program, except that it is limited in the size of the projects it can analyze and it allows only 60 min of operation per session. Users can, however, run and save an unlimited number of sessions, making this software very useful for educational purposes. It is recommended that students using this textbook who seek more in-depth knowledge of the material presented in Chapter 9 on *stormwater management* make use of this software. A tutorial is available to illustrate use of the software, and a library of hydrology information is provided. A brief overview of the software is given below.

HydroCAD is a computer-aided design system for modeling the hydrology and hydraulics of stormwater runoff. It is based largely on the hydrology techniques developed by the Soil Conservation Service (SCS/NRCS), combined with other hydrology and hydraulics calculations; it also includes an option to use the rational method.

For a given rainfall event, these techniques are used to generate hydrographs throughout a watershed. Typically, this allows the engineer to verify that a given drainage system is adequate for the area under consideration or to predict where flooding or erosion is likely to occur.

HydroCAD takes this capability one step further by maintaining a complete database for the watershed and drainage system. With this database, HydroCAD becomes a working model for the entire system where changes can easily be made and their effects viewed. With HydroCAD this takes just minutes, not hours, so the engineer can interact with the watershed model in a way not previously possible. This lets the engineer evaluate numerous possible designs and choose the one most suitable, not just from a safety and environmental standpoint, but based on cost and other considerations as well.

HydroCAD frees the engineer to concentrate on creative design, a goal that is often sacrificed when each alternative requires hours or days of tedious calculations. No program can replace human creativity, but HydroCAD can free that creativity by providing a working model that leaves the engineer free to ask "what if...?"

The HydroCAD *routing diagram* shows the individual "nodes" that make up the project. Arrows that indicate how their outflows are routed are used to connect the nodes. Multiple inflows are summed automatically. Based on the routing diagram, HydroCAD is able to determine the correct sequence of calculations and then calculate the flows throughout the project.

Each routing diagram is composed of the following types of nodes: *subcatchments, reaches, ponds,* and *links.* A *subcatchment* models the effect of rainfall on a specific section of the watershed, and produces a runoff hydrograph. Subcatchments are described by a number of parameters, such as area, curve number, and time of concentration. A *reach* models the effect of a hydrograph being routed through a uniform stream, channel, or pipe under open-channel flow conditions. This results in attenuation and delay of the peak flow due to the storage and travel time of the reach. Reaches are described by a number of parameters, such as shape, slope, length, and roughness coefficient. A *pond* is used to model the storage effects of any retention or detention area, from a small catch basin to a large pond. This includes analysis of the outlet devices that control the pond's discharge, such as culverts and weirs. A *link* is used to enter a hydrograph generated outside HydroCAD or to interconnect several routing diagrams. An example routing diagram is shown in Figure D.1.

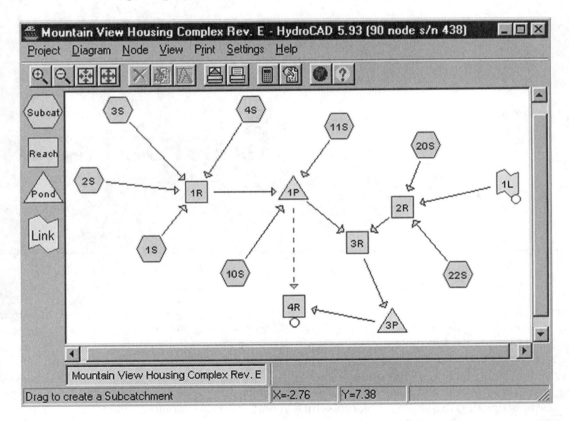

**FIGURE D.1**

*A typical HydroCAD routing diagram, showing subcatchments, reaches, ponds, and links.* (Screen image courtesy of Applied Microcomputer Systems. www.hydrocad.net)

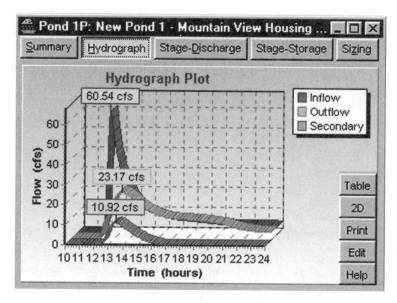

**FIGURE D.2**

*A typical HydroCAD hydrograph.*(Screen image courtesy of Applied Microcomputer Systems. www.hydrocad.net)

In Figure D.1, the subcatchments are represented as hexagons, the reaches as squares, and the ponds as triangles. Clicking and dragging the nodes into the project window easily constructs the diagram. Each node is automatically numbered and each can also be described or annotated with text. Hydraulic calculations can be done using SI metric or U.S. Customary units. The results can be depicted in tabular form or in the form of hydrographs, as illustrated in Figure D.2.

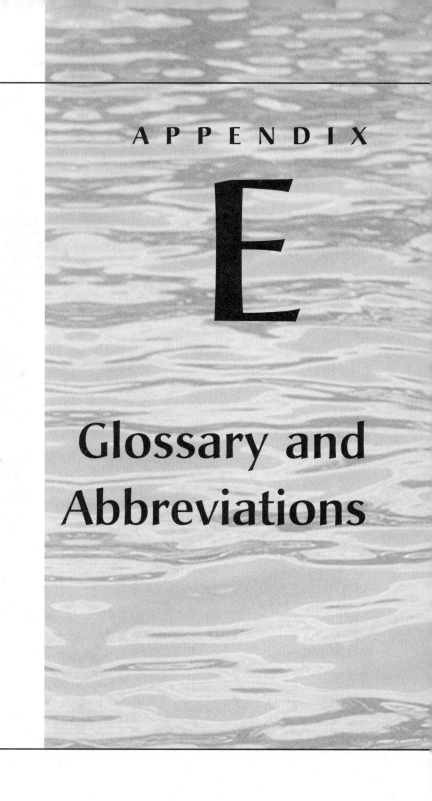

# Glossary and Abbreviations

Many of the technical terms and abbreviations related to the subject of *environmental technology* are new to the student. For the most part, these are defined when they are first introduced in the text. In some cases, though, a student may read certain sections or chapters out of sequence. This glossary will then be useful for getting a quick and brief definition of a new term or abbreviation that is unfamiliar. It can also serve as a review and study aid.

The definitions here are intentionally brief; further discussion of each item can be found in the appropriate section of the text. A list of commonly used environmental abbreviations follows the glossary definitions.

Additional definitions can be found at the EPA's *Terms of Environment* Web site at http://www.epa.gov/students/terms_of_environment.htm. Also, the *Environmental Engineering Dictionary and Directory,* edited by Thomas Pankratz (published by Lewis Publishers/CRC Press, Boca Raton, Florida, 2000), provides more than 8000 terms and abbreviations, as well as a guide for environmental products, services, and companies.

# GLOSSARY

**ABSOLUTE PRESSURE** Pressure measured with reference to a total vacuum as a zero or starting point.

**ABSORPTION** A process by which one substance is trapped throughout the volume of another, usually a liquid, by solution or chemical reaction.

**ABSORPTION FIELD** See *leaching field.*

**ACID** A substance that causes an increase in the hydrogen ion concentration of a solution and that can react to neutralize a base or alkaline substance.

**ACID RAIN** Precipitation with higher than normal acidity, caused primarily by sulfur and nitrogen dioxide air pollution.

**ACOUSTIC MATERIAL** A material that absorbs or reflects sound energy and is used for noise control.

**ACRE–FOOT** A unit used for expressing large quantities of water, as in conservation reservoirs, equivalent to the volume that would cover 1 ac of land to a depth of 1 ft.

**ACTIVATED CARBON** A very porous material, which, after being subjected to intense heat to drive off impurities, can then be used to adsorb pollutants from air or water.

**ACTIVATED SLUDGE** The suspended solids in an aeration tank or at the bottom of a secondary clarifier in a sewage treatment plant, consisting mostly of living microorganisms.

**ACTIVATED SLUDGE PROCESS** A biological sewage treatment system in which living microbes, suspended in a mixture of sewage and air, absorb the organic pollutants and convert them to stable substances.

**ADIABATIC LAPSE RATE** The decrease in ambient temperature with elevation, in dry air, that represents the boundary between a stable and an unstable atmosphere; it is equal to $-1°C/100$ m ($-5.4°F/1000$ ft).

**ADSORPTION** A physical process involving the contact and trapping of water pollutants or air pollutants on the surface of a solid substance, usually activated carbon.

**ADVANCED TREATMENT** Purification processes used after or during secondary wastewater treatment to remove nutrients or additional solids and dissolved organics; also called tertiary treatment.

**AERATION** A physical treatment process in which air is thoroughly mixed with water or wastewater for purification.

**AEROBE** A microorganism that requires an aerobic environment to live and reproduce.

**AEROBIC** In the presence of air or available molecular oxygen.

**AEROSOL** A suspension of small solid or liquid particles in air.

**AIR CLASSIFIER** A mechanical device used to separate paper and other light materials from solid waste in a municipal recycling facility.

**AIR QUALITY INDEX** A number on a uniform scale between 0 and 500, published daily by the EPA, to inform the public about local air quality conditions for each of the five major air pollutants; the higher the index number, the worse is the air pollution. (Formerly called Pollutant Standards Index.)

**ALGAE** Microscopic single-cell plants suspended in water; phytoplankton.

**ALGAL BLOOM** Visible overgrowth of algae in a lake due to eutrophication.

**ALKALINITY** A property of water containing dissolved salts, characterized by the ability to neutralize acids.

**ALUM** Aluminum sulfate, one of the most commonly used chemical coagulants used for water treatment.

**AMBIENT SAMPLE** An air sample collected from the outdoor or surrounding air after pollutants from various sources have been dispersed.

**ANAEROBE** A microorganism that lives under anaerobic conditions, without free oxygen.

**ANAEROBIC** In the absence of air or available molecular oxygen.

**ANTHROPOGENIC** Caused by human activities.

**AQUATIC ORGANISM** An organism that lives in water.

**AQUICLUDE** An underground layer of relatively impermeable soil or rock that does not yield appreciable quantities of groundwater.

**AQUIFER** An underground layer of soil or rock that is porous enough to yield significant amounts of groundwater for public supply.

**AREA METHOD** One method by which solid waste is placed, compacted, and covered in a sanitary landfill, without excavation.

**ARTESIAN AQUIFER** An aquifer that is enclosed or sandwiched between two impermeable layers of soil or rock; also called a confined aquifer.

**ASH** Incombustible residue left after incineration of fuel or solid waste.

**ATOM** The smallest part of an element that can exist and still retain the same chemical characteristics.

**AUTOTROPHIC ORGANISMS** Self-nourishing green plants that obtain food from photosynthesis; the beginning link of the food chain.

**BACKWASH**  The washing cycle for a rapid filter in a water treatment plant in which clean water flows up through the filter.

**BACKWATER ANALYSIS**  Computation of upstream water surface elevations in gradually varied flow where the flow depth is above the normal depth on a mild slope.

**BACTERIA**  Microscopic single-celled organisms that do not contain chlorophyll and do not nourish themselves by photosynthesis.

**BAGHOUSE FILTER**  An air-cleaning device that removes very small particles from dirty flue gases as the gas stream passes through a special filter fabric shaped like a long, inverted bag.

**BALING**  A mechanical process in which municipal solid waste is compressed under high pressure and compacted into rectangular blocks or bales for subsequent land disposal.

**BASE**  A substance that causes an increase in the hydroxyl radical concentration of a solution and that can react to neutralize an acid.

**BASE FLOW**  Dry-weather flow in a stream, fed by groundwater seeping out of the ground and into the stream channel.

**BATTER BOARDS**  Wooden boards placed across a trench during construction to help in establishing line and grade for a sewer pipeline.

**BIOAUGMENTATION**  Addition of bacterial mixtures to contaminated soil to facilitate bioremediation.

**BIOCHEMICAL OXYGEN DEMAND**  The amount of oxygen required by microorganisms to decompose organic waste in water; a measure of the amount of organic pollution; BOD.

**BIODEGRADABLE**  Readily broken down or decomposed into simpler substances by biological action of microbes.

**BIOLOGICAL TREATMENT**  See *secondary treatment.*

**BIOREACTOR**  A vessel or lagoon used for biological treatment of hazardous waste.

**BIOREMEDIATION**  Use of microorganisms to convert harmful chemical compounds to less harmful compounds in order to clean up or remediate a contaminated site.

**BIOSOLIDS**  Treated sewage sludge; a primarily organic solid product, produced by wastewater treatment processes, that can be beneficially recycled.

**BIOVENTING**  See *sparging.*

**BOTTOM ASH**  Residue remaining on the grate or hearth after incineration.

**BRACKISH**  Inland water or groundwater containing more than 1000 mg/L of dissolved salts from natural mineral sources.

**BROWNFIELDS**  Abandoned, idle, or underused industrial and commercial facilities where expansion or redevelopment is complicated by real or perceived environmental contamination.

**BUBBLER**  A device used to collect gaseous air pollutants for analysis by absorption in an appropriate liquid.

**BULKING SLUDGE**  Activated sludge that does not settle properly in a secondary clarifier, causing high effluent BOD and TSS.

**CARBONACEOUS MATERIAL**  Carbon compounds in sewage and garbage.

**CARCINOGENIC**  Capable of causing cancer.

**CATALYST**  A substance that modifies a chemical reaction (makes it slower or faster) without being consumed.

**CATCHMENT AREA**  The area that contributes runoff to a stream or urban drainage system; also, watershed or drainage basin.

**CENTRIFUGAL PUMP**  A mechanical device that adds energy to a liquid, using a rapidly rotating impeller in a specially shaped casing; most common type of pump used for water treatment and distribution.

**CESSPOOL**  A covered pit for disposal of sanitary sewage, usually prohibited now in the United States.

**CHANNEL FLOW TIME**  Time of flow in a stream or pipeline to a drainage basin outlet; part of the time of concentration used in stormwater computations.

**CHEMICAL OXYGEN DEMAND**  The amount of oxygen needed to oxidize all the organics in a wastewater sample, a measure of the level of organic pollution; COD.

**CHEMISTRY**  The study of the composition, properties, and reactions of substances: atoms, elements, molecules, and compounds.

**CHLORINATION**  The process of adding chlorine to water or wastewater, primarily for disinfection.

**CHLORINE RESIDUAL**  The small amount of chlorine compounds that remains in water or wastewater after disinfection, providing continued sanitary protection in the distribution system.

**CHLOROFLUOROCARBONS**  Synthetic organic compounds that contribute to global air pollution.

**CLARIFIER**  A sedimentation basin or settling tank in which suspended solids settle to the bottom and the clarified water or wastewater is drawn off the top.

**COAGULATION**  The addition to water or wastewater of certain chemicals that allow very small suspended particles to collide, stick together, and form settleable flocs.

**COGENERATION**  Energy recovery by production of both steam and electricity at a municipal solid waste incineration facility.

**COLIFORMS** A group of mostly harmless bacteria that live in the intestinal tract of warm-blooded animals and are used as a biological indicator of water pollution.

**COLLOID** Extremely small particles suspended in water or wastewater that cannot be removed by plain sedimentation or filtration without coagulation.

**COMBINED SEWER** A pipeline that may carry a mixture of sanitary, storm, and industrial sewage, common in older cities, but not used in modern construction.

**COMBUSTIBLE MATERIAL** Substances that can be burned or incinerated.

**COMETABOLISM** A biological process used to treat chlorinated hydrocarbons in contaminated soil at abandoned hazardous waste sites.

**COMMINGLED WASTE** A variety of recyclable solid wastes stored in one container.

**COMMINUTOR** A mechanical cutting or shredding device often used for preliminary treatment of sewage.

**COMMUNICABLE DISEASE** A contagious disease that can be transmitted from person to person in a family or community.

**COMPOSITE RUNOFF COEFFICIENT** A weighted average of the runoff coefficients for a large area, used in the rational method of stormwater drainage computations; accounts for different types of land use in the watershed.

**COMPOSITE SAMPLE** A water or wastewater sample obtained by mixing individual grab samples taken at regular time intervals over the sampling period.

**COMPOST** The end product of the composting process, consisting of an inoffensive material resembling potting soil; also called humus.

**COMPOSTING** A biological process for treating garbage and/or sewage sludge, involving aerobic decomposition of organic waste under controlled conditions.

**COMPOUND** A substance made up of a combination of elements.

**CONE OF DEPRESSION** The shape of the groundwater table around a well from which water is being withdrawn.

**CONFINED AQUIFER** See *artesian aquifer.*

**CONSERVATION RESERVOIR** A large open reservoir that serves primarily to store excess wet-weather streamflow for later use during periods of dry weather or drought.

**CONTACT STABILIZATION** A variation of the conventional activated sludge process, used for sewage treatment.

**CONTAGIOUS DISEASE** See *infectious disease.*

**CORROSIVE WASTE** A hazardous waste with pH less than 2 or more than 12.5.

**COVER MATERIAL** Soil used to cover compacted solid waste in a sanitary landfill waste disposal lift or cell.

**CRITICAL FLOW** Open channel flow that occurs at minimum specific energy.

**CROSS-CONNECTION** An unintentional connection between a potable water system and any nonpotable system, through which backflow and contamination of the potable system can occur.

**CROWN** The top inside wall of a pipe.

**CRYPTOSPORIDIUM** A microscopic organism (protozoan) transmitted through ingestion of contaminated drinking water; causes gastrointestinal disease.

**CULLET** Crushed glass recycled for use in the manufacture of new glass products.

**CURIE** A measure of the level of radioactivity.

**CYCLONE** A mechanical air pollution control device used to remove particulates.

**DARCY'S LAW** A formula that expresses the velocity of groundwater flow as a function of the slope of the water table and the soil permeability in the aquifer.

**DAY–NIGHT SOUND LEVEL** A 24-h average sound level with a 10-dB penalty for nighttime noises.

**DECIBEL** A unit of noise measurement that uses a logarithmic scale and that is referenced to the lowest audible sound.

**DECOMPOSITION** The process by which complex organic and inorganic substances are broken down into simpler substances by biological or physical processes; also called decay.

**DEEP WELL** A relatively deep and narrow vertical excavation drilled to penetrate an aquifer for water supply.

**DENITRIFICATION** Conversion of nitrates to nitrogen gas in wastewater treatment systems.

**DESALINATION** The process of separating freshwater from seawater or brackish water; also called desalting.

**DETENTION BASIN** A relatively small reservoir constructed to slow down or temporarily detain surface runoff from a storm.

**DETENTION TIME** The average amount of time water or sewage remains in a treatment tank or basin.

**DEWATERING** Removal of water from sludge.

**DIGESTION** The decomposition of organic waste by microbes under controlled conditions in a sewage treatment plant or garbage compost facility.

**DISCHARGE** The volume rate of flow in a stream, river, or pipeline.

**DISINFECTION** The destruction of disease-causing microbes in water or sewage effluent, usually by the addition of chlorine, ozone, or UV radiation.

**DISPERSED SOURCE**   A broad and unconfined area from which pollutants enter a body of water.

**DISTRIBUTION RESERVOIR**   A water storage tank connected directly to a distribution system, providing about 1 d of capacity.

**DIVERSITY INDEX**   A measure of the variety and population density of different species in an ecosystem.

**DRAINAGE BASIN**   See *catchment area.*

**DRAINAGE DIVIDE**   A line sketched on a topographic map that separates adjacent drainage basins; also called a ridgeline.

**DRAWDOWN**   The vertical distance between the static water level and the pumping water level in a well.

**DROUGHT**   A long period of dry weather that causes low flows in streams and rivers and that affects water supplies adversely.

**DROUGHT FLOW**   Usually the minimum average flow in a stream or river over a 7-consecutive-day period with a recurrence interval of 10 years.

**DUG WELL**   A shallow excavation that penetrates an unconfined aquifer; usually no longer allowed for public drinking water supply.

**DUMP**   An uncontrolled area where wastes have been placed in an environmentally unsound manner.

**DUST**   Suspended solid particles in air larger than 1 $\mu$m in size.

**DUSTFALL BUCKET**   A simple device used to collect and measure settleable particulate levels in the atmosphere.

**ECOLOGY**   The study of living organisms and how they interact with their physical environment.

**ECOSYSTEM**   An identifiable ecological system containing plants and animals and the air, water, and minerals necessary for their survival.

**EFFLUENT**   Waste or wastewater that flows out from a treatment plant or individual treatment process.

**EFFLUENT STANDARDS**   Limitations on the maximum amounts of pollutants that can be discharged from a sewage treatment plant.

**ELECTROSTATIC PRECIPITATOR**   An air-cleaning device that removes very small particulates from flue gases in an electric field.

**ELEMENT**   A substance that cannot be divided into simpler substances by ordinary chemical change.

**EMISSIONS SAMPLING**   See *source sampling.*

**ENVIRONMENT**   Our physical surroundings: air, water, and land.

**EPHEMERAL STREAM**   A stream that becomes completely dry during a drought; also called an intermittent stream.

**EPIDEMIC**   The temporary, but above-average occurrence and spread of a particular disease in a community.

**EPILIMNION**   The uppermost layer of a stratified lake, in which mixing occurs.

**EPISODE**   The temporary occurrence of high air pollution concentrations and acute public health problems, usually during periods of temperature inversion.

**EQUIVALENT PIPE**   A computed or theoretical diameter and/or length of pipe that would have the same hydraulic characteristics as an actual series and/or parallel pipe network.

**EQUIVALENT SOUND LEVEL**   An energy-averaged sound pressure level over a specific period of time.

**EROSION AND SEDIMENT CONTROL**   The use of temporary grass cover, mulch, hay bales, diversion channels, or detention basins at a construction site to prevent the washing away of exposed soil and the clogging of nearby streams with silt and sand.

**ESTUARY**   A brackish water ecosystem between rivers and nearshore ocean waters, where freshwater mixes with salt water and shelter is provided for marine life, birds, and wildlife.

**EUTROPHICATION**   The natural aging of a lake, characterized by high nutrient levels, excessive plant growth, and accumulation of bottom sediments.

**EUTROPHIC LAKE**   A relatively shallow, warm, and turbid lake, with excessive growths of weeds and algae.

**EVAPORATION**   Change of phase of a liquid into a vapor.

**EVAPOTRANSPIRATION**   A part of the hydrologic cycle involving the combined processes of evaporation and transpiration of water by vegetation.

**EXFILTRATION TEST**   A method of testing sewer lines for watertightness, usually as part of an I/I survey.

**EXTENDED AERATION**   A modification of the conventional activated sludge process for sewage treatment.

**FACULTATIVE BACTERIA**   Bacteria that can thrive in either an aerobic or an anaerobic environment.

**FECAL COLIFORMS**   Coliform bacteria from the intestines of warm-blooded animals.

**FECAL STREP**   Fecal *Streptococcus* bacteria, which live in the intestines of warm-blooded animals; used along with fecal coliforms to determine the source of water pollution, whether of animal or human origin.

**FERROUS METAL**   Metal alloy containing mostly iron.

**FILTRATION**   The removal of suspended particles from water or air using a porous material that allows the fluid to pass through, but traps and retains the particles.

**FLEXIBLE MEMBRANE LINER**   Plastic sheet used as an impermeable liner at a landfill; also called a geomembrane; FML.

**FLOC** A particle large enough to settle out of water or wastewater, formed during the coagulation–flocculation process; also, settleable particles of activated sludge.

**FLOCCULATION** Gentle stirring of water or sewage after the addition of coagulation chemicals, which aid in the formation of settleable flocs.

**FLOODPLAIN** The land along a river that would be covered by water during a 100-year flood.

**FLUE GAS DESULFURIZATION** An air pollution control process that removes sulfur oxides from the emissions from coal-fired power plants; FGD.

**FLY ASH** Finely divided suspended particles carried out of a furnace in the combustion gas.

**FOOD CHAIN** An interrelated series of living organisms that feed on each other in a one-way pattern or direction; the producers, the consumers, and the decay organisms.

**FOOD WEB** A complex food chain involving interactions among many different species of organisms.

**FORCE MAIN** A pipeline through which sewage is pumped under pressure, usually to a sewage treatment plant.

**FREEBOARD** The vertical distance between the top of a tank wall and the water surface in the tank.

**FREQUENCY ANALYSIS** A statistical method for determining the recurrence interval of storms, floods, or droughts; also, a method for analyzing complex noises.

**FRESHWATER** Water containing less than 1000 mg/L of dissolved minerals.

**FUME** Very small solid particles suspended in air, usually formed during high-temperature chemical reactions.

**GAGE PRESSURE** Pressure measured with reference to atmospheric pressure as a zero or starting point.

**GARBAGE** Food wastes in refuse, usually originating in the kitchens of homes or restaurants and in food-processing plants.

**GEOGRAPHIC INFORMATION SYSTEM** A computerized mapping system capable of storing, manipulating, and displaying geographically referenced information; GIS.

**GEOMEMBRANE** See *flexible membrane liner.*

**GLOBAL WARMING** Gradual increase in average atmospheric temperature, attributed by many scientists to the greenhouse effect.

**GRAB SAMPLE** A single sample of water, wastewater, or air collected within a short time span for analysis of pollutants.

**GRADUALLY VARIED FLOW** Open channel flow where the liquid depth changes from one end of the channel reach to the other and the slope of the liquid surface is not parallel to the slope of the channel bottom.

**GRAVITY FLOW** Open channel flow in a pipe, ditch, or stream bed, characterized by a free liquid surface at atmospheric pressure.

**GREENHOUSE EFFECT** The gradual warming of the atmosphere due to increasing levels of carbon dioxide and other gases, and the trapping of radiated heat energy.

**GRIT CHAMBER** One of the preliminary processes in a sewage treatment plant, which serves to remove sand and other inert gritty material from the sewage by gravity settling

**GROUNDWATER** Underground water that occupies the pore spaces in soil or fissures in rock.

**GROUNDWATER TABLE** The interface between the zone of aeration and the zone of saturation in the soil, or the level at which the pore spaces in the soil are saturated with water, but are still at atmospheric pressure.

**HAMMER MILL** A mechanical device used to shred and pulverize municipal solid waste prior to composting and/or disposal.

**HARDNESS** A property of water characterized by soap curdling and scale deposits in hot-water systems, caused primarily by the presence of dissolved calcium and magnesium salts.

**HARDY CROSS ANALYSIS** A computational method for analyzing flows and pressures in interconnected water distribution networks.

**HAZARDOUS WASTE** Dangerous waste material that can cause serious illness, injury, or death, and environmental damage.

**HAZEN–WILLIAMS EQUATION** A formula used to compute major pressure losses in water distribution mains and to design the mains.

**HEADWORKS** The beginning of a wastewater treatment system, typically including flow measurement, screening, and grit removal processes.

**HEAVY METALS** Metals such as mercury or lead that have high molecular weights and are toxic to living organisms at trace levels.

**HERTZ** A term that indicates the frequency of a wave, such as a sound wave, standing for cycles per second.

**HETEROTROPHIC ORGANISM** An organism that cannot manufacture its own food by photosynthesis and must consume plants or animals for energy.

**HIGH-VOLUME SAMPLER** A filtration device used for collecting and measuring the amount of suspended particulates in a relatively large sample of air; also called a hi-vol sampler.

**HUMUS** The end product of garbage and/or sludge composting.

**HYDRAULIC GRADE LINE**   A graph of pressure head in a hydraulic system, usually comprising a series of sloping straight lines that show a drop in pressure in the direction of flow.

**HYDRAULIC JUMP**   Rapidly varied flow which occurs when flow passes from supercritical to subcritical conditions, characterized by a relatively sudden increase in depth and turbulent flow.

**HYDRAULICS**   The study of water at rest and in motion in tanks, reservoirs, pipelines, and pumping systems.

**HYDROCARBON**   An organic substance that contains only hydrogen and carbon atoms.

**HYDROGRAPH**   A graph of stream or river discharge versus time.

**HYDROLOGIC CYCLE**   The cycle of water moving through the environment as rainfall, surface and subsurface flow, and vapor.

**HYDROLOGY**   The study of the occurrence and distribution of water on and under Earth's surface.

**HYDROSTATIC PRESSURE**   The force per unit area on the walls of a tank, dam, or pipe caused by the action of a stationary liquid such as water or sewage.

**HYPOLIMNION**   The bottommost layer of a stratified lake, in which little or no mixing occurs.

**IGNEOUS ROCK**   Rock that has cooled and solidified from an original hot, molten condition.

**IGNITABLE WASTE**   A hazardous waste that can readily cause a fire during its storage, treatment, or disposal.

**I/I SURVEY**   A field survey for measuring the extent of infiltration and inflow in a sanitary sewer system.

**IMHOFF CONE**   A graduated transparent glass cone used in laboratories for volumetric determination of settleable solids content in sewage.

**IMHOFF TANK**   An early type of primary settling and sludge storage tank used in wastewater treatment; in effect, a two-storied septic tank.

**INCINERATION**   An engineered process using controlled combustion to burn solid waste and/or sewage sludge for volume reduction and disposal.

**INDUSTRIAL SEWAGE**   Used water from industrial or manufacturing facilities that carries primarily chemical waste products.

**INERTIAL SEPARATOR**   A device used in a materials recycling facility plant to separate and sort solid waste based on its relative weight or density.

**INFECTIOUS DISEASE**   Illness caused by pathogenic microbes and spread from person to person.

**INFILTRATION**   In hydrology, a term referring to the penetration of water from precipitation into the ground. In sanitary sewer systems, a term referring to the seepage of groundwater into the sewer line through poorly constructed joints or cracks.

**INFLOW**   Unwanted runoff that gets into a sanitary sewer from illegal connections to roof drains or basement sump pumps.

**INFLUENT**   Liquid that flows into a water or wastewater treatment plant or purification process.

**INFRASTRUCTURE**   Constructed public works facilities that allow human communities to function and thrive productively.

**INJECTION WELL**   A well into which liquid hazardous waste is pumped for disposal deep underground.

**INORGANIC**   Mineral substances, usually not containing carbon.

**IN SITU TREATMENT**   Treatment of contaminated soil or groundwater at the site of an old waste dump.

**INTERCEPTOR**   A large sewer that collects wastewater from smaller sewer lines and conveys it to a treatment plant or pumping station.

**INVERSION**   See *temperature inversion*.

**INVERT**   The bottom inside wall surface of a pipe.

**ION**   An electrically charged fragment of an atom or molecule.

**IONIZATION**   The process by which molecules dissociate into charged fragments called ions or radicals.

**ISOKINETIC SAMPLE**   A sample drawn through a probe in a smokestack at the same velocity as the gas in the stack.

**JAR TEST**   A lab procedure used to determine the optimum coagulant dose in a water treatment plant.

**LAGOON**   A pond used for biological stabilization of wastewater or for the storage of hazardous waste.

**LANDFILL**   See *sanitary landfill* and *secure landfill*.

**LAND TREATMENT**   The controlled spreading of wastewater, sludge, or hazardous waste on selected land parcels for waste treatment and/or disposal.

**LAPSE RATE**   The decrease in air temperature with increasing altitude; also called prevailing or environmental lapse rate.

**LASER**   An instrument that projects an intense narrow beam of light, used to establish line and grade during pipeline construction.

**LATERAL**   A relatively small sewer in a public right-of-way that collects wastewater directly from homes or buildings.

**LEACHATE**   Highly contaminated liquid generated in and tending to flow out of a sanitary landfill, thereby causing water pollution.

**LEACHING FIELD**   An area comprising several trenches and buried perforated pipes for the distribution and absorption of septic tank effluent in soil; also called an absorption field.

**LIFT STATION** A pumping facility for lifting sewage from a low point and moving it in a force main to a higher elevation, usually to a treatment plant.

**LOW-PRESSURE AIR TEST** A method for determining the degree of watertightness of a sanitary sewer system during an I/I study.

**MAJOR LOSSES** Energy loss in a pipeline, manifested as a pressure drop due to friction between the layers of flowing water and the pipe wall.

**MANHOLE** A structure that provides access to a sewer system for inspection, cleaning, maintenance, sampling, or flow measurement.

**MANIFEST SYSTEM** The federally mandated procedure for monitoring or tracking hazardous waste material from cradle to grave.

**MANNING'S FORMULA** An equation used for analyzing and designing gravity-flow storm or sanitary sewer pipelines.

**MARSTON'S FORMULA** An equation used to estimate the external load acting on a buried pipe for the purpose of designing its bedding.

**MASS BURNING** The incineration of raw or unprocessed municipal solid waste in an energy recovery facility.

**MASS DIAGRAM** See *summation hydrograph*.

**MATERIALS RECYCLING FACILITY** A facility in which commingled recyclable materials are separated from refuse, sorted, and prepared for marketing; MRF.

**MAXIMUM CONTAMINANT LEVEL** The highest concentration of a substance allowed in public drinking water supplies; MCL.

**MAXIMUM CONTAMINANT LEVEL GOAL** A level of a drinking water contaminant not expected to cause any harmful health effect, published as an MCL goal, but not as an enforceable standard; MCLG.

**MEMBRANE FILTER METHOD** A technique for testing the bacteriological quality of water, using a very fine paper filter to trap and collect the bacteria.

**METABOLISM** Biochemical process by which living organisms produce energy to sustain themselves.

**METAMORPHIC ROCK** Rock formed from igneous or sedimentary rock due to the action of extreme heat and pressure over long time periods.

**METEOROLOGY** The study of weather patterns and events in the troposphere.

**METHANE** A gaseous hydrocarbon product from the anaerobic decomposition of garbage or sewage sludge.

**MICROBAR** A unit for expressing sound pressures, equal to one millionth of standard atmospheric pressure.

**MICROBE** A tiny single-celled living organism seen with the aid of a microscope.

**MICROSTRAINER** A physical treatment device in which water or wastewater flows through a revolving drum covered with a finely woven metal fabric that traps suspended solids.

**MINOR LOSSES** Energy loss in flowing water manifested as a pressure drop that occurs as the water flows through vales, bends, and other pipeline fittings.

**MIST** Very small liquid droplets suspended in the atmosphere.

**MIXED LIQUOR SUSPENDED SOLIDS** The contents of an activated sludge aeration tank at a secondary sewage treatment plant; MLSS.

**MOLECULE** The smallest fragment of a compound that can exist and still retain the same chemical properties.

**MOST PROBABLE NUMBER** A statistical estimate of coliform bacteria concentration in a sample of water or wastewater, based on the results of the multiple-tube fermentation test; MPN.

**MULCH** Material applied on top of the soil surface to limit evaporation.

**MULTIPLE-TUBE FERMENTATION TEST** A method for estimating the concentration of coliform bacteria in water or wastewater.

**MULTIPURPOSE RESERVOIR** A large reservoir built to satisfy two or more needs, including flood control, water supply, power generation, irrigation, and recreation.

**MUNICIPAL SOLID WASTE** Nonhazardous refuse generated by individual and community activities; MSW.

**MUTAGENIC** Causing harmful health effects in the next generation.

**NAPPE** A sheet of water that flows freely over a dam or weir.

**NATURAL ATTENUATION** Using natural physical, chemical, and biological processes to break down contaminants.

**NATURAL SUCCESSION** A process by which healthy ecosystems gradually age and change form as time passes.

**NEWTON** A unit of force necessary to give an acceleration of 1 meter per second squared to 1 kilogram of mass.

**NITRATES** Inorganic compounds that can enter water supplies from fertilizer runoff and sewage; associated with "blue baby syndrome."

**NITRIFICATION** The conversion of ammonia into nitrates by bacterial action, causing a decrease in dissolved oxygen levels in water or sewage.

**NOMOGRAPH** A chart used to solve equations graphically.

**NONFERROUS METAL** Metals or metal alloys that do not contain iron.

**NONINFECTIOUS DISEASE**   A disease that is not transmitted from person to person.

**NORMAL DEPTH**   The depth of steady, uniform flow in an open channel, when the slope of the liquid surface is parallel to the slope of the channel bottom.

**NUTRIENT**   A mineral substance that is essential to life.

**OLIGOTROPHIC LAKE**   A relatively deep, cold, clear young lake, with little aquatic life.

**ON-SITE DISPOSAL**   Disposal of sewage at the location where it is generated, usually using a subsurface septic tank and leaching field.

**OPEN CHANNEL FLOW**   Gravity flow in a pipe or open conduit with a free surface at atmospheric pressure.

**ORGANIC COMPOUND**   A substance usually made up of complex molecules that comprise carbon with hydrogen, oxygen, and other elements.

**OVERLAND FLOW**   Runoff that has not yet reached a well-defined stream channel or ditch; also called sheet flow.

**OVERLAND FLOW TIME**   The time it takes sheet flow to reach a stream or a stormwater inlet; one part of the time of concentration.

**OVERTURN**   The mixing of water in a stratified lake due to a change of season and air temperature.

**OXIDATION**   A chemical reaction involving combination with oxygen and/or loss of electrons.

**OXIDATION DITCH**   A concrete-lined basin that uses mechanical aerators to propel sewage around the basin and provide secondary treatment.

**OXYGEN SAG CURVE**   A graph that shows the decrease in dissolved oxygen concentration in a polluted stream as a function of time or distance; also called the oxygen profile.

**PAPER-TAPE SAMPLER**   A filtration-type device used to collect and measure suspended particulates in air.

**PARSHALL FLUME**   A constricted section in an open channel, for the purpose of measuring flow rate.

**PARTICULATE**   A very small fragment of a solid or liquid substance that is suspended in the atmosphere.

**PASCAL**   A unit of pressure, equal to 1 newton per square meter.

**PATHOGEN**   A type of microorganism that can cause disease.

**PERCOLATION**   The flow of water through the pore spaces of soil, due to gravity.

**PERC TEST**   A field test to determine the rate at which water seeps into the ground at a given site, used for septic system design.

**PERENNIAL STREAM**   A stream that has flow all year, despite periods of drought.

**PERMEABILITY**   The ability of porous soil or rock to allow the flow of water through voids and fissures.

**PERSISTENT ORGANIC POLLUTANTS**   A variety of harmful substances, including pesticides like DDT and chlordane, and industrial chemicals such as PCBs and dioxin, which may contaminate food and water supplies.

**PHON**   Sound-level units representing constant loudness over a wide range of frequencies.

**PHOTOCHEMICAL SMOG**   Air pollution caused by the action of sunlight on nitrogen oxides and hydrocarbons.

**PHOTOSYNTHESIS**   The natural process by which green plants convert carbon dioxide, water, nutrients, and sunlight energy into basic food substances.

**PHREATIC SURFACE**   See *water table*.

**pH SCALE**   A logarithmic scale used to indicate the strength of an acidic or basic solution.

**PHYTOPLANKTON**   Tiny autotrophic plants (that is, algae) that live in water.

**PHYTOREMEDIATION**   Use of plants to naturally absorb soil contaminants.

**PICOCURIE**   One trillionth ($10^{-12}$) of a curie.

**PIEZOMETRIC SURFACE**   An imaginery surface or line that represents the height to which water would rise in an artesian well or in a water main.

**PITCH**   A perceived characteristic of sound that depends on frequency; a shrill whistle is high-pitched and a fog horn is low-pitched.

**PLANIMETER**   A device used to measure the area enclosed within an irregular boundary, such as a ridgeline, on a scaled map.

**PLANKTON**   Tiny free-floating plants and animals (algae and protozoa) that live in river, lake, and ocean water.

**PLUME**   A visible or measurable discharge of a pollutant from a point source.

**POINT OF CONCENTRATION**   A point in a watershed for which runoff quantities are calculated.

**POINT OF CONFLUENCE**   A point where a tributary stream feeds into a larger stream or river.

**POINT SOURCE**   A pipe, channel, or chimney from which pollutants are discharged directly into a body of water or the air.

**POLLUTANT STANDARDS INDEX**   See *Air Quality Index*.

**POROSITY**   The percentage of rock or soil volume occupied by spaces or voids.

**POTABLE WATER**   Freshwater that is crystal clear, safe, and pleasant to drink.

**PRELIMINARY TREATMENT**   The first steps in sewage treatment, including the physical processes of screening, comminution, and grit removal.

**PRESSURE HEAD** The height of a column of liquid, usually water, that a given hydrostatic pressure in a water distribution system could support.

**PRIMARY POLLUTANT** A substance that is emitted directly into the environment and that causes harm in its original form.

**PRIMARY STANDARDS** Air and water quality standards that protect public health.

**PRIMARY SLUDGE** A concentrated suspension or slurry of solids that accumulates at the bottom of a primary settling tank in a sewage treatment plant.

**PRIMARY TREATMENT** The removal of floating and settleable solids from wastewater by screening and gravity settling, preceding secondary treatment processes.

**PROTOZOA** Microscopic single-celled animals that consume bacteria and algae for food.

**PUMP HEAD CURVE** A graph that shows the relationship between flow rate and pressure head on the discharge side of a pump.

**PUTREFACTION** Anaerobic decay of protein compounds.

**PYROLYSIS** A high-temperature thermal conversion process using little or no oxygen for processing municipal solid waste.

**RADICAL** An electrically charged group of atoms that act together as a unit in chemical reactions.

**RADIONUCLIDES** Elements that emit potentially harmful radiation as they undergo a process of natural decay.

**RADIUS OF INFLUENCE** The horizontal distance from a well to the area where the water table elevation is not affected by pumping.

**RAINFALL CURVES** A set of graphs that shows the relationships among rainfall intensity, duration, and frequency for a given region of the country.

**RAINFALL INTENSITY** The rate of rainfall expressed in terms of inches per hour or millimeters per hour.

**RAMP METHOD** One of the methods used for placing, compacting, and covering refuse in a sanitary landfill.

**RAPID FILTER** A water purification system that removes suspended solids as the water flows through a granular bed of sand or other material and that is cleaned by backwashing the filter bed.

**RAPIDLY VARIED FLOW** Open channel flow where the liquid depth changes significantly over a relatively short distance, such as in a hydraulic jump.

**RATING CURVE** A graph that shows the relationship between the stage of a stream and its discharge; also, stage–discharge curve.

**RATIONAL METHOD** A common procedure for estimating peak stormwater runoff rates.

**RAW SEWAGE** Wastewater that has not yet been treated to remove pollutants.

**REACTIVE WASTE** Hazardous waste material that is explosive, flammable, or highly corrosive.

**REAERATION** A natural process that occurs in flowing streams, by which air is mixed in the water, thereby increasing the dissolved oxygen level; also, part of contact stabilization.

**RECHARGE AREA** A region where water infiltrates the ground surface and percolates to the underlying groundwater aquifer.

**RECHARGE BASIN** A reservoir built specifically to collect stormwater runoff and allow it to percolate to an underlying aquifer.

**RECURRENCE INTERVAL** The average number of years between storms of specific intensities and durations; also, return period.

**RECYCLING** The recovery, reprocessing, and reuse of certain discarded materials as an alternative to final waste disposal.

**REDUCTION** A chemical reaction involving the removal of oxygen from and/or the addition of electrons to a compound.

**REFORMULATED GASOLINE** An oxygenated form of gasoline that enhances cleaner burning in internal combustion engines; RFG.

**REFUSE** All the solid waste from a community that requires collection and hauling to a disposal or processing site, including garbage, rubbish, and trash.

**REFUSE-DERIVED FUEL** The combustible portion of solid waste burned for energy in an incinerator; RDF.

**REMEDIATION** Cleanup of an abandoned waste disposal site.

**RESPIRATION** The process by which organic material is oxidized inside the cells of living organisms, providing energy for growth and reproduction.

**RETENTION BASIN** A small reservoir holding a permanent pool of water, constructed to retain stormwater runoff.

**RETURN PERIOD** See *recurrence interval.*

**REVULCANIZATION** The process by which waste rubber is processed for reuse as part of a solid waste recycling program.

**RIDGELINE** A line sketched on a topographic map to show the separation of adjacent watersheds; also, drainage divide line.

**RINGLEMANN CHART** A set of five standard shades of gray used for visual measurement of smoke plume density.

**RIVER BASIN** A large watershed encompassing a major river and all its tributary streams.

**RUBBISH** The dry, nonbiodegradable portion of solid waste.

**RUNOFF** Water from rain or snowmelt that flows overland to lakes, streams, and rivers.

**SALTWATER INTRUSION** Gradual displacement of fresh groundwater by seawater or brackish water, due to excessive pumping of groundwater.

**SANITARY LANDFILL** An engineered facility for disposal of municipal solid waste on land, without endangering public health or causing environmental damage.

**SANITATION** The promotion of cleanliness for the prevention of disease and for public health protection.

**SCREENING** A physical treatment process for water or wastewater in which relatively large floating objects are removed as the liquid passes through a coarse bar screen or wire mesh screen.

**SCRUBBER** An air-cleaning device that traps particulates or gases in a spray of water; also called a wet collector or spray tower.

**SCS METHOD** A procedure for estimating the volume and rate of stormwater runoff using soil type as a major criterion.

**SECONDARY POLLUTANT** A pollutant that is not emitted directly into the atmosphere, but is formed after emission by chemical reactions with other substances.

**SECONDARY STANDARDS** Air and water quality standards that relate to esthetic impacts rather than health impacts.

**SECONDARY TREATMENT** Biological treatment of wastewater designed to remove at least 85 percent of the suspended solids and biochemical oxygen demand.

**SECURE LANDFILL** A landfill constructed with a double impermeable bottom liner, a double leachate collection system, an impermeable cover or cap, and a groundwater monitoring system for the disposal of hazardous waste.

**SEDIMENTARY ROCK** Compacted and consolidated soil particles that have become cemented together naturally over a long period of time.

**SEDIMENTATION** The slow settling and separation of suspended solids from a liquid under the force of gravity.

**SELF-PURIFICATION** The processes by which a stream or river assimilates waste and cleanses itself naturally of organic pollutants.

**SEPTAGE** The contents of a septic tank.

**SEPTIC** Anaerobic, or without oxygen.

**SEPTIC TANK** A buried steel or concrete tank that serves for primary settling, sludge digestion, and storage in an on-site sanitary sewage disposal system.

**SETTLEABLE SOLIDS** The coarser fraction of suspended particles in water, wastewater, or air that settle out because of gravity under relatively quiescent (quiet or still) conditions.

**SETTLING CHAMBER** An enlarged compartment or section of a flue in which airstream velocity is reduced so as to allow relatively coarse particulates or dust to settle out by gravity.

**SETTLING TANK** A steel or concrete basin in which settleable solids are allowed to separate from water or wastewater under the force of gravity; also called a clarifier.

**SEWAGE** Used water from domestic, commercial, or industrial establishments carrying sanitary or industrial waste material; also, wastewater.

**SEWERAGE** The entire system of sewage collection, treatment, and disposal infrastructure.

**SHEET FLOW** Runoff that has not yet reached a well-defined stream channel or drainage ditch; also, overland flow.

**SHORT-CIRCUITING** A term referring to the condition in which water or wastewater flows through a treatment tank in less than the theoretical detention time based on the tank volume and flow rate.

**SHUTOFF HEAD** The pressure head developed by a centrifugal pump that operates against a closed discharge valve.

**SICK BUILDING SYNDROME** Indoor air pollution in an office building, characterized by unspecific illness among occupants.

**SIDE WATER DEPTH** Actual depth of water or sewage in a treatment tank; SWD.

**SLUDGE** A slurry or concentrated suspension of solids that accumulates at the bottom of a settling tank or clarifier.

**SLUDGE DEWATERING** The process of drying liquid sludge, thereby changing its condition to that resembling potting soil.

**SLUDGE DIGESTION** Biological stabilization of organic sludge to reduce its volume, destroy pathogens, and prepare it for drying.

**SLUDGE THICKENING** A process that increases the solids concentration of sludge in order to reduce its overall volume.

**SLUDGE VOLUME INDEX** A measure or indicator of the settling behavior of activated sludge in a secondary clarifier.

**SLURRY** A suspension of solid particles in a liquid.

**SLURRY TRENCH** A method used to construct subsurface cutoff walls for the purpose of containing buried hazardous waste at an illegal dump site.

**SMOKE** Very small airborne solid particulates, less than 1 $\mu m$ in size, formed during incomplete combustion.

**SMOKE READING** A visual evaluation of a smoke plume emitted from a stack, made by comparing the smoke density to standard shades of gray on a Ringlemann chart.

**SMOKE TESTING** A method for testing a sanitary sewer system for watertightness during an I/I survey.

**SOFTENING** A treatment process that reduces the hardness of water by removing much of the dissolved calcium and magnesium.

**SOIL EROSION** The wearing away of the land surface, particularly topsoil, by the natural action of wind or water.

**SOIL SERIES** A group of related soils that have developed from similar rocks and that have similar characteristics.

**SOIL SURVEY MAP** A map prepared by the Natural Resource Conservation Service showing the soil series in different areas of a county.

**SOLID WASTE** Useless, unwanted, discarded material including garbage, rubbish, and trash; also called refuse.

**SOLID WASTE MANAGEMENT** The planning, design, construction, and operation of facilities for the collection, transport, processing, and disposal of solid waste.

**SOUND EXPOSURE LEVEL** A unit of sound measurement that accounts for varying durations of different sound events or noises.

**SOUND LEVEL** A measure of noise, using a meter that weights or adjusts the readings so as to respond approximately as the human ear perceives sound, usually expressed in units of dBA.

**SOUND PRESSURE LEVEL** A measure of noise, expressed in decibels, dB.

**SOURCE REDUCTION** See *waste minimization.*

**SOURCE SAMPLING** Air samples collected from the pollutant source; also called emissions sampling.

**SOURCE SEPARATION** Segregation of paper and other recyclables from garbage at the point of waste generation.

**SPARGING** A site remediation method; injection of air below ground to stimulate the biodegradation of soil contaminants.

**SPECIFIC ENERGY** The sum of the depth of flow and the velocity head in open channel flow.

**STACK SAMPLING** Collecting samples of flue gas by a probe that is inserted directly into the stack, for emission analysis.

**STAGE** The depth or elevation of water in a stream or lake.

**STAGE DISCHARGE CURVE** See *rating curve.*

**STANDARD METHODS** An important reference and guide for water and wastewater sampling and analysis, published jointly by the APHA, AWWA, and the WPCF.

**STATIC LEVEL** The position or elevation of the water surface in a well that is not being pumped at the time.

**STORM SEWAGE** Runoff caused by rainfall and collected in a system of surface channels or buried pipes; also called stormwater.

**STORM SEWERS** Pipelines that are designed to carry only stormwater to a point of storage or disposal.

**STORMWATER MANAGEMENT** The planned control of surface runoff in natural and urban systems to prevent flooding and pollution.

**STRATOSPHERE** A stable layer of the atmosphere located just above the troposphere.

**SUBCRITICAL FLOW** Open channel flow where the depth is higher than the critical depth; typically deep flow at low velocity on mild slopes.

**SUBMAIN** A relatively large sanitary sewer that intercepts the flow from smaller lateral sewers; also called a collector sewer.

**SUMMATION HYDROGRAPH** A graph that shows cumulative flow versus time and is used to determine required reservoir storage volumes; also called a mass diagram.

**SUPERCRITICAL FLOW** Open channel flow where the depth is lower than the critical depth; typically shallow flow at high velocity on steep slopes.

**SUPERFUND** A large fund of money set aside by the federal government to help pay for cleaning up abandoned hazardous waste dump sites.

**SUPERNATANT** The water that remains above the sludge layer in a settling tank or digester.

**SURCHARGED SEWER** A gravity sewerline flowing full with wastewater backed up in manholes and with the hydraulic grade line above the pipe crown.

**SURFACE WATER** Water in lakes, streams, and rivers.

**SUSPENDED SOLIDS** Solids carried in water or sewage that would be retained on a glass-fiber filter in a standard lab test.

**SYSTEM HEAD CURVE** A graph that shows the relationship between the flow rate and the total dynamic head in a water distribution system.

**TAILWATER ANALYSIS** Computation of open channel water elevation when the depth of flow is between critical depth and normal depth.

**TEMPERATURE INVERSION** An atmospheric condition in which air temperature increases with altitude, instead of decreasing as normal.

**TERRESTRIAL** Growing or living on land.

**TERTIARY TREATMENT** Processes used after or during secondary wastewater treatment to remove nutrients and/or additional solids and organics; also called advanced treatment or effluent polishing.

**THERMAL POLLUTION** The change in water temperature of a river or lake, usually caused by cooling water discharge from power plants.

**THERMAL STRATIFICATION** The natural process by which separate layers form in lakes because of water temperature differences.

**THERMOCLINE** A layer of water that separates the upper, epilimnion, from the lower, hypolimnion, layer in a stratified lake.

**TIME OF CONCENTRATION** The time it takes a drop of water to flow from the most distant point of a watershed to the outlet; also, the storm duration used in the rational method for runoff calculations.

**TIPPING FLOOR** Unloading area for solid waste at an incinerator or transfer station.

**TOTAL COLIFORM** Bacteria that are used as indicators of pollution in drinking water.

**TOTAL DYNAMIC HEAD** The total pressure head in a system against which a centrifugal pump operates; TDH.

**TOTAL STATIC HEAD** The vertical distance between the free water surfaces on the suction side and the discharge side of a pump.

**TOXIC WASTE** Poisonous waste that causes acute illness, death, or chronic health problems.

**TRANSFER STATION** A centrally located facility where refuse from individual collection trucks is transferred into larger vehicles for transport to a more distant disposal site.

**TRANSPIRATION** A process whereby water is used by vegetation and then returned to the atmosphere through openings in the leaves.

**TRASH** Community refuse that requires special collection, such as an old mattress.

**TREATMENT TECHNIQUE** A specific treatment process (for example, filtration) required to remove certain contaminants from drinking water, used in lieu of an MCL for substances that are very difficult or costly to measure; TT.

**TRIBUTARY** A small stream that feeds into a larger stream or river.

**TRICKLING FILTER** A biological sewage treatment unit in which dissolved organics are absorbed from the settled sewage as it flows over a bed of slime-covered rocks.

**TROMMEL SCREEN** A rotary-drum-type classifier for sorting solid waste on the basis of size and density characteristics.

**TROPHIC LEVEL** A level in the food chain.

**TROPOSPHERE** The lowermost layer of the atmosphere, in which life is sustained, weather patterns develop, and most air pollution problems occur.

**TRUNK LINE** A large sewer that collects wastewater from submains and conveys it to a treatment plant or pumping station; also called an interceptor.

**TUBERCULATION** The formation of nodules of rust on the inside wall of a pipe, increasing pressure loss and decreasing flow capacity.

**TUBE SETTLER** A prefabricated unit of inclined, nested tubes that is installed in a settling tank to increase its efficiency.

**TURBIDITY** A measure of the light-scattering effect caused by finely divided suspended particles in water.

**TURNOVER** The natural destratification and mixing of the water in a lake or reservoir due to the seasonal change in ambient air temperature during the fall and spring of the year.

**UNDERGROUND INJECTION** A method for disposing of hazardous waste by pumping it through deep wells into confined porous aquifers; also called deep well injection.

**VACUUM FILTER** A mechanical device used at a sewage treatment plant for sludge dewatering.

**VADOSE ZONE** See *zone of aeration.*

**VECTOR-BORNE** A mode of disease transmission through insects, rodents, or other living animal carriers of pathogenic organisms.

**VEHICLE-BORNE** A mode of disease transmission from contaminated inanimate objects, including food and water.

**VIRUS** An extremely small pathogenic parasite, roughly $\frac{1}{50}$ the size of bacteria.

**VOLATILE** Easily converted into vapor.

**VOLATILE ORGANIC COMPOUND** Any organic compound that readily evaporates and participates in atmospheric photochemical reactions to form smog; VOC.

**WASTE MINIMIZATION** Reduction of waste at the source by process changes or recycling.

**WASTEWATER** See *sewage.*

**WATERSHED** The land area that contributes runoff to a stream or lake; also called drainage basin and catchment area.

**WATER TABLE** The top of the zone of saturation, where all soil voids are filled with groundwater at atmospheric pressure; static water surface in a well.

**WEATHERING** The gradual process by which solid rock becomes soil from the action of wind and water, as well as from chemical and temperature changes in the environment.

**WEIR** An obstruction or small dam placed in a flowing stream or channel, usually to measure the flow rate.

WETLAND   A land area frequently submerged by surface or groundwater, and which supports vegetation adapted to saturated conditions.

WINDROW   A long, narrow pile of garbage and/or sewage sludge in an open-field composting facility.

WORKING LEVEL   A unit of measure for exposure to radon gas radiation.

ZERO-EMISSION VEHICLE   An alternative to vehicles powered by internal combustion engines, such as solar-powered or electric vehicles.

ZONE OF AERATION   The upper layer of soil, in which the voids or spaces between the soil particles are not completely filled with groundwater.

ZONE OF SATURATION   The layer of soil or rock in which all the voids or fissures are filled with groundwater.

ZOOPLANKTON   Tiny, free-floating animals in water, which serve as food for fish and other organisms.

## ABBREVIATIONS

| | |
|---|---|
| **AC pipe** | asbestos cement pipe |
| **ach** | air changes per hour |
| **APHA** | American Public Health Association |
| **AQI** | Air Quality Index |
| **AWWA** | American Water Works Association |
| **BOD** | biochemical oxygen demand |
| **BMP** | best management practice |
| **CAA** | *Clean Air Act* |
| **CERCLA** | *Comprehensive Environmental Response, Compensation, and Liability Act* |
| **CFC** | chlorofluorocarbons |
| **cfs** | cubic feet per second |
| **CI pipe** | cast-iron pipe |
| **cmd** | cubic meters per day |
| **COD** | chemical oxygen demand |
| **COH** | coefficient of haze |
| **cps** | cycles per second |
| **CSO** | combined sewer overflow |
| **CSS** | combined sewer system |
| **CWA** | *Clean Water Act* |
| **dB** | decibel |
| **dBA** | decibel, A-weighted |
| **DNAPL** | dense nonaqueous-phase liquids |
| **DNL** | day–night level |
| **DO** | dissolved oxygen |
| **EIA** | environmental impact assessment |
| **EIS** | environmental impact statement |
| **EPA** | Environmental Protection Agency |
| **ETS** | Environmental tobacco smoke |
| **FBRR** | Filter Backwash Recycling Rule |

| | |
|---|---|
| **FGD** | flue-gas desulfurization |
| **F/M** | food-to-microorganism ratio |
| **FML** | flexible membrane liner |
| **GAC** | granular activated carbon |
| **GCM** | global circulation model |
| **GIS** | geographic information system |
| **gpcd** | gallons per capita per day |
| **gpg** | grains per gallon |
| **gpm** | gallons per minute |
| **HDPE** | high-density polyethylene |
| **HGL** | hydraulic grade line |
| **HMTA** | *Hazardous Materials Transportation Act* |
| **HWM** | hazardous waste management |
| **Hz** | hertz (cps) |
| **I/I** | infiltration and inflow |
| **JTU** | Jackson turbidity unit |
| **LNAPL** | light nonaqueous-phase liquids |
| **MA7CD10** | minimum average 7-consecutive-day 10-year low flow |
| **MCL** | maximum contaminant level |
| **MCLG** | maximum contaminant level goal |
| **mgd** | million gallons per day |
| **MLSS** | mixed liquor suspended solids |
| **MPN** | most probable number |
| **MRF** | materials recycling facility |
| **MSW** | municipal solid waste |
| **MSWLF** | municipal solid waste landfill |
| **NAAQS** | national ambient air quality standards |
| **NAPL** | nonaqueous-phase liquids |
| **NEPA** | *National Environmental Policy Act* |
| **NFIP** | National Flood Insurance Program |
| **NIMBY** | not in my backyard |
| **NPDES** | National Pollutant Discharge Elimination System |
| **NRC** | noise reduction coefficient |
| **NRCS** | Natural Resource Conservation Service |
| **NSPS** | new source performance standards |
| **NTU** | nephelometric turbidity unit |
| **OCC** | old corrugated cardboard |
| **ONP** | old newspaper |
| **OSHA** | *Occupational Safety and Health Act* |
| **PCB** | polychlorinated biphenol |
| **PET** | polyethylene teraphthalate |
| **POP** | persistent organic pollutant |
| **POTW** | publically owned treatment works |
| **PPL** | pits, ponds, lagoons |
| **ppm** | parts per million |
| **PSI** | Pollutant Standards Index |
| **psi** | pounds per square inch |
| **PURPA** | *Public Utilities Regulatory Act* |
| **RBC** | rotating biological contactor |
| **RBCA** | risk-based corrective action |
| **RCP** | reinforced concrete pipe |

| | | | |
|---|---|---|---|
| **RCRA** | *Resource Conservation and Recovery Act* | **TCLP** | toxicity characteristics leaching procedure |
| **RDF** | refuse-derived fuel | | |
| **RFG** | Reformulated gasoline | **TDH** | total dynamic head |
| **RI/FS** | remedial investigation/feasibility study | **TDS** | total dissolved solids |
| **rms** | root mean square | **TGIF** | Thank Goodness It's Finished |
| **SARA** | *Superfund Amendments and Reauthorization Act* | **TLV** | threshold limit value |
| | | **TOC** | total organic carbon |
| **SCBA** | self-contained breathing apparatus | **TPH** | total petroleum hydrocarbons |
| **SCS** | Soil Conservation Service | **TS** | total solids |
| **SDWA** | *Safe Drinking Water Act* | **TSCA** | *Toxic Substances Control Act* |
| **SEL** | sound exposure level | **TSDF** | treatment, storage, and disposal facility |
| **SIP** | state implementation plan | **TSP** | total suspended particulates |
| **SOC** | synthetic organic chemical | **TSS** | total suspended solids |
| **SPL** | sound pressure level | **TT** | treatment technique |
| **SPS** | source performance standards | **TTHM** | total trihalomethanes |
| **SS** | suspended solids | **USDA** | United States Department of Agriculture |
| **STC** | sound transmission class | **USGS** | United States Geological Survey |
| **STP** | sewage treatment plant | **UST** | underground storage tank |
| **SVI** | sludge volume index | **UV** | ultraviolet |
| **SWD** | sidewater depth | **VOC** | volatile organic compound |
| **SWMM** | Stormwater Management Model | **WEF** | Water Environment Federation |
| **SWTR** | Surface Water Treatment Rule | **WL** | working level |
| **TARP** | Tunnel and Reservoir Project | **WTP** | water treatment plant |

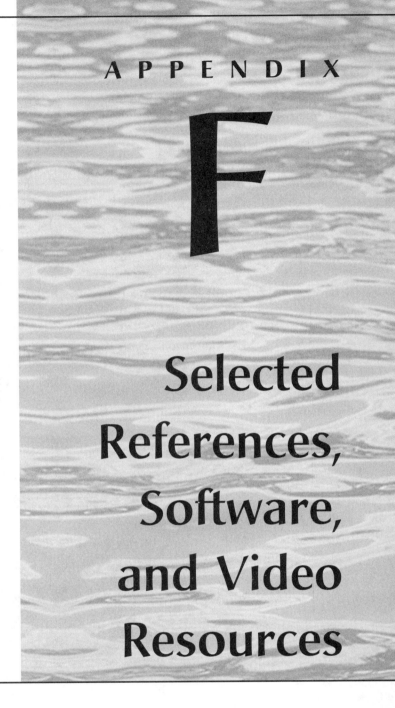

# APPENDIX

# F

# Selected References, Software, and Video Resources

## TEXTBOOKS

### General

Davis, M. L., and Cornwell, D. A., *Introduction to Environmental Engineering*, McGraw-Hill, Inc., New York, 2nd Edition, 1991.

Gupta, Ram S., *Environmental Engineering and Science: An Introduction*, Government Institutes, Rockville, MD, 1997.

Manahan, Stanley E., *Environmental Science and Technology*, Lewis Publishers, Boca Raton, FL, 1998.

Masters, G. M., *Introduction to Environmental Engineering and Science*, 2nd Edition, Prentice Hall, Englewood Cliffs, NJ, 1997.

Ray, B. T., *Environmental Engineering*, PWS Publishing Company, Boston, 1995.

Salvato, J. A., *Environmental Engineering and Sanitation*, 4th Edition, John Wiley & Sons, Inc., New York, 1992.

## Water and Wastewater

Hammer, M. J. Sr., and Hammer, M. J. Jr., *Water and Wastewater Technology*, 4th ed., Prentice-Hall, Inc., Upper Saddle River, New Jersey, 2001.

Metcalf & Eddy, Inc., *Wastewater Engineering: Treatment, Disposal, Reuse*, 3rd Edition, McGraw-Hill, Inc., New York, 1991.

Viessman, W., and Hammer, M. J., *Water Supply and Pollution Control*, 6th Edition, Addison-Wesley Publishing Co., Reading, MA, 1998.

## Air and Noise Pollution

Beranek, L., *Noise and Vibration Control Engineering*, John Wiley & Sons, Inc., New York, 1992.

Cooper, C. D., and Alley, F. C., *Air Pollution Control: A Design Approach*, 2nd Edition, Waveland Press, Inc., Prospect Heights, IL, 1994.

DeNevers, N., *Air Pollution Control Engineering*, McGraw-Hill, Inc., New York, 1995.

Eagleman, J. R., *Air Pollution Meteorology*, Trimedia Publications, Lenexa, KS, 1991.

## Solid and Hazardous Waste

Blackman, W. C., Jr., *Basic Hazardous Waste Management*, Lewis Publishers/CRC Press, Inc., Boca Raton, FL, 1993.

Krieth, F. (ed), *Handbook of Solid Waste Management*, McGraw-Hill, Inc., New York, 1994.

LaGrega, M. D., *et al.*, *Hazardous Waste Management*, McGraw-Hill, Inc., New York, 1994.

Tchobanoglous, G., *et al.*, *Integrated Solid Waste Management*, McGraw-Hill, Inc., New York, 1993.

## World Wide Web

Leshin, C., *Internet Investigations in Environmental Technology*, Prentice-Hall, Inc., Englewood Cliffs, NJ, 1997.

Briggs-Erickson, C., and Murphy, T., *Environmental Guide to the Internet*, 4th Edition, Government Institutes, Inc., Rockville, MD, 1997.

## ENVIRONMENTAL SOFTWARE

A wide variety of public-domain software (shareware) programs related to many of the topics covered in this textbook are available. The EPA Region 5 and Purdue University developed these interactive programs. They are available on CDs from Purdue University (http://pasture.ecn.purdue.edu/%7Ecttpp/programs/cd rom.htm). Downloadable copies are also available from the EPA (http://www.epa.gov/seahome). Program titles and their descriptions, along with ordering and downloading information, can be obtained from the aforementioned Web sites.

## VIDEO RESOURCES

A wide variety of videos focusing on many topics in the areas of environmental engineering, environmental ethics, and waste management are available from Insight Media, 2162 Broadway, New York, NY 10024-0621. Titles, descriptions, and ordering information can be obtained at http://www.insight-media.com/engineering/enter.html.

# APPENDIX

# G

# Answers to Practice Problems

Answers have been rounded off to reflect the precision of the data and/or the accuracy of assumed factors in the problems.

## CHAPTER 2

**1.** 22 psi; 8.6 psi  **2.** 5 m  **3.** 290 kPa  **4.** 115 ft; 32 psi  **5.** 40 kPa; 4 m  **6.** 1.4 m/s  **7.** 12 in.  **8.** 8 m/s  **9.** 0.67 m/s  **10.** 48 psi  **11.** 660 kPa  **12.** 13 kPa  **13.** 10.5 in.; use 12-in. standard size  **14.** 100 L/s

**15.** 2 cfs  **16.** 63 L/s  **17.** 560 L/s; 1.17 m/s  **18.** 1800 gpm; 2.3 ft/s  **19.** 450 mm  **20.** 0.05 percent  **21.** 130 mm; 1.8 m/s  **22.** 1900 gpm; 17 in.  **23.** 1 m/s; 580 L/s; 740 mm  **24.** 100 L/s; 1.3 m/s  **25.** 500 L/s  **26.** 0.16 ft³/s  **27.** 12 L/s  **28.** 120 L/s

## CHAPTER 3

**1.** 50 mm/h; 375 ML  **2.** 2 in./h; 8 ac–ft; 350,000 ft³  **3.** 4 in./h; 1.7 in./h; 2 in./h  **4.** 1 percent  **5.** 27 mm/h

523

**6.** 5 percent **7.** 35 m³/s **8.** See instructor for solution. **9.** $40 \times 10^6$ gal **10.** 0.9 mm/h **11.** sand **12.** 30 m³/h; 33 m³/h

## CHAPTER 4

**1.** 275 mg/L; 16.1 gpg **2.** 68 mg/L **3.** 1250 lb/d **4.** 0.6 mg/L **5.** 10 kg **6.** 100 lb/d **7.** 25 ppb **8.** 8 mg/L **9.** >14 mg/L **10.** 300 mg/L **11.** 270 mg/L; 340 mg/L **12.** 5.8 m/L **13.** 500 mg/L **14.** 350 mg/L; 43 percent **15.** 220/100 mL **16.** 120

## CHAPTER 5

**1.** 180 mg/L **2.** 20 ML/d **3.** 6.4 mg/L **4.** 5.5 mg/L; 6.6 km **5.** 8.8 mg/L **6.** 7.9 mg/L; 2.2 mi

## CHAPTER 6

**1.** 8.8 d **2.** 1250 m³, 5 m **3.** 0.5 in./min **4.** 56.4 ft, 13.4 ft **5.** 20 m by 10 m by 4 m **6.** 4.6 gpm/ft², 7.4 in./min **7.** 6 m by 6 m **8.** 2.6 percent **9.** 19 cylinders; yes **10.** 0.6 mg/L **11.** 50 ML/d

## CHAPTER 7

**1.** 35 ML/d or 9.5 mgd **2.** 3200 L/s **3.** 77 L **4.** 0.62 m² **5.** 625 gpm at 156 ft **6.** 500 gpm at 180 ft **7.** 610 gpm at 195 ft **8.** 185 L/s at 27 m; pump b alone, 110 L/s at 30 m **9.** 23 hp; 30 hp **10.** 65 percent **11.** 4100 m³ **12.** 8 L/s; filling at 34 L/s **13.** 260 mm **14.** 320 mm **15.** 14 in. **16.** AB: 300 L/s; BC: 100 L/s; CA: 200 L/s **17.** 170 kPa **18.** AB: 53 L/s; BC: 11 L/s; DB: 18 L/s; DA: 47 L/s; DC: 29 L/s

## CHAPTER 8

**1.** 830 gpm **2.** 60 L/s **3.** 350 mm at 2.5%; 303.20, 300.70 **4.** 200 mm at 0.0213; 300 mm at 0.00725 **5.** 2000 lb/ft **6.** Class B **7.** Class C **8.** 14 L/min **9.** 870 gal/d/in./mi

## CHAPTER 9

**1.** 18 cfs **2.** 750 L/s **3.** 0.30 **4.** 0.34 **5.** 0.26 to 0.45 **6.** 44 cfs **7.** 3 m³/s **8.** 4 m³/s **9.** 15 m³/s **10.** 1400 cfs; 2400 cfs **11.** 11 cfs, 24 in.; 23 cfs, 30 in.; 35 cfs, 36 in. **12.** 220 L/s, 600 mm; 400 L/s, 650 mm; 600 L/s, 650 mm **13.** 1.3 m³/s **14.** 12,000 m³ **15.** 1.2 ac

## CHAPTER 10

**1.** 18 mg/L; 1500 lb/d **2.** 96 percent; 150 kg/d **3.** 70 percent **4.** 140 ppm **5.** 6 mgd **6.** 19 m³/m² d; 0.5 kg/m³ · d **7.** 31 mg/L; 86 percent **8.** 0.21 **9.** 6 m **10.** 250; no **11.** 50 min/25 mm **12.** 20 min/in. **13.** 7 at 20 m long **14.** 5 at 100 ft; 24 ft wide **15.** 75 tons **16.** 6000 kg; 100 000 L **17.** Approximately 9 m³

## CHAPTER 11

**1.** 80%; 5:1; 20:1 **2.** 88%; 20 yd³ **3.** 90%; 20 m³ **4.** 89%; 9:1; 20:1 **5.** 3.8 ft **6.** 2.7 m **7.** 14 loads/day **8.** 25 loads/day **9.** 15 years **10.** 35 years **11.** 4 m **12.** 21 ft **13.** 30 ac **14.** 23 ha

## CHAPTER 13

**1.** (a) superadiabic (b) superadiabatic (c) 100 m to 150 m (d) yes at 50 m **2.** (a) subadiabatic (b) superadiabatic (c) 0 to 150 m (d) no at 100 m and 200 m **3.** 0.031 ppm **4.** 0.05 ppm; 0.000005% **5.** 10 mg/m³ **6.** 35 ppm; 0.0035% **7.** 17 g/m³ **8.** 1.5 g/m³ **9.** 39 tons/mi²/mo; 141 kg/ha/mo **10.** 120 $\mu$g/m³ **11.** 98.9%; 99.6%; 99.99% **12.** 99.2%; 95.9%; 99.99%

## CHAPTER 14

**1.** 12 s **2.** 1.67 km **3.** 0.1 m **4.** 6800 Hz **5.** 118 dB **6.** 11 $\mu$bar **7.** 98 dBA **8.** 79.8 dBA; 81 dBA **9.** 90 dB **10.** 90.5 dB **11.** 30 phons; 57 dB **12.** 50 phons; 75 dB **13.** 70 dBA; 79 dBA **14.** 72.8 dBA; 75.8 dBA **15.** $L_{10} = 84$; $L_{50} = 78$; $L_{90} = 73$ dBA **16.** $L_{10} = 81$; $L_{50} = 75$; $L_{90} = 69$ dBA

# INDEX

# H

## Color Photographs

**Figure 1**
View of a 36 ML/d (9.3 mgd) advanced wastewater treatment plant in Leominster, Massachusetts. Major process equipment includes aerated grit chambers, spiral lift pumps, rapid mix and flocculation chambers, primary settling tanks, aeration tanks, final settling tanks, chlorination equipment and contact tanks, sludge storage tanks, and vacuum filters. Text reference for wastewater treatment, p. 260. *(Photo courtesy USFilter.)*

**Figure 2**

An aerial view of the Joint Meeting of Essex and Union Counties, New Jersey, 280 ML/d (75 mgd) secondary wastewater treatment facilities. Major process equipment includes four mechanically aerated tanks, final settling tanks, and gravity sludge thickeners. Employing the step aeration modification of the conventional activated sludge process, these facilities are designed for an average of 85 percent BOD removal. Text reference for activated sludge treatment, p. 268. *(Photo courtesy of Joint Meeting of Essex and Union Counties.)*

**Figure 3**

View of a 13 ML/d (3.5 mgd) advanced wastewater treatment plant serving the boroughs of Madison and Chatham, New Jersey. Major process equipment includes comminutors and grit facilities, conventional activated sludge and final clarifiers, an aerated pond that provides three days detention, chlorination-dechlorination facilities, sludge digesters, and mechanical sludge dewatering facilities that supplement the sludge drying beds. Text reference for oxidation ponds, p. 279. *(Photo courtesy of Killam Associates.)*

**Figure 4**
A close-up view of a circular clarifier dewatered for maintenance. The inlet pipe is seen supported from the walkway above it. The v-notch effluent weir is seen on the perimeter of the tank. Text reference for circular clarifiers, p. 148. *(Photo courtesy Killam Associates.)*

**Figure 5**
Waste composition survey at a municipal solid waste (MSW) landfill. Text reference for MSW landfills, p. 311. *(Photo courtesy of Killam Associates.)*

**Figure 6**
Collection of a sample for laboratory testing at a hazardous waste site. The technicians are wearing Level B protective clothing. Text reference for sampling, p. 366. *(Photo courtesy of Killam Associates.)*

**Figure 7**
At Fred Hervey Water Reclamation Plant in El Paso, Texas, wastewater is treated to drinking water standards and is injected into aquifer for reuse. Clarifiers follow a two-stage system that uses biological treatment combined with adsorption on powdered activated carbon to achieve high levels of treatment. Design flow is 38 ML/d (10 mgd). Text reference for clarifiers, p. 148; text reference for biologically activated carbon, p. 165. *(Photo courtesy of USFilter.)*

**Figure 8**
View of open-bay compost system. Sludge mixture is stored and agitated in concrete walled bays. Agitators ride on rails atop bay walls, mixing, aerating, and moving compost to end of bay where it is removed to storage areas. Thermocouples measure temperature to assure pathogen kill. Text reference for sludge composting, p. 325. *(Photo courtesy of USFilter.)*

**Figure 9**
A hydrogen sulfide oxidation system removes $H_2S$ from landfill gas in Broward County, Florida, where the Central Sanitary Landfill and Recycling Center operates the nation's largest landfill gas-to-energy plant. Text reference for landfill gas, p. 341; see case study, p. 343. *(This illustration originally appeared in the May 1997 issue of* Public Works®, *published by Public Works Journal Corporation, 200 South Broad Street, Ridgewood, NJ 07450. ©1997 Public Works Journal Corporation. All rights reserved. Used by permission.)*

**Figure 10**
Pressure filtration vessels at the 64-mgd Lake Casitas, California, water treatment plant operate at filtration rates as high as 8.5 liters per square meter per second (12 gallons per minute per square foot). Pressure filtration was chosen as an economical technique for complying with the Surface Water Treatment Rule (SWTR). Text reference for pressure filtration, p. 157; text reference for SWTR, p. 144. *(This illustration originally appeared in the February 1997 issue of* Public Works®, *in the article "Water District Moves Innovation into Practice," by S. Wickstrum and T. Rao, published by Public Works Journal Corporation, 200 South Broad Street, Ridgewood, NJ 07450. ©1997 Public Works Journal Corporation. All rights reserved. Used by permission.)*

**Figure 11**
A circular clarifier tank at the Atwater, California, wastewater treatment facility. Text reference for wastewater clarifiers, p. 264. *(This illustration originally appeared in the April 1998 issue of* Public Works®, *in the article "Wastewater Treatment Efficiency," by J. Haug and M. Hamamoto, published by Public Works Journal Corporation, 200 South Broad Street, Ridgewood, NJ 07450. ©1998 Public Works Journal Corporation. All rights reserved. Used by permission.)*

**Figure 12**
These reverse osmosis membrane elements have a salt rejection capability as high as 99 percent of total dissolved solids. Text reference for reverse osmosis process, p. 166. *(This illustration originally appeared in the November 1997 issue of* Public Works®*, in the article "Membrane Technology, Part I," by William Suratt, et al., published by Public Works Journal Corporation, 200 South Broad Street, Ridgewood, NJ 07450. ©1997 Public Works Journal Corporation. All rights reserved. Used by permission.)*

**Figure 13**
An aerator (the silver box) atop the water storage tank in the community of Franklin, Indiana, is used to remove iron and manganese from the water. Photo inset shows high-lift pumps at Franklin. Text reference for iron and manganese removal, p. 164; text reference for high-lift pumps, p. 181. *(This illustration originally appeared in the June 1998 issue of* Public Works®*, in the article "Design–Build Construction," by P. Spence, published by Public Works Corporation, 200 South Broad Street, Ridgewood, NJ 07450. ©1998 Public Works Journal Corporation. All rights reserved. Used by permission.)*

**Figure 14**
Tertiary filter and reuse tank. Pulsed bed, shallow bed fine sand filter manufactured by USFilter Zimpro Products group, consists of six cells in concrete tanks for flow of 1.32 ML/d (0.35 mgd). Filtered wastewater is stored in recycle tank for reuse as irrigation water on golf courses at Julington Creek residential development near Jacksonville, Florida. Text reference for wastewater filters, p. 282. *(Photo courtesy of USFilter.)*

**Figure 15**
Deer Island chain and scraper. Manufactured by USFilter's Envirex Products group, 72 chain and flight sludge collectors and motorized scum collectors collect floatable and settleable solids from 18 stacked rectangular tanks at Deer Island Wastewater Treatment Plant, serving the greater Boston area. Text reference for wastewater clarifiers, p. 264. *(Photo courtesy of USFilter.)*